Lecture Notes in Computer Science

Commenced Publication in 1973
Founding and Former Series Editors:
Gerhard Goos, Juris Hartmanis, and Jan van Leeuwen

Editorial Board

Jerzy Marcinkowski   Andrzej Tarlecki (Eds.)

Computer Science Logic

18th International Workshop, CSL 2004
13th Annual Conference of the EACSL
Karpacz, Poland, September 20-24, 2004
Proceedings

 Springer

Volume Editors

Jerzy Marcinkowski
Wrocław University
Institute of Computer Science
Przesmyckiego 20, 51-151 Wrocław, Poland
E-mail: jma@ii.uni.wroc.pl

Andrzej Tarlecki
Warsaw University
Institute of Informatics
Banacha 2, 02-097 Warsaw, Poland
E-mail: tarlecki@mimuw.edu.pl

Library of Congress Control Number: 2004111518

CR Subject Classification (1998): F.4.1, F.4, I.2.3-4, F.3

ISSN 0302-9743
ISBN 3-540-23024-6 Springer Berlin Heidelberg New York

Springer is a part of Springer Science+Business Media

springeronline.com

© Springer-Verlag Berlin Heidelberg 2004
Printed in Germany

Typesetting: Camera-ready by author, data conversion by Scientific Publishing Services, Chennai, India
Printed on acid-free paper   SPIN: 11315872   06/3142   5 4 3 2 1 0

Preface

This volume contains papers selected for presentation at the 2004 Annual Conference of the European Association for Computer Science Logic, held on September 20–24, 2004 in Karpacz, Poland.

The CSL conference series started as the International Workshops on Computer Science Logic, and then, after five meetings, became the Annual Conference of the European Association for Computer Science Logic. This conference was the 18th meeting, and the 13th EACSL conference.

Altogether 99 abstracts were submitted, followed by 88 papers. Each of these papers was refereed by at least three reviewers. Then, after a two-week electronic discussion, the Programme Committee selected 33 papers for presentation at the conference. Apart from the contributed papers, the Committee invited lectures from Albert Atserias, Martin Hyland, Dale Miller, Ken McMillan and Paweł Urzyczyn.

We would like to thank all PC members and the subreferees for their excellent work.

The electronic PC meeting would not be possible without good software support. We decided to use the GNU CyberChair system, created by Richard van de Stadt, and we are happy with this decision. We also would like to thank Michał Moskal who installed and ran CyberChair for us. Finally, we would like to thank ToMasz Wierzbicki, who helped with the preparation of this volume.

We gratefully acknowledge financial support for the conference received from the Polish Committee for Scientific Research, and Wrocław University.

July 2004   Jerzy Marcinkowski and Andrzej Tarlecki

Organization

CSL 2004 was organized by the Institute of Computer Science, Wrocław University.

Programme Committee

Krzysztof Apt (*Amsterdam*)
Hubert Comon (*Cachan*)
Anuj Dawar (*Cambridge*)
Thomas Ehrhard (*Marseille*)
Javier Esparza (*Stuttgart*)
Kousha Etessami (*Edinburgh*)
Erich Graedel (*Aachen*)
Robert Harper (*CMU*)
Johann Makowsky (*Technion*)
Jan Małuszyński (*Linkoping*)

Jerzy Marcinkowski (*Wrocław*, Chair)
Eugenio Moggi (*Geneva*)
Till Mossakowski (*Bremen*)
Hans de Nivelle (*Saarbrücken*)
Luke Ong (*Oxford*)
Andrew Pitts (*Cambridge*)
Michael Rusinowitch (*Nancy*)
David Schmidt (*Kansas*)
Jean-Marc Talbot (*Lille*)
Andrzej Tarlecki (*Warsaw*, Co-chair)

Referees

P. Abdullah
J.-M. Andreoli
J. Avigad
A. Avron
V. Balat
M. Bednarczyk
M. Benedikt
N. Benton
M. Bezem
P. Blain Levy
A.-J. Bouquet
O. Bournez
A. Bove
J. Bradfield
P. Bruscoli
A. Bucciarelli
W. Charatonik
B. Coecke
S. Conchon
V. Cortier
S. Dantchev

P. de Groote
D. Deharbe
G. Delzanno
N. Dershowitz
R. Di Cosmo
C. Dixon
W. Drabent
A. Edalat
A. Filinski
M. Fiore
C. Fuhrman
D. Galmiche
R. Gennari
G. Ghelli
D. Ghica
O. Grumberg
S. Guerrini
H. Herbelin
D. Hutter
N. Immerman
R. Impagliazzo

D. Janin
S. Kahrs
Y. Kazakov
E. Kieroński
C. Kirchner
B. König
B. Konikowska
J. Krajíček
S. Kreutzer
J.-L. Krivine
A. Kučera
G. Lafferriere
J. Laird
D. Larchey-Wendling
C. Lüth
M. Luttenberger
P. Madhusudan
C. Marché
J.-Y. Marion
R. Mayr
C. McBride

D. Miller	C. Ringeissen	S. Tison
O. Miś	E. Ritter	L. Tortora de Falco
M. Müller-Olm	M. Roggenbach	T. Touili
J. Niehren	G. Rosolini	T. Truderung
R. Nieuwenhuis	H. Ruess	J. Tyszkiewicz
U. Nilsson	C. Russo	C. Urban
D. Niwiński	A. Sabry	P. Urzyczyn
A. Nowack	H. Schlingloff	T. Uustalu
D. Nowak	R. Schmidt	L. Van Begin
N.N. Vorobjov	L. Schröder	J. van de Pol
J. Obdrzalek	A. Schubert	J. van Oosten
M. Otto	S. Schwoon	V. van Oostrom
L. Pacholski	N. Segerlind	R. Vestergaard
P. Panangaden	V. Simonet	M. Virgile
G. Pappas	K. Sohr	T. Walsh
W. Pawłowski	L. Straßburger	J. Wells
N. Peltier	T. Streicher	B. Werner
F. Pfenning	J. Stuber	T. Wierzbicki
L. Regnier	A. Szałas	S. Wöhrle
D. Richerby	H. Thielecke	M. Zawadowski

Table of Contents

Invited Lectures

Regular Papers

Notions of Average-Case Complexity for Random 3-SAT

Albert Atserias*

Universitat Politècnica de Catalunya, Barcelona, Spain

Abstract. By viewing random 3-SAT as a distributional problem, we go over some of the notions of average-case complexity that were considered in the literature. We note that for dense formulas the problem is polynomial-time on average in the sense of Levin. For sparse formulas the question remains widely open despite several recent attempts.

1 Introduction

The satisfiability problem for propositional logic is central to computational complexity. The work of Cook [2] showed that the problem is NP-complete, even when restricted to 3-CNF formulas, and is thus hard in the worst-case unless P = NP. Later on, the optimization versions of 3-SAT were also considered and showed hard. Namely, Hastad [4] culminated the monumental work of the 1990s on PCPs by showing that the number of clauses that can be satisfied simultaneously in a 3-CNF formula cannot be approximated within a ratio better than $7/8$ in polynomial-time, unless P = NP. The current decade is perhaps time for studying the *average-case complexity* of 3-SAT. Is it hard on average as well, unless P = NP, or is it easier?

The program comes also motivated from the fact that a fairly natural probability distribution on 3-CNF formulas has attracted the attention of many different communities, from AI to statistical physics, through combinatorics and mathematical logic. Our aim here is to review the background for a complexity-theoretic approach to the average-case complexity of random 3-SAT. In this short note we focus on the different definitions of average-case complexity that were introduced in the literature and their relationship. In the talk we will overview some of the partial results towards settling the main open questions.

2 Notions of Average Case Complexity

For every $n \geq 0$, let $f_n : \{0,1\}^n \to \{0,1\}$ be a Boolean function. In order to simplify notation, we will use f instead of f_n, and we will write $f = f_n$ to emphasize the fact that f is actually a sequence of functions parameterized by n. It will be understood from context that f denotes the sequence of functions $\{f_n\}$ in some cases, and the particular function f_n in others. We adopt the framework of *ensemble of distributions* suggested by Impagliazzo [5], where a different probability distribution is considered

---

* Supported in part by CICYT TIC2001-1577-C03-02 and the Future and Emerging Technologies programme of the EU under contract number IST-99-14186 (ALCOM-FT).

J. Marcinkowski and A. Tarlecki (Eds.): CSL 2004, LNCS 3210, pp. 1–5, 2004.

for each input size. So let $\mu = \mu_n$ be a probability distribution on $\{0,1\}^n$. We should think of μ as a sequence of distributions, one for each $n \geq 0$. The pair (f, μ) is called a *distributional problem*. Informally, the problem reads as follows: given a random input $x \in \{0,1\}^n$ drawn according to distribution μ_n, compute $f_n(x)$.

Levin's Average Case. Let (f, μ) be a distributional problem. Consider an algorithm A computing f, and let $T = T_n : \{0,1\}^n \rightarrow \mathbb{N}$ be its running time on inputs of length n. What should it mean that the running time T of A be polynomial on average with respect to μ? The obvious immediate candidate definition would be this: there exists a $k \geq 1$ such that $E_\mu[T] = O(n^k)$ where E_μ denotes expectation with respect to μ. Unfortunately, this simple definition suffers from a serious problem: the class of functions that are polynomial on average under this definition is not closed under polynomial images. Indeed, if we let μ be the uniform distribution on $\{0,1\}^n$, and let T be such that $T(x) = n$ for all but one string x_0 in $\{0,1\}^n$ for which $T(x_0) = 2^n$, then $E_\mu[T] = O(n)$ while $E_\mu[T^2] = \Omega(2^n)$. This lack of robustness would spoil any attempt to build a theory of polynomial reducibilities among distributional problems. A satisfactory remedy to this was discovered by Levin and reformulated by Impagliazzo for ensembles of distributions: we say that T *is polynomial on average with respect to μ* if there exists a $k \geq 1$ such that $E_\mu[T^{1/k}] = O(n)$. It is now immediate from this definition that the class of functions that is polynomial on average is closed under polynomial functions. We say that a distributional problem (f, μ) has a *polynomial-time on average algorithm* if there exists an algorithm A for f whose running time is polynomial on average with respect to μ.

Impagliazzo's Benign Algorithms. Let $f = f_n : \{0,1\}^n \rightarrow \{0,1\}$ be a Boolean function. A *prudent algorithm* for f is one that, for every input $x \in \{0,1\}^n$, outputs either $f(x)$ or ?. We should think of ? as a "don't know answer". Clearly, a prudent algorithm is useful only if it rarely outputs ?. We say that a distributional problem (f, μ) has a *polynomial-time benign algorithm* if there exists a prudent algorithm $A(x, \delta)$ for f that is polynomial-time in $|x|$ and $1/\delta$, and such that $\Pr_\mu[A(x, \delta) = ?] \leq \delta$ where $\Pr_\mu$ denotes probability with respect to μ. The last clause of this definition formalizes the idea that the algorithm "rarely outputs ?".

Impagliazzo [5] showed that the two notions introduced so far coincide, from which we conclude that the concept is fairly robust. We reproduce the proof since it is informative.

Theorem 1 (Impagliazzo). *Let (f, μ) be a distributional problem. Then, the following are equivalent:*

1. *(f, μ) has a polynomial-time on average algorithm.*
2. *(f, μ) has a polynomial-time benign algorithm.*

Proof: We start by 1. implies 2.: Let $E_\mu[T^{1/k}] \leq cn$. By Markov's inequality, we have $\Pr_\mu[T(x) > (tcn)^k] \leq 1/t$. Thus, for building a benign algorithm, it suffices to run the polynomial-time on average algorithm for $(cn/\delta)^k$ steps, and if it does not terminate, output ?. Next we show that 2. implies 1.: Suppose the benign algorithm runs in $(n/\delta)^k$ steps. Run the benign algorithm with parameter $\delta = 1/2, 1/4, 1/8, \ldots$ until an output different from ? is returned. Then, the expectation of the $2k$-th root of the running time of this algorithm is bounded by $(2n)^{1/2} + 1/2(4n)^{1/2} + 1/4(8n)^{1/2} + \cdots = O(n^{1/2})$ since at most $1/2$ of the inputs return ? in the first round, at most $1/4$ in the second round, and so on. □

Certification Algorithms. Let (f, μ) be a distributional problem. A *sound algorithm* for f is one that, for every input $x \in \{0, 1\}^n$, if it outputs 1, then indeed $f(x) = 1$. Clearly, a sound algorithm A is useful only if it outputs 1 on a large fraction of the "yes" instances, in other words, only if $|\Pr_\mu[f(x) = 1] - \Pr_\mu[A(x) = 1]|$ is small. In such cases we say that it is *almost complete*. We say that a distributional problem (f, μ) has a *polynomial-time certification algorithm* if there exists a sound algorithm $A(x, \delta)$ for f that is polynomial in $|x|$ and $1/\delta$, and such that $|\Pr_\mu[f(x) = 1] - \Pr_\mu[A(x) = 1]| \leq \delta$. The last clause of this definition formalizes the idea of almost completeness. The relationship is now very easy to see:

Lemma 1. *Let (f, μ) be a distributional problem. Then, if (f, μ) has a polynomial-time benign algorithm, then (f, μ) has a polynomial-time certification algorithm.*

Proof: Let $A(x, \delta)$ be the benign algorithm. Consider the following algorithm $B(x, \delta)$: run the benign algorithm $A(x, \delta)$, and if it outputs ?, output 0. Clearly, $B(x, \delta)$ is sound. Moreover, by soundness, we have $|\Pr_\mu[f(x) = 1] - \Pr_\mu[B(x, \delta) = 1]| = \Pr_\mu[f(x) \neq B(x, \delta)]$ which in turn is bounded by $\Pr_\mu[A(x, \delta) = ?] \leq \delta$. □

If we put Theorem 1 and Lemma 1 together we see that if (f, μ) has a polynomial-time on average algorithm, then (f, μ) has a polynomial-time certification algorithm. In the contrapositive form, if (f, μ) is hard to certify, then (f, μ) is hard on average. Although we do not know whether the converse relationship holds in general, we note in the next section that it holds for the particular case of random 3-SAT.

3   Random 3-SAT

Let $x_1, \ldots, x_n$ be n propositional variables. A literal is a variable or its negation. A 3-clause is a tuple of three literals. A 3-CNF formula is a set of 3-clauses. Note that the number of 3-clauses is exactly $(2n)^3$. Thus, a 3-CNF formula can be encoded by a binary string of length $(2n)^3$ denoting which clauses are present and which are not.

There are several probability distributions that have been considered in the literature. The one we adopt here is inspired from the theory of random graphs. The distribution $\mu = \mu_n$ is parameterized by a real number $p = p_n$ in $(0, 1)$ and consists in choosing each clause with independent probability p. This probability model is sometimes referred to as the model A. The model B considers the number of clauses m as fixed and chooses the formula uniformly within that set. Both these models have several variants according to whether clauses are ordered tuples or sets, and may, or may not, have repeated and complementary literals. As in the random graph model, which model to use is often a matter of convenience, and rarely an important issue as far as the results are concerned.

We are interested in the distributional problem $(UNSAT, \mu)$, where $UNSAT = UNSAT_n$ is simply the unsatisfiability problem on 3-CNF formulas with n variables, and $\mu = \mu_n$ is the probability distribution that we just described. Notice that here n is not exactly the length of the input, but it is polynomially related. Notice also that μ is parameterized by $p = p_n$, and the complexity of the distributional problem may very well depend on p. As a matter of fact, when p is large, it can be seen that $(UNSAT, \mu)$ has a benign polynomial-time algorithm. Before proving that, we first show that for all values of p that guarantee unsatisfiability of a random formula with overwhelming probability, the three notions of average-case complexity considered so far coincide.

Theorem 2. *Let $p \geq (\ln 2 + \epsilon)/n^2$, with $\epsilon > 0$. Then, the following are equivalent:*

1. *$(UNSAT, \mu)$ has a polynomial-time on average algorithm.*
2. *$(UNSAT, \mu)$ has a polynomial-time benign algorithm.*
3. *$(UNSAT, \mu)$ has a polynomial-time certification algorithm.*

Proof: By Theorem 1 and Lemma 1, it suffices to show that 3. implies 1. Let $A(x, \delta)$ be the certification algorithm. Assume its running time is $(n/\delta)^k$. Consider the following algorithm. Run the certification algorithm with parameter $\delta = 1/2, 1/4, 1/8, \ldots$ until either "unsatisfiable" is returned, in which case we return "unsatisfiable" as well, or the parameter becomes smaller than 2^{-n}, in which case we run through all 2^n truth assignments to check whether F is satisfiable or not. By soundness of the certification algorithm, it is clear that this algorithm is correct. Let us estimate the expectation of the r-th root of its running time for a constant r to be determined later.

When $p \geq (\ln 2 + \epsilon)/n^2$, the probability that a random formula is satisfiable is $2^{-\gamma n}$ for some constant $\gamma > 0$, as it is easy to see. Let us consider the set of "satisfiable" instances. For those instances, the running time of the algorithm can only be bounded by

$$\sum_{i=1}^{n} (2^i n)^k + 2^{cn}$$

for some constant $c \geq 1$, which is time 2^{dkn} for some other constant $d \geq c$. Hence, the "satisfiable" instances contribute to the expectation of the r-th root of the running time by at most $2^{-\gamma n} 2^{dkn/r}$. Let us now turn to the contribution to the expectation of the "unsatisfiable" instances. The expectation of the r-th root of the running time for those instances is bounded by

$$(2n)^{k/r} + 2^{-1}(4n)^{k/r} + 2^{-2}(8n)^{k/r} + \cdots + 2^{-n+1}(2^n n)^{k/r} + 2^{-n}(2^{cn})^{1/r}$$

since at most $1/2$ of the "unsatisfiable" instances miss the first round, at most $1/4$ of those miss the second round, and so on, until at most $1/2^n$ of the instances miss all rounds in which case the algorithm goes on to cycle through all 2^n truth assignments in time 2^{cn}. It is now straightforward to see that if we take r large enough, say $r > dk/\gamma$, then the total expectation of the r-root of the running time is $O(n)$. $\square$

In general, the proof technique of this result applies to any distributional problem for which the "no" instances represent a fraction that is inversely polynomial with respect to the worst-case running time that it is required to solve the problem. Let us conclude this paper with the promised benign algorithm when p is large. The reader will notice that the proof below resembles the arguments in [1].

Theorem 3. Let $p = \omega(1/n)$. Then $(UNSAT, \mu)$ has a polynomial-time benign algorithm.

Proof Sketch: Consider the following algorithm. Let F be the input 3-CNF formula and let δ be the error parameter. Let $\gamma > 0$ be a small constant to be determined later. If $\delta < 2^{-\gamma n}$, simply run through all 2^n truth assignments to check whether F is satisfiable or not. If $\delta \geq 2^{-\gamma n}$, find the most popular variable x in F. Consider the set of 2-clauses in $F|_{x=0}$ and $F|_{x=1}$, and run a polynomial-time 2-SAT algorithm on the resulting 2-CNF formulas. If both are unsatisfiable, report "unsatisfiable". Otherwise, output ?.

It should be clear from its definition that the algorithm is prudent. It is also clear that the running time of the algorithm is polynomial in n and $1/\delta$. Indeed, when $\delta < 2^{-\gamma n}$, the running time is $2^{O(n)}$ which is polynomial in $1/\delta$, and in the other case the running time is polynomial in n. Let us argue that the probability that it outputs ? is smaller than δ. When $\delta < 2^{-\gamma n}$, the algorithm never outputs ?. So let us assume that $\delta \geq 2^{-\gamma n}$. Each variable appears in $\Theta(n^2)$ clauses. Hence, the expected number of occurrences of each variable is $\Theta(n^2 p)$, which is $\omega(n)$ since $p = \omega(1/n)$. It follows from concentration bounds that the probability that a particular variable appears less than half this number of times is $e^{-\omega(n)}$. Thus, by Markov's inequality, the probability that some variable appears $\omega(n)$ times is at least $1 - ne^{-\omega(n)}$. The number of 2-clauses in $F|_{x=0}$ and $F|_{x=1}$ is thus $\omega(n)$ with at least that much probability. Moreover, the resulting 2-CNF formulas are random, so the probability that one of them is satisfiable is bounded by $2^{-\Omega(n)}$, as is well-known. All in all, the probability that the algorithm does not report "unsatisfiable" is bounded by $2^{-\Omega(n)}$. Thus, the probability that the algorithm outputs ? is bounded by δ since $\delta \geq 2^{-\gamma n}$. Here γ is chosen to be the hidden constant in the $2^{-\Omega(n)}$ bound. □

It follows from this result and Theorem 1 that when $p = \omega(1/n)$, the distributional problem $(UNSAT, \mu)$ is solvable in polynomial-time on average in the sense of Levin. For $p = O(1/n^{3/2-\epsilon})$, recent work has focused on certification algorithms [3]. For $p = O(1/n^2)$, the problem is widely open.

References

1. P. Beame, R. Karp, T. Pitassi, and M. Saks. The efficiency of resolution and Davis-Putnam procedures. *SIAM Journal of Computing*, pages 1048–1075, 2002.
2. S. Cook. The complexity of theorem proving procedures. In *3rd Annual ACM Symposium on the Theory of Computing*, pages 151–158, 1971.
3. J. Friedman and A. Goerdt. Recognizing more unsatisfiable random 3-SAT instances efficiently. In *28th International Colloquium on Automata, Languages and Programming*, volume 2076 of *Lecture Notes in Computer Science*, pages 310–321. Springer-Verlag, 2001.
4. J. Hastad. Some optimal inapproximability results. *Journal of the ACM*, 48:798–859, 2001.
5. R. Impagliazzo. A personal view of average-case complexity. In *10th IEEE Structure in Complexity Theory*, pages 134–147, 1995.

Abstract Interpretation of Proofs: Classical Propositional Calculus

Martin Hyland

DPMMS, Centre for Mathematical Sciences,
University of Cambridge, England

Abstract. Representative abstract interpretations of the proof theory of the classical propositional calculus are described. These provide invariants for proofs in the sequent calculus. The results of calculations in simple cases are given and briefly discussed.

Keywords: classical propositional calculus, proof theory, category theory.

1   Introduction

1.1   Background

The Curry-Howard isomorphism suugests the connection between proofs in intuitionistic propositional logic, simply typed lambda calculus and cartesian closed categories. This set of ideas provides a context in which constructive proofs can be analysed in a direct fashion. For a treatment in which the category theoretic aspect does not dominate see [13]. By contrast analyses of classical proof theory tend to be indirect: typically one reduces to the contructive case via some form of double-negation translation. (Of course there is also work constructing measures of complexity of classical proofs, but that is not a structural analysis in the sense that there is one for constructive proofs.)

In [16], I sketched a proposal to analyse classical proofs in a direct fashion with the intention inter alia of providing some kind of Curry-Howard isomorphism for classical proof. This is currently the focus of an EPSRC project with principals Hyland (Cambridge), Pym (Bath) and Robinson (Queen Mary). Developments have been interesting. While we still lack natural mathematical semantics for an analysis along the lines of [16], the flaws in the detail proposed there are now ironed out (see [1]). The proof net proposal of Robinson [30] was a response to the difficulties of that approach; it has been considered in depth by Fürhmann and Pym [11]. This leads to more familiar semantics and we have a clear idea as to how this resulting semantics departs from the conception of proof embodied in the sequent calculus. But we are far from understanding the full picture.

One motivation for the project on classical proof was a desire for a systematic approach to the idea of invariants of proofs more flexible than that of complexity analyses. In this paper I try further to support this basic project by describing two abstract interpretations of the classical propositional calculus. One should

J. Marcinkowski and A. Tarlecki (Eds.): CSL 2004, LNCS 3210, pp. 6–21, 2004.

regard these as akin to the abstract interpretations used in strictness analysis. The point is to define and compute interesting invariants of proofs. The abstract interpretations considered here are intended to be degenerate in the same way that the (so-called) relational model is a degenerate model of Linear Logic. There the tensor and par of linear logic are identified; our abstract interpretations identify classical conjunction and disjunction. (The notion of degenerate model for Linear Logic is discussed at greater length in [19].)

In joint work with Power I have tried to put the theory of abstract interpretations on a sound footing. That involves categorical logic of a kind familiar to rather few, so here I leave that aside and simply consider some case studies in the hope of provoking interest. These cases studied can be regarded as representative: they arise from free constructions of two different kinds. I give some calculations (in a bit of a rush - I hope they are right) but do not take the analysis very far. A systematic study even of the interpretations given here would be a major undertaking; but the calculations complement those in [12] which are for special cases of the second class of interpretations considered here.

One should observe that in this paper I get nowhere near the complexities considered by Carbone in [4], [5] and [6]. Carbone's work can also be regarded as a study of abstract interpretations: it is nearest to being a precursor of the approach taken here.

I hope that by and large the notation of the paper will seem standard. However I follow some computer science communities by using diagrammatic motation for composition:

$$f : A \longrightarrow B \quad \text{and} \quad g : B \longrightarrow C$$

compose to give

$$f; g : A \longrightarrow C.$$

1.2 Abstract Interpretations

We start with some general considerations concerning the semantics of proofs in the sequent calculus for the classical propositional calculus. The basic idea, which goes back to Szabo, is to take the CUT rule as giving the associative composition in some polycategory. If we simplify (essentially requiring representability of polymaps) along the Fürhmann-Pym-Robinson lines we get the following.

Definition 1. *A* model of classical propositional proof satisfying the Fürhmann-Pym-Robinson equations *consists of the following data.*

- *A $*$-autonomous category $\mathbb{C}$: here the tensor is the conjunction $\wedge$, and its unit is the true $\top$; dually the par is the disjunction $\vee$, and its unit is the false $\bot$.*
- *The equipment on each object A of $\mathbb{C}$ of the structure of a commutative comonoid with respect to tensor.*
- *The equipment on each object A of $\mathbb{C}$ of the structure of a commutative monoid with respect to par.*

One requires in addition

- *that the commutative comonoid structure is compatible with the tensor structure (so ⊤ has the expected comonoid structure and comonoid structure is closed under ∧);*
- *that the commutative monoid structure is compatible with the par structure (so ⊥ has the expected monoid structure and monoid structure is closed under ∨);*
- *and that the structures correspond under the duality (so that the previous two conditions are equivalent).*

There are further categorical nuances which we do not discuss here.

The interpretation of classical proofs in such a structure is a straightforward extension of the interpretation of multiplicative linear proofs in a ∗-autonomous category. The algebraic structure deals with the structural rules of the sequent calculus. The several requirements added are natural simplifying assumptions. They do not really have much proof theoretic justification as things stand.

As indicated above we take a notion of abstract interpretation which arises by the identification of the conjunction ∧ and disjunction ∨.

Definition 2. *By an* abstract interpretation of classical proof *we mean a compact closed category in which each object A is equipped with*

- *the structure $t : A \to I, d : A \to A \otimes A$ of a commutative comonoid,*
- *the structure $e : I \to A, m : A \otimes A \to A$ of a commutative monoid,*

with the structures

- *compatible with the monoidal structure $(I, \otimes)$, and*
- *and interchanged under the duality $(-)^*$.*

One should note that the optical graphs of Carbone [4] are in effect abstract interpretations, but in a more general sense than that considered here.

We gloss the definition a little. According to it, each object is equipped with commutative monoid structure to model the structural rules for ∨ and with commutative comonoid structure to model the structural rules for ∧. Naturally we expect the structural rules to be interchanged by the duality $(-)^*$. Modulo natural identifications we have

$$(t_A)^* = e_{A^\square} : I \to A^*, \qquad (e_A)^* = t_{A^\square} : A^* \to I,$$
$$(d_A)^* = m_{A^\square} : A^* \otimes A^* \to A^*, \quad (m_A)^* = d_{A^\square} : A^* \to A^* \otimes A^*.$$

In addition we ask that the structure be compatible with the monoidal structure. This means first that I should have the expected structure

$$t_I = \mathrm{id}_I : I \to I, \ d_I = \tilde{l}_I = \tilde{r}_I : I \to I \otimes I,$$
$$e_I = \mathrm{id}_I : I \to I, \ m_I = l_I = r_I : I \otimes I \to I,$$

derived from the unit structure

$$l_A : I \otimes A \to A \ r_A : A \otimes I \to A$$
$$\tilde{l}_A : A \to I \otimes A \ \tilde{r}_A : A \to A \otimes I$$

for the tensor unit I. In addition it means that the structures are preserved by tensor: that is, modulo associativities we have

$$d_{A \otimes B} = d_A \otimes d_B; \mathrm{id}_A \otimes c_{A,B} \otimes \mathrm{id}_B : A \otimes B \to A \otimes B \otimes A \otimes B,$$
$$m_{A \otimes B} = \mathrm{id}_A \otimes c_{A,B} \otimes \mathrm{id}_B; m_A \otimes m_B : A \otimes B \otimes A \otimes B \to A \otimes B.$$

For the moment it is best to regard these requirements as being justified by the models which we are able to give.

1.3   Strictness

Any honest consideration of categorical structure should address questions of strictness. In particular one has the distinction between functors preserving structure on the nose and functors preserving structure up to (coherent) natural isomorphism. A setting in which such issues can be dealt with precisely is laid out in [2]. The only issue which need concern us here is that of the strictness of the structure in our definition of abstract interpretation.

We shall largely deal with structures freely generated by some data. So it will be simplest for us to take the monoidal structure to be strictly associative. Similarly we shall be able to take the duality to be strictly involutive so that

$$(f : A \to B)^{**} = (f : A \to B)$$

and to respect the monoidal structure, so that on objects

$$I^* = I \quad \text{and} \quad (A \otimes B)^* = B^* \otimes A^*$$

on the nose, and similarly for maps. Note further that duality in a compact closed category provides adjunctions for all the 1-cells of the corresponding one object bicategory. That is very much choice of structure: so for us every object A is equipped with a left (say) dual A^* with explicit unit and counit

$$I \longrightarrow A \otimes A^* \quad \text{and} \quad A^* \otimes A \longrightarrow I.$$

This is all as explained in [21]. But one should go further: in general there should be natural coherence diagrams connecting the adjunction for $A \otimes B$ with the adjunctions for A and B. (In a sense these conditions parallel the assumption that the comonoid and monoid structures are preserved under tensor product. The relevent coherence theorem extending [21] is not in principle hard, but we do not need it here.)

For the purposes of this paper one can take the definition of abstract interpretation in the strict sense indicated. But not much depends on that: the critical issue for the background theory is simply that the notion is given by algebraic structure over Cat in the sense of [22]. A reader to whom all this is foreign should still understand the examples.

1.4   Miscellaneous Examples

Before turning to the interpretations which are our chief concern, we give a few natural examples of abstract interpretations. For this section we ignore strictness issues.

1. Consider the category **Rel** of sets and relations equipped with the set theoretic product $\times$ as tensor product. **Rel** is compact closed; it is contained in the compact closed core of **SupLat** the category of complete lattices and sup-preserving maps. The duality is

$$(-)^* \quad : (A \xrightarrow{F} B) \longrightarrow (B \xrightarrow{F^{op}} A)$$

in particular is the identity on objects. For each object $A \in$ **Rel** there is a natural choice of commutative comonoid structure arising from product in **Set**. By duality that gives a choice of commutative monoid structure on all objects, and by definition the structures are interchanged by the duality. This gives a simple abstract interpretation.

2. We can extend the above example to one fundamental to Winskel's Domain Theory for Concurrency (see [27] for example). Following [27] write **Lin** (after Linear Logic) for the category with objects preordered sets and maps profunctors between them. We can regard this also as being within the compact closed core of **SupLatt**. We equip the preordered set $\mathbb{P}$ with comonoid structure via the counit

$$t_{\mathbb{P}} : \mathbb{P} \longrightarrow 1 \qquad t_{\mathbb{P}}(a, \star) = \text{true}$$

and the comultiplication

$$d_{\mathbb{P}} : \mathbb{P} \longrightarrow \mathbb{P} \times \mathbb{P} \qquad d_{\mathbb{P}}(a, (b, c)) = a \geq b \text{ and } a \geq c,$$

extending the definition for **Rel**. The duality is

$$(-)^* : (\mathbb{P} \xrightarrow{F} \mathbb{Q}) \longrightarrow (\mathbb{Q}^{op} \xrightarrow{F^{op}} \mathbb{P}^{op})$$

and this is no longer the identity on objects. The duality induces the monoid structure from the comonoid structure so the structures are automatically interchanged by duality.

3. Let **FVec** be the category of finite dimensional k-vector spaces (and k-linear maps) for a field k. A *commutative bialgebra* is an object A of **FVec** equipped with the structure

$$t : A \rightarrow I, \quad d : A \rightarrow A \otimes A$$

of a commutative comonoid and the structure

$$e : I \rightarrow A, m : A \otimes A \rightarrow A$$

of a commutative monoid, satisfying the equations

$$e_A; t_A = \mathrm{id}_I$$
$$m_A; t_A = t_A \otimes t_A; m_I$$
$$e_A; d_A = d_I; e_A \otimes e_A$$
$$m_A; d_A = d_A \otimes d_A; \mathrm{id}_A \otimes c_{A,A} \otimes \mathrm{id}_A; m_A \otimes m_A$$

(Hopf algebras (that is, bialgebras with antipode) are amongst the staples of representation theory. There is a plentiful supply of such: the standard group algebra kG of a finite group G is a Hopf algebra.) If we take the category whose objects are bialgebras, with maps linear maps of the underlying vector spaces, we get an abstract interpretation in our sense.

2   Frobenius Algebras

2.1   Frobenius Abstract Interpretations

Definition 3. *A commutative Frobenius algebra in a symmetric monoidal category is an object A equipped with the structure $A, t : A \to I, d : A \to A \otimes A$ of a commutative comonoid and the structure $A, e : I \to A, m : A \otimes A \to A$ of a commutative monoid, satisfying the equation*

$$A \otimes d; m \otimes A = m; d = d \otimes A; A \otimes m.$$

Note that an algebra is a module over itself (on the left and on the right), and a coalgebra a comodule over itself (again on both sides). We can write the Frobenius equations in diagrams as

$$
\begin{array}{ccccc}
A \otimes A \otimes A & \xleftarrow{\ d \otimes A\ } & A \otimes A & \xrightarrow{\ A \otimes d\ } & A \otimes A \otimes A \\[2mm]
\Big\downarrow{\scriptstyle A \otimes m} & & \Big\downarrow{\scriptstyle m} & & \Big\downarrow{\scriptstyle M \otimes A} \\[2mm]
A \otimes A & \xleftarrow{\ \ d\ \ } & A & \xrightarrow{\ \ d\ \ } & A \otimes A
\end{array}
$$

and we see that they say that d is a map of right and left modules. Equivalently (and by symmetry) they say that m is a map of right and left comodules.

In mathematics algebras with a Frobenius structure have played a role in representation theory for a century, certainly since Frobenius [10]. The condition is explicitly identified in work of T. Nakayama and C. Nesbitt from the late 1930s. Sources for this early history are mentioned in [23]. An important conceptual understanding of the Frobenius condition or structure was suggested by Lawvere [24].

Definition 4. *We say that an abstract interpretation is Frobenius if all the comonoid and monoid structures satisfy the equations of a Frobenius algebra.*

We note that for an object A in **Rel** one readily calculates that

$$A \otimes d; m \otimes A = m; d = d \otimes A; A \otimes m$$

is the relation $A \times A \to A \times A$ identifying equal elements of the diagonal. So **Rel** is a Frobenius abstract interpretation. In view of remarks below, one could regard this as explaining why the objects in **Rel** are self-dual! One the other hand it is easy to see that the Frobenius condition fails for **Lin**. This is related to the fact that we do not generally have $\mathbb{P}^{op} \cong \mathbb{P}$ for posets $\mathbb{P}$.

Since the comonoid structure of a Frobenius algebra is not natural with respect to the monoid structure (and dually not vice-versa either), we are not dealing with a commutative sketch in the sense of [17]: rather one needs the more general theory of [18]. As a consequence the identification of the free Frobenius algebra, given in the next section, is non-trivial. However a simplifying feature of Frobenius algebras is that they carry their own duality with them. In fact Frobenius algebras are self dual: one has the unit

$$I \xrightarrow{\ e_A\ } A \xrightarrow{\ d_A\ } A \otimes A\,,$$

and the counit

$$A \otimes A \xrightarrow{\ m_A\ } A \xrightarrow{\ t_A\ } I\,.$$

By straightforward calculation one has

$$
\begin{aligned}
(e_A; d_A) \otimes \mathrm{id}_A; \mathrm{id}_A \otimes (m_A; t_A) &= e \otimes \mathrm{id}_A; d \otimes \mathrm{id}_A; \mathrm{id}_A \otimes m; \mathrm{id}_A \otimes t \\
&= e \otimes \mathrm{id}_A; m; d; \mathrm{id}_A \otimes t \\
&= \mathrm{id}_A; \mathrm{id}_A = \mathrm{id}_A
\end{aligned}
$$

giving one of the triangle identities; And symmetrically one has

$$\mathrm{id}_A \otimes (e_A; d_A); (m_A; t_A) \otimes \mathrm{id}_A$$

which is the other. This shows that in any symmetric monoidal closed category the Frobenius objects live in the compact closed core. Moreover it is easy to see that the intrinsic duality interchanges the comonoid and monoid structures on a Frobenius algebra. So the abstract interpretation aspect is also automatic. So overall to give abstract interpretations it suffices to find Frobenius algebras in some symmetric monoidal closed category.

2.2   The Free Frobenius Algebra

In recent times the study of Frobenius algebras has become compelling following the identification of 2-dimensional Topological Quantum Field Theories (TQFT) with commutative Frobenius algebras [9]. A readable intuitive explanation is given in [23]. In essence this arises from an identification of the free symmetric monoidal category generated by a Frobenius algebra. We state this in the customary rough and ready way: though we make some of the ideas more precise in a moment, there is a limit to what it is useful to do here.

Proposition 1. *The free symmetric monoidal category generated by a Frobenius algebra can be described in the following equivalent ways.*

1. It is the category of topological Riemann Surfaces. Objects are finite disjoint sums of the circle and maps (from n circles to m circles) are homeomorphism classes of surfaces with boundary consisting of $n + m$ marked circles.

2. It is the category of one dimensional finitary topology up to homology. The objects are finite discrete sets of points and the maps from n points to m points are homology classes of one dimensional simplicial complexes with $n + m$ marked points as boundary.

To make things more precise, we might as well engage at once with the strict version of the above. In that view the free symmetric monoidal category generated by a Frobenius algebra has objects

$$0, 1, 2, ..., n, ...$$

which should be regarded as representatives of finite sets. The maps from n to m are determined by an equivalence relation on $n + m$, which one can think of as giving connected components topologically, together with an association to each of these connected components of a natural number (the genus).

Dijkgraaf's identification of two-dimensional TQFT has been independentently established more or less precisely by a number of people. I not unnaturally like the account in Carmody [7] which already stresses the wiring diagrams in the sense of [17] and [18], as well as rewriting in the style of the identification of the simplicial category by generators and relations [26]. We shall show that the TQFT aspect of Frobenius algebras runs parallel to a simple topological idea of abstract interpretation.

2.3   Representative Calculations

We consider here the obvious interpretation of classical proofs in the symmetric monoidal category generated by a Frobenius algebra. In this interpretation all atomic propositions are interpreted by the generating Frobenius algebra, so are not distinguished. Also the interpretation is not sensitive to negation. Despite that the intrepretation does detect some structural features of proofs. We already explained the data for a map in the free symmetric monoidal category generated by a Frobenius algebra: it consists of a collection of connected components and a genus attached to each. A proof π in classical propositional logic gives rise to its interpretation $V(\pi)$ which is thus a map of this kind. One loses just a lttle information if one considers only the invariants given by the homology

$$H_0(\pi) = H_0(V(\pi), \mathbb{Q}) \quad \text{and} \quad H_1(\pi) = H_1(V(\pi), \mathbb{Q})$$

of a proof π. We set

$$h_0(\pi) = \dim H_0(\pi) \quad \text{and} \quad h_1(\pi) = \dim H_1(\pi).$$

Usually we ensure there will be just one connected component so that one loses no information in passing to homology: the invariant reduces to the genus usually written $g = h_1(\pi)$.

1. **Proofs in MLL.** The Frobenius algebra interpretation of proofs in Multiplicative Linear Logic should be regarded as one of the fundamentals of the subject. For simplicity we follow [14] in dealing with a one-sided sequent calculus.

   - For an axiom

$$A \vdash^{\alpha} A$$

   we have

$$h_0(\alpha) = 1 \quad \text{and} \quad h_1(\alpha) = 0 .$$

   - For the ∧-R rule

$$\frac{\vdash^{\pi_1} \Gamma, A \quad \vdash^{\pi_2} \Delta, B}{\vdash^{\pi} \Gamma, \Delta, A \wedge B}$$

   we have

$$h_0(\pi) = h_0(\pi_1) + h_0(\pi_2) + 1 \quad \text{and} \quad h_1(\pi) = h_1(\pi_1) + h_1(\pi_2) .$$

   - For the ∨-L rule

$$\frac{\vdash^{\pi^{\bullet}} \Gamma, A, B}{\vdash^{\pi} \Gamma, A \vee B}$$

   we have

$$h_0(\pi) = h_0(\pi') \quad \text{and} \quad h_1(\pi) = h_1(\pi') + 1 .$$

   The final claims need to be justified inductively using the fact that we always have one connected component, that is, we always have $h_0 = 1$.

   We deduce from the above that for a proof π, in multiplicative linear logic, the genus counts the number of pars (that is for us occurences of ∨) in the conclusions. Thus the Frobenius algebra interpretation points towards the Danos-Regnier correctness criterion. (My student Richard Garner has given a full analysis along these lines.)

2. **Distributive Law.** Perhaps the simplest interesting non-linear proofs are those of the distributive laws. Consider first the proof

$$\frac{\dfrac{\dfrac{A \vdash A \quad B \vdash B}{A, B \vdash A \wedge B} \quad \dfrac{A \vdash A \quad C \vdash C}{A, C \vdash A \wedge C}}{\dfrac{A, A, B \vee C \vdash A \wedge B, A \wedge C}{\dfrac{A, B \vee C \vdash A \wedge B, A \wedge C}{A \wedge (B \vee C \vdash (A \wedge B) \vee (A \wedge C)}}}}{}$$

From the proof net one readily sees that one has $h_0 = 1$, that is one has one connected component, and that $h_1 = 3$. There are just two occurences of ∨, so a linear proof would have $h_1 = 2$. Thus the invariant does detect the non-linearity.

I do not give here the most natural proof of the converse distributive law: $(A \wedge B) \vee (A \wedge C) \vdash A \wedge (B \vee C)$. It seems necessarily more complex, in that the most obvious proof has $h_0 = 1$ but $g = h_1 = 8$. The proof does not just reverse.

3. **The Natural Numbers.** Recall that (up to $\beta\eta$-equivalence) the constructive proofs of $(A \Rightarrow A) \Rightarrow (A \Rightarrow A)$ correspond to the Church numerals in the lambda calculus. (Implicitly we should rewrite $(A \Rightarrow A) \Rightarrow (A \Rightarrow A)$ as $(A \wedge \neg A) \vee (\neg A \vee A)$; but there are obvious corresponding proofs.) Let π_n be the proof given by the nth Church numeral $\lambda f, x. f^n(x)$. We compute the invariants for these proofs π_n. With just one conclusion we have forced $h_0(\pi_n) = 1$, so we just look at the genus. The proof net picture immediately show us that $h_1(\pi_n) = n + 1$. So the invariant readily distingushes these proofs.

Now consider what it is to compose proofs π_n and π_m with the proof $\mu = \lambda a, b, f. a(b(f))$ of

$$(A \Rightarrow A) \Rightarrow (A \Rightarrow A), (A \Rightarrow A) \Rightarrow (A \Rightarrow A) \vdash (A \Rightarrow A) \Rightarrow (A \Rightarrow A)$$

corresponding to multiplication on the Church numerals. This gives a proof $\mu|\pi_n|\pi_m$ with cuts. We compute the invariants for the interpretation $V(\mu|\pi_n|\pi_m)$ in our model. This is just an exercise in counting holes. Generally we find that

$$h_0(\mu|\pi_n|\pi_m) = 1 \quad \text{and} \quad h_1(\mu|\pi_n|\pi_m) = n + m.$$

However the case $n = m = 0$ is special. We get

$$h_0(\mu|\pi_0|\pi_0) = 2 \quad \text{and} \quad h_1(\mu|\pi_0|\pi_0) = 1.$$

(Note that the Euler characteristic is consistent!)
Of course if we reduce $\mu|\pi_n|\pi_m$ to normal form we get π_{nm} with

$$h_0(\pi_{nm}) = 1 \quad \text{and} \quad h_1(\pi_{nm}) = nm + 1.$$

So the interpretation distinguishes proofs from their normal forms. The need to think this way about classical proof was stressed in [16].

3   Traced Monoidal Categories

3.1   Background

With our Frobenius Algebra interpretation we got the compact closed aspect of our abstract interpretation for free. For our second example we exploit a general method for contructing compact closed categories from traced monoidal categories. We recall the basic facts concerning traced monoidal categories. We do not need the subtleties of the braided case explained in the basic reference [20]. So for us a *traced monoidal category* is a symmetric monoidal category equipped with a trace operation

$$\frac{f : A \otimes U \to B \otimes U}{\operatorname{tr}(f) : A \to B}$$

satisfying elementary properties of feedback. A useful perspective and diagrams without the braidings in [20] is provided by Hasegawa [15]. It is a commonplace amongst workers in Linear Logic that traced monoidal categories provide a backdrop to Girard's Geometry of Interaction.

If $\mathbf{C}$ is a traced monoidal category, then its integral completion $\mathrm{Int}(\mathbf{C})$ is defined as follows.

– The objects of $\mathrm{Int}(\mathbf{C})$ are pairs (A_0, A_1) of objects of $\mathbf{C}$.
– Maps $(A_0, A_1) \to (B_0, B_1)$ in $\mathrm{Int}(\mathbf{C})$ are maps $A_0 \otimes B_1 \to B_0 \otimes A_1$ of $\mathbf{C}$.
– Composition of $f : (A_0, A_1) \to (B_0, B_1)$ and $g : (B_0, B_1) \to (C_0, C_1)$ is given by taking the trace $\mathrm{tr}(\sigma; f \otimes g; \tau)$ of the composite of $f \otimes g$ with the obvious symmetries

$$A_0 \otimes C_1 \otimes B_0 \otimes B_1 \xrightarrow{\sigma} A_0 \otimes B_1 \otimes B_0 \otimes C_1 \,,$$

and

$$B_0 \otimes A_1 \otimes C_0 \otimes B_1 \xrightarrow{\tau} C_0 \otimes A_1 \otimes B_0 \otimes B_1 \,.$$

– Identities $(A_0, A_1) \to (A_0, A_1)$ are given by the identity $A_0 \otimes A_1 \to A_0 \otimes A_1$.

The basic result from [20] is the following.

Theorem 1. *(i) Suppose that $\mathbf{C}$ is a traced monoidal category. Then $\mathrm{Int}(\mathbf{C})$ is a compact closed category.*
(ii) Int extends to a 2-functor left biadjoint to the forgetful 2-functor from compact closed categories to traced monoidal categories.

3.2   Abstract Interpretations via Traces

Suppose that we have a traced monoidal category $\mathbf{C}$ in which every object A is equipped with the structure of a commutative comonoid

$$I \xleftarrow{w} A \xrightarrow{d} A \otimes A$$

and of a commutative monoid

$$I \xrightarrow{e} A \xleftarrow{m} A \otimes A \,.$$

Consider the compact closed category $\mathrm{Int}(\mathbf{C})$. Given an object (A_0, A_1), we have maps

$$(A_0, A_1) \longrightarrow (I, I) \quad \text{given by} \quad A_0 \otimes I \xrightarrow{w \otimes e} I \otimes A_1$$

$$(A_0, A_1) \longrightarrow (A_0 \otimes A_0, A_1 \otimes A_1) \quad \text{given by} \quad A_0 \otimes A_1 \otimes A_1 \xrightarrow{d \otimes m} A_0 \otimes A_0 \otimes A_1$$

which clearly equip it with the structure of a commutative comonoid; and dually we have maps

$$(I, I) \longrightarrow (A_0, A_1) \quad \text{given by} \quad I \otimes A_1 \xrightarrow{e \otimes w} A_0 \otimes I$$

$$(A_0 \otimes A_0, A_1 \otimes A_1) \longrightarrow (A_0, A_1) \quad \text{given by} \quad A_0 \otimes A_0 \otimes A_1 \xrightarrow{m \otimes d} A_0 \otimes A_1 \otimes A_1$$

which equip it with the structure of a commutative monoid. These structures are manifestly interchanged (on the nose) by the duality. Thus such situations will always lead to abstract interpretations.

3.3   Traced Categories with Biproducts

We consider the special case where the tensor product in a traced monoidal category is a biproduct (see for example [25]). Under these circumstances one has a canonical choice of commutative comonoid and monoid structure, and so a natural abstract interpretation.

We recall that a category $\mathcal{C}$ with biproducts is enriched in commutative monoids. More concretely each hom-set $\mathcal{C}(A, B)$ is equipped with the structure of a commutative monoid (which we write additively) and composition is bilinear in that structure. It follows that for each object A its endomorphisms $\mathrm{End}_{\mathcal{C}}(A) = \mathcal{C}(A, A)$ has the structure of what is now called a rig, that is to say a (commutative) ring without negatives. One can explain in these terms what it is to equip a category with biproducts with a trace. Here we concentrate on the one object case, which is the only case considered in the main reference [3].

We recall the notion of Conway Algebra (essentially in Conway [8]) as articulated in [3]

Definition 5. *A Conway Algebra is a rig A equipped with a unary operation*

$$(-)^* : A \longrightarrow B \, ; \, a \to a^*$$

satisfying the two equations

$$(ab)^* = 1 + a(ba)^*b$$
$$(a + b)^* = (a^*b)^*a^*$$

It is immediate that in a traced monoidal category $\mathbf{C}$ whose tensor product is a biproduct each $\mathrm{End}_{\mathbf{C}}(A)$ is a Conway Algebra, the operation $(-)^*$ being given by

$$a^* = \mathrm{tr}\begin{pmatrix} 0 & 1 \\ 1 & a \end{pmatrix}.$$

In the case of a category generated by a single object U, the requirement that $\mathrm{End}_{\mathbf{C}}(U)$ be a Conway algebra is in fact sufficient. Generally one takes the trace of a map $A \oplus C \longrightarrow B \oplus C$ given by the matrix

$$\begin{pmatrix} a & b \\ c & d \end{pmatrix}$$

with $a \in \mathcal{C}(A, B)$, $b \in \mathcal{C}(C, B)$, $c \in \mathcal{C}(A, C)$ $d \in \mathcal{C}(C, C)$ using the natural formula

$$\mathrm{tr}\begin{pmatrix} a & b \\ c & d \end{pmatrix} = a + bd^*c$$

So to identify the free traced monoidal category with biproducts on an object U it suffices to identify the free Conway algebra on no generators.

Fortunately that is already known. In [8] Conway effectively identifies the elements of the free Conway algebra on no generators: the distinct elements are those of the form

$$\{n \mid n \geq 0\} \cup \{n(1^*)^m \mid n, m \geq 1\} \cup \{1^{**}\}.$$

The algebraic structure can be deduced from the following absorbtion rules.

$$1 + (1^*)^n = (1^*)^n$$
$$(1^*)^n + 1^{**} = 1^{**} + 1^{**} = 1^{**}$$
$$n.1^{**} = 1^*.1^{**} = 1^{**}.1^{**} = 1^{**}$$
$$1^{***} = 2^* = 1^{**}$$

3.4   Representative Calculations

The objects in the free traced monoidal category with biproducts generated by a single object are (as we had earlier)

$$0, 1, 2, \ldots, n, \ldots$$

representatives of finite sets. But now the maps from n to m are given by $m \times n$ matrices with entries in the free Conway algebra just described.

Taking Int gives us objects of the form (n, m) with n and m finite cardinals. We consider the interpretation which arises when each atomic proposition A is interpreted by the object $(1, 0)$ with $\neg A$ therefore interpreted by $(0, 1)$. Proofs π will have interpretations $V(\pi)$ which will be suitably sized matrices as above.

1. **Proofs in MLL.** The data in an intereptation of a proof in multiplicative linear logic is familiar. Again we follow [14] by considering only one sided sequents. Suppose we have $\vdash^\pi \Gamma$. There will be some number, n say, of occurences of atomic propositions (literals) and the same number of the corresponding negations. So Γ will be interpreted by the object (n, n), and π by an $n \times n$ matrix. For MLL this matrix will always be a permutation matrix giving the information of the axiom links in π. (Of course the permutation is just a construct of the order in which the literals and their negations are taken.)

2. **Distributive Laws.** For simple proofs like those of the distributive laws the interpretation continues just to give information akin to that of axiom links. Consider first the proof of

$$A \wedge (B \vee C \vdash (A \wedge B) \vee (A \wedge C)$$

which we gave earlier. The interpretation is a map from $(3, 0$ to $(4, 0)$, and so is given by a 4×3 matrix: it is

$$\begin{pmatrix} 1 & 0 & 0 \\ 0 & 1 & 0 \\ 1 & 0 & 0 \\ 0 & 0 & 1 \end{pmatrix}.$$

The natural proof of the converse distributive law,

$$(A \wedge B) \vee (A \wedge C) \vdash A \wedge (B \vee C),$$

may seem more complicated, but our current interpretation does not notice that. One gets

$$\begin{pmatrix} 1 & 0 & 1 & 0 \\ 0 & 1 & 0 & 0 \\ 0 & 0 & 0 & 1 \end{pmatrix},$$

which is just the transpose of the previous matrix.

3. **Natural Numbers.** The interpretation of $(A \Rightarrow A) \Rightarrow (A \Rightarrow A)$ is $(2,2)$ so the natural number proofs π_n are interpreted as 2×2 matrices. We get

$$V(\pi_0) = \begin{pmatrix} 0 & 0 \\ 0 & 1 \end{pmatrix}, \qquad V(\pi_{n+1}) = \begin{pmatrix} n & 1 \\ 1 & 0 \end{pmatrix}.$$

As before we consider what it is to compose proofs π_n and π_m with the proof $\mu = \lambda a, b, f. \, a(b(f))$ of multiplication. This is interpreted by the (obvious permutation) matrix

$$\begin{pmatrix} 0 & 0 & 1 & 0 & 0 & 0 \\ 0 & 0 & 0 & 0 & 0 & 1 \\ 1 & 0 & 0 & 0 & 0 & 0 \\ 0 & 0 & 0 & 0 & 1 & 0 \\ 0 & 0 & 0 & 1 & 0 & 0 \\ 0 & 1 & 0 & 0 & 0 & 0 \end{pmatrix}$$

where the first two rows and columns come from the codomain. Essentially we have to compose and take a trace. At first sight this is not very exciting and things seem much as before. We find

$$V(\mu|\pi_{n+1}|\pi_{m+1}) = \begin{pmatrix} n+m & 1 \\ 1 & 0 \end{pmatrix}.$$

But the connectivity of π_0 introduces an unexpected nuance. To compute $V(\mu|\pi_0|\pi_{m+1})$, we can compose one way to get the matrix

$$\begin{pmatrix} 0 & 0 & 0 & 0 & 0 & 0 \\ 0 & 0 & 0 & 0 & 1 & 0 \\ 1 & 0 & 0 & 0 & 0 & 0 \\ 0 & 0 & 0 & 0 & m & 1 \\ 0 & 0 & 0 & 1 & 0 & 0 \\ 0 & 1 & 0 & 0 & 0 & 0 \end{pmatrix}$$

and then we need to take a not so obvious trace. We end up with

$$V(\mu|\pi_0|\pi_{m+1}) = \begin{pmatrix} 0 & 0 \\ 0 & m^* \end{pmatrix}.$$

For $V(\mu|\pi_0|\pi_0)$, the calculations are marginally simpler and we end up with

$$V(\mu|\pi_0|\pi_0) = \begin{pmatrix} 0 & 0 \\ 0 & 1^* \end{pmatrix} .$$

One sees that even simple cuts can produce cycles in a proof of a serious kind, and these are detected by our second interpretation.

4   Summary

In this paper I hope to have presented evidence that there are mathematical interpretations of classical proof which produce what can be regarded as invariants of proofs. Clearly there are many more possibilities than those touched on here. It seems worth making a couple of concluding comments.

First while the interpretations given do handle classical proofs, they do not appear to detect any particular properties of them. All examples given concern (very simple) familiar constructive proofs. There would have been no special interest for example in treating Pierce's Law.

Secondly, these interpretations are sensitive to cut elimination. This appears to be a necessary feature of any mathematical theory of classical proof respecting the symmetries. Even for constructive prrofs it suggests a quite different criterion for the identity of proofs than that given by equality of normal form. This criterion would have the merit of being sensitive inter alia to the use of Lemmas in mathematical practice.

References

1. G. Bellin and M. Hyland and E. Robinson and C. Urban. Proof Theory of Classical Logic. In preparation.
2. R. Blackwell and G. M. Kelly and A. J. Power. Two-dimensional monad theory. *Journal of Pure and Applied Algebra* **59** (1989), 1–41.
3. S. L. Bloom and Z. Esik. *Iteration Theories*. Springer-Verlag, 1993.
4. A. Carbone. Duplication of directed graphs and exponential blow up of proofs. *Annals of Pure and Applied Logic* **100** (1999), 1–67.
5. A. Carbone. Asymptotic cyclic expansion and bridge groups of formal proofs. *Journal of Algebra*, 109–145, 2001.
6. A.Carbone, Streams and strings in formal proofs. *Theoretical Computer Science*, 288(1):45–83, 2002.
7. S. Carmody. *Cobordism Categories*. PhD Dissertation, University of Cambridge, 1995.
8. J. H. Conway. *Regular algebra and finite machines.* Chapman and Hall, 1971.
9. R. Dijkgraaf. *A geometric approach to two dimensional conformal field theory.* Ph.D. Thesis, Univeristy of Utrecht, 1989.
10. G. Frobenius. Theorie der hyperkomplexen Grössen. *Sitzungsbereich K. Preuss Akad. Wis.* **24** 1903, 503–537, 634–645.

11. C. Führmann and D. Pym. Order-enriched Categorical Models of the Classical Sequent Calculus. Submitted.
12. C. Führmann and D. Pym. On the Geometry of Interaction for Classical Logic (Extended Abstract). To appear in Proceedings LICS 04, IEEE Computer Society Press, 2004.
13. J.-Y. Girard. *Proofs and Types*. Cambridge University Press, 1989.
14. J.-Y. Girard. Linear Logic. *Theoretical Computer Science* **50**, (1987), 1–102.
15. M. Hasegawa. *Models of sharing graphs. (A categorical semantics for* Let *and* Letrec.*)* Distinguished Dissertation in Computer Science, Springer-Verlag, 1999.
16. J. M. E. Hyland. Proof Theory in the Abstract. *Annals of Pure and Applied Logic* **114** (2002), 43–78.
17. M. Hyland and J. Power. Symmetric monoidal sketches. *Proceedings of PPDP 2000*, ACM Press (2000), 280–288.
18. M. Hyland and J. Power. Symmetric monoidal skecthes and categories of wiring diagrams. *Proceedings of CMCIM 2003*, to appear.
19. J. M. Hyland and A. Schalk. Glueing and Orthogonality for Models of Linear Logic. *Theoretical Computer Science* **294** (2003) 183–231.
20. A. Joyal and R. Street and D. Verity. Traced monoidal categories. *Math. Proc Camb Phil. Soc.* **119** (1996), 425–446.
21. G. M. Kelly and M. Laplaza. Coherence for compact closed categories. *Journal of Pure and Applied Algebra* **19** (1980), 193–213.
22. G. M. Kelly and A. J. Power. Adjunctions whose counits are equalizers, and presentations of finitary enriched monads. *Journal of Pure and Applied Algebra* **89** (1993), 163–179.
23. J. Kock. *Frobenius Algebras and 2D Topological Quantum Field Theories*. Cambridge University Press, 2003.
24. F. W. Lawvere. Ordinal sums and equational doctrines. In *Seminar on Triples and Categorical Homology Theory*, LNM **80** (1969), 141–155.
25. S. Mac Lane. *Categories for the working mathematician*. Graduate Texts in Mathematics **5**. Springer (1971).
26. J. P. May. *Simplicial objects in algebraic topology*. Van Nostrand, Princeton. 1967.
27. M. Nygaard and G. Winskel. Domain Theory for Concurrency. *Theoretical Computer Science*, to appear.
28. A. J. Power and E. P. Robinsons. A characterization of PIE limits. *Math. Proc. Camb. Phil. Soc.* **110**, (1991), 33–47.
29. D. Prawitz. Ideas and results in proof theory. In *Proceedings of the Second Scandinavian Logic Symposium*. J.-E. Fenstad (ed), North-Holland 1971, 237–309.
30. E. P. Robinson. Proof Nets for Classical Logic. *Journal of Logic and Computation*. **13**, 2003, 777–797.
31. M. E. Szabo. Polycategories. *Comm. Alg.* **3** (1997), 663–689.

Applications of Craig Interpolation to Model Checking

Kenneth McMillan

Cadence Berkeley Labs

A Craig interpolant [1] for a mutually inconsistent pair of formulas (A,B) is a formula that is (1) implied by A, (2) inconsistent with B, and (3) expressed over the common variables of A and B. It is known that a Craig interpolant can be efficiently derived from a refutation of $A \land B$, for certain theories and proof systems. For example, interpolants can be derived from resolution proofs in propositional logic, and for "cutting planes" proofs for systems of linear inequalities over the reals [5, 3]. These methods have been extended to the theory of linear inequalities with uninterpreted function symbols [4].

The derivation of interpolants from proofs has a number of applications in model checking. For example, interpolation can be used to construct an inductive invariant for a transition system that is strong enough to prove a given property. In effect, we can use interpolation to construct an abstract "image operator" that can be iterated to a fixed point to obtain an invariant. This invariant contains only information actually deduced by a prover in refuting counterexamples to the property of a fixed number of steps. Thus, in a certain sense, we abstract the invariant relative to a given property. This avoids the complexity of computing the strongest inductive invariant (i.e., the reachable states) as is typically done in model checking.

This approach gives us a complete procedure for model checking temporal properties of finite-state systems that allows us to exploit recent advances in SAT solvers for the proof generation phase. Experimentally, this method is found to be quite robust for verifying properties of industrial hardware designs, relative to other model checking approaches.

The same approach can be applied to infinite-state systems, such as programs and parameterized protocols, although there is no completeness guarantee in this case. Alternatively, interpolants derived from proofs can be used to infer predicates that are useful for predicate abstraction [6]. This approach has been used in a software model checking to verify properties of C programs with in excess of 100K lines of code [2].

References

1. W. Craig. Linear reasoning: A new form of the Herbrand-Gentzen theorem. *J. Symbolic Logic*, 22(3):250–268, 1957.

J. Marcinkowski and A. Tarlecki (Eds.): CSL 2004, LNCS 3210, pp. 22–23, 2004.

2. T. A. Henzinger, R. Jhala, Rupak Majumdar, and K. L. McMillan. Abstractions from proofs. In *ACM Symp. on Principles of Prog. Lang. (POPL 2004)*, 2004. to appear.
3. J. Krajíček. Interpolation theorems, lower bounds for proof systems, and independence results for bounded arithmetic. *J. Symbolic Logic*, 62(2):457–486, June 1997.
4. K. L. McMillan. An interpolating theorem prover. In *TACAS 2004*, pages 16–30, 2004.
5. P. Pudlák. Lower bounds for resolution and cutting plane proofs and monotone computations. *J. Symbolic Logic*, 62(2):981–998, June 1997.
6. Hassen Saïdi and Susanne Graf. Construction of abstract state graphs with PVS. In Orna Grumberg, editor, *Computer-Aided Verification, CAV '97*, volume 1254, pages 72–83, Haifa, Israel, 1997. Springer-Verlag.

Bindings, Mobility of Bindings, and the ∇-Quantifier: An Abstract

Dale Miller

INRIA-Futurs & École Polytechnique, France

We present a meta-logic that contains a new quantifier ∇ (for encoding "generic judgments") and inference rules for reasoning within fixed points of a given specification. We then specify the operational semantics and bisimulation relations for the finite π-calculus within this meta-logic. Since we restrict to the finite case, the ability of the meta-logic to reason within fixed points becomes a powerful and complete tool since simple proof search can compute the unique fixed point. The ∇ quantifier helps with the delicate issues surrounding the scope of variables within π-calculus expressions and their executions (proofs). We shall illustrate several merits of the logical specifications we write: they are natural and declarative; they contain no-side conditions concerning names of bindings while maintaining a completely formal treatment of such bindings; differences between late and open bisimulation relations are easy to see declaratively; and proof search involving the application of inference rules, unification, and backtracking can provide complete proof systems for both one-step transitions and for bisimulation. This work is joint with Alwen Tiu and is described in more detail in the following papers.

References

1. Miller, D., Tiu, A.: A proof theory for generic judgments: An extended abstract. In: Proceedings of LICS 2003, IEEE (2003) 118–127
2. Miller, D., Tiu, A.: A proof theory for generic judgments. ACM Transactions on Computational Logic (To appear.)
3. Tiu, A., Miller, D.: A proof search specification of the π-calculus. Submitted. (2004)

My (Un)Favourite Things

Paweł Urzyczyn

Warsaw University

Abstract. The talk will be devoted to four open problems I was unsuccessfully trying to solve in the past. These problems concern:

- Regular and Context-Free Dynamic Logic;
- The question of polymorphic collapse;
- Higher-order push-down stores and procedures;
- Polymorphic definability of recursive data types.

Each of these problems addresses a basic issue in the logical foundations of computer science, and each has been open for a long time. This talk aims at bringing back the challenge, and sorting out the related confusions.

J. Marcinkowski and A. Tarlecki (Eds.): CSL 2004, LNCS 3210, p. 25, 2004.

On Nash Equilibria in Stochastic Games[*]

Krishnendu Chatterjee[1], Rupak Majumdar[2], and Marcin Jurdziński[3]

[1] EECS, University of California, Berkeley, USA
[2] CS, University of California, Los Angeles, USA
[3] CS, University of Warwick, UK
{c_krish,mju}@eecs.berkeley.edu, rupak@cs.ucla.edu

Abstract. We study infinite stochastic games played by n-players on a finite graph with goals given by sets of infinite traces. The games are *stochastic* (each player simultaneously and independently chooses an action at each round, and the next state is determined by a probability distribution depending on the current state and the chosen actions), *infinite* (the game continues for an infinite number of rounds), *nonzero sum* (the players' goals are not necessarily conflicting), and undiscounted. We show that if each player has a reachability objective, that is, if the goal for each player i is to visit some subset R_i of the states, then there exists an ϵ-Nash equilibrium in memoryless strategies, for every $\epsilon > 0$. However, exact Nash equilibria need not exist. We study the complexity of finding such Nash equilibria, and show that the payoff of some ϵ-Nash equilibrium in memoryless strategies can be ϵ-approximated in NP.

We study the important subclass of n-player *turn-based probabilistic* games, where at each state at most one player has a nontrivial choice of moves. For turn-based probabilistic games, we show the existence of ϵ-Nash equilibria in pure strategies for games where the objective of player i is a *Borel* set B_i of infinite traces. However, exact Nash equilibria may not exist. For the special case of ω-regular objectives, we show exact Nash equilibria exist, and can be computed in NP when the ω-regular objectives are expressed as parity objectives.

1 Introduction

The interaction of several agents is naturally modeled as non-cooperative games [19, 21]. The simplest, and most common interpretation of a non-cooperative game is that there is a single interaction among the players ("one-shot"), after which the payoffs are decided and the game ends. However, many, if not all, strategic endeavors occur over time, and in stateful manner. That is, the games progress over time, and the current game is decided based on the history of the interactions. Infinite *stochastic games* [24, 9, 6] form a natural model for such interactions. A stochastic game is played over a finite *state space*, and is played in rounds. In each round, each player chooses an available action out of a finite

[*] This research was supported in part by the AFOSR MURI grant F49620-00-1-0327, ONR grant N00014-02-1-0671, NSF grants CCR-9988172 and CCR-0225610.

J. Marcinkowski and A. Tarlecki (Eds.): CSL 2004, LNCS 3210, pp. 26–40, 2004.

set of actions simultaneously with and independently from all other players, and the game moves to a new state under a possibly probabilistic transition relation based on the current state and the joint actions. For the verification and control of reactive systems, such games are infinite: play continues for an infinite number of rounds, giving rise to an infinite sequence of states, called the *outcome* of the game. The players receive a payoff based on a payoff function mapping infinite outcomes to a real in $[0, 1]$.

Payoffs are generally Borel measurable functions [18]. For example, the payoff set for each player is a Borel set B_i in the Cantor topology on S^ω (where S is the set of states), and player i gets payoff 1 if the outcome of the game is a member of B_i, and 0 otherwise. In verification, payoff functions are usually index sets of ω-*regular languages*. ω-regular sets occur in low levels of the Borel hierarchy (they are in $\Sigma_3 \cap \Pi_3$), but they form a robust and expressive language for determining payoffs for commonly used specifications [16]. The simplest ω-regular games correspond to safety (closed sets) or reachability (open sets) objectives.

Games may be *zero sum*, where two players have directly conflicting objectives and the payoff of one player is one minus the payoff of the other, or *nonzero sum*, where each player has a prescribed payoff function based on the actions of all the other players. The fundamental question for games is the existence of equilibrium values. For zero sum games, this involves showing a *determinacy* theorem that states that the expected optimum value obtained by player 1 is exactly one minus the expected optimum value obtained by player 2. For one-shot zero sum games, this is von Neumann's minmax theorem [30]. For infinite games, the existence of such equilibria is not obvious, in fact, by using the axiom of choice, one can construct games for which determinacy does not hold. However, a remarkable result by Martin [18] shows that all stochastic zero sum games with Borel payoffs are determined.

For nonzero sum games, the fundamental equilibrium concept is a *Nash equilibrium* [11], that is, a strategy profile such that no player can gain by deviating from the profile, assuming all other players continue playing their strategies in the profile. Again, for one-shot games, the existence of such equilibria is guaranteed by Nash's theorem [11]. However, the existence of Nash equilibria in infinite games is not immediate: Nash's theorem holds for finite bimatrix games, but in case of stochastic games, the possible number of strategies is infinite. The existence of Nash equilibria is known only in very special cases of stochastic games. In fact, Nash equilibria may not exist, and the best one can hope for is an ϵ-Nash equilibrium for all $\epsilon > 0$, where an ϵ-Nash equilibrium is a strategy profile where unilateral deviation can only increase the payoff of a player by at most ϵ. Exact Nash equilibria do exist in discounted stochastic games [10], and other special cases [26, 27]. For limit average payoffs, exact Nash equilibria need not exist even for two-player games [1]. Recently the existence of ϵ-Nash equilibria for all $\epsilon > 0$ was proved in [28, 29] for the two-player case, and the general case remains an important open question. For games with payoffs defined by Borel sets, surprisingly little is known. Secchi and Sudderth [23] showed that exact Nash equilibria do exist when all players have payoffs defined by closed sets ("safety games"), where the objective of each player is to stay within a certain set of good states. Formally, each player i has a subset of states S_i as their safe states, and gets a payoff 1 if the play never leaves the set S_i and gets payoff 0 otherwise. This result was generalized to general state and action spaces [23, 15], where only ϵ-equilibria exist.

However, not much more is known: for example, the existence of ϵ-Nash equilibria when the payoff for each player is given by an open set ("reachability games") remained open. In the open (or reachability) game, each player i has a subset of states R_i as reachability targets. Player i gets payoff 1 if the outcome visits some state from R_i at some point, and 0 otherwise. In this case, an ϵ-equilibrium is the best one can hope for: there exist two-player reachability games for which no exact Nash equilibria exist [13]. In this paper, we answer this question in the affirmative: we show that every n-player reachability game has an ϵ-Nash equilibrium, for every $\epsilon > 0$. Moreover, there is an ϵ-Nash equilibrium profile of *memoryless strategies* (a strategy is memoryless if it only depends on the current state of the game, and not on the history). However, strategies in general may require randomization. We achieve our result by going through discounted games. Our proof has three main technical ingredients: the existence of equilibria in certain discounted games, the approximability of such equilibria with simple strategies, and the approximability of certain undiscounted payoffs in a Markov decision process (MDP) by discounted payoffs. First, we use the existence of exact Nash equilibria in memoryless strategies in discounted reachability games, that is, reachability games where play stops at each round with a probability $\beta < 1$, and continues to the next round with probability $1 - \beta$. Second, we show that the exact Nash equilibrium in discounted games can be approximated by *simple* strategy profiles, where each player plays uniformly over a multi-set of actions, and the size of the multi-set can be bounded in terms of the number of players and the size of the game. Third, using a result of Condon [3], we show that for an MDP, for any $\epsilon > 0$, we can choose the discount factor β such that the difference between the payoffs in the original MDP and the discounted MDP is bounded by ϵ. It follows that on the MDP obtained by fixing these simple strategies for all but player i, player i cannot significantly increase his payoff by switching to a different strategy. This construction yields an ϵ-Nash equilibrium in memoryless strategies. In contrast, the proof for safety games [23] proceeds by induction on the number of players, and does not give a Nash equilibrium in memoryless strategies. It is not known if exact Nash equilibria in memoryless strategies exist for safety games.

Computing the values of a Nash equilibrium, when it exists, is another challenging problem [20, 31]. For one-shot zero sum games, equilibrium values and strategies can be computed in polynomial time (by reduction to linear programming) [19]. For zero-sum stochastic games with ω-regular payoffs, doubly exponential time algorithms can be obtained to approximate the value to within ϵ by expressing the value as nested fixpoints, and then reducing to the validity of a sentence in the theory of reals [6]. For one-shot nonzero sum games, no polynomial time algorithm to compute an exact Nash equilibrium in a two-player game is known [20].

We study the complexity of approximating Nash equilibria in reachability games. We give an NP algorithm to approximate the value of some ϵ-Nash equilibrium in memoryless strategies, for any constant $\epsilon > 0$. We show that the problem of finding a Nash equilibrium in a reachability game in *pure* (i.e., not requiring randomization) memoryless strategies, if it exists, where each player gets at least some (specified) payoff is NP-complete. In contrast, for matrix games, finding a *pure* Nash equilibrium, if it exists, can always be achieved in polynomial time in size of the input. Related NP-hardness results appear in [4], but our results do not follow from theirs as their hardness proof does not hold for pure strategy Nash equilibria. For two-player zero-sum games with reachability objective, values can be algebraic and there are simple examples when they are irrational [6]. Hence one can only hope to compute it to an ϵ-precision.

Interestingly, the techniques we develop for the nonzero sum games can be used to get better complexities for the zero sum case. In particular, we show an improved NP $\cap$ co-NP upper bound to approximate the valued for two-player zero sum reachability games within ϵ-tolerance for any constant ϵ. This improves the previously best known EXPTIME upper bound for reachability games [6]. This also generalizes a result of Condon [3]. Open and closed sets form the lowest level of the Borel hierarchy, and together with [23], this paper answers positively the existence of (ϵ) Nash equilibria in such games. We leave the generalization of these results to higher levels of the Borel hierarchy as an interesting open problem.

An important class of stochastic games are *turn-based probabilistic games* [27, 22, 9], where at each stage, at most one player has a nontrivial choice of actions. In verification, such games model asynchronous interactions among components. For turn-based probabilistic games, we prove the existence of ϵ-Nash equilibria for payoffs specified by arbitrary Borel sets. Moreover, we show that there is an ϵ-equilibrium strategy profile in *pure* strategies, that is, strategies that do not require randomization. Our proof proceeds in two steps. First, we prove a pure strategy determinacy theorem for turn-based probabilistic zero sum games and Borel payoff functions. The proof is a specialization of Martin's determinacy proof for stochastic games with Borel payoffs [18]. Second, using this and a general construction of *threat strategies* [19], we show that ϵ-Nash equilibria exist for all turn based probabilistic nonzero sum games with arbitrary Borel set payoffs. We show this result is optimal: there exist turn-based probabilistic games with Borel payoffs for which exact Nash equilibria do not exist. As a special case, we study turn-based probabilistic games with *parity* winning objectives [7]. Parity objectives form a canonical form for all ω-regular winning objectives [7]. Using existence of pure memoryless optimal strategy for zero sum turn-based probabilistic parity games, we show the existence of pure strategy exact Nash equilibria for parity payoffs. This proves that (exact) Nash equilibria exist for turn based probabilistic games with ω-regular payoffs. We give corresponding complexity results to compute Nash equilibrium solutions. Using an NP $\cap$ co-

Model	Type	Payoffs	Existence	Strategy	Complexity
General	Z	Borel	[18]		2EXPTIME to ϵ-approximate for parity payoffs [6]
TB	Z	Borel	[18]	(*) pure	NP $\cap$ co-NP for parity payoffs [2]
General	NZ	Safety	NE [23]	?	?
General	NZ	Reachability	(*) ϵ-NE	(*) memoryless	(*) NP to ϵ-approximate
General	NZ	Borel	?	?	?
General	Z	Reachability	[8]	memoryless ϵ-optimal [8]	(*) NP $\cap$ co-NP to ϵ-approximate
TB	NZ	Borel	(*) ϵ-NE	(*) pure	?
TB	NZ	$\Sigma_3 \cap \Pi_3$	(*) NE	(*) pure	(*) NP for parity payoffs

Fig. 1. Summary of known and new/our (asterisk (*)) results. "General" is for stochastic games, "TB" for turn-based probabilistic games. "Z" denotes zero sum, "NZ" denotes nonzero sum. A '?' denotes open problem. The "Strategy" column indicates existence of an (exact or ϵ) Nash equilibrium with that class of strategies, for example, there exists an ϵ-Nash equilibrium profile in memoryless strategies for n-player nonzero sum reachability games

NP strategy construction algorithm for probabilistic turn-based zero sum parity games [2], we get an NP algorithm to find an exact Nash equilibrium in these games. Figure 1 gives a summary of known and new (our) results in this area.

2 Definitions

An *n-player stochastic game* G consists of a finite, nonempty set of states S, n players $1, 2, \ldots, n$, finite sets of action sets $A_1, A_2, \ldots, A_n$ for the players, a conditional probability distribution p on $S \times (A_1 \times A_2 \times \cdots \times A_n)$ called the law of motion, and bounded, real valued payoff functions $\phi_1, \phi_2, \ldots, \phi_n$ defined on the history space $H = S \times A \times S \times A \cdots$, where $A = A_1 \times A_2 \times \cdots A_n$. The game is called a *n-player deterministic game* if for all states $s \in S$ and action choices $a = (a^1, a^2, \ldots, a^n)$ there is a unique state s' such that $p(s'|s, a) = 1$.

Play begins at an initial state $s_0 = s \in S$. Each player independently and concurrently selects a mixed action a_i^1 with a probability distribution $\sigma_i(s)$ belonging to $\mathcal{P}(A_i)$, the set of probability measures on A_i. Given s_0 and the chosen mixed actions $a^1 = (a_1^1, a_2^1, \ldots a_n^1) \in A$, the next state s_1 has the probability distribution $p(\cdot|s_0, a^1)$. Then again each player i independently selects a_i^2 with a distribution $\sigma_i((s_0, a^1, s_1))$ and given $a^2 = (a_1^2, a_2^2, \ldots, a_n^2)$, the next state s_2 has the probability distribution $p(\cdot|s_1, a^2)$. Play continues in this way for an infinite number of steps and generates a random history $h = (s_0, a^1, s_1, a^2, \ldots) \in H$. The payoff is decided based on the infinite history.

A function π_i that specifies for each partial history $h' = (s_0, a^1, s_1, a^2, \ldots, s_k)$ the conditional distribution $\pi_i(h') \in \mathcal{P}(A_i)$ for player i's next action a_i^{k+1} is called a *strategy* for player i. A strategy profile $\pi = (\pi_1, \pi_2, \ldots, \pi_n)$ consists of a strategy π_i for each player i. A *selector* for a player i is a mapping

$\sigma_i : S \to \mathcal{P}(A_i)$. A selector profile $\sigma = (\sigma_1, \sigma_2, \ldots, \sigma_n)$ consists of a selector σ_i for each player i. A selector σ_i defines a *memoryless* strategy σ_i^∞ for player i: σ_i^∞ chooses the mixed action $\sigma_i(s')$ each time the play visits s'. A strategy profile $\sigma^\infty = (\sigma_1^\infty, \sigma_2^\infty, \ldots, \sigma_n^\infty)$ is a memoryless strategy profile if all the strategies $\sigma_1^\infty, \sigma_2^\infty, \ldots, \sigma_n^\infty$ are memoryless. Given a memoryless strategy profile $\sigma^\infty = (\sigma_1^\infty, \sigma_2^\infty, \ldots, \sigma_n^\infty)$ we write $\sigma = (\sigma_1, \sigma_2, \ldots, \sigma_n)$ to denote the corresponding selector profile for the players. An initial state s and a strategy profile $\pi = (\pi_1, \pi_2, \ldots, \pi_n)$ together with the law of motion p determine a probability distribution $P_{s,\pi}$ on the history space. We write $E_{s,\pi}$ for the expectation operator associated with $P_{s,\pi}$. A strategy π_i for player i is *pure* if for every history $h = (s_0, a^1, s_1, \ldots, a^k, s_k)$ there is a action $a_k \in A_{is_k}$ such that $\pi_i(a) = 1$. In other words, a strategy is pure if for every history the strategy chooses one action rather than a probability distribution over the action set. A strategy profile is pure if all the strategies of the profile are pure.

Given a strategy profile $\tau = (\tau_1, \tau_2, \ldots, \tau_n)$ the strategy profile $\tau_{-i} = (\tau_1, \ldots, \tau_{i-1}, \tau_{i+1}, \ldots, \tau_n)$ is the strategy profile obtained by deleting the strategy τ_i from τ whereas for any strategy μ_i of player i, $\rho(\tau_{-i}, \mu_i) = (\tau_1, \ldots, \tau_{i-1}, \mu_i, \tau_{i+1}, \ldots, \tau_n)$ denotes the strategy profile where player i follows μ_i and the other players follows the strategy of τ_{-i}. Similar definitions hold for selector profiles as well.

Assume now that the payoff functions $\phi_i : H \to \mathbb{R}$ are bounded and measurable, where $\mathbb{R}$ is the set of reals. If the initial state of the game is s and each player i choses a strategy π_i, then the payoff to each player i is the expectation $E_{s,\pi}\phi_i$, where π is the strategy profile $\pi = (\pi_1, \pi_2, \ldots, \pi_n)$. For $\epsilon \geq 0$, an ϵ-equilibrium at the initial state s is a profile $\pi = (\pi_1, \pi_2, \ldots, \pi_n)$ such that, for all $i = 1, 2, \ldots, n$ we have $E_{s,\pi}\phi_i \geq \sup_{\mu_i} E_{s,\rho(\pi_{-i}, \mu_i)}\phi_i - \epsilon$, where μ_i ranges over the set of all strategies for player i. In other words, each π_i guarantees an expected payoff for player i which is within ϵ of the best possible expected payoff for player i when every other player $j \neq i$ plays π_j. A 0-equilibrium is called a *Nash equilibrium* and for every $\epsilon > 0$ an ϵ-equilibrium is called an ϵ-*Nash equilibrium* [11]. A strategy profile π for an ϵ-Nash equilibrium is called an ϵ-equilibrium profile. A strategy profile π for a Nash equilibrium is called a Nash equilibrium profile.

Let $R_1, R_2, \ldots, R_n$ be subsets of the state space S. The subset of states R_i is referred as the *target set* for player i. Then let $R_1^\infty, R_2^\infty, \ldots, R_n^\infty$ be the subsets of H defined by $R_i^\infty = \{h = (s_0, a^1, s_1, a^2, \ldots) \mid \exists k \in \mathbb{N}.s_k \in R_i\}$ and take the payoff function $\phi_i^{R_i}$ to be the indicator function of R_i^∞ for $i = 1, 2, \ldots, n$. Thus each player receives a payoff of 1 if the process of states $s_0, s_1, \ldots$ reaches a state in R_i and receives payoff 0 otherwise. We call stochastic games with the payoff functions of this form *reach-a-set games*.

A two-player zero sum reachability game (also called *concurrent reachability game* [5]) G is a two-player stochastic game with $R_1 \subseteq S$ as a target set of states for player 1. Given a random history $h = (s_0, a^1, s_1, a^2, \ldots)$ player 1 gets a payoff 1 if the history contains a state in R_1, else the player 2 gets a payoff 1. In other words, player 1 plays a reachability game with target set R_1 and player 2 has a strictly competing objective of keeping the game out of R_1 forever. Notice that this is not a two-player reach-a-set game.

3   Existence of ϵ-Nash Equilibria

The main result of this section is the existence of ϵ-Nash equilibria in n-player reach-a-set games for every $\epsilon > 0$.

Theorem 1 (ϵ-equilibrium). *An n-player reach-a-set game G with finite state and action spaces has an ϵ-Nash equilibrium at every initial state $s \in S$ for every $\epsilon > 0$. Moreover, there is a memoryless ϵ-equilibrium strategy profile for every $\epsilon > 0$.*

Already for two-player zero sum reachability games optimal strategies need not exist [13, 6]. The example can be easily adapted to show that Nash equilibria need not exist even for 2-player reach-a-set games. Hence an ϵ-Nash equilibrium for all $\epsilon > 0$, is the best one can achieve for n-player reach-a-set games.

Memoryless Nash Equilibrium in Discounted Games. We first prove the existence of a Nash equilibrium in memoryless strategies in a *discounted* n-player reach-a-set game. Given a n-player game G we use G^β to denote a β-discounted version of the game G. The game G^β at each step halts with probability β (goes to a special sink state *halt* which has a reward 0 for every player) and continues as the game G with probability $1 - \beta$. We refer to β as the discount-factor. The proof of the next lemma uses Kakutani's Fixed point theorem to show the existence of Nash equilibria in discounted games [25, 23].

Lemma 1. *For every β-discounted n-player reach-a-set game G^β there exist selectors $\sigma_i : S \to \mathcal{P}(A_i), i = 1, 2, \ldots, n$, such that the memoryless profile $\sigma^\infty = (\sigma_1^\infty, \sigma_2^\infty, \ldots, \sigma_n^\infty)$ is a Nash equilibrium profile in G^β for every $s \in S$.*

k-uniform Strategies. Next, we show that a Nash equilibrium in the discounted game can be approximated using "simple" strategy profiles. We start with some definitions. An MDP is a 1-player stochastic game. A MDP reach-a-set game is a 1-player stochastic reach-a-set game. Fix an n-player stochastic game G. Let $\sigma^\infty = (\sigma_1^\infty, \sigma_2^\infty, \ldots, \sigma_n^\infty)$ be a memoryless strategy profile and $\sigma = (\sigma_1, \sigma_2, \ldots, \sigma_n)$ be the corresponding selector profile. Then the game G_σ is a Markov chain whose law of motion p_σ is defined by the functions in the selector profile σ and the law of motion p of the game G. Similarly, $G_{\sigma_{\square i}}$ is an MDP where the mixed action of each player $j \neq i$ at a state s is fixed according to the selector function $\sigma_j(s)$. The law of the motion $p_{\sigma_{\square i}}$ of the MDP is determined by the selectors in σ_{-i} and law of motion p of G. Given a MDP reach-a-set game G with the target R, the *value* of the game at state s is denoted by $v(s) = \sup_\pi E_{s,\pi} \phi_1^R$, where π ranges over all strategies and ϕ_1^R is the reach-a-set game payoff for the player in the game G. Similarly, we use $v^\beta(s) = \sup_\pi E_{s,\pi}^\beta \phi_1^R$, to denote the value at state s in the game G^β, where G^β is the β-discounted version of the game G.

Given a n-player reach-a-set game G let $|S|$ denote the size of the state space and $l = \max_i |A_i|$ denote the maximum number of actions available to any player at any state of G. A selector function σ_i for player i is *pure* if for all states $s \in S$ there is an action $a_i \in A_i$ such that $\sigma_i(s)(a_i) = 1$. A selector function σ_i^k for player i is k-*uniform* if for all states $s \in S$ there exists a multiset M of pure selectors with $|M| \leq k$ such that σ_i^k is the uniform distribution over M. A selector profile $\sigma^k = (\sigma_1^k, \sigma_2^k, \ldots, \sigma_n^k)$ is k-uniform if all the selectors σ_i^k are k-uniform for all $i \in \{1, 2, \ldots, n\}$. A memoryless strategy profile $\sigma^{k,\infty} = (\sigma_1^{k,\infty}, \sigma_2^{k,\infty}, \ldots, \sigma_n^{k,\infty})$ is k-uniform if the selector profile $\sigma^k = (\sigma_1^k, \sigma_2^k, \ldots, \sigma_n^k)$ corresponding to the strategy profile $\sigma^{k,\infty}$ is k-uniform. We use a technical lemma by Lipton et. al. [14] for matrix games (Lemma 2), and a Lipschitz continuity property for MDPs.

Lemma 2 ([14]). *Let J be a matrix game with n-players and each player has at most l actions. Let π be a Nash-equilibrium strategy profile. Then for every $\epsilon > 0$, for every $k \geq \frac{3n^2 \ln n^2 l}{\epsilon^2}$, there exists a k-uniform strategy profile π' such that: (a) for every action a, if $\pi(a) = 0$ then $\pi'(a) = 0$; (b) for every action a, if $\pi(a) > 0$ then $|\pi(a) - \pi'(a)| \leq \epsilon$.*

Let G_1 and G_2 be two MDP's defined on the same state space S and action space A with laws of motion p_1 and p_2 respectively. The *difference* of the two MDP's, denoted $err(G_1, G_2)$, is defined as: $err(G_1, G_2) = \sum_{s,s' \in S^2, a \in A} |p_1(s|s', a) - p_2(s|s', a)|$.

Lemma 3. *Let G^β be a β-discounted n-player reach-a-set game. Let σ^∞ be a memoryless Nash equilibrium profile with selector profile $\sigma = (\sigma_1, \sigma_2, \ldots, \sigma_n)$. For every $\epsilon > 0$, for every $k \geq \frac{3n^2 \ln n^2 l}{(\frac{\epsilon}{n \cdot |S|^2 \cdot l})^2}$, there exists a k-uniform memoryless strategy profile $\sigma^{k,\infty}$ (with selector profile σ^k) such that for all players i, the MDPs $G_{\sigma_{\blacksquare i}}$ and $G_{\sigma_{\blacksquare i}^k}$ satisfy that $err(G_{\sigma_{\blacksquare i}}, G_{\sigma_{\blacksquare i}^k}) \leq \epsilon$.*

Proof. Consider any player i and a player $j \neq i$. Consider the selector profile σ_j for player j. It follows from Lemma 2 that there is a k-uniform selector profile σ^k, where $k \geq \frac{3n^2 \ln n^2 l}{(\frac{\epsilon}{n \cdot |S|^2 \cdot l})^2}$, such that for player j the following conditions hold: (a) every action that is played with probability 0 in σ_j is played with probability 0 in σ_j^k, (b) for every action a that is played with positive probability in σ_j, the difference in probability $|\sigma_j(s)(a) - \sigma_j^k(s)(a)|$ for any $s \in S$ is at most $\frac{\epsilon}{n \cdot |S|^2 \cdot l}$. Consider the MDPs $G_{\sigma_{\blacksquare i}}$ and $G_{\sigma_{\blacksquare i}^k}$, with laws of motion p_1 and p_2 respectively. Since there are n players, for any pair of states s, s', and action a, the difference in probabilities $|p_1(s' | s, a) - p_2(s' | s, a)|$ is at most $\frac{\epsilon}{|S|^2 \cdot l}$. Since size of the state space in S and the number of actions of player i is at most l, there can be at most $|S|^2 \cdot l$ edges. Hence the result follows. ∎

Lemma 3 and Lipschitz continuity of values of MDP reach-a-set game with respect to err [9] gives the following lemma.

Lemma 4. *Let G^β be a β-discounted n-player reach-a-set game. For every $\epsilon > 0$, for every $k \geq \frac{3n^4 |S|^4 l^2 \ln n^2 l}{\epsilon^2}$, there exists a k-uniform memoryless strategy* *profile* $\sigma^{k,\infty} = (\sigma_1^{k,\infty}, \sigma_2^{k,\infty}, \ldots, \sigma_n^{k,\infty})$ *such that* $\sigma^{k,\infty}$ *is an* ϵ*-Nash equilibrium profile in the game* G^β.

Given any game G we denote by $p_{\min}$ the minimum non-zero transition probability in the law of motion p of G; formally $p_{\min} = \min\{p(t \mid s, a_1 \ldots a_n) \mid p(t \mid s, a_1 \ldots a_n) > 0\}$. The following results follows from Lemma 1 and Lemma 4.

Lemma 5. *Given a n-player β-discounted stochastic reach-a-set game G^β, for every $\epsilon > 0$ there exists a memoryless strategy profile $\pi = (\pi_1, \pi_2, \ldots, \pi_n)$ such that π is an ϵ-Nash equilibrium profile in the game G^β and for every player i the minimum transition probability in the MDP $G_{\sigma_{\blacksquare}^k i}$ is at least*

$$p_{\min} \cdot \left(\frac{\epsilon^2}{3n^4 |S|^4 l^2 \ln n^2 l} \right)^n.$$

Our final ingredient is a result by Condon [3] that relates $p_{\min}$ with the discount factor.

Lemma 6. *Let G be a MDP reach-a-set game such that the minimum nonzero transition probability $p_{\min}$ of the MDP is at least α. For every $\epsilon > 0$ there is a discount factor $\beta = \epsilon \alpha^{O(|S|)}$, such that for all states $s \in S$ of the β-discounted game G^β we have $v(s) - v^\beta(s) \leq \epsilon$.*

Nash Equilibrium. We now prove Theorem 1 using the results on discounted reach-a-set games above. Given the game G and $\epsilon > 0$ let α be $p_{\min} \left(\frac{\frac{\epsilon}{2}}{3n^4 |S|^4 l^2 \ln n^2 l} \right)^n$. We construct a game G^β which is a discounted version of G with discount-factor β, such that $\beta = \left(\frac{\epsilon}{2} \right) (\alpha)^{O(|S|)}$. Let $\sigma^{k,\infty} = (\sigma_1^{k,\infty}, \sigma_2^{k,\infty}, \ldots, \sigma_n^{k,\infty})$ be a memoryless strategy profile such that $\sigma^{k,\infty}$ is an $\frac{\epsilon}{2}$-Nash equilibrium profile in the game G^β and for every player i the minimum transition probability in the MDP $G_{\sigma_{\blacksquare}^k i}$ is at least α (existence follows from Lemma 5). Consider any player i and the strategy profile $\sigma_{-i}^{k,\infty}$. The game $G_{\sigma_{\blacksquare}^k i}$ is a MDP where the mixed actions of all the other players are fixed according to the σ_{-i}^k. Also, $G_{\sigma_{\blacksquare}^k i}^\beta$ is the MDP which is the β-discounted version of the game $G_{\sigma_{\blacksquare}^k i}$. Hence for every player i we have

$$\sup_{\mu_i} E_{s, \rho(\sigma_{\blacksquare}^k i, \mu_i)} \phi_i^{R_i} \leq v^\beta(s) + \tfrac{\epsilon}{2} \text{ (from Lemma 6)}$$
$$\leq \left(E_{s, \sigma^{k,\blacksquare}}^\beta \phi_i^{R_i} + \tfrac{\epsilon}{2} \right) + \tfrac{\epsilon}{2}$$
$$\leq E_{s, \sigma^{k,\blacksquare}} \phi_i^{R_i} + \epsilon$$

Hence $\sigma^{k,\infty} = (\sigma_1^{k,\infty}, \sigma_2^{k,\infty}, \ldots, \sigma_n^{k,\infty})$ is an ϵ-Nash equilibrium profile in G.

4   Computational Complexity

Let π be an ϵ-equilibrium profile. Let $v_i^\pi(s) = E_{s,\pi} \phi_i^{R_i}$ denote the value at state s for player i for the strategy profile π. The value of an ϵ-equilibrium profile π at a state s is the value vector $\boldsymbol{v}^\pi(s) = (v_1^\pi(s), v_2^\pi(s), \ldots, v_n^\pi(s))$. Our main results about the computational complexity of computing the value of any ϵ-equilibrium profile within a tolerance of ϵ are summarized below.

Theorem 2 (Computing Values of a Memoryless Equilibrium Profile).
Let G be an n-player reach-a-set game, let s be a state in G, and let $v = (v_1, v_2, \ldots, v_n) \in [0, 1]^n$ be a value vector.

(1) For any constant $\epsilon > 0$, the complexity of computing if there is an ϵ-equilibrium profile π in memoryless strategies such that for all player i we have $v_i^\pi(s) \geq v_i - \epsilon$ is NP.
(2) The complexity of determining whether there is a pure Nash equilibrium strategy profile π such that the value $v_i^\pi(s)$ for every player i from the state s for the profile π is greater than equal to v_i is NP-complete. The problem is NP-hard already for deterministic reach-a-set games.

Approximating in NP. We will prove that for every fixed $\epsilon > 0$ the value of some ϵ-Nash equilibrium can be approximated in NP.

   The following Lemma follows by approximating a memoryless ϵ-Nash equilibrium (existence follows from Theorem 1) by k-uniform memoryless equilibrium using arguments similar to Lemmas 3 and 4.

Lemma 7. *Given a n-player stochastic reach-a-set game G for every $\epsilon > 0$, there exists, for every $k \geq \frac{12n^4|S|^4l^2 \ln n^2 l}{\epsilon^2}$ a k-uniform memoryless strategy profile $\sigma^{k,\infty}$ such that $\sigma^{k,\infty}$ is a ϵ-equilibrium profile for the game G.*

Lemma 8. *Given a constant ϵ the value of an ϵ-equilibrium with a memoryless strategy profile of a n-player stochastic reach-a-set game can be approximated within ϵ tolerance by an NP algorithm.*

Proof. The NP algorithm guesses a k-uniform selector σ^k for a k-uniform memoryless ϵ-equilibrium strategy profile $\sigma^{k,\infty}$. It then verifies that the value for the MDP's $G_{\sigma_{-i}^k}$ for every state $s \in S$ and each player i is within ϵ-tolerance as compared to the value of the Markov chain define by G_{σ^k}. Since the computation of values of a MDP can be achieved in polynomial time (using linear programming) it follows that the approximation within ϵ tolerance can be achieved by a NP-algorithm. ∎

NP-completeness. We first prove it is NP-hard to compute a pure, memoryless Nash equilibrium profile for n-player deterministic reach-a-set games by reduction from 3-SAT. Given a 3-SAT formula ψ with n-clauses and m-variables we will construct a n-player deterministic reach-a-set game G_ψ. Let the variables in the formula ψ be $x_1, x_2, \ldots, x_m$ and the clauses be $C_1, C_2, \ldots, C_n$. In the game G_ψ each clause is a player. The state space S, the law of motion and the target states are defined as follows. There is a state i for each variable x_i, two sink states $sink$ and $m + 1$, and states (i, j) for $i = 1, \ldots, m$ and $j = 0, 1$ for an assignment j to the variable x_i: $S = \{1, 2, \ldots, m, m + 1\} \cup \{(i, j) \mid i = 1, \ldots, m, \text{ and } j = 0, 1\} \cup \{sink\}$. For any state (i, j) the game always moves to the state $i + 1$. Let $C_i = \{c_{i_1}, c_{i_2}, \ldots, c_{i_k}\}$ be the set of clauses in which variable x_i occurs. Then, in state i players $i_1, i_2, \ldots, i_k$ have a choice of moves between $\{0, 1\}$. If all the players chose move 0 the game proceeds to state $(i, 0)$, if all the players chose move 1 the game proceeds to state $(i, 1)$, else the game goes to the sink state. Once the game reaches the sink state or the state $m + 1$ it remains there for ever. Let $C_i^0 = \{c_{k_1}^0, c_{k_2}^0, \ldots, c_{k_l}^0\}$ be the set of clauses that are satisfied assigning $x_i = 0$, then the state $(i, 0)$ is a target state for players $k_1, k_2, \ldots, k_l$. Similarly, let $C_i^1 = \{c_{k_1^{\blacksquare}}^1, c_{k_2^{\blacksquare}}^1, \ldots, c_{k_j^{\blacksquare}}^1\}$ be the set of clauses that are satisfied by assigning the variable $x_i = 1$ then the state $(i, 1)$ is a target state for players $k_1', k_2', \ldots, k_j'$. States $1, 2, \ldots, m + 1$ and the *sink* state is not a target state for any player.

We reduce the 3-SAT problem ψ to the problem of determining whether there is an equilibrium in G_ψ such that each player has a value $\geq v_i$ at state s. Each player gets a value 1 at state 1 iff the formula ψ is satisfiable. If the formula is satisfiable then consider a satisfying assignment to the variables. Then at each state i all the players chose the move as specified by the satisfying assignment and hence every player get a payoff 1. If there is a pure memoryless strategy profile such that all the players get a payoff 1 in the game G_ψ then it follows from the construction of G_ψ that there is an assignment such that every clause is satisfied and hence the 3-SAT formula ψ is satisfiable.

Inclusion in NP is proved by guessing a pure memoryless strategy profile (i.e., at each state, guess an action for each player), and verifying that in the resulting Markov chain, each player i gets at least the payoff v_i, and also that in the Markov decision process obtained by fixing the strategies of all but player i, player i cannot improve his expected payoff. This proves Theorem 2(2).

Concurrent Reachability Games. The techniques developed for n-player nonzero sum games can be used to find better bounds in two-player zero sum reachability games. Our proof goes through a special case of two-player (nonzero sum) reach-a-set games, namely two-player *constant-sum* games. For a two-player reach-a-set game G let Π denote the set of all ϵ-equilibrium strategy profiles for $\epsilon \geq 0$. We use the following notation: $v_1(s) = \sup_{\pi \in \Pi} E_{s,\pi} \phi_1^{R_1}$ and $v_2(s) = \sup_{\pi \in \Pi} E_{s,\pi} \phi_2^{R_2}$. A two-player reach-a-set game is constant-sum if for all states $s \in S$ we have (a) $v_1(s) + v_2(s) = 1$ and (b) for all $\pi \in \Pi$, $E_{s,\pi} \phi_1^{R_1} + E_{s,\pi} \phi_1^{R_2} = 1$. For this special case, we prove a NP $\cap$ co-NP bound to approximate the value of a ϵ-equilibrium profile, for any fixed ϵ.

Lemma 9. *Let G be a two-player constant-sum reach-a-set game, s an initial state and v^1 and v^2 be two values. For a fixed ϵ it can be determined in NP $\cap$ co-NP whether $v_1(s) \geq v^1 - \epsilon$ and $v_2(s) \geq v^2 - \epsilon$.*

Proof. Inclusion in NP follows as a special case of Lemma 8. To prove inclusion in co-NP, consider the case when $v_1(s) < v^1 - \epsilon$. The following NP algorithm determines if $v_1(s) < v^1 - \epsilon$. It guesses the k-uniform selector σ_2^k for player 2 and verifies that the value of player 1 in the state s in the MDP $G_{\sigma_{\blacksquare 1}^k}$ is less than $v^1 - \epsilon$. From Lemma 7 we know that there is a k-uniform memoryless

ϵ-equilibrium profile $\sigma^{k,\infty} = (\sigma_1^{k,\infty}, \sigma_2^{k,\infty})$ with selector profile $\sigma^k = (\sigma_1^k, \sigma_2^k)$. Since the value of a MDP at any state can be computed in polynomial time (using linear programming) the required result follows. ∎

We now reduce two-player zero sum reach-a-set games to two player constant-sum reach-a-set games in the following way. First, we compute the set of states where the players have value 1 (i.e., for each player $i \in \{1, 2\}$, the set of states W_i where player i can ensure that she gets payoff arbitrarily close to 1). This can be done in polynomial time [5]. Second, consider the constant-sum reach-a-set game G' constructed from the concurrent reachability game G where the set of states W_1 and W_2 are converted to *sink* states and the objective for player i is to reach the set W_i. Then, we can show the value obtained by player 1 in the game G is equal to the value $v_1(s)$ obtained by player 1 in G'. This gives an NP ∩ co-NP algorithm for two player zero sum reachability games, improving the previously known EXPTIME bound [6].

Theorem 3. *The value of a two-player concurrent reachability game can be approximated within ϵ-tolerance in NP ∩ co-NP, for a fixed $\epsilon > 0$.*

The natural question at this point is whether there is a polynomial time algorithm for concurrent zero sum reachability games. Since simple stochastic games [3] can be easily reduced to concurrent reachability games, a polynomial time algorithm for this problem will imply a polynomial time algorithm for simple stochastic games. This is a long standing open problem.

5   Games with Turns

An n-player stochastic game is *turn-based* (or perfect information) if at each state, there is exactly one player who determines the next state. Formally, we extend the action sets A_i for $i = 1, \dots, n$ to be state dependent, that is, for each state $s \in S$, there are action sets A_{is} for $i = 1, \dots, n$, and we restrict the action sets so that for any $s \in S$, there is at most one $i \in \{1, \dots, n\}$ such that $|A_{is}| > 1$.

We consider payoff functions that are index sets of *Borel sets* (see e.g., [12] for definitions), that is, given a Borel set B, we consider a payoff function χ_B that assigns a payoff 1 to a play that is in the set B, and 0 to a play that is not in the set B. With abuse of notation, we identify the set B with the payoff function χ_B. We consider turn based games in which each player is given a Borel payoff B_i. A two-player Borel game is *zero sum* if the payoff set B of one player is the complement $S^\omega \setminus B$ of the other player, that is, the players have strictly opposing objectives. A deep result by Martin shows that two player zero sum infinite stochastic games with Borel payoffs have a value [18]. The proof constructs, for each real $v \in (0, 1]$ a zero sum turn-based deterministic infinite-state game with Borel payoff such that a (pure) winning strategy for player 1 in this game can be used to construct a (mixed) winning strategy in the original game that assures player 1 a payoff of at least v. From the determinacy of turn-based deterministic games with Borel payoffs [17], the existence of value

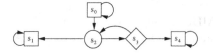

Fig. 2. A turn-based probabilistic game

in zero sum stochastic games with Borel payoffs follows. Moreover, the proof constructs ϵ-optimal mixed winning strategies. A careful inspection of Martin's proof in the special case of turn-based probabilistic games shows that the ϵ-optimal strategies of player 1 are pure. This is because the mixed strategies are derived from solving certain one-shot concurrent games at each round. In our special case these one-shot games have pure winning strategies since only one player has a choice of actions.

Lemma 10. *For each $\epsilon > 0$ there is a pure strategy π_1 of player 1 such that for all strategies π_2 of player 2 we have $E_s^{\pi_1,\pi_2}\{f\} \geq v - \epsilon$.*

Theorem 4. *For each $\epsilon > 0$ there exists an ϵ-Nash equilibrium in every n-player turn based probabilistic game with Borel payoffs.*

Proof. Our construction is based on a general construction from repeated games. The basic idea is that player i plays optimal strategies in the zero sum game against all other players, and any deviation by player i is punished indefinitely by the other players by playing ϵ-optimal spoiling strategies in the zero sum game against player i (see, e.g., [19, 28]). Let player i have the payoff set B_i, for $i = 1, \ldots, n$. Consider the n zero sum games played between i and the team $[n] \setminus \{i\}$, with the winning objective B_i for i. By Lemma 10 here is a pure ϵ-optimal strategy π_i^i for player i in this game, and a pure ϵ-optimal spoiling strategy for players $j \neq i$. This spoiling strategy induces a strategy π_i^j for each player $j \neq i$. Now consider the strategy τ^i for player i as follows. Player i plays the strategy π_i^i as long as all the other players j play π_j^i and switch to π_j^i as soon as some player j deviates. Since the strategies are pure, any deviation is immediately noted. The strategies τ^i for $i = 1, \ldots, n$ form an ϵ-Nash equilibrium. ∎

Note that the construction above for probabilistic Borel games guarantees only ϵ-optimality. Example 1 shows that there are two player turn-based probabilistic zero sum games where only ϵ-optimal strategies exist, and optimal strategies do not exist. Hence Theorem 4 cannot be strengthened to Nash equilibrium.

Example 1. Consider the turn-based probabilistic game shown in Figure 2. At state s_0 player 1 chooses between two actions: $a_1 = s_0 \to s_1$ and $a_2 = s_0 \to s_2$. At state s_1 the play actions to state s_2 and s_3 with probability $\frac{1}{2}$. At state s_3 player 2 chooses between two actions: $b_1 = s_3 \to s_2$ and $b_2 = s_3 \to s_4$. The game is zero sum, and the objective B for player 1 is:

$$B = \{\pi_1 \pi_2 \ldots \mid \pi_1 = s_0 \wedge \exists k. \ \pi_k = s_1\}$$
$$\cup \{\pi_1 \pi_2 \ldots \mid \pi_1 = s_0 \wedge \exists k. \ \pi_1 = \pi_2 = \ldots = \pi_k = s_0 \ \wedge \ \pi_{k+1} \neq s_0$$
$$\wedge \ \exists j. \ k + 1 \leq j \leq 3k. \ \pi_j = s_4\}$$

Informally, the winning condition for player 1 is as follows. Player 1 wins if either the game reaches the state s_1, or the game reaches s_4 and the number of visits to s_0 is greater than the number of visits to s_3 before s_4 is hit. Player 1 can get payoff arbitrarily close to 1 in the following way. For any $\epsilon > 0$, consider the strategy σ_1 for player 1 that chooses the action $a_1 = s_0 \rightarrow s_0$ for k times and then chooses the action $a_2 = s_0 \rightarrow s_2$, where $\epsilon > \frac{1}{2^k}$. The strategy σ_1 ensures that player 1 wins with probability at least $1 - \epsilon$. On the other hand, player 1 has no optimal strategy. Consider any strategy σ_1 for player 1 that chooses the action $a_2 = s_0 \rightarrow s_2$ after k steps. The strategy for player 2 is as follows: choose action $b_1 = s_3 \rightarrow s_2$ for $k+1$ times and then choose the action $b_2 = s_3 \rightarrow s_4$. The probability that player 1 wins is at most $1 - \frac{1}{2^{k+1}}$. Hence, player 1 has ϵ-optimal strategies for every $\epsilon > 0$, but no optimal strategy. ∎

Exact Nash equilibria do exist in special cases. First, the determinacy result for zero sum turn based deterministic games with payoffs corresponding to Borel sets [17] shows the existence of optimal winning strategies. Hence, the construction of Theorem 4 gives exact Nash equilibria.

Corollary 1. *Every n-player turn-based deterministic game with Borel payoffs has a Nash equilibrium with a pure strategy profile.*

Second, in the special case of turn-based probabilistic games with parity winning conditions, pure and memoryless optimal winning strategies exist for two player zero-sum case [2]. Moreover, the pure memoryless optimal strategies can be computed in NP ∩ co-NP. (Notice that the winning condition in Example 1 is not ω-regular.) Therefore we have the following.

Theorem 5. *There exists a Nash equilibrium with pure strategy profile in every turn-based probabilistic game with parity payoff conditions. The value profile of some Nash equilibrium can be computed in FNP.*

Acknowledgments. We are grateful to Prof. William Sudderth for pointing out an error in the earlier version of the paper. We thank Antar Bandyopadhyay and Tom Henzinger for many insightful discussions.

References

1. D. Blackwell and T.S. Ferguson. The big match. *Annals of Mathematical Statistics*, 39:159–163, 1968.
2. K. Chatterjee, M. Jurdziński, and T.A. Henzinger. Quantitative stochastic parity games. In *SODA 04*, pages 114–123, 2004.
3. A. Condon. The complexity of stochastic games. *Information and Computation*, 96(2):203–224, 1992.
4. V. Conitzer and T. Sandholm. Complexity results about Nash equilibria. In *IJCAI 03*, pages 765–771, 2003.

5. L. de Alfaro, T.A. Henzinger, and O. Kupferman. Concurrent reachability games. In *FOCS 98*, pages 564–575. IEEE Computer Society Press, 1998.
6. L. de Alfaro and R. Majumdar. Quantitative solution of omega-regular games. In *STOC 01*, pages 675–683. ACM Press, 2001.
7. E.A. Emerson and C. Jutla. Tree automata, mu-calculus and determinacy. In *FOCS 91*, pages 368–377. IEEE Computer Society Press, 1991.
8. H. Everett. Recursive games. In *Contributions to the Theory of Games III*, volume 39 of *Annals of Mathematical Studies*, pages 47–78, 1957.
9. J. Filar and K. Vrieze. *Competitive Markov Decision Processes*. Springer-Verlag, 1997.
10. A.M. Fink. Equilibrium in a stochastic n-person game. *Journal of Science of Hiroshima University*, 28:89–93, 1964.
11. J.F. Nash Jr. Equilibrium points in n-person games. *Proceedings of the National Academy of Sciences USA*, 36:48–49, 1950.
12. A. Kechris. *Classical Descriptive Set Theory*. Springer, 1995.
13. P.R. Kumar and T.H. Shiau. Existence of value and randomized strategies in zero-sum discrete-time stochastic dynamic games. *SIAM J. Control and Optimization*, 19(5):617–634, 1981.
14. R.J. Lipton, E. Markakis, and A. Mehta. Playing large games using simple strategies. In *EC 03*, pages 36–41. ACM Press, 2003.
15. A. Maitra and W.D. Sudderth. Borel stay-in-a-set games. *International Journal of Game Theory*, 32:97–108, 2003.
16. Z. Manna and A. Pnueli. *The Temporal Logic of Reactive and Concurrent Systems: Specification*. Springer-Verlag, 1992.
17. D.A. Martin. Borel determinacy. *Annals of Mathematics*, 102(2):363–371, 1975.
18. D.A. Martin. The determinacy of Blackwell games. *The Journal of Symbolic Logic*, 63(4):1565–1581, 1998.
19. G. Owen. *Game Theory*. Academic Press, 1995.
20. C.H. Papadimitriou. On the complexity of the parity argument and other inefficient proofs of existence. *JCSS*, 48(3):498–532, 1994.
21. C.H. Papadimitriou. Algorithms, games, and the internet. In *STOC 01*, pages 749–753. ACM Press, 2001.
22. T.E.S. Raghavan and J.A. Filar. Algorithms for stochastic games — a survey. *ZOR — Methods and Models of Operations Research*, 35:437–472, 1991.
23. P. Secchi and W.D. Sudderth. Stay-in-a-set games. *International Journal of Game Theory*, 30:479–490, 2001.
24. L.S. Shapley. Stochastic games. *Proc. Nat. Acad. Sci. USA*, 39:1095–1100, 1953.
25. M. Sobel. Noncooperative stochastic games. *Ann. Math. Stat.*, 42:1930–1935, 1971.
26. F. Thuijsman. *Optimality and Equilibria in Stochastic Games*. CWI-Tract 82, CWI, Amsterdam, 1992.
27. F. Thuijsman and T.E.S. Raghavan. Perfect information stochastic games and related classes. *International Journal of Game Theory*, 26:403–408, 1997.
28. N. Vieille. Two player stochastic games I: a reduction. *Israel Journal of Mathematics*, 119:55–91, 2000.
29. N. Vieille. Two player stochastic games II: the case of recursive games. *Israel Journal of Mathematics*, 119:93–126, 2000.
30. J. von Neumann and O. Morgenstern. *Theory of games and economic behavior*. Princeton University Press, 1947.
31. B. von Stengel. Computing equilibria for two-person games. *Chapter 45, Handbook of Game Theory*, 3:1723–1759, 2002. (editors R.J. Aumann and S. Hart).

A Bounding Quantifier

Mikołaj Bojańczyk[*]

Uniwersytet Warszawski, Wydział MIM, Banacha 2, Warszawa, Poland

Abstract. The logic MSOL+$\mathbb{B}$ is defined, by extending monadic second-order logic on the infinite binary tree with a new *bounding quantifier* $\mathbb{B}$. In this logic, a formula $\mathbb{B}X.\varphi(X)$ states that there is a finite bound on the size of sets satisfying $\varphi(X)$. Satisfiability is proved decidable for two fragments of MSOL+$\mathbb{B}$: formulas of the form $\neg\mathbb{B}X.\varphi(X)$, with φ a $\mathbb{B}$-free formula; and formulas built from $\mathbb{B}$-free formulas by nesting $\mathbb{B}$, $\exists$, $\vee$ and $\wedge$.

1 Introduction

Using monadic second-order logic over infinite trees one cannot express properties such as: "there exists bigger and bigger sets such that..." or "there is a bound on the size of sets such that...". In this paper we present decision procedures for an extension of MSOL where such properties are definable.

The need for such cardinality constraints occurs naturally in applications. For instance, a graph that is interpreted in the full binary tree using monadic second-order logic (MSOL) is known to have bounded tree-width if and only if it does not contain bigger and bigger complete bipartite subgraphs [1]. Another example: a formula of the two-way μ-calculus [12] has a finite model if and only if it has a tree model in which there is a bound on the size of certain sets [2]. Sometimes boundedness is an object of interest in itself, cf. [4], where pushdown games with the bounded stack condition are considered.

In light of these examples, it seems worthwhile to consider the logic MSOL+$\mathbb{B}$ obtained from MSOL by adding two new quantifiers $\mathbb{B}$ and $\mathbb{U}$, which express properties like the ones just mentioned. Let $\psi(X)$ be a formula expressing some property of a set X in a labeled infinite tree. The formula $\mathbb{B}X.\psi(X)$ is satisfied in those trees t where there is a finite bound – which might depend on t – on the size of sets F such that the tree $t[X := F]$ satisfies $\psi(X)$. We also consider the dual quantifier $\mathbb{U}$, which states that there is no finite bound.

Adding new constructions to MSOL has a long history. A notable early example is a paper of Elgot and Rabin [6], where the authors investigated what predicates P can be added to MSOL over $\langle \mathbb{N}, \leq \rangle$ while preserving decidability of the theory. Among the positive examples they gave are monadic predicates representing the sets $\{i! : i \in \mathbb{N}\}$, $\{i^k : i \in \mathbb{N}\}$ and $\{k^i : i \in \mathbb{N}\}$. This line of

---

[*] Supported by the European Community Research Training Network GAMES and Polish KBN grant No. 4 T11C 042 25.

J. Marcinkowski and A. Tarlecki (Eds.): CSL 2004, LNCS 3210, pp. 41–55, 2004.
© Springer-Verlag Berlin Heidelberg 2004

research was recently continued by Carton and Thomas in [5], where the list was extended by so called *morphic* predicates.

A construction similar to our bounding quantifier can be found in [7], where Klaedtke and Ruess consider extending MSOL on trees and words with cardinality constraints of the form:

$$|X_1| + \cdots + |X_r| < |Y_1| + \cdots + |Y_s|.$$

Although MSOL with these cardinality constraints is in general undecidable, the authors show a decision procedure for a fragment of the logic, where, among other restrictions, quantification is allowed only over finite sets. Interestingly, MSOL+$\mathbb{B}$ is definable using cardinality constraints, although it does fall outside the aforementioned fragment and cannot be described using the techniques in [7]:

$$\mathbb{B}X.\psi \quad \text{iff} \quad \exists Y.\text{Finite}(Y) \wedge \forall X.(\psi(X) \to |X| < |Y|).$$

Finally, a quantifier that also deals with cardinality can be formulated based on the results of Niwiński in [8]. It is not however the size of sets satisfying $\psi(X)$, but the number of such sets that is quantified. More precisely, a binary tree t satisfies $\exists^{\mathfrak{c}}X.\psi(X)$ if there are continuum sets F such that $t[X := F]$ satisfies $\psi(X)$. This quantifier, it turns out, is definable in MSOL, and thus its unrestricted use retains decidability.

Our bounding quantifier $\mathbb{B}$, however, is *not* definable in MSOL. Using the bounding quantifier, one can define nonregular languages and hence the question: is satisfiability of MSOL+$\mathbb{B}$ formulas decidable? In this paper we investigate this question and, while being unable to provide an exhaustive answer, we present decision procedures for two nontrivial fragments of MSOL+$\mathbb{B}$.

This investigation leads us to identify a class of tree languages, new to our knowledge, which we call quasiregular tree languages. A set of infinite trees is *L-quasiregular* if it coincides with the regular language L over the set of regular trees and, moreover, is the sum of some family of tree regular languages. The intuition behind an L-quasiregular language is that it is a slight non-regular variation over the language L, yet in most situations behaves the same way as L.

On the one hand, quasiregular languages are simple enough to have decidable emptiness: an L-quasiregular language is nonempty iff L is nonempty. On the other hand, quasiregular languages are powerful enough to allow nontrivial applications of the bounding quantifier: they are closed under bounding quantification, existential quantification, conjunction and disjunction. This yields the decidability result:

Theorem 43
The satisfiability problem for existential bounding formulas, i. e. ones built from an MSOL core by application of $\mathbb{B}$, $\exists$, $\wedge$ and $\vee$ is decidable.

Unfortunately quasiregular languages do not capture all of MSOL+$\mathbb{B}$. For instance, they are not closed under complementation, hence Theorem 43 gives no insight into properties that use the dual quantifier $\mathbb{U}$.

For this reason, we also conduct a separate analysis of the $\mathbb{U}$ quantifier (note that satisfiability for $\mathbb{U}$ is related to *validity* for $\mathbb{B}$). By inspection of an underlying automaton, we prove:

Theorem 47
Satisfiability is decidable for formulas of the form $\mathbb{U}X.\psi$, *where* ψ *is MSOL.*

We are, however, unable to extend this result in a fashion similar to Theorem 43, by allowing for non-trivial nesting.

The plan of the paper is as follows. In Section 2, we briefly survey possible applications of the bounding quantifier. After the preliminaries in Section 3, we introduce the quantifier in Section 4. In Sections 4.1 and 4.2 we prove decidability for bounding existential formulas, while in Section 4.3, we prove decidability for formulas which use the unbounding quantifier outside an MSOL formula.

Acknowledgements. I would like to thank Igor Walukiewicz, Damian Niwiński and Thomas Colcombet for their valuable suggestions.

2 Applications

In this section we briefly and informally overview three possible applications. We would like to emphasize that in none of these cases does using the bounding quantifier give *new* results, it only simplifies proofs of existing ones.

The first application comes from graph theory. Sometimes a graph $G = (V, E)$ can be interpreted in the unlabeled full binary tree $\{0, 1\}^*$ via two formulas: a formula $\alpha(x)$ true for the vertices used to represent a vertex from V and a formula $\beta(x, y)$ representing the edge relation E. From [1], it follows that such a graph $G(\alpha, \beta)$ is of bounded tree-width if and only if there is a fixed bound N on the size n of full bipartite subgraphs $K_{n,n}$ of $G(\alpha, \beta)$. Given two sets $F, G \subseteq \{0, 1\}^*$ one can express using MSOL that these sets represent the left and right parts of a bipartite subgraph. The property that there exist bigger and bigger sets F, G encoding a bipartite graph can then, after some effort, be expressed as a formula of the form $\mathbb{U}Z.\psi(Z)$, where the unboundedness of only a single set Z is required. The validity of such a formula in the unlabeled tree can be verified using either one of the Theorems 43 and 47, hence we obtain conceptually simple decidability proof for the problem: "does a graph represented in the full binary tree have bounded tree-width?" [1]

Another application is in deciding the winner in a certain type of pushdown game. A *pushdown game* is a two-player game obtained from a *pushdown graph*. The vertices of the graph are the configurations $(q, \gamma) \in Q \times \Gamma^*$ of a pushdown automaton of state space Q and stack alphabet Γ, while the edges represent the transitions. The game is obtained by adding a partition of Q into states Q_0 of player 0 and states Q_1 of player 1, along with a *winning condition*, or set of plays in $(Q \times \Gamma^*)^{\mathbb{N}}$ that are winning for the player 0. In [4], the authors consider the *bounded stack* winning condition, where a play is winning for player 0 if there is a fixed finite bound on the size of the stacks appearing in it. Using a natural interpretation of the pushdown game in a binary tree, the fact that player 0 wins the game from a fixed position v is equivalent to the satisfiability of a formula

$$\exists S_0\ \forall S_1\ \mathbb{B}X.\ \psi(S_0, S_1, X, v)$$

in which $\psi(S_0, S_1, X, v)$ says that X represents a stack appearing in the unique play starting in vertex v and concordant with the strategies S_0 and S_1. We are able to quantify over strategies due to memoryless determinacy of the relevant game. Moreover, by a closer inspection of the game, one can show that $\forall S_1$ can be shifted inside the $\mathbb{B}$ quantifier, yielding an existential bounding formula whose satisfiability is decidable by Theorem 43.

Finally, the bounding quantifier can be applied to the following decision problem [3, 2]: "Is a given formula ψ of the modal μ-calculus with backward modalities satisfiable in some finite structure?" In [3] it is shown that the answer is yes iff a certain nonregular language L_ψ of infinite trees is nonempty. This language expresses the property that certain paths in a tree are of bounded length, and can easily be expressed using an existential quasiregular formula.

3   Preliminaries

In this section we define the basic notions used in the paper: infinite trees, regular languages of infinite trees and regular trees.

A *finite A-sequence* is a function $\boldsymbol{a} : \{0, \dots, n\} \to A$, while an *infinite A-sequence* is a function $\boldsymbol{a} : \mathbb{N} \to A$. We use boldface letters to denote sequences. Given a function $f : A \to B$ and an A-sequence $\boldsymbol{a}$, $f \circ \boldsymbol{a}$ is a well defined B-sequence. Often we will forsake the functional notation and write $\boldsymbol{a}_i$ instead of $\boldsymbol{a}(i)$. The length $|\boldsymbol{a}| \in \mathbb{N} \cup \{\infty\}$ of a sequence is the size of its domain. We use A^* to denote the set of finite A-sequences and $A^{\mathbb{N}}$ for the set of infinite A-sequences. The concatenation of two sequences $\boldsymbol{a}$ and $\boldsymbol{b}$, denoted by $\boldsymbol{a} \cdot \boldsymbol{b}$, is defined in the usual fashion.

Let Σ be some finite set, called the *alphabet*. An *infinite Σ-tree* is a function $t : \{0, 1\}^* \to \Sigma$. Therefore, all infinite trees have the same domain. We denote the set of infinite Σ-trees by $\mathrm{Trees}^{\infty}(\Sigma)$. An *infinite tree language over Σ* is any subset of $\mathrm{Trees}^{\infty}(\Sigma)$. Since we will only consider infinite trees in this paper and the next one, we will omit the word infinite and simply write Σ-tree and tree language. A node is any element of $\{0, 1\}^*$. We order nodes using the prefix relation $\leq$. Given $v \in \{0, 1\}^*$, the *subtree of t rooted in v* is the tree $t|_v$ defined by:

$$t|_v(w) = t(v \cdot w).$$

A *regular tree* is a tree with finitely many distinct subtrees; the class of all regular trees is denoted by REG. An *infinite path* is any infinite sequence of nodes $\boldsymbol{\pi}$ such that:

$$\boldsymbol{\pi}_0 = \varepsilon \quad \boldsymbol{\pi}_1 = \boldsymbol{\pi}_0 \cdot a_0 \quad \boldsymbol{\pi}_2 = \boldsymbol{\pi}_1 \cdot a_1 \quad \cdots \qquad a_i \in \{0, 1\}\ .$$

Given two nodes $v < w$, we define the set $\mathrm{Bet}(v, w)$ of elements between v and w as $v \cdot \{0, 1\}^* \setminus w \cdot \{0, 1\}^*$.

Let Σ be an alphabet and $*$ a letter outside Σ. A Σ-*context* is any $\Sigma \cup \{*\}$-tree C where the label $*$ occurs only once, in a position called the *hole* of C. We don't require this position to be a leaf, since there are no leaves in an infinite tree, but all nodes below the hole are going to be irrelevant to the context. The *domain* dom(C) of a context C is the set of nodes that are not below or equal to the hole. Given a Σ-tree t and a context $C[]$ whose hole is v, the tree $C[t]$ is defined by:

$$C[t](w) = \begin{cases} t(u) & \text{if } w = v \cdot u \text{ for some } u \in \{0,1\}^*; \\ C(w) & \text{otherwise.} \end{cases}$$

The composition of two contexts C and D is the unique context $C \cdot D$ such that $(C \cdot D)[t] = C[D[t]]$ holds for all trees t. We do not use multicontexts for infinite trees.

3.1   Nondeterministic Tree Automata and Regular Tree Languages

As in the case of finite trees, regular languages of infinite trees can be defined both using automata and monadic second-order logic. The two approaches are briefly described in this section.

Definition 31.   [Parity condition] A sequence $\boldsymbol{a} \in A^{\mathbb{N}}$ of numbers belonging to some finite set of natural numbers A is said to satisfy the *parity condition* if the smallest number occurring infinitely often in $\boldsymbol{a}$ is even.

A *nondeterministic tree automaton with the parity condition* is a tuple

$$\mathcal{A} = \langle Q, \Sigma, q_I, \delta, \Omega \rangle$$

where Q is a finite set of *states*, Σ is the finite *input alphabet*, $q_I \in Q$ is the *initial state*, $\delta \subseteq Q \times \Sigma \times Q \times Q$ is the *transition relation* and $\Omega : Q \to \mathbb{N}$ is the *ranking function*. Elements of the finite image $\Omega(Q)$ are called *ranks*. A *run* of $\mathcal{A}$ over a Σ-tree t is any Q-tree ρ such that

$$\langle \rho(v), t(v), \rho(v \cdot 0), \rho(v \cdot 1) \rangle \in \delta \qquad \text{for every } v \in \{0,1\}^*.$$

The run ρ is *accepting* if for every infinite path $\boldsymbol{\pi}$, the sequence of ranks $\Omega \circ \rho \circ \boldsymbol{\pi}$ satisfies the parity condition. The automaton *accepts a tree t from state* $q \in Q$ if there is some accepting run with state q labeling the root. A tree is *accepted* if it is accepted from the initial state q_I. The *language of* $\mathcal{A}$, denoted $L(\mathcal{A})$, is the set of trees accepted by $\mathcal{A}$; such a language is said to be *regular*. An automaton is *nonempty* if and only if its language is.

We say two trees s and t are *equivalent for an automaton* $\mathcal{A}$, which is denoted $s \simeq_{\mathcal{A}} t$, if for every state q of $\mathcal{A}$, the tree s is accepted from q if and only if the tree t is. If the trees s and t are equivalent for $\mathcal{A}$, then they cannot be distinguished by a context, i.e. for every context $C[]$, the tree $C[s]$ is accepted by $\mathcal{A}$ if and only if the tree $C[t]$ is.

We now proceed to define the logical approach to regular languages of infinite trees. Consider an alphabet $\Sigma = \{\sigma_1, \ldots, \sigma_n\}$. As in the finite tree case, with a Σ-tree, we associate a relational structure

$$\underline{t} = \langle \{0,1\}^*, S_0, S_1, \leq, \underline{\sigma}_1^t, \ldots, \underline{\sigma}_n^t \rangle.$$

The relations are interpreted as follows: S_0 is the set of left sons $\{0,1\}^*0$, S_1 is the set of right sons $\{0,1\}^*1$, $\leq$ is the prefix ordering, while $\underline{\sigma}_i^t$ is the set of nodes that are labeled by the letter σ_i.

With a sentence ψ of monadic second-order logic we associate the language $L(\psi)$ of trees t such that $\underline{t}$ satisfies ψ. Such a language is said to be *MSOL-definable*. A famous result of Rabin [9] says that a language of infinite trees is MSOL-definable if and only if it is regular.

4   The Bounding Quantifier

The logic MSOL+$\mathbb{B}$ is obtained from MSOL by adding two quantifiers: the *bounding quantifier* $\mathbb{B}$, and its dual *unbounding quantifier* $\mathbb{U}$, which we define here using infinitary disjunction and conjunction:

$$\mathbb{B}^i X.\varphi := \forall X.(\varphi(X) \Rightarrow |X| < i) \qquad \mathbb{U}^i X.\varphi := \exists X.(\varphi(X) \wedge |X| \geq i)$$

$$\mathbb{B}X.\ \varphi := \bigvee_{i \in \mathbb{N}} \mathbb{B}^i X.\varphi \qquad\qquad \mathbb{U}X.\ \varphi := \bigwedge_{i \in \mathbb{N}} \mathbb{U}^i X.\varphi$$

MSOL+$\mathbb{B}$ defines strictly more languages than MSOL (see Fact 44), hence it is interesting to consider decidability of the following problem:

Is a given formula of MSOL+$\mathbb{B}$ satisfiable in some infinite tree?

The remainder of this paper is devoted to this question. Although unable to provide a decision procedure for the whole logic, we do identify two decidable fragments. The first, existential bounding formulas, is proved decidable in Sections 4.1 and 4.2, while the second, formulas of the form $\mathbb{U}X.\psi$ with ψ in MSOL, is proved decidable in Section 4.3.

4.1   Quasiregular Tree Languages

Before we proceed with the proof of Theorem 43, we define the concept of a quasiregular tree language. We then demonstrate some simple closure properties of quasiregular tree languages and, in Section 4.2, show that quasiregular tree languages are closed under bounding quantification. These closure properties, along with the decidable nonemptiness of quasiregular languages, yield the decision procedure for existential bounding formulas found in Theorem 43.

For technical reasons, we will find it henceforth convenient to work on trees where the alphabet is the powerset $P(\Sigma)$ of some set Σ. The same results would hold for arbitrary alphabets, but the notation would be more cumbersome. By $\mathrm{Val}(\Sigma)$ we denote the set of $P(\Sigma)$-trees. Elements of the set Σ will be treated as set variables, the intuition being that a tree in $\text{Val}(\Sigma)$ represents a valuation of the variables in Σ. Given a tree $t \in \text{Val}(\Sigma)$ and a set $F \subseteq \{0, 1\}^*$, the tree

$$t[X := F] \in \text{Val}(\Sigma \cup \{X\})$$

is defined by adding the element X to the labels of all nodes in F and removing it, if necessary, from all the other nodes.

Bounding quantification for an arbitrary tree language $L \subseteq \text{Val}(\Sigma)$ is defined as follows. A tree $t \in \text{Val}(\Sigma \setminus \{X\})$ belongs to the language $\mathbb{B}X.L$ if there is some finite bound on the size of sets F such that the tree $t[X := F]$ belongs to L.

We now give the key definition of a quasiregular tree language.

Definition 41 (Quasiregular Language). Let L be a regular tree language. A tree language K is L-*quasiregular* if

 - $K \cap \text{REG} = L \cap \text{REG}$, and
 - K is the union of some family of regular tree languages

A tree language is *quasiregular* if it is L-quasiregular for some regular language L. For the rest of Section 4.1 we will use the letter L for regular languages and the letter K for quasiregular ones.

Lemma 1. *If K is L-quasiregular, then $K \subseteq L$.*

Proof
Let $\{L_i\}_{i \in I}$ be the family of regular tree languages whose union is K. We will show that each language L_i is a subset of L. Indeed, over regular trees L_i is a subset of L, since K and L agree over regular trees. This implies the inclusion $L_i \subseteq L$ for arbitrary trees, since otherwise the regular language $L_i \setminus L$ would be nonempty and therefore, by Rabin's Basis Theorem [10], contain a regular tree. $\square$

The following easy fact shows that emptiness is decidable for quasiregular tree languages given an appropriate presentation:

Fact 42. If K is L-quasiregular, then K is nonempty iff L is nonempty.

Proof
If L is nonempty, then it contains by Rabin's Basis Theorem a regular tree and hence K must contain this same tree. The other implication follows from Lemma 1. $\square$

In particular, every nonempty quasiregular language contains a regular tree. For a variable X we define the *projection function* Π_X which given a tree returns the tree with X removed from all the labels. Projection is the tree language operation corresponding to existential quantification, as testified by the following equation:

$$L(\exists X.\psi) = \Pi_X(L(\psi)).$$

A set $F \subseteq \{0,1\}^*$ is *regular* if the unique tree $t[X := F] \in \mathrm{Val}(\{X\})$ is regular. Equivalently, F is regular if it is a regular word language. The following is a standard result:

Lemma 2. *If a regular tree t belongs to the projection $\Pi_X(L)$ of a regular language L, then $t[X := F]$ belongs to L for some regular set F.*

Proof

Since t is a regular tree, the set $\{t\}$ is a regular tree language and so is $\Pi_X^{-1}(\{t\})$. Therefore the intersection $L \cap \Pi_X^{-1}(\{t\})$ is regular and nonempty and, by Rabin's Basis Theorem, contains some regular tree. Obviously, the X component in this tree must be a regular set. □

Now we are ready to show some basic closure properties of quasiregular languages:

Lemma 3. *Quasiregular languages are closed under projection, intersection and union.*

Proof

The cases of intersection and union are trivial; we will only do the proof for projection. Let K be L-quasiregular. We will show that the projection $\Pi_X(K)$ is $\Pi_X(L)$-quasiregular. First we prove that $\Pi_X(K)$ is the union of a family of regular languages. By assumption, K is the union some family of regular tree languages $\{L_i\}_{i \in I}$. But then

$$\Pi_X(K) = \Pi_X\left(\bigcup_{i \in I} L_i\right) = \bigcup_{i \in I} \Pi_X(L_i)$$

and, since regular tree languages are closed under projection, $\Pi_X(K)$ is the union of some family of regular tree languages.

We also need to show that for every regular tree t,

$$t \in \Pi_X(K) \quad \text{iff} \quad t \in \Pi_X(L).$$

The left to right implication follows from Lemma 1. The right to left implication follows from the fact that if $t \in \Pi_X(L)$ then, by Lemma 2, for some regular set F, $t[X = F] \in L$. Since the tree $t[X = F]$ is regular, it also belongs to K and hence t belongs to $\Pi_X(K)$.

□

4.2   Closure Under Bounding Quantification

In this section, we show that quasiregular tree languages are closed under application of the bounding quantifier. This, together with the closure properties described in Lemma 3, yields a decision procedure for the fragment of MSOL+$\mathbb{B}$ that nests $\mathbb{B}$ along with existential quantification, conjunction and disjunction.

Recall that a *chain* is any set of nodes that is linearly ordered by $\leq$. We say that a chain C is a *trace path* of a set of nodes F if the set $F \cap \mathrm{Bet}(v, u)$ is nonempty for all nodes $v < w$ in C.

Lemma 4. *A set of at least 3^n nodes has a trace path of size n. An infinite set has an infinite trace path.*

Let $t \in \mathrm{Val}(\Sigma)$ and $L \subseteq \mathrm{Val}(\Sigma \cup \{X\})$. An (L, X)-*bad chain* in the tree t is an infinite chain C whose every finite subset is a trace path of some set F such that $t[X := F] \in L$. The set of trees containing no (L, X)-bad chain is denoted by $\mathbb{C}X.L$. Bad chains have the desirable property of being MSOL definable, as testified by:

Lemma 5. *If L is regular then $\mathbb{C}X.L$ is regular.*

Lemma 6. *If K is L-quasiregular then $\mathrm{REG} \cap \mathbb{C}X.K = \mathrm{REG} \cap \mathbb{C}X.L$.*

Proof
This follows from the fact that for a finite (and therefore regular) node set F and a regular tree t, the tree $t[X := F]$ is regular, and hence belongs to L if and only if it belongs to K. □

Lemma 7. *Let L be a regular tree language. A regular tree that belongs to $\mathbb{C}X.L$ also belongs to $\mathbb{B}X.L$.*

Proof
Consider a regular tree t with m distinct subtrees. Let $\mathcal{A}$ being some automaton recognizing L with k being the index of the relation $\simeq_{\mathcal{A}}$. Setting n to be $3^{k \cdot m + 1}$, we will show that if t does not belong to the language $\mathbb{B}^n X.L$, then an (L, X)-bad chain must exist.

Consider indeed a set F of at least n nodes such that the tree $t[X := F]$ belongs to L. By Lemma 4, this set has a trace path with more than $k \cdot m$ nodes. Let $v < w$ be two nodes on this trace path such that

$$t[X := F]|_v \simeq_{\mathcal{A}} t[X := F]|_w \qquad \text{and} \qquad t|_v = t|_w .$$

Such two nodes exist by virtue of the trace path's size. Moreover, since v and w are on the trace path, the intersection $F \cap \mathrm{Bet}(v, w)$ is nonempty. Let $u \in \{0, 1\}^*$ be such that $w = v \cdot u$.

We claim that the chain $\{v \cdot u^i : i \in \mathbb{N}\}$ is a bad chain. For this, we will show that for every $i \in \mathbb{N}$, the subchain $C_i = \{v \cdot u^j : j \leq i\}$ can be expanded to a set F_i satisfying $t[X := F_i] \in L$.

This is done by pumping i times the part of the set F between v and w. Consider the following partition of F:

$$F_1 = \{u' : u' < v\} \cap F \qquad F_2 = \mathrm{Bet}(v, w) \cap F \qquad F_3 = \{u' : u' > w\} \cap F$$

One can easily check that the following set F_i contains the subchain C_i:

$$F_i \quad = \quad F_1 \quad \cup \quad \bigcup_{j \in \{0, \ldots, i\}} v \cdot u^j \cdot v^{-1} \cdot F_2 \quad \cup \quad v \cdot u^i \cdot v^{-1} \cdot F_3.$$

Moreover, since for all $j \in \{0, \ldots, i\}$, the equivalence

$$t[X := F_j]_{v \cdot u^j} \simeq_{\mathcal{A}} t[X := F]|_w$$

holds, the tree $t[X := F_i]$ belongs to L. $\qquad\square$

Using the Lemma 7 above, we can show that quasiregular tree languages are closed under application of the bounding quantifier.

Lemma 8. *If K is quasiregular, then so is $\mathbb{B}X.K$.*

Proof

If K is quasiregular, then it is a union $\bigcup_{i \in I} L_i$ of some family of regular tree languages. Therefore $\mathbb{B}X.$ K is also a union regular tree languages:

$$\mathbb{B}X.K = \bigcup_{i \in I} \bigcup_{j > 0} \mathbb{B}^j X. \ L_i.$$

Let L be such that K is L-quasiregular. We will show:

$$\mathbb{C}X.L \cap \mathrm{REG} = \mathbb{B}X.K \cap \mathrm{REG}.$$

The right to left inclusion follows from Lemma 6 and the simple inclusion $\mathbb{B}X.K \subseteq \mathbb{C}X.K$. The left to right inclusion follows from Lemma 7. $\qquad\square$

Putting together the closure properties of quasiregular tree languages proved in this and the previous section, we obtain:

Theorem 43
The satisfiability problem for existential bounding formulas, i.e. ones built from arbitrary MSOL formulas by application of $\mathbb{B}$, $\exists$, $\wedge$ and $\vee$ is decidable.

Proof

By Lemmas 3 and 8, the language $L(\psi)$ of an existential bounding formula is L-quasiregular for some effectively obtained regular tree language L. By Fact 42, the emptiness of $L(\psi)$ is equivalent to the emptiness of L. $\qquad\square$

Unfortunately, we cannot hope to extend the quasiregular tree language approach to decide all possible nestings of the bounding quantifier, as certified by the following Fact:

Fact 44. *Even for regular L, $\neg\mathbb{B}X.L$ is not necessarily quasiregular.*

Proof

The language L in question is obtained from a formula ψ with free variables X and Y. This formula states that Y contains no infinite subchains and that X is a subchain of Y.

In a regular tree $t \in \mathrm{Val}(\{X, Y\})$ with n distinct subtrees, a subchain of Y can be of size at most n – otherwise Y has an infinite subchain and ψ does not hold. Therefore $\neg\mathbb{B}X.L(\psi)$ is a nonempty language without a regular tree and cannot be quasiregular. $\qquad\square$

4.3   The Unbounding Quantifier

In this section we present a procedure which, given a regular language $L \subseteq \mathrm{Val}(\Sigma)$ and a variable $X \in \Sigma$, decides whether the language $\mathbb{U}X.L$ is nonempty. This implies that satisfiability is decidable for formulas of the form $\mathbb{U}X.\psi(X)$, where ψ is in MSOL. Unfortunately, we are unable to extend this decision procedure to accommodate nesting, the way we did in Theorem 43. On the other hand though, the procedure runs in polynomial time in the size of an input parity automaton.

In order to help the reader's intuition a bit, we will begin our analysis by debunking a natural, yet false, idea: for every regular language L there is some $n \in \mathbb{N}$ such that the language $\mathbb{U}X.L$ is nonempty if and only if the language $\mathbb{U}^n X.L$ is.

The intuition behind this idea would be that a pumping process should inflate arbitrarily a set F satisfying $t[X := F] \in L$ once it has reached some threshold size. The problem, however, is that a tree may contain labels which are not part of the set F, and the pumping might violate this labeling. A suitable counterexample is the following language $L \subseteq \mathrm{Val}(\{X, Y\})$:

X is a subset of Y and Y is a finite set.

Obviously the language $\mathbb{U}X.L$ is empty, yet for every $n \in \mathbb{N}$, the language $\mathbb{U}^n X.L$ is nonempty. We will have to bear such issues in mind in the proofs below, taking care that we pump only the part of the labeling corresponding to X.

Let us fix a set Σ, a regular language $L \subseteq \mathrm{Val}(\Sigma)$ and a variable $X \in \Sigma$ for the rest of this section. We will use $\hat{\Sigma}$ to denote the set $\Sigma \setminus \{X\}$. Analogously to the "language" definition of $\mathbb{B}X.L$ in Section 4.1, we say a tree $t \in \mathrm{Val}(\hat{\Sigma})$ belongs to $\mathbb{U}X.L$ if there is *no* finite bound on the size of sets F such that $t[X := F] \in L$.

An infinite sequence of nodes $\boldsymbol{v}$ is *increasing* if $\boldsymbol{v}_i < \boldsymbol{v}_{i+1}$ holds for all $i \in \mathbb{N}$. A family of node sets $\mathcal{F}$ is *traced* by an increasing sequence $\boldsymbol{v}$ of nodes if for all $i \in \mathbb{N}$,

$$|F \cap \mathrm{Bet}(\boldsymbol{v}_i, \boldsymbol{v}_{i+1})| \geq i \qquad \text{for some } F \in \mathcal{F} \ .$$

Lemma 9. *A family that contains sets of unbounded size is traced.*

We fix now some nondeterministic parity automaton recognizing L:

$$\mathcal{A} = \langle Q, \Sigma, A, q_I, \delta, \Omega \rangle.$$

Without loss of generality, we assume that every state $q \in Q$ is used in some accepting run. The rest of this section is devoted to an analysis this automaton and to establishing a structural property equivalent to the nonemptiness of the language $\mathbb{U}X.L$.

A *descriptor* is any element of $Q \times \Omega(Q) \times Q$. With a $P(\Sigma)$-context C we associate the set $\mathit{Trans}(C)$ consisting of those descriptors (q, m, r) such that there is a run of $\mathcal{A}$ that starts in the root of C in state q and:

– The (finite) path of the run that ends in the hole of C uses states of rank at least m and ends in the state r.
– All the (infinite) paths of the run that do not go through the hole of C satisfy the parity condition.

The intuition is that $Trans(C)$ describes the possible runs of $\mathcal{A}$ which go through the context C. The compositions of two descriptors and then of two sets of descriptors are defined below (descriptors which do not agree on the state p do not compose):

$$(q, n, p) \cdot (p, m, r) = (q, \min(n, m), r) \qquad \text{(descriptor)}$$
$$X \cdot Y = \{x \cdot y : x \in X, y \in Y\} \qquad \text{(set of descriptors)}.$$

The descriptor set of the context composition $C \cdot D$ can be computed from the composition of the descriptor sets of the contexts C and D:

$$Trans(C \cdot D) = Trans(C) \cdot Trans(D). \tag{1}$$

We will also be using descriptors of $P(\hat{\Sigma})$-contexts. For a $P(\hat{\Sigma})$-context C and $k \in \mathbb{N}$, we define $Trans_k(C)$ to be the set

$$\bigcup_{F : |F| \geq k} Trans(C[X := F]).$$

A *schema* is a pair $R = (R^\bullet, R^\circ)$ of descriptor sets. A schema is meant to describe a $P(\hat{\Sigma})$-context, the intuition being that the R° descriptors can be obtained from any sets F, while the $R^\bullet$ descriptors are obtained from "large" sets. A $P(\hat{\Sigma})$-context C is said to k-*realize* a schema R if $R^\circ \subseteq Trans_0(C)$ and $R^\bullet \subseteq Trans_k(C)$. The composition $R \cdot S$ of two schemas $R = (R^\bullet, R^\circ)$ and $S = (S^\bullet, S^\circ)$ is defined to be the schema

$$R \cdot S = (R^\bullet \cdot S^\circ \cup R^\circ \cdot S^\bullet, R^\circ \cdot S^\circ).$$

The following obvious fact describes how composition of schemas corresponds to composition of $P(\hat{\Sigma})$-contexts.

Fact 45. Let R and S be schemas which are respectively k-realized by $P(\hat{\Sigma})$-contexts C and D. The schema $R \cdot S$ is k-realized by the context $C \cdot D$.

We now proceed to define the notion of an infinitary sequence. The intuition here is that an infinitary sequence exhibits the existence of a tree belonging to $\mathbb{U}X.L$, which is obtained by composing all the contexts in $\boldsymbol{C}$:

Definition 46 A sequence of schemas $\boldsymbol{R}$ is *infinitary* if both

– There is a sequence of $P(\hat{\Sigma})$-contexts $\boldsymbol{C}$ such that for every $n \in \mathbb{N}$ the schema $\boldsymbol{R}_n$ is n-realized by the context $\boldsymbol{C}_n$; and

- For some fixed state $r \in Q$ and all $n \in \mathbb{N}$, there is a state sequence $\boldsymbol{q}$ with $\boldsymbol{q}_0 = r$ such that:
  1. For $i < n$, $(\boldsymbol{q}_i, m, \boldsymbol{q}_{i+1}) \in R_i^{\circ}$ for some rank m;
  2. For $i = n$, $(\boldsymbol{q}_i, m, \boldsymbol{q}_{i+1}) \in R_i^{\bullet}$ for some rank m;
  3. For $i > n$, $(\boldsymbol{q}_i, m, \boldsymbol{q}_{i+1}) \in R_i^{\circ}$ for some even rank m.

Lemma 10. $\mathbb{U}X.L$ *is nonempty iff there exists an infinitary sequence.*

Proof
Consider first the right to left implication. Let $\boldsymbol{R}$ be the infinitary sequence with C and $r \in Q$ being the appropriate sequence of contexts and starting state from Definition 46. Let $t \in \mathrm{Val}(\hat{\Sigma})$ be the infinite composition of all successive contexts in C:

$$t = \boldsymbol{C}_0 \cdot \boldsymbol{C}_1 \cdot \boldsymbol{C}_2 \cdots$$

Let D be a context such that $(q_I, n, r) \in \mathit{Trans}(D)$ for some $n \in \Omega(Q)$. This context exists by our assumption on $\mathcal{A}$ not having useless states. Using the properties of the sequence $\boldsymbol{R}$ postulated in Definition 46, one can easily verify that the tree $D[t]$ belongs to $\mathbb{U}X.L$.

For the left to right implication, consider a tree t in $\mathbb{U}X.L$. From this tree we will extract an infinitary sequence. Consider the family of node sets

$$\{F \subseteq \{0, 1\}^* : t[X := F] \in L\}.$$

By assumption on t, this family contains sets of unbounded size. Therefore, by Lemma 9, it is traced by some increasing sequence $\boldsymbol{v}$. Consider the sequence C of contexts, where C_i is obtained from the tree $t|_{\boldsymbol{v}_i}$ by placing the hole in the node corresponding to $\boldsymbol{v}_{i+1}$. One can verify that the sequence of schemas

$$\boldsymbol{R}_i = (\mathit{Trans}(C_i), \mathrm{Trans}_i(C_i)),$$

along with $r = q_I$, is infinitary. $\qquad\square$

Although infinitary sequences characterize the unboundedness of L, they are a little hard to work with. That is why we use a special type of infinitary sequence, which nonetheless remains equivalent to the general case (cf. Lemma 12). Consider a very simple schema R which consists of two loops in R° and a connecting descriptor in $R^{\bullet}$:

$$R = (R^{\circ}, R^{\bullet}) \quad \text{where} \quad R^{\circ} = \{(q, k, q), (p, m, p)\} \quad \text{and} \quad R^{\bullet} = \{(q, n, p)\}.$$

We say the pair of states (q, p) used above is *inflatable* if the sequence $\boldsymbol{R}$ constantly equal R is infinitary, for some choice of ranks k, m and n. Note that in this case, the rank m must be even.

Lemma 11. *The set of inflatable pairs can be computed in polynomial time.*

Proof
For $i \in \mathbb{N}$, consider the set A_i of triples

$$\langle (q_1, n_1, p_1), (q_2, n_2, p_2), (q_3, n_3, p_3) \rangle \in (Q \times \Omega(Q) \times Q)^3$$

such that for some $P(\hat{\Sigma})$-context C:

- $(q_1, n_1, p_1), (q_3, n_3, p_3) \in \mathrm{Trans}_0(C)$;
- $(q_2, n_2, p_2) \in \mathrm{Trans}_i(C)$.

Using a dynamic algorithm, the set A_i can be computed in time polynomial on i and the size of the state space Q. By a pumping argument, one can show that the pair (q, p) is inflatable if and only if

$$\langle (q, n_1, q), (q, n_2, p), (p, n_3, p) \rangle \in A_{|Q|+1} \qquad \text{for some even } n_3.$$

$\square$

We now proceed to show Lemma 12, which shows that one can consider inflatable pairs instead of arbitrary infinitary sequences.

Lemma 12. *There is an infinitary sequence iff there is an inflatable pair.*

Proof
An inflatable pair is by definition obtained from an infinitary sequence, hence the right to left implication. For the other implication, consider an infinitary sequence $\boldsymbol{R}$ along with the appropriate sequence of contexts $\boldsymbol{C}$. With every two indices $i < j$, we associate the schema $R[i, j]$ obtained by composing the schemas $\boldsymbol{R}_i \cdots \boldsymbol{R}_{j-1}$. Since there is a finite number of schemas, by Ramsey's Theorem [11] there is a schema R and a set of indices $I = \{i_1 < i_2 < \cdots \} \subseteq \mathbb{N}$ such that $R[i, j] = R$ for every $i < j$ in I. Naturally, in this case $R \cdot R = R$ and, by Fact 45, the sequence constantly equal R is an infinitary sequence which realizes the sequence of contexts $\boldsymbol{D}$ defined by

$$\boldsymbol{D}_j = \boldsymbol{C}_{i_j} \cdots \boldsymbol{C}_{i_{j+1}-1}.$$

We will now show how to extract an inflatable pair from this sequence. Let $q \to p$ be the relation holding for those states $q, p \in Q$ such that (q, m, p) belongs to R° for some rank m. Since $R \cdot R = R$, the relation $\to$ is transitive. Let $(q, m, p) \in R^\bullet$ be a descriptor used for infinitely many n in clause 2 of Definition 46. We claim:

- $r \to q'$, $q' \to q'$ and $q' \to q$, for some $q' \in Q$. This follows from transitivity of $\to$ if we take n in Definition 46 to be big enough to find a loop.
- $p \to p'$ and $(p', m, p') \in R^\circ$ for some $p' \in Q$ and even rank m. This is done as above.

Consider finally the sequence of contexts $\boldsymbol{E}$ defined by

$$\boldsymbol{E}_i = \boldsymbol{D}_i \cdot \boldsymbol{D}_{i+1} \cdot \boldsymbol{D}_{i+2}.$$

By Fact 45, for all $i \in \mathbb{N}$ the schema $R \cdot R \cdot R = R$ is i-realized by the context $\boldsymbol{E}_i$. One can easily verify that the sequence $\boldsymbol{E}$ witnesses the fact that (q', p') is an inflatable pair.

$\square$

From Lemmas 10, 11 and 12 we immediately obtain:

Theorem 47
Satisfiability is decidable for formulas of the form $\mathbb{U}X.\psi$, where ψ is MSOL.

Note that the appropriate algorithm is in fact polynomial in the size of a parity automaton recognizing ψ.

5   Closing Remarks

The results in this paper can only be thought of as initiating research regarding of the bounding quantifier: we have not shown satisfiability decidable for the whole logic. The Theorems 43 and 47 can thus be improved by showing satisfiability decidable (or undecidable) for larger fragments than the ones considered above. Moreover, a better complexity assessment for Theorem 43 would be welcome.

References

1. K. Barthelmann. When can an equational simple graph be generated by hyperedge replacement? In *MFCS*, volume 1450 of *Lecture Notes in Computer Science*, pages 543–552, 1998.
2. M. Bojańczyk. Two-way alternating automata and finite models. In *International Colloquium on Automata, Languages and Programming*, volume 2380 of *Lecture Notes in Computer Science*, pages 833–844, 2002.
3. M. Bojańczyk. The finite graph problem for two-way alternating automata. *Theoretical Computer Science*, 298(3):511–528, 2003.
4. A. Bouquet, O. Serre, and I. Walukiewicz. Pushdown games with unboundedness and regular conditions. In *Foundations of Software Technology and Theoretical Computer Science*, volume 2914 of *Lecture Notes in Computer Science*, pages 88–99, 2003.
5. O. Carton and W. Thomas. The monadic theory of morphic infinite words and generalizations. In *Mathematical Foundations of Computer Science*, volume 1893 of *Lecture Notes in Computer Science*, pages 275–284, 2000.
6. C. C. Elgot and M. O. Rabin. Decidability and undecidability of extensions of second (first) order theory of (generalized) successor. *Journal of Symbolic Logic*, 31:169–181, 1966.
7. F. Klaedtke and H. Ruess. Parikh automata and monadic second–order logics with linear cardinality constraints. Technical Report 177, Institute of Computer Science at Freiburg University, 2002.
8. D. Niwiński. On the cardinality of sets of infinite trees recognizable by infinite automata. In *Mathematical Foundations of Computer Science*, volume 520 of *Lecture Notes in Computer Science*, 1991.
9. M. O. Rabin. Decidability of second-order theories and automata on infinite trees. *Transactions of the AMS*, 141:1–23, 1969.
10. M. O. Rabin. *Automata on Infinite Objects and Church's Problem*. American Mathematical Society, Providence, RI, 1972.
11. F. P. Ramsey. On a problem of formal logic. *Proceedings of the London Mathematical Society*, 30:264–285, 1930.
12. M. Vardi. Reasoning about the past with two-way automata. In *International Colloquium on Automata, Languages and Programming*, number 1443 in Lecture Notes in Computer Science, pages 628–641, 1998.

Parity and Exploration Games
on Infinite Graphs*

Hugo Gimbert

Université Paris 7, LIAFA, case 7014
2, place Jussieu
75251 Paris Cedex 05, France
Hugo.Gimbert@liafa.jussieu.fr

Abstract. This paper examines two players' turn-based perfect-information games played on infinite graphs. Our attention is focused on the classes of games where winning conditions are boolean combinations of the following two conditions: (1) the first one states that an infinite play is won by player 0 if during the play infinitely many different vertices were visited, (2) the second one is the well known parity condition generalized to a countable number of priorities.

We show that, in most cases, both players have positional winning strategies and we characterize their respective winning sets. In the special case of pushdown graphs, we use these results to show that the sets of winning positions are regular and we show how to compute them as well as positional winning strategies in exponential time.

1 Introduction

Two-player games played on graphs have attracted a lot of attention in computer science. In verification of reactive systems it is natural to see the interactions between a system and its environment as a two-person game [19, 9], in control theory the problem of controller synthesis amounts often to finding a winning strategy in an associated game [1].

Depending on the nature of the examined systems various types of two-player games are considered. The interactions between players can be turn-based [23, 19] or concurrent [7, 8], finite like in reachability games or infinite like in parity or Muller games, the players can have perfect or imperfect information about the play. Moreover, the transitions may be deterministic or stochastic [6, 8] and finally the system itself can be finite or infinite.

Another source of diversity comes from players' objectives, i.e. winning conditions.

Our work has as a framework turn-based perfect information infinite games on pushdown graphs. The vertices of such graphs correspond to configurations

---

* This research was supported by European Research Training Network: Games and Automata for Synthesis and Validation.

of a pushdown automaton and edges are induced by push-down automaton transitions. The interest in such games comes, at least in part, from practical considerations, pushdown systems can model, to some extent, recursive procedure calls. On the other hand, pushdown graphs constitute one of the simplest class of infinite graphs that admit non trivial positive decidability results and since the seminal paper of Muller and Schupp [14] many other problems are shown to be decidable for this class [2, 13, 5, 18, 3, 22, 4, 17].

Let us describe briefly a play of such a game. The set of vertices is partitioned into two sets: vertices belonging to player 0 and vertices belonging to his adversary 1. Initially, a pebble is placed on a vertex. At each turn the owner of the vertex with the pebble chooses a successor vertex and moves the pebble onto it. Then the owner of this new vertex proceeds in the same way, and so on. The successive pebble positions form an infinite path in the graph, this is the resulting play.

In this framework, different objectives have been studied. Such an objective is described in general as the set of infinite plays that are winning for player 0, and it is called a winning condition. A lot of attention has been given to the case where this set is regular, which gives rise to Müller and parity winning conditions [23, 22, 19] which lie on the level Δ_2 of the Borel hierarchy. However, recently Cachat et al. [5], presented a new winning condition of Borel complexity Σ_3 which still remains decidable. This Σ_3-condition specifies that player 0 wins a play if there is no vertex visited infinitely often. Yet another condition, *unboundedness*, was introduced by Bouquet et al. [3]. The unboundedness condition states that player 0 wins a play if the corresponding sequence of stack heights is unbounded. Obviously the conditions of [5] and [3] are tightly related, if no configuration of the push-down system is visited infinitely often then the stack is unbounded. The converse can be established as well if the winning strategies are memoryless, i.e. do not depend on the past.

In this paper, we first transfer the condition of [3] to arbitrary infinite graphs of finite degree. In the context of arbitrary infinite graphs we examine *Exploration* condition which states that a play is won by player 0 if the pebble visits an infinite number of different vertices. Obviously for the particular case of pushdown graphs this gives the same condition as [3]. In fact we go a step further and consider the games whose winning conditions are boolean combinations of Exploration condition and of the classical parity condition. We note respectively *Exp ∪ Parity* and *Exp ∩ Parity* the games obtained by taking the union and the intersection of Exploration and Parity conditions.

We also consider a particular extension of the classical Parity condition to the case with an infinite number of priorities and denote it $Parity_\infty$ (see also [11] for another approach to parity games with an infinity of priorities).

We prove the following results in the context of the games over any infinite graphs:

- Both players have positional winning strategies for the game with the winning condition *Exp ∪ Parity*, including the case where there is an infinite number of priorities.

– In the case where there are finitely many priorities, player 1 has also a winning positional strategy in the game where the winning condition for player 0 is of type $Exp \cap Parity$. Moreover, we can easily characterize the set of winning positions of player 0.

Even if general results concerning winning strategies over arbitrary infinite graphs are of some interest we are much more interested in decidability results for the special case of pushdown graphs. In the case where the game graph is a pushdown graph, we prove for both types of games $Exp \cup Parity$ and $Exp \cap Parity$ that the sets of winning configurations (positions) for player 0 (and also for player 1) are regular subsets of $Q\Gamma^*$ where Q is the set of states of pushdown system and Γ is the stack alphabet. We provide also an algorithm for computing a Büchi automaton with $2^{\mathcal{O}(d^2|Q|^2+|\Gamma|)}$ states recognizing those winning sets, where d is the number of priorities of the underlying parity game and Q and Γ are as stated above. Moreover, we show that for both games and both players, the set of winning positional strategies is regular and recognized by an alternating Büchi automaton. In the case of the $Exp \cup Parity_d$ game, this automaton has $\mathcal{O}(d|Q|^2 + |\Gamma|)$ states whereas in the case of the $Exp \cap Parity_d$ game, it has $\mathcal{O}(d^2|Q|^2 + d|\Gamma|)$ states

These results constitute an extension of the results of [5, 3, 22, 18, 20]: The papers [22, 20, 18] examine only *Parity* conditions with a finite number of priorities for pushdown games. Bouquet et al. [3] were able to extend the decidability results to the games with the winning condition of the form $Exp \cup Buchi$ or $Exp \cap Buchi$, i.e. union and intersections of Büchi condition with Exploration condition. However this class of conditions is not closed under boolean operations (intersecting Büchi and co-Büchi conditions with an Exploration condition is not in this class). In our paper we go even further since we allow boolean combinations of Exp conditions with parity conditions. Since parity conditions, after appropriate transformations, are closed under boolean operations we show in fact that it is decidable to determine a winner for the smallest class of conditions containing Exploration and Büchi conditions and closed under finite boolean operations.

For computing the winning sets and the winning strategies, we make use of tree automata techniques close to the one originated in the paper of Vardi [20] and applied in [16, 12]. This is a radical departure from the techniques applied in [21, 22, 3, 18] which are based on game-reductions.

This paper is organized as follows. In the first part, we introduce some basic definitions and the notions of Exploration and Parity games. In the second part, we prove the results concerning the winning strategies for the games $Exp \cup Parity_\infty$ and $Exp \cap Parity$, and make some comments about the $Parity_\infty$ game. In the third part, we describe the construction of automata computing the winning sets and the winning positional strategies. Due to space limitation, most proofs are omitted and can be found in the full version [10].

2   Parity and Exploration Games

In this section, we present basic notions about games and we define different winning conditions.

2.1   Generalities

The games we study are played on oriented graphs of finite degree, with no dead-ends, whose vertex set is partitioned between the vertices of player 0 and the vertices of player 1. Such a graph is called an arena. At the beginning of a play, a pebble is put on a vertex. During the play, the owner of the vertex with the pebble moves it to one of the successors vertices. A play is the infinite path visited by the pebble. A winning condition determines which player is the winner. Here follows the formal description of these notions.

Notations. Let $G = (V, E)$ be an oriented graph with the set $E \subset V \times V$ of edges. Given a vertex v, vE denotes the set of successors of v, $vE = \{w \in V : (v, w) \in E\}$, whereas Ev is the set of predecessors of v. For a set $H \subseteq E$ of edges, $Dom(H)$, the domain of H, denotes the set of the vertices adjacent to edges of H.

Parity Arenas. An arena is a tuple (V, V_0, V_1, E), where (V, E) is a graph of finite degree with no dead-ends and (V_0, V_1) is a partition of V. Let $i \in \{0, 1\}$ be a player. V_i is the set of vertices of player i. We will often say that $G = (V, E)$ itself is an arena, when the partition (V_0, V_1) is obvious. An infinite path in G is called a play, whereas a finite path in G is called a finite play. When the vertices of G are labeled with natural numbers with a map $\phi : V \to \mathbb{N}$, G is said to be a parity arena.

Winning Conditions and Games. A winning condition determines the winner of a play. Formally, it is a subset Vic $\subseteq V^\omega$ of the set of infinite plays. A game is a couple (G, Vic) made of an arena and a winning condition. Often, when the arena G is obvious, we will say that Vic itself is a game. A play $p \in V^\omega$ is won by player 0 if $p \in$ Vic. Otherwise, if $p \notin$ Vic, it is said to be won by player 1. Vic is said to be *concatenation-closed* if $V^* \text{Vic} = \text{Vic}$.

Strategies, Winning Strategies and Winning Sets. Depending on the finite path followed by the pebble, a strategy allows a player to choose between a restricted number of successor vertices. Let $i \in \{0, 1\}$ be a player. Formally, a strategy for player i is a map σ, which associates to any finite play $v_0 \cdots v_n$ such that $v_n \in V_i$ a nonempty subset $\sigma(v_0...v_n) \subseteq v_n E$. A play $p = (v_n)_{n \in \mathbb{N}} \in V^\omega$ is said to be consistent with σ if, for any n such that $v_n \in V_i$, $v_{n+1} \in \sigma(v_0 \cdots v_n)$. Given a subset $X \subseteq V$ of the vertices, A strategy for player i is said to be winning the game (G, Vic) on X if any infinite play starting in X and consistent with this strategy is won by player i. If there exists such a strategy, we say that player i wins (G, Vic) on X. If $X = V$, we simply say that i wins (G, Vic). The winning set of player i is the greatest set of vertices such that i wins Vic on this set.

Positional Strategies. With certain strategies, the choices advised to the player depend only on the current vertex. Such a strategy can be simply described by the set of edges it allows the players to use. $\sigma \subseteq E$ is a positional strategy for player i in the arena G if there is no dead-end in the subgraph $(Dom(\sigma), \sigma)$ induced by σ and σ does not restrict the moves of the adversary: if $v \in V_{1-i} \cap Dom(\sigma)$ then $\{v\} \times vE \subseteq \sigma$. Let $X \subseteq V$ be a subset of vertices. If $Dom(\sigma) = X$, σ is said to be defined on X. We say that a player wins positionally a game Vic on X if he has a positional strategy winning on X.

Subarenas and Traps. Let $X \subseteq V$ be a subset of vertices and $F \subseteq E$ a subset of edges. $G[X]$ denotes the graph $(X, E \cap X^2)$ and $G[X, F]$ denotes the graph $(Dom(F) \cap X, F \cap X^2)$. When $G[X]$ or $G[X, F]$ is an arena, it is said to be a subarena of G. X is said to be a trap for player i in G if $G[X]$ is a subarena and player i can't move outside of X, i.e. $\forall v \in X \cap V_i, vE \subseteq X$.

2.2   Winning Conditions

Let $G = (V, V_0, V_1, E)$ be an arena and $X \subseteq V$. We define various winning conditions.

Attraction Game to X. Player 0 wins if the pebble visits X at least once. The corresponding winning condition is $Attraction(X) = V^* X V^\omega$. The winning set for player 0 is denoted by $Att_0(G, X)$ or $Att_0(X)$, when G is obvious. Symmetrically, we define $Att_1(G, X)$ and $Att_1(X)$, the sets of vertices where player 1 can attract the pebble to X. Note that for this game, both players have positional winning strategies.

Trap Game and Büchi Game to X. Player 0 wins the trap game in X if the pebble stays ultimately in X. The winning condition is $TrapX = V^* X^\omega$. The dual game is the Büchi game to X, where player 0 wins if the pebble visits X infinitely often. The winning condition is $Buchi(X) = (V^* X)^\omega$.

Exploration Game. This is a game over an infinite graph, where player 0 wins a play if the pebble visits infinitely many different vertices. The winning condition is $Exp = \{v_0 v_1 \cdots \in V^\omega \mid \text{the set } \{v_0, v_1, \ldots\} \text{ is infinite}\}$.

    The exploration condition is an extension of the *Unboundedness* condition introduced in [3]. The Unboundedness condition concerns games played on the configuration graph of a pushdown system. On such a graph, the set of plays is exactly the set of runs of the underlying pushdown automaton, and 0 wins a play if the height of the stack is unbounded, which happens if and only if infinitely many different configurations of the pushdown automaton are visited.

    The exploration condition is also closely related to the Σ_3-condition considered in [5], which states that 0 wins a play if every vertex is visited finitely often. Notice that such a play is necessarily also winning for the exploration condition, but the converse is not true. However, given an arena, it is easy to see that each player has the same positional winning strategies for both games. Since the Exploration game is won positionally by both players (cf. Proposition 1), it implies both games have the same winning positions. Hence, in that sense, the Explosion game and the Σ_3-game introduced in [5] are equivalent.

Parity Game. G is a parity arena equipped with a priority mapping $\phi : V \rightarrow \mathbb{N}$. Player 0 wins a play if there exists a highest priority visited infinitely often and this priority is even, or if the sequence of priorities is unbounded. Thus, the winning condition is

$$Parity_\infty = \{(v_i)_{i \in \mathbb{N}} : \overline{\lim_{i \in \mathbb{N}}} \ \phi(v_i) \in \{0, 2, \ldots, +\infty\}\}$$

where $\overline{\lim}_{i \in \mathbb{N}} \ \phi(v_i) = \lim_{i \in \mathbb{N}} \ \sup_{j \geq i} \ \phi(v_j)$ denotes the limit sup of the infinite sequence of visited priorities. If G is labeled by a finite number of priorities, i.e. if there exists $d \in \mathbb{N}$ such that $\phi : V \rightarrow [0, d]$, we write also the winning condition as $Parity_d$. In this case, a classical result [9, 19, 23] states that both players win this game positionally.

3   Playing the Games $Exp \cup Parity_\infty$ and $Exp \cap Parity_d$

In this section we study the winning strategies for the games $Exp \cup Parity_\infty$ and $Exp \cap Parity_d$. In the case of the game $Exp \cup Parity_\infty$, we show that each player has a positional strategy, winning on the set of his winning posiitons. Concerning the game $Exp \cap Parity_d$, we show that this remains true for player 1, and we exhibit an arena where no winning strategy of player 0 is positional. However, we give a characterization of the winning set of player 0.

3.1   The Game $Exp \cup Parity_\infty$

G is a parity arena equipped with $\phi : V \rightarrow \mathbb{N}$.

Proposition 1. *Each player wins positionally the game* $Exp \cup Parity_\infty$ *on his winning set.*

Proof. It is crucial to observe that $Exp \cup Parity_\infty$ can be expressed as the limit of a decreasing sequence of winning conditions:

$$Exp \cup Parity_\infty = \bigcap_{n \in \mathbb{N}} Vic_n \ ,$$

where

$$Vic_n = Attraction(\{n + 1, n + 2, \ldots\}) \cup Parity_\infty \ .$$

Moreover, each game (G, Vic_n) is won positionally by players 0 and 1 on their winning sets X_n and $V \backslash X_n$. It is easy to establish that player 1 wins positionally

$(G, \bigcap_n Vic_n)$ on $\bigcup_n V \setminus X_n = V \setminus \bigcap_n X_n$. For winning positionally $\bigcap_n Vic_n$ on $\bigcap_n X_n$, player 0 can manage to play in such a way that, as long as the pebble stays in $\{0, 1, \ldots, n\}$, the play is consistent with a winning strategy for Vic_n. Then, if the pebble stays bounded in some set $\{n, n+1, \ldots\}$, the play is won for condition $\bigcap_{m \geq n} Vic_m \subset Parity_\infty$. If the pebble leaves every set $\{0, \ldots, n\}$, then the play visits infinitely many different vertices and the play is won for Exp by player 0.   □

Since the Exp game is a special case of the $Exp \cup Parity_\infty$ game where all the vertices are labeled with priority 1, we get the following corollary.

Corollary 1. *Each player wins positionally the game* Exp *on his winning set.*

3.2   The Game $Parity_\infty$

A natural question that arises is whether the players have some positional winnign strategies for the $Parity_\infty$ game. Notice that $Exp \subseteq Parity_\infty$ in the special case where, for every priority d, $\phi^{-1}(d)$ is finite. Indeed, any play visiting infinitely many different vertices will visit infinitely many different priorities.

Hence, in this special case, by Proposition 1, the game $Parity_\infty$ is won positionally by both players. This is not true anymore if $\phi^{-1}(d)$ is infinite for some d. Consider the example given on Fig. 1. The circles are the vertices of player 0 and the squares those of player 1. Player 0 wins $Parity_\infty$ from everywhere but has no positional winning strategy.

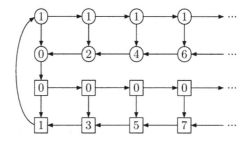

Fig. 1. Player 0's strategy must recall the highest odd vertex reached by player 1 in the lower row in order to answer with a higher even vertex in the second row

It is interesting to note that if the winning player is determined by the lowest priority visited infinitely often rather than by the greatest one, then both players have positional winning strategies, even if infinitely many priorities are assumed [11].

3.3   The Game $Exp \cap Parity_d$

The analysis of the $Exp \cap Parity_d$ game extends the results of [23]. In this section, G is a parity arena equipped with $\phi : V \to [0, d]$.

Proposition 2. *Player* 1 *wins positionally the game* $\text{Exp} \cap \text{Parity}_d$ *on his winning set.*

Proof. Without loss of generality, we can assume that player 1 wins everywhere. The proof is by induction on d.

If $d = 0$, it is impossible for player 1 to win any play and his winning set is empty. If d is odd and $d \neq 0$, let W be the attractor for player 1 in the set of vertices coloured by the maximal odd priority d. Since $V \backslash W$ is a trap for player 1 coloured from 0 to $d - 1$, and by inductive hypothesis, player 1 can win positionally $(G[V \backslash W], Exp \cap Parity_{d-1})$ with some strategy $\sigma_{V \backslash W}$. To win, player 1 shall use $\sigma_{V \backslash W}$ inside $V \backslash W$ and shall attract the pebble to a vertex of colour d when it reaches the set W. That way, either the play stays ultimately in $V \backslash W$ and some suffix is consistent with $\sigma_{V \backslash W}$ or it reaches the odd priority d infinitely often. In both cases, player 1 is the winner.

The case where d is even is less trivial. It is easy to prove that there exists the greatest subarena of G where player 1 wins positionally. It remains to prove that this subarena coincides with the whole arena. $\square$

It may happen that player 0 has a winning strategy from every vertex but he has no positional winning strategy. Such an example is given by Fig. 2.

Fig. 2. To win the $Exp \cap Parity_2$ game, player 0 has to visit new vertices arbitrarily far to the right hand side of the arena and has also to visit the unique vertex of color 2 infinitely often

Nevertheless, we can characterize the arenas in which player 0 wins the game $Exp \cap Parity_d$ from every position:

Proposition 3. *Let* $G = (V, E)$ *be an arena, coloured from 0 to $d > 0$. Let D be the set of vertices coloured by d. Player* 0 *wins the game* $(G, \text{Exp} \cap \text{Parity}_d)$ *on V if and only if there exists a subarena $G[W]$, coloured from 0 to $d - 1$ such that one of the following conditions holds:*

- **Case d even:** *Player* 0 *wins the games* $(G[W], \text{Exp} \cap \text{Parity}_{d-1})$ *and* (G, Exp) *everywhere and she wins the game* $(G, Attraction(D))$ *on $V \backslash W$.*
- **Case d odd:** *Player* 0 *wins the game* $(G, Trap(W))$ *with a positional strategy* $\sigma_{Trap(W)}$ *and she wins the game* $(G[W, \sigma_{Trap(W)}], \text{Exp} \cap \text{Parity}_{d-1})$.

The conditions of Proposition 3 are illustrated on Fig. 3.

Remark 1. Note that winning the game $(G[W, \sigma_{Trap(W)}], Exp \cap Parity_{d-1})$ means that player 0 has a strategy σ_W winning the game $(G[W], Exp \cap Parity_{d-1})$ which advises player 0 to play moves **consistent with the positional strategy** $\sigma_{\mathbf{Trap(W)}}$.

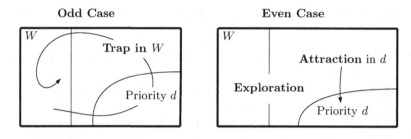

Fig. 3. Conditions of Proposition 3

Proof. We sketch the proof of the direct implication. In the case where d is even this proof is simple. Consider $W = V \backslash Att_0(D)$. Since $V \backslash W$ is a trap, player 0 wins $(G[V \backslash W], Exp \cap Parity_d)$. The other claims are trivially true.

The case where d is odd is more tricky. We establish first that the family of subarenas of G where Proposition 3 holds is closed by arbitrary union, then we prove that the maximal arena of this family is necessarily G itself.

We sketch the proof of the converse implication. We shall construct a strategy σ_G for player 0 winning the game $(G, Exp \cap Parity_d)$. This construction depends on the parity of d.

d Odd: By hypothesis, player 0 has a positional strategy $\sigma_{Trap(W)}$ winning the game $(G, Trap(W))$ and a strategy σ_{Sub} winning the game $(G[W, \sigma_{Trap(W)}], Exp \cap Parity_{d-1})$. The strategy σ_G is constructed in the following way:

- If the pebble is not in W, player 0 plays according to her positional strategy $\sigma_{Trap(W)}$.
- If the pebble is in W, player 0 uses her strategy σ_{Sub} in the following way: Let p be the sequence of vertices visited up to now and let p' be the longest suffix of p consisting of vertices of W. Player 0 takes a move according to $\sigma_{Sub}(p')$.

The strategy σ_G is winning for the game $(G, Exp \cap Parity_d)$. Indeed, since σ_{Sub} is a strategy in the arena $G[W, \sigma_{Trap(W)}]$, all moves consistent with σ_G are consistent with $\sigma_{Trap(W)}$. Hence, the play is ultimately trapped in W and is ultimately consistent with σ_W, thus won by player 0.

d Even: By hypothesis and by Corollary 1, player 0 has a positional strategy $\sigma_{Exp} \subseteq E$ winning (G, Exp). She has also a positional strategy $\sigma_{Att} \subseteq E$ winning $(G, Attraction(D))$ on $V \backslash W$ and a strategy σ_{Sub} winning the game $(G[W], Exp \cap Parity_{d-1})$.

σ_G is constructed in the following way. At a a given moment player 0 is in one of the three playing modes: *Attraction, Sub* or *Exploration*. It can change the mode when the pebble moves to a new vertex. Player 0 begins to play in *Exploration* mode. Here follows the precise description of the strategy σ_G, summarized by Fig. 4.

- The playing mode *Exploration* can occur wherever the pebble is. Player 0 plays according to her positional strategy σ_{Exp}. When a new vertex v is visited for the first time the mode is changed either to *Sub* mode if $v \in W$ or to *Attraction* mode if $v \notin W$.
- The playing mode *Attraction* can occur only if the pebble is not in W. Player 0 plays according to her positional strategy σ_{Att}. When a vertex of priority d is eventually visited, the playing mode is switched to *Exploration*.
- The playing mode *Sub* can occur only if the pebble is in W. Player 0 plays using her strategy σ_{Sub} in the following way. Let p be the sequence of vertices visited up to now and p' the longest suffix of p consisting of vertices of W. Then 0 takes a move according to $\sigma_{Sub}(p')$. If the pebble leaves W, the playing mode is switched to *Exploration*.

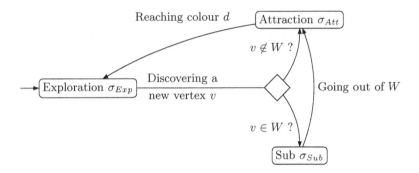

Fig. 4. Rules of transition between playing modes

Notice that, by definition of σ_{Att} and σ_{Exp}, it is not possible that an infinite play consistent with σ_G stays forever in the playing modes *Attraction* or *Exploration*. Hence, such a play can be of two different types. Either the pebble stays ultimately in the playing mode *Sub* or it goes infinitely often in the modes *Exploration* and *Attraction*. In the first case, it stays ultimately in W and the play is ultimately consistent with σ_{Sub}. In the second case, the pebble visits infinitely often the even priority d and discovers infinitely often a new vertex. In both cases, this play is won by player 0 for the $Exp \cap Parity_d$ condition. □

4   Computation of the Winning Sets and Strategies on Pushdown Arenas

In this section, we apply our results to the case where the infinite graph is the graph of the configurations of a pushdown automaton. And we get an algorithm to compute the winning sets. Moreover, in all cases except for player 0 in the game $Exp \cap Parity_d$, we can also compute winning positional strategies.

Definitions. A pushdown system is a tuple $\mathcal{P} = (Q, \Gamma, \Delta, \perp)$ where Q is a finite set of control states, Γ is a finite stack alphabet, $\perp$ is a special letter called the stack bottom, $\perp \notin \Gamma$ and $\Delta \subseteq Q \times (\Gamma \cup \{\perp\}) \times (\Gamma \cup \{-1\}) \times Q$ is the set of transitions.

The transition $(q, \alpha, \beta, r) \in \Delta$ is said to be a *push transition* if $\beta \in \Gamma$ and a *pop transition* if $\beta = -1$. In both cases, it is said to be an α-transition and a (q, α)-transition. Concerning $\perp$, we impose the restriction that there exists no pop $\perp$-transition. Moreover, we work only with complete pushdown systems, in the sense that, for every couple $(q, \alpha) \in Q \times (\Gamma \cup \{\perp\})$, there exists at least one (q, α)-transition.

Notice that, in the sense of language recognition, any pushdown automaton is equivalent to one of this kind, and the reduction is polynomial.

A configuration of $\mathcal{P}$ is a sequence $q\gamma$, where $q \in Q$ and $\gamma \in \Gamma^*$. Intuitively, q represents the current state of $\mathcal{P}$ while γ is the stack content above the bottom symbol $\perp$. We assume that the symbols on the right of γ are at the top of the stack. Note that $\perp$ is assumed implicitly at the bottom of the stack, i.e. actually the complete stack content is always $\perp\gamma$.

The set of all configurations of $\mathcal{P}$ is denoted by $V_\mathcal{P}$. Transition relation $E_\mathcal{P}$ over configurations is defined in the usual way: Let $q\gamma\alpha$, where $q \in Q, \gamma \in \Gamma^*$ and $\alpha \in \Gamma$, be a configuration.

- $(q\gamma\alpha, r\gamma) \in E_\mathcal{P}$ if there exists a pop transition $(q, \alpha, -1, r) \in \Delta$,
- $(q\gamma\alpha, r\gamma\alpha\beta) \in E_\mathcal{P}$ if there exists a push transition $(q, \alpha, \beta, r) \in \Delta$.

Let $q\epsilon$ be a configuration with empty stack. Then

- $(q\epsilon, r\beta) \in E_\mathcal{P}$ if there exists a push transition $(q, \perp, \beta, r) \in \Delta$.

We shall write $q\gamma \xrightarrow{\delta} r\gamma'$ to express that a transition $\delta \in \Delta$ of the pushdown automaton corresponds to an edge $(q\gamma, r\gamma') \in E_\mathcal{P}$ between two configurations. The graph $G_\mathcal{P} = (V_\mathcal{P}, E_\mathcal{P})$ is called the *pushdown graph* of $\mathcal{P}$.

If Q is partitioned in (Q_0, Q_1), this partition extends naturally to the set of configurations of $\mathcal{P}$ and we $G_\mathcal{P}$ is an arena. Moreover, when the control states Q are labeled by priorities with a map $\phi : Q \to [0, d]$, this labeling extends naturally to $V_\mathcal{P}$ by setting $\phi(q\gamma) = \phi(q)$. $G_\mathcal{P}$ is then a parity arena.

Subgraph Trees and Strategy Trees. With any subset $\sigma \subseteq E_\mathcal{P}$ of the edges of a pushdown arena we associate a tree $T_\sigma : \Gamma^* \to 2^\Delta$ with vertices labeled by sets of transition of $\mathcal{P}$. This construction is illustrated by Fig. 5.

A vertex of the tree is a stack content of $\mathcal{P}$. A transition $\delta \in \Delta$ belongs to the set labeling a vertex $\gamma \in \Gamma^*$ if there exists a state $q \in Q$ and a configuration $r\gamma'$ such that $q\gamma \xrightarrow{\delta} r\gamma'$ and $(q\gamma, r\gamma') \in \sigma$. Such a tree is called the *coding tree* of σ. Notice that the transformation $\sigma \to T_\sigma$ is one-to-one. If σ is a strategy for player i, we call T_σ a *strategy tree* for player i.

The next theorem states that the languages of positional winning strategies is regular. Thus, we can build a Büchi alternating automaton of size $\mathcal{O}(d|Q|^2 + |\Gamma|)$ which recognizes the language of couples (σ_0, σ_1) such that σ_i is a winning positional strategy for player i and the domains of σ_0 and σ_1 constitute a partition of

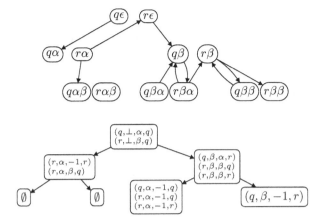

Fig. 5. A finite subset of $E_{\mathcal{P}}$ and its coding tree. Only the labels of the vertices $\{\epsilon, \alpha, \beta, \alpha\alpha, \alpha\beta, \beta\alpha, \beta\beta\}$ are represented. Other vertices of the coding tree are labeled with $\emptyset$

$V_{\mathcal{P}}$. In the case of the $Parity_d$ and $Exp \cup Parity_d$ games, Proposition 1 establishes that this language is non-empty. Hence it is possible to compute a regular tree (σ_0, σ_1) of size $2^{\mathcal{O}(d|Q|^2 + |\Gamma|)}$. This regular tree can be seen as the description of a couple of winning stack strategies for both players. This kind of strategy has been defined in [21].

Theorem 1. *Let i be a player and $\mathrm{Vic} \in \{Parity_d, Exp \cup Parity_d, Exp \cap Parity_d\}$. The language of strategy trees which correspond to winning positional strategies for player i is regular. One can effectively construct an alternating Büchi automaton $\mathcal{A}_{\mathrm{Vic},i}$ with $\mathcal{O}(d|Q|^2 + |\Gamma|)$ states which recognizes it.*

Proof. The construction of $\mathcal{A}_{\mathrm{Vic},i}$ uses techniques close to the one of [20, 16]. Unfortunately, we couldn't manage to use directly the results of those papers about two-way tree automata, because we don't know how to use a two-way automata to detect a cycle in a strategy tree.

Our aim is to construct a tree automaton recognizing a tree $t : \Gamma^* \to 2^\Delta$ iff there exists a winning positional strategy σ such that $t = T_\sigma$. In fact we shall rather construct a Büchi alternating automaton recognizing the complement of the set $\{T_\sigma | \sigma$ winning positional strategy $\}$. First of all it is easy to implement an alternating automaton verifying if the tree t is or is not a strategy tree. It is less trivial to construct the automaton checking if a positional strategy $\sigma \subset E_{\mathcal{P}}$ is winning or not. However, it can be expressed by a simple criterion concerning the cycles and the exploration paths of the graph $(Dom(\sigma), \sigma)$ induced by σ. Those criteria are summarized in Table 1.

We have to construct automata checking each condition of Table 1. They are derived from an automaton detecting the existence of a special kind of finite path called a *jump*. A jump between two vertices with the same stack γ is a path between those vertices, that never pops any letter of γ (see Fig. 4).

Table 1. Characterization of winning positional strategies

Winning Condition	i	Condition on cycles	Condition on exploration paths
Parity	0	Even	Even
	1	Odd	Odd
$Exp \cup Parity_d$	0	Even	No condition
	1	Odd	No exploration path
$Exp \cap Parity_d$	0	No cycle	Even
	1	No condition	Odd

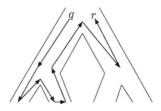

Fig. 6. A jump from $q\gamma$ to $r\gamma$ in a strategy tree

This kind of path is interesting since a cycle is simply a jump from a vertex to itself, and because the existence of an exploration path of priority c is equivalent to the existence of one of the two kinds of paths illustrated on Fig. 7.

Due to the high computational power of alternation, it is possible to construct automata checking the existence of jumps and detecting the kinds of paths of Fig. 7, with only $\mathcal{O}(d|Q^2| + |\Gamma|)$ control states.   □

Computation of Winning Sets. Using the automata recognizing languages of winning positional strategies, it is possible to recognize the language of winning positions. For each player i, Theorem 2 leads to an EXPTIME procedure to compute a regular tree $\Gamma^* \to 2^Q$ of exponential size that associates with a stack γ the set $\{q \in Q : q\gamma$ is winning for player $i\}$. Once computed, deciding if a given position is winning for player i can be done in linear time.

Theorem 2. *For each player i and each winning condition* Vic $\in \{$Parity$_d$, Exp $\cup$ Parity$_d$, Exp $\cap$ Parity$_d\}$, *the tree* $\Gamma^* \to 2^Q$ *which associates with a stack* γ *the set* $\{q \in Q : q\gamma$ *is winning for* $i\}$ *is regular and one can compute a nondeterministic Büchi automaton recognizing it. Such an automaton has* $2^{\mathcal{O}(d|Q|^2+|\Gamma|)}$ *states if* Vic $\in \{$Parity$_d$, Exp$\cup$Parity$_d\}$ *and* $2^{\mathcal{O}(d^2|Q|^2+d|\Gamma|)}$ *if* Vic $=$ Exp $\cap$ Parity$_d$.

Proof. For the games *Parity$_d$* and *Exp $\cup$ Parity$_d$*, this Theorem is a direct corollary of Theorem 1. In fact, we can build a Büchi alternating automaton which recognizes the language of couples (σ_0, σ_1) such that σ_i is a winning positional strategy for player i and the domains of σ_0 and σ_1 are a partition of $V_\mathcal{P}$. The winning sets are then obtained by projection, which requires to transform this alternating automaton to a non-deterministic one and leads to an exponential blowup of the state space [15].

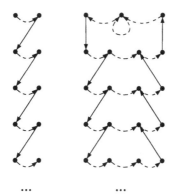

···   ···

Fig. 7. The dotted arrows are jumps of priority less than c. The top-down regular arrows are push transitions, while the down-top ones are pop-transitions. On the left hand side, infinitely many jumps have priority c. On the right hand side, the upper jump, that is a loop, has priority c

In the case of the $Exp \cap Parity_d$ game, we use also the characterization of the winning sets given by Proposition 3. We define the notion of a winning-proof, which is a tree on Γ^* labeled by tuples of subsets of Δ, and is defined such that the existence of a winning-proof in an arena is equivalent to the conditions of Proposition 3. Here follows the definition of a winning-proof in a subarena G of a pushdown arena $G_\mathcal{P}$.

In the case where $d = 0$, it is a strategy tree $T_{\sigma_{Exp}}$ winning the game (G, Exp).

In the case where $d > 0$ and is even, it is a tuple $T_d = (T', T_{\sigma_{Exp}}, T_{\sigma_{Att}}, T_{d-1})$ where T' is the coding tree of a subarena G' of G, $T_{\sigma_{Exp}}$ is a strategy tree winning the game (G, Exp), $T_{\sigma_{Att}}$ is a strategy tree winning the game $(G, Attraction(D))$ on $Dom(G')$ and T_{d-1} is a $(d-1)$-winning proof in G'.

In the case where d is odd, it is a tuple $T_d = (T', T_{\sigma_{Trap}}, T_{d-1})$ where T' is the coding tree of a subarena G' of G, $T_{\sigma_{Trap}}$ is a strategy tree winning the game $(G, Trap(Dom(G')))$ and T_{d-1} is a $(d-1)$-winning proof in G'.

Each one of those $\mathcal{O}(d)$ conditions can be verified with an alternating automaton with $\mathcal{O}(d|Q|^2 + |\Gamma|)$ states. The corresponding automata constructions are very close to the ones of Theorem 1. Hence, the language of d-winning proofs is regular and recognized by an alternating automaton with $\mathcal{O}(d^2|Q|^2 + d|\Gamma|)$ states.

As in the positional case, by projection, we obtain the desired non-deterministic automaton with $2^{\mathcal{O}(d^2|Q|^2 + d|\Gamma|)}$ states. $\square$

Acknowledgements. We thank Wiesław Zielonka and Olivier Serre for some enlightening discussions on games on pushdown graphs, and the anonymous referee for their careful comments.

References

1. A. Arnold, A. Vincent, and I. Walukiewicz. Games for synthesis of controlers with partial observation. *Theoretical Computer Science*, 303(1):7–34, 2003.
2. A. Bouajjani, J. Esparza, and O. Maler. Reachability analysis of pushdown automata: Applications to model checking. In *CONCUR'97, LNCS*, volume 1243, pages 135–150, 1997.
3. A. Bouquet, O. Serre, and I. Walukiewicz. Pushdown games with unboudedness and regular conditions. In *Proc. of FSTTCS'03, LNCS*, volume 2914, pages 88–99, 2003.
4. T. Cachat. Symbolic strategy for games on pushdown graphs. In *Proc. of 29th ICALP, LNCS*, volume 2380, pages 704–715, 2002.
5. T. Cachat, J. Duparc, and W. Thomas. Pushdown games with a σ_3 winning condition. In *Proc. of CSL 2002, LNCS*, volume 2471, pages 322–336, 2002.
6. K. Chatterejee, M. Jurdziński, and T.A. Henzinger. Simple stochastic parity games. In *CSL'03*, volume 2803 of *LNCS*, pages 100–113. Springer, 2003.
7. L. de Alfaro and T.A. Henzinger. Concurrent ω-regular games. In *LICS'00*, pages 142–154. IEEE Computer Society Press, 2000.
8. L. de Alfaro, T.A. Henzinger, and O. Kupferman. Concurrent reachability games. In *FOCS'98*, pages 564–575. IEEE Computer Society Press, 1998.
9. E. A. Emerson and C. S. Jutla. Tree automata, mu-calculus and determinacy. In *Proc. of 32th FOCS*, pages 368–377. IEEE Computer Society Press, 1991.
10. H. Gimbert. Parity and exploration games on infinite graphs, full version. www.liafa.jussieu.fr/~hugo.
11. E. Grädel. Positional determinacy of infinite games. In *STACS 2004, LNCS*, volume 2996, pages 4–18, 2004.
12. O. Kupferman, N. Piterman, and M. Vardi. Pushdown specifictaions. In *Proc. of LPAR'02, LNCS*, October 2002.
13. O. Kupferman and M.Y. Vardi. An automata-theoretic approach to reasoning about infinite state systems. In *Proc. of CAV'00, LNCS*, volume 1855, pages 36–52, 2000.
14. D. E. Muller and P. E. Schupp. The theory of ends, pushdown automata and second order logic. *Thoretical Computer Science*, 37:51–75, 1985.
15. D. E. Muller and P. E. Schupp. Simulating alternating tree automata by nondeterministic automata. *Theoretical Computer Science*, 141:69–107, 1995.
16. N. Piterman and M.Vardi. From bidirectionnality to alternation. *Theoretical Computer Science*, 295:295–321, 2003.
17. O. Serre. Games with winning conditions of high borel complexity, to appear in proc. of icalp'04.
18. O. Serre. Note on winning positions on pushdown games with omega-regular conditions. *Information Processing Letters*, 85(6):285–291, 2003.
19. W. Thomas. On the synthesis of strategies in infinite games. In *Proc. of STACS'95,LNCS*, volume 900, pages 1–13, 1995.
20. M. Vardi. Reasoning about the past with two-way automata. In *Proc. of ICALP'98, LNCS*, volume 1443, pages 628–641, 1998.
21. I. Walukiewicz. Pushdown processes: Games and model checking. In *Proc. of CAV'96, LNCS*, volume 1102, pages 62–74, 1996.
22. I. Walukiewicz. Pushdown processes: Games and model checking. *Information and Computation*, 164(2):234–263, 2001.
23. W. Zielonka. Infinite games on finitely coloured graphs with applications to automata on infinite trees. *Theoretical Computer Science*, 200:135–183, 1998.

Integrating Equational Reasoning into Instantiation-Based Theorem Proving

Harald Ganzinger and Konstantin Korovin

MPI für Informatik, Saarbrücken

Abstract. In this paper we present a method for integrating equational reasoning into instantiation-based theorem proving. The method employs a satisfiability solver for ground equational clauses together with an instance generation process based on an ordered paramodulation type calculus for literals. The completeness of the procedure is proved using the the model generation technique, which allows us to justify redundancy elimination based on appropriate orderings.

1 Introduction

The basic idea of instantiation-based theorem proving is to combine clever generation of instances of clauses with propositional satisfiability checking. Thus, it seems to be promising to exploit the impressive performance of modern propositional SAT technology in the more general context of first-order theorem proving. Accordingly, we have seen several attempts recently at designing new first-order prover architectures combining efficient propositional reasoning into instance generation scenarios, cf. [4, 11, 5, 15, 3, 12, 9, 13] among others.

Integration of efficient equational reasoning into such systems has been a challenging problem, important for many practical applications. In this paper we show how to integrate equational reasoning into the instantiation framework developed in [8]. In [8] we presented instance generation inference systems based on selection from propositional models, together with a notion of redundancy based on closure orderings, and showed their refutational completeness.

Our approach of integrating equational reasoning into this framework aims to preserve attractive properties of the instantiation process, in particular:

1. no recombination of clauses,
2. the length of clauses does not grow,
3. optimal efficiency in the ground case,
4. semantic selection,
5. redundancy criteria.

As in our previous work, we will use in a modular fashion a satisfiability solver for ground clauses. Let us remark that in the presence of equality such solvers have received considerable attention and very efficient implementations are available, see e.g., [7]. We also use selection based on models of ground clauses to guide the theorem proving process. Another ingredient of our procedure is a

J. Marcinkowski and A. Tarlecki (Eds.): CSL 2004, LNCS 3210, pp. 71–84, 2004.
© Springer-Verlag Berlin Heidelberg 2004

paramodulation-based calculus for reasoning with sets of literals. As we show, with the help of such a calculus it is possible to generate suitable instances of clauses, witnessing unsatisfiability of the selected literals on the ground level. For the completeness proofs we use the model generation technique (see [1, 2, 14]) which allows us to justify redundancy elimination based on entailment from smaller clauses, where "smaller" refers to suitable closure orderings.

Let us briefly compare our method with two other approaches, we aware of, that deal with equational reasoning in the context of instantiation-based theorem proving. In [13] and [16] an equational version of the disconnection calculus is presented, however in their method, literals from different clauses are recombined into a new clause when (superposition-type) equational steps are done. Our instance generation inference systems entirely avoid that recombination which, according to [11], can be a major source of inefficiency in resolution- and superposition-type inference systems. In [15], Plaisted and Zhu consider an extension of their OSHL calculus with equality. It is based on paramodulation with unit clauses, but for non-unit clauses it requires the generally less efficient Brand's transformation method. Our method is applicable for arbitrary first-order clauses with equality.

2   Preliminaries

We shall use standard terminology for first-order clause logic with equality. The symbol "$\simeq$" is used to denote formal equality. By "$\models$" we denote entailment in first-order logic with equality. A *clause* is a possibly empty multiset of literals L_i, usually written $L_1 \vee \ldots \vee L_n$; a *literal* being either an equation $s \simeq t$ or a disequations $u \not\simeq v$ built from terms s, t, u, and v over the given signature. We consider $\simeq$ (and $\not\simeq$) as symmetric syntactically, identifying $s \simeq t$ with $t \simeq s$. We say that C is a sub-clause of D, and write $C \subseteq D$, if C is a sub-multiset of D. The empty clause, denoted by $\square$, denotes falsum. If L is a literal, $\overline{L}$ denotes the complement of L.

A substitution is called a *proper instantiator* of an expression (a term, literal, or clause) if at least one variable of the expression is mapped to a non-variable term. We call D *more specific* than C if $C\tau = D$ for some proper instantiator τ of C. *Renamings* are injective substitutions, sending variables to variables. Two clauses are *variants* of each other if one can be obtained from the other by applying a renaming.

Instance-based theorem proving requires us to work with a refined notion of instances of clauses that we call closures. A *closure* is a pair consisting of a clause C and a substitution σ written $C \cdot \sigma$. We work modulo renaming, that is, do not distinguish between closures $C \cdot \sigma$ and $D \cdot \tau$ for which C is a variant of D and $C\sigma$ is a variant of $D\tau$. Note the distinction between the two notations $C\sigma$ and $C \cdot \sigma$. The latter is a closure *representing* the former which is a clause. A closure is called *ground* if it represents a ground clause. A (ground) closure $C \cdot \sigma$ is called a *(ground) instance* of a set of clauses S if C is a clause in S, and then we say that the closure $C \cdot \sigma$ is a *representation (of the clause $C\sigma$) in S.*

Inference systems and completeness proofs will be based on orderings on ground clauses and closures. Let $\succ_{gr}$ be a total simplification ordering on ground terms. We can assume that $\succ_{gr}$ is defined on ground clauses by a total, well-founded and monotone extension of the order from terms to clauses, as defined, e.g., in [14]. We will extend $\succ_{gr}$ in two ways to orderings on ground closures. The first ordering is $\succ_l$, defined in Section 4, and will be used in reasoning with the unit paramodulation calculus UP. The second is $\succ_{cl}$, defined in Section 5, and will be used in reasoning about instantiations of clauses.

The (Herbrand) *interpretations* we deal with are sometimes partial, given by consistent sets I of ground literals. (As usual, I is called *consistent* if, and only if, $I \not\models \Box$.) A ground literal L is called *undefined* in I if neither $I \models L$ nor $I \models \overline{L}$. I is called *total* if no ground literal is undefined in I. A ground clause C is called *true* (or valid) in a partial interpretation I if $I \models C$. This is the same as saying that $J \models C$ for each total extension J of I. C is called *false* in I if $I \models \neg C$, or, equivalently, if $J \not\models C$ for each total extension J of I. Truth values for closures are defined from the truth values of the clauses they represent.

3   An Informal Description of the Procedure

Let us first informally describe our instantiation-based inference process for equational reasoning. We assume that a satisfiability solver for ground equational clauses is given.

Let S be a given set of first-order clauses. We start by mapping all variables in all clauses in S to a distinguished constant $\bot$, obtaining a set of ground clauses $S\bot$. If $S\bot$ is unsatisfiable then S is first-order unsatisfiable and we are done. Otherwise, we non-deterministically select a literal from each clause in S, obtaining a set of literals Lit.

The next natural step, similar to the case without equality, would be to consider applicable ordered paramodulation inferences, but instead of generating paramodulants to generate corresponding instances of clauses. Unfortunately, adding these instances is not sufficient for a solver on ground clauses to detect unsatisfiability, as the following example shows. Consider the unsatisfiable set of literals $S = \{f(h(x)) \simeq c, h(y) \simeq y, f(a) \not\simeq c\}$. The only applicable ordered paramodulation inference is between the first and the second equation, but the resulting instances are the given equations themselves. On the other hand, the set of ground literals $S\bot$ is satisfiable, so a solver for ground literals can not detect the unsatisfiability of S.

Our approach to this problem is to apply separate first-order reasoning with the selected literals Lit. If Lit is first-order satisfiable, then S is satisfiable and we are done. Otherwise, we generate relevant instances of clauses from S witnessing unsatisfiability of Lit at the ground level. This is done using a paramodulation-based system on literals. In particular, relevant instances can be generated by propagating substitutions from proofs of the empty clause in such a system. Finally, we add obtained instances to S, and repeat the procedure.

Let us now modify the previous example and assume that the literals above are among the selected ones in some clauses, e.g.,

$$S = \{f(h(x)) \simeq c \vee h(h(x)) \not\simeq a, \ h(y) \simeq y, \ f(a) \not\simeq c\}$$
$$Lit = \{f(h(x)) \simeq c, \ h(y) \simeq y, \ f(a) \not\simeq c\}.$$

We can derive the empty clause from Lit by first paramodulating the second literal into the first literal, followed by paramodulation of the result into the third literal. Now, from this paramodulation proof we can extract a relevant substitution σ, which maps x and y to a. Then, the new set of clauses is obtained by applying σ to the old clauses: $S' = S \cup \{f(h(a)) \simeq c \vee h(h(a)) \not\simeq a, h(a) \simeq a\}$. Now, the set $S'\bot$ can be shown to be unsatisfiable by a solver for ground clauses, so we conclude that the original set S is first-order unsatisfiable. In the case if $S'\bot$ is satisfiable we would continue the procedure with this new set of clauses. Let us note that usually the search for the proof of the empty clause from the set of literals is done via some kind of saturation process which can generate a lot of inferences. But the proof itself usually involves only some of them, and as we have seen, we need to propagate only substitutions used in the proof.

We use a solver for ground clauses not only for testing unsatisfiability of $S\bot$. In addition, in the case of satisfiable $S\bot$, the instantiation process can be guided by a model $I_\bot$ of $S\bot$. For this, we restrict the selection of literals to the literals L, such that $L\bot$ is true in $I_\bot$.

Now we overview how we are going to prove completeness of such instantiation process. First, in Section 4 we introduce a calculus UP for ground closures of literals based on ordered paramodulation. We will use this calculus to obtain relevant instantiations of clauses. Then, in Section 5 we show that if a set of clauses is saturated enough, then either it is satisfiable, or otherwise its unsatisfiability can be detected by a ground solver. In the subsequent Section 6 we show how to obtain a saturated set as a limit of a fair saturation process. The problem of how to ensure that a saturation process is fair is considered in Section 7. Up to this point we are working with the UP calculus defined on ground closures. Ground closures allow us to present completeness proofs and fine grained notions of redundancy. Nevertheless, from the practical point of view it is infeasible to work with each ground closure separately, and therefore in Section 8 we present the UPL calculus which is a lifted version of UP. Finally, in Section 9 we consider the issue of how to propagate information on redundant closures to the UPL calculus. This is done via dismatching constraints.

4   Unit Paramodulation on Literal Closures

In this section we introduce an inference system on ground closures of literals, based on ordered paramodulation. This system (and its lifted versions) will be used to guide our instantiation process as shown in the following sections.

Unit-Paramodulation calculus (UP)

$$\frac{(l \simeq r) \cdot \sigma \quad L[l'] \cdot \sigma'}{L[r]\theta \cdot \rho} \ (\theta) \qquad\qquad \frac{(s \not\simeq t) \cdot \tau}{\Box} \ (\mu)$$

where (i) $l\sigma \succ_{gr} r\sigma$; (ii) $\theta = mgu(l, l')$;   where (i) $s\tau = t\tau$;
(iii) $l\sigma = l'\sigma' = l'\theta\rho$; (iv) l' is not a variable.   (ii) $\mu = mgu(s, t)$.

An inference in UP is called proper if the substitution θ, (μ) is a proper instantiator and non-proper otherwise. Let us note that a set of literal closures can be contradictory, yet the empty clause is not derivable in UP.

Example 1. Consider a set of literal closures $\mathcal{L} = \{(f(x) \simeq b) \cdot [a/x], a \simeq b, f(b) \not\simeq b\}$ and assume that $a \succ_{gr} b$. Then, $\mathcal{L}$ is inconsistent but the empty clause is not derivable by UP from $\mathcal{L}$.

UP-Redundancy. Let R be an arbitrary ground rewrite system and $\mathcal{L}$ be a set of literal closures, we denote $irred_R(\mathcal{L})$ the set of closures $L \cdot \sigma \in \mathcal{L}$ with irreducible σ w.r.t. R. In order to introduce the notion of UP-redundancy we need the following ordering on literal closures. Let $\succ_l$ be an arbitrary total well-founded extension of $\succ_{gr}$ from ground literals to ground closures of literals, such that if $L\sigma \succ_{gr} L'\sigma'$ then $L \cdot \sigma \succ_l L' \cdot \sigma'$.

 Let $\mathcal{L}$ be a set of literal closures. We say that $L \cdot \sigma$ is UP-*redundant* in $\mathcal{L}$ if for every ground rewrite system R oriented by $\succ_{gr}$, and such that σ is irreducible w.r.t. R we have $R \cup irred_R(\mathcal{L}_{L \cdot \sigma \succ_l}) \models L\sigma$. Here, $\mathcal{L}_{L \cdot \sigma \succ_l}$ denotes the set of all closures in $\mathcal{L}$ less than $L \cdot \sigma$ w.r.t. $\succ_l$. We denote the set of all UP-redundant closures in $\mathcal{L}$ as $\mathcal{R}_{UP}(\mathcal{L})$. With the help of this redundancy notion we can justify the following simplification rule.

Non-proper Demodulation

$$\frac{(l \simeq r) \cdot \sigma \quad L[l'] \cdot \sigma'}{L[r]\theta \cdot \sigma'}$$

where (i) $l' = l\theta$, (ii) $l\sigma \succ_{gr} r\sigma$, (iii) θ is a non-proper instantiator,
(iv) $l\sigma = l'\sigma'$, (v) $Var(r) \subseteq Var(l)$, (vi) $L[l']\sigma' \succ_{gr} (l \simeq r)\sigma$.

Let us show that non-proper demodulation is a simplification rule, i.e., after adding the conclusion of this rule the right premise becomes UP-redundant.

Lemma 1. *Non-proper demodulation is a simplification rule.*

Proof. Indeed, let $L[r]\theta \cdot \sigma'$ be the conclusion of an application of the non-proper demodulation rule with the premise $(l \simeq r) \cdot \sigma$, $L[l'] \cdot \sigma'$. Now let R be a rewrite system orientable by $\succ_{gr}$ such that σ' is irreducible w.r.t. R. Since θ is a non-proper instantiator, $l' = l\theta$, and $Var(r) \subseteq Var(l)$ we have that σ is also irreducible. Therefore, $L[l']\sigma'$ follows from the smaller closures $(l \simeq r) \cdot \sigma$ and $L[r]\theta \cdot \sigma'$.

Let us show that demodulation with proper unifiers can not be used as a simplification rule in general.

Example 2. Consider the following closures.

$$(1) \quad (g(x) \simeq c) \cdot [f(d)/x] \qquad (3) \quad (f(d) \simeq m) \cdot []$$
$$(2) \quad (g(f(x)) \simeq c) \cdot [d/x] \qquad (4) \quad (g(m) \not\simeq c) \cdot []$$

We can derive the empty clause by UP inferences from (2), (3) and (4). But if we simplify (2) by demodulation with (1) we obtain a tautological closure and the empty clause would not be derivable by UP. The reason for this is that the substitution $[f(d)/x]$ in (1) is reducible (by (3)), whereas the substitution $[d/x]$ in (2) is not.

An UP-*saturation process* is a finite or infinite sequence of sets of closures of literals $\{\mathcal{L}_i\}_{i=1}^{\infty}$ where each set $\mathcal{L}_i$ is obtained from $\mathcal{L}_{i-1}$ by either adding a conclusion of an UP-inference with premises from $\mathcal{L}_{i-1}$ or by removing an UP-redundant w.r.t. $\mathcal{L}_{i-1}$ closure. Let us denote by $\mathcal{L}^{\infty}$ the set of *persisting closures*, that is, the lower limit of the sequence $\mathcal{L}_i$. An UP-saturation process $\{\mathcal{L}_i\}_{i=1}^{\infty}$ is called UP-*fair* if for every UP-inference with premises in $\mathcal{L}^{\infty}$, the conclusion is redundant w.r.t. $\mathcal{L}_i$ for some i.

Lemma 2. *For any two UP-fair saturation processes $\{\mathcal{L}_i\}_{i=1}^{\infty}$ and $\{\mathcal{L}'_i\}_{i=1}^{\infty}$ with $\mathcal{L}_1 = \mathcal{L}'_1$, we have $\mathcal{L}^{\infty} \setminus \mathcal{R}_{UP}(\mathcal{L}^{\infty}) = \mathcal{L}'^{\infty} \setminus \mathcal{R}_{UP}(\mathcal{L}'^{\infty})$.* [1]

This lemma allows us to introduce for every set of literal closures $\mathcal{L}$ its unique UP-saturation $\mathcal{L}^{sat}$ where $\mathcal{L}^{sat} = \mathcal{L}^{\infty} \setminus \mathcal{R}_{UP}(\mathcal{L}^{\infty})$, for some UP-fair saturation process.

5   Completeness for Saturated Sets of Clauses

In this section we prove that if a set of clauses S is saturated enough, then either it can be shown to be unsatisfiable by a ground solver, i.e., $S\perp$ is unsatisfiable, or otherwise S is first-order satisfiable. In the later sections we show how to achieve saturated sets.

First we introduce the notion of Inst-redundancy which will be used to extract relevant closures form clause sets, and also to measure progress in the instantiation process. For this we extend the order $\succ_{gr}$ from ground clauses to ground closures as follows. We say that $C \cdot \tau \succ'_{cl} D \cdot \rho$ if either $C\tau \succ_{gr} D\rho$ or $C\tau = D\rho$ and $C\theta = D$ for a proper instantiator θ. It is straightforward to see that $\succ'_{cl}$ is a well-founded order, so we define $\succ_{cl}$ to be any total well-founded extension of $\succ'_{cl}$.

Let S be a set of clauses and C a ground closure. C is called Inst-*redundant* in S if there exist closures $C_1, \ldots, C_k$ that are ground instances of S such that, (i) for each i, $C \succ_{cl} C_i$, and (ii) $C_1, \ldots, C_k \models C$. A clause C (possibly non-ground)

---
[1] See www.cs.man.ac.uk/~korovink for the proof, omitted due to the lack of space.

is called Inst-redundant in S if each ground closure $C \cdot \sigma$ is Inst-redundant in S. We denote the set of Inst-redundant closures in S as $\mathcal{R}_{\text{Inst}}(S)$.

Consider a set of clauses S, a model $I_\perp$ of $S\perp$. A *selection function* sel based on $I_\perp$ is a function mapping clauses to literals such that for each $C \in S$, $\text{sel}(C) \in C$ and $\text{sel}(C)\perp$ is true in $I_\perp$. Let us consider a selection function sel based on I. Define a set of S-*relevant* instances of literals $\mathcal{L}_S$ as the set of all literal closures $L \cdot \sigma$ such that

1. $L \vee C \in S$,
2. $(L \vee C) \cdot \sigma$ is not Inst-redundant in S,
3. $L = \text{sel}(L \vee C)$.

Let $\mathcal{L}_S^{sat}$ denote the UP-saturation of $\mathcal{L}_S$. We say that the set of clauses S is Inst-*saturated* w.r.t. a selection function sel, if $\mathcal{L}_S^{sat}$ does not contain the empty clause. The following auxiliary lemma about UP is obvious.

Lemma 3. *Let R be a ground rewrite system and UP is applicable to $(l \simeq r)\cdot\sigma$, $L[l'] \cdot \sigma'$ with the conclusion $L[r]\theta \cdot \rho$. Then if σ and σ' are irreducible w.r.t. R then ρ is also irreducible.*

Now we are ready to prove our main completeness theorem. Let us remark that in the proof we will use both orderings $\succ_l$ and $\succ_{cl}$. In fact, the model construction will be done in $\succ_l$, but the counterexample reduction in $\succ_{cl}$.

Theorem 1. *If a set S of clauses is Inst-saturated, and $S\perp$ is satisfiable, then S is also satisfiable.*

Proof. Let S be an Inst-saturated set of clauses, such that $S\perp$ is satisfiable in a model $I_\perp$, and sel is a selection function based on $I_\perp$. Let $\mathcal{L}_S$ be S-relevant instances of literals. We have that $\mathcal{L}_S^{sat} \perp$ does not contain the empty clause.

By induction on $\succ_l$ we construct a candidate model to S based on $\mathcal{L}_S^{sat}$. Suppose, as an induction hypothesis, that sets of literals ϵ_M have been defined for the ground closures $M \in \mathcal{L}_S^{sat}$ smaller than L in $\succ_l$, and let I_L denote the set $\bigcup_{L \succ_l M} \epsilon_M$. Let R_L denote the ground rewrite system obtained by orienting all positive equations in I_L w.r.t. $\succ_l$. Suppose that $L = L' \cdot \sigma$. Then define $\epsilon_L = \{L'\sigma\}$, if

1. $L'\sigma$ is irreducible by R_L, and
2. $L'\sigma$ is undefined in I_L (i.e. neither $I_L \models L'\sigma$ nor $I_L \models \overline{L}'\sigma$).

In this case we say that L is *productive*. Otherwise, we define $\epsilon_L = \emptyset$. Define I_S to be the set $\bigcup_{L \in \mathcal{L}_S^{sat}} \epsilon_L$ and $R_S = \bigcup_{L \in \mathcal{L}_S^{sat}} R_L$. It is easy to see that I_S is consistent and R_S is a convergent interreduced rewrite system and every $L\sigma \in I_S$ is irreducible by R_S. Let I be an arbitrary total consistent extension of I_S. Now we show that I is a model to all ground instances of S. Assume otherwise.

Let $D = D' \cdot \sigma$ be the minimal w.r.t. $\succ_{cl}$ ground instance of S that is false in I. Let us show that for every variable x in D', $x\sigma$ is irreducible by R_S. Otherwise, let $(l \to r)\tau \in R_L$ and $x\sigma = x\sigma[l\tau]_p$ for some variable x in D'. Then, we can define a substitution σ' by changing σ on x with $x\sigma' = x\sigma[r\tau]_p$. We have that $I \not\models D'\sigma'$ and $D \succ_{cl} D' \cdot \sigma'$, which contradicts to the minimality of the counterexample.

Now we note that D is not Inst-redundant in S. Otherwise it would follow from smaller, w.r.t. $\succ_{cl}$, closures $D_1, \ldots, D_n$. Hence, one of D_i is false in I contradicting to the minimality of the counterexample.

Since D is not Inst-redundant, we have that for some literal L, $D' = L \vee D''$ and $L \cdot \sigma \in \mathcal{L}_S$. And also $L\sigma$ is false in I.

Assume that $L \cdot \sigma$ is UP-redundant in $\mathcal{L}_S^{sat}$. Then, since σ is irreducible by R_S we have

$$R_S \cup irred_{R_S}(\{L' \cdot \sigma' \in \mathcal{L}_S^{sat} | L \cdot \sigma \succ_l L' \cdot \sigma'\}) \models L\sigma.$$

Therefore, there is $L' \cdot \sigma \in irred_{R_S}(\mathcal{L}_S^{sat})$ false in I, (if $L \cdot \sigma$ is not UP-redundant in $\mathcal{L}_S^{sat}$ we take $L' \cdot \sigma = L \cdot \sigma$). Let $M \cdot \tau$ be the minimal w.r.t. $\succ_l$ closure in $irred_{R_S}(\mathcal{L}_S^{sat})$ which is false in I. Let us show that $M \cdot \tau$ is irreducible by R_S. Otherwise, assume that $M \cdot \tau$ is reducible by $l \to r \in R_S$ and $(l' \to r') \cdot \rho \in \mathcal{L}_S^{sat}$ is the closure producing $l \to r$ in R_S. Since τ is irreducible by R_S, UP-inference is applicable to $(l' \to r') \cdot \rho$ and $M[l''] \cdot \tau$ with the conclusion $M[r']\theta \cdot \mu$, where $l'\rho = l''\tau = l''\theta\mu$ and $\theta = mgu(l', l'')$. We have that $M[r']\theta \cdot \mu$ is false in I. Now we show that $M[r']\theta \cdot \mu$ is not UP-redundant in $\mathcal{L}_S^{sat}$. Assume otherwise. From Lemma 3 follows that μ is irreducible by R_S. From definition of UP-redundancy, we have

$$R_S \cup irred_{R_S}(\{M' \cdot \tau' \in \mathcal{L}_S^{sat} | M[r']\theta \cdot \mu \succ_l M' \cdot \tau'\}) \models M[r']\theta\mu.$$

Therefore, there is $M' \cdot \tau' \in irred_{R_S}(\mathcal{L}_S^{sat})$ such that $M \cdot \tau \succ_l M[r']\theta \cdot \mu \succ_l M' \cdot \tau'$ and $M'\tau'$ false in I. This contradicts to the minimality of $M \cdot \tau$. But, if $M[r']\theta \cdot \mu$ is not UP-redundant we have $M[r']\theta \cdot \mu \in \mathcal{L}_S^{sat}$, and since μ is irreducible by R_S, $M[r']\theta \cdot \mu \in irred_{R_S}(\mathcal{L}_S^{sat})$, we again obtain a contradiction to the minimality of $M \cdot \tau$. We conclude that $M \cdot \tau$ is irreducible by R_S.

Now we have that $M \cdot \tau$ is in $\mathcal{L}_S^{sat}$, irreducible by R_S, and not productive. Therefore $I_{M\cdot\tau} \models \overline{M\tau}$. Consider all possible cases. Let $M \cdot \tau$ be an equation $(s \simeq t) \cdot \tau$. We have that $I_{M\cdot\tau} \models (s \not\simeq t)\tau$. Since, all literals in $I_{M\cdot\tau}$ and $s\tau, t\tau$ are irreducible by $R_{M\cdot\tau}$, and $R_{M\cdot\tau}$ is a convergent rewrite system we have $(s \not\simeq t)\tau \in I_{M\cdot\tau}$. Therefore $(s \not\simeq t)\tau$ is produced to $I_{M\cdot\tau}$ by some $(s' \not\simeq t') \cdot \tau'$. But this is impossible since $(s' \not\simeq t')\tau' \succ_{gr} (s \simeq t)\tau = M\tau$, and hence $(s' \not\simeq t') \cdot \tau' \succ_l M \cdot \tau$. Now assume that $M \cdot \tau$ is a disequation $(s \not\simeq t) \cdot \tau$. We have $I_{M\cdot\tau} \models (s \simeq t)\tau$ and since $s\tau$ and $t\tau$ are irreducible by $R_{M\cdot\tau}$ we have $s\tau = t\tau$. But then equality resolution is applicable to $M \cdot \tau$, contradicting that $\mathcal{L}_S^{sat}$ does not contain the empty clause.

Finally we conclude that I is an model for S.

6   Effective Saturation Strategies

In this section we shall investigate how saturation of a set of clauses can be achieved effectively. First we show how saturation is done on closures and later we show how saturation process can be lifted to general clauses.

An Inst-*saturation process* is a sequence of triples $\{\langle S^i, I^i_\perp, \mathsf{sel}^i \rangle\}^\infty_{i=1}$, where S^i is a set of clauses, $I^i_\perp$ a model of $S^i\perp$ and sel^i a selection function based on that model. Given $\langle S^i, I^i_\perp, \mathsf{sel}^i \rangle$, a *successor state* $\langle S^{i+1}, I^{i+1}_\perp, \mathsf{sel}^{i+1} \rangle$ is obtained by one of these steps: (i) $S^{i+1} = S^i \cup N$, where N is a set of clauses such that $S^i \models N$; or (ii) $S^{i+1} = S^i \setminus \{C\}$, where C is Inst-redundant in S^i. If $S^{i+1}\perp$ is unsatisfiable, the process terminates with the result "unsatisfiable". Let us denote by S^∞ the set of persisting clauses, that is, the lower limit of $\{S^i\}^\infty_{i=1}$. In order to ensure that we always reach an Inst-saturated set in the limit of the saturation process we need the notion of Inst-fair saturation.

Consider a finite set of closures $K = \{(L_1 \vee C_1) \cdot \sigma, \ldots, (L_n \vee C_n) \cdot \sigma\}$ of clauses from S^∞. We denote $\mathcal{L} = \{L_1 \cdot \sigma, \ldots, L_n \cdot \sigma\}$. The pair $(K, \mathcal{L})$ is called a *persistent conflict* if $\mathcal{L}^{sat}$ contains the empty clause and for infinitely many i we have $\mathsf{sel}^i(L_j \vee C_j) = L_j$ for $1 \leq j \leq n$.

We call an Inst-saturation process Inst-*fair* if for every persistent conflict $(K, \mathcal{L})$, at least one of the closures in K is Inst-redundant in S_i for some i.

Now our goal is to show that for the limit S^∞ of an Inst-fair saturation process, such that $S^\infty\perp$ is satisfiable, we can build a model $I_\perp$ and a selection function sel, based on $I_\perp$ such that S^∞ is Inst-saturated w.r.t. sel. The main problem here is that when we use selection functions based on truth in propositional models, these models change when we add more instances. Note that it is possible that the limit S^∞ of an Inst-fair saturation process is not Inst-saturated for some model $I_\perp$ of $S^\infty\perp$, likewise it is possible that $I^i_\perp \cup I^j_\perp$ is inconsistent for every $i \neq j$ (so, for example, we can not take union of $I^i_\perp$ for $I_\perp$).

Lemma 4. *Let S^∞ be a set of persistent clauses of an* Inst-*fair saturation process* $\{\langle S^i, I^i_\perp, \mathsf{sel}^i \rangle\}^\infty_{i=1}$, *and* $S^\infty\perp$ *is satisfiable. Then, there exists a model* $I\perp$ *of* $S^\infty\perp$ *and a selection function* sel *based on* $I\perp$ *such that* S^∞ *is* Inst-*saturated w.r.t.* sel.

Proof. Let $\{C_i\}^\infty_{i=1}$ be an enumeration of clauses in S^∞. For each n we construct a model J^n of $\{C_i\perp\}^{i=n}_{i=1}$ and a selection function sel^n_J based on J^n, by induction on n. For each n the following invariants will be satisfied.

1. J^n is consistent and sel^n_J is a selection function for clauses $\{C_i\}^{i=n}_{i=1}$ based on J^n.
2. $J^{n-1} \subseteq J^n$ and sel^n_J coincides with sel^{n-1}_J on clauses $\{C_i\}^{i=n-1}_{i=1}$.
3. There are infinitely many k such that $J^n \subseteq I^k$ and for all $1 \leq l \leq n$, $\mathsf{sel}^k(C_l) = \mathsf{sel}^n_J(C_l)$.

If $n = 1$ then we have that there exists $L \in C_1$ such that $L \in \mathsf{sel}^k$ for infinitely many k. We take $J^1 = \{L\perp\}$ and $\mathsf{sel}^1_J(C_1) = \{L\}$. Trivially, all invariants (1–3) are satisfied.

Let $n \geq 1$ and assume that we have a model J^n and sel^n_J for $\{C_i\perp\}^{i=n}_{i=1}$ such that invariants (1–3) are satisfied. Since $C_{n+1} \in S^\infty$ we have that for some m and every $p \geq m$, $C_{n+1} \in S^p$. From this and invariant (3) follows that for some $L \in C_{n+1}$ there are infinitely many k such that $J^n \subseteq I^k$, and $\mathsf{sel}^k(C_l) = \mathsf{sel}^n_J(C_l)$ for all $1 \leq l \leq n$, and $\mathsf{sel}^k(C_{n+1}) = L$. Define $J^{n+1} = J^n \cup \{L\perp\}$ and

$\mathsf{sel}^{n+1}_J(C_l) = \mathsf{sel}^n_J(C_l)$ for $1 \le l \le n$, $\mathsf{sel}^{n+1}_J(C_{n+1}) = L$. It is easy to see that all invariants (1–3) are satisfied for $J^{n+1}, \mathsf{sel}^{n+1}_J$.

We define $I_\perp = \cup^\infty_{i=1} J_i$ and $\mathsf{sel}(C_i) = \mathsf{sel}^i_J(C_i)$ for $i \ge 1$. From compactness follows that $I_\perp$ is consistent, and sel is a selection function based on $I_\perp$.

Now we need to show that S^∞ is saturated w.r.t. sel. Assume otherwise. Then, there is a finite subset $\mathcal{L}$ of $\mathcal{L}_{S^\infty}$, such that $\mathcal{L}^{sat}$ contains the empty clause. Let $K = \{(L_1 \vee C_1) \cdot \sigma, \ldots, (L_n \vee C_n) \cdot \sigma\}$ be the set of closures of clauses from S^∞, producing $\mathcal{L}$ to $\mathcal{L}_{S^\infty}$. Then, from the construction of $I_\perp$ and in particular from the invariant (3) follows that there are infinitely many i such that $\mathsf{sel}^i(L_j \vee C_j) = L_j$ for $1 \le j \le n$. Hence, $(K, \mathcal{L})$ is a persistent conflict. Since the saturation process is Inst-fair we have that at least one of the closures in K is Inst-redundant in $\mathcal{L}_{S^\infty}$. But this is impossible since all closures in $\mathcal{L}$ are S^∞-relevant and can not be produced by Inst-redundant closures.

Corollary 1. *Let $\{\langle S^i, I^i_\perp, \mathsf{sel}^i \rangle\}^\infty_{i=1}$ be an Inst-fair saturation process. Then, either (1) for some i we obtain an unsatisfiable $S^i\perp$ and therefore S^1 is unsatisfiable, or (2) for all i, $S^i\perp$ is satisfiable and therefore, (by Lemma 4 and Theorem 1) S^1 is satisfiable, moreover if for some i, S^i is Inst-saturated then at this step we can conclude that S^1 is satisfiable.*

In the next sections we consider the issue of how to ensure that an Inst-saturation process is Inst-fair.

7   Relevant Instances from Proofs

In order to obtain an Inst-fair saturation we need to make closures in persistent conflicts Inst-redundant. A uniform way to make a closure $C \cdot \sigma$ of a clause $C \in S$, Inst-redundant in S, is to add to S a proper (possible nonground) instance of C, which generalises $C\sigma$. Next we will study how to find instantiations which are relevant to the persisting conflicts.

Let us consider a persistent conflict $(K, \mathcal{L})$, where $K = \{(L_1 \vee C_1) \cdot \sigma, \ldots, (L_n \vee C_n) \cdot \sigma\}$ and $\mathcal{L} = \{L_1 \cdot \sigma, \ldots, L_n \cdot \sigma\}$. Since $\mathcal{L}^{sat}$ contains the empty clause we have that there is a proof of the empty clause in UP from closures in $\mathcal{L}$. Our next goal is to show that in any proof at least one inference is a proper UP-inference. To speak more formally about the proofs we assume that proofs are represented as binary trees with nodes labelled by closures together with substitutions from the corresponding inferences. We assume that at each node of a proof, left subproof is variable disjoint from the right subproof.

Example 3. A proof of the empty clause in UP from literal closures.

$$\frac{\dfrac{f(x) \simeq g(x) \cdot [h(a)/x] \quad f(y) \simeq h(y) \cdot [h(a)/y]}{g(x) \simeq h(x) \cdot [h(a)/x]} \, [x/y] \quad \dfrac{g(h(u)) \not\simeq h(h(u)) \cdot [a/u]}{h(h(u)) \not\simeq h(h(u)) \cdot [a/u]}}{\dfrac{h(h(u)) \not\simeq h(h(u)) \cdot [a/u]}{\square}} \, [h(u)/x]$$

Let us consider a proof P and a leaf of this proof with a closure $L \cdot \sigma$. Let $\theta_1, \ldots, \theta_n$ be substitutions along the branch from this leaf to the root of the proof. We call the composition $\theta = \theta_1 \cdots \theta_n$ as a *P-relevant instantiator* and the closure $L\theta \cdot \tau$ as a *P-relevant instance* of $L \cdot \sigma$, where $L\theta\tau = L\sigma$. If we consider the left most leaf of the proof in the example above, then the P-relevant instance will be $(f(h(u)) \simeq g(h(u))) \cdot [a/u]$ with the P-relevant instantiator $[h(u)/x]$.

Lemma 5. *Let P be a proof of the empty clause, and PI be the set of P-relevant instances of all leafs of P. Then, $PI\bot$ is unsatisfiable.*

Corollary 2. *Let $(K, \mathcal{L})$ be a persistent conflict and P is a proof of the empty clause from $\mathcal{L}$ in UP, then at least one of P-relevant instantiator is proper.*

For a persistent conflict $(K, \mathcal{L})$, this corollary allows us to make closures in K Inst-redundant by adding their P-relevant proper instances. Let us continue with Example 3. Assume that literal closures at the leafs of P are in $\mathcal{L}$ for a persistent conflict $(K, \mathcal{L})$, so $\mathcal{L} = \{(f(x) \simeq g(x)) \cdot [h(a)/x], \ldots\}$ and, e.g., $K = \{(f(x) \simeq g(x) \vee h(g(x)) \simeq c) \cdot [h(a)/x], \ldots\}$, then we can add a P-relevant proper instance $f(h(u)) \simeq g(h(u)) \vee h(g(h(u))) \simeq c$ to the clause set, making the first closure in K Inst-redundant. Thus, to make an Inst-saturation process Inst-fair we need to UP-saturate literal closures from the persisting conflicts and add proper instantiations of clauses with substitutions that can be obtained from the proofs of the empty clause.

8   From Literal Closures to Literal Clauses

So far we have been considering closures as the basic entities for persistent conflicts and the UP calculus. Of course, working with each ground closure separately is of little practical use. This motivates our next step of lifting UP calculus from literal closures to literals.

Unit paramodulation for literals (UPL)

$$\frac{(l \simeq r) \quad L[l']}{L[r]\theta} \; (\theta) \qquad\qquad \frac{s \not\simeq t}{\Box} \; (\mu)$$

where (i) $\theta = mgu(l, l')$; (ii) l' is not a variable; (iii) $l\sigma \succ_{gr} r\sigma$ for some grounding substitution σ; $\mu = mgu(s, t)$;

Proofs in UPL can be represented in the same way as proofs in UP (see Section 7). And in the same way we can define notions of a P-relevant instantiator and a P-relevant instance.

By a simple lifting argument we can prove the following lemma, connecting UPL with UP calculus.

Lemma 6. *Let Lit be a set of literals such that $Lit\bot$ is satisfiable and $\mathcal{L}$ be a set of ground closures of literals from Lit such that the empty clause is derivable in UP from $\mathcal{L}$. Then, there is a proof P of the empty clause in UPL from Lit*

such that for at least one closure $L \cdot \sigma \in \mathcal{L}$, P-relevant instance of L is $L\theta$ where θ is a proper instantiator and $L\sigma = L\theta\tau$ for some ground substitution τ.

This lemma implies that an Inst-saturation process $\{\langle S^i, I^i_\perp, \mathsf{sel}^i \rangle\}_{i=1}^\infty$ is Inst-fair if the following holds. Consider a finite set $K = \{(L_1 \vee C_1), \ldots, (L_n \vee C_n)\}$ of clauses from S^∞, such that for infinitely many i we have $\mathsf{sel}^i(L_j \vee C_j) = L_j$ for $1 \leq j \leq n$. Let P be an UPL proof of the empty clause from $\{L_1, \ldots, L_n\}$ and $L_i\theta$ be a proper P-relevant instance. Then, for some step j, all ground closures $(L_i \vee C_i) \cdot \theta\sigma$ are Inst-redundant in S^j.

We can observe that since θ is a proper instantiator, to make all closures $(L_i \vee C_i) \cdot \theta\sigma$ Inst-redundant, we can just add $(L_i \vee C_i)\theta$ to the clause set.

9   Representation of Closures via Dismatching Constraints

We have seen that in the process of obtaining a saturated set we make certain closures Inst-redundant by proper instantiations. It might be desirable to discard these redundant closures when we consider UPL calculus. In this section we show how it can be done with the help of dismatching constraints, defined below. We remark that in the context of resolution and paramodulation various kinds of constraints have been considered (see e.g. [14, 10, 6]).

A *simple dismatching constraint* is a formula $ds(\bar{s}, \bar{t})$, where $\bar{s}, \bar{t}$ are two variable disjoint tuples of terms, with the following semantics. A solution to a constraint $ds(\bar{s}, \bar{t})$ is a substitution σ such that for every substitution γ, $\bar{s}\gamma \neq \bar{t}\sigma$, (here $=$ is the syntactic equality). It is easy to see that a constraint $ds(\bar{s}, \bar{t})$ is satisfiable if and only if for all substitutions γ, $\bar{s}\gamma \neq \bar{t}$. In other words, a dismatching constraint $ds(\bar{s}, \bar{t})$ is not satisfiable if and only if there is a substitution μ such that $\bar{s}\mu = \bar{t}$, which is a familiar matching problem. We will use conjunctions of simple dismatching constraints, called just *dismatching constraints*, $\wedge_{i=1}^n ds(\bar{s}_i, \bar{t}_i)$, where $\bar{s}_i$ is variable disjoint from all $\bar{t}_j$, and $\bar{s}_k$, for $i \neq k$. Let us note that there is a polynomial time algorithm for testing satisfiability of the dismatching constraints. To check whether a constraint $\wedge_{i=1}^n ds(\bar{s}_i, \bar{t}_i)$ is (un)satisfiable, we just need to solve n matching problems.

A *constrained clause* $C \mid [\, D \,]$ is a clause C together with a dismatching constraint D. We will assume that in a constrained clause $C \mid [\, \wedge_{i=1}^n ds(\bar{s}_i, \bar{t}_i) \,]$, the clause C is variable disjoint from all s_i, $1 \leq i \leq n$. A *constrained clause* $C \mid [\, D \,]$ represents the set of all ground closures $C \cdot \sigma$, denoted as $Cl(C \mid [\, D \,])$, such that σ is a solution to D. For a set S of constrained clauses, $Cl(S)$ denotes the set of all ground closures represented by constrained clauses from S.

Now if we consider a set of clauses S such that $C \in S$ and $C\theta \in S$ for some proper instantiator θ, then we can discard all Inst-redundant ground closures $C \cdot \theta\sigma$, by adding a dismatching constraint to C, obtaining $C \mid [\, ds(\bar{x}\theta, \bar{x}) \,]$. In the general case, when a constrained clause $C \mid [\, D \,]$ is in S and we add $C\theta$ to S for some proper instantiator θ, then we can discard all Inst-redundant ground closures $C \cdot \theta\sigma$, by extending the dismatching constraint D, obtaining $C \mid [\, D \wedge ds(\bar{x}\theta, \bar{x}) \,]$. We can always assume that all variables in $\bar{x}\theta$ are disjoint from variables in C and D.

The notion of Inst-redundancy can be adapted from clauses to constrained clauses, by saying that a constrained clause $C \mid [\ D\ ]$ is Inst-redundant if all closures in $Cl(C \mid [\ D\ ])$ are Inst-redundant.

Let S be a set of constrained clauses, then $Unc(S)$ denotes the set of all unconstrained clauses obtained from S by dropping all constraints. We say that a set of constrained clauses S is *well-constrained* if $Cl(S) \setminus \mathcal{R}_{\mathrm{Inst}}(Cl(S)) = Cl(Unc(S)) \setminus \mathcal{R}_{\mathrm{Inst}}(Cl(Unc(S)))$. Thus, constraints in well-constrained sets of clauses is just a tool of discarding Inst-redundant closures.

Next we can replace, UPL calculus with the calculus on constrained literals.

Unit paramodulation with dismatching constraints (UPD)

$$\frac{(l \simeq r) \mid [\ D_1\ ] \quad L[l'] \mid [\ D_2\ ]}{L[r]\theta \mid [\ (D_1 \wedge D_2)\theta\ ]}\ (\theta) \qquad\qquad \frac{s \not\simeq t \mid [\ D\ ]}{\square}\ (\mu)$$

where (i) $\theta = mgu(l, l')$; (ii) l' is not a vari-   where (i) $\mu = mgu(s, t)$; (ii)
able; (iii) for some grounding substitution σ, $D\mu$ is satisfiable.
satisfying $(D_1 \wedge D_2)\theta$, $l\sigma \succ_{gr} r\sigma$;

Naturally we can define the notion of UPD-redundancy, saying that a constrained literal $L \mid [\ D\ ] \in LD$ is UPD-*redundant* in LD if all closures in $Cl(L \mid [\ D\ ])$ are UP-redundant in $Cl(LD)$. And in the same way as for UP we can define UPD-saturation process and LD^{sat}.

Now in the place of S-relevant literal closures $\mathcal{L}_S$ we define the set of S-relevant constrained literals LD_S as the set of all constrained literals $L \mid [\ D\ ]$ such that

1. $(L \vee C) \mid [\ D\ ] \in S$,
2. $(L \vee C) \mid [\ D\ ]$ is not Inst-redundant in S,
3. $L = \mathsf{sel}(L \vee C)$.

We say that the set of constrained clauses is Inst-saturated if LD_S^{sat} does not contain the empty clause.

The following lemma can be proved by a simple lifting argument.

Lemma 7. *Let LD be a set of constrained literals. If $Cl(LD)^{sat}$ (saturation w.r.t. UP) contains the empty clause, then LD^{sat} (saturation w.r.t. UPD) also contains the empty clause.*

From Lemma 7 a lifted version of Theorem 1 from Section 5 follows.

Theorem 2. *If a well-constrained set of clauses S is Inst-saturated and $Unc(S)\perp$ is satisfiable, then $Unc(S)$ is also satisfiable.*

References

1. L. Bachmair and H. Ganzinger. Equational reasoning in saturation-based theorem proving. In W. Bibel and P.H. Schmitt, editors, *Automated Deduction — A Basis for Applications*, volume I, chapter 11, pages 353–397. Kluwer, 1998.

2. L. Bachmair and H. Ganzinger. Resolution theorem proving. In A. Robinson and A. Voronkov, editors, *Handbook of Automated Reasoning*, volume 1, pages 19–100. Elsevier, 2001.

3. P. Baumgartner. FDPLL – a first-order Davis-Putnam-Logeman-Loveland Procedure. In *Proc. CADE*, volume 1831 of *LNAI*, pages 200–219, 2000.

4. P. Baumgartner and C. Tinelli. The model evolution calculus. In F. Baader, editor, *Proc. CADE-19*, number 2741 in LNAI, pages 350–364. Springer, 2003.

5. J.-P. Billon. The disconnection method: a confluent integration of unification in the analytic framework. In *Tableaux 1996*, volume 1071 of *LNAI*, pages 110–126, 1996.

6. R. Caferra and N. Zabel. A method for simultaneous search for refutations and models by equational constraint solving. *J. of Symbolic Computation*, 13(6):613–641, 1992.

7. H. Ganzinger, G. Hagen, R. Nieuwenhuis, A. Oliveras, and C. Tinelli. DPLL(T): fast decision procedures. In *16th Int. Conf. on Computer Aided Verification*, LNCS, 2004. to appear.

8. H. Ganzinger and K. Korovin. New directions in instantiation-based theorem proving. In *Proc. 18th IEEE Symposium on Logic in Computer Science*, pages 55–64. IEEE, 2003.

9. J.N. Hooker, G. Rago, V. Chandru, and A. Shrivastava. Partial instantiation methods for inference in first order logic. *J. of Automated Reasoning*, 28:371–396, 2002.

10. C. Kirchner, H. Kirchner, and M. Rusinowitch. Deduction with symbolic constraints. *Revue Francaise d'Intelligence Artificielle*, 4(3):9–52, 1990. Special issue on automated deduction.

11. S.J. Lee and D. Plaisted. Eliminating duplication with the Hyper-linking strategy. *J. of Automated Reasoning*, 9:25–42, 1992.

12. R. Letz and G. Stenz. Proof and model generation with disconnection tableaux. In *Proc. LPAR 2001*, volume 2250 of *LNAI*, pages 142–156, 2001.

13. Reinhold Letz and Gernot Stenz. Integration of equality reasoning into the disconnection calculus. In *Tableaux 2002*, volume 2381 of *LNAI*, pages 176–190, 2002.

14. R. Nieuwenhuis and A. Rubio. Paramodulation-based theorem proving. In A. Robinson and A. Voronkov, editors, *Handbook of Automated Reasoning*, volume I, pages 371–443. Elsevier, 2001.

15. D. Plaisted and Y. Zhu. Ordered semantic hyper-linking. *J. of Automated Reasoning*, 25(3):167–217, 2000.

16. G. Stenz. *The Disconnection Calculus*. Logos, 2002. Dissertation, TU München.

Goal-Directed Methods for Łukasiewicz Logic

George Metcalfe[1], Nicola Olivetti[2], and Dov Gabbay[1]

[1] Department of Computer Science, King's College London,
Strand, London WC2R 2LS, UK
{metcalfe,dg}@dcs.kcl.ac.uk
[2] Department of Computer Science, University of Turin,
Corso Svizzera 185, 10149 Turin, Italy
olivetti@di.unito.it

Abstract. In this paper we present goal-directed deduction methods for Łukasiewicz infinite-valued logic **Ł**, giving logic programming style algorithms which both have a logical interpretation and provide a suitable basis for implementation. We begin by considering a basic version with connections to calculi for other logics, then make refinements to obtain greater efficiency and termination properties, and to deal with further connectives and truth constants. We finish by considering applications of these algorithms to fuzzy logic programming.

Keywords: Łukasiewicz Logics, Fuzzy Logics, Goal-Directed Methods.

1 Introduction

Łukasiewicz logics were introduced by Jan Łukasiewicz in the 1920s [9], and are currently studied intensively in connection with several areas of research. Firstly, in fuzzy logic where the infinite-valued Łukasiewicz logic **Ł**, along with Gödel logic **G** and Product logic **Π**, emerges as one of the fundamental "t-norm based" fuzzy logics [7]. Also in algebra, where Chang's MV-algebras for **Ł** have applications to many fields of mathematics, see e.g. [3], and in geometry where formulae of **Ł** are related to particular geometric functions via McNaughton's representation theorem [10]. Finally, various semantic interpretations of Łukasiewicz logics have been given, most importantly using Ulam's game with errors/lies which has applications to adaptive error-correcting codes [15].

From the automated reasoning perspective, a variety of proof methods have been advanced for **Ł**. These fall into three main categories:

1. *Gentzen-style calculi*: sequent and hypersequent calculi are provided by the authors in [12], and a many-placed sequent calculus (via a reduction to finite-valued logics) is given by Aguzzoli and Ciabattoni in [1].
2. *Tableaux systems*: both Hähnle [6] and Olivetti [17] have given tableaux for **Ł** that are co-NP via reductions to mixed integer programming problems.
3. *Resolution methods*: Wagner [21] (using hyperplanes), and Mundici and Olivetti [16] have given resolution calculi for **Ł**.

We note also that connectives from **Ł** have been used as the basis for fuzzy logic programming methods, see e.g. [19, 8, 20].

J. Marcinkowski and A. Tarlecki (Eds.): CSL 2004, LNCS 3210, pp. 85–99, 2004.

In this work our aim is to make use of the theoretical insights provided by the Gentzen-style calculi of [12] to give proof calculi for both **Ł** and an important extension of **Ł** with constants called *Rational Pavelka Logic* **RPL** (see e.g. [7]), that are not only geared towards automated reasoning, but also have intuitive logical and algorithmic interpretations. For this purpose we develop *goal-directed methods*, a generalization of the logic programming style of deduction particularly suitable for proof search, which decompose goal formulae according to their structure and database formulae according to the goals, thereby eliminating much non-determinism and avoiding decomposing irrelevant formulae. Goal-directed systems have been given for a wide range of logics including classical, intuitionistic, intermediate, modal, substructural and many-valued logics [13, 4, 5, 11], and have been used as the basis for various non-classical logic programming languages, see e.g. [18].

We proceed as follows. We begin in Sections 2 and 3 by introducing **Ł** and the goal-directed methodology respectively. In Section 4 we then give a basic goal-directed algorithm for **Ł**, pointing out similarities with calculi for other logics. In Section 5 we turn our attention to improving efficiency, giving a variety of terminating and non-terminating versions, and to showing how other connectives and truth constants for **RPL** may be treated. Finally in Section 6 we explore applications of these algorithms to fuzzy logic programming in **Ł**.

2   Łukasiewicz Logic

Łukasiewicz logic **Ł** is defined as follows:

Definition 1 (Ł). **Ł** *is based on a language with a binary connective $\rightarrow$ and a constant $\perp$, and consists of the following axioms together with modus ponens.*

$$(\text{Ł}1) \quad A \rightarrow (B \rightarrow A)$$
$$(\text{Ł}2) \quad (A \rightarrow B) \rightarrow ((B \rightarrow C) \rightarrow (A \rightarrow C))$$
$$(\text{Ł}3) \quad ((A \rightarrow B) \rightarrow B) \rightarrow ((B \rightarrow A) \rightarrow A)$$
$$(\text{Ł}4) \quad ((A \rightarrow \perp) \rightarrow (B \rightarrow \perp)) \rightarrow (B \rightarrow A)$$

Other connectives are defined as follows:

$$\neg A =_{def} A \rightarrow \perp \qquad A \odot B =_{def} \neg(A \rightarrow \neg B) \qquad A \wedge B =_{def} A \odot (A \rightarrow B)$$
$$\top =_{def} \neg\perp \qquad A \oplus B =_{def} \neg A \rightarrow B \qquad A \vee B =_{def} (A \rightarrow B) \rightarrow B$$

Appropriate algebras for **Ł** were introduced by Chang in [2].

Definition 2 (MV-Algebra). *An MV-algebra is an algebra $\mathcal{A} = \langle L, \oplus, \neg, \perp \rangle$ with a binary operation $\oplus$, a unary operation $\neg$ and a constant $\perp$, satisfying:*

$$(mv1) \perp \oplus a = a \qquad\qquad (mv2)\ a \oplus b = b \oplus a$$
$$(mv3)\ (a \oplus b) \oplus c = a \oplus (b \oplus c) \qquad (mv4)\ \neg\neg a = a$$
$$(mv5)\ a \oplus \neg\perp = \neg\perp \qquad\qquad (mv6)\ \neg(\neg a \oplus b) \oplus b = \neg(\neg b \oplus a) \oplus a$$

We also define: $a \rightarrow b =_{def} \neg a \oplus b$, $a \odot b =_{def} \neg(\neg a \oplus \neg b)$, $a \wedge b =_{def} a \odot (a \rightarrow b)$, $a \vee b =_{def} \neg(\neg a \wedge \neg b)$, $\top =_{def} \neg\perp$.

A valuation *for $\mathcal{A}$ is a function v from the set of propositional variables to L extended to formulae by the condition $v(\#(A_1,\ldots,A_n)) = \#(v(A_1),\ldots,v(A_n))$ for each connective $\#$. We say that a formula A is* valid in $\mathcal{A}$ *iff for all valuations v for $\mathcal{A}$, $v(A) = \top$, and write $\models_L A$ iff A is valid in all MV-algebras.*

A formula A is a *logical consequence of a set of formulae Γ in $\mathcal{A}$ iff for any valuation v for $\mathcal{A}$ such that $v(B) = \top$ for all $B \in \Gamma$, also $v(A) = \top$, and we write $\Gamma \models_L A$ iff A is a logical consequence of Γ in all MV-algebras.*

The usual standard MV-algebra for **L** is based on the real unit interval $[0,1]$; however it will be more convenient in this paper for us to consider a "knocked down" version based instead on the real interval $[-1,0]$.

Definition 3 ($[-1,0]_L$). *Let $x \oplus y = min(0, x+y+1)$ and $\neg x = -1 - x$, and define $[-1,0]_L = \langle [-1,0], \oplus, \neg, -1 \rangle$.*

We now state Chang's algebraic completeness theorem (slightly revised) for **L**.

Theorem 1 (Chang [2]). *The following are equivalent: (1) A is derivable in* **L**. *(2) $\models_L A$. (3) A is valid in $[-1,0]_L$.*

We also mention a deduction theorem for **L**, writing A^n for $\overbrace{A \odot \ldots \odot A}^{n}$.

Theorem 2. $\Gamma, A \models_L B$ *iff* $\Gamma \models_L A^n \to B$ *for some $n \geq 0$.*

Complexity issues have been settled for **L** by Mundici in [14].

Theorem 3 (Mundici [14]). *The tautology problem for* **L** *is co-NP-complete.*

3   Goal-Directed Methods

In this paper we adopt the following "logic programming style" goal-directed paradigm of deduction. For a given logic denote by $\Gamma \vdash^? A$ the query "does A follow from Γ?" where Γ is a database (collection) of formulae and A is a goal formula. The deduction is *goal-directed* in the sense that the next step in a proof is determined by the form of the current goal: a complex goal is decomposed until its atomic constituents are reached, an atomic goal q is matched (if possible) with the "head" of a formula $G \to q$ in the database, and its "body" G asked in turn. This can be viewed as a sort of resolution step, or generalized Modus Tollens.

This model of deduction can be refined in several ways: (1) by putting constraints/labels on database formulae, restricting those available to match an atomic goal, (2) by adding more control to ensure termination, either by loop-checking or by "diminishing resources" i.e. removing formulae "used" to match an atomic goal, (3) by re-asking goals previously occurring in the deduction using *restart rules*. Note however that for applications such as deductive databases and logic programming, a terminating proof procedure is not always essential. We might want to get a proof of the goal from the database quickly if one exists, but be willing to have no answer otherwise (and preempt termination externally).

Goal-directed procedures have been proposed for a variety of logics, in some cases as refinements of sequent calculi called "Uniform proof systems", see e.g. [13]. Here we illustrate the goal-directed methodology by presenting a diminishing resources with bounded restart algorithm given in [4] for the implicational fragment of intuitionistic logic, henceforth adopting the convention of writing $\{A_1, \ldots, A_n\} \to q$ for $A_1 \to (A_2 \to \ldots (A_n \to q) \ldots)$.

Definition 4 (GDLJ$^\to$). *Queries have the form $\Gamma \vdash^? G; H$ where Γ is a multiset of formulae called the database, G is a formula called the goal, and H is a sequence of atomic goals called the history. The rules are as follows (where $H * (q)$ is q appended to H):*

(success) $\Gamma \vdash^? q; H$ *succeeds if* $q \in \Gamma$

(implication)   *From* $\Gamma \vdash^? \Pi \to q; H$ *step to* $\Gamma, \Pi \vdash^? q; H$

(reduction)   *From* $\Gamma, \Pi \to q \vdash^? q; H$ *step to* $\Gamma \vdash^? A; H * (q)$ *for all* $A \in \Pi$

(bounded restart) *From* $\Gamma \vdash^? q; H$ *step to* $\Gamma \vdash^? p; H * (q)$ *if* p *follows* q *in* H

Example 1. Consider the following proof, observing that *(bounded restart)* is needed at (2) to compensate for the removal of $p \to q$ at (1):

$$
\begin{array}{ll}
\vdash^? [(p \to q) \to p, p \to q] \to q; \emptyset & (implication) \\
(p \to q) \to p, p \to q \vdash^? q; \emptyset \quad (1) & (reduction) \\
(p \to q) \to p \vdash^? p; (q) & (reduction) \\
\vdash^? p \to q; (q, p) & (implication) \\
p \vdash^? q; (q, p) \quad (2) & (bounded\ restart) \\
p \vdash^? p; (q, p, q) & (success)
\end{array}
$$

Moreover a goal-directed calculus for the implicational fragment of *classical logic* is obtained by simply liberalising the *(bounded restart)* rule to allow restarts from *any* previous goal [4]. Also by modifying the history to allow states of the database to be recorded we can give goal-directed calculi for *Gödel logics* [11].

4   A Basic Goal-Directed System

In this section our aim will be to define a basic goal-directed system for **Ł** based on a language with connectives $\to$ and $\bot$, recalling that other standard connectives are definable from this pair. We start by defining *goal-directed queries* for **Ł**, similar to those given for intuitionistic and classical logic above, which consist of a database together with a multiset of goals, and a history of previous states of the database with goals.

Definition 5 (Goal-Directed Query). *A goal-directed query for* **GDŁ** *(query for short) is a structure of the form:*

$$Q = \Gamma \vdash^? \Delta; H \quad where$$

- *Γ is a multiset of formulae called the* database.
- *Δ is a multiset of formulae called the* goals.

– H is a multiset of pairs of multisets of formulae (states of the database with goals), written $\{(\Gamma_1 \vdash^? \Delta_1), \dots, (\Gamma_n \vdash^? \Delta_n)\}$, called the history.

Q is atomic if all formulae occurring in Q are atomic. Note also that we will frequently write nx for the multiset containing n copies of x.

A query may be understood intuitively as asking if either the goals "follow from" the database or there is an alternative state in the history such that the goals there follow from the appropriate database. Formally we determine the validity of queries using the model $[-1, 0]_{\text{Ł}}$ as follows:

Definition 6 (Validity). $Q = \Gamma_1 \vdash^? \Delta_1; \{(\Gamma_2 \vdash^? \Delta_2); \dots; (\Gamma_n \vdash^? \Delta_n)\}$ is valid in **Ł**, written $\models^*_L Q$, iff for all valuations v for $[-1, 0]_L$ for some i, $1 \leq i \leq n$:

$$\sum v(\Gamma_i) \leq \sum v(\Delta_i)$$

Equivalently we can interpret a query Q as a formula of *Abelian logic* **A**, the logic of lattice ordered abelian groups, and identify the validity of Q with the validity of this formula in **A**. Details are provided in [12]; here we simply emphasize that for formulae our interpretation agrees with the usual one for **Ł**.

Lemma 1. $\models^*_L \vdash^? A; \emptyset$ iff $\models_L A$.

It is easy to see that a query is valid if every goal A matches with an occurrence of either A or $\perp$ in the database. We express this using the following relation:

Definition 7 ($\subseteq^*$). (1) $\Delta \subseteq^* \Gamma$ if $\Delta \subseteq \Gamma$. (2) $\Delta \cup \{A\} \subseteq^* \Gamma \cup \{\perp\}$ if $\Delta \subseteq^* \Gamma$.

We now give our first goal-directed calculus for **Ł**.

Definition 8 (GDŁ). GDŁ has the following rules:

(*success*) $\Gamma \vdash^? \Delta; H$ succeeds if $\Delta \subseteq^* \Gamma$

(*implication*) From $\Gamma \vdash^? \Pi \to q, \Delta; H$ step to $\Gamma \vdash^? \Delta; H$ and $\Gamma, \Pi \vdash^? q, \Delta; H$

(*l-reduction*) From $\Gamma, \Pi \to q \vdash^? q, \Delta; H$ step to $\Gamma \vdash^? \Pi, \Delta; H \cup \{(\Gamma \vdash^? q, \Delta)\}$

(*r-reduction*) From $\Gamma \vdash^? q, \Delta; H \cup \{(\Gamma', \Pi \to q \vdash^? \Delta')\}$ step to:
    $\Gamma', q \vdash^? \Pi, \Delta'; H \cup \{(\Gamma' \vdash^? \Delta'), (\Gamma \vdash^? q, \Delta)\}$

(*n-reduction*) From $\Gamma, \Pi \to \perp \vdash^? \Delta; H$ step to $\Gamma, \perp \vdash^? \Pi, \Delta; H \cup \{(\Gamma \vdash^? \Delta)\}$

(*mingle*)   From $\Gamma \vdash^? q, \Delta; H \cup \{(\Gamma', q \vdash^? \Delta')\}$ step to:
    $\Gamma, \Gamma' \vdash^? \Delta, \Delta'; H \cup \{(\Gamma', q \vdash^? \Delta'), (\Gamma \vdash^? q, \Delta)\}$

(*restart*)   From $\Gamma \vdash^? \Delta; H \cup \{(\Gamma' \vdash^? \Delta')\}$ step to $\Gamma' \vdash^? \Delta'; H \cup \{(\Gamma \vdash^? \Delta)\}$

Note that the (*implication*) rule means that a query with an implicational goal $\Pi \to q$ steps to two further queries: one where this goal is removed, and one where Π is added to the database and $\Pi \to q$ is replaced by q as a goal. We can also define a version where $\Pi \to q$ is the only goal in which case only one premise is needed, i.e.:

($implication'$) From $\Gamma \vdash^? \Pi \to q; H$ step to $\Gamma, \Pi \vdash^? q; H$

Note also that for convenience of presentation we have split the reduction process into three "reduction rules": *local reduction* (*l-reduction*) and *remote reduction* (*r-reduction*) which treat the cases where a goal matches the head of a formula in the database and in the database of a state in the history respectively, and *negation reduction* (*n-reduction*) which decomposes a formula in the database with head $\perp$. Finally, observe that (*mingle*) combines two states in the case where one has a goal occurring in the database of the other, and that (*restart*) allows the computation to switch to a state recorded in the history.

GDŁ is easily adapted to provide goal-directed calculi for other logics. For example a calculus for *classical logic* is obtained by changing all mention in Definition 5 of multisets to sets, and replacing the restart rule with:

($restart'$) From $\Gamma \vdash^? \Delta; H \cup \{(\Gamma' \vdash^? \Delta')\}$ step to $\Gamma', \Gamma \vdash^? \Delta'; H \cup \{(\Gamma \vdash^? \Delta)\}$

Example 2. We illustrate **GDŁ** with a proof of the axiom (Ł4), using (1) and (2) to mark separate branches:

$$
\begin{array}{rll}
& \vdash^? [(p \to q) \to q, q \to p] \to p; \emptyset & (implication') \\
(p \to q) \to q, q \to p \vdash^? & p; \emptyset & (l\text{-}reduction) \\
(p \to q) \to q \vdash^? & q; \{((p \to q) \to q \vdash^? p)\} & (l\text{-}reduction) \\
& \vdash^? p \to q; \{(\vdash^? q), ((p \to q) \to q \vdash^? p)\} & (implication') \\
p \vdash^? & q; \{(\vdash^? q), ((p \to q) \to q \vdash^? p)\} & (r\text{-}reduction) \\
q \vdash^? & p, p \to q; \{(\vdash^? q), (p \vdash^? q)\} & (implication) \\
(1) \quad q \vdash^? & p; \{(\vdash^? q), (p \vdash^? q)\} & (mingle) \\
q \vdash^? & q; \{(\vdash^? q), (p \vdash^? q), (q \vdash^? p)\} & (success) \\
(2) \quad q, p \vdash^? & p, q; \{(\vdash^? q), (p \vdash^? q), (q \vdash^? p)\} & (success)
\end{array}
$$

We now show that **GDŁ** is *sound* with respect to the interpretation of queries given in Definition 6.

Theorem 4 (Soundness of GDŁ). *If Q succeeds in* **GDŁ** *then* $\models_{\text{Ł}}^* Q$.

Proof. We proceed by induction on the height of a derivation in **GDŁ** showing that each rule preserves validity. For example for (*implication*), given a valuation v for $[-1,0]_{\text{Ł}}$ we can ignore the history H as it is repeated in both premise and conclusion, and we have $\sum v(\Gamma) \le \sum v(\Delta)$ and $\sum v(\Gamma) + \sum v(\Pi) \le \sum v(q) + \sum v(\Delta)$. If $\sum v(\Pi) \le v(q)$ then $v(\Pi \to q) = 0$ and we use the first premise, otherwise $v(\Pi \to q) = v(q) - \sum v(\Pi)$ and we use the second. $\qquad\square$

To prove *completeness* we distinguish two functions of **GDŁ**: (1) to decompose complex formulae using rules that take us from valid queries to valid queries, and (2) to determine the validity of atomic hypersequents. We show that (1) allows us to reduce the derivability of a hypersequent to the derivability of hypersequents containing only atoms and irrelevant (in the sense that they do not affect the validity of the hypersequent) formulae, and that (2) allows us to prove all valid atomic hypersequents. We begin with the former.

Definition 9 (Complete). *A rule is* complete *if whenever its conclusion is valid then also all of its premises are valid.*

Lemma 2. *(implication), the reduction rules, and (restart) are complete.*

If we apply the *(implication)* and reduction rules exhaustively to a query using *(restart)* to move between different states of the history, then we end up with queries where the states are atomic goals that fails to match the head of a non-atomic formula in the database of *any* state. We call such queries *irreducible*.

Definition 10 (Irreducible). $\Gamma_1 \vdash^? \Delta_1; \{(\Gamma_2 \vdash^? \Delta_2), \ldots, (\Gamma_n \vdash^? \Delta_n)\}$ *is irreducible iff:*

1. Δ_i *is atomic for* $i = 1, \ldots, n$.
2. $\Gamma_i = \Pi_i \cup \Sigma_i$ *for* $i = 1, \ldots, n$ *where* Σ_i *is atomic.*
3. $Head(\Pi_1 \cup \ldots \cup \Pi_n) \cap (\Delta_1 \cup \ldots \cup \Delta_n \cup \{\bot\}) = \emptyset$.

where $Head([A_1, \ldots, A_n] \to q) = q$ *and* $Head(\Gamma) = \{Head(A) \mid A \in \Gamma\}$.

Lemma 3. *For a valid query Q we can apply the rules of* **GDŁ** *to step to valid irreducible queries.*

If an irreducible query is valid then by removing non-atomic formulae we obtain an atomic query that is also valid.

Lemma 4. *If the following conditions hold:*

1. $\models^*_{\text{Ł}} \Gamma_1, \Pi_1 \vdash^? \Delta_1; \{(\Gamma_2, \Pi_2 \vdash^? \Delta_2), \ldots, (\Gamma_n, \Pi_n \vdash^? \Delta_n)\}$
2. Γ_i *and* Δ_i *are atomic for* $i = 1, \ldots, n$
3. $Head(\Pi_1 \cup \ldots \cup \Pi_n) \cap (\Delta_1 \cup \ldots \cup \Delta_n \cup \{\bot\}) = \emptyset$

Then: $\models^*_{\text{Ł}} \Gamma_1 \vdash^? \Delta_1; \{(\Gamma_2 \vdash^? \Delta_2), \ldots, (\Gamma_n \vdash^? \Delta_n)\}$

Proof. Assume $\not\models^*_{\text{Ł}} \Gamma_1 \vdash^? \Delta_1; \{(\Gamma_2 \vdash^? \Delta_2), \ldots, (\Gamma_n \vdash^? \Delta_n)\}$. We get that there is a valuation v for $[-1, 0]_{\text{Ł}}$ such that $\sum v(\Gamma_i) > \sum v(\Delta_i)$ for $i = 1, \ldots, n$. We define v' as follows:

$$v'(q) = \begin{cases} 0 & \text{if } q \in Head(\Pi_1 \cup \ldots \cup \Pi_n) \\ v(q) & \text{otherwise} \end{cases}$$

Since $Head(\Pi_1 \cup \ldots \cup \Pi_n) \cap (\Delta_1 \cup \ldots \cup \Delta_n) = \emptyset$ we have that $\sum v'(\Delta_i) = \sum v(\Delta_i)$ for $i = 1, \ldots, n$. We also have get that $\sum v'(\Pi_i) = 0$ and $\sum v'(\Gamma_i) \geq \sum v(\Gamma_i)$ for $i = 1, \ldots, n$. Hence $\sum v'(\Pi_i) + \sum v'(\Gamma_i) > \sum v'(\Delta_i)$ for $i = 1, \ldots, n$ and $\not\models^*_{\text{Ł}} \Gamma_1, \Pi_1 \vdash^? \Delta_1; \{(\Gamma_2, \Pi_2 \vdash^? \Delta_2), \ldots, (\Gamma_n, \Pi_n \vdash^? \Delta_n)\}$, a contradiction. □

The next step is to show that all valid irreducible queries succeed in **GDŁ**.

Lemma 5. *All valid atomic queries succeed in* **GDŁ**.

Proof. Let $Q = \Gamma_1 \vdash^? \Delta_1; \{\Gamma_2 \vdash^? \Delta_2, \ldots, \Gamma_n \vdash^? \Delta_n\}$ be atomic. If $\models_{\text{Ł}} Q$ then the set $\{\sum \Gamma_i > \sum \Delta_i \mid 1 \leq i \leq n\}$ is inconsistent over $[-1, 0]$, and hence there exists $\lambda_1, \ldots, \lambda_n \in \mathbb{N}$ such that $\lambda_i > 0$ for some $1 \leq i \leq n$ and:

$$\bigcup_{i=1}^{n} \lambda_i \Delta_i \subseteq^* \bigcup_{i=1}^{n} \lambda_i \Gamma_i$$

We show that Q succeeds in **GDŁ** by induction on $\lambda = \sum_{i=1}^{n} \lambda_i$. If $\lambda = 1$ then Q succeeds by an application of (*restart*) if necessary and (*success*). For $\lambda > 1$ we have two cases. First suppose that for some $i \neq j$, $1 \leq i, j \leq n$, we have $\lambda_i, \lambda_j > 0$ and there exists $q \in \Delta_i \cap \Gamma_j$. Since we can always apply (*restart*) to change the state of the database, we can assume without loss of generality that $i = 1$ and $j = 2$. Now by applying (*mingle*) we obtain a query:

$$Q' = \Gamma_1, \Gamma_2 - \{q\} \vdash^? \Delta_1 - \{q\}, \Delta_2; \{\Gamma_1 \vdash^? \Delta_1, \ldots, \Gamma_n \vdash^? \Delta_n\}$$

If $\lambda_1 \geq \lambda_2$ then we have:

$$(\lambda_1 - \lambda_2)\Delta_1 \cup \lambda_2(\Delta_1 - \{q\} \cup \Delta_2) \cup \bigcup_{i=3}^{n} \lambda_i \Delta_i \subseteq^* (\lambda_1 - \lambda_2)\Gamma_1 \cup \lambda_2(\Gamma_1 \cup \Gamma_2 - \{q\}) \cup \bigcup_{i=3}^{n} \lambda_i \Gamma_i$$

Moreover $(\lambda_1 - \lambda_2) + \lambda_2 + \sum_{i=3}^{n} \lambda_i < \lambda$ and hence by the induction hypothesis Q' succeeds in **GDŁ** so we are done. The case where $\lambda_2 \geq \lambda_1$ is very similar.

Alternatively if for all $q \in \Delta_i$ for $i = 1, \ldots, n$ where $\lambda_i > 0$ there is no j such that $\lambda_j > 0$ and $q \in \Gamma_j$, then in at least one state there must be more occurrences of $\perp$ than goals, so the query succeeds by (*restart*) and (*success*). $\quad\square$

Lemma 6. *If* $\Gamma_1 \vdash^? \Delta_1; \{(\Gamma_2 \vdash^? \Delta_2), \ldots, (\Gamma_n \vdash^? \Delta_n)\}$ *succeeds in* **GDŁ** *then* $\Gamma_1, \Pi_1 \vdash^? \Delta_1; \{(\Gamma_2, \Pi_2 \vdash^? \Delta_2), \ldots, (\Gamma_n, \Pi_n \vdash^? \Delta_n)\}$ *succeeds in* **GDŁ**.

We are now able to conclude that **GDŁ** is complete.

Theorem 5 (Completeness of GDŁ). *If* $\models_{\text{Ł}}^* Q$ *then* Q *succeeds in* **GDŁ**.

Proof. We apply the complete rules to Q terminating with valid irreducible queries by Lemmas 2 and 3. The atomic part of these queries must be valid by Lemma 4, and hence succeed by Lemma 5, but then by Lemma 6, the whole of each query succeeds. $\quad\square$

5 Refinements

In this section we make a number of refinements to **GDŁ**; in particular we use the introduction of new propositional variables to give more efficient reduction rules, we give revised (*mingle*) and (*success*) rules to obtain terminating calculi, and finally we treat connectives other than $\to$ and $\perp$, and truth constants.

5.1 More Efficient Reduction

One significant problem for the efficiency of **GDŁ** is that the reduction rules place multiple copies of the database in the history, meaning that each formula may be reduced an exponential number of times. For example notice that in Example 2 the formula $(p \to q) \to q$ occurs (3 lines down) twice: once in the database and once again in the history. The solution we propose here is to introduce *new propositional variables* which ensure that formulae are not duplicated, and may be thought of as keeping track of different options in the history.

Definition 11 (GDŁ$_e$). *We define a new calculus* **GDŁ$_e$** *by replacing the reduction rules in* **GDŁ** *with the following (recalling that $2x$ stands for $\{x, x\}$):*

(a-reduction) From $\Gamma, \Sigma \vdash^? \Sigma, \Delta; H$ step to $\Gamma \vdash^? \Delta; H$

(l-reduction) From $\Gamma, \Pi \to q' \vdash^? q, \Delta; H$ where $q' \in \{q, \bot\}$ step to:
$q' \vdash^? \Pi, 2x; H \cup \{(\Gamma, 2x \vdash^? q, \Delta)\}$ *where x is new.*

(r-reduction) From $\Gamma \vdash^? q, \Delta; H \cup \{\Gamma', \Pi \to q \vdash^? \Delta'\}$ step to:
$q \vdash^? \Pi, 2x; H \cup \{(\Gamma', 2x \vdash^? \Delta'), (\Gamma \vdash^? q, \Delta)\}$ *where x is new.*

Note that for **GDŁ$_e$** we have given an alternative breakdown of the reduction process. In particular goals matching atoms in the database are removed by *(a-reduction)* while *(l-reduction)* takes care of the cases where a goal matches the head of a formula in the database or the head is $\bot$. In both *(l-reduction)* and *(r-reduction)* the formula reduced is replaced by two copies of a new propositional variable x, which act as a marker allowing the formula to be decomposed independently from the database.

Example 3. We illustrate **GDŁ$_e$** with the following proof:

$$\vdash^? [(p \to q) \to r, (q \to p) \to r] \to r; \emptyset \qquad (implication)$$

$$\begin{array}{ll} (p \to q) \to r, \\ (q \to p) \to r \vdash^? r; \emptyset & (l\text{-}reduction) \end{array}$$

$$r \vdash^? p \to q, 2x; \{(2x, (q \to p) \to r \vdash^? r)\} \qquad (implication)$$

$$(1) \quad r \vdash^? 2x, \{(2x, (q \to p) \to r \vdash^? r)\} \qquad (mingle)$$

$$r, x, (q \to p) \to r \vdash^? x, r, \{(r \vdash^? 2x), (2x, (q \to p) \to r \vdash^? r)\} \qquad (success)$$

$$(2) \quad r, p \vdash^? q, 2x; \{(2x, (q \to p) \to r \vdash^? r)\} \qquad (restart)$$

$$2x, (q \to p) \to r \vdash^? r; \{(r, p \vdash^? q, 2x)\} \qquad (l\text{-}reduction)$$

$$r \vdash^? q \to p, 2y; \{(r, p \vdash^? q, 2x), (2x, 2y \vdash^? r)\} \qquad (implication)$$

$$(2.1) \quad r \vdash^? 2y; \{(r, p \vdash^? q, 2x), (2x, 2y \vdash^? r)\} \qquad (mingle)$$

$$2x, r, y \vdash^? r, y; \{(r, p \vdash^? q, 2x), (2x, 2y \vdash^? r), (r \vdash^? 2y)\} \quad (success)$$

$$(2.2) \quad r, q \vdash^? p, 2y; \{(r, p \vdash^? q, 2x), (2x, 2y \vdash^? r)\} \qquad (mingle)$$

$$2r, q \vdash^? q, 2x, 2y; \{(r, p \vdash^? q, 2x), (2x, 2y \vdash^? r),$$
$$(r, q \vdash^? p, 2y)\} \qquad (mingle)$$

$$2r, q, x, 2y \vdash^? q, x, 2y, r; \{(r, p \vdash^? q, 2x), (2x, 2y \vdash^? r),$$
$$(r, q \vdash^? p, 2y), (2r, q \vdash^? q, 2x, 2y)\} \quad (success)$$

Theorem 6. *Q succeeds in* **GDŁ$_e$** *iff $\models^*_{\text{Ł}} Q$.*

Proof. Soundness and completeness for **GDŁ$_e$** are proved in exactly the same manner as the corresponding results for **GDŁ**, the main issue being to check that the new reduction rules are both sound and complete. As an example we check the soundness of *(l-reduction)* assuming as before that $H = \emptyset$. Suppose that we have a valuation v for $[-1, 0]_{\text{Ł}}$. If $\sum v(\Pi) < v(q)$ then $v(\Pi \to q) = 0$ and extending v with $v(x) = 0$ we get from the premise that $v(\Gamma) = v(\Gamma) + 2v(x) \leq v(q) + v(\Delta)$ as required. If $\sum v(\Pi) \geq v(q)$ then suppose that $v(\Gamma) + v(\Pi \to q) = v(\Gamma) + v(q) - \sum v(\Pi) > v(q) + v(\Delta)$. This means that $v(\Gamma) - \sum v(\Pi) + \epsilon > v(\Delta)$ for some $\epsilon < 0$. We now define:

$$v(x) = \frac{v(q) - \sum v(\Pi) + \epsilon}{2}$$

noting that dividing by two ensures that $0 \leq v(x) \leq -1$, which gives that $v(q) > \sum v(\Pi) + 2v(x)$ and $v(\Gamma) + 2v(x) > v(\Delta)$, a contradiction.   □

5.2   Terminating Calculi

There are several options for developing terminating calculi for **Ł**. Perhaps the simplest approach is to feed irreducible queries to a linear programming problem solver, which with the improved efficiency reduction rules of the previous section gives a co-NP algorithm. Alternatively if we want to maintain a *logical* interpretation to each step we can adapt the *(mingle)* rule, giving terminating strategies with either queries of an exponential size or involving non-deterministic choices.

A goal-directed calculus using linear programming is as follows, noting that the *(success)* rule transforms a query into a set of equations over $[-1, 0]$; validity of the query being implied by the inconsistency of this set:

Definition 12 (GDŁ$_{lp}$). GDŁ$_{lp}$ *consists of the (implication), reduction and (restart) rules of* **GDŁ$_e$** *together with the following success rule:*

   (success) $\Gamma_1 \vdash^? \Delta_1; \{\Gamma_2 \vdash^? \Delta_2, \ldots, \Gamma_n \vdash^? \Delta_n\}$ *succeeds if:*
$$\{\sum atom(\Gamma_i) > \sum atom(\Delta_i)\}_{1 \leq i \leq n} \text{ is inconsistent over } [-1, 0].$$

where $atom(\Gamma) = \{q \mid q \in \Gamma, \ q \text{ atomic}\}$.

Theorem 7. $\models^*_L Q$ *iff* Q *succeeds in* **GDŁ$_{lp}$**.

Theorem 8. **GDŁ$_{lp}$** *is co-NP.*

Proof. To show that a query with length l fails in **GDŁ$_{lp}$**, we apply *(implication)* and the reduction rules exhaustively using *(restart)* to move between different states in the history, choosing a branch non-deterministically. It is easy to see that applying each rule takes polynomial time in l, and that since each of the rules except *(restart)* strictly reduces the complexity of the query, the length of the branch is polynomial in l. Moreover both the number of propositional variables, and the number of different states are polynomial in l, so since linear programming is polynomial, checking *(success)* is also polynomial in l. Hence derivability in **GDŁ$_{lp}$** is co-NP.   □

Note that **GDŁ$_{lp}$** is also easily adapted to cope with *finite-valued* Łukasiewicz logics, since we can just change the *(success)* rule to check for inconsistency over the set $[-1, -\frac{n-2}{n-1}, \ldots, -\frac{1}{n-1}, 0]$ for the n-valued logic **Ł$_n$**.

We get "logical methods" for solving linear programming problems by changing the *(mingle)* rule of **GDŁ$_e$** to allow matching of multiple occurrences of atoms, the idea being to obtain a terminating procedure by removing all occurrences of one particular propositional variable at a time. We give two versions:

Definition 13 (GDŁ$_t^i$). GDŁ$_t^i$ *for* $i = 1, 2$ *consists of the same rules as* **GDŁ$_e$** *except that* (*mingle*) *is replaced by* (*mingle$_i$*):

(*mingle$_1$*) *From* $\Gamma \vdash^? \lambda q, \Delta; H \cup \{(\Gamma', \mu q \vdash^? \Delta')\}$ *step to:*
$$\mu\Gamma, \lambda\Gamma' \vdash^? \mu\Delta, \lambda\Gamma'; H \cup \{(\Gamma', \mu q \vdash^? \Delta'), (\Gamma \vdash^? \lambda q, \Delta)\}$$

(*mingle$_2$*) *From* $\Gamma \vdash^? \lambda q, \Delta; H \cup \{(\Gamma', \mu q \vdash^? \Delta')\}$ *step to:*
either $\mu\Gamma, \lambda\Gamma' \vdash^? \mu\Delta, \lambda\Gamma'; H \cup \{(\Gamma \vdash^? \lambda q, \Delta)\}$
or $\mu\Gamma, \lambda\Gamma' \vdash^? \mu\Delta, \lambda\Gamma'; H \cup \{(\Gamma', \mu q \vdash^? \Delta'\}$

Note that **GDŁ$_t^1$** keeps all previous states of the database in the history, giving an exponential blow-up in the size of the query. **GDŁ$_t^2$** on the other hand keeps the size of queries linear but requires non-deterministic choices.

Example 4. We illustrate **GDŁ$_t^2$** with a proof of an atomic query:

$$
\begin{array}{ll}
2p \vdash^? 3q, r; \{(q, r \vdash^? p), (2q \vdash^? p)\} & (mingle_2) \\
2p, 3r \vdash^? r, 3p; \{(2q \vdash^? p), (2p \vdash^? 3q, r)\} & (a\text{-}reduction) \\
2r \vdash^? p; \{(2q \vdash^? p), (2p \vdash^? 3q, r)\} & (switch) \\
2p \vdash^? 3q, r; \{(2q \vdash^? p), (2r \vdash^? p)\} & (mingle_2) \\
4p \vdash^? 2r, 3p; \{(2q \vdash^? p), (2r \vdash^? p)\} & (a\text{-}reduction) \\
p \vdash^? 2r; \{(2q \vdash^? p), (2r \vdash^? p)\} & (mingle_2) \\
2p \vdash^? 2p; \{(2q \vdash^? p), (2r \vdash^? p)\} & (success)
\end{array}
$$

Theorem 9. Q *succeeds in* **GDŁ$_t^i$** *iff* $\models_L^* Q$ *for* $i = 1, 2$.

Theorem 10. **GDŁ$_t^i$** *terminates with a suitable control strategy for* $i = 1, 2$.

Proof. We sketch suitable control strategies. The first step is to reduce formulae in the query to obtain irreducible queries. Since the reduction rules all reduce the complexity of the query it is sufficient here to ensure that (*switch*) is not applied ad infinitum e.g. using a split history to show which states have already been considered. For the second step we require that (*a-reduction*) is applied eagerly i.e. whenever possible, and that (*mingle$_i$*) is applied with the maximum matching propositional variables i.e. we add the condition $q \notin \Gamma \cup \Gamma' \cup \Delta \cup \Delta'$. Moreover we must insist that (*mingle$_i$*) is applied exhaustively to just one propositional variable at a time, using (*switch*) to move between states. □

5.3   Rules for Other Connectives

Although the standard connectives of **Ł** are all definable in a language with just the connectives $\rightarrow$ and $\bot$, this treatment is often unsatisfactory for two reasons. First, the definitions may multiply the size of formulae exponentially e.g. $A \vee B =_{def} (A \rightarrow B) \rightarrow B$, and second, it may introduce occurrences of $\bot$ which are not very good for proof search e.g. $A \wedge B =_{def} ((A \rightarrow \bot) \vee (B \rightarrow \bot)) \rightarrow \bot$. We therefore illustrate here one possible approach to dealing with other well known connectives of **Ł** in an efficient fashion. We begin by defining normal forms for goal and database formulae in a language with connectives $\rightarrow$, $\odot$, $\wedge$, $\vee$ and $\bot$, where every formula A in this language is equivalent to both a goal-formula and a strong conjunction of database formulae, with complexity linear in the complexity of A.

Definition 14 (Normal Forms). Goal-formulae G *and* database-formulae D *are defined by mutual induction as follows:*

$$G = q|\bot|G \vee G|G \wedge G|D \to G \qquad H = q|\bot|H \vee H \qquad D = H|G \to D$$

Lemma 7. *For a formula A there exists a goal-formula G, and database-formulae $D_1, \ldots, D_n$, such that G and $D_1 \odot \ldots \odot D_n$ have complexity linear in the complexity of A, and $A \equiv_L G \equiv_L D_1 \odot \ldots \odot D_n$, where $B \equiv_L C$ iff $v(B) = v(C)$ for all $[-1,0]_L$ valuations v.*

We now provide a goal-directed calculus for **L** using these normal forms.

Definition 15 (GDL_f). GDL_f *consists of the rules (success), (a-reduction), (mingle) and (restart) of* **GDL_e** *together with:*

(implication) From $\Gamma \vdash^? \Pi \to G, \Delta; H$ step to $\Gamma \vdash^? \Delta; H$ and $\Gamma, \Pi \vdash^? G, \Delta; H$

(or) From $\Gamma \vdash^? A \vee B, \Delta; H$ step to:
$\Gamma, y \vdash^? x, \Delta; H \cup \{(x \vdash^? y, A), (x \vdash^? y, B)\}$ where x and y are new

(and) From $\Gamma \vdash^? A \wedge B, \Delta; H$ step to $\Gamma \vdash^? A, \Delta; H$ and $\Gamma \vdash^? B, \Delta; H$

(l-reduction) From $\Gamma, \Pi \to (A \vee q') \vdash^? q, \Delta; H$ where $q' \in \{q, \bot\}$ step to:
$\Gamma, \Pi \to A \vdash^? q, \Delta; H$ and
$q' \vdash^? \Pi, 2x; H \cup \{(\Gamma, 2x \vdash^? q, \Delta)\}$ where x is new.

(r-reduction) From $\Gamma \vdash^? q, \Delta; H \cup \{\Gamma', \Pi \to (A \vee q) \vdash^? \Delta'\}$ step to:
$\Gamma \vdash^? q, \Delta; H \cup \{(\Gamma', \Pi \to A \vdash^? \Delta')\}$ and
$q \vdash^? \Pi, 2x; H \cup \{(\Gamma', 2x \vdash^? \Delta'), (\Gamma \vdash^? q, \Delta)\}$ where x is new.

Theorem 11. *Q succeeds in* **GDL_f** *iff* $\models_L^* Q$.

5.4   Adding Truth Constants

All the algorithms we have given so far have dealt with proving *theorems* of **L** i.e. formulae which always take the value 0 in the model $[-1,0]_L$. However, from the point of view of fuzzy logic, it is natural to ask for proofs of partially true conclusions from partially true premises. The key fact here is that in **L** for any valuation v for $[-1,0]_L$ if for a formula A, $v(A) = r$, then for any formula B, $v(B) \geq r$ iff $v(A \to B) = 0$. Hence by adding a truth constant r to our language i.e. a constant where $v(r) = r$ for all valuations v for $[-1,0]_L$, we can express that a formula has value greater than or less than r. Now it is possible here to add constants for any and indeed all real numbers between -1 and 0; however it is more usual (see e.g. [7]) to consider just the rational numbers (giving a countable language) thereby getting *Rational Pavelka Logic* **RPL**.

Obtaining a calculus for **RPL** is straightforward, the key step being to redefine the relation in the *(success)* rule to deal with extra constants.

Definition 16 $(\subseteq_2^*)$. $\Delta \subseteq_2^* \Gamma$ *iff:* (1) $\Gamma = \Gamma_1 \cup \{a_1, \ldots, a_n\}$. (2) $\Delta = \Delta_1 \cup \Delta_2 \cup \{b_1, \ldots, b_m\}$. (3) $\Delta_1 \subseteq \Gamma_1$. (4) $|\Delta_2| + \sum_{i=1}^n a_i \le \sum_{i=1}^m b_i$.

The only other change we need to make is to allow reduction of database formulae with any truth constant as a head. In the general case this is bad for proof search since it means that all such formulae can be reduced no matter what the goal; however for fuzzy reasoning it is often sufficient to have constants only in the body of database formulae.

Definition 17 (GDŁ$_c$). GDŁ$_c$ *consists of the rules of* GDŁ$_e$ *with the condition in the success rule changed to* $\Delta \subseteq_2^* \Gamma$ *and the condition* $q' \in \{q, \perp\}$ *in* (*l-reduction*) *changed to "$q' = q$ or $q' = a$ for some truth constant a".*

Theorem 12. Q *succeeds in* GDŁ$_c$ *iff* $\models_{\mathit{Ł}}^* Q$.

6   Application to Fuzzy Logic Programming

In this section we show that our algorithms can be used as the basis for fuzzy logic programming applications, illustrating the potential of our approach by considering an example for Ł taken from [8]. In this case a database consists of a number of fuzzy statements (Horn clauses) such as "students are young" with associated lower bounds for truth values, and queries involve deriving lower bounds for (non-implicational) *logical consequences* of such databases like "Mary is young". Observe however that although the statements are first order, the use of a restricted function-free language means that for a finite domain we can translate such statements into formulae of propositional logic by taking the lattice conjunction of all possible instances.[1] Moreover whereas in [8] truth values representing lower bounds are treated as separated entities, here following the previous section, we can represent "A has truth value greater than equal to a" by the formulae $a \to A$. Below we list the statements considered in [8] translated into a set of propositional formulae of Ł for each individual i in the domain (e.g. yng_i has the meaning "i is young"), noting that in [8] the disjunctive clause $s3$ is written as two separate clauses.

Label	Statement	Database Entries
s1	Students are young.	$[-\frac{1}{6}, sdnt_i] \to yng_i$
s2	Young people are single.	$[-\frac{1}{3}, yng_i] \to sng_i$
s3	Students who have children are married or cohabitants.	$[sdnt_i, pnt_i] \to (mrd_i \vee chbt_i)$
s4	Cohabitants are young.	$[-\frac{1}{3}, chbt_i] \to yng_i$
s5	Single, married and cohabitant are mutually exclusive.	$[sng_i, mrd_i] \to \perp$, $[sng_i, chbt_i] \to \perp$, $[mrd_i, chbt_i] \to \perp$

We now want to derive *logical consequences* from such a database. By the deduction theorem for Ł (Theorem 2) we can simply allow as many occurrences

---

[1] We leave finding a more efficient approach using unification for future work.

of a formula as we want in the database and use (for example) the algorithm $\mathbf{GDL_c}$ to prove the desired consequence. This approach is obviously not terminating but although the refinements required to obtain a terminating algorithm for logical consequence are not overly difficult, we again leave this task for future work. In [8] a number of queries are considered, all of which can be treated in our approach. For example, to find a lower bound for the truth value of "Lea is single" given that "Lea is a student" we can prove $c \to sng_i$ from $s1$, $s2$ and $sdnt_i$ as follows (where c is a to-be-determined truth constant).

$$s1, s2, sdnt_i \vdash^? c \to sng_i; \emptyset$$
$$s1, s2, sdnt_i, c \vdash^? sng_i; \emptyset$$
$$sng_i \vdash^? -\tfrac{1}{3}, yng_i, 2x; \{(s1, 2x, sdnt_i, c \vdash^? sng_i)\}$$
$$yng_i \vdash^? -\tfrac{1}{6}, sdnt_i, 2y; \{(sng_i \vdash^? -\tfrac{1}{3}, yng_i, 2x), (2y, 2x, sdnt_i, c \vdash^? sng_i)\}$$
$$2x, 2y, c, yng_i \vdash^? -\tfrac{1}{6}, 2y, sng_i; \{(sng_i \vdash^? -\tfrac{1}{3}, yng_i, 2x), (2y, 2x, sdnt_i, c \vdash^? sng_i)\}$$
$$2x, 2y, c, yng_i \vdash^? -\tfrac{1}{6}, 2y, -\tfrac{1}{3}, yng_i, 2x; \{(sng_i \vdash^? -\tfrac{1}{3}, yng_i, 2x)$$
$$(2y, 2x, sdnt_i, c \vdash^? sng_i)\}$$

We observe that the computation succeeds at the second line iff $c = -1$ and at the last iff $c \leq -\tfrac{1}{2}$, giving that $c \to sng_i$ is derivable from the given database for all $c \leq -\tfrac{1}{2}$. In general we have to consider all possible proofs to obtain the greatest lower bound. We note further that unlike the algorithms presented in [8] we can also cope either with embedded implications e.g. "very young people are single" could be represented as $[-\tfrac{1}{3}, -\tfrac{1}{6} \to yng_i] \to sng_i$, or clauses with disjunctive bodies e.g. "students and young people are single" could be represented as $(sdnt_i \vee yng_i) \to sng_i$. Hence our algorithms provide the basis for a far more expressive approach to fuzzy logic programming in $\mathbf{L}$.

7   Concluding Remarks

In this paper we have presented a basic goal-directed calculus for Łukasiewicz logic $\mathbf{L}$ with a purely logical intepretation, subsequently refined to obtain more efficient and terminating reduction methods, and extensions to Rational Pavelka Logic $\mathbf{RPL}$. These calculi are a significant improvement on other automated reasoning methods for $\mathbf{L}$ proposed in the literature. Unlike the tableaux calculi of [6, 17], and the resolution calculi of [21, 16], each step in our calculi has an intuitive logical interpretation. Moreover the goal-directed methodology both gives an algorithmic reading of the logic, and ensures that, rather than decomposing all formulae (as in the cited approaches), only formulae relevant to the current proof are treated. We note also that as for the mentioned tableaux calculi we obtain a co-NP decision procedure for $\mathbf{L}$ using linear programming. Finally, a promising direction for applications of these calculi is as the basis for fuzzy logic programming algorithms. Using our techniques we are able not only to derive lower bounds for logical consequences in $\mathbf{L}$ from Horn clauses with lower bounds, as in [8, 20], (see also the "quantitative" variant of Prolog introduced by Van Emden, and developed by Subrahmanian in [19]), but also to deal with the full range of propositional formulae of $\mathbf{L}$, giving a far more expressive logic programming language. We intend to investigate issues regarding the implementation of such a language in future work.

References

1. S. Aguzzoli and A. Ciabattoni. Finiteness in infinite-valued logic. *Journal of Logic, Language and Information*, 9(1):5–29, 2000.
2. C. C. Chang. Algebraic analysis of many-valued logics. *Transactions of the American Mathematical Society*, 88:467–490, 1958.
3. R. Cignoli, I. M. L. D'Ottaviano, and D. Mundici. *Algebraic Foundations of Many-Valued Reasoning*, volume 7 of *Trends in Logic*. Kluwer, 1999.
4. D. Gabbay and N. Olivetti. *Goal-directed Proof Theory*. Kluwer, 2000.
5. D. Gabbay and N. Olivetti. Goal oriented deductions. In D. Gabbay and F. Guenthner, editors, *Handbook of Philosophical Logic*, volume 9, pages 199–285. Kluwer, second edition, 2002.
6. R. Hähnle. *Automated Deduction in Multiple-Valued Logics*. Oxford University Press, 1993.
7. P. Hájek. *Metamathematics of Fuzzy Logic*. Kluwer, Dordrecht, 1998.
8. F. Klawonn and R. Kruse. A Łukasiewicz logic based Prolog. *Mathware & Soft Computing*, 1(1):5–29, 1994.
9. J. Łukasiewicz and A. Tarski. Untersuchungen über den Aussagenkalkül. *Comptes Rendus des Séances de la Societé des Sciences et des Lettres de Varsovie, Classe III*, 23, 1930.
10. R. McNaughton. A theorem about infinite-valued sentential logic. *Journal of Symbolic Logic*, 16(1):1–13, 1951.
11. G. Metcalfe, N. Olivetti, and D. Gabbay. Goal-directed calculi for Gödel-Dummett logics. In M. Baaz and J. A. Makowsky, editors, *Proceedings of CSL 2003*, volume 2803 of *LNCS*, pages 413–426. Springer, 2003.
12. G. Metcalfe, N. Olivetti, and D. Gabbay. Sequent and hypersequent calculi for abelian and Łukasiewicz logics. To appear in ACM TOCL, 2004.
13. D. Miller, G. Nadathur, F. Pfenning, and A. Scedrov. Uniform proofs as a foundation for logic programming. *Annals of Pure and Applied Logic*, 51:125–157, 1991.
14. D. Mundici. Satisfiability in many-valued sentential logic is NP-complete. *Theoretical Computer Science*, 52(1-2):145–153, 1987.
15. D. Mundici. The logic of Ulam's game with lies. In C. Bicchieri and M.L. Dalla Chiara, editors, *Knowledge, belief and strategic interaction*, pages 275–284. Cambridge University Press, 1992.
16. D. Mundici and N. Olivetti. Resolution and model building in the infinite-valued calculus of Łukasiewicz. *Theoretical Computer Science*, 200(1–2):335–366, 1998.
17. N. Olivetti. Tableaux for Łukasiewicz infinite-valued logic. *Studia Logica*, 73(1):81–111, 2003.
18. D. J. Pym and J. A. Harland. The uniform proof-theoretic foundation of linear logic programming. *Journal of Logic and Computation*, 4(2):175–206, 1994.
19. V. S. Subrahmanian. Intuitive semantics for quantitative rule sets. In R. Kowalski and K. Bowen, editors, *Logic Programmin, Proceedings of the Fifth International Conference and Symposium*, pages 1036–1053, 1988.
20. P. Vojtás. Fuzzy logic programming. *Fuzzy Sets and Systems*, 124:361–370, 2001.
21. H. Wagner. A new resolution calculus for the infinite-valued propositional logic of Łukasiewicz. In R. Caferra and G. Salzer, editors, *Int. Workshop on First-Order Theorem Proving*, pages 234–243, 1998.

A General Theorem on Termination of Rewriting

Jeremy E. Dawson and Rajeev Goré

Logic and Computation Program, NICTA* and
Automated Reasoning Group
Australian National University, Canberra, ACT 0200, Australia
Jeremy.Dawson@nicta.com.au Rajeev.Gore@anu.edu.au

Abstract. We re-express our theorem on the strong-normalisation of display calculi as a theorem about the well-foundedness of a certain ordering on first-order terms, thereby allowing us to prove the termination of systems of rewrite rules. We first show how to use our theorem to prove the well-foundedness of the lexicographic ordering, the multiset ordering and the recursive path ordering. Next, we give examples of systems of rewrite rules which cannot be handled by these methods but which can be handled by ours. Finally, we show that our method can also prove the termination of the Knuth-Bendix ordering and of dependency pairs.

Keywords: rewriting, termination, well-founded ordering, recursive path ordering

1   Introduction

The traditional method for proving the termination of a rewriting systems uses well-founded orderings [7]. The traditional method for proving strong-normalisation of λ-calculi is to use structural induction on lambda terms and their types, backed by auxiliary well-founded inductions. Recently, structural induction on derivations have been used to prove cut-elimination [12] and to prove strong-normalisation of the generalised sequent framework of display calculi [3].

We re-express our theorem on the strong-normalisation of display calculi as a theorem about the well-foundedness of a certain ordering on first-order terms, thereby allowing us to prove the termination of systems of rewrite rules. We then show how to use our theorem to prove the well-foundedness of the lexicographic ordering, the multiset ordering and the recursive path ordering. Next, we give examples of systems of rewrite rules which cannot be handled by these methods but which can be handled by ours. Finally, we show that our method can also prove the termination of the Knuth-Bendix ordering and of dependency pairs.

The results in this paper have been machine-checked using the logical framework Isabelle, see http://web.rsise.anu.edu.au/~jeremy/isabelle/snabs/

* National ICT Australia is funded by the Australian Government's Department of Communications, Information Technology and the Arts and the Australian Research Council through Backing Australia's Ability and the ICT Centre of Excellence program

J. Marcinkowski and A. Tarlecki (Eds.): CSL 2004, LNCS 3210, pp. 100–114, 2004.

Related Work: Important methods for proving termination of rewrite systems often use simplification orderings. These include the recursive path orderings ([7], [11]) and the Knuth-Bendix ordering. In §3.7 and §3.8 we use our theorem to prove the well-foundedness of these orderings.

These orderings have been generalised by Ferreira & Zantema [9], Dershowitz & Hoot [6] and Borralleras, Ferreira & Rubio [2]. Their theorems, like ours, are not limited to simplification orderings, but they prove the well-foundedness of larger relations which satisfy the subterm property, so if they were closed under context (rewrite relations), they would also be simplification orderings. Jean Goubault-Larrecq [10] has proved results which are more general in that they replace the notion of subterm with an arbitrary well-founded relation. He also generalises methods for dealing with the λ-calculus, where the notion of substitution creates difficulties for techniques such as ours.

Arts & Giesl [1] describe a method of proving termination using "dependency pairs". They give key theorems and extensive development of resulting techniques. Their method does not require a simplification ordering or the subterm property. In §3.9 we use our theorem to prove one of their key theorems.

1.1   Notation and Terminology

Assume that we have fixed some syntax for defining "terms" like r, s and t. We deliberately leave this notion vague for now.

For an irreflexive binary relation ρ on terms, we will write $(r, t) \in \rho$, $(r, t) \in <_\rho$, $r <_\rho t$ or $t >_\rho r$ interchangeably. Relations will be assumed irreflexive unless the contrary is stated, but are not assumed transitive even when written $<_\rho$. We say r is *strongly normalising*, or is $\in SN$, (with respect to ρ) if there is no infinite descending sequence $r = r_0 >_\rho r_1 >_\rho r_2 >_\rho \ldots$ of terms, and ρ is *well-founded* (or *Noetherian*) if every $r \in SN$. We write $\leq_\rho$, $<_\rho^+$ and $<_\rho^*$ for the reflexive closure, the transitive closure and the reflexive transitive closure, respectively, of $<_\rho$. We write $\sigma \circ \rho$ for the relational composition of relations σ and ρ, in the sense that $(r, s) \in \sigma \circ \rho$ if there exists t such that $(r, t) \in \rho$ and $(t, s) \in \sigma$.

Often (see our examples) such a relation is described by giving a finite set of "rewrite rules" $l_i \to r_i$, where l_i and r_i are terms containing (meta-)variables for which terms may be substituted.

When a relation is defined by a set of rewrite rules, it is also usually taken that it is *closed under context*, ie, that where $C[\_]$ is a context (a term with a "hole"), and l rewrites to r, then $C[l]$ rewrites to $C[r]$. Often such a relation is called a *reduction*, such as the "β-reduction" of the lambda calculus.

The setting of our main theorem is that we are given a binary relation ρ which may or may not consist of the substitution instances of a finite set of rewrite rules, and which is not necessarily closed under context. Then we define *ctxt* ρ to be the closure under context of ρ and prove that *ctxt* ρ is well-founded. When ρ consists of the substitution instances of a finite set of rewrite rules the well-foundedness of *ctxt* ρ proves that the associated rewrite system *terminates*.

To formalise "closure under context", we must specify a language of terms. This is a first-order language, with a fixed set of function symbols, or term constructors, of fixed or variable (finite) arity, whose arguments and results are terms. This language may or may not contain term variables. The meta-language used to express rewrite rules and to discuss rewrites *does* contain term variables.

Given a term like $t = f(a, b, g(c, d))$, its *immediate* subterms are a, b and $g(c, d)$, its *proper* subterms are these and c and d, and its subterms are these and also t itself. We write $\bar{s}$ for a sequence of terms $s_1, \ldots, s_m$, such that these are the immediate subterms of $f(\bar{s})$. We define the "closure of ρ under context", *ctxt* ρ: for example, if $(c', c) \in \rho$, then $(f(a, g(c')), f(a, g(c))) \in ctxt \, \rho$.

Definition 1 (Closure Under Context). *Given an irreflexive relation ρ and terms t_0 and t_1, the pair $(t_1, t_0) \in ctxt \, \rho$ (say "t_0 reduces to t_1") if either* (a) *$(t_1, t_0) \in \rho$, or* (b) *t_0 and t_1 are identical except that exactly one immediate subterm of t_0 reduces to the corresponding immediate subterm of t_1.*

We also define, inductively, the set *SN* of strongly normalising terms. This definition is equivalent, at least in classical logic, to that given in §1.1.

Definition 2 (Strongly Normalising). *The set SN of strongly normalising terms is the smallest set of terms such that if every term t_1 to which t_0 reduces is in SN then $t_0 \in SN$. It follows that if t_0 cannot be reduced then $t_0 \in SN$.*

It is to be understood that concepts such as "strongly normalising", "reduction", etc, relate to *ctxt* ρ, which is, itself, closed under context. It follows therefore that if $t \in SN$ then all subterms of t are in *SN*.

2   A Proof of Termination

We now re-express our theorem about strong-normalisation for display calculi from [3] to make it applicable to term rewriting.

2.1   Various Well-Founded Orderings

To prove that *ctxt* ρ is well-founded, we use a binary relation $<_{dt}$ on terms, and show that it is well-founded. The relation $<_{dt}$ depends on a relation $<_{cut}$, which we have the freedom to define. The relation $<_{cut}$ must be well-founded, and invariably it is a superset of ρ. It will normally, but not necessarily, depend on the parts of the terms at or near its head (ie the roots of the corresponding abstract syntax trees).

Definition 3 ($<_{sn1}$, $<_{sn2}$, $<_{dt}$). *Given ρ and $<_{cut}$ we define three further binary relations on terms, $<_{sn1}$, $<_{sn2}$ and $<_{dt}$:*

(a) *$t_1 <_{sn1} t_0$ if t_0 and t_1 are the same except that one of the immediate subterms of t_0 is in SN and reduces to the corresponding immediate subterm of t_1.*

(b) *$t_1 <_{sn2} t_0$ if t_0 and t_1 are the same except that a proper subterm of t_0 is in SN and reduces to the corresponding proper subterm of t_1.*

(c) *$t_1 <_{dt} t_0$ iff either $t_1 <_{cut} t_0$ or $t_1 <_{sn1} t_0$.*

Note that $t_1 <_{sn1} t_0$ implies $t_1 <_{sn2} t_0$, and our main theorem uses only $<_{sn1}$. However $<_{sn2}$ is sometimes easier to work with because it is closed under context. We also define an auxiliary function *fwf* for "from well-founded" which maps a binary relation ρ to a binary relation *fwf* ρ as below:

Definition 4. *Given a relation ρ, the pair $(r, t) \in$ fwf ρ if and only if $(r, t) \in \rho$ and t is strongly normalising (for ρ).*

Clearly *fwf* ρ is well-founded, regardless of whether ρ is. We then prove

Theorem 1. *The relations $<_{sn1}$ and $<_{sn2}$ are each well-founded [3].*

Despite the notation, these relations need not be transitive. Intuitively, $t_1 <_{dt} t_0$ means that t_1 is closer to being "normalised" (in some sense) than is t_0.

Recall that we require $<_{cut}$ to be well-founded. We also need $<_{dt}$ to be well-founded. Given that $<_{cut}$ and $<_{sn1}$ are well-founded, to prove that $<_{dt}$ is well-founded, we need one of a number of sufficient conditions on the interaction between $<_{cut}$ and $<_{sn1}$.

Lemma 1. *Assume that τ and σ are well-founded relations. Then each of the following conditions is sufficient for $\tau \cup \sigma$ to be well-founded:*

(a) $\tau \circ \sigma \subseteq \sigma^* \circ \tau$,
(b) $\tau \circ \sigma \subseteq \sigma \circ \tau^*$,
(c) $\tau \circ \sigma \subseteq \tau \cup \sigma$,
(d) $\tau \circ \sigma \subseteq (\sigma \circ (\tau \cup \sigma)^*) \cup \tau$.

Of these, the last is from Doornbos & von Karger [8], and is implied by each of the others, which are in earlier results discussed and cited in [8].

Clearly $<_{sn1} \subseteq <_{sn2}$, so to prove that $<_{dt}$ is well-founded it is enough to prove that $<_{cut} \cup <_{sn2}$ is well-founded. Sometimes it is easier to prove one of the above conditions for $<_{sn2}$ than for $<_{sn1}$.

2.2   Strong Normalisation Induced from Immediate Subterms

We next define the set *ISN*, for "inductively strongly normalising", as the set of terms that are in *SN* if their immediate subterms are in *SN*. Clearly, $SN \subseteq ISN$.

Definition 5 (*ISN*). *A term $t \in$ ISN iff: if all the immediate subterms of t are in SN then $t \in SN$.*

The next lemma follows from this definition. We use only the $\Longleftarrow$ part.

Lemma 2. *A term $t \in$ SN iff every subterm of t is in ISN.*

Proof. $\Longrightarrow$: Assume $t \in SN$ and let u be a subterm of t, where $t = C[u]$. Consider any infinite sequence $u = u_0, u_1, \ldots$ of reductions starting with u. Since the reduction relation is closed under context, $t = C[u_0], C[u_1], \ldots$ is also an infinite sequence of reductions, contradicting $t \in SN$. Therefore $u \in SN$ and so $u \in ISN$.

$\Longleftarrow$: By induction on the structure of term t — assume that the result holds for each immediate subterm t' of t. Assume that every subterm of t, including t itself, is in *ISN*. Therefore each immediate subterm t' of t has the property that every subterm of t' is in *ISN*, and so, by the inductive hypothesis, $t' \in SN$. As each $t' \in SN$, and $t \in ISN$, so $t \in SN$.   □

2.3   Properties We Require of ρ

Finally, given a rewrite system, our result requires that the relation ρ satisfy certain properties. The most general version of these is

Condition 1. *For all $(r, l) \in \rho$, if all proper subterms of l are in SN then, for all subterms r' of r, either*

(a) $r' \in SN$ or
(b) $r' <^+_{dt} l$.

In practice we use a simpler condition which implies Condition 1, such as

Condition 2. *For all $(r, l) \in \rho$, for all subterms r' of r, either*

(a) r' is a proper subterm of l or is a subterm of a reduction of a proper subterm of l
(b) $r' <_{cut} l$
(c) r' is obtained from l by reduction of l at a proper subterm.

For if we assume that all proper subterms of l are in *SN* as required in the precondition of Condition 1, then Condition 2(a) implies that $r' \in SN$, and 2(c) implies that $r' <_{sn1} l$, which implies that $r' <^+_{dt} l$.

Note that sometimes, as in Example 3.3, we *enlarge* the relation ρ to satisfy Conditions 1 and 2.

2.4   Strong-Normalisation

Lemma 3. *Suppose ρ satisfies Condition 1, and let t_0 be any term. If all terms $t' <^+_{dt} t_0$ are in ISN, then $t_0 \in ISN$.*

Proof. Given t_0, assume that ρ satisfies Condition 1 and that
    (a): all terms t' such that $t' <^+_{dt} t_0$ are in *ISN*.
We need to show $t_0 \in ISN$, so we assume that
    (b): all immediate subterms of t_0 are in *SN*,
and we show that $t_0 \in SN$.

To show that $t_0 \in SN$, we show that every term t_1 that can be obtained from t_0 by a single reduction is in *SN*. We consider the two possible cases: whether the reduction occurs in a proper subterm of t_0, or is a reduction of t_0 as a whole.

Firstly, consider a reduction in a proper subterm of t_0: that is, $(t_1, t_0) \in ctxt \, \rho \setminus \rho$. Then the reduction is in an immediate (strongly normalising) subterm of t_0. So $t_1 <_{sn1} t_0$ and hence $t_1 <_{dt} t_0$ by Definition 3(c). Therefore $t_1 \in ISN$ by assumption (a). Now each immediate subterm of t_1 is equal to, or is a reduction of, an immediate subterm of t_0, which itself is in *SN* by assumption (b). All immediate subterms of t_1 are therefore in *SN*. Therefore, as $t_1 \in ISN$, $t_1 \in SN$.

Secondly, consider a reduction of the whole term t_0: that is, $(t_1, t_0) \in \rho$. Thus we get the new term t_1 whose subterms are either in SN because of Condition 1(a) or are subterms which are $<_{dt}^+$-smaller than t_0, by Condition 1(b). That is, every subterm t_1^s of t_1 (including t_1 itself) satisfies one of the following:

$$\text{(c): } t_1^s \in SN \qquad\qquad \text{(d): } t_1^s <_{dt}^+ t_0$$

Then $t_1^s \in ISN$, in case (c) because $SN \subseteq ISN$, and in case (d) by assumption (a). Since t_1^s is an arbitrary subterm of t_1, Lemma 2 implies that $t_1 \in SN$.

In either case $t_1 \in SN$. Since t_1 was obtained via an arbitrary reduction from t_0, it follows that $t_0 \in SN$. Thus we have $t_0 \in ISN$. $\qquad\square$

Theorem 2. *If ρ satisfies Condition 1 and $<_{dt}$ is well-founded, then every term is strongly normalising.*

Proof. As $<_{dt}$ and hence $<_{dt}^+$ are well-founded, it follows from Lemma 3 by well-founded induction that every term is in ISN; now use Lemma 2. $\qquad\square$

We now have a way of proving termination of a suitable rewrite system. Given the relation ρ, normally the rewrite rules and their substitutional instances, we define a well-founded relation $<_{cut}$ such that, when $<_{sn1}$ and $<_{dt}$ are defined as in Definition 3, the resulting $<_{dt}$ is well-founded and Condition 1 is satisfied. This is enough to show the well-foundedness of $ctxt\ \rho$. In the following section we show some examples using this procedure.

3   Examples Using the Theorem

As explained above, the crux is to find an appropriate definition of $<_{cut}$.

3.1   Multiset Order

Given an irreflexive relation ρ on a set E, we can define the *multiset order* derived from ρ on multisets of elements of E. We use A, B and C for finite multisets of elements of E, by which we mean both that they contain only finitely many distinct elements and that they contain only finitely many copies of each such element. We use $A \sqcup B$ to stand for the multiset union. We consider the irreflexive relation $<_{m1}$ defined on finite multisets:

$\leq_{m1}$: $\forall C, \forall b \in E$. if, for all $c \in C$, $c <_\rho b$, then $C \sqcup A <_{m1} \{b\} \sqcup A$

If ρ is a strict order (an irreflexive, transitive relation), then $<_{m1}^+$ is equal to the multiset order derived from ρ.

We represent a multiset as a tree, with two sorts of node, "inner" nodes (I) and "leaf" nodes (L). Viewing such a tree as a term, the function symbols are I and, for each $e \in E$, $L(e)$, where I has arbitrary arity and each $L(e)$ is nullary. The "leaf multiset" of a tree is the multiset of its leaf nodes, but with each $L(e)$ changed to e. Note that different trees can have the same leaf multiset.

We define a rewrite relation on such trees (terms) as follows. For every (finite) multiset $C = [c_1, c_2, \ldots, c_k]$ and every element $b \in E$ such that for all $c_i \in C$, $c_i <_\rho b$ (as in the definition of $<_{m1}$) we have a rule

$$L(b) \longrightarrow I(L(c_1), L(c_2), \dots, L(c_k))$$

Theorem 3. *Given a well-founded order ρ, the derived multiset order (on finite multisets) is well-founded.*

Proof. Clearly, whenever $A <_{m1} B$, any tree whose leaf multiset is B can be reduced to a tree whose leaf multiset is A using the rewrite relation defined above. Since for a given finite multiset B there always is a tree whose corresponding multiset is B, when we show that rewriting terminates we have shown that the multiset order is well-founded.

We prove that this rewrite system terminates. We define $<_{cut}$ by the rules

$$L(x) >_{cut} I(y) \qquad\qquad L(x) >_{cut} L(y) \text{ iff } x >_{\rho} y$$

It is clear that $<_{cut}$ is well-founded when ρ is. To show that $<_{cut} \cup <_{sn1}$ is well-founded, we use Lemma 1, by showing that in fact $<_{sn1} \circ <_{cut} = \emptyset$. For suppose $t >_{sn1} u >_{cut} v$. Then u must be of the form $L(x)$, and a proper subterm of t must reduce to a proper subterm of u – but u has no proper subterms.

To show that Condition 2 is satisfied, when $L(b) \longrightarrow I(L(c_1), \dots, L(c_k))$, we have $L(b) >_{cut} I(L(c_1), \dots, L(c_k))$ by the first rule for $<_{cut}$, and, for every subterm $L(c_i)$ of the reduced subterm, $L(b) >_{cut} L(c_i)$ by the second rule.

3.2   A Non Simplification Ordering

Example 5 of [5], with the single rule $f(f(x)) \longrightarrow f(g(f(x)))$ is one for which a simplification ordering cannot be used, because a simplification ordering would take $g(f(x))$ to $f(x)$ and so $f(g(f(x)))$ to $f(f(x))$, giving a cycle.

But Theorem 2 is not limited to simplification orderings. We define $<_{cut}$ according to the number of consecutive f symbols starting from the head of a term. Alternatively, we could use the total number of pairs of adjacent f symbols, as suggested in [5]. Thus $f(f(x)) >_{cut} f(g(y))$, $f(f(x)) >_{cut} g(y)$, and $f(f(x)) >_{cut} f(x)$. Finally, any subterm of x is a proper subterm of $f(f(x))$. Thus Condition 2 is satisfied. Clearly also, rewriting a subterm cannot increase the number of consecutive f symbols, so $<_{sn1} \circ <_{cut} \subseteq <_{cut}$ (and likewise $<_{cut} \circ <_{sn1} \subseteq <_{cut}$). Thus $<_{dt} = <_{cut} \cup <_{sn1}$ is well-founded by Lemma 1. Therefore the system terminates, by Theorem 2.

3.3   A Recursive Path Ordering Example

We now consider a typical example whose termination is shown by the recursive path ordering (see, for example, [6] or [7]).

$$D(x + y) \longrightarrow Dx + Dy \qquad\qquad (x \times y) \times z \longrightarrow x \times (y \times z)$$
$$D(x \times y) \longrightarrow x \times Dy + Dx \times y \qquad (x + y) + z \longrightarrow x + (y + z)$$
$$x \times (y + z) \longrightarrow x \times y + x \times z$$

The recursive path ordering is a simplification ordering. To prove termination we must actually add the following simplification rules as rewrite rules:

$$x + y \longrightarrow x \qquad x + y \longrightarrow y \qquad x \times y \longrightarrow x \qquad x \times y \longrightarrow y$$

Clearly, if additional rules R' are added to a rewrite system R, and the resulting system $R \cup R'$ terminates, then the original system R terminates.

We define $<_{cut}$ as below by first defining the relation $<_h$, which depends on the head symbol, and then defining the relations $<_\times$ and $<_+$ which are for the rewrite rules capturing associativity:

- $D(x) >_h y \times z$ $D(x) >_h y + z$ $x \times y >_h z + w$
- if x' is an immediate subterm of x, then $x + y >_+ x' + z$ and $x \times y >_\times x' \times z$
- $<_{cut} = (<_h \cup <_\times \cup <_+)$.

We now prove that this rewrite system terminates. Clearly $<_{cut}$ is well-founded, since the immediate subterm relation is well-founded.

To show that $<_{cut} \cup <_{sn1}$ is well-founded, we in fact show that $<_{cut} \cup <_{sn2}$ is well-founded, using Lemma 1(a). First we show that $<_{cut} \circ <_{sn2} \subseteq <_{sn2}^* \circ <_{cut}$. For suppose $t >_{cut} u >_{sn2} v$. If $t >_h u$ then clearly $t >_h v$. Suppose $t >_\times u$, say $t = x \times y$, $u = x' \times z$ and x' is an immediate subterm of x. There are two cases for $u >_{sn2} v$, namely that a strongly normalising subterm of either x' or z reduces to a corresponding subterm of v. Firstly, if $z \longrightarrow w$, then we also have $v = x' \times w$ and $t >_{cut} v$. Secondly, if $x' \longrightarrow w'$, then $v = w' \times z$. As x' is a subterm of x, let w be the term obtained from x by rewriting its subterm x' to w'. Then $t = x \times y >_{sn2} w \times y$: but note that this step of the argument would not hold for $>_{sn1}$. Then w' is an immediate subterm of w, so we have $w \times y >_{cut} w' \times z = v$. That is, $(v, t) \in <_{sn2} \circ <_{cut}$.

So, in each case, $(v, t) \in <_{sn2}^* \circ <_{cut}$, and so $<_{cut} \cup <_{sn2}$ is well-founded, by Lemma 1(a).

Finally, we need to consider all pairs (r', l), where $l \rightarrow r$ is a rewrite rule and r' is a subterm of r. In many cases $l >_h r'$: for example $D(x + y) >_h Dx + Dy$. In other cases, r' is a proper subterm of l: for example whenever a (meta-)variable appears on the right-hand side of a rule, and r' is the term the variable stands for, or any subterm of it. In the case of the associative rewrite rule $l = (x \times y) \times z \longrightarrow r = x \times (y \times z)$, we have that $l >_\times r$, by definition; the rule for the associativity of $+$ is similar.

Finally, we have cases for (r', l) such as $(D(x), D(x+y))$ or $(y \times z, (x \times y) \times z)$. Here, since a proper subterm l' of l is assumed to be strongly normalising, and $x + y \longrightarrow x$ (likewise $x \times y \longrightarrow y$) we have $D(x + y) >_{sn1} D(x)$ and $(x \times y) \times z >_{sn1} y \times z$.

Thus in all cases either r' is a proper subterm of l or (assuming proper subterms of l are in SN) $l >_{dt} r'$, so the system terminates by Theorem 2.

3.4   Ackermann's Function

Ackermann's function on the natural numbers can be defined by the following rewrite rules [5, Example 29]

$$A(0, y) \longrightarrow S(y)$$
$$A(S(x), 0) \longrightarrow A(x, S(0))$$
$$A(S(x), S(y)) \longrightarrow A(x, A(S(x), y))$$

It can be shown to terminate using the lexicographic path ordering. This is reflected in the relation $>_{cut}$ we use, which is defined by the following cases:

$$A(x, y) >_{cut} S(z) \tag{1}$$

$$A(S(x), y) >_{cut} A(x, z) \tag{2}$$

$$A(x, S(y)) >_{cut} A(x, y) \tag{3}$$

We now prove that this rewrite system terminates. It is clear that $>_{cut}$ is well-founded using the lexicographic ordering on arguments. It is also clear that for each (r', l), where $l \to r$ is a rewrite rule and r' is a subterm of r, either $l >_{cut} r'$ or r' is a proper subterm of l.

It remains to show that $<_{cut} \cup <_{sn1}$ is well-founded. Again we show that $<_{cut} \cup <_{sn2}$ is well-founded, using Lemma 1(a). We show that $<_{cut} \circ <_{sn2} \subseteq <_{sn2}^* \circ <_{cut}$. For suppose $t >_{cut} u >_{sn2} v$. If $t >_{cut} u$ by rule (1), ie $t = A(x, y)$ and $u = S(z)$, then $v = S(z')$ and so $t >_{cut} v$. If $t >_{cut} u$ by rule (2), then $t = A(S(x), y)$ and $u = A(x, z)$. There are two cases for $u >_{sn2} v$: $v = A(x, z')$ where $z \longrightarrow z'$, in which case $t >_{cut} v$, or $v = A(x', y)$ where $x \longrightarrow x'$ by reducing a strongly normalising subterm of x, in which case $t = A(S(x), y) >_{sn2} A(S(x'), y) >_{cut} A(x', z) = v$. The case for rule (3) is similar.

So in all cases $(v, t) \in <_{sn2}^* \circ <_{cut}$, and so $<_{cut} \cup <_{sn2}$ is well-founded, by Lemma 1(a). Therefore the system terminates by Theorem 2.

3.5   Insertion Sort

This example [5, Example 32] is more difficult than the previous two, though the approach is similar. The rules are

$$sort(nil) \longrightarrow nil$$
$$sort(cons(x, y)) \longrightarrow insert(x, sort(y))$$
$$insert(x, nil) \longrightarrow cons(x, nil)$$
$$insert(x, cons(v, w)) \longrightarrow choose(x, cons(v, w), x, v)$$
$$choose(x, cons(v, w), y, 0) \longrightarrow cons(x, cons(v, w))$$
$$choose(x, cons(v, w), 0, s(z)) \longrightarrow cons(v, insert(x, w))$$
$$choose(x, cons(v, w), s(y), s(z)) \longrightarrow choose(x, cons(v, w), y, z)$$

To define $>_{cut}$, we start by defining an order $>_h$, contained in $>_{cut}$, which depends on the head symbol, using this (transitive) order on symbols: $sort > \{insert, choose\} > cons$. We notice that the rules (considered as definitions) define $insert$ and $choose$ in terms of each other, so we continue by defining

$$insert(x, w) >_{cut} choose(y, w, a, b) >_{cut} insert(x, w')$$
$$choose(y, w, a, b) >_{cut} choose(y', w, a', b')$$

where $>_{cut}$ is transitive and w', a' are immediate subterms of w, a. It is easy to see that $>_{cut}$ is well-founded. As in §3.3, we add a simplification rule

$$cons(x, y) \longrightarrow y$$

Then, for every (r', l) where $l \to r$ is a rewrite rule and r' is a subterm of r, either $l >_{cut} r'$, $l >_{sn1} r'$ or r' is a proper subterm of l.

Again we show that $<_{cut} \circ <_{sn2} \subseteq <^*_{sn2} \circ <_{cut}$, using the same technique as in the previous two examples. Therefore the system terminates by Theorem 2.

The ordering $<_{cut}$ is similar to that used in [5, Example 32], and [6].

3.6   The Factorial Example

Example 21 of [5] is almost the usual definition of the factorial function, but modified so that we can not use a simplification ordering. The rules are

$$P(S(x)) \longrightarrow x \qquad\qquad F(0) \longrightarrow 0$$
$$F(S(x)) \longrightarrow S(x) \times F(P(S(x))) \qquad 0 \times y \longrightarrow 0$$
$$S(x) \times y \longrightarrow x \times y + y \qquad\qquad x + 0 \longrightarrow x$$
$$x + S(y) \longrightarrow S(x + y)$$

As usual we define a (transitive) ordering $<_h$ of terms based on the following ordering of head symbol: $F > \times > + > S$. But we need to define $<_{cut}$ to be the union of $<_h$ and the following additional cases

$$F(S(x)) >_{cut} F(P(S(x))) \qquad F(S(x)) >_{cut} F(P(x))$$

We can not use a simplification ordering because if we allowed $P(x) \longrightarrow x$ the system would not terminate, but would "cycle" between terms containing $F(S(x))$ and terms containing $F(P(S(x)))$. We do, however, need to add the rule $S(x) \longrightarrow x$. The proofs of termination given in [5, Examples 21, 25] are based on interpreting arguments to the function symbols as natural numbers.

Now for each (r', l) where $l \to r$ is a rewrite rule and r' is a subterm of r, it is reasonably easy to see that we have one of the following cases:

– r' is a proper subterm of l
– $r' <_h l$
– r' is l, but with S removed from a subterm (to give $r' <_{sn1} l$)
– r' is $F(P(S(x)))$ and l is $F(S(x))$ (so $r' <_{cut} l$).

It is easy to see that $<_{cut}$ is well-founded. To show that $<_{cut} \cup <_{sn2}$ is well-founded we need to use a more general case of Lemma 1 than hitherto. In fact we show $<_{cut} \circ <_{sn2} \subseteq (<^*_{sn2} \circ <_{cut}) \cup <^+_{sn2}$, which implies (d) of Lemma 1.

Suppose $t >_{cut} u >_{sn2} v$. If $t >_h u$ then clearly $t >_h v$. Suppose $t = F(S(x))$ and $u = F(P(S(x)))$ or $u = F(P(x))$. If $u >_{sn2} v$ by way of reducing the x in u to x' (reducing a strongly normalising subterm of x), then $t = F(S(x)) >_{sn2} F(S(x')) >_{cut} v$. Otherwise we need to consider specific cases:

– $u = F(P(S(x)))$, $S(x) \longrightarrow x$ where $S(x) \in SN$, $v = F(P(x))$: then $t >_{cut} v$
– $u = F(P(S(x)))$, $P(S(x)) \longrightarrow x$ where $P(S(x)) \in SN$, $v = F(x)$: here, as $S(x) \longrightarrow x$ and $S(x) \in SN$, we have $t >_{sn2} v$
– $u = F(P(x))$, where $x = S(y)$, $P(S(y)) \longrightarrow y$ and $P(S(y)) \in SN$, $v = F(y)$: here, as $S(y) \longrightarrow y$ and $S(y) \in SN$, we have $t = F(S(S(y))) >_{sn2} F(S(y)) >_{sn2} F(y) = v$.

Thus $<_{cut} \cup <_{sn2}$ is well-founded and the system terminates by Theorem 2.

3.7   Recursive Path Orderings

To show that our theorem is at least as general as the recursive path orderings we now derive their termination.

Given a well-founded ordering $<_\sigma$, we use the derived lexicographic ordering *lex* σ, which is restricted to fixed-length sequences, and so is well-founded.

Given a well-founded ordering $<_\rho$ on function symbols, the lexicographic path ordering *lpo* ρ, also written $<_{lpo}$ [7, Defn 4.22], is then defined as below (omitting reference to *lpo*-equivalent terms, and where $<_{lex}$ is *lex* (*lpo* ρ)

$$\frac{s_i \geq_{lpo} t}{f(s_1,\ldots,s_m) >_{lpo} t} \qquad \frac{f >_\rho g \qquad \forall i \in \{1,\ldots,n\}.\ f(s_1,\ldots,s_m) >_{lpo} t_i}{f(s_1,\ldots,s_m) >_{lpo} g(t_1,\ldots,t_n)}$$

$$\frac{(s_1,\ldots,s_m) >_{lex} (t_1,\ldots,t_m) \qquad \forall i \in \{1,\ldots,m\}.\ f(s_1,\ldots,s_m) >_{lpo} t_i}{f(s_1,\ldots,s_m) >_{lpo} f(t_1,\ldots,t_m)}$$

We note that the definition of the multiset path ordering, and the proof that it is well-founded, correspond closely to what follows.

The following lemma is actually an easy consequence of the transitivity of $>_{lpo}$, but that result is much more difficult to prove.

Lemma 4. *If* $s >_{lpo} t$ *and* t' *is a subterm of* t *then* $s >_{lpo} t'$.

Proof. It is enough to show that if $s >_{lpo} t = g(t_1,\ldots,t_n)$ then $s >_{lpo} t_j$ where $j \in \{1,\ldots,n\}$. We show this by induction on the size (or structure) of s. If $s >_{lpo} t$ by the second or third rules, then it is immediate that $s >_{lpo} t_j$. If $s = f(s_1,\ldots,s_m) >_{lpo} t$ by the first rule, then for some s_i, either $s_i >_{lpo} t$ and so $s_i >_{lpo} t_j$ by induction, or $s_i = t >_{lpo} t_j$ by the first rule, and then $s >_{lpo} t_j$ by the first rule.   □

Theorem 4. *For a well-founded ordering* $<_\rho$ *on function symbols, the derived lexicographic path ordering is well-founded.*

Proof. We first define $<_{cut}$ using $<_h$ as the ordering of terms according to the head symbol and defining another relation $<_{lw1}$ as below:

- $f(\bar{s}) >_h g(\bar{t})$ iff $f >_\rho g$
- if $((\bar{t}),(\bar{s})) \in$ *lex* (*fwf* (*lpo* r)), then $f(\bar{t}) <_{lw1} f(\bar{s})$
- $<_{cut} = (<_h \cup <_{lw1})$.

The idea behind the definition is similar to the way the definition of $<_{sn1}$ requires the $<_{sn1}$-greater term to be strongly normalising. As *fwf* (*lpo* r) is well-founded, *lex* (*fwf* (*lpo* r)) and so $<_{lw1}$ are well-founded. Clearly also $<_h \circ <_{lw1} \subseteq <_h$, so $<_{cut}$ is well-founded, by Lemma 1. Further, $<_{sn1} \subseteq <_{lw1}$, so $<_{cut} \cup <_{sn1} = <_{cut}$ is well-founded.

To show that the rewrite relation $<_{lpo}$ satisfies Condition 1, suppose (r',l) is given, where $r <_{lpo} l$ and r' is a subterm of r. Then, by Lemma 4, $r' <_{lpo} l$.

If $l = f(s_1,\ldots,s_m) >_{lpo} r'$ by the first rule, then some $s_i >_{lpo} r'$. Now assuming that $s_i \in SN$, we have $r' \in SN$.

Now suppose $l >_{lpo} r'$ by the second or third rules. Again, we are assuming that all proper subterms of l are strongly normalising, and then $r' <_{cut} l$.   □

It may be noted that using our theorem as above actually proves that $ctxt\,(lpo\ \rho)$ is well-founded — in fact, since $lpo\ \rho$ is closed under context this point is of no significance. The proof that the multiset path ordering $<_{mpo}$ [7, Defn 4.24] is well-founded is very similar. In the "status-based" recursive path ordering, lists of arguments to a function symbol are compared using the lexicographic, multiset or other derived ordering $\phi(\_)$, depending on the function symbol. This is well-founded also, by a similar proof. The necessary properties for such an ordering $\phi(\_)$ are the following: that $\phi(\sigma)$ is well-founded if σ is, and that if $s'_i <_\sigma s_i$ then $(s_1, \ldots, s'_i, \ldots, s_m) <_{\phi(\sigma)} (s_1, \ldots, s_i, \ldots, s_m)$. This second property is used above in the step $<_{sn1} \subseteq <_{lw1}$.

3.8   The Knuth-Bendix Ordering

For a rewrite system (with rules containing variables), the Knuth-Bendix ordering $<_{kb}$ is based on a strict well-founded order $<$ on function symbols and, additionally, a weight function w on function symbols and variables. Since our approach is based on a relation ρ (which amounts to the rewrite rules after all possible substitutions for variables), we describe the Knuth-Bendix ordering in this context. Weights are natural numbers, and the weight of any constant or object language variable is positive: thus every subterm has positive weight. At most one unary function symbol (call it k) can have zero weight, and then $k > f$ for any other function symbol f. The weight of a term is the sum of the weights of the function symbols and variables in it. Then we have that $s >_{kb} t$ iff $w(s) \geq w(t)$ and one of

$$w(s) > w(t) \tag{4}$$
$$s = k^n(t) \text{ for some } n > 0 \tag{5}$$
$$s = f(\bar{s}) \text{ and } t = g(\bar{t}) \text{ where } f > g \tag{6}$$
$$s = f(\bar{s}), t = f(\bar{t}) \text{ and } (\bar{t}, \bar{s}) \in lex\ (<_{kb}) \tag{7}$$

Theorem 5. *The Knuth-Bendix ordering is well-founded.*

Proof. The proof that $<_{kb}$ is well-founded is similar to that in §3.7, and so is somewhat abbreviated here. We define $s >_{cut} t$ iff $w(s) \geq w(t)$ and either one of the rules (4), (5), (6) holds or $s >_{kw1} t$ holds, where $<_{kw1}$ is defined by:

$$f(\bar{t}) <_{kw1} f(\bar{s}) \text{ iff } (\bar{t}, \bar{s}) \in lex\ (fwf\ (<_{kb})) \tag{8}$$

To show $<_{cut}$ is well-founded, we have that each individual rule provides a well-founded relation and it is easy to apply Lemma 1 to show that their union is well-founded, noting that if (6) applies then $g \neq k$. As in the case of the lexicographic path order $<_{sn1} \subseteq <_{kw1}$.

To show that Condition 1 is satisfied, suppose that $l >_{kb} r$ and that r' is a subterm of r. Assume that all proper subterms of l are in SN, in which case, if $l >_{kb} r'$ then $l >_{cut} r'$. We show by induction on the structure of l, that $r' <_{kb} l'$ for some subterm l' of l. Then if l' is a proper subterm of l, Condition 2(a) is satisfied; if $l' = l$ then $r' <_{kb} l$ and so $r' <_{cut} l$, so Condition 2(b) is satisfied.

If $w(r') < w(r)$ or $w(r) < w(l)$ then $w(l) > w(r')$ and $l >_{kb} r'$. If $r' = r$ then $l >_{kb} r'$. If $w(l) = w(r) = w(r')$ and r' is a proper subterm of r, then $r = k^n(r')$ for some $n > 0$, and so $l >_{kb} r$ by either (5) or (7) (as the g of (6) cannot be k). If $l >_{kb} r$ by (5) then $l >_{kb} r'$ by (5). If $l >_{kb} r$ by (7), then, for some l_1, r_1, $l = k(l_1)$, $r = k(r_1)$, $l_1 >_{kb} r_1$ and r' is a subterm of r_1, so by induction we have that $r' <_{kb} l'$ for some l' which is a subterm of l_1 and so of l. □

3.9   Dependency Pairs

Arts & Giesl [1] describe a method of establishing termination using "dependency pairs". They distinguish function symbols which appear at the head of the left-hand side of a rewrite rule ("defined symbols") and those which do not ("constructor symbols"). They follow the convention that for a rule $l \to r$ of a rewrite system, l is not a lone variable, and any variable in r is also in l. For each defined symbol d they introduce a new corresponding "tuple symbol" $d^\sharp$. From a term t we obtain a term $t^\sharp$ by changing the head symbol of t to the corresponding tuple symbol.

Previously we have considered a "rule" $l \to r$ after substitution, and the variables in our analyses have been metavariables where, for example, we might have considered a rule $g(x) \to x$ with $l = g(x)$, $r = x$, and r' a proper subterm of x. This approach no longer holds: $l \to r$ will mean a rule before substitution, and we will use σ for a substitution.

For a rewrite rule $l \to r$, and subterm r' of r, if the head of r' is a defined symbol then $l^\sharp \to r'^\sharp$ is a *dependency pair*. We now state and prove the "sufficiency" part of Theorem 7 of [1].

Theorem 6. *If there is a well-founded quasi-ordering $\leq$ which is closed under context and where both $\leq$ and $<$ are closed under substitution, such that*

(a) $l \geq r$ for all rules $l \to r$, and
(b) $s > t$ for all dependency pairs $s \to t$

then the rewrite system terminates.

Proof. Assume $>$ is minimal such that the conditions hold. Then there is no instance of $c^\sharp(x) > d^\sharp(y)$ or $c(x) > d(y)$ where c is a constructor symbol and d is a defined symbol, as neither (a) nor (b) nor the requirement that $\leq$ be closed under context require it, whence the transitivity of $\leq$ cannot require it.

We define $<_{cut}$ by

$$s^\sharp < t^\sharp \implies s\sigma <_{cut} t\sigma \tag{9}$$

$$c(x) <_{cut} d(y) \tag{10}$$

for any substitution σ, constructor symbol c and defined symbol d.

Then $<_{cut}$ is well-founded because $<$ is and because, as remarked above, there is no instance of $c^\sharp(x) > d^\sharp(y)$ or $c(x) > d(y)$.

To show that $<_{cut} \cup <_{sn1}$ is well-founded, we use Lemma 1(a), showing that $<_{cut} \circ <_{sn1} \subseteq <_{cut}$. Suppose that $t >_{cut} u >_{sn1} v$. If $t >_{cut} u$ by rule (10) then

$t >_{cut} v$ by the same rule. If $t >_{cut} u$ by rule (9) then $t^\sharp > u^\sharp$, where a proper subterm u' of u is rewritten to a corresponding subterm v' of v. Since $u' \geq v'$ by assumption (a) because $\leq$ is closed under substitution, we have $u^\sharp \geq v^\sharp$ since $\leq$ is closed under context, and so $t^\sharp > v^\sharp$ and $t >_{cut} v$.

Finally, we show that Condition 2 is satisfied. For a rule $l \to r$ and substitution σ, so $(r\sigma, l\sigma) \in \rho$, and subterm r' of $r\sigma$, there are three cases for r':

- the head of r' is a constructor symbol in r, in which case $l\sigma >_{cut} r'$ by (10)
- the head of r' is a defined symbol in r, in which case $r' = r_1\sigma$ for some subterm r_1 of r, $l^\sharp \to r_1^\sharp$ is a dependency pair, and $l\sigma >_{cut} r_1\sigma = r'$ by (9)
- r' is a subterm of $x\sigma$ for some variable x in r, in which case r' is a proper subterm of $l\sigma$, since any variable in r appears as a proper subterm of l.

4 Observations and Conclusion

In §3.3 we had the somewhat paradoxical situation that it was necessary to add additional rewrite rules to enlarge the relation ρ to prove termination. An interesting question is whether we can reformulate Theorem 2 to avoid this need since, the larger ρ is, the more difficult it should be to prove termination.

It is interesting to consider further questions of a similar nature. We prove the well-foundedness of $ctxt \, \rho$ based on certain conditions on ρ. Since $ctxt \, (ctxt \, \rho) = ctxt \, \rho$ and assuming we can apply Theorem 2 to ρ, can we also apply it to $ctxt \, \rho$ using the same or a different choice of $<_{cut}$? In fact, we can, and with the same choice of $<_{cut}$. Suppose ρ satisfies Condition 1. Consider $(C[r], C[l]) \in ctxt \, \rho \setminus \rho$, where $(r, l) \in \rho$, and let r' be a subtree of $C[r]$. If $r' = C[r]$, then $r' <_{sn1} C[l]$ under the assumption that all proper subtrees of $C[l]$ are in SN. If r' is a subtree of r, then Condition 2(a) holds for $(r', C[l])$ because l is a proper subtree of $C[l]$, l reduces to r, and r' is a subtree of r. Thus l, r and r' are in SN. Finally, r' could be $C'[r]$, where $C'[l]$ is a proper subterm of $C[l]$. If $C'[r] = C'[l]$ then l is in a different part of $C[l]$ from $C'[l]$ so $r' = C'[l]$ is a proper subterm of $C[l]$; otherwise r' is the reduction of $C'[l]$, and Condition 2(a) holds.

Likewise, rewriting with ρ terminates if and only if rewriting with ρ^+ terminates: this is because $ctxt \, \rho \subseteq ctxt \, \rho^+ \subseteq (ctxt \, \rho)^+$ and $ctxt \, \rho$ is well-founded if and only if $(ctxt \, \rho)^+$ is. Now, when we can apply Theorem 2 to ρ, can we also apply it to ρ^+ ? So far we have shown that if $ctxt \, \rho$ satisfies Condition 2, then so does $(ctxt \, \rho)^+$ (with the same $<_{cut}$), provided that $ctxt \, \rho$ is well-founded. The proof is a complex triple induction.

If we ask how powerful Theorem 2 is, that is, which terminating rewrite systems can it handle, then we find that it can handle all of them, but this result is unhelpful. For given ρ, where $ctxt \, \rho$ is well-founded, then, writing $\leq_{sub}$ for the subterm relation, we can define $<_{cut} = \rho \circ <_{sub}$. Then Condition 2(a) holds, trivially. Further, it can be shown that $ctxt \, \rho \cup <_{sub}$ is well-founded: though, as we have seen, $ctxt \, (\rho \cup <_{sub})$ need not be. Since $<_{dt} \subseteq (ctxt \, \rho \cup <_{sub})^+$, we have $<_{dt}$ is well-founded. That is, where $ctxt \, \rho$ is well-founded, we can always find an appropriate relation $<_{cut}$. But it may be no easier to show $<_{cut}$ well-founded than to show $ctxt \, \rho$ well-founded.

We have described a proof of the termination of a rewrite system (or any relation closed under context) which provides simple proofs of the termination of a range of rewrite systems. It provides a reasonably easy proof of the well-foundedness of the lexicographic and multiset path orderings (which are simplification orderings), but it is not limited to simplification orderings. It can also be used to prove the termination of Knuth-Bendix orderings and a key theorem for the method of dependency pairs.

There are several termination results for orderings which are not necessarily closed under context, but contain a rewrite ordering, such as the results of Ferreira & Zantema [9], Dershowitz & Hoot [6] and Borralleras, Ferreira & Rubio [2]. We are currently exploring the linkage between our results and these termination results.

Acknowledgements. We wish to thank Linda Buisman for investigating some of the examples, and researching the criteria for well-foundedness of a union of well-founded orderings. We also wish to thank Jean Goubault-Larrecq and Hubert Comon for pointers to the literature. Finally, we thank some anonymous referees for very helpful comments.

References

1. Thomas Arts and Jürgen Giesl. Termination of Term Rewriting Using Dependency Pairs. Theoretical Computer Science 236 (2000), 133-178, 2000.
2. Cristina Borralleras, Maria Ferreira and Albert Rubio. Complete Monotonic Semantic Path Orderings. In Proc. 17th Conference on Automated Deduction (CADE-17), LNCS 1831, 346–364, Springer, 2000.
3. Jeremy E. Dawson and Rajeev Goré. A New Machine-checked Proof of Strong Normalisation for Display Logic. In Computing: The Australasian Theory Symposium, 2003, Electronic Notes in Theoretical Computer Science 78, 16–35, Elsevier.
4. Jeremy E. Dawson and Rajeev Goré. Formalised cut admissibility for display logic. In Proc. 15th International Conference on Theorem Proving in Higher Order Logics (TPHOLs02), LNCS 2410, 131–147, Springer, 2002.
5. Nachum Dershowitz. 33 Examples of Termination. Proc. French Spring School of Theoretical Computer Science (1993), LNCS 909, 16–25, Springer 1995.
6. Nachum Dershowitz & Charles Hoot. Natural Termination. Theoretical Computer Science 142 (1995), 179-207.
7. Nachum Dershowitz & David A. Plaisted. Rewriting. In Alan Robinson & Andrei Voronkov, eds, Handbook of Automated Reasoning, 535–610, Elsevier, 2001.
8. Henk Doornbos & Burghard Von Karger. On the Union of Well-Founded Relations. L. J. of the IGPL 6 (1998), 195-201.
9. M C F Ferreira and H Zantema. Well-foundedness of Term Orderings. In Conditional Term Rewriting Systems (CTRS-94) ed N. Dershowitz, LNCS 968, 106–123, 1995.
10. Jean Goubault-Larrecq. Well-founded recursive relations. In Computer Science Logic (CSL'2001), LNCS 2142, 484–497, 2001.
11. Sam Kamin & Jean-Jacques Lévy. Two generalizations of the recursive path ordering. Unpublished, Department of Computer Science, University of Illinois, 1980.
12. Frank Pfenning. Structural Cut Elimination, Proc. LICS 94, 1994.

Predicate Transformers and Linear Logic: Yet Another Denotational Model

Pierre Hyvernat[1,2]

[1] Institut mathématique de Luminy, Marseille, France
[2] Chalmers Institute of Technology, Göteborg, Sweden
hyvernat@iml.univ-mrs.fr

Abstract. In the refinement calculus, monotonic predicate transformers are used to model specifications for (imperative) programs. Together with a natural notion of simulation, they form a category enjoying many algebraic properties.

We build on this structure to make predicate transformers into a denotational model of full linear logic: all the logical constructions have a natural interpretation in terms of predicate transformers (*i.e.* in terms of specifications). We then interpret proofs of a formula by a safety property for the corresponding specification.

Introduction

The first denotational model for linear logic was the category of *coherent spaces* ([1]). In this model, formulas are interpreted by graphs; and proofs by *cliques* (complete subgraphs). This forms a special case of domain à *la* Scott.

From a conceptual point of view, the construction of interfaces is a little different: first, the model looks a little more dynamic; then, *seeds* —the notion corresponding to cliques— are not closed under substructures; and finally, they are closed under arbitrary unions (usually, only directed unions are allowed).

What was a little unexpected is that the interpretation of linear proofs used in the relational model can be lifted directly to this structure to yield a denotational model of full linear logic in the spirit of _/hyper/multi-coherence or finiteness spaces.

A promising direction for further research is to explore the links between the model presented below and non-determinism as it appears both in the differential lambda-calculus ([2, 3]) and different kind of process calculi. We expect such a link because of the following remarks: this model comes from the semantics of imperative languages; it can be extended to a model of the differential lambda calculus (which can be seen as a variant of "lambda calculus with resource") and there is a completely isomorphic category in which predicate transformers are replaced by (two-sided) transition systems. In particular, all of the logical operations presented below have natural interpretations in terms of processes...

J. Marcinkowski and A. Tarlecki (Eds.): CSL 2004, LNCS 3210, pp. 115–129, 2004.

1   Relations and Predicate Transformers

Definition 1. *A* relation *r* between two sets is a subset of their cartesian product. *We write $r^\sim$ for the* converse relation: *$r^\sim = \{(b,a) \mid (a,b) \in r\}$.*

The composition of two relations $r \subseteq A \times B$ and $r' \subseteq B \times C$ is defined by $r' \cdot r = \{(a,c) \mid (\exists b \in B)\ (a,b) \in r \land (b,c) \in r'\}$.

If X is a set, $\mathbf{Id}_X$ denotes the identity on X, i.e. $\mathbf{Id}_X = \{(a,a) \mid a \in X\}$.

There seems to be three main notions of morphisms between sets. These give rise to three important categories in computer science:

- **Set**, where morphisms are functions;
- **Rel**, where morphisms are (binary) relations;
- **Pow**, where morphisms are monotonic *predicate transformers.*

One can go from **Set** to **Rel** and from **Rel** to **Pow** using the same categorical construction ([4]) which cannot be applied further.

Definition 2. *A* predicate transformer *from A to B is a function from $\mathcal{P}(A)$ to $\mathcal{P}(B)$. A predicate transformer P is* monotonic *if $x \subseteq x'$ implies $P(x) \subseteq P(x')$.*

From now on, we will consider only monotonic predicate transformers. The adjective "monotonic" is thus implicit everywhere.

The term "predicate" might not be the most adequate but the terminology was introduced by E. Dijkstra some decades ago, and has been used extensively by computer scientists since then. Formally, a predicate on a set A can be identified with a subset of A by the separation axiom of ZF set theory; the confusion is thus harmless.

Definition 3. *If r is a relation between A and B, we write $\langle r \rangle : \mathcal{P}(A) \to \mathcal{P}(B)$ for the following predicate transformer:*   *(called the* direct image of r*)*

$$\langle r \rangle(x) = \{b \in B \mid (\exists a \in A)\ (a,b) \in r \land a \in x\} \ .$$

Note that in the traditional version of the refinement calculus ([5]), our $\langle r \rangle$ is written $\{r^\sim\}$, but this notation clashes with set theoretic notation and would make our formulas very verbose with $\_^\sim$ everywhere. pt

2   Interfaces

Several denotational models of linear logic can be seen as "refinements" of the relational model. This very crude model interprets formulas by sets; and proofs by subsets. It is degenerate in the sense that any formula is identified with its linear negation! Coherent spaces ([1]), hypercoherent spaces ([6]), finiteness spaces ([7]) remove (part of) this degeneracy by adding structure on top of the relational model. We follow the same approach:

Definition 4. *An* interface X *is given by a set $|X|$ (called the* state space*) and a predicate transformer P_X on $|X|$ (called the* specification*).*

The term "specification" comes from computer science, where a specification usually takes the form:

if the program is started in a state satisfying ϕ, it will terminate; and the final state will satisfy ψ.

Such a specification can be identified with the (monotonic) predicate transformer $\psi \mapsto$ "biggest such ϕ". This point of view is that of the **wp** calculus, introduced by Dijkstra ("**wp**" stands for "weakest precondition"). Note that the specification "goes backward in time": it associates to a set of final states (which we want to reach) a set of initial states (which guarantee that we will reach our goal).[1]

For a complete introduction to the field of predicate transformers in relation to specifications, we refer to [5].

In the coherence semantics, a "point" is a complete subgraph,[2] called a *clique*. Since the intuitions behind our objects are quite different, we change the terminology.

Definition 5. *Let X be an interface, a subset $x \subseteq |X|$ is called a* seed *of X if $x \subseteq P_X(x)$. We write $\mathcal{S}(X)$ for the collection of seeds of X.*

More traditional names for seeds are safety properties, or P-invariant properties: if some initial state is in x, no matter what, after each execution of a program satisfying specification P, the final state will still be in x. In other words, P maintains an invariant, namely "staying in x". In particular, there can be no program deadlock when starting from x.

The collection of cliques in the (hyper)coherent semantics forms a c.p.o.: the sup of any directed family exists. The collection of seeds in an interface satisfies the stronger property:

Lemma 1. *For any interface X, $\bigl(\mathcal{S}(X), \subseteq\bigr)$ is a complete sup-lattice.*

Proof. $\emptyset$ is trivially a seed; and by monotonicity of P, a union of seeds is a seed. $\square$

The fact that seeds are closed under union may seem counter-intuitive at first; but one possible interpretation is that we allow for non-deterministic data. For example, all denotational models of linear logic have an object for the booleans: its state space is $\{t, f\}$, and the cliques are always $\emptyset$, $\{t\}$ and $\{f\}$. The union of $\{t\}$ and $\{f\}$ is usually not itself a clique because "one cannot get both true and false". However, if one interprets union as a non-deterministic sum, then $\{t, f\}$ is a perfectly sensible set of data.

However, nothing guarantees that a seed is the unions of all its finite subseeds; a given seed needs not even contain any finite seed!. (The canonical example being $P_X(x) = X$, with X infinite.)

---

[1] In a previous version, interfaces also had to enjoy the property $P(\emptyset) = \emptyset$ and $P(|X|) = |X|$. This condition doesn't interact well with second order interpretation and has thus been dropped.

[2] The intuition is that a set of data is coherent iff it is pairwise coherent.

3   Constructions on Interfaces

A denotational model interprets formulas as objects in a category (and proofs as morphisms). We thus need to define all the constructions of linear logic at the level of interfaces. The most interesting cases are the linear negation and the tensor product (and the exponentials, but they will be treated in section 6).

Note that there will always be an "ambient" set A for predicates. We write $\overline{x}$ for the A-complement of x.

Let $X = (|X|, P_X)$ and $Y = (|Y|, P_Y)$ be two interfaces;

Definition 6. *The* dual *of X is defined as $(|X|, P_X^\perp)$ where $P_X^\perp(x) = \overline{P_X(\overline{x})}$. We write it $X^\perp$. An* antiseed *of X is a seed in $X^\perp$.*

In terms of specifications, $a \in P^\perp(x)$ means "*if the program is started in a, and if execution terminates, the final state will not be in x*". If P is concerned with **wp** calculus, then $P^\perp$ is more concerned with **wlp** calculus. (Weakest liberal precondition, also introduced by Dijkstra: we are not interested in termination, only the states which we will never reach.)

This operation of "negation" is the reason we do not ask for any properties on the predicate transformer. It respects neither continuity nor commutation properties! In many respects, this operation is not very well-behaved.

Definition 7. *The* tensor *of X and Y is the interface $(|X| \times |Y|, P_X \otimes P_Y)$ where $P_X \otimes P_Y(r)$ is the predicate transformer*

$$r \mapsto \bigcup_{x \times y \subseteq r} P_X(x) \times P_Y(y) .$$

We write it $X \otimes Y$.

$P_X \otimes P_Y$ is the most natural transformer to construct on $|X| \times |Y|$. It was used in [8] to model parallel execution of independent pieces of programs. The intuition is the following: a program satisfies $P_X \otimes P_Y$ if, when you start it in the pair $(a_i, b_i) \in P_X \otimes P_Y(r)$ of initial states, the two final states will be related through r. In particular, this means that execution is *synchronous*: both executions need to terminate.

Definition 8. *The* with *of X and Y is the interface $(|X| + |Y|, P_X \& P_Y)$ where $P_X \& P_Y(x, y) = \big(P_X(x), P_Y(y)\big).$[3] We write it $X \& Y$.*

This operation is not very interesting from the specification point of view: it is a kind of disjoint union.

Definition 9. *The other connectives are defined as usual:*

- $0 = (\emptyset, \mathbf{Id})$; $\top = 0^\perp$; $1 = (\{*\}, \mathbf{Id})$; $\perp = 1^\perp$;
- $X \oplus Y$ *(plus) is the interface $\big(X^\perp \& Y^\perp\big)^\perp$;*

---

[3] It uses implicitly the fact that $\mathcal{P}(|X| + |Y|) \simeq \mathcal{P}(|X|) \times \mathcal{P}(|Y|)$.

- $X \bindnasrepma Y$ *(par) is the interface* $\left(X^{\perp} \otimes Y^{\perp}\right)^{\perp}$;
- $X \multimap Y$ *is the interface* $X^{\perp} \bindnasrepma Y$.

We have:

Lemma 2. $\perp = 1; \top = 0$ *and* $X \oplus Y = X \mathbin{\&} Y$.

The proof is immediate. The first two equalities are satisfied in several of the denotational models of LL; the second one is a little less common. (For example, it is satisfied in finiteness spaces, but in no ...-coherence spaces.)

As an application of the definitions, let's massage the definition of $A \multimap B$ into something readable:

$(a, b) \in A \multimap B(r)$
$\quad \Leftrightarrow \; \{ \text{ definition } \}$
$(a, b) \in \left(A^{\perp} \bindnasrepma B\right)(r)$
$\quad \Leftrightarrow \; \{ \text{ definition, involutivity of } \_^{\perp} \}$
$(a, b) \in \left(A \otimes B^{\perp}\right)^{\perp}(r)$
$\quad \Leftrightarrow \; \{ \text{ definition of } \_^{\perp} \}$
$(a, b) \notin A \otimes B^{\perp}(\bar{r})$
$\quad \Leftrightarrow \; \{ \text{ definition of } \otimes \}$
$\neg\left((\exists x \times y \subseteq \bar{r}) \, a \in A(x) \wedge b \in B^{\perp}(y)\right)$
$\quad \Leftrightarrow \; \{ \text{ logic } \}$
$(\forall x \times y \subseteq \bar{r}) \, a \notin A(x) \vee b \notin B^{\perp}(y)$
$\quad \Leftrightarrow \; \{ \text{ logic } \}$
$(\forall x \times y \subseteq \bar{r}) \, a \in A(x) \Rightarrow b \in B(\bar{y})$
$\quad \Leftrightarrow \; \{ \text{ lemma: } x \times y \subseteq \bar{r} \text{ iff } \langle r \rangle x \subseteq \bar{y} \}$
$(\forall \langle r \rangle x \subseteq \bar{y}) \, a \in A(x) \Rightarrow b \in B(\bar{y})$
$\quad \Leftrightarrow \; \{ \text{ change of variable: } y \mapsto \bar{y} \}$
$(\forall \langle r \rangle x \subseteq y) \, a \in A(x) \Rightarrow b \in B(y)$.

From this, we derive:

Lemma 3. $(a, b) \in A \multimap B(r)$ *iff* $a \in A(x) \Rightarrow b \in B(\langle r \rangle x)$ *for all* $x \subseteq |X|$. *For any interface* X, $\mathbf{Id}_{|X|} \in \mathcal{S}(X \multimap X)$.

The shapes of images along $X \bindnasrepma Y$ are usually difficult to visualize, but we have the following on "rectangles":

Lemma 4. *Let* X *and* Y *be interfaces; then for all* $x \subseteq |X|$ *and* $y \subseteq |Y|$ *we have:* $P_X \otimes P_Y(x \times y) = P_X(x) \times P_Y(y) \subseteq P_X \bindnasrepma P_Y(x \times y)$.

Proof. That $P_X \otimes P_Y(x \times y) = P_X(x) \times P_Y(y)$ is straightforward.

Suppose now $a \in P_X(x)$ and $b \in P_Y(y)$, let's show that $(a, b) \in P_X \bindnasrepma P_Y(x \times y)$: suppose $x' \times y' \subseteq \overline{x \times y}$
$\quad \Rightarrow \; \{ \text{ claim (see below) } \}$
$x \subseteq \overline{x'} \vee y \subseteq \overline{y'}$
$\quad \Rightarrow \; \{ \text{ monotonicity } \}$
$a \in P_X(\overline{x'}) \vee b \in P_Y(\overline{y'})$.

Claim: $x' \times y' \subseteq \overline{x \times y} \Rightarrow x \subseteq \overline{x'} \vee y \subseteq \overline{y'}$

Proof of claim: suppose $\neg(x \subseteq \overline{x'}) \wedge \neg(y \subseteq \overline{y'})$
$$\Rightarrow x \cap x' \neq \emptyset \wedge y \cap y' \neq \emptyset$$
$$\Rightarrow x \times y \cap x' \times y' \neq \emptyset$$
$$\Rightarrow \neg(x' \times y' \subseteq \overline{x \times y}).$$ □

Furthermore, seeds in A and B are related to seeds in $A \otimes B$ and $A \,\mathfrak{N}\, B$ in the following way:

Lemma 5. *Let A and B be interfaces. We have:*

(i) if $x \in \mathcal{S}(A)$ and $y \in \mathcal{S}(B)$ then $x \times y \in \mathcal{S}(A \otimes B)$;
(ii) if $x \in \mathcal{S}(A)$ and $y \in \mathcal{S}(B)$ then $x \times y \in \mathcal{S}(A \,\mathfrak{N}\, B)$.

Proof. The first point is obvious; the second point is a direct consequence of Lemma 4. □

4   Linear Proofs and Seeds

The previous section gave a way to interpret any linear formula F by a interface F^*. (When no confusion arises, F^* is written F.) We now interpret linear proofs of F as subsets of the state space of F^*.[4] We refer to [1] or the abundant literature on the subject for the motivations governing those inference rules.

(1) If π is $\dfrac{}{\vdash 1}$ then $\pi^* = \{*\}$;

(2) if π is $\dfrac{}{\vdash \Gamma, \top}$ then $\pi^* = \emptyset$;

(3) if π is $\dfrac{\pi_1 \vdash \Gamma}{\vdash \Gamma, \bot}$ then $\pi^* = \{(\gamma, *) \mid \gamma \in \pi_1^*\}$;

(4) if π is $\dfrac{\pi_1 \vdash \Gamma, A, B}{\vdash \Gamma, A \,\mathfrak{N}\, B}$ then $\pi^* = \{(\gamma, (a, b)) \mid (\gamma, a, b) \in \pi_1^*\}$;

(5) if π is $\dfrac{\pi_1 \vdash \Gamma, A \qquad \pi_2 \vdash \Delta, B}{\vdash \Gamma, \Delta, A \otimes B}$

then $\pi^* = \pi_1^* \otimes \pi_2^* = \{(\gamma, \delta, (a, b)) \mid (\gamma, a) \in \pi_1^* \wedge (\delta, b) \in \pi_2^*\}$;

(6) if π is $\dfrac{\pi_1 \vdash \Gamma, A}{\vdash \Gamma, A \oplus B}$ then $\pi^* = \{(\gamma, (1, a)) \mid (\gamma, a) \in \pi_1^*\}$;

---

[4] Recall that a sequent $A_1, \ldots A_n$ is interpreted by $A_1 \,\mathfrak{N}\, \ldots A_n$ and the notation $\pi \vdash \Gamma$ means "π is a proof of sequent Γ".

(7) if π is $\dfrac{\pi_1 \vdash \Gamma, B}{\vdash \Gamma, A \oplus B}$ then $\pi^* = \{(\gamma, (2, b)) \mid (\gamma, b) \in \pi_1^*\}$;

(8) if π is $\dfrac{\pi_1 \vdash \Gamma, A \qquad \pi_2 \vdash \Gamma, B}{\vdash \Gamma, A \,\&\, B}$

then π^* is $\{(\gamma, (1, a)) \mid (\gamma, a) \in \pi_1^*\} \cup \{(\gamma, (2, b)) \mid (\gamma, b) \in \pi_2^*\}$;

(9) if π is $\dfrac{\pi_1 \vdash \Gamma, A \qquad \pi_2 \vdash \Delta, A^\perp}{\vdash \Gamma, \Delta}$

then $\pi^* = \{(\gamma, \delta) \mid (\exists a)\ (\gamma, a) \in \pi_1^* \wedge (\delta, a) \in \pi_2^*\}$.

This interpretation is correct in the following sense:

Proposition 1. *If π a proof of F, then π^* is a seed in F^*.*

Proof. By induction on the structure of π: we will check that seeds propagate through the above constructions. It is mostly trivial computation, except for two interesting cases:

(5): suppose that π_1 is a seed in $\Gamma \,\invamp\, A$ and that π_2 is a seed in $\Delta \,\invamp\, B$. We need to show that $\pi_1 \otimes \pi_2 = \{(\gamma, \delta, (a, b)) \mid (\gamma, a) \in \pi_1 \wedge (\delta, b) \in \pi_2\}$ is a seed in the sequent $\Gamma \,\invamp\, \Delta \,\invamp\, (A \otimes B)$.

Let $(\gamma, \delta, (a, b)) \in \pi_1 \otimes \pi_2$

$\Leftrightarrow$

$(\gamma, a) \in \pi_1$ and $(\delta, b) \in \pi_2$

$\Rightarrow$ $\{\pi_1$ and π_2 are seeds in Γ, A and $\Delta, B\}$

$(\gamma, a) \in \Gamma, A(\pi_1)$ and $(\delta, \pi_2) \in \Delta, B(\pi_2)$.

By contradiction, let $(\gamma, \delta, (a, b)) \notin \Gamma, \Delta, A \otimes B(\pi_1 \otimes \pi_2)$

$\Rightarrow$

$(\gamma, \delta, (a, b)) \in \Gamma^\perp \otimes \Delta^\perp \otimes (A \otimes B)^\perp \overline{(\pi_1 \otimes \pi_2)}$

$\Rightarrow$ $\{$ for some $u \times v \times r \subseteq \overline{\pi_1 \otimes \pi_2}$: $\}$

$\gamma \in \Gamma^\perp(u) \wedge \delta \in \Delta^\perp(v) \wedge \underbrace{(a, b) \in (A \otimes B)^\perp(r)}$

$\Rightarrow$

$\ldots \wedge \Big((\forall x \times y \subseteq \overline{r})\ a \in A^\perp(\overline{x}) \vee b \in B^\perp(\overline{y})\Big).$

In particular, define $x = \langle \pi_1 \rangle u$ and $y = \langle \pi_2 \rangle v$; it is easy to show that $x \times y \subseteq \overline{r}$, so that we have $a \in A^\perp(\overline{x})$ or $b \in B^\perp(\overline{y})$.

Suppose $a \in A^\perp(\overline{x})$: we have $\gamma \in \Gamma^\perp(u)$ and $u \times \overline{x} \subseteq \overline{\pi_1}$ (easy lemma); so by definition, $(\gamma, a) \in \Gamma^\perp \otimes A^\perp(\overline{\pi_1})$, i.e. $(\gamma, a) \notin \Gamma, A(\pi_1)$! This is a contradiction.

Similarly, one can derive a contradiction from $b \in B^\perp(\overline{y})$.

This finishes the proof that $\pi_1 \otimes \pi_2$ is a seed of $\Gamma, \Delta, A \otimes B$.

(9): let π_1 be a seed in $\Gamma, A = \Gamma^\perp \multimap A$ and π_2 a seed in $\Delta, A^\perp$, i.e. $\pi_2^{\sim}$ is a seed in $A \multimap \Delta$. Let's show that $\pi = \{(\gamma, \delta) \mid (\exists a)\ (\gamma, a) \in \pi_1 \wedge (\delta, a) \in \pi_2\} = \pi_2^{\sim} \cdot \pi_1$ is a seed in Γ, Δ.

Suppose $(\gamma, \delta) \in \pi_2^{\sim} \cdot \pi_1$, i.e. that $(\gamma, a) \in \pi_1$ and $(a, \delta) \in \pi_2^{\sim}$ for some a. We will prove that (γ, δ) is in $\Gamma, \Delta(\pi) = \Gamma^\perp \multimap \Delta(\pi)$. According to Lemma 3, we need to show that if $\gamma \in \Gamma^\perp(u)$ then $\delta \in \Delta(\langle \pi \rangle u)$.

Let $\gamma \in \Gamma^\perp(u)$
$\quad \Rightarrow \quad \{ (\gamma, a) \in \pi_1 \subseteq \Gamma^\perp \multimap A(\pi_1) \}$
$a \in A(\langle \pi_1 \rangle u)$
$\quad \Rightarrow \quad \{ (a, \delta) \in \pi_2^\sim \subseteq A \multimap \Delta(\pi_2^\sim) \}$
$\delta \in \Delta(\langle \pi_2^\sim \rangle \langle \pi_1 \rangle u)$
$\quad \Leftrightarrow$
$\delta \in \Delta(\langle \pi \rangle u).$ $\square$

5   Morphisms, Categorical Structure

To complete the formal definition of a category of interfaces, we need to define morphisms between interfaces. This is done in the usual way:

Definition 10. *A linear arrow from X to Y is a seed in $X \multimap Y$.*

Here is a nicer characterization of linear arrows from X to Y:

Lemma 6. $r \in \mathcal{S}(X \multimap Y)$ *iff* $\langle r \rangle \big(P_X(x) \big) \subseteq P_Y \big(\langle r \rangle(x) \big)$ *for all* $x \subseteq |X|$.

Proof. Suppose r is a seed in $X \multimap Y$, let $b \in \langle r \rangle P_X(x)$
$\quad \Rightarrow$
there is some a s.t. $(a, b) \in r$ and $a \in P_X(x)$
$\quad \Rightarrow \quad \{ r$ is a seed in $X \multimap Y \}$
$(a, b) \in P_X \multimap P_Y(r)$
$\quad \Rightarrow \quad \{$ definition of $\multimap \}$
$b \in P_Y(\langle r \rangle x).$

Conversely, suppose $\langle r \rangle P_X(x) \subseteq P_Y \langle r \rangle(x)$; let $(a, b) \in r$, and $a \in P_X(x)$. We have $b \in \langle r \rangle P_X(x)$, and by hypothesis, $b \in P_Y(\langle r \rangle x)$. $\square$

Lemma 7. *If* $r \in \mathcal{S}(X \multimap Y)$ *and* $r' \in \mathcal{S}(Y \multimap Z)$ *then* $r' \cdot r \in \mathcal{S}(X \multimap Z)$.

Proof. This is the essence of point *(9)* from Proposition 1; or a simple corollary to Lemma 6. $\square$

Taken together with Lemma 3, this makes interfaces into a category:

Definition 11. *We write* **Int** *for the category with interfaces as objects and linear arrows as morphisms.*

This category is an enrichment of the usual category **Rel**. The construction can be summarized in the following way:

Lemma 8. Int *is obtained by lifting* **Rel** *through the following specification structure ([9]):*

- *if X is a set,* $\mathrm{Pr}_X \equiv \mathcal{P}(X) \to \mathcal{P}(X)$;
- *if $r \subseteq X \times Y$, $P \in \mathrm{Pr}_X$ and $Q \in \mathrm{Pr}_Y$, then $P\{r\}Q$ iff $\langle r \rangle \cdot P \subseteq Q \cdot \langle r \rangle$.*

Let's now turn our attention to the structure of this category:

Lemma 9. *In* **Int**, $\top$ *is terminal and* $\&$ *is the cartesian product.*

Proof. This is immediate. □

Lemma 10. $\_^{\perp}$ *is an involutive contravariant functor.*

Proof. Involutivity is trivial; contravariance is only slightly trickier:
r is a seed in $A \multimap B$
$\quad \Leftrightarrow \quad \{\text{ Lemma 6 }\}$
$\forall x \; \langle r \rangle A(x) \subseteq B \langle r \rangle x$
$\quad \Leftrightarrow$
$\forall x \; \overline{B \langle r \rangle x} \subseteq \overline{\langle r \rangle A(x)}$
$\quad \Leftrightarrow \quad \{\text{ lemma: } y \subseteq \overline{\langle r \rangle x} \text{ iff } \langle r^{\sim} \rangle y \subseteq \overline{x} \}$
$\forall x \; \langle r^{\sim} \rangle \overline{B \langle r \rangle x} \subseteq \overline{A(x)}$
$\quad \Rightarrow \quad \{\text{ in particular, for } x \text{ of the form } \overline{\langle r^{\sim} \rangle x}; \text{ we have } \overline{x} \subseteq \langle r \rangle \overline{\langle r^{\sim} \rangle x} \text{ (lemma) }\}$
$\forall x \; \langle r^{\sim} \rangle B^{\perp}(x) \subseteq A^{\perp}\big(\langle r^{\sim} \rangle x\big)$
i.e. $r^{\sim}$ is a seed in $B^{\perp} \multimap A^{\perp}$. The action of $\_^{\perp}$ on morphisms is just $\_^{\sim}$. □

Corollary 1. **Int** *is autodual through* $\_^{\perp}$; $\mathbf{0}$ *is initial; and* $\oplus$ *is the coproduct.*

It is now easy to see that linear arrows transform seeds into seeds, and, in the other direction, antiseeds into antiseeds:

Proposition 2. *Suppose* r *is a linear arrow from* X *to* Y:

(i) $\langle r \rangle$ *is a sup-lattice morphism from* $\mathcal{S}(X)$ *to* $\mathcal{S}(Y)$;
(ii) $\langle r^{\sim} \rangle$ *is a sup-lattice morphism from* $\mathcal{S}(Y^{\perp})$ *to* $\mathcal{S}(X^{\perp})$.

Proof. Let $r \in \mathcal{S}(X \multimap Y)$ and $x \subseteq X(x)$; we want to show that $\langle r \rangle x \subseteq Y(\langle r \rangle x)$. Let $b \in \langle r \rangle x$
$\quad \Leftrightarrow$
$(\exists a) \; (a, b) \in r \wedge a \in x$
$\quad \Rightarrow \quad \{ r \text{ is a seed in } X \multimap Y \}$
$(\exists a) \; (a, b) \in X \multimap Y(r) \wedge a \in x$
$\quad \Rightarrow \quad \{ \text{ definition of } X \multimap Y \text{ with the fact that } \langle r \rangle x \subseteq \langle r \rangle x \}$
$b \in Y(\langle r \rangle x)$.
Showing that $\langle r \rangle$ commutes with sups is immediate: it commutes with arbitrary unions, even when the argument is not a seed.
The second point follows because $r^{\sim} \in \mathcal{S}(Y^{\perp} \multimap X^{\perp})$. □

Lemma 11. $\otimes$ *[$\mathfrak{N}$] is a categorical tensor product with neutral element* $\mathbf{1}$ *[$\perp$].*

Proof. We need to show the bifunctoriality of $\otimes$. This was actually proved in the previous section (Proposition 1, point *(5)*). The bifunctoriality of $\mathfrak{N}$ follows by duality; and the rest is immediate. □

As a summary of this whole section, we have:

Proposition 3. **Int** *is a* $*$-*autonomous category. (In particular,* **Int** *is symmetric monoidal closed.)*

Proof. This amounts to checking trivial equalities, in particular, that the following diagram commutes: (where d is the natural isomorphism $X \simeq X^{\perp\perp}$)

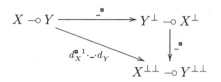

It is immediate because $d = \mathbf{Id}$ and $\_^{\perp} = \_^{\sim}$.   □

6   Exponentials

The category **Int** is thus a denotational model for multiplicative additive linear logic. Let's now add the exponentials $!X$ and $?X$.

Unsurprisingly, we will use finite multisets; here are the necessary definitions and notations:

Definition 12. *Let S be a set;*

- *if $(s_i)_{i \in I}$ and $(t_j)_{j \in J}$ are finite families on S, say $(s_i) \simeq (t_j)$ iff there is a bijection σ from I to J such that $s_i = t_{\sigma(i)}$ for all i in I.*
- *A finite multiset over S is an equivalence class of $\simeq$. We write $[s_i]$ for the equivalence class containing (s_i).*
- *$\mathcal{M}_f(S)$ is the collection of finite multisets over S.*
- *Concatenation of finite families[5] can be lifted to multisets; it is written $+$.*
- *If x and y are two subsets of S, we write $x * y$ for the set $\{[a, b] \mid a \in x \land b \in y\}$. Its indexed version is written $\prod_{i \in I} x_i$; it is a kind of commutative product.*
- *If U and V are two subsets of $\mathcal{M}_f(A)$, the set $\{u + v \mid u \in U \land v \in V\}$ is written $U * V$ (same symbol, but no confusion arises).*

Definition 13. *For $X = (|X|, P)$, define $!X = (\mathcal{M}_f(|X|), !P)$ where*

$$[a_1, \ldots a_n] \in !P(U) \quad \Leftrightarrow \quad (\exists (x_i)_{1 \leq i \leq n}) \prod_i x_i \subseteq U \land (\forall i = 1, \ldots n)\, a_i \in P(x_i)$$

Let $?X = \left(!(X^{\perp})\right)^{\perp}$.

Recall that a multiset $[a_i]$ is in $\prod x_i$ iff there is a bijection σ s.t. $\forall i, a_i \in x_{\sigma(i)}$.

A useful intuition is that $[a_1, \ldots] \in !P(U)$ iff $[a_1, \ldots]$ is in a "weak infinite tensor" $\bigoplus_n X^{\otimes n}(U)$. In terms of specifications and programs, it suggests multi-threading: for an initial state $[a_1, \ldots a_n]$, start n occurrences of the program in the states $a_1, \ldots a_n$; the final state is nothing but the multiset of all the n final states.[6] The "weak" part means that we forget the link between a particular final state and a particular initial state.

---

[5] Defined on the disjoint sum of the different index sets.
[6] The interpretation of $!$, like that of $\otimes$ is a synchronous operation.

Note that this is a "non-uniform" model in the sense that the web of $!X$ contains all finite multisets, not just those whose underlying set is a seed. It is thus closer to non-uniform (hyper)coherence semantics (see [10] or [11]) than to the traditional (hyper)coherence semantics.

Let's prove a simple lemma about the exponentials:

Lemma 12. *Suppose $U \subseteq \mathcal{M}_f(|A|)$:*

(i) $[a] \in !A(U)$ iff there is some x "included" in U (i.e. $\forall a \in x \; [a] \in U$) s.t. $a \in A(x)$;

(ii) $l + l' \in !A(U)$ iff there are $V * V' \subseteq U$ s.t. $l \in !A(V)$ and $l' \in !A(V')$;

(iii) $[a] \in ?A(U)$ iff for all $\overline{x}$ "included" in $\overline{U}$, $a \in A(x)$;

(iv) $l + l' \in ?A(U)$ iff for all $\overline{V} * \overline{V'} \subseteq \overline{U}$, $l \in ?A(V)$ or $l' \in ?A(V')$.

Proof. The first point is immediate and the second is left as an exercise. The third and last point are consequences of the definition of ? in terms of !. □

Define now the interpretation of proofs with exponentials:

(10) if π is $\quad \dfrac{\pi_1 \vdash \Gamma, A}{\vdash \Gamma, ?A} \quad$ then $\pi^* = \big\{ (\gamma, [a]) \mid (\gamma, a) \in \pi_1^* \big\}$;

(11) if π is $\quad \dfrac{\pi_1 \vdash \Gamma}{\vdash \Gamma, ?A} \quad$ then $\pi^* = \big\{ (\gamma, []) \mid \gamma \in \pi_1^* \big\}$;

(12) if π is $\quad \dfrac{\pi_1 \vdash \Gamma, ?A, ?A}{\vdash \Gamma, ?A} \quad$ then $\pi^* = \big\{ (\gamma, l + l') \mid (\gamma, l, l') \in \pi_1^* \big\}$;

(13) if π is $\quad \dfrac{\pi_1 \vdash ?\Gamma, A}{\vdash ?\Gamma, !A}$

then we define $(\gamma_1, \ldots \gamma_l, [a_1 \ldots a_n]) \in \pi^*$ if for each $j = 1, \ldots l$, there is a partition $\gamma_j = \sum_{1 \leq i \leq n} \gamma_j^i$ and the following holds: for each $i = 1, \ldots n$, $(\gamma_1^i, \ldots \gamma_l^i, a_i) \in \pi_1^*$.

Proposition 4. *If π a proof of $\vdash \Gamma$, then π^* is a seed of Γ.*

Proof. Points *(10)* and *(11)* are immediate.

(12): suppose π_1 is a seed $\Gamma, ?A, ?A$ and let $(\gamma, l + l')$ be an element of π. By contradiction, suppose that $(\gamma, l + l') \notin \Gamma, ?A(\pi)$

$\Leftrightarrow$

$(\gamma, l + l') \in \Gamma^\perp \otimes !A^\perp(\overline{\pi})$

$\Leftrightarrow$ { for some $u \times U \subseteq \overline{\pi}$ }

$\gamma \in \Gamma^\perp(u) \wedge l + l' \in !A^\perp(U)$

$\Leftrightarrow$ { Lemma 12 }

$\gamma \in \Gamma^\perp(u) \wedge (\exists V * V' \subseteq U) \; l \in !A^\perp(V) \wedge l' \in !A^\perp(V')$

$\Rightarrow$ { lemma: $u \times V \times V' \subseteq \overline{\pi_1}$ }

$\gamma \in \Gamma^\perp(u) \wedge l \in \ !A^\perp(V) \wedge l' \in \ !A^\perp(V')$

$\Rightarrow$

$(\gamma, l, l') \in \Gamma^\perp \otimes \ !A^\perp \otimes \ !A^\perp(\overline{\pi_1})$

$\Leftrightarrow$

$(\gamma, l, l') \notin \Gamma, ?A, ?A(\pi_1)$, which contradicts the fact that π_1 is a seed in $\Gamma, ?A, ?A$.

(13): suppose that Γ contains only one formula B. The general case will follow from a lemma proved below (Lemma 13). Suppose that π_1 is a seed in $?B, A$; let $(l, [a_1, \ldots a_n])$ be in π, *i.e.* $(l_i, a_i) \in \pi_1$ for $i = 1, \ldots n$, for some partition $(l_1, \ldots l_n)$ of l.

Suppose by contradiction that $(l, [a_1 \ldots a_n]) \notin \ ?B, !A(\pi)$

$\Leftrightarrow$

$(l, [a_1, \ldots a_n]) \in \ !B^\perp \otimes \ ?A^\perp(\overline{\pi})$

$\quad \Leftrightarrow \ \{$ for some $U \times V \subseteq \overline{\pi} \ \}$

$l \in \ !B^\perp(U) \wedge [a_1, \ldots a_n] \in \ ?A^\perp(V)$

$\quad \Rightarrow \ \{$ definition of $?A \ \}$

$l \in \ !B^\perp(U) \wedge \left(\left(\forall(x_i) \right) \prod x_i \subseteq \overline{V} \Rightarrow (\exists i) \ a_i \in A(\overline{x_i}) \right)$

$\quad \Leftrightarrow \ \{$ Lemma 12 for l: for some (U_i) s.t. $\prod_i U_i \subseteq U \ \}$

$(\forall i) \ l_i \in \ !B^\perp(U_i) \wedge \left(\left(\forall(x_i)_i \right) \prod_i x_i \subseteq \overline{V} \Rightarrow (\exists i) \ \ldots \right.$

$\quad \Rightarrow \ \{$ define $x_i = \langle \pi_1 \rangle U_i$; lemma: $\prod_i x_i \subseteq \overline{V} \ \}$

$\left((\forall i) \ l_i \in \ !B^\perp(U_i) \right) \wedge \left((\exists i) \ a_i \in A(\overline{x_i}) \right)$

$\quad \Rightarrow \ \{$ lemma: $U_i \times x_i \subseteq \overline{\pi_1} \ \}$

$(\exists i) \ (\exists U_i \times x_i \subseteq \overline{\pi_1}) \ l_i \in \ !B^\perp(U_i) \wedge a_i \in A(\overline{x_i})$

$\Leftrightarrow$

$(l_i, a_i) \in \ !B^\perp \otimes A^\perp(\overline{\pi_1})$

$\Leftrightarrow$

$(l_i, a_i) \notin \ ?B, A(\pi_1)$, which contradicts the fact that π_1 is a seed in $?B, A$.   □

Lemma 13. *For all interfaces X and Y, we have $!(X \ \& \ Y) = \ !X \otimes \ !Y$.*

Proof. The state spaces are isomorphic via $\mathcal{M}_f(|X|+|Y|) \simeq \mathcal{M}_f(|X|) \times \mathcal{M}_f(|Y|)$. We will use this transparently, for example $l_X + l_Y \in R$ iff $(l_X, l_Y) \in R$. This is possible because the sets are disjoint: we can always split a multiset in $x * y$ into two multisets in x and y. (In other words: if $x \cap y = \emptyset$ then $x * y \simeq x \times y$.)

Notice also that $(1, a) \in X \ \& \ Y(x, y) \Leftrightarrow a \in X(x)$ so that when considering a particular element of $X + Y(x, y)$, only one part of the argument (x, y) is really important; the other can be dropped (or replaced with $\emptyset$).

$\subseteq$: suppose $[a_1, \ldots a_n] + [b_1, \ldots b_m] \in \ !(X \ \& \ Y)(R)$

$\quad \Leftrightarrow \ \{$ for some $(x_i)_{i=1 \ldots n}$ and $(y_j)_{j=1 \ldots m} \ \}$

$\prod_i x_i * \prod_j y_j \subseteq R \wedge (\forall i) \ a_i \in X(x_i) \wedge (\forall j) b_j \in Y(y_j)$

$\quad \Rightarrow \ \{$ define $U' = \prod_i x_i$ and $V' = \prod_j y_j \ \}$

$(\exists U' \times V' \subseteq R) \ [a_i] \in \ !X(U') \wedge [b_j] \in \ !Y(V')$

$\Leftrightarrow$

$([a_1, \ldots a_n], [b_1, \ldots b_m]) \in \ !X \otimes \ !Y(R)$.

$\supseteq$: suppose $([a_1, \ldots a_n], [b_1, \ldots b_m]) \in \ !X \otimes \ !Y(R)$

$\Leftrightarrow$ { for some $U' \times V' \subseteq R$ }
$[a_1, \ldots a_n] \in !X(U') \wedge [b_1, \ldots b_m] \in !Y(V')$
$\Leftrightarrow$ { for some (x_i) s.t. $\bigcap x_i \subseteq U'$ and (y_j) s.t. $\bigcap y_j \subseteq V'$ }
$(\forall i)\ a_i \in X(x_i) \wedge (\forall j)\ b_j \in Y(y_j)$
$\Rightarrow$ { $\bigcap_i x_i \times \bigcap_j y_j \subseteq U \times V$ and thus $\bigcap_i x_i * \bigcap_j y_j \subseteq R$ }
$[a_1, \ldots a_n] + [b_1, \ldots b_m] \in !(X \& Y)(R).$ $\qquad\qquad\square$

This allows us to transform any sequent $?\Gamma = ?B_1 \,\invamp\, \ldots ?B_n$ into $?(B_1 \oplus \ldots B_n)$, and thus, formally ends the proof of Proposition 4 point *(13)*.

7   Linear Interfaces and Linear Seeds

What is the structure of those interfaces that come from a linear formula? The answer is unfortunately trivial:

Proposition 5. *If F is a linear formula, then $P_F = \mathbf{Id}_{\mathcal{P}|F|}$.*

Proof. Immediate induction. Let's treat the case of the exponentials: suppose $F(x) = x$; suppose moreover that $[a_1, \ldots a_n] \in U$
$\Rightarrow$
$a_i \in F(\{a_i\})$ for all i and $\bigcap\{a_i\} = \{[a_1, \ldots a_n]\} \subseteq U$
$\Rightarrow$
$[a_1, \ldots a_n] \in !F(U)$
Similarly, suppose $[a_1, \ldots a_n] \in !F(U)$
$\Rightarrow$
each $a_i \in F(x_i) = x_i$ for some (x_i) s.t. $\bigcap x_i \subseteq U$
$\Rightarrow$ { $[a_1, \ldots a_n] \in \bigcap x_i$ }
$[a_1, \ldots a_n] \in U.$ $\qquad\qquad\square$

In particular, every subset of $|F|$ is a clique and an anticlique: the situation is thus quite similar to the purely relational model. In the presence of atoms however, interfaces become much more interesting.

Adding atoms is sound because the proof of Proposition 1 doesn't rely on the particular properties of interfaces. Note that we need to introduce a general axiom rule and its interpretation:

(14) if π is $\dfrac{}{\vdash X, X^\perp}$ then $\pi^* = \mathbf{Id}_{|X|} = \{(a, a) \mid a \in |X|\}.$

This is correct in the sense that π^* is always a clique in $X \,\invamp\, X^\perp$.

With such atoms, the structure of linear interfaces gets non trivial.[7] For example, let's consider the following atom $X = (\{-, +\}, P)$ defined by:

- $P(\emptyset) = \emptyset$ and $P(|X|) = |X|$;
- $P(\{+\}) = \{-\}$ and $P(\{-\}) = \{+\}$.

---

[7] We can extend this to a model for $\square^1$ logic, and even to full second order, see [12].

This is the simplest example of an interesting interface, and corresponds to a "switch" specification. (Interpret − as "off" and + as "on".)

Lemma 14. *if P is the above specification:*

(i) $P^\perp = P$;
(ii) $P \cdot P = \mathbf{Id}$;
(iii) $\mathcal{S}(X) = \{\emptyset, \{+, -\}\}$;
(iv) $\{(+, -), (-, +)\} \in \mathcal{S}(X \otimes X)$.

Proof. This is just trivial computation...   □

Point *(iv)* shows in particular that a seed in $X \otimes Y$ needs not contain a product of seeds in X and Y. (Compare with Lemma 5.)

The hierarchy generated from this single interface is however still relatively simple: call a specification *deterministic* if it commutes with non-empty unions and intersections.

Lemma 15. *Let F be any specification constructed from the above P and the linear connectives. Then F is deterministic. Moreover, F is of the form $\langle f \rangle$ where f is an obvious bijection on the state space of F.*[8]

A less trivial (in the sense that it is not deterministic) specification is the following: if X is a set, $\mathsf{magic}_X(x) = X$. In terms of programming, the use of the magic command allows to reach any predicate, even the empty one!

Lemma 16. $\mathbf{Id}_{|X|} \subsetneq \mathsf{magic}_X \multimap \mathsf{magic}_X(\mathbf{Id}_{|X|})$ *if* $X \neq \emptyset$.

Thus we cannot strengthen the definition of seeds to read "$x = P(x)$" without imposing further constraints on our specifications. It is still an open question to find a nice class of predicate transformers for which it would be possible. (However, considerations about second order seem to indicate that strengthening the definition of seeds in such a way is not a good idea.)

In the case with atoms, because the structure of seeds (sup-lattice) is quite different from the structure of cliques in the ...-coherent model (domain), it is difficult to relate seeds and cliques. In particular, a seed needs not be a clique (since the union of arbitrary cliques is not necessarily a clique); and a clique needs not be a seed (since a subset of a seed is not necessarily a seed).

Conclusion

One aspect which was not really mentioned here is the fact that linear arrows from A to B are equivalent to the notion of *forward data refinement* (Lemma 6) from the refinement calculus. In particular, a linear proof of $A \multimap B$ is a proof that specification B *implements* specification A. It would interesting to see if any

---

[8] Where $\langle f \rangle(x) = \{f(a) \mid a \in x\}$.

application to the refinement calculus could be derived from this work. In the same direction, trying to make sense of the notions of *backward data refinement*, or of *general data refinement* in terms of linear logic could prove interesting.[9]

The fact that this model is degenerate in the propositional case is disappointing, but degeneracy disappear when we consider Π^1 logic, and *a fortiori* when we consider full second-order (see [12]). The point of extending this propositional model to Π^1 is to remove the dependency on specific valuations for the atoms present in a formula.

One the interesting consequences of this work is that a a proof of a formula F gives a guarantee that the system specified by the formula F can avoid deadlocks seems to point toward other fields like process calculi and similar models for "real" computations. This direction is currently being pursued together with the following link with the differential lambda-calculus ([2]): one property of this model which doesn't reflect any logical property is the following; we have a natural transformation $A \multimap !A$ called *co-dereliction*, which has a natural interpretation in terms of differential operators on formulas (see [3]). Note that such a natural transformation forbids any kind of completeness theorem, at least as far as "pure" linear logic is concerned.

References

1. Girard, J.Y.: Linear logic. Theoretical Computer Science **50** (1987)
2. Ehrhard, T., Regnier, L.: The differential lambda calculus. Theoretical Computer Science **309** (2003) 1–41
3. Ehrhard, T., Regnier, L.: Differential interaction nets. unpublished note (2004)
4. Gardiner, P.H.B., Martin, C.E., de Moor, O.: An algebraic construction of predicate transformers. Science of Computer Programming **22** (1994) 21–44
5. Back, R.J., von Wright, J.: Refinement Calculus: a systematic introduction. Graduate texts in computer science. Springer-Verlag, New York (1998)
6. Ehrhard, T.: Hypercoherences: a strongly stable model of linear logic. Mathematical Structures in Computer Science **3** (1993) 365–385
7. Ehrhard, T.: Finiteness spaces. to appear in Mathematical Structures in Computer Science (2004)
8. Back, R.J., von Wright, J.: Product in the refinement calculus. Technical Report 235, Turku Center for Computer Science (1999)
9. Abramsky, S., Gay, S.J., Nagarajan, R.: A specification structure for deadlock-freedom of synchronous processes. Theoretical Computer Science **222** (1999) 1–53
10. Bucciarelli, A., Ehrhard, T.: On phase semantics and denotational semantics: the exponentials. Annals of Pure and Applied Logic **109** (2001) 205–241
11. Boudes, P.: Non-uniform hypercoherences. In Blute, R., Selinger, P., eds.: Electronic Notes in Theoretical Computer Science. Volume 69., Elsevier (2003)
12. Hyvernat, P.: Predicate transformers and linear logic: second order. unpublished note (2004)

---

[9] A data refinement from specification F to specification G is a predicate transformer P s.t. $P \cdot F \subseteq G \cdot P$; a forward [resp. backward] data refinement is a data refinement which commutes with arbitrary unions [resp. arbitrary intersections].

Structures for Multiplicative Cyclic Linear Logic: Deepness vs Cyclicity*

Pietro Di Gianantonio

dipartimento di Matematica e Informatica, Università di Udine
via delle Scienze 206 I-33100, Udine – Italy
`digianantonio@dimi.uniud.it`

Abstract. The aim of this work is to give an alternative presentation for the multiplicative fragment of Yetter's cyclic linear logic. The new presentation is inspired by the calculus of structures, and has the interesting feature of avoiding the cyclic rule. The main point in this work is to show how cyclicity can be substituted by deepness, i.e. the possibility of applying an inference rule at any point of a formula. We finally derive, through a new proof technique, the cut elimination property of the calculus.

Keywords: proof theory, linear logic, cyclic linear logic, calculus of structures, Lambek calculus.

1   Introduction

A non-commutative version of linear logic appeared as soon as linear logic was published [1]; in 1987 Jean Yves Girard, in a series of lectures, suggested a version of linear logic containing non-commutative connectives. This logic was later fully developed by Yetter [2] and named Cyclic Linear Logic (CyLL). This immediate interest for a non-commutative logic can be explained by the fact that linear logic puts great emphasis on structural rules, and so it was natural to consider the commutativity rule and check whether it is possible to define a proof system without it. Looking at the subject from a semantic point of view, non-commutative connectives are present in the "logic of quantum mechanics" [3] a logic aiming to model empirical verification and containing a non commutative connective "and then" (&). In this logic, the formula $A\&B$ is interpreted as "we have verified A and then we have verified B". Non-commutative connectives are present also in Lambek's syntactic calculus [4], a calculus modeling linguistic constructors. Both these calculi are strictly related with cyclic linear logic, in particular, Lambek's calculus can be seen as a fragment of the multiplicative cyclic linear logic. Later on the cyclic linear logic has been extended by the introduction of a commutative version of the multiplicative connectives [5, 6], leading to the definition of non-commutative logic (NL), a logic that encompasses both cyclic linear logic and standard linear logic.

---

* Supported by Italian MIUR Cofin "Protocollo", and EEC Working Group "Types".

J. Marcinkowski and A. Tarlecki (Eds.): CSL 2004, LNCS 3210, pp. 130–144, 2004.

The multiplicative fragment of cyclic linear logic can be simply obtained by taking the multiplicative LL, and substituting the structural rule of Exchange:

$$\frac{\Delta}{\Gamma}$$ the list Γ is a permutation of the list Δ

with the Cycling rule:

$$\frac{\Delta, \Gamma}{\Gamma, \Delta} \text{ Cycling}$$

Cycling rule is considered so crucial that the whole logic is named after it. However this rule still misses a natural explanation, and it is often explained in terms of necessity:

> The reader should note that in terms of the semantics we will develop, the seemingly unnatural Cycling rule is forced by having a system with a single negation ... [2]

More recently Guglielmi proposed the calculus of structures (CoS) [7] as a calculus for defining logics, alternative to sequent calculus and whose main feature is deep inference, that is the possibility of applying inference rules arbitrarily deep inside formulae. This greater liberty in applying inference rules can be used to treat logics whose formalization in the sequent calculus is not completely satisfactory (as modal logic [8]), or to ensure structural properties for the derivation, properties that are not present in sequent calculus derivations (as locality [9, 10]). Moreover there are examples of logics that cannot be treated at all in sequent calculus [7].

In this work, we present cyclic linear logic using the CoS. We show that the cycling rule can be avoided in the CoS formulations, namely, we show that if one takes the formulation of multiplicative linear logic in the CoS and then simply drops the commutative rules for par and tensor, one immediately obtains a formulation of cyclic linear logic (with no cycling rule present). This fact gives an explanation for the cycling rule, i.e. the cycling rule is a rule that recover the lack of deep inference in the sequent calculus. More in detail, deep inference can be, informally, described as follows:

> for any formulae A, B and positive context S, from $A \Rightarrow B$ and $S[A]$ derive $S[B]$.

Deep inference as a rule is normally not present in proof systems but one can argue, after having defined positive contexts, that deep inference has to be an admissible rule. In our work, we show that, with respect to admissibility, the cycling rule and deep inference are equivalent, i.e., for any proof system containing all the remaining rules of cyclic linear logic, the cycling rule is admissible if and only if deep inference is admissible. Therefore, the CoS gives a method for substituting, in Yetter's words, a *seemingly unnatural rule*: cycling, with a natural one: deep inference. The proof transformation between the two systems does not add complexity cost: given a CyLLproof in sequents calculus, it is possible to define a corresponding proof in the CoS containing the same order of applied rules.

As a further result, in this work we present a new technique for proving cut elimination that can be usefully employed in the treatment of other logics inside the CoS. The CoS is a very recent formalism and so there is the need of developing a bunch of techniques for obtaining meta-theoretical results. Our original proof of cut elimination is a step in this direction.

The article is organized as follows: Section 2 gives a short explanation of the calculus of structures, Section 3 formalizes the multiplicative linear logic in the CoS, Section 4 presents cyclic linear logic together with a proof of cut elimination, Section 5 gives a short account for possible future works.

2   Calculus of Structures

The CoS is characterized by two main features, the possibility of applying inference rules at any point in a formula (deep inference) and the idea to consider formulae up to an equivalence relation equating formulae provable equivalent by some elementary arguments. The equivalence classes of this relation are called *structures*. In this article we retain the first feature of the CoS but drop the second one, i.e. we do not use structures and work directly on formulae. The main reason for this choice is the fact that the cut elimination proof, given in following, needs to consider a proof system where also formulae belonging to the same equivalence class are kept distinct. As a consequence, we do not present here the true CoS but a slightly different formalism using a different syntax and having a different treatment of the equality. The reader not familiar with the CoS will gain a more direct presentation: we use only formulae and we avoid the syntactic overhead caused by structures. The reader familiar with the CoS should have no problem in relating the two presentations.

Before introducing our formalism, we want to present an alternative view of the sequent calculus. In constructing a derivation for a formula A, in sequent calculus, one reduces the derivability of a formula A to the derivability of a set of sequents. That is, in the intermediate steps of the construction of a bottom up derivation of A, one reduces the problem of deriving A to the problem of deriving a set of sequents. The intuitive meaning of this set of sequents (let it be $\vdash B_{1,1}, \ldots, B_{1,n_1} \quad \vdash B_{2,1}, \ldots, B_{2,n_2} \quad \ldots \quad \vdash B_{m,1}, \ldots, B_{m,n_m}$) is the formula $(B_{1,1} \otimes \ldots \otimes B_{1,n_1}) \otimes (B_{2,1} \otimes \ldots \otimes B_{2,n_2}) \otimes \ldots \otimes (B_{m,1} \otimes \ldots \otimes B_{m,n_m})$. In fact one can easily map a set of derivations for the sequents $\vdash B_{1,1}, \ldots, B_{1,n_1}$ $\vdash B_{2,1}, \ldots, B_{2,n_2} \quad \ldots \quad \vdash B_{m,1}, \ldots, B_{m,n_m}$, into a derivation for the formula and vice-versa. One can see the above set of sequents as a different writing of the previous formula. This alternative writing of a formula is a way of marking the main connectives to which inference rules can be applied. We argue that sequent calculus is a formalism for writing derivations where the main connectives have, at most, syntactic deepness two. In this respect, the most remarkable difference between sequent calculus and the CoS lies on the fact that in the CoS rules can be applied at an arbitrary deepness inside a formula.

In the CoS there is more freedom in applying rules, and as a consequence, derivations loose some of their internal structures but on the other hand there are many examples of logics where the use of deep rules gives some advantages from the point of view of the proof theory [9, 7, 8, 10].

3   Multiplicative Linear Logic

As a first step in presenting cyclic linear logic, we present multiplicative linear logic (MLL) [1] in the CoS. We use the standard syntax of multiplicative linear logic, i.e. our formulae are given by the syntax:

$$A := \perp \mid 1 \mid a \mid \bar{a} \mid (A \otimes A) \mid (A \otimes A)$$

Formulae in the form $a, \bar{a}$ are called atomic. We denote by $\bar{a}$ the *negation* of a. The negation of an arbitrary formula $\bar{A}$ is syntactically defined by the following (De Morgan) rules:

$$\overline{(A \otimes B)} \triangleq \bar{B} \otimes \bar{A}$$
$$\overline{A \otimes B} \triangleq \bar{B} \otimes \bar{A}$$
$$\overline{\bar{a}} \triangleq a$$
$$\overline{1} \triangleq \perp$$
$$\overline{\perp} \triangleq 1$$

3.1   Equivalence Between Formulae

As we already remarked, the CoS introduces the notion of structures which are equivalence classes of formulae. Formulae contained in the same equivalence class are considered to be elementary logical equivalence. In a derivation there is always the freedom to choose the most suitable representative of a structure. In this way, it is possible to omit what is considered bureaucracy, and so better highlight the important steps in a derivation. Here we follow a different approach, we do not use structures and work directly on formulae. An abstract motivation for our choice is the idea that working with structures, hence with equivalence classes, it is possible to hide some interesting aspects of the proof theory. A more concrete argument against the use of structures is the fact our cut elimination proof relies on the distinction between formulae belonging to the same equivalence class and cannot be presented in term of structures.

Having decide to work directly on formulae, we need to introduce some rules, not explicitly present in the CoS, allowing to substitute, in a derivation, a formulae with by an elementary equivalent one. This can be obtained by introducing a set of rules, each rule stating a particular property of a particular connective. However, in order to have a more compact presentation, we group together these "equivalence" rules in a single rule. To this end we introduce a relation ~ between formulae; ~ related formulae that can be shown equivalent by a *single* application of the commutativity, associativity and identity laws for the connectives par and tensor. Note that in definingthe relation $\sim$ we do not use any symmetric or transitive closure, hence $\sim$ is *not* a equivalence relation. The relation $\sim$ is defined by the following set of schemata:

$$A \,\invamp\, B \;\sim\; B \,\invamp\, A \qquad \text{Par commutative}$$
$$A \otimes B \;\sim\; B \otimes A \qquad \text{Times commutative}$$
$$A \,\invamp\, (B \,\invamp\, C) \;\sim\; (A \,\invamp\, B) \,\invamp\, C \qquad \text{Par associative}$$
$$A \otimes (B \otimes C) \;\sim\; (A \otimes B) \otimes C \qquad \text{Times associative}$$

$$\bot \,\invamp\, A \;\sim\; A \qquad \text{Par unit L}$$
$$A \;\sim\; A \,\invamp\, \bot \qquad \text{Par unit R}$$
$$1 \otimes A \;\sim\; A \qquad \text{Times unit L}$$
$$A \;\sim\; A \otimes 1 \qquad \text{Times unit R}$$

3.2   Proof System

We take full advantage, in presenting our calculus, from the fact that logical rules are closed by positive contexts. Positive contexts are generated by the grammar:

$$S ::= \circ \mid (A \,\invamp\, S) \mid (S \,\invamp\, A) \mid (A \otimes S) \mid (S \otimes A)$$

we denote by $S[A]$ the formula obtained by replacing, in the structural context S, the place holder $\circ$ by the formula A.

The proof system, for multiplicative linear logic, is given by the following set of inference rules:

$$\frac{}{1} \;\text{Empty (Emp)}$$

$$\frac{S[B]}{S[A]} \;\text{if } A \sim B \qquad \text{Equivalence (Eq)}$$

$$\frac{S[1]}{S[A \,\invamp\, \overline{A}]} \;\text{Interaction (Int)}$$

$$\frac{S[A \otimes \overline{A}]}{S[\bot]} \;\text{Cut}$$

$$\frac{S[(A \,\invamp\, B) \otimes C]}{S[A \,\invamp\, (B \otimes C)]} \;\text{Switch (Sw)}$$

As we already remark, the Equivalence rule can be seen as a compact way to represent a set of inference rules, namely the 8 rules obtain by considering, one by one, the 8 schemata defining the relation $\sim$.

Definition 1. *We denote with* ⊢ *A the derivability of the formula A by the above defined set of rules.*

The article [11] contains a presentation of multiplicative exponential linear logic (MELL) in the calculus of structures. Apart from the differences in the syntax and in the equivalence rule, we use the same rules presented in [11] for the multiplicative connectives. Similarly to what has been done in [11] we can prove that our calculus satisfies the cut-elimination property and is equivalent to multiplicative linear logic.

Proposition 1. *For any formula A,*

(i) ⊢ *A if and only if A is provable in* MLL*;*
(ii) *if* ⊢ *A then A is provable without using the Cut rule.*

We omit proofs since they are already present in [11] and can be easily derived by the corresponding proofs for cyclic linear logic given in the next sections.

4 Multiplicative Cyclic Linear Logic

Cyclic Linear Logic can be optained by simply removing the commutative rules from MLL, that is we consider a new relation $\sim_N$ that is equal to the relation $\sim$, given in Section 3.1, except for the omission of the commutative rules for the par and tensors. However, to have a coherent proof system, we need to substitute the equivalence rule with the a new one having the following form:

$$\frac{S[B]}{S[A]} \text{ if } A \sim_N \text{ or } B \sim_N A \qquad \text{EquivalenceN (EqN)}$$

In the commutative calculus, this reflexive formulation of the equivalence rule is not necessary. In fact, through commutativity, it is possible to derive each extra case given by the EquivalenceN rule by (at most three) consecutive applications of the Equivalence rule. Similarly to the Equivalence rule, the EquivalenceN rules can be seen as a compact way to represent a set of inference rules, namely the 12 rules obtain by considering, in the both directions, the 6 schemata defining the relation $\sim_N$.

We need to add also a mirror image version of the Switch rule:

$$\frac{S[A \otimes (B \mathbin{\mathgrave{} } C)]}{S[(A \otimes B) \mathbin{\mathgrave{} } C]} \text{ Switch Mirror (SwM)}$$

which is not present in the MLL since is there derivable through commutativity.

Definition 2. *We call system* NLS *the inference calculus formed by the rules: Empty, EquivalenceN, Switch, Switch Mirror, Interaction, and Cut. We denote with* ⊢N *A the fact that the formula A is derivable in system* NLS*.*

Our proof system is equivalent to multiplicative CyLL.

Theorem 1. *For any formula A, $\vdash_N A$ if and only if A is provable in multiplicative* CyLL.

Proof. In order to prove this theorem we consider the presentation of multiplicative CyLL given in [2]. We start by proving the left to right implication, that is, everything provable in our system is provable in multiplicative CyLL. The implication follows immediately from two properties of CyLL. The first one is that derivation is closed by positive context. That is, if S is a positive context not containing negation, and the formulae $S[A]$ and $\overline{A}\,\mathbin{\text{⅋}} B$ are derivable, then also the formula $S[B]$ is derivable. This fact can be proved in the following way. Let Π be a derivation, in multiplicative CyLL, for $S[A]$, since Π cannot examine the formula of A until the formula A appears as an element of a sequent, looking at derivations bottom-up-wise, the derivation Π is "independent" from A until it builds a sequent $\Phi(A)$ containing the formula A. From the sequent $\Phi(A)$, using the Cyclic rule it is then possible to derive a sequent Φ', A (having A as last formula) from which, by the Cut rule, one derives Φ', B and, by the Cyclic rule, $\Phi(B)$. From $\Phi(B)$, by following the pattern in Π, one can finally derive $S[B]$.

The second property is that for any rule

$$\frac{S[A]}{S[B]}$$

contained in system NLS, the formula $\overline{A}\,\mathbin{\text{⅋}} B$ is derivable in multiplicative CyLL. This fact can be checked straightforwardly.

The other implication is also simple. First, we define a translation, $\_{}_S$, from sequents and sets of sequents, into formulae:

$$\underline{A_1,\ldots,A_m}_S \triangleq \underline{A_1}_S \mathbin{\text{⅋}} \ldots \mathbin{\text{⅋}} \underline{A_m}_S$$

$$\underline{\{\Gamma_1,\ldots,\Gamma_n\}}_S \triangleq \underline{\Gamma_1}_S \otimes \ldots \otimes \underline{\Gamma_n}_S$$

It is then easy to check that, any CyLL rule different from the Cycling rule is derivable, that is for any rule in the form:

$$\frac{\vdash \Gamma_1 \ldots \vdash \Gamma_n}{\vdash \Delta},$$

from the formula $\underline{\{\Gamma_1,\ldots,\Gamma_n\}}_S$, using the rules in NLS, it is possible to derive the formula $\underline{\Delta}_S$.

Finally, we prove that the Cycling rule is admissible. The Cycling rule has form

$$\frac{\Delta, \Gamma}{\Gamma, \Delta} \quad \text{Cycling}$$

Its admissibility in system NLS can be express in the following way: if $\vdash A \mathbin{\text{⅋}} B$ then $\vdash B \mathbin{\text{⅋}} A$. The proof works as follows:

$$\frac{\overline{}}{1} \text{ Empty}$$
$$\frac{}{B \,\vartheta\, \overline{B}} \text{ Interaction}$$
$$\frac{}{B \,\vartheta\, (1 \otimes \overline{B})} \text{ EquivalenceN}$$
$$\vdots \text{ Hypothesis}$$
$$\frac{B \,\vartheta\, ((A \,\vartheta\, B) \otimes \overline{B})}{B \,\vartheta\, (A \,\vartheta\, (B \otimes \overline{B}))} \text{ Switch}$$
$$\frac{}{B \,\vartheta\, (A \,\vartheta\, \bot)} \text{ Cut}$$
$$\frac{}{B \,\vartheta\, A} \text{ EquivalenceN} \qquad\qquad \square$$

4.1   Cut Elimination

A fundamental feature of every logical system is the cut-elimination property, which can also be proved for NLS. If we consider the different logics so far presented in the CoS, [9, 7, 8, 10] and we compare, for these logics, the proofs of cut-elimination in the sequent calculus, and in the CoS, normally we have that the latter proofs are lengthier. This fact can be explained by remarking that, in its complete formulation, the CoS gives more freedom in constructing derivations. It follows that derivations can be quite an anarchic object, and the standard proof technique of structural induction on the complexity of derivations is more difficult to use. A standard technique, for proving cut elimination in the CoS, is to use of the so-called splitting lemma [7]. The splitting lemma states that one can consider just derivations of a particular shape, i.e. a particular subset of derivations is sufficient to derive any provable judgment. We think that the splitting lemma is applicable also to this case, however here we prefer to use a different proof technique. There are two reasons for this choice. The first one is that when we conceive our proof, the splitting lemma was not discovered yet. The second reason is that our proof enlighten some interesting aspect of the CoS, namely the admissibility of some instances of the Equivalence rules. The main idea in our proof is to use formulae instead of structures. Once this choice has been made, the cut elimination proof is obtained by standard techniques. In more detail, we define a restricted calculus with a minimal set of derivations rules. We then prove that all the omitted rules are admissible in the minimal calculus. As immediate consequence, since the cut rule is an omitted one, we have a proof cut elimination. Since we need to show the admissibility of the omitted rules by taking each rule at a time our proof of is quite lengthy also if the single steps, and the general structure, are quite simple.

As a first step, we need to present a restricted calculus having a minimal set of rules. To motivate this restricted calculus we need to introduce the concept of duality between rules. Given of a rule S

$$\frac{A}{B} \text{ S}$$

the dual of S is the rule

$$\frac{\overline{B}}{\overline{A}} \; \mathrm{dS}$$

For example the rule Interaction is dual to the rule Cut and Switch, Switch Mirror are dual to themselves. It is also easy to observe that from any rule S it is possible to derive its dual, using the Interaction, the Switch, and the Cut rules. In the following, we are going to prove that for any pair of dual rules one of them can be eliminated. Duality between rules is a standard concept in the CoS, and it also standard result the fact that for each each pair of dual rules one of them can be eliminated. What makes our approach different from the previous ones is the fact that we have an explicit rule for equivalence and that we apply the notion of duality also to it. In particular, as we already remark, EquivalenceN rule is a compact way of expressing a set of rules: a rule saying that par is associative, another saying that par is commutative etc. Considering this underlying set of rules, one can observe that it contains pairs of dual rules. For example the rule stating associativity of par is dual to the rule stating associativity of times, the rule for the introduction of the times unit is dual to the rule for the elimination of the par unit, and so on. In the restricted calculus, we insert just one single instance for each pair of dual rules. In doing so we depart from the main stream of the CoS; not only we make the application of the equivalence rule explicit but in the restricted system we do not allow the application of some equivalences. In particular we show that the rules stating the associativity of times are admissible. With this aim, we define a restricted version of the Equivalence rule. This restricted version of the Equivalence considers a new relation, $\rightsquigarrow$, on formulae.

Definition 3. *The relation on formulae $\rightsquigarrow$ is defined as follows:*

$$
\begin{array}{lll}
(A \mathbin{\text{⅋}} B) \mathbin{\text{⅋}} C & \rightsquigarrow A \mathbin{\text{⅋}} (B \mathbin{\text{⅋}} C) & \textit{Par associative L} \\
A \mathbin{\text{⅋}} (B \mathbin{\text{⅋}} C) & \rightsquigarrow (A \mathbin{\text{⅋}} B) \mathbin{\text{⅋}} C & \textit{Par associative R} \\
\bot \mathbin{\text{⅋}} A & \rightsquigarrow A & \textit{Par unit L} \\
A \mathbin{\text{⅋}} \bot & \rightsquigarrow A & \textit{Par unit R} \\
\mathbf{1} \otimes A & \rightsquigarrow A & \textit{Times unit L} \\
A \otimes \mathbf{1} & \rightsquigarrow A & \textit{Times unit R}
\end{array}
$$

The Restricted Equivalence rules is:

$$\frac{S[A]}{S[B]} \;\; \textit{if } B \rightsquigarrow A \qquad\qquad \textit{Restricted Equivalence (REq)}$$

Moreover, it is useful to consider a restricted version of the Interaction rule. In fact, it is possible to reduce the Interaction rule to its atomic version. We call Atomic Interaction the Interaction rule restricted to atomic formulae.

$$\frac{S[\mathbf{1}]}{S[A \mathbin{\text{⅋}} \overline{A}]} \;\; \text{with } A \text{ atomic formula} \qquad\qquad \text{Atomic Interaction}$$

Definition 4. *We call system* NLS$_r$ *the inference calculus formed by the rules: Empty, Restricted Equivalence, Atomic Interaction, Switch, and Switch Mirror. We write $\vdash_r A$ to indicate that the formula A is provable in system* NLS$_r$.

We aim to prove that system NLS$_r$ is equivalent to the NLS. In particular, we will prove that all the rules in NLS are admissible in NLS$_r$. The proof proceeds by several steps, each step proving the admissibility of one missing rule.

Lemma 1. *The Interaction rule is derivable in* NLS$_r$, *that is, for any formula A, and context S, from $S[\mathbf{1}]$, using the* NLS$_r$ *rules it is possible to derive $S[A \otimes \overline{A}]$.*

Proof. By induction on the complexity of the formula A. If A is a unit, then the thesis follows from the Restricted Equivalence rule. In the case where A is an atom, the thesis follows from the Atomic Interaction rule. In the case where A is in the form $A' \otimes A''$ we have the following chain of implications:

$\vdash_r S[\mathbf{1}] \;\Rightarrow\;$ (by inductive hypothesis)
$\vdash_r S[A' \otimes \overline{A'}] \;\Rightarrow\;$ (by Restricted Equivalence rule)
$\vdash_r S[(A' \otimes \mathbf{1}) \otimes \overline{A'}] \;\Rightarrow\;$ (by inductive hypothesis)
$\vdash_r S[(A' \otimes (A'' \otimes \overline{A''})) \otimes \overline{A'}] \;\Rightarrow\;$ (by Switch Mirror rule)
$\vdash_r S[((A' \otimes A'') \otimes \overline{A''}) \otimes \overline{A'}] \;\Rightarrow\;$ (by Restricted Equivalence rule)
$\vdash_r S[(A' \otimes A'') \otimes (\overline{A''} \otimes \overline{A'})].$

The case where A is in the form $A' \otimes A''$, is perfectly equivalent (mirror image) to the previous one. $\square$

Next, we prove admissibility of the Equivalence rule. The proof will be done by induction on the structures of the derivation. To make the induction working, we need to take a stronger, and more involved, inductive hypothesis. A preliminary definition and a lemma are here necessary. We start by defining new classes of contexts.

Definition 5. *(i) A* left context, *T_l, is defined by the following grammar:*

$$T_l ::= \circ \;\mid\; A \otimes T_l \mid A \otimes T_l$$

symmetrically, a right context, *T_r, is defined by the grammar:*

$$T_r ::= \circ \;\mid\; T_r \otimes A \mid T_r \otimes A$$

(ii) A left par-context, *V_l, is defined by the following grammar:*

$$V_l ::= \circ \;\mid\; A \otimes V_l$$

symmetrically, a right par-context, *V_r, is defined by the grammar:*

$$V_r ::= \circ \;\mid\; V_r \otimes A$$

Lemma 2. *For any formulae A, B, C, context S, left par-context V_l and right par-context V_r the following points hold:*

(i) *if* $\vdash_r S[V_l[A \otimes B] \otimes C]$ *then* $\vdash_r S[V_l[A \otimes (B \otimes C)]]$, *and symmetrically if* $\vdash_r S[A \otimes V_r[B \otimes C]]$ *then* $\vdash_r S[V_r[(A \otimes B) \otimes C]]$,

(ii) *if* $\vdash_r S[A \otimes V_r[\mathbf{1}]]$ *then* $\vdash_r S[V_r[A]]$, *and symmetrically, if* $\vdash_r S[V_l[\mathbf{1}] \otimes A]$ *then* $\vdash_r S[V_l[A]]$,

(iii) *if* $\vdash_r S[\bot \mathbin{\text{⅋}} A]$ *then* $\vdash_r S[A]$ *and symmetrically, if* $\vdash_r S[A \mathbin{\text{⅋}} \bot]$ *then* $\vdash_r S[A]$

Proof. All three points are proved by structural induction of the derivation Δ of the judgment in the premise. Here we present just the proof of point (i) which is the most involved one, having the larger number of cases to consider. The other points can be treated with perfectly similar arguments. For each point, the cases to consider concern the last rule, R, applied in the derivation Δ. The simple cases are the ones where R modifies the derived formula only inside one of the contexts S, V_l, V_r, or inside one of the formulae A, B, C: in these cases the thesis simply derives by inductive hypothesis and by an application of R.

For the remaining cases, we schematically present each case by a pair of derivations, the one on left is the application of R considered, while the one on right shows how the case can be treated, i.e. by, in case, applying the inductive hypothesis, and by the given derivation.

(i.a) The rule R generates one of the formulae A, B, C. This can only happen if one of the formulae is a times unit. The case where A is generated is described and treated as follows:

$$\frac{S[V_l[B] \otimes C]}{S[V_l[\mathbf{1} \otimes B] \otimes C]} \text{ REq} \qquad \frac{\begin{array}{c} S[V_l[B] \otimes C] \\ \vdots \ (\text{Sw})^* \\ S[V_l[B \otimes C]] \end{array}}{S[V_l[\mathbf{1} \otimes (B \otimes C)]]} \text{ REq}$$

The cases where B or C are generated can be dealt in a similarl way.

(i.b) The last rule R is a Switch rule involving the context (V_l) and the formula $(A \otimes B)$. This interaction can occur in two forms, the first one is the case where $V_l \equiv V_l'[D \mathbin{\text{⅋}} \circ]$, and:

$$\frac{S[V_l'[(D \mathbin{\text{⅋}} A) \otimes B] \otimes C]}{S[V_l'[D \mathbin{\text{⅋}} (A \otimes B)] \otimes C]} \text{ Sw} \qquad \frac{S[V_l'[(D \mathbin{\text{⅋}} A) \otimes (B \otimes C)]]}{S[V_l'[D \mathbin{\text{⅋}} (A \otimes (B \otimes C))]]} \text{ Sw}$$

(i.c) A second form of interaction between the context (V_l) and the formula $(A \otimes B)$, is given by case where $V_l \equiv V_l'[(D \otimes E) \mathbin{\text{⅋}} \circ]$, and:

$$\frac{S[V_l'[D \otimes (E \mathbin{\text{⅋}} (A \otimes B))] \otimes C]}{S[V_l'[(D \otimes E) \mathbin{\text{⅋}} (A \otimes B)] \otimes C]} \text{ SwM} \qquad \frac{\dfrac{\dfrac{S(V_l'[D \otimes ((E \mathbin{\text{⅋}} (A \otimes B)) \otimes C)]]}{S(V_l'[D \otimes (E \mathbin{\text{⅋}} ((A \otimes B) \otimes C))]]} \text{ SwM}}{S[V_l'[(D \otimes E) \mathbin{\text{⅋}} ((A \otimes B) \otimes C)]]} \text{ Sw}}$$

(i.d) The context S interacts with the formula C, in this case $S \equiv S'[\circ \wp D]$ and:

$$\frac{S'[V_l[A \otimes (B \otimes (C \wp D))]]}{\vdots \quad \text{SwM} \cdot \text{SwM}}$$
$$S'[V_l[(A \otimes (B \otimes C)) \wp D]]$$
$$\vdots \quad (\text{REq})^*$$
$$S'[V_l[A \otimes (B \otimes C)] \wp D]$$

$$\frac{S'[V_l[A \otimes B] \otimes (C \wp D)]}{S'[(V_l[A \otimes B] \otimes C) \wp D]} \text{ SwM}$$

(i.e) The context S interacts with the context V_l, in this case $S \equiv S'[D \wp \circ]$ and the last inference rule is:

$$\frac{S'[(D \wp V_l[A \otimes B]) \otimes C]}{S'[D \wp (V_l[A \otimes B] \otimes C)]} \text{ SwM}$$

for this case, it is sufficient to simply apply the inductive hypothesis.

(i.f) Finally we need to consider the interaction between the context S and the formula $V_l[A \otimes B] \otimes C$, in this case $S \equiv S'[\circ \wp (D \otimes E)]$ and:

$$\frac{S'[((V_l[A \otimes B] \otimes C) \wp D) \otimes E))]}{S'[(V_l[A \otimes B] \otimes C) \wp (D \otimes E)]} \text{ SwM}$$
$$\frac{S'[(V_l[A \otimes (B \otimes C)] \wp D) \otimes E]}{S'[(V_l[A \otimes (B \otimes C)] \wp (D \otimes E)]} \text{ SwM}$$

The case where $S \equiv S'[(D \otimes E) \wp \circ]$ is equally easy. □

Notice that as a special case of point (i) and (ii) of the above lemma we have that

(i) if $\vdash_r S[(A \otimes B) \otimes C]$ then $\vdash_r S[A \otimes (B \otimes C)]$, and symmetrically if $\vdash_r S[A \otimes (B \otimes C)]$ then $\vdash_r S[(A \otimes B) \otimes C]$,

(ii) if $\vdash_r S[A \otimes 1]$ then $\vdash_r S[A]$, and symmetrically, if $\vdash_r S[1 \otimes A]$ then $\vdash_r S[A]$,

it follows:

Proposition 2. *The EquivalenceN rule is admissible in* NLS$_r$.

Next we proof the admissibility of the atomic cuts, also for this case the proof is done by induction on the structures of the derivation, and also for this case to make the induction working, we need to take a stronger, and more involved, inductive hypothesis. The following lemma implies the admissibility of atomic cuts.

Lemma 3. *For any atom a, context S, left context T_l and right context T_r, if $\vdash_r S[T_l[a] \otimes T_r[\overline{a}]]$ (or $\vdash_r S[T_l[\overline{a}] \otimes T_r[a]]$) then $\vdash_r S[T_l[\bot] \wp T_r[\bot]]$.*

Proof. The proof is by structural induction on the derivation Δ of $\vdash_r S[T_l[a] \otimes T_r[\overline{a}]]$, and it is quite similar to the proof of Lemma 2. The different cases considered by the structural induction can be split in two groups. The simple cases are the ones where the last rule R in Δ works internally to one of the contexts S, T_l, T_r; in all these cases the thesis follows by inductive hypothesis and by an application of the rule R. The other cases are the ones where the rule R modifies more than one context, or generates one of the atoms a, $\overline{a}$. In detail:

(a) an Atomic Interaction rule generates one of the atoms. This case can be described and treated as follows:

$$\cfrac{S[T_l[a] \otimes T'_r[\mathbf{1}]]}{S[T_l[\bot \mathbin{\mathscr{G}} a] \otimes T'_r[\mathbf{1}]]} \text{ REq}$$
$$\vdots \ (\text{REq} + \text{SwM})^*$$
$$\cfrac{S[(T_l[\bot] \mathbin{\mathscr{G}} a) \otimes T'_r[\mathbf{1}]]}{S[T_l[\bot] \mathbin{\mathscr{G}} (a \otimes T'_r[\mathbf{1}])]} \text{ Sw}$$
$$\vdots \ (\text{EqN} + \text{SwM})^*$$
$$\cfrac{S[T_l[\bot] \mathbin{\mathscr{G}} T'_r[(a \otimes \mathbf{1})]]}{S[T_l[\bot] \mathbin{\mathscr{G}} T'_r[a]]} \text{ EqN}$$
$$\cfrac{}{S[T_l[\bot] \mathbin{\mathscr{G}} T'_r[\bot \mathbin{\mathscr{G}} a]]} \text{ REq}$$

$$\cfrac{S[T_l[a] \otimes T'_r[\mathbf{1}]]}{S[T_l[a] \otimes T'_r[\overline{a} \mathbin{\mathscr{G}} a]]} \text{ AInt}$$

One should remark that the one of the right is not a true derivation in $\mathsf{NLS_r}$, in fact the EquivalenceN rule is just admissible in $\mathsf{NLS_r}$. The right diagram should be interpreted as a schematic proof that the formula $S[T_l[\bot] \mathbin{\mathscr{G}} T'_r[\bot \mathbin{\mathscr{G}} a]]$ is derivable in $\mathsf{NLS_r}$.

(b) a Switch rule makes the contexts S and T interact. This case can be described and treated as follows: $S \equiv S'[A \mathbin{\mathscr{G}} \circ]$ and

$$\cfrac{S'[(A \mathbin{\mathscr{G}} T_l[a]) \otimes T_r[\overline{a}]]}{S'[A \mathbin{\mathscr{G}} (T_l[a] \otimes T_r[\overline{a}])]} \text{ Sw} \qquad\qquad \cfrac{S'[(A \mathbin{\mathscr{G}} T_l[\bot]) \mathbin{\mathscr{G}} T_r[\bot]]}{S'[A \mathbin{\mathscr{G}} (T_l[\bot] \mathbin{\mathscr{G}} T_r[\bot])]} \text{ REq}$$

It remains to consider the cases where the context S interacts with the whole formula $T_l[a] \otimes T_r[\overline{a}]$, where the contexts T_l interact with formula a and where the contexts T_r interact with formula $\overline{a}$. All these cases are immediate. $\qquad\square$

Lemma 4. *The Cut rule is admissible in* $\mathsf{NLS_r}$, *that is, for any formula A, context S, if $\vdash_r S[A \otimes \overline{A}]$ then $\vdash_r S[\bot]$.*

Proof. This lemma can be seen as a sort of dual of Lemma 1 By structural induction on the formula A. If A is a unit then the thesis follows from the admissibility of the EquivalenceN rule.

In the case where A is an atom, the thesis follows from the previous lemma and from the admissibility of the EquivalenceN rule.

The case where A is in the form $A' \mathbin{\invamp} A''$ can be treated as follows:

$$\cfrac{\cfrac{\cfrac{S[(A' \mathbin{\invamp} A'') \otimes (\overline{A''} \otimes \overline{A'})]}{S[((A' \mathbin{\invamp} A'') \otimes \overline{A''}) \otimes \overline{A'}]} \text{ EqN}}{S[(A' \mathbin{\invamp} (A'' \otimes \overline{A''})) \otimes \overline{A'}]} \text{ Sw}}{}$$

$$\vdots \text{ Inductive hypothesis}$$

$$\cfrac{S[(A' \mathbin{\invamp} \bot) \otimes \overline{A'}]}{S[A' \otimes \overline{A'}]} \text{ EqN}$$

$$\vdots \text{ Inductive hypothesis}$$

$$S[\bot]$$

Note that the above is not a true derivation but just a schematic proof of the derivability of $S[\bot]$. The case where A is in the form $A' \otimes A''$ is perfectly equivalent to this one. □

Having proved that all the rules in NLS are admissible in NLS$_r$ we can finally state:

Proposition 3. *For every formula A if $\vdash_N A$ the $\vdash_r A$.*

That is, the restricted system NLS$_r$ is as powerful as the complete one NLS and from this we have:

Theorem 2. *The system NLS satisfies the cut-elimination property.*

5   Further Works

A natural question to consider is whether the above treatment for the multiplicative cycling logic can be extended to richer logics. In particular, one should consider the complete system of the cycling linear logic [2] and the multiplicative non-commutative logic of Abrusci and Ruet [5]. The complete cyclic linear logic extend the multiplicative part, considered here, by adding the missing linear logic connectives. While the multiplicative non-commutative logic contains both commutative and non-commutative multiplicative connectives.

Without giving any proof we claim that the first extension can be carried out quite smoothly; it is sufficient to consider the presentation of LL in the CoS given in [10] and modify it by removing the commutativity rules. In this way we obtain a proof system for cyclic linear logic. In this system cut elimination can be proved using the technique presented in this article, with the only extra difficulty of using a more complex induction to deal with the exponential and additive connectives. In fact these connectives can multiply the occurrences of a formula in the premises.

The treatment of the multiplicative non-commutative linear logic [5] is still an open problem. We remark that it is possible to formulate, in the CoS, all the rules of the non-commutative logic, given, in the sequent calculus formulation, in [6]. However this presentation will be an obvious and uninteresting result. In this way deep inference will not play any role. A more interesting application of the CoS in this setting would be a proof that the Seesaw rule (the non commutative logic equivalent to the Cycling rule) can be substituted by deep inference. However so far we were not able to find a nice formulation for the non-commutative logic, in the CoS. Our difficulties can be explained by the fact that deep inference alone, in the non-commutative calculus, is not able to reduce the Interaction rule to the Atomic Interaction rule, and dually Cut to and atomic form of cut. These reductions are possible instead in the other logics and are a key ingredient in the cut elimination proofs.

Acknowledgments

We thank the anonymous referees for their useful suggestions and Alessio Guglielmi for many useful suggestions and stimulating discussions.

References

1. Girard, J.Y.: Linear logic. Theoretical Computer Science **50** (1987) 1–102
2. Yetter, D.N.: Quantales and (non-commutative) linear logic. J. Symbolic Logic **55** (1990)
3. Mulvey, C.J.: &. In: Second topology conference. Number 12 in Rediconti del Circolo Matematico di Palermo (1986) 94–104
4. Lambek, J.: The mathematics of sentece structures. Amer. Math. Mon. **65** (1958) 154–170
5. Abrusci, V.M., Ruet, P.: Non-commutative logic I: the multiplicative fragment. Annals of Pure and Applied Logic **101** (2000) 29–64
6. Ruet, P.: Non-commutative logic II: sequent calculus and phase semantics. Mathematical Structures in Computer Science **10** (2000) 277–312
7. Guglielmi, A.: A system of interaction and structure. Technical Report WV-02-10, Dresden University of Technology (2002) Conditionally accepted by ACM Transactions on Computational Logic, http://pikas.inf.tu-dresden.de/~guglielm/paper.html.
8. Stewart, C., Stouppa, P.: A systematic proof theory for several modal logics. Technical Report WV-03-08, Technische Universität Dresden (2003)
9. Brunnler, K.: Atomic cut elimination for classical logic. In Baaz, M., Makowsky, J.A., eds.: CSL 2003. Volume 2803 of Lecture Notes in Computer Science., Springer-Verlag (2003) 86–97
10. Straßburger, L.: A local system for linear logic. In: LPAR 2002. Volume 2514 of Lectures Notes in Artificial Intelligence., Springer-Verlag (2002) 388–402
11. Guglielmi, A., Straßburger, L.: Non-commutativity and MELL in the calculus of structures. In Fribourg, L., ed.: CSL 2001. Volume 2142 of Lecture Notes in Computer Science., Springer-Verlag (2001) 54–68

On Proof Nets for Multiplicative Linear Logic with Units

Lutz Straßburger and François Lamarche

INRIA-Lorraine, Projet Calligramme
615, rue du Jardin Botanique — 54602 Villers-lès-Nancy — France
{Lutz.Strassburger, Francois.Lamarche}@loria.fr

Abstract. In this paper we present a theory of proof nets for full multiplicative linear logic, including the two units. It naturally extends the well-known theory of unit-free multiplicative proof nets. A linking is no longer a set of axiom links but a tree in which the axiom links are subtrees. These trees will be identified according to an equivalence relation based on a simple form of graph rewriting. We show the standard results of sequentialization and strong normalization of cut elimination. Furthermore, the identifications enforced on proofs are such that the proof nets, as they are presented here, form the arrows of the free (symmetric) *-autonomous category.

1 Introduction

For a long time formal logicians have been aware of the need to determine, given a formal system $\mathcal{S}$ and two proofs of a formula A in that system, when these two proofs are "the same" proof. As a matter of fact this was already a concern of Hilbert when he was preparing his famous lecture in 1900 [Thi03]. This problem has taken more importance during the last few years, because many logical systems permit a close correspondence between proofs and programs.

In a formalism like the sequent calculus (and to a lesser degree, natural deduction), it is oftentimes very easy to see that two derivations π_1 and π_2 should be identified because π_1 can be transformed in to π_2 by a sequence of rule permutations that are obviously trivial. It is less immediately clear *in general* what transformations can be effected on a proof without changing its essence. But here category theory is very helpful, providing criteria for the identification of proofs that are simple, general and unambiguous, if sometimes too strong [Gir91].

The advent of linear logic marked a significant advance in that quest. In particular the multiplicative fragment of linear logic comes equipped with an extremely successful theory of proof identification: not only do we know exactly when two sequent proofs should be identified (the allowed rule permutations are described in [Laf95]), but there is a class of simple formal objects that precisely represent these equivalence classes of sequent proofs. These objects are called proof nets, and they have a strong geometric character, corresponding to additional graph structure ("axiom links") on the syntactical forest of the sequent. More precisely, given a sequent $\Gamma = A_1, \ldots, A_n$ and a proof π of that sequent,

J. Marcinkowski and A. Tarlecki (Eds.): CSL 2004, LNCS 3210, pp. 145–159, 2004.

then the proof net that represents π is simply given by the syntactical forest of Γ decorated with additional edges (shown in thick lines) that represent the identity axioms that appeared in the proof:

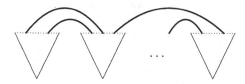

Moreover proof nets are vindicated by category theory, since the category of two-formula sequents and proof nets is precisely the free *-autonomous category [Bar79] (without units) on the set of generating atomic formulas [Blu93]. As a matter of fact axiom links were already visible, under the name of *Kelly-Mac Lane graphs* in the early work [KL71] that tried to describe free autonomous categories; Girard's key insights [Gir87] here were in noticing that there was an inherent symmetry that could be formalized through a negation (thus the move from autonomous to *-autonomous), and that the addition of the axiom links to the sequent's syntactic forest were enough to completely characterize the proof.

The theory of proof nets has been extended to larger fragments of linear logic; when judged from the point of view of their ability to identify proofs that should be identified, these extensions can be shown to have varying degrees of success. One of these extensions, which complies particularly well with the categorical ideal, is the inclusion of additive connectives presented in [HvG03], in which the additives correspond exactly to categorical product and coproduct.

In this paper we give a theory of proof nets for the full multiplicative fragment. That is, our theory of proof nets includes the multiplicative units. We prove that it allows us to construct the free *-autonomous category with units on a given set of generating objects, thus getting full validation from the categorical imperative.

There are only two other presentations for multiplicative units that we are aware of. In [KO99], the authors provide an internal language for autonomous and *-autonomous categories based on the $\lambda\mu$-calculus, and in [BCST96], a non-standard version of two-sided proof nets for a weaker logic is developed from which the authors also claim to have constructed free *-autonomous categories. Our approach is different in the following way: By making full use of the symmetry given by the combination of an involutive negation and a one-sided sequent calculus, we get a notion of proof net which is considerably simpler than the one provided in [BCST96].

The Main Problem

We assume that the reader is familiar with the sequent calculus for classical multiplicative linear logic.

The theory of *-autonomous categories tells us that whenever a proof contains a rule instance r which appears after a $\perp$-introduction rule and which does

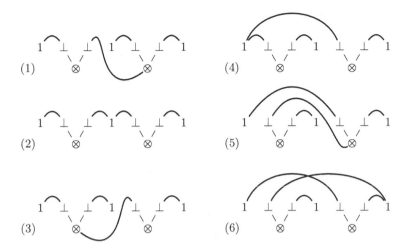

Fig. 1. Different representations of the same proof

not introduce a connective under that $\bot$, then r can be pushed above that $\bot$-introduction without changing the proof:

$$\cfrac{\cfrac{\Gamma}{\bot,\Gamma}\bot}{\bot,\Gamma'}r \quad \cdots \quad \longleftrightarrow \quad \cfrac{\cfrac{\Gamma \quad \cdots}{\Gamma'}r}{\bot,\Gamma'}\bot$$

This seemingly trivial permutation actually has deep consequences. Supposing that rule r was a $\otimes$-introduction, there is now a choice of two branches on which to do the $\bot$-introduction.

$$\cfrac{\cfrac{\cfrac{\Gamma,A}{\bot,\Gamma,A}\bot \quad B,\Delta}{\bot,\Gamma,A\otimes B,\Delta}}{\ }\otimes \quad \longleftrightarrow \quad \cfrac{\cfrac{\Gamma,A \quad B,\Delta}{\Gamma,A\otimes B,\Delta}\otimes}{\bot,\Gamma,A\otimes B,\Delta}\bot \quad \longleftrightarrow \quad \cfrac{\Gamma,A \quad \cfrac{B,\Delta}{\bot,B,\Delta}\bot}{\bot,\Gamma,A\otimes B,\Delta}\otimes$$

Ordinary proof nets for multiplicative linear logic are characterized by the presence of *links*, which connect the atoms of the syntactical forest of the sequent. When extending them to multiplicative units, the first impulse is probably to try to attach the $\bot$s that are present on the sequent forest on other atomic formulas. This is what is done in [BCST96] and corresponds, in the sequent calculus, to doing the $\bot$-introductions as early as possible, that is, as high up on the sequent tree as can be done. The paragraph above shows that an arbitrary choice has to be made because of tensor introductions: in a $\otimes$-intro one branch of the sequent proof tree or the other has to be chosen for doing the $\bot$-intro. In such a situation correct identification of proofs can only be achieved by considering equivalence classes of graphs, and the theory of proof nets involves an equivalence relation on a set of "correct" graphs.

Another possibility is to attach these ⊥s "as low as possible" on the forest, corresponding to the idea that in the sequent calculus deduction the ⊥-intro would be done as late as possible, for example just before the ⊥ instance gets a connective introduced under it. One way of implementing this is linking the ⊥ instance to the last connective that was introduced above it. This is not the only way of doing things, for example we could imagine links that attach that ⊥ instance to several subformulas of the sequent forest, corresponding to the several conclusions of the sequent that existed above the ⊥-introduction.

But whatever way we choose to "normalize" proofs, we claim that if the conventional notion of "link" is used for ⊥s (i.e., if we consider a proof π on the sequent Γ as the sequent forest of Γ decorated with special edges that encode information about the essence of π) we still need to use equivalence classes of such graphs, and there is no hope of having a normal form in that universe of enriched sequent graphs. For instance, the six graphs in Figure 1 are easily seen to represent equivalent proofs, because going from an odd-numbered example to its successor is just sliding a ⊥-intro up in one of the ⊗-intro branches, and going from an even-numbered example to its successor is just doing the reverse transformation. But notice that examples (3) and (5) are *distinct but isomorphic* graphs, since one can be exactly superposed on the other *by only using the Exchange rule.* Thus it is impossible, given the information at our disposal, to choose one instead of the other to represent the abstract proof they both denote. The only way this could be done would be by using arbitrary extra information, like the order of the formulas in the sequent, a strategy that only replaces the overdeterminism of the sequent calculus by another kind of overdeterminism.

The same can be said of Examples (2) and (6), which are also isomorphic modulo Exchange. But notice that these two comply to the "as early as possible" strategy, while the previous two were of the "as late as possible" kind. So for neither strategy can there be a hope a graphical normal form. The interested reader can verify that the six examples above are part of a "ring" of 24 graphs that are all equivalent from the point of view of category theory.

Thus there is one aspect of our work that does not differ from [BCST96], which is our presentation of abstract proofs as equivalence classes of graphs. But some related aspects are significantly different:

- The graphs that belong to our equivalence classes are *standard multiplicative proof nets,* where the usual notions, like correctness criteria and the empire of a tensor branch, will apply. It is just that some ⅋ and ⊗ links are used in a particular fashion to deal with the units. (The readers can choose their favorite correctness criterion since they are all equivalent; in this paper we will use the one of [DR89] because of its popularity.)
- The equivalence relation we will present is based on a very simple set of rewriting rules on proof graphs. As a matter of fact, there is only *one* non-trivial rule, since the other rules have to do with commutativity and associativity of the connectives and can be dispensed with if we use, for example, n-ary connectives.

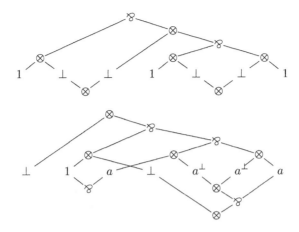

Fig. 2. Two examples of proof graphs

2   Cut Free Proof Nets for MLL

Let $\mathcal{A} = \{a, b, \ldots\}$ be an arbitrary set of atoms, and let $\mathcal{A}^\perp = \{a^\perp, b^\perp, \ldots\}$. The set of MLL *formulas* is defined as follows:

$$\mathcal{F} ::= \mathcal{A} \mid \mathcal{A}^\perp \mid 1 \mid \perp \mid \mathcal{F} \otimes \mathcal{F} \mid \mathcal{F} \otimes \mathcal{F} \quad.$$

Additionally, we will define the set of MLL *linkings* (which can be seen as a special kind of formulas) as follows:

$$\mathcal{L} ::= 1 \mid a \otimes a^\perp \mid a^\perp \otimes a \mid \perp \otimes \mathcal{L} \mid \mathcal{L} \otimes \perp \mid \mathcal{L} \otimes \mathcal{L} \quad.$$

Here, a stands for any element of $\mathcal{A}$. We will use $A, B, \ldots$ to denote formulas, and $P, Q, \ldots$ to denote linkings. *Sequents* (denoted by $\Gamma, \Delta, \ldots$) are finite lists of formulas (separated by comma).

In the following, we will always consider both formulas and linkings as binary trees (and sequents as forests), whose leaves are decorated by elements of $\mathcal{A} \cup \mathcal{A}^\perp \cup \{1, \perp\}$, and whose inner nodes are decorated by $\otimes$ or $\otimes$. We can also think of the nodes being decorated by the whole subformula above that node.

2.1   Definition.   A *pre-proof graph* is a graph consisting of a linking P and a sequent Γ, both of which share the same set of leaves. It will be denoted as $P \triangleright \Gamma$.

Following the tradition, we will draw these graphs such that the roots of the formula trees are at the bottom, the root of the linking tree is at the top, and the leaves are in between. Figure 2 shows two examples. The first of them corresponds to the first graph in Figure 1. A more compact notation for this is

$$(1_1 \otimes \perp_2) \otimes (\perp_3 \otimes ((1_4 \otimes \perp_5) \otimes (\perp_6 \otimes 1_7)))$$
$$\triangledown$$
$$1_1, \perp_2 \otimes \perp_3, 1_4, \perp_5 \otimes \perp_6, 1_7$$

and

$$\perp_1 \otimes ((1_2 \otimes \perp_4) \bindnasrepma ((a_3 \otimes a_5^\perp) \bindnasrepma (a_6^\perp \otimes a_7)))$$
$$\triangledown$$
$$\perp_1, 1_2 \bindnasrepma a_3, \perp_4 \otimes ((a_5^\perp \otimes a_6^\perp) \bindnasrepma a_7)$$.

Here, the indices are used to show how the leaves of the linking and the leaves of the sequent are identified. In this way we will, throughout this paper, use indices on atoms to distinguish between different occurrences of the same atom (i.e. a_3 and a_7 do not denote different atoms). In the same way, indices on the units 1 and $\perp$ are used to distinguish different occurrences.

2.2   Definition.   A *switching* of a pre-proof graph $P \triangleright \Gamma$ is a graph G that is obtained from $P \triangleright \Gamma$ by omitting for each $\bindnasrepma$-node one of the two edges that connect the node to its children. [DR89]

2.3   Definition.   A pre-proof graph $P \triangleright \Gamma$ is called *correct* if all its switchings are connected and acyclic. A *proof graph* is a correct pre-proof graph.

The examples in Figure 2 are proof graphs.

Let $P \triangleright \Gamma$ be a pre-proof graph where one $\perp$ is selected. Let it be indexed as $\perp_i$. Now, let G be a switching of $P \triangleright \Gamma$, and let G' be the graph obtained from G by removing the edge between $\perp_i$ and its parent in P (which is always a $\otimes$). Then G' is called an *extended switching* of $P \triangleright \Gamma$ with respect to $\perp_i$. Observe that, if $P \triangleright \Gamma$ is correct, then every extended switching is a graph that has exactly two connected components.

We will use the notation $P\{Q\} \triangleright \Gamma$ to distinguish the subtree Q of the linking tree of the graph. Then $P\{ \}$ is the context of Q.

2.4   Equivalence on Pre-proof Graphs.   On the set of pre-proof graphs we will define the relation $\sim$ to be the smallest equivalence relation satisfying

$$P\{Q \bindnasrepma R\} \triangleright \Gamma \ \sim \ P\{R \bindnasrepma Q\} \triangleright \Gamma$$
$$P\{(Q \bindnasrepma R) \bindnasrepma S\} \triangleright \Gamma \ \sim \ P\{Q \bindnasrepma (R \bindnasrepma S)\} \triangleright \Gamma$$
$$P\{Q \otimes R\} \triangleright \Gamma \ \sim \ P\{R \otimes Q\} \triangleright \Gamma$$
$$P\{\perp_i \otimes (Q \otimes \perp_j)\} \triangleright \Gamma \ \sim \ P\{(\perp_i \otimes Q) \otimes \perp_j)\} \triangleright \Gamma$$
$$P\{Q \bindnasrepma (R \otimes \perp_i)\} \triangleright \Gamma \ \overset{(*)}{\sim} \ P\{(Q \bindnasrepma R) \otimes \perp_i)\} \triangleright \Gamma \quad ,$$

where the last equation only holds if the following side condition is fulfilled:
(∗) In each extended switching of $P\{Q \bindnasrepma (R \otimes \perp_i)\} \triangleright \Gamma$ with respect to $\perp_i$ no node of the subtree Q is connected to $\perp_i$.

The following proof graph is equivalent to the second one in Figure 2:

$$(((\perp_1 \otimes 1_2) \otimes \perp_4) \bindnasrepma (a_3 \otimes a_5^\perp)) \bindnasrepma (a_6^\perp \otimes a_7)$$
$$\triangledown$$
$$\perp_1, 1_2 \bindnasrepma a_3, \perp_4 \otimes ((a_5^\perp \otimes a_6^\perp) \bindnasrepma a_7) \quad .$$

$$\mathrm{id} \, \frac{}{a \otimes a^\perp \, \triangleright \, a, a^\perp} \qquad \mathrm{ex} \, \frac{P \, \triangleright \, \Gamma, A, B, \Delta}{P \, \triangleright \, \Gamma, B, A, \Delta}$$

$$1 \, \frac{}{1 \, \triangleright \, 1} \qquad \bot \, \frac{P \, \triangleright \, \Gamma}{\bot \otimes P \, \triangleright \, \bot, \Gamma}$$

$$\mathbin{\text{⅋}} \, \frac{P \, \triangleright \, A, B, \Gamma}{P \, \triangleright \, A \mathbin{\text{⅋}} B, \Gamma} \qquad \otimes \, \frac{P \, \triangleright \, \Gamma, A \quad Q \, \triangleright \, B, \Delta}{P \mathbin{\text{⅋}} Q \, \triangleright \, \Gamma, A \otimes B, \Delta}$$

Fig. 3. Translation of cut free sequent calculus proofs into pre-proof graphs

2.5   Definition.   A *pre-proof net*[1] is an equivalence class $[P \triangleright \Gamma]_\sim$. A pre-proof net is *correct* if one of its elements is correct. In this case it is called a *proof net*.

In the following, we will for a given proof graph $P \triangleright \Gamma$ write $[P \triangleright \Gamma]$ to denote the proof net formed by its equivalence class (i.e. we will omit the $\sim$ subscript).

2.6   Lemma.   *If $P \triangleright \Gamma$ is correct and $P \triangleright \Gamma \sim P' \triangleright \Gamma$, then $P' \triangleright \Gamma$ is also correct.*

Proof: That the first four equations preserve correctness is obvious. If in the last equation there is a switching that makes one side disconnected, then it also makes the other side disconnected. For acyclicity, we have to check whether there is a switching that produces a cycle on the right-hand side of the equation and not on the left-hand side. This is only possible if the cycle contains some nodes of Q and the $\bot_i$. But this case is ruled out by the side condition $(*)$.   □

Lemma 2.6 ensures that the notion of proof net is well-defined, in the sense that all its members are proof graphs, i.e. correct.

3   Sequentialization

Figure 3 shows how cut free sequent proofs of MLL can be inductively translated into pre-proof graphs.

We will call a pre-proof net *sequentializable* if one of its representatives can be obtained from a sequent calculus proof via this translation.

3.1   Theorem.   *A pre-proof net is sequentializable iff it is a proof net.*

For the proof we will need the observation that any proof graph is an ordinary unit-free proof net, and the well-known fact that there is always a splitting tensor in such a net.

3.2   Observation   Every proof graph $P \triangleright \Gamma$ is an ordinary unit-free proof net in the style of [DR89]. To make this precise, define for the linking P the *linking formula* $P^\star$ inductively as follows:

---

[1] What we call *pre-proof net* is in the literature often called *proof structure*.

$$a^{\perp\star} = a \qquad 1^\star = \perp \qquad (A \otimes B)^\star = A^\star \otimes B^\star$$
$$a^\star = a^\perp \qquad \perp^\star = 1 \qquad (A \mathbin{⅋} B)^\star = A^\star \mathbin{⅋} B^\star \qquad .$$

In other words, $P^\star$ is obtained from P by replacing each leaf by its dual and by leaving all inner nodes unchanged. We now connect the leaves of $P^\star$ and Γ by ordinary axiom links according to the leaf identification in $P \triangleright \Gamma$. If we forget the fact that $\perp$ and 1 are the units and think of them as ordinary dual atoms, then we have an ordinary unit-free proof net[2].

3.3   Lemma. *If in a unit-free proof net all roots are $\otimes$-nodes, then one of them is splitting, i.e. by removing it the net becomes disconnected.* [Gir87]

Proof of Theorem 3.1 (Sketch): It is easy to see that the rules 1 and id give proof graphs and that the rules $\perp$, $\mathbin{⅋}$, and $\otimes$ preserve the correctness. Therefore every sequentializable pre-proof net is correct.

For the other direction pick one representative $P \triangleright \Gamma$ of the proof net and proceed by induction on the sum of the number of $\otimes$-nodes in the graph and the number of $\mathbin{⅋}$-nodes in Γ. We now interpret $P \triangleright \Gamma$ as an ordinary unit-free proof net (according to Observation 3.2), and remove all $\mathbin{⅋}$-roots (for those inside Γ apply the $\mathbin{⅋}$ rule and proceed by induction hypothesis). Then apply Lemma 3.3. If the splitting $\otimes$ is inside Γ, we can apply the $\otimes$-rule and proceed by induction hypothesis; if it is inside P, it must come from an axiom link or a bottom link. In both cases we can obtain two smaller proof graphs, to which we can apply the induction hypothesis to get two sequent proofs, which can be composed by plugging one into a leaf of the other.   □

4   Cut and Cut Elimination

A *cut* is a formula $A \oplus A^\perp$, where $\oplus$ is called the *cut connective*, and where the function $(-)^\perp$ is defined on formulas as follows (with a little abuse of notation):

$$a^{\perp\perp} = a \qquad 1^\perp = \perp \qquad (A \otimes B)^\perp = A^\perp \mathbin{⅋} B^\perp$$
$$a^\perp = a^\perp \qquad \perp^\perp = 1 \qquad (A \mathbin{⅋} B)^\perp = A^\perp \otimes B^\perp \qquad .$$

A *sequent with cuts* is a sequent where some of the formulas are cuts. But cuts are not allowed to occur inside formulas, i.e. all $\oplus$-nodes are roots. A *pre-proof graph with cuts* is a pre-proof graph $P \triangleright \Gamma$, where Γ may contain cuts. The $\oplus$-nodes have the same geometric behavior as the $\otimes$-nodes. Therefore the correctness criterion stays literally the same, and we can define *proof graphs with cuts* and *proof nets with cuts* accordingly. In the translation from sequent proofs containing the cut rule into pre-proof graphs with cuts, the cut is treated as follows:

$$\text{cut } \frac{\Gamma, A \quad A^\perp, \Delta}{\Gamma, \Delta} \quad \rightsquigarrow \quad \text{cut } \frac{P \triangleright \Gamma, A \quad Q \triangleright A^\perp, \Delta}{P \mathbin{⅋} Q \triangleright \Gamma, A \oplus A^\perp, \Delta} \qquad .$$

---

[2] If Γ consists of only one formula, then we have an object which is in [BC99] called a *bipartite proof net*. In fact, two proof graphs (in our sense) are equivalent if and only if the two linkings (seen as formulas) are isomorphic (in the sense of [BC99]).

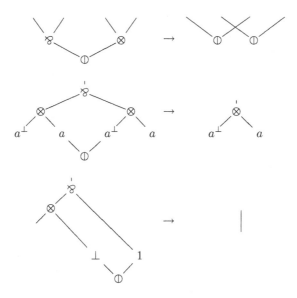

Fig. 4. Cut elimination reduction steps

Since the $\mathbb{1}$ behaves in the same way as the $\otimes$, we immediately have the generalization of the sequentialization:

4.1   Theorem.   *A pre-proof net with cuts is sequentializable if and only if it is correct, i.e. it is a proof net with cuts.*

On the set of cut pre-proof graphs we can define the cut reduction relation $\rightarrow$ as follows:

$$\begin{array}{ccc}
P & & P \\
\triangledown & \rightarrow & \triangledown \\
(A \,\mathbin{\text{⅋}}\, B) \oplus (A^\perp \otimes B^\perp), \Gamma & & A \oplus A^\perp, B \oplus B^\perp, \Gamma
\end{array}$$

$$\begin{array}{ccc}
P\{(a_h^\perp \otimes a_i) \,\mathbin{\text{⅋}}\, (a_j^\perp \otimes a_k)\} & & P\{a_h^\perp \otimes a_k\} \\
\triangledown & \rightarrow & \triangledown \\
a_i \oplus a_j^\perp, \Gamma & & \Gamma
\end{array}$$

$$\begin{array}{ccc}
P\{(Q \otimes \perp_i) \,\mathbin{\text{⅋}}\, 1_j\} & & P\{Q\} \\
\triangledown & \rightarrow & \triangledown \\
\perp_i \oplus 1_j, \Gamma & & \Gamma
\end{array}$$

These reduction steps are shown in graphical notation in Figure 4.

4.2   Lemma.   *If $P \rhd \Gamma$ is correct and $P \rhd \Gamma \rightarrow P' \rhd \Gamma'$, then $P' \rhd \Gamma'$ is also correct.*

Proof: It is impossible that a cut reduction step introduces a cycle in a switching or makes it disconnected. $\qquad\square$

Observe that it can happen that in a proof graph no reduction is possible, although there are cuts present in the sequent. For example, in

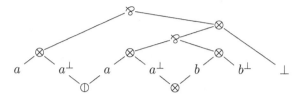

the cut cannot be reduced.

In a given proof graph $P \triangleright \Gamma$, a $①$-node that can be reduced will be called *ready*. Obviously, a cut on a $⊗$-$⅋$-pair is always ready, but for a cut on atoms or units this is not necessarily the case, as the example above shows. However, we have the following theorem:

4.3   Theorem. *Given a proof graph $P \triangleright \Gamma$ and a $①$-node in Γ, there is an equivalent proof graph $P' \triangleright \Gamma$, in which that $①$-node is ready, i.e. can be reduced.*

This is an immediate consequence of the following two lemmas.

4.4   Lemma. *For every proof graph $P \triangleright a_i ① a_j^\perp, \Gamma$ that contains an atomic cut, there is an equivalent proof graph $P'\{(a_h^\perp \otimes a_i) ⅋ (a_j^\perp \otimes a_k)\} \triangleright a_i ① a_j^\perp, \Gamma$.*

4.5   Lemma. *For every proof graph $P \triangleright \perp_i ① 1_j, \Gamma$ that contains a cut on the units, there is an equivalent proof graph $P'\{(Q \otimes \perp_i) ⅋ 1_j\} \triangleright \perp_i ① 1_j, \Gamma$.*

For proving them, we will use the following three lemmas.

4.6   Lemma. *Let $P\{(\perp_k \otimes R\{x_i\}) ⅋ (S\{x_j^\perp\} \otimes \perp_h)\} \triangleright x_i ① x_j^\perp, \Gamma$ be a proof graph, where x is an arbitrary atom or a unit, and $x^\perp$ its dual. Then at least one of $P\{\perp_k \otimes (R\{x_i\} ⅋ (S\{x_j^\perp\} \otimes \perp_h))\} \triangleright x_i ① x_j^\perp, \Gamma$ and $P\{((\perp_k \otimes R\{x_i\}) ⅋ S\{x_j^\perp\}) \otimes \perp_h\} \triangleright x_i ① x_j^\perp, \Gamma$ is equivalent to it.*

4.7   Lemma. *Let $P\{(\perp_k \otimes R\{x_i\}) ⅋ (x_j^\perp \otimes Q)\} \triangleright x_i ① x_j^\perp, \Gamma$ be a proof graph, where x is an arbitrary atom or a unit, and $x^\perp$ its dual. Then $P\{\perp_k \otimes (R\{x_i\} ⅋ (x_j^\perp \otimes Q))\} \triangleright x_i ① x_j^\perp, \Gamma$ is equivalent to it.*

4.8   Lemma. *Let $P\{(\perp_k \otimes R\{x_i\}) ⅋ x_j^\perp\} \triangleright x_i ① x_j^\perp, \Gamma$ be a proof graph, where x is an arbitrary atom or a unit, and $x^\perp$ its dual. Then $P\{\perp_k \otimes (R\{x_i\} ⅋ x_j^\perp)\} \triangleright x_i ① x_j^\perp, \Gamma$ is equivalent to it.*

Proof of Lemma 4.4 (Sketch): Since the proof graph is correct, the linking P must be of the shape $P''\{R\{a_h^\perp \otimes a_i\} ⅋ S\{a_j^\perp \otimes a_k\}\}$ for some contexts $P''\{\ \}$, $R\{\ \}$ and $S\{\ \}$. The contexts $R\{\ \}$ and $S\{\ \}$ can be reduced to $\{\ \}$ by applying Lemma 4.6 and Lemma 4.7 repeatedly.

Proof of Lemma 4.5 (Sketch): Similar to Lemma 4.4, but in this case we also need Lemma 4.8.

Let us now extend the relation $\rightarrow$ to proof nets as follows: $[P \triangleright \Gamma] \rightarrow [Q \triangleright \Delta]$ if an only if there are proof graphs $P' \triangleright \Gamma$ and $Q' \triangleright \Delta$ such that

$$P \triangleright \Gamma \sim P' \triangleright \Gamma \rightarrow Q' \triangleright \Delta \sim Q \triangleright \Delta \quad .$$

4.9   Lemma. *There is no infinite sequence*

$$[P \triangleright \Gamma] \rightarrow [P' \triangleright \Gamma'] \rightarrow [P'' \triangleright \Gamma''] \rightarrow \cdots$$

Proof: In each reduction step the size of the sequent (i.e. the number of $\otimes$, $\otimes$ and $\oplus$-nodes) is reduced.   □

4.10   Lemma. *Let* $P \triangleright \Gamma \sim P' \triangleright \Gamma$, *and let* $P \triangleright \Gamma \rightarrow Q \triangleright \Delta$ *and* $P' \triangleright \Gamma \rightarrow Q' \triangleright \Delta$, *i.e. in both reductions the same cut is reduced. Then we have* $Q \triangleright \Delta \sim Q' \triangleright \Delta$.

Proof: Easy case analysis.   □

4.11   Lemma. *If* $Q \triangleright \Delta \leftarrow P \triangleright \Gamma \rightarrow R \triangleright \Sigma$, *then either* $Q \triangleright \Delta = R \triangleright \Sigma$, *or there is a proof graph* $S \triangleright \Phi$ *such that* $Q \triangleright \Delta \rightarrow S \triangleright \Phi \leftarrow R \triangleright \Sigma$.

4.12   Lemma. *If* $[Q \triangleright \Delta] \leftarrow [P \triangleright \Gamma] \rightarrow [R \triangleright \Sigma]$, *then either* $[Q \triangleright \Delta] = [R \triangleright \Sigma]$, *or there is a proof net* $[S \triangleright \Phi]$ *such that* $[Q \triangleright \Delta] \rightarrow [S \triangleright \Phi] \leftarrow [R \triangleright \Sigma]$.

Proof (Sketch): Let $\oplus_1$ denote the cut that is reduced in Γ to obtain Δ and $\oplus_2$ the one that is reduced to obtain Σ. The basic idea is to apply Theorem 4.3 in order to make both cuts ready at the same time and then apply Lemma 4.11 and Lemma 4.10. There is essentially only one case in which it is not possible to make both cuts ready at the same time, namely, when they use the same axiom link. In other words, $P \triangleright \Gamma$ is of the following shape:

$$P'\{(P''\{a_h^\perp \otimes a_i\} \otimes P'''\{a_j^\perp \otimes a_k\}) \otimes P''''\{a_l^\perp \otimes a_m\}\}$$
$$\triangledown$$
$$a_i \oplus_1 a_j^\perp, a_k \oplus_2 a_l^\perp, \Phi$$

But whatever order of reduction is used, in both cases we get something of the shape $S'\{a_h^\perp \otimes a_m\} \triangleright \Phi$.   □

4.13   Theorem. *The cut elimination reduction* $\rightarrow$ *on proof nets is strongly normalizing. The normal forms are cut free proof nets.*

Proof: Termination is provided by Lemma 4.9 and confluence follows from Lemma 4.12. That the normal form is cut free is ensured by Theorem 4.3.   □

5   *-Autonomy

For any formula A, we can provide an identity proof net $\mathrm{id}_A = [I_A \triangleright A^\perp, A]$, where I_A is called the *identity linking* which is defined inductively on A as follows:

$$\begin{aligned}
I_a &\quad = I_{a^\square} &&= a \otimes a^\perp \\
I_\perp &\quad = I_1 &&= \perp \otimes 1 \\
I_{A \otimes B} &\quad = I_{A \otimes B} &&= I_A \otimes I_B
\end{aligned}$$

Observe that we can have that $I_A = I_{A^\square}$ because changing the order of the arguments of a $\otimes$ or $\otimes$ in the linking of a proof graph does not change the proof net (see 2.4).

Furthermore, for any two proof nets $f = [P \triangleright A^\perp, B]$ and $g = [Q \triangleright B^\perp, C]$, we can define their composition $g \circ f$ to be the result of the cut elimination procedure to $[P \mathbin{⅋} Q \rhd A^\perp, B \oplus B^\perp, C]$. That this is well-defined and associative follows almost immediately from the strong normalization of cut elimination. We also have that $f \circ \mathrm{id}_A = f = \mathrm{id}_B \circ f$.

This gives rise to a category $\mathbf{PN}(\mathcal{A})$ whose objects are the MLL formulas built over $\mathcal{A} \cup \mathcal{A}^\perp \cup \{\perp, 1\}$, and whose arrows are the proof nets. More precisely, the arrows between two objects A and B are the (cut-free) proof nets $[P \rhd A^\perp, B]$. The operation $\otimes$ on formulas can be extended to a bifunctor $\otimes : \mathbf{PN}(\mathcal{A}) \times \mathbf{PN}(\mathcal{A}) \to \mathbf{PN}(\mathcal{A})$ by defining for two arrows $f = [P \rhd A^\perp, B]$ and $g = [Q \rhd C^\perp, D]$ the arrow $f \otimes g = [P \mathbin{⅋} Q \rhd A^\perp \mathbin{⅋} C^\perp, B \otimes D]$. It can easily be seen that this bifunctor makes our category symmetric monoidal (with unit 1): The basic natural isomorphisms demanded by the definition (associativity, right unit, left unit, symmetry) are

$$\alpha_{A,B,C} = [I_A \mathbin{⅋} I_B \mathbin{⅋} I_C \rhd A^\perp \mathbin{⅋} (B^\perp \mathbin{⅋} C^\perp), (A \otimes B) \otimes C]$$
$$\rho_A = [\perp \otimes I_A \rhd A^\perp \mathbin{⅋} \perp, A]$$
$$\lambda_A = [\perp \otimes I_A \rhd \perp \mathbin{⅋} A^\perp, A]$$
$$\sigma_{A,B} = [I_A \mathbin{⅋} I_B \rhd A^\perp \mathbin{⅋} B^\perp, B \otimes A]$$

It is easy to check these are indeed proof nets, that α, ρ, λ, and σ are natural isomorphisms for all formulas A, B, and C, and that the corresponding diagrams (see [BW99]) commute.

Furthermore, we can exhibit the (contravariant) duality functor $(-)^\perp$ whose object function has already been defined. For an arrow $f = [P \rhd A^\perp, B] : A \to B$ let $f^\perp = [P \rhd B, A^\perp] : B^\perp \to A^\perp$. This determines a symmetric *-autonomous category structure [Bar79, BW99]. In particular, we define the bifunctor $- \mathbin{⅋} -$ as $A \mathbin{⅋} B = (A^\perp \otimes B^\perp)^\perp$ and its unit object as $\perp = 1^\perp$. The last thing to check is that we have the natural bijection

$$\mathrm{Hom}(A \otimes B, C) \cong \mathrm{Hom}(A, B^\perp \mathbin{⅋} C)$$
$$[P \rhd A^\perp \mathbin{⅋} B^\perp, C] \mapsto [P \rhd A^\perp, B^\perp \mathbin{⅋} C] \quad .$$

6   The Free *-Autonomous Category

In this section we will show that the category of proof nets is the free symmetric *-autonomous category. Let $\mathcal{A}$ be a set and let $\eta_{\mathcal{A}} : \mathcal{A} \to \mathrm{Obj}(\mathbf{PN}(\mathcal{A}))$ be the function that maps every element of $\mathcal{A}$ to itself seen as atomic formula. To say that $\mathbf{PN}(\mathcal{A})$ is the *free *-autonomous category generated by* $\mathcal{A}$ amounts to saying that

6.1   Theorem. *For any *-autonomous category*[3] $(\mathcal{C}, \otimes, 1_{\mathcal{C}}, (-)^\perp)$ *and any map* $G^\circ : \mathcal{A} \to \mathrm{Obj}(\mathcal{C})$, *there is a unique functor* $G : \mathbf{PN}(\mathcal{A}) \to \mathcal{C}$, *preserving the *-autonomous structure, such that* $G^\circ = \mathrm{Obj}(G) \circ \eta_{\mathcal{A}}$, *where* $\mathrm{Obj}(G)$ *is the restriction of* G *on objects.*

The remainder of this section is devoted to a sketch the proof of this theorem. For this we will introduce the following notation.

---

[3] For simplicity we assume that for every object C of $\mathcal{C}$ we have $C^{\perp\perp} = C$. This can be relaxed to a natural isomorphism by standard trickery.

Let I be an index set. A *bracketing* of I is given by a total order $I = \{i_1, \ldots, i_k\}$ and a binary tree structure whose set of leaves is I, such that the order is respected. We will denote bracketings of I also by I. The whole point of this is, given an I-indexed family $(C_i)_{i \in I}$ of objects of $\mathcal{C}$, that we can write $\bigotimes_I \{C_{i_1}, \ldots, C_{i_k}\}$ to denote the object of $\mathcal{C}$ that is obtained by applying the functor $- \otimes -$ according to the bracketing I. By a standard theorem of symmetric monoidal categories, any two objects obtained from different bracketings of the same set have a unique "coherence" isomorphism between them. Notice that this will involve the symmetry only if the order differs on the bracketings. Similarly, $\bigotimes_I \{C_{i_1}, \ldots, C_{i_k}\}$ is defined. For empty I, let $\bigotimes_\emptyset \emptyset = 1_{\mathcal{C}}$ and $\bigotimes_\emptyset \emptyset = \bot_{\mathcal{C}} = 1_{\mathcal{C}}^{\bot}$. The purpose of this notation is to state the following property of *-autonomous categories.

6.2   Proposition. *Let $\mathcal{C}$ be a *-autonomous category, and let $C_1, \ldots, C_n$ be objects of $\mathcal{C}$. Let $I, J \subseteq \{1, \ldots, n\}$, and let $\complement I = \{1, \ldots, n\} \setminus I$ and $\complement J = \{1, \ldots, n\} \setminus J$ be their complements. Then for all bracketings of $I, J, \complement I, \complement J$, we have a natural bijection between $\mathrm{Hom}_{\mathcal{C}} \left(\bigotimes_I \{C_i^{\bot} \mid i \in I\}, \bigotimes_{\complement I} \{C_i \mid i \in \complement I\} \right)$ and $\mathrm{Hom}_{\mathcal{C}} \left(\bigotimes_J \{C_i^{\bot} \mid i \in J\}, \bigotimes_{\complement J} \{C_i \mid i \in \complement J\} \right)$.*

Proof: The proof is done by repeatedly applying the associativity and commutativity of the two functors $- \otimes -$ and $- \bigotimes -$, the natural isomorphisms for the units, and the natural bijection $\mathrm{Hom}_{\mathcal{C}}(A \otimes B^{\bot}, C) \cong \mathrm{Hom}_{\mathcal{C}}(A, B \bigotimes C)$, which is imposed by the *-autonomous structure.   □

Let now the *-autonomous category $\mathcal{C}$ and the embedding $G^{\circ} : \mathcal{A} \to \mathrm{Obj}(\mathcal{C})$ be given. We will exhibit the functor $G : \mathbf{PN}(\mathcal{A}) \to \mathcal{C}$ which has the desired properties. On the objects, this functor is uniquely determined as follows:

$$G(a) = G^{\circ}(a) \qquad G(\bot) = \bot_{\mathcal{C}} \qquad G(A \bigotimes B) = G(A) \bigotimes G(B)$$
$$G(a^{\bot}) = G^{\circ}(a)^{\bot} \qquad G(1) = 1_{\mathcal{C}} \qquad G(A \otimes B) = G(A) \otimes G(B)$$

There is no other choice since the objects $1_{\mathcal{C}}$ and $\bot_{\mathcal{C}}$ in $\mathcal{C}$, as well as the functors $(-)^{\bot}$, $- \otimes -$, and $- \bigotimes -$ are uniquely determined by the *-autonomous structure on $\mathcal{C}$.

For defining G on the morphisms, the situation is not as simple. We will first ignore the fact that the units are units and interpret a proof graph (with cuts) $P \triangleright \Gamma$ as an ordinary unit-free proof net with conclusions $A_0, \ldots, A_n, B_1 \oplus B_1^{\bot}, \ldots, B_m \oplus B_m^{\bot}$, where $A_0 = P^{\star}$ (see Observation 3.2), $A_1, \ldots, A_n$ are the formulas in Γ that are not cuts, and $B_1 \oplus B_1^{\bot}, \ldots, B_m \oplus B_m^{\bot}$ are the cuts in Γ. To each such object we will uniquely assign a family of morphisms

$$\bigotimes_I \{G(A_i)^{\bot} \mid i \in I\} \to \bigotimes_{\complement I} \{G(A_i) \mid i \in \complement I\}$$

indexed by the bracketings on the subsets $I \subseteq \{0, \ldots, n\}$ and their complements. Proposition 6.2 ensures that every member of such a family of morphisms determines the others uniquely. The construction is done by induction on the size of the proof graph, using Lemma 3.3. (In fact, it is quite similar to the sequentialization.)

Observe that in particular this construction gives us for each proof graph $P \triangleright A^{\bot}, B$ a unique arrow $\psi_{P \triangleright A^{\bullet}, B} : G(P^{\star})^{\bot} \to G(A^{\bot}) \bigotimes G(B)$. Furthermore, observe that for every linking P, the object $G(P^\star)^\perp$ in $\mathcal{C}$ is isomorphic to $\bigotimes_{a \otimes a^\bullet} \{G(a) \,\mathbin{\rotatebox[origin=c]{180}{$\&$}}\, G(a)^\perp\}$, where $a \otimes a^\perp$ ranges over the axiom links in P. This means that the *-autonomous structure on $\mathcal{C}$ uniquely determines a morphism $\phi_P : 1_\mathcal{C} \to G(P^\star)^\perp$. This can be composed with $\psi_{P \triangleright A^\bullet, B}$ to get $\xi_{[P \triangleright A^\bullet, B]} : 1_\mathcal{C} \to G(A^\perp) \,\mathbin{\rotatebox[origin=c]{180}{$\&$}}\, G(B)$. That this is well-defined, is ensured by the following lemma (in which we no longer ignore the fact that the units are units).

6.3   Lemma.   *If* $Q \triangleright A^\perp, B \sim P \triangleright A^\perp, B$, *then* $\xi_{[P \triangleright A^\bullet, B]} = \xi_{[Q \triangleright A^\bullet, B]}$.

Consequently, to each proof net $f = [P \triangleright A^\perp, B]$, we can uniquely assign the arrow $G(f) : G(A) \to G(B)$ that is determined by $\xi_{[P \triangleright A^\bullet, B]}$ via Proposition 6.2.

It remains to show that $G : \mathbf{PN}(\mathcal{A}) \to \mathcal{C}$ is indeed a functor (i.e. identities and composition are preserved). That for each formula A, the proof $[I_A \triangleright A^\perp, A]$ is mapped to identity $\mathrm{id} : G(A) \to G(A)$ is an easy induction on the structure of A and left to the reader. The preservation of composition is ensured by the following lemma.

6.4   Lemma.   *Let* $T \triangleright \Gamma \to S \triangleright \Delta$, *i.e. the proof graph* $S \triangleright \Delta$ *is obtained from* $T \triangleright \Gamma$ *by applying a single cut reduction step. Then* $\xi_{[T \triangleright \Gamma]}$ *and* $\xi_{[S \triangleright \Delta]}$ *denote the same morphism* $1_\mathcal{C} \to \mathbin{\rotatebox[origin=c]{180}{$\&$}} \{G(A_1), \ldots, G(A_n)\}$, *where* $A_1, \ldots, A_n$ *are the formulas in* Γ *(resp.* Δ*) that are not cuts.*

It might be worth mentioning, that Theorem 6.1 provides a decision procedure for the equality of morphisms in the free symmetric *-autonomous category, which is in our opinion simpler than the ones provided in [BCST96] and [KO99].

7   Conclusion

We think we made a convincing case for the the cleanest approach yet to proof nets with the multiplicative units. There is always the possibility that another "ideology" than category theory will arise and will tell us to identify sequent proofs in a different way, perhaps collapsing fewer proofs, and help us construct more rigid proof objects. But we doubt very much that such a thing exists, given that the permutation rules that category theory imposes on the sequent calculus are so natural and so hard to weaken.

There are some issues that are left open and that we want to explore in the future:

- The relation with the new proof formalism called the calculus of structures [GS01, BT01]. We should mention that the idea behind our approach originates from the new viewpoints that are given by the calculus of structures.
- The addition of additives to our theory. This should not be very hard, given the work done in [HvG03]. The true challenge is to include also the additive units.
- The development of a theory of proof nets for classical logic. The problem is finding the right extension of the axioms of a *-autonomous category, such that on the one hand classical proofs are identified in a natural way, and on the other hand there is no collapse into a boolean algebra.

– The search for meaningful invariants. It is very probable that the equivalence classes of graphs we define have a geometric meaning, and can be related to more abstract invariants like those given by homological algebra. We are convinced that the work in in [Mét94] is only the tip of the iceberg.

References

[Bar79]   Michael Barr. *-Autonomous Categories*, volume 752 of *Lecture Notes in Mathematics*. Springer-Verlag, 1979.

[BC99]   Vincent Balat and Roberto Di Cosmo. A linear logical view of linear type isomorphisms. In *Computer Science Logic, CSL 1999*, volume 1683 of *LNCS*, pages 250–265. Springer-Verlag, 1999.

[BCST96]   Richard Blute, Robin Cockett, Robert Seely, and Todd Trimble. Natural deduction and coherence for weakly distributive categories. *Journal of Pure and Applied Algebra*, 113:229–296, 1996.

[Blu93]   Richard Blute. Linear logic, coherence and dinaturality. *Theoretical Computer Science*, 115:3–41, 1993.

[BT01]   Kai Brünnler and Alwen Fernanto Tiu. A local system for classical logic. In R. Nieuwenhuis and A. Voronkov, editors, *LPAR 2001*, volume 2250 of *Lecture Notes in Artificial Intelligence*, pages 347–361. Springer-Verlag, 2001.

[BW99]   Michael Barr and Charles Wells. *Category Theory for Computing Science*. Les Publications CRM, Montréal, third edition, 1999.

[DR89]   Vincent Danos and Laurent Regnier. The structure of multiplicatives. *Annals of Mathematical Logic*, 28:181–203, 1989.

[Gir87]   Jean-Yves Girard. Linear logic. *Theoretical Computer Science*, 50:1–102, 1987.

[Gir91]   Jean-Yves Girard. A new constructive logic: Classical logic. *Mathematical Structures in Computer Science*, 1:255–296, 1991.

[GS01]   Alessio Guglielmi and Lutz Straßburger. Non-commutativity and MELL in the calculus of structures. In Laurent Fribourg, editor, *Computer Science Logic, CSL 2001*, volume 2142 of *LNCS*, pages 54–68. Springer-Verlag, 2001.

[HvG03]   Dominic Hughes and Rob van Glabbeek. Proof nets for unit-free multiplicative-additive linear logic. In *18'th IEEE Symposium on Logic in Computer Science (LICS 2003)*, 2003.

[KL71]   Gregory Maxwell Kelly and Saunders Mac Lane. Coherence in closed categories. *Journal of Pure and Applied Algebra*, 1:97–140, 1971.

[KO99]   Thong-Wei Koh and Chih-Hao Luke Ong. Internal languages for autonomous and *-autonomous categories. In Martin Hofmann, Giuseppe Rosolini, and Dusko Pavlovic, editors, *Proceedings of the 8th Conference on Category Theory and Computer Science, Edinburgh, September 1999*, volume 29 of *Electronic Notes in Theoretical Computer Science*. Elsevier, 1999.

[Laf95]   Yves Lafont. From proof nets to interaction nets. In J.-Y. Girard, Y. Lafont, and L. Regnier, editors, *Advances in Linear Logic*, volume 222 of *London Mathematical Society Lecture Notes*, pages 225–247. Cambridge University Press, 1995.

[Mét94]   François Métayer. Homology of proof nets. *Archive of Mathematical Logic*, 33:169–188, 1994.

[Thi03]   Rüdiger Thiele. Hilbert's twenty-fourth problem. *American Mathematical Monthly*, 110:1–24, 2003.

The Boundary Between Decidability and Undecidability for Transitive-Closure Logics

Neil Immerman[1,*], Alex Rabinovich[2], Tom Reps[3,**], Mooly Sagiv[2,**], and Greta Yorsh[2,***]

[1] Dept. of Comp. Sci. Univ. of Massachusetts
immerman@cs.umass.edu
[2] School of Comp. Sci., Tel Aviv Univ.,
{rabinoa,msagiv,gretay}@post.tau.ac.il
[3] Comp. Sci. Dept., Univ. of Wisconsin
reps@cs.wisc.edu

Abstract. To reason effectively about programs, it is important to have some version of a transitive-closure operator so that we can describe such notions as the set of nodes reachable from a program's variables. On the other hand, with a few notable exceptions, adding transitive closure to even very tame logics makes them undecidable.

In this paper, we explore the boundary between decidability and undecidability for transitive-closure logics. Rabin proved that the monadic second-order theory of trees is decidable, although the complexity of the decision procedure is not elementary. If we go beyond trees, however, undecidability comes immediately.

We have identified a rather weak language called $\exists\forall(\mathrm{DTC}^+[E])$ that goes beyond trees, includes a version of transitive closure, and is decidable. We show that satisfiability of $\exists\forall(\mathrm{DTC}^+[E])$ is NEXPTIME complete. We furthermore show that essentially any reasonable extension of $\exists\forall(\mathrm{DTC}^+[E])$ is undecidable.

Our main contribution is to demonstrate these sharp divisions between decidable and undecidable. We also compare the complexity and expressibility of $\exists\forall(\mathrm{DTC}^+[E])$ with related decidable languages including MSO(trees) and guarded fixed point logics.

We mention possible applications to systems some of us are building that use decidable logics to reason about programs.

1 Introduction

To reason effectively about programs, it is important to have some version of a transitive-closure operator so that we can describe such notions as the set of nodes reachable from a program's variables. On the other hand, with a few notable exceptions, adding transitive closure to even very tame logics makes them undecidable.

In this paper, we explore the boundary between decidability and undecidability for transitive-closure logics. Rabin [13] proved that the monadic second-order theory of trees is decidable, although the complexity of the decision procedure is not elementary. If we go beyond trees, however, undecidability comes immediately.

---

* Supported by NSF grant CCR-0207373 and a Guggenheim fellowship.
** Supported by ONR contract N00014-01-1-0796 and the von Humboldt Foundation.
*** Supported by the Israel Science Foundation.

Modal logics and their extension to the μ calculus have proved quite useful. The μ calculus has an EXPTIME-complete satisfiability problem [3] and the same has been shown true even for the more expressive guarded fixed-point logic, as long as the vocabulary remains of bounded arity [6]. Guarded fixed-point logic can express reachability from a specific constant, or from some point of a specific color, and it can restrict this reachability to be along paths specified, for example, by a regular expression. What it cannot express is a reachability relation between a pair of variables, i.e., that there is a path from u to v.

We have identified a rather weak language, called $\exists\forall(\mathrm{DTC}^+[E])$, that goes beyond trees, includes a version of the latter sort of transitive closure, and is decidable. We show that satisfiability of $\exists\forall(\mathrm{DTC}^+[E])$ is NEXPTIME complete. We furthermore show that **essentially any reasonable extension of** $\exists\forall(\mathrm{DTC}^+[E])$ **is undecidable**.

The main contribution of this paper is to demonstrate the above sharp divisions between decidable and undecidable. We also compare the complexity and expressibility of $\exists\forall(\mathrm{DTC}^+[E])$ with related decidable languages, including MSO(trees) and guarded fixed-point logics.

The main application we have in mind is for the static-analysis methods that we are pursuing. Very generally, we model the properties of an infinite set of data structures that can be generated by the program we are analyzing, using a bounded set of first-order, three-valued structures [14]. In [15], it is shown that this modeling can be improved so that it computes the most precise possible transformation summarizing each program step, through the use of decidable logics.

Furthermore, in [9] we show that we can use a method we call "structure simulation" to significantly extend the sets of data structures that we can model with decidable logics over trees (monadic second-order logic) or graphs ($\exists\forall(\mathrm{DTC}^+[E])$). In the latter case, transitive-closure information must be restricted to deterministic paths.

The advantage of $\exists\forall(\mathrm{DTC}^+[E])$ compared with MSO(trees) is that while the latter is usually much more expressive, we can go beyond trees in the former. As an example, to express reachability in dynamic, undirected graphs, as in [2], we need not only a spanning forest, but a record of all the remaining edges in the undirected graph [9].

Fig. 1 summarizes results concerning the decidability and complexity of satisfiability for relevant logics. All the languages will be defined precisely in the next two sections. For previously known results we include a reference, and for results new to this paper we include the number of the relevant theorem.

Decidable	Complexity	Citation
μ calculus	EXPTIME complete	[3]
Guarded Fixed Point	EXPTIME complete	[6]
MSO(trees)	non-elementary	[13]
FO^2	NEXPTIME complete	[11,4]
$\exists\forall$	Σ_2^p complete	[1]
$\exists\forall(\mathrm{TC}^-)$	Σ_2^p complete	Prop 2
$\exists\forall(\mathrm{DTC}^+[E])$	NEXPTIME complete	Th 4, 5
$\exists\forall(\mathrm{TC}, f)$	NEXPTIME complete	Cor 6

Undecidable	Citation
$\mathrm{FO}^2(\mathrm{TC})$	[5]
$\mathrm{FO}^2(\mathrm{DTC})$	[5]
$\forall(\mathrm{TC}^+[E])$	Cor 9
$\forall(\mathrm{DTC}^+)$	Th 8
$\forall(\mathrm{DTC}^-[E])$	Th 13

Fig. 1. Summary of the decidability and complexity, and the undecidability of the logics we study. The arity of all relation symbols is bounded. The results are the same for $\forall$ and $\exists\forall$, and they are the same for the satisfiability and finite-satisfiability problems

2   Background and Tiling

As we have mentioned, being able to express reachability is crucial for our applications. However, adding a transitive-closure operator tends to make even very tame logics undecidable. We use $TC_{u,u'}[\varphi]$ to denote the reflexive, transitive closure of binary relation $\varphi(u, u')$ [8]. **Note:** In this paper, we confine our attention to applications of $TC[\varphi]$ for which φ is quantifier-free and TC-free. Furthermore, we assume throughout that the arity of all relation symbols is bounded.[1]

For example, consider the simple, decidable logic FO^2. This is first-order logic restricted to having only two variables, x, y. Grädel et al. [5] prove that if we add the transitive-closure operator (TC) to FO^2 then the resulting logic is undecidable. In fact, they prove that even $FO^2(DTC)$ is undecidable. Here DTC — deterministic transitive closure — is the restriction of transitive closure to paths that have no choices. For the binary relation $E(x, y)$, define $E_d(x, y)$ as follows:

$$E_d(x, y) \overset{\text{def}}{=} E(x, y) \ \wedge \ \forall z(E(x, z) \to z = y) \ .$$

That is, if vertex v has more than one outgoing E-edge, then it has no outgoing E_d-edges. Then define DTC as follows: $DTC[E] \overset{\text{def}}{=} TC[E_d]$.

It is surprising that $FO^2(DTC)$ is undecidable, but the proof is that even this seemingly very weak language is strong enough to express tilings.

Definition 1. *Define a* tiling problem, $\mathcal{T} = \langle T, R, D \rangle$, *to consist of a finite list of tile types,* $T = [t_0, \ldots t_k]$, *together with horizontal and vertical adjacency relations,* $R, D \subseteq T^2$. *Here* $R(a, b)$ *means that tiles of type b fit immediately to the right of tiles of type a, and* $D(a, b)$ *means that tiles of type b fit one step down from those of type a. A* solution *to a tiling problem is an arrangement of instances of the tiles in a rectangular grid such that a* t_0 *tile occurs in the top left position, and a* t_k *tile occurs in the bottom right position, and all adjacency relationships are respected.*

Given a Turing machine, M, and an input, w, we can build a tiling problem, $\mathcal{T}$, of size $O(|M| + |w|)$, such that $\mathcal{T}$ has a solution iff M on input w eventually halts. Here any correct tiling solution would represent an accepting computation of M on input w. Think of t_0 as representing the initial state and t_k as representing the final accepting state. Thus, as is well known, any logic that can express tilings has undecidable finite satisfiability – and general satisfiability – problems.

(Standard definitions of tiling problems only require t_0 at the top left, and do not also ask for t_k at the lower right. This minor change does not affect the undecidability and complexity results, but makes some of our constructions slightly simpler.) See [1] for a nice treatment of tiling problems, as well as discussions of many relevant decidable and undecidable logics.

3   Decidability of $\exists\forall(DTC^+[E])$

We start with the first-order logic $\exists\forall$, consisting of first-order formulas in prenex form with all existential quantifiers preceding all universal quantifiers. The vocabulary has

---

[1] For our intended applications, arity 2 is sufficient and arity 3 is a luxury. In theory, an unbounded arity can significantly increase some of the complexity bounds.

no function symbols. It is well known and easy to see that the satisfiability problem for $\exists\forall$ is decidable: Let $\varphi \in \exists\forall$. Form the Skolemization, φ_S, by replacing the existential quantifiers, $\exists x_1, \ldots, x_k$, by new constants, $c_1, \ldots, c_k$. Suppose $\mathcal{A} \models \varphi_S$. Let $\mathcal{C}$ be the substructure of $\mathcal{A}$ whose universe consists of the constant symbols appearing in φ_S. Since φ_S is universal, we have that $\mathcal{C} \models \varphi_S$. Thus, φ has a model iff it has a small model, i.e., one of size less than $|\varphi|$. We say that $\exists\forall$ has the *small-model property*, in this case with models of at most linear size. To test if a universal formula, φ_S, is satisfiable, we would guess a structure, $\mathcal{A}$, of size at most $n = |\varphi_S|$ and then check that $\mathcal{A} \models \varphi_S$. Testing whether a given structure satisfies an input universal first-order formula is co-NP complete. Thus satisfiability of $\exists\forall$ formulas is in, and in fact complete for, Σ_2^p, the second-level of the polynomial-time hierarchy.

Since the existential quantifiers in $\exists\forall$ formulas can be eliminated by adding constants, we limit our discussion to universal formulas. Let $\forall(\text{DTC})$ consist of universal formulas in which DTC may occur. Unfortunately, as we will see, satisfiability of $\forall(\text{DTC})$ and $\forall(\text{TC})$ are undecidable (Theorem 8).

It is the positive occurrences of TC that cause the satisfiability of $\forall(\text{TC})$ to be undecidable. Let $\exists\forall(\text{TC}^-)$ consist of formulas in prenex form in which TC only occurs negatively.

Proposition 2. *Satisfiability and finite satisfiability of $\exists\forall(\text{TC}^-)$ are decidable with complexity complete for Σ_2^p.*

Proof: The above argument for $\exists\forall$ continues to work. If $\varphi \in \exists\forall(\text{TC}^-)$ is satisfiable, let $\mathcal{A} \models \varphi_S$, where φ_S is the Skolemization of φ. As above, let $\mathcal{C}$ be the substructure of $\mathcal{A}$ whose universe consists of the constant symbols appearing in φ_S. Then $\mathcal{C} \models \varphi_S$ because if a path did not exist in $\mathcal{A}$ then it still does not exist in $\mathcal{C}$. (Recall that we only apply TC to quantifier-free formulas.) Furthermore, we can test in polynomial time whether such a path exists in $\mathcal{C}$. Thus, the complexity of satisfiability remains Σ_2^p complete. $\square$

Definition 3. *Define $\exists\forall(\text{DTC}^+[E])$ to be the restriction of $\exists\forall(\text{DTC})$ in which the language has only one binary relation symbol, E, (plus unary relation symbols and constants), and all applications of DTC are positive occurrences of the form $\text{DTC}[E]$. In addition, we include in $\exists\forall(\text{DTC}^+[E])$ **arbitrary negative occurrences** of $\text{TC}[\varphi]$ for φ quantifier-free.[2] However, it is very important that there are **no negative occurrences** of DTC, for otherwise the language would become undecidable (Theorem 13).*

Theorem 4. *$\exists\forall(\text{DTC}^+[E])$ has the small-model property, with models of size at most $2^{O(n^2)}$, where n is the size of the formula. Thus, satisfiability and finite satisfiability of $\exists\forall(\text{DTC}^+[E])$ are decidable, with complexity at most NEXPTIME.*

Proof: Using Skolemization, it suffices to prove these results for $\forall(\text{DTC}[E])$. Let $\varphi \in \forall(\text{DTC}[E])$ be satisfiable and let $\mathcal{A} \models \varphi$. We will show that there exists a model $\mathcal{B} \models \varphi$ such that $\|\mathcal{B}\| \leq 2^{O(n^2)}$. Here $\|\mathcal{B}\|$ denotes the cardinality of the universe of the structure $\mathcal{B}$, and $n = |\varphi|$,

---

[2] A more accurate name for $\exists\forall(\text{DTC}^+[E])$ would really be $\exists\forall(\text{DTC}^+[E], \text{TC}^-)$, but this is a mouthful, and all bounds remain the same whether or not the negative occurrences of TC are allowed.

Let $c_1 \ldots c_k$ be the constants occurring in φ. For each pair of constants, c_i, c_j, such that $\mathcal{A} \models \text{DTC}[E](c_i, c_j)$, there is a unique path p_{ij} from c_i to c_j in $\mathcal{A}$. Let $\mathcal{A}'$ be the substructure of $\mathcal{A}$ whose universe consists of the constants, plus all vertices that lie on any of the paths p_{ij}.

We claim that $\mathcal{A}' \models \varphi$. To see this, first observe that for any two elements a, b of the universe of $\mathcal{A}'$ we have

$$\mathcal{A} \models \text{DTC}[E](a, b) \Rightarrow \mathcal{A}' \models \text{DTC}[E](a, b) \tag{1}$$

(The proof of Theorem 12 exploits the fact that the converse need not hold.) Since a and b occur on paths p_{ij}, if $\mathcal{A} \models \text{DTC}[E](a, b)$ then the path from a to b must be along the paths p_{ij}. Thus $\mathcal{A}' \models \text{DTC}[E](a, b)$ holds as well.

Since $\mathcal{A}'$ is a substructure of $\mathcal{A}$ and φ is a universal formula with only positive occurrences of DTC, it follows from Equation (1) that $\mathcal{A}' \models \varphi$. (Note that the negative occurrences of $\text{TC}[\varphi]$ with φ quantifier-free do not cause problems: since $\mathcal{A}'$ is a substructure of $\mathcal{A}$ it follows that if $\mathcal{A} \models \neg\text{TC}[\varphi](a, b)$, then $\mathcal{A}' \models \neg\text{TC}[\varphi](a, b)$ as well.)

Structure $\mathcal{A}'$ consists of a set of "trees" directed from leaf to root, all of whose leaves and roots are constants; however, (1) some of the "trees" may end in a cycle rather than a root; and (2) multiple edges may occur from some of the roots to other vertices. Note that if there is more than one edge from vertex v, then v does not occur on any DTC path, except perhaps as the last vertex. For this reason, if there are multiple edges in $\mathcal{A}$ from constant c_i, then we can remove all such edges and replace them by a new unary relation symbol Q_i true of all the vertices that had edges from c_i; as long as we modify φ accordingly. (In particular, we would change all occurrences of "$E(x, y)$" to "$E(x, y) \vee (x = c_i \wedge Q_i(y))$".) Because we can eliminate issue (2), we henceforth assume that the graph $\mathcal{A}'$ has outdegree at most one.

Note that some of the paths, $p_{ij}, p_{i'j'}$ may intersect. If so, for simplicity we identify the first point of intersection for each pair of paths as a new constant. Observe that there are a total of at most $k - 1$ such new constant symbols. Thus from now on we will only consider *direct paths* p_{ij} containing no intermediate constants. See Fig. 2 for an example graph where constants c_7, c_8, and c_9 have been added.

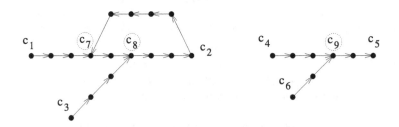

Fig. 2. Example $\mathcal{A}'$ from proof of Theorem 4 after constants c_7, c_8, c_9 have been added

After these normalization steps, $\mathcal{A}'$ consists of k' constants and at most k' direct paths, p_{ij}, where $k' \leq 2k - 1$. Let r be the number of unary relation symbols, and m be the number of (universal) quantifiers in φ. We claim that no direct path p_{ij} need have length greater than $2^{rm} + m + 1$. Suppose on the contrary that the length of p_{12} is greater than $2^{rm} + m + 1$. Let the color of a vertex be the set of unary relation symbols that it satisfies. There are 2^r possible colors and 2^{rm} possible m-tuples of colors; consequently there must be at least two identically colored consecutive m-tuples, $u_1, \ldots, u_m$, and $v_1, \ldots, v_m$, in the interior of p_{12}. (By "consecutive" we mean the m-tuple is a path.) Form the structure $\mathcal{B}$ from $\mathcal{A}'$ by deleting vertices u_2 through v_1 and adding an edge from u_1 to v_2.

We claim that $\mathcal{B} \models \varphi$. It suffices to show that for any m-tuple of vertices from $\mathcal{B}$, $b_1, b_2, \ldots, b_m$, there is a corresponding, isomorphic[3] m-tuple from $\mathcal{A}'$, $a_1, a_2, \ldots, a_m$. Note that every vertex in $\mathcal{B}$ is in $\mathcal{A}'$, and furthermore, the only difference between $\mathcal{B}$ and $\mathcal{A}'$ concerning these vertices is that $E(u_1, v_2)$ holds in $\mathcal{B}$ but not in $\mathcal{A}'$.

If any b_i is not on the path p_{12}, then we let a_i be the identical vertex in $\mathcal{A}'$. We may thus confine our attention to the most difficult case, namely, that $b_1, b_2, \ldots, b_m$ are all in the path p_{12}. Assume for simplicity that they occur in order. Our only problem is if for some ℓ, $b_\ell = u_1$ and $b_{\ell+1} = v_2$. In this case, we let $a_t = b_t$ for $t \leq \ell$, but we let $a_{\ell+1} = u_2$. Similarly, if $b_{\ell+i-1} = v_i$ for all $i \in \{2, \ldots s\}$, then we must let $a_{\ell+i-1} = u_i$. Consider the first gap (if any), i.e., b_i and b_{i+1} are not consecutive. We have that $b_i = v_z$ and $a_i = u_z$, for some z. We can let $a_j = b_j$ for $j > i$, see Fig. 3. Note that we have replaced some v_i's by u_i's but all unary relations, edge relations and connectivity have been preserved. Thus, as desired, $a_1, a_2, \ldots, a_m$ is isomorphic to $b_1, b_2, \ldots, b_m$.

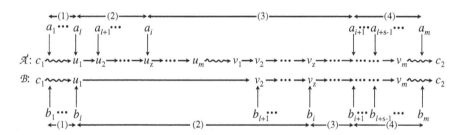

Fig. 3. Illustration of how for every m-tuple of vertices $b_1, \ldots, b_m$ from $\mathcal{B}$ there is a corresponding isomorphic m-tuple of vertices $a_1, \ldots, a_m$ from $\mathcal{A}'$. In region (2) of $\mathcal{B}$, $b_l, \ldots, b_i$ are assigned consecutive vertices; similarly, in region (2) of $\mathcal{A}'$, $a_l, \ldots, a_i$ are assigned consecutive vertices. Because b_i and b_{i+1} are separated by two or more E edges in region (3) of $\mathcal{B}$ (i.e., there a "gap"), the assignments for $a_{i+1}, \ldots, a_m$ in region (4) of $\mathcal{A}'$ can match those for $b_{i+1}, \ldots, b_m$ in region (4) of $\mathcal{B}$ exactly

Thus $\mathcal{B} \models \varphi$ as desired. We can continue shortening any remaining paths of length greater than $2^{rm} + m + 1$. It follows that there is a model $\mathcal{B}$ of φ and $\|\mathcal{B}\| \leq (2k - 1)(2^{rm} + m + 1) \leq 2^{|\varphi|^2}$, as desired. $\qquad\square$

It follows from Theorem 4 that the satisfiability of $\exists\forall(\text{DTC}^+[E])$ formulas can be checked in NEXPTIME. We next show that this cannot be improved.

Theorem 5. *The satisfiability of $\exists\forall(\text{DTC}^+[E])$ formulas is NEXPTIME-complete.*

---

[3] More explicitly, we mean that the map taking each b_i to a_i is an isomorphism of the induced substructures of $\mathcal{B}$ and $\mathcal{A}'$ generated by $b_1, \ldots, b_m$ and $a_1, \ldots, a_m$, respectively. This may be thought of as an Ehrenfeucht-Fraïssé game in which the spoiler chooses the b_i's and the duplicator answers with the a_i's [8].

Proof: Let $\mathcal{T}$ be a tiling problem as in Definition 1, and let n be a natural number. It is an NEXPTIME-complete problem to test on input $(\mathcal{T}, 1^n)$ whether there is a $\mathcal{T}$-tiling of a square grid of size 2^n by 2^n [12].

We will define a formula φ_n that expresses exactly a solution to this tiling problem. There will be two constants: s, denoting the cell in the upper-left corner, and t, denoting the cell in the lower-right corner. The desired model will consist of 2^{2n} tiles:

$$s = [1,1,t_0] \cdots [1,2^n,t]; \quad [2,1,t'] \cdots [2,2^n,t'']; \quad \cdots [2^n,1,t'''] \cdots [2^n,2^n,t_k] = t$$

The binary relation E will hold between each pair of consecutive tiles, including, for example, $[1,2^n,t]$ and $[2,1,t']$. We will include the following unary relation symbols: $H_1,\ldots H_n$, indicating the horizontal position as an n-bit number; $V_1,\ldots V_n$, indicating the vertical position; and $T_0,\ldots T_k$, indicating the tile type.

The formula φ_n is the conjunction of the following assertions:

1. $T_0(s) \;\wedge\; \bigwedge\limits_{i=1}^{n}\big(\neg H_i(s) \wedge \neg V_i(s)\big) \;\wedge\; T_k(t) \;\wedge\; \bigwedge\limits_{i=1}^{n}\big(H_i(t) \wedge V_i(t)\big)$

2. $\forall x \;\bigwedge\limits_{0 \le i < j \le k} \neg(T_i(x) \wedge T_j(x))$

3. $\forall x,y\big((\mathrm{Suc}_v(x,y) \to \mathrm{Vert}(x,y)) \;\wedge\; (\mathrm{Suc}_h(x,y) \to \mathrm{Hor}(x,y))\big)$

4. $\mathrm{DTC}[E](s,t) \;\wedge\; \forall x,y\big(E(x,y) \to \mathrm{Next}(x,y)\big)$

Here (1) says that s is the first tile, has tile type t_0, and t is the last tile and has tile type t_k. We have chosen for simplicity to encode the tile types in unary so we need (2), which says that tile types are mutually exclusive.

Conjunct (3) says that the arrangement of tiles honors $\mathcal{T}$'s adjacency requirements. The abbreviation $\mathrm{Suc}_h(x,y)$ means that x and y have the same vertical position and y's horizontal position is one more than that of x. $\mathrm{Suc}_v(x,y)$ means that x and y have the same horizontal position and y's vertical position is one more than that of x. The abbreviations $\mathrm{Hor}(x,y)$ and $\mathrm{Vert}(x,y)$ are disjunctions over the tile types asserting that the tiles in positions x and y are horizontally, respectively vertically, compatible; for example,

$$\mathrm{Hor}(x,y) \;\equiv\; \bigvee\limits_{R(t_i,t_j)} (T_i(x) \wedge T_j(y)) \tag{2}$$

Finally, (4) says that there is a path from s to t. The abbreviation $\mathrm{Next}(x,y)$ means $\mathrm{Suc}_h(x,y)$ or x has horizontal position 2^n, y has horizontal position 1, and y's vertical position is one more than that of x. $\qquad\square$

The formula φ_n described in the above proof can be written in length $O(n)$ using only two variables. When satisfiable, it has a minimal model of size $2^{\Omega(n)}$. In Corollary 16 we extend the above argument, showing that the $2^{O(n^2)}$ bound of Theorem 4 is in fact optimal. For this we need a variant of the above φ_n that uses n variables.

4   Logics with One Function Symbol

We next discuss the language $\forall(\mathrm{TC}, f)$, which consists of universal first-order logic with a transitive-closure operator and one unary function symbol, plus arbitrary unary relation symbols and constants. This is closely related to the language $\exists\forall(\mathrm{DTC}^+[E])$.

One important difference is that in $\forall(f)$ we may write a formula that has only infinite models.[4]

It is well known that the satisfiability and finite-satisfiability problems for monadic second-order logic with a single unary function symbol are decidable,[5] although their complexities are not elementary, even when restricted to first-order quantification [10, 13, 1, 7].

It is not hard to modify the proofs of Theorems 4 and 5 to apply to $\forall(\mathrm{TC}, f)$. (For functions, the implication of Equation (1) is a biimplication, and thus the result goes through for positive and negative DTC's.)

Corollary 6. *The finite satisfiability problem for* $\forall(\mathrm{TC}, f)$ *is NEXPTIME complete.*[6]

Proof: If a formula $\varphi \in \forall(\mathrm{TC}, f)$ has a finite model $\mathcal{A}$, then it must have a model of the form $\mathcal{A}'$ as in the proof of Theorem 4. The only difference is that since f must be a total function, there are no roots; that is, all trees end in cycles. The size of the smallest model is still $2^{O(n^2)}$. The difference in counting is slight, namely, applications of the function symbol f can extend the apparent number of constant symbols: $f(c_i)$ behaves like a new constant symbol c_i', and $f(x)$ behaves like a new universally quantified variable y, such that $E(c_i, c_i')$ and $E(x, y)$, respectively, must hold. Thus, the proof of Theorems 4 and 5 go through if we replace k and m by qk and qm, respectively, where q is the number of occurrences of f in φ. $\qquad\square$

5   Undecidability of Related Logics

We next show that most reasonable extensions of the language $\exists\forall(\mathrm{DTC}^+[E])$ can express the solution to tiling problems, and thus are undecidable. In this section we show that any of the following changes cause undecidability: the use of TC; the presence of more than one binary relation symbol; or a single positive use of $\mathrm{DTC}[\sigma]$, where σ is quantifier-free. In the next section, we show that $\forall(\mathrm{DTC}^-[E])$ is undecidable. To begin, we first show

Theorem 7. *Satisfiability and finite satisfiability of* $\forall(\mathrm{DTC}^+[V], \mathrm{DTC}^+[H])$ — *universal logic with two binary relations, V and H, and their positive deterministic transitive closure — are undecidable.*

Proof: Let $\mathcal{T}$ be a tiling problem (Definition 1). We show how to write a formula $\varphi \in \forall(\mathrm{DTC}^+[V], \mathrm{DTC}^+[H])$ such that φ is satisfiable iff $\mathcal{T}$ has a solution.

Formula φ contains four constant symbols, a, b, c, and d, representing the four corners of the solution to $\mathcal{T}$; see Fig. 4.

We assert that every element satisfies exactly one of the tile relations, $T_0, \ldots, T_k$. We assert $T_0(a) \wedge T_k(d)$, i.e., the upper left tile is t_0 and the lower right is t_k. We assert

---

[4] For example: $\forall x, y(c \neq f(x) \ \wedge \ (f(x) = f(y) \to x = y))$.

[5] This is equivalent to the MSO theory of trees with multiple successor functions.

[6] This holds as well for the general satisfiability problem. For infinite structures there is a similar "small model" except that from some constants there is an infinite chain that intersects no other vertices of the structure. The infinite chain must repeat an m-tuple of colors and can from thereafter repeat exactly. Thus it has a representation of size $2^{O(n^2)}$.

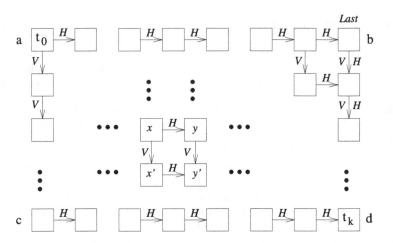

Fig. 4. A tiling as expressed in Theorem 7

that H and V paths exist between the four corners: $\text{DTC}[H](a, b) \wedge \text{DTC}[H](c, d) \wedge \text{DTC}[V](a, c) \wedge \text{DTC}[V](b, d)$.

We add a unary predicate, *Last*, and assert the conjunction of the universal closure of the following formulas: $Last(b)$, $\neg V(x, b)$, $V(x, y) \rightarrow (Last(x) \leftrightarrow Last(y))$, and $(H(x, y) \wedge \neg V(x, y)) \rightarrow \neg Last(x)$. These assure that *Last* is true exactly of the tiles in the rightmost column. In this column, we make the H-edges go down along the V-edges, i.e., $Last(x) \wedge Last(y) \rightarrow (H(x, y) \leftrightarrow V(x, y))$. This allows us to express the fact that H-edges continue all the way to the right in every row, i.e., we assert: $\forall x \text{DTC}[H](x, d)$.

We assert that H and V edges satisfy the corresponding horizontal and vertical tiling constraints, using the formulas Hor and Vert as in Equation (2). $\forall x, y((H(x, y) \wedge \neg Last(x)) \rightarrow \text{Hor}(x, y)) \wedge (V(x, y) \rightarrow \text{Vert}(x, y)))$.

We assert that the intermediate rows are filled in: $\forall x, y, x', y'((H(x, y) \wedge V(x, x') \wedge V(y, y')) \rightarrow H(x', y'))$.

Finally, we assert that the columns are filled in and line up: $\forall x, y, x', y'((\neg Last(x) \wedge H(x, y) \wedge V(x, x') \wedge H(x', y')) \rightarrow V(y, y'))$.

It is not hard to see that the conjunction of the above assertions is equivalent to the existence of a solution to the tiling problem, $\mathcal{T}$. Thus satisfiability of $\forall(\text{DTC}^+[V], \text{DTC}^+[H])$ is undecidable. □

Theorem 7 shows that a second binary relation over which we can take DTC causes undecidability. We can modify the proof to show that even if there is only one (positive) occurrence of DTC, the logic is still undecidable if a second binary relation is allowed, or if DTC is allowed to be taken not just over the relation E, but over a formula that also involves unary relation symbols.

Theorem 8. *Satisfiability and finite satisfiability of $\forall(\text{DTC}^+)$ are undecidable. This holds even if there is only one occurrence of DTC and only one binary relation symbol. Also, if there is a second binary relation symbol, then the single occurrence of DTC can be restricted to the form $\text{DTC}[E]$.*

Proof: We modify the proof of Theorem 7 so that the path from a to d through the tiled rectangle is along a single snake-like path of the edge predicate, E, as in Fig. 5.

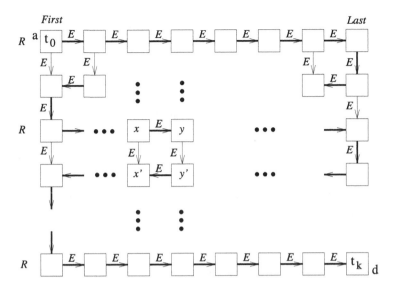

Fig. 5. A tiling expressed with a single occurrence of DTC as in Theorem 8

We do this by adding unary relation *First* denoting the first column of the tiling rectangle, plus the relation R true of the tiles in the odd-number rows. We then make the E-path go left-to-right on the rows satisfying R and right-to-left on the other rows.

Define the edges along the snake-like path, $\sigma(x, y) \equiv E(x, y) \wedge ((R(x) \leftrightarrow R(y)) \vee (First(x) \wedge \neg R(x) \wedge R(y)) \vee (Last(x) \wedge R(x) \wedge \neg R(y)))$.

The single use of DTC is the assertion $\text{DTC}[\sigma](a, d)$. We also assert the completion of squares (see Fig. 5),

$(E(x, y) \wedge E(y, y') \wedge E(y', x') \wedge (R(x) \leftrightarrow R(y)) \wedge (R(x') \leftrightarrow R(y')) \wedge (R(y) \leftrightarrow \neg R(y'))) \rightarrow E(x, x')$.

Finally, we add the following assertions, which together make sure that all models must be valid tilings:

1. $T_0(a) \wedge T_k(d) \wedge First(a) \wedge Last(d) \wedge \neg(First(x) \wedge Last(x))$

2. $\displaystyle\bigvee_{i=0}^{k} T_i(x) \wedge \bigwedge_{0 \leq i < j \leq k} \neg(T_i(x) \wedge T_j(x))$

3. $(E(x, y) \wedge (R(x) \leftrightarrow \neg R(y))) \rightarrow ((First(x) \leftrightarrow First(y)) \wedge (Last(x) \leftrightarrow Last(y)))$

4. $E(x, y) \rightarrow \neg\big((R(x) \wedge R(y) \wedge (Last(x) \vee First(y))) \vee (\neg R(x) \wedge \neg R(y) \wedge (Last(y) \vee First(x)))\big)$

5. $((E(x, y) \wedge R(x) \wedge R(y)) \vee (E(y, x) \wedge \neg R(x) \wedge \neg R(y))) \rightarrow \text{Hor}(x, y)$

6. $(E(x, y) \wedge (R(x) \leftrightarrow \neg R(y))) \rightarrow \text{Vert}(x, y)$

Again formulas Hor and Vert are as in Equation (2). The conjunction of the universal closure of all the above assertions thus asserts a solution to the tiling problem, $\mathcal{T}$, as desired. To prove the last assertion in the statement of the theorem: with a second relation symbol, W, we can let E correspond to σ, and W correspond to $E \wedge \neg\sigma$.   □

We remark that if in the proof of Theorem 8 we reverse the edges that are not σ edges, then we can use $TC[E]$ in lieu of $DTC[\sigma]$ and the proof goes through. Thus we have,

Corollary 9. *Satisfiability and finite satisfiability of $\forall(TC^+[E])$ are undecidable. This holds even if there is only a single occurrence of TC (it occurs as TC[E]) and E is the only binary relation symbol.*

Note that the formulas in Theorems 7, 8, and Corollary 9 use only two variables except in the completion-of-squares formula. In fact, using an extra occurrence of TC, we can write equivalent formulas with only two variables. We do this by reversing the vertical edges in the even columns. We then assert that each non-boundary edge, $\langle x, y \rangle$ is in an appropriate cycle, i.e., $TC[E](y, x)$ or $DTC[\gamma](y, x)$ holds, for appropriate γ.

Corollary 10. *If we allow a second occurrence of a transitive-closure operator, the undecidability results of Theorems 7, 8, and Corollary 9 all remain true for the corresponding languages with only two variables.*

6   Undecidability of $\forall(DTC^-[E])$

We were quite surprised to find that although $\forall(TC^-)$ is decidable, $\forall(DTC^-[E])$ is not. We give the somewhat subtle proof in this section. First we show that $\forall(DTC^-[E])$ has an infinity axiom.

Proposition 11. *There is a sentence in $\forall(DTC^-[E])$ that is satisfiable, but only in an infinite model.*

Proof: The idea is that we know that if $E(c_0, c_1)$ and $\neg DTC[E](c_0, c_1)$ both hold, then there must be another edge from c_0. We can use this observation to write an infinity axiom that essentially expresses the existence of a successor function. We write the conjunction of the following formulas:

1. $\forall v (v \neq c_1 \rightarrow (E(v, c_1) \wedge \neg DTC[E](v, c_1)))$
2. $\forall v u_1 u_2 (v \neq c_1 \wedge E(u_1, v) \wedge E(u_2, v) \rightarrow u_1 = u_2)$
3. $c_0 \neq c_1 \wedge \forall v \neg E(v, c_0)$

(1) says that every vertex besides c_1 has an edge to c_1 but not a DTC path to c_1, so it must have outdegree greater than 1; (2) says that every vertex besides c_1 has in-degree at most one; and (3) says that c_0 has in-degree 0. Thus, there must be an infinite chain of edges starting at c_0.

These formulas are satisfied by a model that contains the natural numbers plus a new point called c_1, with edges $E(n, c_1)$ and $E(n, n + 1)$, for $n = 0, 1, \ldots$. □

Theorem 12. *Satisfiability and finite satisfiability of $\forall(DTC[E])$ are undecidable.*

Proof: We take as our starting point the undecidability proof of Theorem 8. Our new idea is to remove all of the non-boldface E's in Fig. 5 and to replace them by a gadget of new green vertices, satisfying the unary relation symbol, G, and associated edges. The existence of the green vertices and their associated edges will be implied by the "not DTC trick" introduced in the proof of Proposition 11, together with some universal first-order statements that make sure that the vertical edges continue to be attached appropriately.

Just as in the proof of Theorem 8, we express the existence of a tiling. Since we have removed the non-boldface E's, we can now simply express the path from the first tile to the last as $\mathrm{DTC}[E](a, d)$.

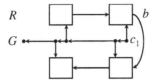

Fig. 6. Gadget used in Theorem 12

To define the gadget, we add two new constants, b, for the top rightmost tile, and c_1 for the top rightmost green vertex, just below it. The green path proceeds in the opposite direction of the non-green, tile path directly above it, see Fig. 6.

We make the following assertions. These all concern the green row below each R, i.e., right-going, row of tiles. For simplicity, we skip the analogous case below each left-going row of tiles.

1. $G(c_1) \wedge E(c_1, b) \wedge \forall ux(E(c_1, x) \wedge G(x) \wedge E(b, u) \rightarrow E(x, u))$
2. $\forall x((\neg G(x) \leftrightarrow \mathrm{DTC}[E](x, d)) \wedge (\neg G(x) \leftrightarrow \mathrm{DTC}[E](a, x)))$
3. $\forall xyz(G(x) \wedge E(x, y) \wedge E(x, z) \wedge y \neq z \rightarrow (G(y) \leftrightarrow \neg G(z)))$
4. $\forall uvxyz\big(\neg G(u) \wedge \neg G(v) \wedge G(x) \wedge G(y) \wedge G(z) \wedge R(u) \wedge R(v) \wedge E(v, u) \wedge E(x, u) \wedge$
   $E(x, y) \wedge E(y, z) \rightarrow E(z, v)\big)$
5. $\forall uvxyz\big(\neg G(u) \wedge \neg G(v) \wedge G(x) \wedge G(y) \wedge G(z) \wedge \neg R(u) \wedge \neg R(v) \wedge E(u, v) \wedge$
   $E(x, u) \wedge E(x, y) \wedge E(y, z) \rightarrow E(z, v)\big)$
6. $\forall u, v, x, y\big(\neg G(u) \wedge \neg G(v) \wedge G(x) \wedge G(y) \wedge R(u) \wedge \neg R(v) \wedge E(x, u) \wedge E(x, y) \wedge$
   $E(y, v) \rightarrow \mathrm{Vert}(u, v)\big)$

(1) starts us out by saying that c_1 is green, has an edge to b, and its green successor has an edge to the tile directly below b. (2) says that green vertices do not have DTC paths to d, but all non-green vertices do; it also says that all the non-green edges occur on the DTC-path from a to d. (3) says that if the outdegree of a green vertex is at least 2, then it has a green and a non-green successor. We will assure later, inductively, that each green vertex has an edge to a non-green vertex. Since the non-green vertex has a DTC-path to d, but the green vertex does not, this assures that the green vertex has outdegree 2. (4) is an inductive condition, which says that if x, y, and z are consecutive green nodes, and if x points up to a non-green node, u, then z points up to u's predecessor, v. (5) is the similar condition for the edges going down.

Finally, condition (6) asserts that these green gadgets transmit the vertical information between the non-green, i.e., tile, nodes as desired.   □

Theorem 12 leaves open the question of the decidability of $\forall(\mathrm{DTC}^-[E])$. It would seem that the positive use of DTC was crucial in the statement $\mathrm{DTC}[E](a, d)$. However, even this can be replaced by the "not DTC trick". (The positive uses of DTC in formula (2) of the proof of Theorem 12 can easily be removed.) The conclusion is that $\forall(\mathrm{DTC}^-[E])$ is undecidable.

Theorem 13. *Satisfiability and finite satisfiability of* $\forall(\text{DTC}^-[E])$ *are undecidable.*

Proof: We modify the proof of Theorem 12 by removing the assertion $\text{DTC}[E](a, d)$ and replacing it using the "not DTC trick". More explicitly, we add another unary predicate B true of the tiles, and we add another constant, c_0. Then we make the following additional assertions:

1. $B(a) \wedge \forall x(B(x) \wedge x \neq d \rightarrow E(x, c_0) \wedge \neg\text{DTC}[E](x, c_0))$
2. $\forall xy(B(x) \wedge y \neq c_0 \wedge E(x, y) \rightarrow B(y))$
3. The in-degree for B-vertices from B-vertices is at most one, and it is zero for a.

(1) and (2) together assert that each B-vertex besides d has an edge to another B-vertex. It follows that either $\text{DTC}[E](a, d)$ holds, or there is an infinite path. Thus, the formula is finitely satisfiable iff the corresponding tiling problem has a solution. (To show that the general satisfiability problem for $\forall(\text{DTC}^-[E])$ is undecidable, we would modify the construction to assert that there is no node d, and thus an infinite path, so that the corresponding Turing machine, when started on blank tape, never halts. The tiling would have to be modified so that the first row has length one, and each successive row has one greater length. This is necessary so that an infinite path corresponds to an infinite computation rather than an infinitely long first row.)   □

7   Complexity of the Decision Procedure

In this section, we study the complexity of the decision procedure for $\exists\forall(\text{DTC}^+[E])$. The first thing we do is look more carefully at the proof of Theorem 5, and show that our lower bound is tight, matching the $2^{O(n^2)}$ upper bound of Theorem 4.

Lemma 14. *The formula* φ_n *used in the proof of Theorem 5 may be written in length* $O(n)$.

Proof: The only difficulty in keeping φ_n to total size $O(n)$ is in writing the formulas $\text{Suc}_h(x, y)$ and $\text{Suc}_v(x, y)$. These are nearly identical and we will restrict our attention to $\text{Suc}_h(x, y)$. Recall that $\text{Suc}_h(x, y)$ means that the horizontal position of y is one greater than the horizontal position of x. (Our convention is that the bit positions are numbered 1 to n, with 1 being the high-order bit, and n the low-order bit.) $\text{Suc}_h(x, y)$ can be written as follows:

$$\text{Suc}_h(x, y) \equiv \bigvee_{i=1}^{n} \Big[\bigwedge_{j>i}(H_j(x) \wedge \neg H_j(y)) \quad \wedge \quad (\neg H_i(x) \wedge H_i(y))$$

$$\wedge \quad \bigwedge_{j<i}(H_j(x) \leftrightarrow H_j(y)) \Big]$$

However, the length of the above formula is $O(n^2)$. We can decrease the size by keeping track of the position i in the above formula. We do this by adding $2n$ more unary relation symbols, $G_j, K_j, 1 \leq j \leq n$. The intuitive meaning of $K_i(x)$ is that it is bit i of the horizontal number that will be incremented as we go from x to its successor. This means that $\neg H_i(x)$, and for all $j > i$, $H_j(x)$; i.e., there is a "0" in position i, and a "1" in each position to the right of i.

The intuitive meaning of $G_j(x)$ is that $j > i$ where $K_i(x)$. We also use the abbreviation $L_j(x) \equiv \neg(K_j(x) \vee G_j(x))$. (The mnemonic is that G holds for elements in positions "greater" than the K position; L holds for elements in "lesser" positions.)

The advantage of having these new relations is that we can now reduce the length of $\text{Suc}_h(x, y)$ as follows:

$$\text{Suc}_h(x, y) \equiv \bigwedge_{j=1}^{n} \Big[(G_j(x) \wedge H_j(x) \wedge \neg H_j(y)) \quad \vee \quad (K_j(x) \wedge \neg H_j(x) \wedge H_j(y))$$

$$\vee \quad (L_j(x) \wedge (H_j(x) \leftrightarrow H_j(y))) \Big]$$

Finally, we must write down several more conditions. The conjunction of the following conditions assures that the new relations G_i and K_i are defined correctly.

1. $\forall x (K_1(x) \vee K_2(x) \vee \cdots \vee K_n(x) \vee (H_1(x) \wedge H_2(x) \cdots H_n(x)))$

2. $\forall x (\bigwedge_{i=1}^{n-1} (K_i(x) \to G_{i+1}(x)) \wedge \bigwedge_{i=1}^{n-1} (K_{i+1}(x) \to L_i(x)))$

3. $\forall x (\bigwedge_{i=1}^{n-1} (L_{i+1}(x) \to L_i(x)) \wedge \bigwedge_{i=1}^{n-1} (G_i(x) \to G_{i+1}(x)))$

4. $\forall x (\bigwedge_{i=1}^{n} \neg(K_i(x) \wedge G_i(x)) \wedge \bigwedge_{i=1}^{n} ((G_i(x) \to H_i(x)) \wedge (K_i(x) \to \neg H_i(x))))$

□

It follows from Lemma 14 and the proof of Theorem 5 that we can write a sequence of formulas $\varphi_n \in \exists\forall(\text{DTC}^+[E])$, $n = 1, 2, \ldots$ such that $|\varphi_n| = O(n)$, φ_n has only two variables, and yet φ_n's smallest model is of size $2^{\Omega(n)}$. This is the best possible with only two variables. To match the $2^{O(n^2)}$ upper bound of Theorem 4, we need a formula with n variables.

We can count up to 2^{n^2} using a sequence of n consecutive vertices, each with a number between 1 and 2^n. We will add n more unary relation symbols, C_i, $1 \leq i \leq n$. A tile will then be encoded by n vertices as follows:

$$[C_1, h_1, v_1, t] \ [C_2, h_2, v_2, t] \ \cdots \ [C_n, h_n, v_n, t]$$
$$[C_1, h'_1, v'_1, t'] \ [C_2, h'_2, v'_2, t'] \ \cdots \ [C_n, h'_n, v'_n, t']$$

That is, the first n vertices hold tile t with its (collective) horizontal and vertical numbers $\langle h_1, \ldots, h_n \rangle$ and $\langle v_1, \ldots, v_n \rangle$ having values between 1 and 2^{n^2}, the next n vertices hold tile t' with the successor number, etc. Using very similar ideas to the proof of Lemma 14 we can prove,

Lemma 15. *Given any tiling problem, $\mathcal{T}$, we can write a sequence of formulas φ'_n of length $O(n)$, $n = 1, 2, \ldots$, such that φ_n is satisfiable iff there is a solution to $\mathcal{T}$ that is a 2^{n^2} by 2^{n^2} square.*

Corollary 16. *The $2^{O(n^2)}$ upper bound of Theorem 4 is optimal.*

8   Conclusions

We have introduced the language $\exists\forall(DTC^+[E])$, which is a decidable transitive-closure logic that goes beyond trees. We have shown that all the reasonable extensions of $\exists\forall(DTC^+[E])$ that we could think of are undecidable. Uses of $\exists\forall(DTC^+[E])$ exist, but how useful it might be remains to be seen.

We showed that the satisfiability of $\exists\forall(DTC^+[E])$ is NEXPTIME complete. The lower bound depended on a formula that describes an exponentially long sequence of colors. We suspect that in practice the formulas one encounters would have much, much shorter sequences of color types. We suspect that techniques related to Ehrenfeucht-Fraïssé games can automatically find the relevant color sequences. These ideas might lead to a satisfiability algorithm that is feasible in practice.

References

1. E. Börger, E. Grädel, and Y. Gurevich. *The Classical Decision Problem*. Springer-Verlag, 1996.
2. G. Dong and J. Su. Space-bounded foies. In *Principles of Database Systems*, pages 139–150. ACM Press, 1995.
3. E. Allen Emerson and Charanjit S. Jutla. The complexity of tree automata and logics of programs. In *Proc. 29th IEEE Symposium on Foundations of Computer Science*, pages 328–337. IEEE Computer Society Press, 1988.
4. E. Grädel, Ph. Kolaitis, and M. Vardi. On the decision problem for two-variable first-order logic. *Bulletin of Symbolic Logic*, 3:53–69, 1997.
5. E. Grädel, M. Otto, and E. Rosen. Undecidability results on two-variable logics. *Archive of Math. Logic*, 38:313–354, 1999.
6. E. Grädel and I. Walukiewicz. Guarded fixed point logic. In *Proc. 14th IEEE Symposium on Logic in Computer Science*, pages 45–54. IEEE Computer Society Press, 1999.
7. J.G. Henriksen, J. Jensen, M. Jørgensen, N. Klarlund, B. Paige, T. Rauhe, and A. Sandholm. Mona: Monadic second-order logic in practice. In *Tools and Algorithms for the Construction and Analysis of Systems, First International Workshop, TACAS '95, LNCS 1019*, 1995.
8. N. Immerman. *Descriptive Complexity*. Springer-Verlag, 1999.
9. N. Immerman, A. Rabinovich, T. Reps, M. Sagiv, and G. Yorsh. Verification via structure simulation. To appear in CAV'04, 2004.
10. Albert R. Meyer. Weak monadic second-order theory of successor is not elementary recursive. In *Logic Colloquium, (Proc. Symposium on Logic, Boston, 1972)*, pages 132–154, 1975.
11. M. Mortimer. On languages with two variables. *Zeitschr. f. math. Logik u. Grundlagen d. Math*, 21:135–140, 1975.
12. C. Papadimitriou. *Computational Complexity*. Addison–Wesley, 1994.
13. M. Rabin. Decidability of second-order theories and automata on infinite trees. *Trans. Amer. Math. Soc.*, 141:1–35, 1969.
14. M. Sagiv, T. Reps, and R. Wilhelm. Parametric shape analysis via 3-valued logic. *Trans. on Prog. Lang. and Syst.*, pages 217–298, 2002.
15. G. Yorsh, T. Reps, and M. Sagiv. Symbolically computing most-precise abstract operations for shape analysis. In *TACAS*, pages 530–545, 2004.

Game-Based Notions of Locality
over Finite Models

Marcelo Arenas, Pablo Barceló, and Leonid Libkin

Department of Computer Science, University of Toronto
{marenas, pablo, libkin}@cs.toronto.edu

Abstract. Locality notions in logic say that the truth value of a formula can be determined locally, by looking at the isomorphism type of a small neighborhood of its free variables. Such notions have proved to be useful in many applications. They all, however, refer to isomorphism of neighborhoods, which most local logics cannot test for. A more relaxed notion of locality says that the truth value of a formula is determined by what the logic itself can say about that small neighborhood. Or, since most logics are characterized by games, the truth value of a formula is determined by the type, with respect to a game, of that small neighborhood. Such game-based notions of locality can often be applied when traditional isomorphism-based locality cannot.

Our goal is to study game-based notions of locality. We work with an abstract view of games that subsumes games for many logics. We look at three, progressively more complicated locality notions. The easiest requires only very mild conditions on the game and works for most logics of interest. The other notions, based on Hanf's and Gaifman's theorems, require more restrictions. We state those restrictions and give examples of logics that satisfy and fail the respective game-based notions of locality.

1 Introduction

Locality is a property of logics that finds its origins in the work by Hanf [13] and Gaifman [10], and that was shown to be very useful in the context of finite model theory. Locality is primarily used in two ways: for proving inexpressibility results, and for establishing normal forms for logical formulae. The former has led to new easy winning strategies in logical games [6, 8, 20], with applications in descriptive complexity (e.g., the study of monadic NP and its relatives [8], or circuit complexity classes [21]), in databases (e.g., establishing bounds on the expressiveness of aggregate queries [16], or on query rewriting in data integration and exchange [7, 1]), and in formal languages (e.g., in characterizing subclasses of star-free languages [27]). Local normal forms like those in [10, 25] have found many applications as well, for example, in the design of low-complexity model-checking algorithms [9, 12, 26], in automata theory [25] and in computing weakest preconditions for database transactions [2].

There are two closely related ways of stating locality of logical formulae. One, originating in Hanf's work [13], says that if two structures $\mathfrak{A}$ and $\mathfrak{B}$ realize the

J. Marcinkowski and A. Tarlecki (Eds.): CSL 2004, LNCS 3210, pp. 175–189, 2004.
© Springer-Verlag Berlin Heidelberg 2004

same multiset of isomorphism types of neighborhoods of radius d, then they agree on a given sentence Φ. Here d depends only on Φ.

The notion of locality inspired by Gaifman's theorem [10] says that if the d-neighborhoods of two tuples $\bar{a}_1$ and $\bar{a}_2$ in a structure $\mathfrak{A}$ are isomorphic, then $\mathfrak{A} \models \varphi(\bar{a}_1) \leftrightarrow \varphi(\bar{a}_2)$. Again, d depends on φ, and not on $\mathfrak{A}$.

If all formulae in a logic are local, it is easy to prove bounds on its expressive power. For example, connectivity violates the Hanf notion of locality, as one cycle of length $2m$ and two disjoint cycles of length m realize the same multiset of isomorphism types of neighborhoods of radius d as long as $m > 2d + 1$. Likewise, the transitive closure of a graph violates the Gaifman notion of locality: in the graph in Fig. 1, one can find two elements a, b such that the radius-d neighborhoods of (a, b) and (b, a) are isomorphic, and yet the transitive closure distinguishes these tuples.

Fig. 1. Locality and transitive closure

These notions of locality, while very useful in many applications, have one deficiency: they all refer to *isomorphism* of neighborhoods, which is a very strong property (typically not expressible in a logic that satisfies one of the locality properties). There are situations when these notions are not applicable simply because structures do not have enough isomorphic neighborhoods! One example was given in [21] which discussed applicability of locality techniques to the study of small parallel complexity classes: consider a directed tree in which all non-leaf nodes have different out-degrees. Then locality techniques cannot be used to derive any results about logics over such trees.

Intuitively, it seems that requiring isomorphism of neighborhoods is too much. Suppose we are dealing with first-order logic FO, which is local in the sense of Gaifman. For a structure $\mathfrak{A}$, it appears that if FO itself cannot see the difference between two large enough neighborhoods of points a and b in $\mathfrak{A}$, then it should not be able to see the difference between elements a and b in $\mathfrak{A}$. That is, for a given formula $\varphi(x)$, if radius-d neighborhoods of a and b cannot be distinguished by sufficiently many FO formulae, then $\mathfrak{A} \models \varphi(a) \leftrightarrow \varphi(b)$.

Gaifman's theorem [10] actually implies that this is the case: if φ is of quantifier rank k, then there exist numbers d and l, dependent on k only, such that if radius-d neighborhoods of a and b cannot be distinguished by formulae of quantifier rank l, then $\mathfrak{A} \models \varphi(a) \leftrightarrow \varphi(b)$.

In fact, it seems that if a logic is local (say, in the sense of Gaifman), then for each formula φ there is a number d such that if the logic cannot distinguish radius-d neighborhoods of $\bar{a}$ and $\bar{b}$, then $\varphi(\bar{a}) \leftrightarrow \varphi(\bar{b})$.

The goal of this paper is to introduce such notions of locality based on logical indistinguishability of neighborhoods, and see if they apply to logics that are known to possess isomorphism-based locality properties. Since logical equivalence is often captured by Ehrenfeucht-Fraïssé-type of games, we shall refer to such new notions of locality as *game-based*.

We shall discover that the situation is more complex than one may have expected, and passing from isomorphism-based locality to game-based is by no means guaranteed for logics known to possess the former.

To be able to talk about general game-based locality notions, we need a unifying framework for talking about logical games that subsumes games for FO, and many of its counting and generalized quantifier extensions. A game is described via an *agreement*, which is a collection of *tactics*, and each tactic is a set of partial functions according to which the game is played. We present this framework in Section 3.

To analyze game-based locality, we then study conditions on agreements that guarantee one of the locality notions. We look at three progressively more complex notions: weak locality, Gaifman-locality, and Hanf-locality, which are described in Section 4. Weak locality is a variation of Gaifman-locality that applies to non-overlapping neighborhoods. In general, establishing some variation of game-based locality for a logic $\mathcal{L}$ does *not* imply that fragments or extensions of $\mathcal{L}$ will possess the same locality property.

Weak locality turns out to require very little and holds for so-called *basic* agreements, as shown in Section 6. While most games of interest are based on basic agreements, we give an example of one unary generalized-quantifier extension of FO which fails weak locality.

In Section 7, we study Hanf-locality under games and show that it holds for a class of agreements that we call matching. These include games for some counting extensions of FO, but game-based Half-locality fails for FO itself (as was already observed in [25]) and some of its generalized-quantifier extensions.

In Section 8, we study Gaifman-locality under games. We show that this notion often implies a normal form result for a logic, similar to Gaifman's theorem for FO. We establish Gaifman-locality for games corresponding to FO and some of its extensions.

Due to space limitations, proofs are not presented in this extended abstract. A full version containing all the proofs can be obtained from the authors.

2   Notation

We work with finite structures, whose universes are subsets of some countable infinite set U. All vocabularies will be finite sequences of relation symbols $\sigma = \langle R_1, \ldots, R_n \rangle$; a σ-structure $\mathfrak{A}$ consists of a finite universe $A \subset U$ and an interpretation of each m-ary relation symbol R_i in σ as a subset of A^m. We adopt the convention that the universe of a structure is denoted by the corresponding Roman letter, that is, the universe of $\mathfrak{A}$ is A, the universe of $\mathfrak{B}$ is B, etc. Isomorphism of structures will be denoted by $\cong$.

For a relation $F \subseteq A \times B$, we use $\mathrm{dom}(F)$ to denote its domain $\{a \in A \mid \exists b \ (a, b) \in R\}$ and $\mathrm{rng}(F)$ to denote the range $\{b \in B \mid \exists a \ (a, b) \in R\}$. We use the same notation dom and rng for the domain and range of a (partial) function. The graph of a function $f : A \to B$ is denoted by $\text{graph}(f) = \{(a, b) \mid b = f(a)\}$.

Given two tuples $\bar{a}_1$ and $\bar{a}_2$, we write $\bar{a}_1\bar{a}_2$ for their concatenation.

Next, we introduce the logics considered in the paper. First-order logic will be denoted by FO. Then, we define *simple unary generalized quantifiers* $\mathbf{Q}_S$ [19, 28]. Let $S \subseteq \mathbb{N}$. We denote by $\text{FO}(\mathbf{Q}_S)$ the extension of FO with the following formation rule: if $\psi(x, \bar{y})$ is a formula, then $\varphi(\bar{y}) = \mathbf{Q}_S x \; \psi(x, \bar{y})$ is a formula. The semantics is as follows: $\mathfrak{A} \models \varphi(\bar{a})$ if $|\{a \mid \mathfrak{A} \models \psi(a, \bar{a})\}| \in S$. One could also define FO extended with a collection of simple unary generalized quantifiers.

We consider one special case of unary quantifiers: *modulo quantifiers* (cf. [23, 24, 28]). If $S = \{n \cdot p \mid n \in \mathbb{N}\}$, then we write $\mathbf{Q}_p$ instead of $\mathbf{Q}_S$.

Finally, we define a powerful counting logic that subsumes most counting extensions of FO, in particular FO extended with arbitrary collections of unary generalized quantifiers. The structures for this logic are two-sorted, being $\mathbb{N}$ the second sort. There is a constant symbol for each $k \in \mathbb{N}$. The logic has *infinitary* connectives $\bigvee$ and $\bigwedge$, and *counting terms*: if φ is a formula and $\bar{x}$ a tuple of free first-sort variables in φ, then $\#\bar{x}.\varphi$ is a term of the second sort, whose free variables are those in φ except $\bar{x}$. Its value is the number of tuples $\bar{a}$ that make $\varphi(\bar{a}, \cdot)$ true. This logic, denoted by $\mathcal{L}_{\infty\omega}(\mathbf{Cnt})$, defines all properties of finite structures.

To restrict it, we use the notion of quantifier rank $\text{qr}(\cdot)$ which is defined as the maximum depth of quantifier nesting (*excluding* quantification over the numerical universe for two-sorted logics). For $\mathcal{L}_{\infty\omega}(\mathbf{Cnt})$, we also define $\text{qr}(\#\bar{x}.\varphi)$ as $\text{qr}(\varphi) + |\bar{x}|$.

We now define $\mathcal{L}^*_{\infty\omega}(\mathbf{Cnt})$ as $\mathcal{L}_{\infty\omega}(\mathbf{Cnt})$ restricted to formulae and terms that have finite rank. This logic subsumes known counting extensions of FO, but cannot express many properties definable, say, in fixed-point logics or fragments of second-order logic [20].

3 Games and Logics

We now present the first way of abstractly viewing games such as Ehrenfeucht-Fraïssé games, as well as games for counting and unary-quantifier extensions of FO. Such games are played by two players, the *spoiler* and the *duplicator*, on two σ-structures $\mathfrak{A}$ and $\mathfrak{B}$. The goal of the spoiler is to show that the structures are different while the duplicator is trying to show that they are the same.

In most games, the spoiler and the duplicator agree on a class of relations before the game starts, that is, for each $A, B \subset U$, they have sets $\mathfrak{F}(A, B) = \{\mathcal{F}_1(A, B), \ldots, \mathcal{F}_s(A, B)\}$, where each $\mathcal{F}_i(A, B)$ is a a family of subsets of $A \times B$. The game starts with a position $(\bar{a}_0, \bar{b}_0)$, where $\bar{a}_0 \in A^l, \bar{b}_0 \in B^l$ (l could be 0). After i rounds, the *position* of the game consists of $(\bar{a}_0, a_1, \ldots, a_i)$ in $\mathfrak{A}$ and $(\bar{b}_0, b_1, \ldots, b_i)$ in $\mathfrak{B}$. Given a position $(\bar{a}_0\bar{a}, \bar{b}_0\bar{b})$ after round i, the game proceeds as follows:

1. The spoiler selects a structure, $\mathfrak{A}$ or $\mathfrak{B}$.
2. The duplicator picks a family of relations $\mathcal{F}(A, B) \in \mathfrak{F}(A, B)$, if the spoiler selected $\mathfrak{A}$, or $\mathcal{F}(B, A) \in \mathfrak{F}(B, A)$, if the spoiler selected $\mathfrak{B}$. Assume that the spoiler chose $\mathfrak{A}$, the other case being completely symmetric.
3. The spoiler chooses one relation $F \in \mathcal{F}(A, B)$ and an element $a \in \mathrm{dom}(F)$.
4. The duplicator responds with an element $b \in \mathrm{rng}(F)$ such that $(a, b) \in F$, and the game continues from the position $(\bar{a}_0 \bar{a} a, \ \bar{b}_0 \bar{b} b)$.

We now present games corresponding to FO, $\mathcal{L}^*_{\infty\omega}(\mathbf{Cnt})$, and FO($\mathbf{Q}_p$).

- If $\mathfrak{F}(A, B) = \{\{A \times B\}\}$ for every $A, B \subset U$, then this is the usual Ehrenfeucht-Fraïssé game: the spoiler is free to choose any point in A, and the duplicator is free to choose any point in B.
- Let $f_1, \ldots, f_r$ enumerate all the bijections $A \to B$. Suppose $\mathfrak{F}(A, B) = \{\{\mathrm{graph}(f_1)\}, \ldots, \{\mathrm{graph}(f_r)\}\}$. Then we have the *bijective* game of [14]. In this game, in each round, the duplicator selects a bijection $f : A \to B$; the spoiler plays $a \in A$ and the duplicator responds by $f(a) \in B$.
- Given $A, B \subset U$, consider sets $\mathcal{F}(A, B)$ of the form $\{C_i \times D_i \mid C_i \subseteq A, D_i \subseteq B, i \in I\}$, where every subset of B occurs as one of the D_i's, and $|C_i| \equiv |D_i| \pmod{p}$ for each i. $\mathfrak{F}(A, B)$ consists of all $\mathcal{F}(A, B)$'s of this form. This is the setting of the game for modulo p quantifiers $\mathbf{Q}_p$ [24]. In each round of this game, the spoiler chooses $D \subseteq B$ and the duplicator selects $C \subseteq A$ with $|C| \equiv |D| \pmod{p}$. Then the spoiler plays $a \in A$ and the duplicator responds with $b \in B$ such that $a \in C$ iff $b \in D$.

The presentation of games given above is standard in the literature. For stating results in the paper, we shall use a slightly different way of presenting games. Suppose we have a position $(\bar{a}_0 \bar{a}, \ \bar{b}_0 \bar{b})$ in the game, and the duplicator chooses a family $\mathcal{F}(A, B) \in \mathfrak{F}(A, B)$. By doing so, the duplicator is certain that, no matter what relation $F \in \mathcal{F}(A, B)$ the spoiler chooses, for every $a \in \mathrm{dom}(F)$, he has a response $b \in \mathrm{rng}(F)$. That is, for every $F \in \mathcal{F}(A, B)$, the duplicator has one or more *functions* $f : A \to B$ with $\mathrm{graph}(f) \subseteq F$, such that if the spoiler plays $a \in A$, he can respond with $f(a) \in B$. From now on, we shall be defining games using such a functional approach.

Definition 1. *An* agreement *is a collection* $\mathfrak{F} = \{\mathfrak{F}(A, B) \mid A, B$ *are finite subsets of* $U\}$, *where each* $\mathfrak{F}(A, B)$ *is of the form* $\{\mathcal{F}_1(A, B), \ldots, \mathcal{F}_m(A, B)\}$, $m \geq 0$, *and each* $\mathcal{F}_i(A, B)$ *is a nonempty collection of partial functions* $f : A \to B$. *We shall call the sets* $\mathcal{F}_i(A, B)$'s tactics.

The $\mathfrak{F}$*-game on* $(\mathfrak{A}, \bar{a}_0)$ *and* $(\mathfrak{B}, \bar{b}_0)$ *is played as follows. Suppose after* i *rounds the position is* $(\bar{a}_0 \bar{a}, \bar{b}_0 \bar{b})$ *(before the game starts, the tuples* $\bar{a}, \bar{b}$ *are empty). Then, in round* $i + 1$:

1. *The spoiler chooses a structure,* $\mathfrak{A}$ *or* $\mathfrak{B}$. *Below we present the moves assuming he chose* $\mathfrak{A}$, *the case of* $\mathfrak{B}$ *is symmetric.*
2. *The duplicator chooses a tactic* $\mathcal{F}(A, B) \in \mathfrak{F}(A, B)$.
3. *The spoiler chooses a partial function* $f \in \mathcal{F}(A, B)$ *and an element* $a \in \mathrm{dom}(f)$; *the game continues from the position* $(\bar{a}_0 \bar{a} a, \ \bar{b}_0 \bar{b} f(a))$.

The duplicator wins after k-rounds if $\mathfrak{F}(A, B) \neq \emptyset$ and $\mathfrak{F}(B, A) \neq \emptyset$, and the position of the game defines a partial isomorphism. If the duplicator has a winning strategy that guarantees a win in k rounds, we write $(\mathfrak{A}, \bar{a}_0) \equiv_k^{\mathfrak{F}} (\mathfrak{B}, \bar{b}_0)$.

One can pass from the usual representation of games to the functional one without loss of generality:

Lemma 1. *Given $\mathfrak{F}' = \{\mathfrak{F}'(A, B)\}$, where each $\mathfrak{F}'(A, B)$ is a collection of families of relations on finite $A, B \subset U$, there is an agreement $\mathfrak{F}$ such that the relations $\equiv_k^{\mathfrak{F}'}$ and $\equiv_k^{\mathfrak{F}}$ coincide for every k.*

We now return to games seen in the previous example, and show how they look under the functional approach. We define three agreements: $\mathfrak{F}(\mathrm{FO})$, $\mathfrak{F}(\mathcal{L}_{\infty\omega}^*(\mathbf{Cnt}))$, and $\mathfrak{F}(\mathrm{FO}(\mathbf{Q}_p))$.

- $\mathfrak{F}(\mathrm{FO})$: a tactic is a singleton set $\{f\}$, where $f : A \to B$ is a total function. Then $\mathfrak{F}(A, B)$, for each pair of finite sets (A, B), contains all possible tactics.
- $\mathfrak{F}(\mathcal{L}_{\infty\omega}^*(\mathbf{Cnt}))$: same as above, except that each tactic is $\{f\}$ where $f : A \to B$ is a bijection (there are no tactics if $|A| \neq |B|$).
- $\mathfrak{F}(\mathrm{FO}(\mathbf{Q}_p))$: given $A, B \subset U$, a tactic is a set $\mathcal{F}$ of partial maps such that for every $D \subseteq B$, there exists $f \in \mathcal{F}$ such that $\mathrm{dom}(f) = A$ and $|\{c \in A \mid f(c) \in D\}| \equiv |D| \pmod{p}$.

Definition 2. *Given an agreement $\mathfrak{F}$, we say that the $\mathfrak{F}$-game is a game for a logic $\mathcal{L}$, if there exists a partition $\{\mathcal{L}_0, \mathcal{L}_1, \ldots\}$ of the formulae in $\mathcal{L}$ such that for every $k \geq 0$, there exists $k' \geq 0$ with the property that*

$$(\mathfrak{A}, \bar{a}_0) \equiv_{k'}^{\mathfrak{F}} (\mathfrak{B}, \bar{b}_0) \quad \text{implies} \quad (\mathfrak{A} \models \varphi(\bar{a}_0) \Leftrightarrow \mathfrak{B} \models \varphi(\bar{b}_0)), \quad \text{for all } \varphi \in \mathcal{L}_k.$$

If the converse holds as well, that is, for every $k' \geq 0$ there exists $k \geq 0$ such that, $(\mathfrak{A}, \bar{a}_0) \equiv_{k'}^{\mathfrak{F}} (\mathfrak{B}, \bar{b}_0)$, whenever $\mathfrak{A} \models \varphi(\bar{a}_0) \Leftrightarrow \mathfrak{B} \models \varphi(\bar{b}_0)$ for every $\varphi \in \mathcal{L}_k$, then we say that $\mathfrak{F}$-games capture $\mathcal{L}$.

Games are usually applied to prove *inexpressibility* results, in which case one only needs the condition that a given game is a game for a logic. In many cases, however, the converse holds too, that is, games completely characterize logics. The following is a reformulation, under our view of games, of standard results on characterizing logics by games [5, 17, 14, 16, 24, 28].

Proposition 1. *If $\mathcal{L}$ is one of FO, $\mathcal{L}_{\infty\omega}^*(\mathbf{Cnt})$, and $\mathrm{FO}(\mathbf{Q}_p)$, then $\mathfrak{F}(\mathcal{L})$-games are games for $\mathcal{L}$, with $\mathcal{L}_k$ being the set of $\mathcal{L}$-formulae of quantifier rank $\leq k$. Furthermore, these games capture the corresponding logic.*

4   Locality

Given a σ-structure $\mathfrak{A}$, its *Gaifman graph*, denoted by $G(\mathfrak{A})$, has A as the set of nodes. There is an edge (a_1, a_2) in $G(\mathfrak{A})$ iff there is a relation symbol R in σ such that for some tuple t in the interpretation of this relation in $\mathfrak{A}$, both a_1, a_2

occur in t. By the *distance* $d(a_1, a_2)$ we mean the distance in the Gaifman graph, with $d(a, a) = 0$. If there is no path from a_1 to a_2 in $G(\mathfrak{A})$, then $d(a_1, a_2) = \infty$. We write $d(\bar{a}, b)$ for the minimum of $d(a, b)$ over a from $\bar{a}$.

Let $\mathfrak{A}$ be a σ-structure, and $\bar{a} = (a_1, \ldots, a_m) \in A^m$. The *radius r ball around* $\bar{a}$ is the set $B_r^{\mathfrak{A}}(\bar{a}) = \{b \in A \mid d(\bar{a}, b) \leq r\}$. The *$r$-neighborhood of $\bar{a} = (a_1, \ldots, a_m)$ in $\mathfrak{A}$* is the structure $N_r^{\mathfrak{A}}(\bar{a})$ of vocabulary σ expanded with n constant symbols, where the universe is $B_r^{\mathfrak{A}}(\bar{a})$; σ-relations are restrictions of σ-relations in $\mathfrak{A}$ to $B_r^{\mathfrak{A}}(\bar{a})$, and the n additional constants are $a_1, \ldots, a_n$.

Since we define a neighborhood around an m-tuple as a structure with additional constant symbols, for any isomorphism h between $N_r^{\mathfrak{A}}(a_1, \ldots, a_m)$ and $N_r^{\mathfrak{B}}(b_1, \ldots, b_m)$, it must be the case that $h(a_i) = b_i$, $1 \leq i \leq m$.

Let $\mathfrak{A}, \mathfrak{B}$ be σ-structures, where σ only contains relation symbols. Let $\bar{a} \in A^m$ and $\bar{b} \in B^m$. We write $(\mathfrak{A}, \bar{a}) \leftrightarrows_d (\mathfrak{B}, \bar{b})$ if there exists a bijection $f : A \to B$ such that $N_d^{\mathfrak{A}}(\bar{a}c)$ and $N_d^{\mathfrak{B}}(\bar{b}f(c))$ are isomorphic, for every $c \in A$. This definition is most commonly used when $m = 0$; then $\mathfrak{A} \leftrightarrows_d \mathfrak{B}$ means that for some bijection $f : A \to B$, $N_d^{\mathfrak{A}}(c) \cong N_d^{\mathfrak{B}}(f(c))$ for all $c \in A$. That is, $\mathfrak{A} \leftrightarrows_d \mathfrak{B}$ iff $\mathfrak{A}$ and $\mathfrak{B}$ realize the same multiset of isomorphism types of d-neighborhoods of points.

We say that a formula $\varphi(\bar{x})$ is *Hanf-local*, if there exists a number $d \geq 0$ such that $\mathfrak{A} \models \varphi(\bar{a}) \Leftrightarrow \mathfrak{B} \models \varphi(\bar{b})$ whenever $(\mathfrak{A}, \bar{a}) \leftrightarrows_d (\mathfrak{B}, \bar{b})$. This concept was first introduced by Hanf [13] for FO over infinite structures, then modified by [8] to work for sentences over finite models.

Gaifman's theorem [10] states that every FO formula $\varphi(\bar{x})$ is equivalent to a Boolean combination of sentences and formulae in which quantification is restricted to $B_r(\bar{x})$, with r determined by φ. In particular, this implies that for every FO formula, we have two numbers, d and k, such that if $\mathfrak{A}$ and $\mathfrak{B}$ agree on all FO sentences of quantifier-rank $\leq k$ and $N_d^{\mathfrak{A}}(\bar{a}) \cong N_d^{\mathfrak{B}}(\bar{b})$, then $\mathfrak{A} \models \varphi(\bar{a}) \Leftrightarrow \mathfrak{B} \models \varphi(\bar{b})$. This concept is normally used when $\mathfrak{A} = \mathfrak{B}$; then it says that a formula $\varphi(\bar{x})$ is *Gaifman-local* if there exists a number $d \geq 0$ such that for every structure $\mathfrak{A}$, if $N_d^{\mathfrak{A}}(\bar{a}_1) \cong N_d^{\mathfrak{A}}(\bar{a}_2)$, then $\mathfrak{A} \models \varphi(\bar{a}_1) \leftrightarrow \varphi(\bar{a}_2)$.

A formula $\varphi(\bar{x})$ is *weakly-local* [21] if the above condition holds for disjoint neighborhoods: that is, there is a number $d \geq 0$ such that for every structure $\mathfrak{A}$, if $N_d^{\mathfrak{A}}(\bar{a}_1) \cong N_d^{\mathfrak{A}}(\bar{a}_2)$ and $B_d^{\mathfrak{A}}(\bar{a}_1) \cap B_d^{\mathfrak{A}}(\bar{a}_2) = \emptyset$, then $\mathfrak{A} \models \varphi(\bar{a}_1) \leftrightarrow \varphi(\bar{a}_2)$.

The following implications are known [15, 21]: Hanf-local $\Rightarrow$ Gaifman-local $\Rightarrow$ weakly-local. Examples of logics in which all formulae are Hanf- (and hence Gaifman and weakly) local are all the logics considered so far: FO, FO($\mathbf{Q}_p$), $\mathcal{L}_{\infty\omega}^*(\mathbf{Cnt})$ [10, 15, 20, 23]. There are examples of formulae that are Gaifman-but not Hanf-local [15] and weakly but not Gaifman-local [21].

We now state the definition that relaxes the concept of locality, by placing requirements weaker than isomorphism of neighborhoods. For $d, l \geq 0$, we use the notation $(\mathfrak{A}, \bar{a}) \leftrightarrows_{d,l}^{\mathfrak{F}} (\mathfrak{B}, \bar{b})$ if there exists a bijection $f : A \to B$ such that $N_d^{\mathfrak{A}}(\bar{a}c) \equiv_l^{\mathfrak{F}} N_d^{\mathfrak{B}}(\bar{b}f(c))$ for every $c \in A$.

Definition 3. *An agreement $\mathfrak{F}$ is Hanf-local if for every $k, m \in \mathbb{N}$, there exists $d, l \in \mathbb{N}$ such that for every two structures $\mathfrak{A}, \mathfrak{B}$, $\bar{a} \in A^m$ and $\bar{b} \in B^m$,*

$$(\mathfrak{A}, \bar{a}) \leftrightarrows_{d,l}^{\mathfrak{F}} (\mathfrak{B}, \bar{b}) \quad \Rightarrow \quad (\mathfrak{A}, \bar{a}) \equiv_k^{\mathfrak{F}} (\mathfrak{B}, \bar{b}).$$

We call $\mathfrak{F}$ Gaifman-local *if for every* $k, m \in \mathbb{N}$, *there exists* $d, l \in \mathbb{N}$ *such that for every two structures* $\mathfrak{A}, \mathfrak{B}$, $\bar{a} \in A^m$ *and* $\bar{b} \in B^m$,

$$\mathfrak{A} \equiv_l^{\mathfrak{F}} \mathfrak{B} \quad and \quad N_d^{\mathfrak{A}}(\bar{a}) \equiv_l^{\mathfrak{F}} N_d^{\mathfrak{B}}(\bar{b}) \quad \Rightarrow \quad (\mathfrak{A}, \bar{a}) \equiv_k^{\mathfrak{F}} (\mathfrak{B}, \bar{b}).$$

Finally, we call $\mathfrak{F}$ weakly-local *if for every* $k, m \in \mathbb{N}$, *there exist* $d, l \in \mathbb{N}$ *such that for every structure* $\mathfrak{A}$, $\bar{a} \in A^m$ *and* $\bar{b} \in A^m$,

$$N_d^{\mathfrak{A}}(\bar{a}) \equiv_l^{\mathfrak{F}} N_d^{\mathfrak{A}}(\bar{b}) \quad and \quad B_d^{\mathfrak{A}}(\bar{a}) \cap B_d^{\mathfrak{A}}(\bar{b}) = \emptyset \quad \Rightarrow \quad (\mathfrak{A}, \bar{a}) \equiv_k^{\mathfrak{F}} (\mathfrak{A}, \bar{b}).$$

Our main question is the following: *When is a logic local under its games?* Or, more precisely: suppose $\mathfrak{F}$-games are games for a logic $\mathcal{L}$; is $\mathfrak{F}$ Hanf-, Gaifman-, or weakly-local?

If a logic is local under its games, we need an assumption *weaker* than isomorphism in order to prove that formulae cannot distinguish some elements of a structure. Consider, for example, the case of Gaifman-locality, applied to one structure $\mathfrak{A}$. Normally, to derive $\varphi(\bar{a}_1) \leftrightarrow \varphi(\bar{a}_2)$, we would need to assume that $N_d(\bar{a}_1) \cong N_d(\bar{a}_2)$ for some appropriate d. But suppose we know that φ comes from a logic Gaifman-local under $\mathfrak{F}$-games. If k is such that $(\mathfrak{A}, \bar{a}_1) \equiv_k^{\mathfrak{F}} (\mathfrak{A}, \bar{a}_2)$ implies $\varphi(\bar{a}_1) \leftrightarrow \varphi(\bar{a}_2)$, then we find $d, l \in \mathbb{N}$ that ensure

$$N_d^{\mathfrak{A}}(\bar{a}_1) \equiv_l^{\mathfrak{F}} N_d^{\mathfrak{A}}(\bar{a}_2) \Rightarrow (\mathfrak{A}, \bar{a}_1) \equiv_k^{\mathfrak{F}} (\mathfrak{A}, \bar{a}_2) \Rightarrow \mathfrak{A} \models \varphi(\bar{a}_1) \leftrightarrow \varphi(\bar{a}_2).$$

Thus, instead of isomorphism of neighborhoods, we have a weaker requirement that they be indistinguishable by the $\mathfrak{F}$-game, in l rounds.

Even though the notion of locality under games is easier to apply, it is harder to analyze than the standard isomorphism-based locality. For example, if a logic $\mathcal{L}$ is local (Hanf-, or Gaifman-, or weakly) under isomorphisms, and $\mathcal{L}'$ is a sublogic of $\mathcal{L}$, then $\mathcal{L}'$ is local as well. The same, however, is *not* true for game-based locality, as we shall see, as properties of games guaranteeing locality need not be preserved if one passes to weaker games.

5   Basic Structural Properties

We now look at some most basic properties of agreements that are expected to hold. Intuitively, they are: (1) the spoiler is free to play any point he wants to; (2) the duplicator can mimic spoiler's moves when they play on the same structure; (3) the games on $(\mathfrak{A}, \mathfrak{B})$ and $(\mathfrak{B}, \mathfrak{C})$ can be composed into a single game on $(\mathfrak{A}, \mathfrak{C})$, and (4) agreements do not depend on a particular choice of elements of U.

Definition 4. *An agreement* $\mathfrak{F}$ *is said to be* admissible *if the following hold:*

(1) For every $\mathcal{F}(A, B) \in \mathfrak{F}$, *we have* $\bigcup \{\mathrm{dom}(f) \mid f \in \mathcal{F}(A, B)\} = A$ *(the spoiler can play any point he wants to);*

(2) For every $A \subset U$, *there exists* $\mathcal{F}(A, A) \in \mathfrak{F}$ *such that every* $f \in \mathcal{F}(A, A)$ *is the identity on* $\mathrm{dom}(f)$ *(the duplicator can repeat spoiler's moves if they play on the same set);*

(3) For every $\mathcal{F}(A,B)$, $\mathcal{F}(B,C) \in \mathfrak{F}$, the composition $\mathcal{F}(A,B) \circ \mathcal{F}(B,C) = \{g \circ f \mid f \in \mathcal{F}(A,B) \text{ and } g \in \mathcal{F}(B,C)\}$ is a tactic in $\mathfrak{F}$ (games compose);

(4) If $\mathcal{F}(A,B)$ is a tactic in $\mathfrak{F}$, and $g : A' \to A, h : B \to B'$ are bijections, then $\{h \circ f \circ g \mid f \in \mathcal{F}(A,B)\}$ is a tactic over A', B' (agreements do not depend on the choice of elements of U).

It is an easy observation that the agreements $\mathfrak{F}(\mathrm{FO})$, $\mathfrak{F}(\mathcal{L}^*_{\infty\omega}(\mathbf{Cnt}))$, and $\mathfrak{F}(\mathrm{FO}(\mathbf{Q}_p))$ are admissible.

Proposition 2. *Given an admissible agreement $\mathfrak{F}$ and $m, k \geq 0$,*

(a) $\equiv^{\mathfrak{F}}_{k}$ *is an equivalence relation on structures $(\mathfrak{A}, \bar{a}), \bar{a} \in A^m$;*

(b) *If $h : \mathfrak{A} \to \mathfrak{B}$ is an isomorphism, then $(\mathfrak{A}, \bar{a}) \equiv^{\mathfrak{F}}_{k} (\mathfrak{B}, h(\bar{a}))$.*

In many logics, the equivalence classes of $\equiv^{\mathfrak{F}}_{k}$ are definable by formulae (they correspond to *types*, or rank-k types, as k typically refers to the quantifier rank). Then definable sets are unions of types. We introduce an abstract notion of definable sets: a set $S \subseteq A^m$ is $(\mathfrak{F}, k)$-*definable* in $\mathfrak{A}$ if it is closed under $\equiv^{\mathfrak{F}}_{k}$: that is, $\bar{a} \in S$ and $(\mathfrak{A}, \bar{a}) \equiv^{\mathfrak{F}}_{k} (\mathfrak{A}, \bar{a}_1)$ imply $\bar{a}_1 \in S$. For admissible agreements, definable sets behave in the expected way.

Proposition 3. *If $\mathfrak{F}$ is an admissible agreement, then $(\mathfrak{F}, k)$-definable sets are closed under Boolean combinations and Cartesian product; furthermore, the projection $A^{m+1} \to A^m$ applied to an $(\mathfrak{F}, k)$-definable set produces an $(\mathfrak{F}, k+1)$-definable set.*

6   Weak Locality

We now move to the first locality condition, weak locality. In many applications of locality, at least for proving expressibility bounds, one actually uses weak locality as it is easier to work with disjoint neighborhoods (see, e.g., Fig. 1). While examples of weakly-local formulae violating other notions of locality exist, they are not particularly natural [21].

To guarantee weak locality, we impose two very mild conditions on $\mathfrak{F}$-games. The first has to do with compositionality. Composition of games is a standard technique that allows one to use $\mathfrak{A} \equiv^{\mathfrak{F}}_{k} \mathfrak{A}'$ and $\mathfrak{B} \equiv^{\mathfrak{F}}_{k} \mathfrak{B}'$ to conclude $\mathcal{H}(\mathfrak{A}, \mathfrak{B}) \equiv^{\mathfrak{F}}_{l} \mathcal{H}(\mathfrak{A}', \mathfrak{B}')$, for some operation $\mathcal{H}$ (see, e.g., [22] for a survey). While in general such compositionality properties depend on the type of games and the operator $\mathcal{H}$, there is one scenario where they almost universally apply: when $\mathcal{H}$ is the disjoint union of structures [22] (in fact, l is usually equal to k in this situation). We want our games to satisfy this property. We use $\sqcup$ for disjoint union of sets and functions.

Definition 5. *An agreement $\mathfrak{F}$ is* compositional, *if for every two tactics $\mathcal{F}(A,B)$ and $\mathcal{G}(C,D)$ in $\mathfrak{F}$ such that $A \cap C = B \cap D = \emptyset$, the tactic $\mathcal{F}(A,B) \sqcup \mathcal{G}(C,D)$ defined as the set of disjoint unions of partial functions $f : A \to B$ from $\mathcal{F}(A,B)$ and $g : C \to D$ from $\mathcal{G}(C,D)$ is in $\mathfrak{F}$.*

The second condition says that if in a game $\mathfrak{A} \equiv_k^{\mathfrak{F}} \mathfrak{B}$, both players play restricted to subsets $C \subseteq A$ and $D \subseteq B$, then such a game may be considered as a game on substructures of $\mathfrak{A}$ and $\mathfrak{B}$ generated by C and D, respectively. Again, this condition is true for practically all reasonable games.

We formalize it as follows. We denote the set of all nonempty restrictions of partial functions from $\mathcal{F}(A, B)$ to $C \subseteq A$ by $\mathcal{F}(A, B)|_C$. Consider a tactic $\mathcal{F}(A, B)$, and nonempty sets $C \subseteq A$ and $D \subseteq B$. We say that $\mathcal{F}(A, B)$ is *shrinkable* to (C, D) if $a \in C \Leftrightarrow f(a) \in D$ for every $f \in \mathcal{F}(A, B)$ and $a \in \mathrm{dom}(f)$.

Definition 6. *An agreement $\mathfrak{F}$ is* shrinkable *if for every $\mathcal{F}(A, B) \in \mathfrak{F}$, and nonempty subsets $C \subseteq A$ and $D \subseteq B$, if $\mathcal{F}(A, B)$ is shrinkable to (C, D), then $\mathcal{F}(A, B)|_C$ is a tactic over (C, D) that belongs to $\mathfrak{F}$.*

An admissible $\mathfrak{F}$ is called basic *if it is both shrinkable and compositional.*

A simple examination of the agreements seen so far in this paper shows:

Proposition 4. *The agreements $\mathfrak{F}(\mathrm{FO})$, $\mathfrak{F}(\mathcal{L}_{\infty\omega}^*(\mathbf{Cnt}))$ and $\mathfrak{F}(\mathrm{FO}(\mathbf{Q}_p))$ are basic.*

Recall that an agreement $\mathfrak{F}$ is weakly-local if for every $k, m \geq 0$, there exist $d, l \geq 0$ such that for every structure $\mathfrak{A}$ and every $\bar{a}, \bar{b} \in A^m$, if $N_d^{\mathfrak{A}}(\bar{a}) \equiv_l^{\mathfrak{F}} N_d^{\mathfrak{A}}(\bar{b})$ and the neighborhoods $N_d^{\mathfrak{A}}(\bar{a})$ and $N_d^{\mathfrak{A}}(\bar{b})$ are disjoint, then $(\mathfrak{A}, \bar{a}) \equiv_k^{\mathfrak{F}} (\mathfrak{A}, \bar{b})$. We define the *weak-locality rank with respect to $\mathfrak{F}$*, denoted by $\mathrm{wlr}_{\mathfrak{F}}(k, m)$, as the minimum d for which the above condition holds.

Theorem 1. *Every basic agreement $\mathfrak{F}$ is weakly-local. Furthermore, $\mathrm{wlr}_{\mathfrak{F}}(k, m) = O(2^k)$.*

Corollary 1. *The agreements $\mathfrak{F}(\mathrm{FO})$, $\mathfrak{F}(\mathcal{L}_{\infty\omega}^*(\mathbf{Cnt}))$ and $\mathfrak{F}(\mathrm{FO}(\mathbf{Q}_p))$ are weakly-local.*

That is, FO, $\mathrm{FO}(\mathbf{Q}_p)$, and $\mathcal{L}_{\infty\omega}^*(\mathbf{Cnt})$ are weakly-local under their games.

Nevertheless, there are extensions of FO with simple unary generalized quantifiers that are not weakly-local under their games.

Let PRIME be the set of primes and $\mathbf{Q}_{\mathrm{PRIME}}$ the corresponding generalized quantifier. That is, $\mathrm{FO}(\mathbf{Q}_{\mathrm{PRIME}})$ extends FO with formulae $\mathbf{Q}_{\mathrm{PRIME}} y \; \varphi(\bar{x}, y)$ such that $\mathfrak{A} \models \mathbf{Q}_{\mathrm{PRIME}} y \; \psi(\bar{a}, y)$ if $|\{a \mid \mathfrak{A} \models \psi(\bar{a}, a)\}|$ is a prime number. We show that $\mathrm{FO}(\mathbf{Q}_{\mathrm{PRIME}})$ is not weakly-local under its games.

We first define the agreement $\mathfrak{F}(\mathrm{FO}(\mathbf{Q}_{\mathrm{PRIME}}))$. For two finite sets $A, B \subset U$, a tactic is a set $\mathcal{F}$ of partial maps such that for every nonempty $D \subseteq B$, there exists $f \in \mathcal{F}$ such that $\mathrm{dom}(f) = A$ and $|f^{-1}(D)| \in \mathrm{PRIME}$ iff $|D| \in \mathrm{PRIME}$. (In terms of the game, in every round the spoiler selects a set $D \subseteq B$; the duplicator selects $C \subseteq A$ such that $|C|$ is prime iff $|D|$ is. Then the spoiler plays $a \in A$ and the duplicator responds with $b \in B$ such that $a \in C$ iff $b \in D$.) Notice that this agreement is not compositional, and hence not basic.

It is known [28] that for every $\mathrm{FO}(\mathbf{Q}_{\mathrm{PRIME}})$-formula $\varphi(\bar{x})$ of quantifier rank k, if $(\mathfrak{A}, \bar{a}) \equiv_k^{\mathfrak{F}(\mathrm{FO}(\mathbf{Q}_{\mathrm{PRIME}}))} (\mathfrak{B}, \bar{b})$, then $\mathfrak{A} \models \varphi(\bar{a})$ iff $\mathfrak{B} \models \varphi(\bar{b})$. Thus, to show that $\mathrm{FO}(\mathbf{Q}_{\mathrm{PRIME}})$ is not weakly-local under its games, it suffices to prove the following:

Proposition 5. $\mathfrak{F}(\mathrm{FO}(\mathbf{Q}_{\mathrm{PRIME}}))$ *is not weakly-local.*

For this, we give a formula $\varphi(x)$ such that for every $d, l \geq 0$, there is a structure $\mathfrak{A}$ and $a, b \in A$ such that $N_d^{\mathfrak{A}}(a) \equiv_l^{\mathfrak{F}(\mathrm{FO}(\mathbf{Q}_{\mathrm{PRIME}}))} N_d^{\mathfrak{A}}(b)$, $B_d^{\mathfrak{A}}(a) \cap B_d^{\mathfrak{A}}(b) = \emptyset$, and yet $\mathfrak{A} \models \varphi(a) \wedge \neg \varphi(b)$.

Let σ be a signature of a unary relation R and a binary relation E, and let $d, l \geq 0$. Consider the structure $\mathfrak{A}$ whose E-relation is shown in Fig. 2 below; the relation R is interpreted as the set of all a_i's, b_i's, and c_i's. Let $\varphi(x)$ be $\mathbf{Q}_{\mathrm{PRIME}} y\, (R(y) \wedge \neg E(x, y))$.

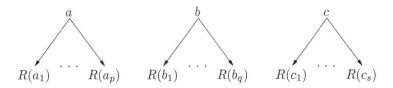

Fig. 2. A structure for proving that $\mathrm{FO}(\mathbf{Q}_{\mathrm{PRIME}})$ is not weakly-local under its games

There are infinitely many primes r such that all the numbers $r - i$ ($i \leq l$) are composite. Choose two sufficiently large p, q ($p \neq q$) from this set so that $N_d^{\mathfrak{A}}(a) \equiv_l^{\mathfrak{F}(\mathrm{FO}(\mathbf{Q}_{\mathrm{PRIME}}))} N_d^{\mathfrak{A}}(b)$ (notice that d can be taken to be 1, without loss of generality). By Dirichlet's Theorem, the arithmetic progression $np + q$ ($n = 0, 1, \ldots$) contains an infinite number of primes. Let $n \geq 1$ be such that $np + q$ is a prime and let $s = np$. Then, $\mathfrak{A} \models \varphi(a)$, since $q + s = np + q$ is prime, and $\mathfrak{A} \not\models \varphi(b)$, since $p + s = (n + 1)p$ is composite. Thus, the agreement $\mathfrak{F}(\mathrm{FO}(\mathbf{Q}_{\mathrm{PRIME}}))$ is not weakly-local.

7   Hanf-Locality

We now present a condition that guarantees Hanf-locality of agreements. While still easy to state, this condition already fails for some logics, notably for FO.

We say that $\mathcal{F}(A, B)$ is a *matching* tactic, if the union $\bigcup_{f \in \mathcal{F}(A,B)} \mathrm{graph}(f)$ is a matching on $A \times B$. That is, the union of all the functions from $\mathcal{F}(A, B)$ is a partial bijection. For example, all the tactics in $\mathfrak{F}(\mathcal{L}_{\infty\omega}^*(\mathbf{Cnt}))$ are matching.

From a tactic $\mathcal{F}(A, B)$ we define a relation $\approx_{\mathcal{F}(A,B)}$ as the minimal relation that contains $\bigcup_{f \in \mathcal{F}(A,B)} \mathrm{graph}(f)$ and satisfies the following: if $a \approx_{\mathcal{F}(A,B)} b'$, $a' \approx_{\mathcal{F}(A,B)} b$ and $f(a') = b'$ for some $f \in \mathcal{F}(A, B)$, then $a \approx_{\mathcal{F}(A,B)} b$.

Another way of looking at this relation is the following: $a \approx_{\mathcal{F}(A,B)} b$ if there is a sequence $\langle a_0, b_1, a_1, b_2, a_2, \ldots, b_{m-1}, a_{m-1}, b_m \rangle$ where $a_0 = a$, $b_m = b$, and for every i, there are $f, f' \in \mathcal{F}(A, B)$ such that $b_i = f(a_{i-1}) = f'(a_i)$, $1 \leq i \leq m-1$, and $b_m = f(a_{m-1})$ for some $f \in \mathcal{F}(A, B)$.

Definition 7. *An agreement $\mathfrak{F}$ is called* matching *if for every tactic $\mathcal{F}(A, B) \in \mathfrak{F}$, there exists a matching tactic $\mathcal{G}(A, B) \in \mathfrak{F}$ such that $\bigcup_{g \in \mathcal{G}(A,B)} \mathrm{graph}(g)$ is contained in $\approx_{\mathcal{F}(A,B)}$.*

If every tactic in an agreement is matching, then the agreement itself is matching. However, some agreements can be matching and have non-matching tactics (examples will be given in the full version of the paper). The following holds trivially:

Proposition 6. $\mathfrak{F}(\mathcal{L}^*_{\infty\omega}(\mathbf{Cnt}))$ *is a matching agreement.*

Recall that an agreement $\mathfrak{F}$ is Hanf-local, if for every $k, m \geq 0$ there exist $d, l \geq 0$ such that, for every two structures $\mathfrak{A}, \mathfrak{B}$ and every $\bar{a} \in A^m$ and $\bar{b} \in B^m$, if $(\mathfrak{A}, \bar{a}) \leftrightarrows^{\mathfrak{F}}_{d,l} (\mathfrak{B}, \bar{b})$, then $(\mathfrak{A}, \bar{a}) \equiv^{\mathfrak{F}}_k (\mathfrak{B}, \bar{b})$. The minimum d for which the above condition holds is called the *Hanf-locality rank with respect to* $\mathfrak{F}$, and is denoted by $\mathrm{hlr}_{\mathfrak{F}}(k, m)$.

Theorem 2. *If an agreement* $\mathfrak{F}$ *is basic and matching, then it is Hanf-local. Furthermore,* $\mathrm{hlr}_{\mathfrak{F}}(k, m) = O(2^k)$.

Corollary 2. $\mathfrak{F}(\mathcal{L}^*_{\infty\omega}(\mathbf{Cnt}))$ *is Hanf-local, and* $\mathrm{hlr}_{\mathfrak{F}(\mathcal{L}^{\bullet}_{\infty\omega}(\mathbf{Cnt}))}(k, m) = O(2^k)$.

Thus, $\mathcal{L}^*_{\infty\omega}(\mathbf{Cnt})$ is Hanf-local under its games. This nice behavior, however, does not extend to other logics known to possess the isomorphism-based Hanf-locality property.

Proposition 7. (see [25])   FO *and* $\mathrm{FO}(\mathbf{Q}_p)$ *are not Hanf-local under their games.*

For FO, this is proved by taking G_1 to be the complete graph with $2N$ vertices, and G_2 to be the disjoint union of two complete graphs with N vertices each. For every d and l, any bijection between the nodes of these graphs witnesses $G_1 \leftrightarrows^{\mathfrak{F}(\mathrm{FO})}_{d,l} G_2$ as long as $N > l$, and yet G_1 and G_2 disagree on $\exists x \exists y \neg E(x, y)$. For $\mathrm{FO}(\mathbf{Q}_p)$, the same proof works, but N is taken to be $p \cdot (l + 1)$.

8   Gaifman-Locality

Recall that $\mathfrak{F}$ is Gaifman-local if for every $k, m \geq 0$ there exist $d, l \geq 0$ such that, for every $\mathfrak{A}$ and $\mathfrak{B}$ and every $\bar{a} \in A^m$ and $\bar{b} \in B^m$, we have $(\mathfrak{A}, \bar{a}) \equiv^{\mathfrak{F}}_k$ $(\mathfrak{B}, \bar{b})$ whenever $\mathfrak{A} \equiv^{\mathfrak{F}}_l \mathfrak{B}$ and $N^{\mathfrak{A}}_d(\bar{a}) \equiv^{\mathfrak{F}}_l N^{\mathfrak{B}}_d(\bar{b})$. The minimum such d is called *Gaifman-locality rank with respect to* $\mathfrak{F}$, and denoted by $\mathrm{lr}_{\mathfrak{F}}(k, m)$,

Our goal is to show that agreements defining games for FO, $\mathrm{FO}(\mathbf{Q}_p)$, and $\mathcal{L}^*_{\infty\omega}(\mathbf{Cnt})$, are all Gaifman-local. The proof of this fact is easier for more expressive logics such as $\mathcal{L}^*_{\infty\omega}(\mathbf{Cnt})$ (this will be explained shortly). In that case, one can show the following:

Lemma 2. *If* $\mathfrak{F}$ *is a basic and matching agreement, then* $\mathfrak{F}$ *is Gaifman-local, and* $\mathrm{lr}_{\mathfrak{F}}(k, m) \leq 3 \cdot \mathrm{hlr}_{\mathfrak{F}}(k, m) + 1$.

This tells us that $\mathcal{L}^*_{\infty\omega}(\mathbf{Cnt})$ is Gaifman-local under their games:

Corollary 3. $\mathfrak{F}(\mathcal{L}^*_{\infty\omega}(\mathbf{Cnt}))$ *is Gaifman-local, and* $\mathrm{lr}_{\mathfrak{F}(\mathcal{L}^{\bullet}_{\infty\omega}(\mathbf{Cnt}))}(k, m) = O(2^k)$.

We next move to Gaifman-locality for FO and FO with modulo quantifiers. Gaifman-locality for them is the hardest of the locality conditions we consider here, mainly because of the following three reasons. First, it requires reasoning about overlapping neighborhoods, which is known to cause complications in the study of locality (see, e.g., [11]). Second, it is a strong notion that implies the existence of normal forms for logical formulae. Such normal forms have been shown for FO [10, 25]. Third, while establishing Gaifman-locality and normal forms, we match the best bound for Gaifman-locality rank for FO, $O(4^k)$. (In Gaifman's original proof, it was $O(7^k)$, the $O(4^k)$ bound is from [18]. For the "one-structure" version, and the isomorphism-based locality, the bound can be further reduced to $O(2^k)$ [20].)

We now show that logics which are Gaifman-local under their games admit a normal form, under the condition that the relations $\equiv_k^{\mathfrak{F}}$ are of finite index (as they are for FO and several other logics). In that case, every formula is equivalent to a Boolean combination of sentences and formulae evaluated in a neighborhood of its free variables. More precisely, for a logic $\mathcal{L}$ that satisfies the basic closure properties of [4] (that is, any reasonable logic, e.g., closed under $\vee, \wedge, \neg$), we can show the following.

Theorem 3. *Let $\mathcal{L}$ be a logic captured by an admissible Gaifman-local agreement $\mathfrak{F}$, where $\mathfrak{F}$ has the property that for every k, the relations $\equiv_k^{\mathfrak{F}}$ are of finite index. Then, for every $\mathcal{L}$-formula $\varphi(\bar{x})$, one can find a number d, a sequence $\Phi_1, \ldots, \Phi_n$ of $\mathcal{L}$-sentences, a sequence $\varphi_1(\bar{x}), \ldots, \varphi_m(\bar{x})$ of $\mathcal{L}$-formulae, and a Boolean function $\beta : \{0,1\}^{n+m} \to \{0,1\}$ such that*

$$\mathfrak{A} \models \varphi(\bar{a}) \quad \Leftrightarrow \quad \beta\Big(\Phi_1(\mathfrak{A}), \ldots, \Phi_n(\mathfrak{A}), \varphi_1(N_d^{\mathfrak{A}}(\bar{a})), \ldots, \varphi_m(N_d^{\mathfrak{A}}(\bar{a}))\Big) = 1$$

where

$$\Phi_i(\mathfrak{A}) = \begin{cases} 1 & \text{if } \mathfrak{A} \models \Phi \\ 0 & \text{if } \mathfrak{A} \models \neg\Phi \end{cases} \quad \text{and} \quad \varphi_j(N_d^{\mathfrak{A}}(\bar{a})) = \begin{cases} 1 & \text{if } N_d^{\mathfrak{A}}(\bar{a}) \models \varphi_j(\bar{a}) \\ 0 & \text{if } N_d^{\mathfrak{A}}(\bar{a}) \models \neg\varphi_j(\bar{a}). \end{cases}$$

Thus, proving Gaifman-locality under games is comparable to proving a result like Gaifman's theorem itself. We now do this for FO and the following generalization of FO($\mathbf{Q}_p$).

If $p_1, \ldots, p_r$ is a sequence of numbers, then FO($\mathbf{Q}_{p_1}, \ldots, \mathbf{Q}_{p_r}$) extends FO with all the generalized quantifiers $\mathbf{Q}_{p_i}$'s. This logic is captured by $\mathfrak{F}(\text{FO}(\mathbf{Q}_{p_1}, \ldots, \mathbf{Q}_{p_r}))$-games, where each tactic in the agreement $\mathfrak{F}(\text{FO}(\mathbf{Q}_{p_1}, \ldots, \mathbf{Q}_{p_r}))$ is simply a union of tactics from each of the $\mathfrak{F}(\text{FO}(\mathbf{Q}_{p_i}))$'s.

Theorem 4. *If $\mathfrak{F}$ is either $\mathfrak{F}(\text{FO})$ or $\mathfrak{F}(\text{FO}(\mathbf{Q}_{p_1}, \ldots, \mathbf{Q}_{p_r}))$, for an arbitrary sequence $p_1, \ldots, p_r$, then $\mathfrak{F}$ is Gaifman-local. Furthermore, $\text{lr}_{\mathfrak{F}}(k, m) = O(4^k)$.*

Note that the bound shown for both FO and FO($\mathbf{Q}_{p_1}, \ldots, \mathbf{Q}_{p_r}$) matches the best bound previously known for FO [18]. Furthermore, since for both FO and $\mathfrak{F}(\text{FO}(\mathbf{Q}_{p_1}, \ldots, \mathbf{Q}_{p_r}))$ the relation $\equiv_k^{\mathfrak{F}}$ is of finite index, the normal form result (Theorem 3) applies to them. For FO, this is of course known and follows from local normal forms of [10, 25]. Our proof, however, is new, and is based entirely on structural properties of games.

9   Conclusions

We looked at the natural extensions of three standard locality notions that use logical equivalence (or equivalence under games) of neighborhoods, as opposed to a much stronger condition of isomorphisms. Such locality notions can be applied in several scenarios where the standard isomorphism-based notions of locality are inapplicable. In fact, their applicability to FO has already been used in data exchange and integration scenarios to help draw the boundary between rewritable and non-rewritable queries [1].

We defined an abstract view of games that let us consider the notions of locality in an abstract setting, independent of a particular logic. This approach is applicable to many logics which are captured by games and whose types are definable in the logic itself (with some exceptions, of course, such as finite variable logics [3], but some of them are non-local). We identified conditions that guarantee the main notions of locality for those games.

The notions for which most questions remain is Gaifman-locality. Unlike others, which admit $O(2^k)$ bounds on locality rank, for Gaifman-locality we could only show a $O(4^k)$ bound, and even that matches the very recently discovered bound for FO, as those previously known were of the order of 7^k. We would like to settle the case of Gaifman-locality completely, by finding natural conditions for it that account for all the known cases, and by precisely calculating the locality rank.

References

1. M. Arenas, P. Barceló, R. Fagin, L. Libkin. Locally consistent transformations and query answering in data exchange. In *PODS 2004*, pages 229–240.
2. M. Benedikt, T. Griffin, L. Libkin. Verifiable properties of database transactions. *Information and Computation* 147(1): 57–88 (1998).
3. A. Dawar, S. Lindell, S. Weinstein. Infinitary logic and inductive definability over finite structures. *Information and Computation* 119(2): 160–175 (1995).
4. H.-D. Ebbinghaus. Extended logics: the general framework. In *Model-Theoretic Logics*, Springer, 1985, pages 25–76.
5. K. Etessami. Counting quantifiers, successor relations, and logarithmic space. *JCSS* 54(3): 400–411 (1997).
6. R. Fagin. Easier ways to win logical games. In N. Immerman and Ph. Kolaitis, eds. *Descriptive Complexity and Finite Models*, AMS, 1997, pages 1–32.
7. R. Fagin, Ph. Kolaitis, R. Miller, L. Popa. Data exchange: semantics and query answering. *TCS*, to appear. Extended abstract in *ICDT 2003*, pages 207–224.
8. R. Fagin, L. Stockmeyer and M. Vardi. On monadic NP vs monadic co-NP. *Information and Computation* 120(1): 78–92 (1994).
9. J. Flum, M. Grohe. Fixed-parameter tractability, definability, and model-checking. *SIAM J. Comput.* 31(1): 113–145 (2001).
10. H. Gaifman. On local and non-local properties. *Logic Colloquium 1981*, North Holland, 1982.
11. M. Grohe and T. Schwentick. Locality of order-invariant first-order formulas. *TOCL* 1(1): 112–130 (2000).

12. M. Grohe, S. Wöhrle. An existential locality theorem. In *CSL 2001*, pages 99–114.
13. W. Hanf. Model-theoretic methods in the study of elementary logic. In J.W. Addison et al., eds., *The Theory of Models*, North Holland, 1965, pages 132–145.
14. L. Hella. Logical hierarchies in PTIME. *Information and Computation* 129(1): 1–19 (1996).
15. L. Hella, L. Libkin and J. Nurmonen. Notions of locality and their logical characterizations over finite models. *J. Symbolic Logic* 64(4): 1751–1773 (1999).
16. L. Hella, L. Libkin, J. Nurmonen and L. Wong. Logics with aggregate operators. *JACM* 48(4): 880–907 (2001).
17. N. Immerman and E. Lander. Describing graphs: A first order approach to graph canonization. In *"Complexity Theory Retrospective"*, Springer Verlag, Berlin, 1990.
18. H.J. Keisler, W. Lotfallah. Shrinking games and local formulas. *Annals of Pure and Applied Logic* 128: 215–225 (2004).
19. Ph. Kolaitis and J. Väänänen. Generalized quantifiers and pebble games on finite structures. *Annals of Pure and Applied Logic* 74: 23–75 (1995).
20. L. Libkin. On counting logics and local properties. *TOCL* 1(1): 33–59 (2000).
21. L. Libkin, L. Wong. Lower bounds for invariant queries in logics with counting. *TCS* 288(1): 153–180 (2002).
22. J. Makowsky. Algorithmic aspects of the Feferman-Vaught Theorem. *Annals of Pure and Applied Logic* 126: 159–213 (2004).
23. J. Nurmonen. On winning strategies with unary quantifiers. *Journal of Logic and Computation* 6(6): 779–798 (1996).
24. J. Nurmonen. Counting modulo quantifiers on finite structures. *Information and Computation* 160(1-2): 62–87 (2000).
25. T. Schwentick and K. Barthelmann. Local normal forms for first-order logic with applications to games and automata. *STACS 1998*, pages 444-454.
26. D. Seese. Linear time computable problems and first-order descriptions. *Mathematical Structures in Computer Science* 6(6): 505–526 (1996).
27. W. Thomas. Languages, automata, and logic. In *Handbook of Formal Languages, Vol. 3*, Springer, 1997, pages 389–455.
28. J. Väänänen. Unary quantifiers on finite models. *J. Logic, Language and Information* 6(3): 275–304 (1997).

Fixed Points of Type Constructors
and Primitive Recursion

Andreas Abel[*,1] and Ralph Matthes[2]

[1] Department of Computer Science, Chalmers University of Technology
abel@cs.chalmers.se
[2] Department of Computer Science, University of Munich
matthes@informatik.uni-muenchen.de

Abstract. For nested or heterogeneous datatypes, terminating recursion schemes considered so far have been instances of iteration, excluding efficient definitions of fixed-point unfolding. Two solutions of this problem are proposed: The first one is a system with equi-recursive non-strictly positive type constructors of arbitrary finite kinds, where fixed-point unfolding is computationally invisible due to its treatment on the level of type equality. Positivity is ensured by a polarized kinding system, and strong normalization is proven by a model construction based on saturated sets. The second solution is a formulation of primitive recursion for arbitrary type constructors of any rank. Although without positivity restriction, the second system embeds—even operationally—into the first one.

1 Introduction

Recently, higher-rank datatypes have drawn interest in the functional programming community [Oka96, Hin01]. Rank-2 non-regular types, so-called *nested datatypes*, have been investigated in the context of the functional programming language Haskell. To define total functions which traverse nested datastructures, Bird et al. [BP99a] have developed *generalized folds* which implement an iteration scheme and are strong enough to encode most of the known algorithms for nested datatypes. In this work, we investigate schemes to overcome some limitations of iteration which we expound in the following.

Since the work of Böhm *et al.* [BB85] it is well-known that iteration for rank-1 datatypes can be simulated in typed lambda calculi. The easiest examples are iterative definitions of addition and multiplication for Church numerals. The iterative definition of the predecessor, however, is inefficient: It traverses the whole numeral in order to remove one constructor. Surely, taking the predecessor should run in constant time.

Primitive recursion is the combination of iteration and efficient predecessor. A typical example for a primitive recursive algorithm is the natural definition

---

[*] The first author gratefully acknowledges the support by both the PhD Programme *Logic in Computer Science* (GKLI) of the *Deutsche Forschungs-Gemeinschaft* and the CoVer project by the *Stiftelsen för Strategisk Forskning* (SSF).

J. Marcinkowski and A. Tarlecki (Eds.): CSL 2004, LNCS 3210, pp. 190–204, 2004.

of the factorial function. It is common belief that primitive recursion cannot be reduced to iteration in a computationally faithful manner. This is because no encoding of natural numbers in the polymorphic lambda-calculus (System F) seems possible which supports a constant-time predecessor operation (see Spławski and Urzyczyn [SU99]).

In this article, we present two approaches to overcome the predecessor dilemma for higher-rank datatypes. A *first* solution, presented in Section 2, is System Fix^ω of non-strictly positive equi-recursive type constructors, which handles folding and unfolding for fixed points on the level of types, trivially yielding an efficient predecessor. Fix^ω is proven strongly normalizing in Section 3. Even though the system has no native means of recursion, a powerful scheme of primitive recursion is definable in Fix^ω. This schema is embodied in our formulation of a *second* system MRec^ω, given in Section 4. In Section 4.2 we give an extensive example of a function which can most naturally be implemented with primitive recursion—redecoration for triangular matrices. Finally, we give the details of the definition of MRec^ω within Fix^ω, hence establishing strong normalization of the primitive recursion scheme as well (Section 5).

2   System Fix^ω

Since Mendler [Men87], it is known that type equations of the form $X = A$ with X a type variable and A a type expression, can only be added to system F in case X only occurs positively in A. Otherwise, strong normalization of typable terms is lost. In this section, we show that the positive part of Mendler's finding, namely strong normalization in the case where X only occurs positively in A, can be extended to equations for type constructors of arbitrary finite kind, hence within the framework F^ω of higher-order parametric polymorphism.

Equations in solved form, i.e., with a constructor variable on the left-hand side, can equivalently be treated by an explicit type constructor for fixed-points [Urz96]. In the case of fixed-points of types, the purported solution of $X = C$ would be written as the type $\mathsf{fix}\,X.C$, with its characteristic equation being $\mathsf{fix}\,X.C = [\mathsf{fix}\,X.C/X]C$. In the case of nested datatypes, we are interested in equations like $X\,A = A + X(A \times A)$, where X now denotes a *type transformation*. This would be solved by $\mathsf{PList} = \mathsf{fix}\,X.\lambda A.\,A + X(A \times A)$. $\mathsf{PList}\,A$ stands for powerlists over A, i.e., lists with 2^n elements of type A, for some (unspecified) n, which can clearly be seen to be the least fixed-point of the above equation. With function kinds at hand, we can pass from $\mathsf{fix}\,X.C$ to $\mathsf{fix}\,F$ with $F := \lambda X.C$, which is a type transformer.

A manifest idea to isolate positive constructors systematically is to distinguish covariant (monotone), contravariant (antitone) and invariant (no information about monotonicity) constructors through the kinding system. Such systems have been found independently by L. Cardelli, B. C. Pierce, the first author and others, but published only by Steffen [Ste98] and Duggan and Compagnoni [DC98]. In both publications, polarized kinds are used to model subtyping of container types like lists and arrays in object-oriented calculi. We are reusing

Polarities	p	$::=$	$+$	covariant
Kinds		$\mid$	$-$	contravariant
		$\mid$	$\circ$	invariant

Kinds	κ	$::=$	$* \mid p\kappa \to \kappa'$
Constructors	A, B, F, G	$::=$	$X \mid \lambda X^{p\kappa}. F \mid F\,G \mid A \to B \mid \forall X^{\kappa}. A \mid \boxdot xF$
Objects (terms)	r, s, t	$::=$	$x \mid \lambda x.t \mid r\,s$
Contexts	Δ	$::=$	$\diamond \mid \Delta, x\!:\!A \mid \Delta, X^{p\kappa}$

Table 1. Language of Fix^{ω}

their ideas to formulate positive recursive constructors in a strongly normalizing language.

Each function kind $\kappa \to \kappa'$ is decorated with a polarity, yielding $p\kappa \to \kappa'$ in Duggan and Compagnoni's notation. For covariant constructors, $p = +$, for contravariant, $p = -$, and $p = \circ$ if the constructor is neither co- nor contravariant or its variance is unknown. Consequently, abstracted variables now carry kinding and polarity information. For instance, we have

$$\lambda X^{+(+*\to*)}\lambda A^{+*}. X\,(X\,A) \;:\; +(+* \to *) \to (+* \to *).$$

The kinding expresses that $X \circ X$ is covariant if X is, and that the "twice" operation $\lambda X.X \circ X$ is itself covariant on covariant arguments, meaning that we may form its fixed point, which would in turn be covariant.[1] We can also classify invariant constructors, e.g., $\lambda A^{\circ *}. A \to X\,(A \times A) : \circ * \to *$ for invariant X that occurs covariantly, indicated by $X^{+(\circ * \to *)}$ in the context, and contravariant constructors like $\lambda X^{-*}. X \to \bot : -* \to *$. Consequently, $\lambda X^{+*}. (X \to \bot) \to \bot :$ $+* \to *$, which hence includes non-strictly positive type transformers.[2]

Language of Fix$^\omega$. Table 1 shows the syntactic entities of System Fix$^\omega$, an extension of F$^\omega$ by polarized kinds and fixed-points of constructors. Typically, the empty context "$\diamond$" will be suppressed. Furthermore we assume all variables in a context Δ to be pairwise distinct. Capture-avoiding substitution of constructor G for variable X in constructor F is written as $[G/X]F$, likewise substitution in terms is denoted by $[s/x]t$. As usual, it is assumed that constructor application and term application associate to the left, e.g., $F\,G\,X$ denotes $(F\,G)\,X$ and $(\lambda x.r)\,s\,t$ denotes $((\lambda x.r)\,s)\,t$. Iterated applications may be "vectorized", i.e., $r\,t_1 \ldots t_n$ will be written as $r\boldsymbol{t}$ with $\boldsymbol{t} := t_1, \ldots, t_n$. Then, $|\boldsymbol{t}| := n$.

While we are using Curry-style objects to express solely the operational behavior, for the type constructors we decided on Church style in order to simplify

---

[1] The kind of $\lambda X.X \circ X$ obtained here is a syntactic approximation and simplification of the more logic-based concept of rank-2 monotonicity introduced in [Mat01].

[2] Most dependently typed systems such as Coq do not allow non-strict positivity for their native fixed points due to the consistency problem reported in [CP88].

the semantics definition in Section 3. The same decision has been taken by Giannini et al. [GHR93], where equivalence with pure Church typing is also established. Note, however, that impredicative systems with dependent types are richer in this mixed style [vBL$^+$97].

Operations on Polarities and Contexts. Negation of a polarity $-p$ is given by the three equations $-(+) = -$, $-(-) = +$ and $-(\circ) = \circ$. We define application $p\Delta$ of a polarity p to a polarized context Δ. Positive polarity is neutral and changes nothing: $+\Delta = \Delta$. The operation $-\Delta$ reverses all polarities in Δ. Furthermore $\circ\Delta$ discards all co- and contravariant type variable bindings.

Kinding. We introduce a judgement $\Delta \vdash F : \kappa$ which combines the usual notions of wellkindedness and positive and negative occurrences of type variables. It assures that fixed-points can only be formed over positive type constructors.

$$\frac{X^{p\kappa} \in \Delta \quad p \in \{+, \circ\}}{\Delta \vdash X : \kappa} \qquad \frac{\Delta, X^{p\kappa} \vdash F : \kappa'}{\Delta \vdash \lambda X^{p\kappa}.\, F : p\kappa \to \kappa'} \qquad \frac{\Delta \vdash F : p\kappa \to \kappa' \quad p\Delta \vdash G : \kappa}{\Delta \vdash FG : \kappa'}$$

$$\frac{-\Delta \vdash A : * \quad \Delta \vdash B : *}{\Delta \vdash A \to B : *} \qquad \frac{\Delta, X^{\circ\kappa} \vdash A : *}{\Delta \vdash \forall X^{\kappa}.\, A : *} \qquad \frac{\Delta \vdash F : +\kappa \to \kappa}{\Delta \vdash \mathsf{fix}\, F : \kappa}$$

Kinding is syntax-directed, and, since we are using Church-style constructors, for given Δ and F, the kind κ of F can be computed by structural recursion on F. As a consequence, all rules are invertible in the strong sense that we can recover the applied rule and all the parts of its premises from a given kinding judgement.

The arrow in kinds and in types is assumed to associate to the right, e.g., $A \to B \to C$ stands for $A \to (B \to C)$ and $+* \to -\kappa \to \kappa'$ stands for $+* \to (-\kappa \to \kappa')$.

Example 1. We can define standard type constructors via the usual impredicative encodings and get more informative kinds:

$$\begin{aligned}
\times &: \quad +* \to +* \to * \\
\times &:= \lambda X^{+*}\lambda Y^{+*}\forall Z^*.\, (X \to Y \to Z) \to Z \\
+ &: \quad +* \to +* \to * \\
+ &:= \lambda X^{+*}\lambda Y^{+*}\forall Z^*.\, (X \to Z) \to (Y \to Z) \to Z \\
\exists^{\kappa} &: \quad +(\circ\kappa \to *) \to * \\
\exists^{\kappa} &:= \lambda F^{+(\circ\kappa \to *)}\forall Z^*.\, (\forall X^{\kappa}.\, F\, X \to Z) \to Z
\end{aligned}$$

Notice that all these examples use non-strict positivity. We will use $+$ and $\times$ infix.

Example 2. The reader is invited to check the examples in the introduction, using $\bot := \forall X^*.\, X$ of kind $*$.

Kinding enjoys the usual properties of weakening and strengthening, as well as substitution which respects polarities:

Lemma 1 (Substitution). *If* $\Delta, X^{p\kappa} \vdash F : \kappa'$ *and* $p\Delta \vdash G : \kappa$ *then* $\Delta \vdash [G/X]F : \kappa'$.

Proof. By induction on $\Delta, X^{p\kappa} \vdash F : \kappa'$.

Constructor Equality. The β-equality $F = F'$ of constructors F, F' is given by the following rules, hence only in the qualified form with contexts:

Computation axioms.

$$\frac{\Delta, X^{p\kappa} \vdash F : \kappa' \qquad p\Delta \vdash G : \kappa}{\Delta \vdash (\lambda X^{p\kappa}.\, F)\, G = [G/X]F : \kappa'} \qquad \frac{\Delta \vdash F : {+}\kappa \to \kappa}{\Delta \vdash \text{fix } F = F\,(\text{fix } F) : \kappa}$$

Congruences.

$$\frac{X^{p\kappa} \in \Delta \qquad p \in \{+, \circ\}}{\Delta \vdash X = X : \kappa} \qquad \frac{\Delta \vdash F = F' : p\kappa \to \kappa' \qquad p\Delta \vdash G = G' : \kappa}{\Delta \vdash F\,G = F'\,G' : \kappa'}$$

$$\frac{\Delta, X^{p\kappa} \vdash F = F' : \kappa'}{\Delta \vdash \lambda X^{p\kappa}.\, F = \lambda X^{p\kappa}.\, F' : p\kappa \to \kappa'} \qquad \frac{\Delta, X^{\circ\kappa} \vdash A = A' : *}{\Delta \vdash \forall X^\kappa.\, A = \forall X^\kappa.\, A' : *}$$

$$\frac{-\Delta \vdash A' = A : * \qquad \Delta \vdash B = B' : *}{\Delta \vdash A \to B = A' \to B' : *} \qquad \frac{\Delta \vdash F = F' : {+}\kappa \to \kappa}{\Delta \vdash \text{fix } F = \text{fix } F'}$$

Symmetry and transitivity.

$$\frac{\Delta \vdash F = F' : \kappa}{\Delta \vdash F' = F : \kappa} \qquad \frac{\Delta \vdash F_1 = F_2 : \kappa \qquad \Delta \vdash F_2 = F_3 : \kappa}{\Delta \vdash F_1 = F_3 : \kappa}$$

Lemma 2 (Reflexivity). *If* $\Delta \vdash F : \kappa$ *then* $\Delta \vdash F = F : \kappa$.

Lemma 3 (Kindedness). *If* $\Delta \vdash F = F' : \kappa$ *then* $\Delta \vdash F : \kappa$ *and* $\Delta \vdash F' : \kappa$.

Wellformed contexts. Δ cxt

$$\frac{}{\diamond \text{ cxt}} \qquad \frac{\Delta \text{ cxt}}{\Delta, X^{p\kappa} \text{ cxt}} \qquad \frac{\Delta \text{ cxt} \qquad \Delta \vdash A : *}{\Delta, x{:}A \text{ cxt}}$$

Welltyped terms. $\Delta \vdash t : A$

$$\frac{(x{:}A) \in \Delta \qquad \Delta \text{ cxt}}{\Delta \vdash x : A} \qquad \frac{\Delta, x{:}A \vdash t : B}{\Delta \vdash \lambda x.t : A \to B} \qquad \frac{\Delta \vdash r : A \to B \qquad \Delta \vdash s : A}{\Delta \vdash r\,s : B}$$

$$\frac{\Delta, X^{\circ\kappa} \vdash t : A}{\Delta \vdash t : \forall X^\kappa.\, A} \qquad \frac{\Delta \vdash t : \forall X^\kappa.\, A \qquad \circ\Delta \vdash F : \kappa}{\Delta \vdash t : [F/X]A} \qquad \frac{\Delta \vdash t : A \qquad \Delta \vdash A = B : *}{\Delta \vdash t : B}$$

Welltyped terms are closed under substitution (as are constructors, cf. Lemma 1).

As opposed to iso-recursive types with "verbose" folding and unfolding, equi-recursive types yield a leaner term language and hence a more succinct semantics.

Lemma 4. *If* $\Delta \vdash t : A$ *then* Δ cxt *and* $\Delta \vdash A : *$.

Proof. By induction on $\Delta \vdash t : A$.

Reduction. The one-step reduction relation $t \longrightarrow t'$ between terms t and t' is defined as the closure of the β-axiom $(\lambda x.t)\, s \longrightarrow_\beta [s/x]t$ under all term constructors. We denote the transitive closure of $\longrightarrow$ by $\longrightarrow^+$. In the next section, we will see that welltyped terms t_0 admit no infinite reduction $t_0 \longrightarrow t_1 \longrightarrow \ldots$

Constant-Time Predecessor. For the type $\mathsf{Nat} := \mathsf{fix}\,\lambda A^{+*}.\,1 + A$ (using $+$, defined above, and $1 := \forall A^{*}.\,A \to A$), it is an easy exercise to define closed terms $\mathsf{O} : \mathsf{Nat}$ and $\mathsf{S} : \mathsf{Nat} \to \mathsf{Nat}$ (using the injections into sums) that represent the natural numbers, and a closed term $\mathsf{P} : \mathsf{Nat} \to \mathsf{Nat}$ (using the definable case analysis construct) such that $\mathsf{P}\,\mathsf{O} \longrightarrow^{+} \mathsf{O}$ and $\mathsf{P}\,(\mathsf{S}\,x) \longrightarrow^{+} x$.

3   Strong Normalization of Fix$^{\omega}$

In this section we prove strong normalization of Fix^{ω} by a model construction where constructors are interpreted as operators on saturated sets. Due to space constraints, the proof necessarily remains sketchy, but all definitions and facts are given which are required to recover the detailed proof.

As is usual for proving (strong) normalization by a model, only the type system has to be reflected in its construction. In System F^{ω}, this is just a simply-typed lambda calculus, namely the (simply-)kinded type constructors. Our system Fix^{ω} additionally has the notions of monotonicity and fixed point. Essentially, we therefore have to give a model of a simply-typed calculus of "syntactically monotone lambda terms". Although the reader will not be surprised by our solution, the authors were surprised that they were not able to find it in the literature.

Following van Raamsdonk and Severi [vRS95, vRS^{+}99] we define the set of strongly normalizing lambda-terms inductively by the following rules (which are implicitly also contained in [Gog95]).

$$\frac{t_i \in \mathsf{SN} \text{ for } 1 \le i \le |t|}{x\,t \in \mathsf{SN}} \qquad \frac{t \in \mathsf{SN}}{\lambda x.t \in \mathsf{SN}} \qquad \frac{[s/x]t\,s \in \mathsf{SN} \qquad s \in \mathsf{SN}}{(\lambda x.t)\,s\,s \in \mathsf{SN}}$$

This characterization is sound, i.e., if $t_0 \in \mathsf{SN}$ then there is no infinite reduction sequence $t_0 \longrightarrow t_1 \longrightarrow \ldots$, for a proof see *loc. cit.* Our aim is to show $t \in \mathsf{SN}$ for each welltyped term t.

3.1   Lattices of Operators on Saturated Sets

A set of terms $\mathcal{A}$ is called *saturated*, $\mathcal{A} \in \mathsf{SAT}^{*}$, if it contains only strongly normalizing terms, $\mathcal{A} \subseteq \mathsf{SN}$, and $\mathcal{A}$ is closed under addition of strongly normalizing neutral terms and strongly normalizing weak head expansion:

$$\frac{t_i \in \mathsf{SN} \text{ for } 1 \le i \le |t|}{x\,t \in \mathcal{A}} \qquad \frac{[s/x]t\,s \in \mathcal{A} \qquad s \in \mathsf{SN}}{(\lambda x.t)\,s\,s \in \mathcal{A}}$$

For sets of terms $\mathcal{A}$, $\mathcal{B}$ we define the function space $\mathcal{A} \to \mathcal{B} := \{r \in \mathsf{SN}\ |\ r\,s \in \mathcal{B} \text{ for all } s \in \mathcal{A}\}$. If $\mathcal{A}$ and $\mathcal{B}$ are saturated, so is $\mathcal{A} \to \mathcal{B}$. Furthermore the function space construction is antitone in the domain and monotone in the codomain: if $\mathcal{A}' \subseteq \mathcal{A}$ and $\mathcal{B} \subseteq \mathcal{B}'$ then $\mathcal{A} \to \mathcal{B} \subseteq \mathcal{A}' \to \mathcal{B}'$.

Given an index set I and a family $\mathcal{A}_i$ ($i \in I$) of saturated sets, the infimum $\bigcap_{i \in I} \mathcal{A}_i$ is also saturated. Formation of the infimum is monotone: Given a second family $\mathcal{A}'_i$ of pointwise greater members, $\mathcal{A}_i \subseteq \mathcal{A}'_i$, the infimum is also greater $\bigcap_{i \in I} \mathcal{A}_i \subseteq \bigcap_{i \in I} \mathcal{A}'_i$. Taking set SN as top element, the saturated sets, together with inclusion, $(\mathsf{SAT}^*, \subseteq)$, constitute a complete lattice.

In our model for Fix^ω, types ($=$ constructors of kind "$*$") will be interpreted as saturated sets. To model constructors of higher kinds κ, we need to define a poset $(\mathsf{SAT}^\kappa, \sqsubseteq^\kappa)$ of (higher-order) operators on saturated sets for each kind κ. For the base kind, let $\mathcal{A} \sqsubseteq^* \mathcal{A}' \; :\Longleftrightarrow \; \mathcal{A}, \mathcal{A}' \in \mathsf{SAT}^*$ and $\mathcal{A} \subseteq \mathcal{A}'$. To require $\mathcal{A}, \mathcal{A}' \in \mathsf{SAT}^*$ is convenient because the reflexive elements of $\sqsubseteq^*$ are now exactly the saturated sets: $\mathcal{A} \in \mathsf{SAT}^* \; \Longleftrightarrow \; \mathcal{A} \sqsubseteq^* \mathcal{A}$. The notion of saturated set $\mathsf{SAT}^{p\kappa \to \kappa^\square}$ and inclusion $\sqsubseteq^{p\kappa \to \kappa^\square}$ for higher kinds is defined by induction on the kind. Let $\mathcal{F}, \mathcal{F}' \in \mathsf{SAT}^\kappa \to \mathsf{SAT}^{\kappa^\square}$ be set-theoretic functions.

$$\mathcal{F} \sqsubseteq^{p\kappa \to \kappa^\square} \mathcal{F}' \; :\Longleftrightarrow \; \mathcal{F}(\mathcal{G}) \sqsubseteq^{\kappa^\square} \mathcal{F}'(\mathcal{G}') \text{ for all } \mathcal{G}, \mathcal{G}' \in \mathsf{SAT}^\kappa \text{ with } \mathcal{G} \sqsubseteq^{p\kappa} \mathcal{G}'$$
$$\mathcal{F} \in \mathsf{SAT}^{p\kappa \to \kappa^\square} \; :\Longleftrightarrow \; \mathcal{F} \sqsubseteq^{p\kappa \to \kappa^\square} \mathcal{F}$$

Here, we used the abbreviations

$$\mathcal{G} \sqsubseteq^{+\kappa} \mathcal{G}' \; :\Longleftrightarrow \; \mathcal{G} \sqsubseteq^\kappa \mathcal{G}',$$
$$\mathcal{G} \sqsubseteq^{-\kappa} \mathcal{G}' \; :\Longleftrightarrow \; \mathcal{G}' \sqsubseteq^\kappa \mathcal{G},$$
$$\mathcal{G} \sqsubseteq^{\circ\kappa} \mathcal{G}' \; :\Longleftrightarrow \; \mathcal{G} \sqsubseteq^\kappa \mathcal{G}' \text{ and } \mathcal{G}' \sqsubseteq^\kappa \mathcal{G}.$$

(An easy induction on κ shows that $\mathcal{G} \sqsubseteq^{\circ\kappa} \mathcal{G}'$ implies $\mathcal{G} = \mathcal{G}'$, but the present definition is more suitable for a uniform treatment of all variances in the proofs to follow.)

Each SAT^κ has a top element and infima: For the base kind, $\top^* = \mathsf{SN}$ and $\bigsqcap^* = \bigcap$; for higher kinds they are defined pointwise: Let $\mathcal{F}_i \in \mathsf{SAT}^{p\kappa \to \kappa^\square}$ for each $i \in I$. Then $\top^{p\kappa \to \kappa^\square} \in \mathsf{SAT}^{p\kappa \to \kappa^\square}$ with $\top^{p\kappa \to \kappa^\square}(\mathcal{G}) := \top^{\kappa^\square}$, and $\bigsqcap_{i \in I}^{p\kappa \to \kappa^\square} \mathcal{F}_i \in \mathsf{SAT}^{p\kappa \to \kappa^\square}$ with $(\bigsqcap_{i \in I}^{p\kappa \to \kappa^\square} \mathcal{F}_i)(\mathcal{G}) := \bigsqcap_{i \in I}^{\kappa^\square} \mathcal{F}_i(\mathcal{G})$. With these definitions, each poset $(\mathsf{SAT}^\kappa, \sqsubseteq^\kappa)$ forms a complete lattice.

By Tarski's fixed-point theorem, each monotone operator $\mathcal{F}$ on a complete lattice has a least fixed point $\mathsf{lfp}\,\mathcal{F}$. Indeed, given $\mathcal{F} \in \mathsf{SAT}^{+\kappa \to \kappa}$, we can define the least fixed point by $\mathsf{lfp}\,\mathcal{F} := \bigsqcap^\kappa \{\mathcal{G} \in \mathsf{SAT}^\kappa \mid \mathcal{F}(\mathcal{G}) \sqsubseteq^\kappa \mathcal{G}\}$, i.e., as the least pre-fixed point of $\mathcal{F}$, which, by the theorem, is indeed a pre-fixed point of $\mathcal{F}$, and also a post-fixed point: $\mathsf{lfp}\,\mathcal{F} \sqsubseteq^\kappa \mathcal{F}(\mathsf{lfp}\,\mathcal{F})$. We will use lfp to interpret fixed points $\mathsf{fix}\,F$ of wellkinded constructors F.

3.2   Interpretation of Constructors

In the following part we will define an interpretation $[\![F]\!]_\theta \in \mathsf{SAT}^\kappa$ for each constructor of kind κ, where θ is a *valuation* for the free constructor variables in F. For convenience, a valuation θ is a set-theoretical object which maps both constructor variables X to sets $\mathcal{F}$ and term variables x to terms t. Update of a valuation is written as $\theta[X \mapsto \mathcal{F}]$ resp. $\theta[x \mapsto t]$. We extend inclusion and saturatedness to valuations by defining:

$$\theta \sqsubseteq^\Delta \theta' \; :\Longleftrightarrow \; \theta(X) \sqsubseteq^{p\kappa} \theta'(X) \text{ for all } X^{p\kappa} \in \Delta$$
$$\theta \in \mathsf{SAT}^\Delta \; :\Longleftrightarrow \; \theta \sqsubseteq^\Delta \theta$$

Lemma 5. *If* $\theta \sqsubseteq^{\Delta} \theta'$, *then* $\theta \sqsubseteq^{+\Delta} \theta'$, $\theta' \sqsubseteq^{-\Delta} \theta$, $\theta \sqsubseteq^{\circ\Delta} \theta'$ *and* $\theta' \sqsubseteq^{\circ\Delta} \theta$.

Proof. By induction on the generation of Δ.

For the following definition and lemma which is the crucial part of this normalization proof, let $\Delta \vdash F : \kappa$. For $\theta \in \mathsf{SAT}^{\Delta}$, we define the interpretation $[\![F]\!]_{\theta} \in \mathsf{SAT}^{\kappa}$ by induction on the structure of F. Simultaneously we need to prove monotonicity of $[\![F]\!]$, the cases for definition and proof are given below.

Lemma 6 (Monotonicity). *If* $\theta \sqsubseteq^{\Delta} \theta'$ *then* $[\![F]\!]_{\theta} \sqsubseteq^{\kappa} [\![F]\!]_{\theta'}$.

For $\theta = \theta'$, immediate consequence of monotonicity is welldefinedness of the interpretation, $[\![F]\!]_{\theta} \in \mathsf{SAT}^{\kappa}$.

Corollary 1 (p-Monotonicity). *Let* $p\Delta \vdash F : \kappa$. *If* $\theta \sqsubseteq^{\Delta} \theta'$ *then* $[\![F]\!]_{\theta} \sqsubseteq^{p\kappa} [\![F]\!]_{\theta'}$.

Proof (of the corollary). In case $p = +$ the corollary just restates monotonicity (Lemma 6). If $p = -$ then $\theta' \sqsubseteq^{-\Delta} \theta$ by Lemma 5. Using monotonicity, $[\![F]\!]_{\theta'} \sqsubseteq^{\kappa} [\![F]\!]_{\theta}$. This is by definition equivalent to $[\![F]\!]_{\theta} \sqsubseteq^{-\kappa} [\![F]\!]_{\theta'}$. If otherwise $p = \circ$, then by Lemma 5 both $\theta \sqsubseteq^{\circ\Delta} \theta'$ and $\theta' \sqsubseteq^{\circ\Delta} \theta$. By monotonicity $[\![F]\!]_{\theta} \sqsubseteq^{\kappa} [\![F]\!]_{\theta'}$ and $[\![F]\!]_{\theta'} \sqsubseteq^{\kappa} [\![F]\!]_{\theta}$ which entail by definition $[\![F]\!]_{\theta} \sqsubseteq^{\circ\kappa} [\![F]\!]_{\theta'}$.

Definition of $[\![F]\!]_{\theta}$ and proof of monotonicity. By induction on the shape of F.

- $\Delta \vdash X : \kappa$. Set $[\![X]\!]_{\theta} := \theta(X)$. By assumption, $X^{p\kappa} \in \Delta$ with $p \in \{+, \circ\}$. The requirement $\theta \sqsubseteq^{\Delta} \theta'$ implies $\theta(X) \sqsubseteq^{\kappa} \theta'(X)$, hence $[\![X]\!]_{\theta} \sqsubseteq^{\kappa} [\![X]\!]_{\theta'}$ by definition.
- $\Delta \vdash \lambda X^{p\kappa}. F : p\kappa \to \kappa'$. The interpretation is a set-theoretic function $[\![\lambda X^{p\kappa}. F]\!]_{\theta} \in \mathsf{SAT}^{\kappa} \to \mathsf{SAT}^{\kappa'}$, $[\![\lambda X^{p\kappa}. F]\!]_{\theta}(\mathcal{G}) := [\![F]\!]_{\theta[X \mapsto \mathcal{G}]}$. To show monotonicity, assume $\mathcal{G}, \mathcal{G}' \in \mathsf{SAT}^{\kappa}$ with $\mathcal{G} \sqsubseteq^{p\kappa} \mathcal{G}'$. By inversion of the typing derivation, $\Delta, X^{p\kappa} \vdash F : \kappa'$, and, since $\theta[X \mapsto \mathcal{G}] \sqsubseteq^{\Delta, X^{p\kappa}} \theta'[X \mapsto \mathcal{G}']$, by induction hypothesis $[\![F]\!]_{\theta[X \mapsto \mathcal{G}]} \sqsubseteq^{\kappa'} [\![F]\!]_{\theta'[X \mapsto \mathcal{G}']}$. Hence, $[\![\lambda X^{p\kappa}. F]\!]_{\theta}(\mathcal{G}) \sqsubseteq^{\kappa'} [\![\lambda X^{p\kappa}. F]\!]_{\theta'}(\mathcal{G}')$ by definition. To conclude, $[\![\lambda X^{p\kappa}. F]\!]$ is monotone.
- $\Delta \vdash F\,G : \kappa'$. Set $[\![F\,G]\!]_{\theta} := [\![F]\!]_{\theta}([\![G]\!]_{\theta})$. Monotonicity and welldefinedness can be seen as follows. By inversion of the kinding derivation, $\Delta \vdash F : p\kappa \to \kappa'$ and $p\Delta \vdash G : \kappa$. Assume $\theta \sqsubseteq^{\Delta} \theta'$. By the first induction hypothesis, $[\![F]\!]_{\theta} \sqsubseteq^{p\kappa \to \kappa'} [\![F]\!]_{\theta'}$. By the second induction hypothesis, with Corollary 1, $[\![G]\!]_{\theta} \sqsubseteq^{p\kappa} [\![G]\!]_{\theta'}$. Putting things together, $[\![F]\!]_{\theta}([\![G]\!]_{\theta}) \sqsubseteq^{\kappa'} [\![F]\!]_{\theta'}([\![G]\!]_{\theta'})$, which by definition entails our goal.
- $\Delta \vdash A \to B : *$. Set $[\![A \to B]\!]_{\theta} := [\![A]\!]_{\theta} \to [\![B]\!]_{\theta}$. By inversion, $-\Delta \vdash A : *$ and $\Delta \vdash B : *$. By induction hypothesis and Corollary 1, $[\![A]\!]_{\theta} \sqsubseteq^{-*} [\![A]\!]_{\theta'}$, hence $[\![A]\!]_{\theta'} \subseteq [\![A]\!]_{\theta}$. Again, by induction hypothesis, $[\![B]\!]_{\theta} \sqsubseteq^{*} [\![B]\!]_{\theta'}$, hence $[\![B]\!]_{\theta} \subseteq [\![B]\!]_{\theta'}$. Together, $[\![A \to B]\!]_{\theta} \subseteq [\![A \to B]\!]_{\theta'}$. Since the functional construction is saturated, we conclude with $[\![A \to B]\!]_{\theta} \sqsubseteq^{*} [\![A \to B]\!]_{\theta'}$.

- $\Delta \vdash \forall X^{\kappa}. A : *.$ Set $[\![\forall X^{\kappa}. A]\!]_{\theta} := \bigcap_{\mathcal{F} \in \mathsf{SAT}^{\kappa}} [\![A]\!]_{\theta[X \mapsto \mathcal{F}]}.$ By inversion, $\Delta, X^{\circ \kappa} \vdash A : *.$ For arbitrary $\mathcal{F} \in \mathsf{SAT}^{\kappa}$, $\theta[X \mapsto \mathcal{F}] \sqsubseteq^{\Delta, X^{\bullet \kappa}} \theta'[X \mapsto \mathcal{F}]$, hence $[\![A]\!]_{\theta[X \mapsto \mathcal{F}]} \sqsubseteq^* [\![A]\!]_{\theta^{\bullet}[X \mapsto \mathcal{F}]}$ by induction hypothesis. This entails $[\![\forall X^{\kappa}. A]\!]_{\theta} \sqsubseteq^* [\![\forall X^{\kappa}. A]\!]_{\theta^{\bullet}}$ by monotonicity and saturatedness of the infimum.
- $\Delta \vdash \mathsf{fix}\, F : \kappa.$ Set $[\![\mathsf{fix}\, F]\!]_{\theta} := \mathsf{lfp}([\![F]\!]_{\theta}).$ By inversion, $\Delta \vdash F : +\kappa \to \kappa$, hence, by induction hypothesis, $\mathcal{F} := [\![F]\!]_{\theta} \in \mathsf{SAT}^{+\kappa \to \kappa}$ is a monotone operator on SAT^{κ}, and by Tarski's theorem the least fixed-point $\mathsf{lfp}\,\mathcal{F} \in \mathsf{SAT}^{\kappa}$ exists. To show monotonicity, we assume $\theta \sqsubseteq^{\Delta} \theta'$ and define $\mathcal{F}' := [\![F]\!]_{\theta^{\bullet}}.$ By monotonicity of $[\![F]\!]$, $\mathcal{F} \sqsubseteq^{+\kappa \to \kappa} \mathcal{F}'.$ In particular, $\mathcal{F}(\mathcal{G}) \sqsubseteq^{\kappa} \mathcal{F}'(\mathcal{G})$ for every $\mathcal{G} \in \mathsf{SAT}^{\kappa}.$ Since $\mathsf{lfp}\,\mathcal{F}$ is a monotone function in its argument $\mathcal{F}$, we are done. $\square$

The interpretation is compatible with substitution and constructor equality, as we show in the following lemmata.

Lemma 7 (Soundness of Substitution). *If* $\Delta, X^{p\kappa} \vdash F : \kappa'$ *and* $p\Delta \vdash G : \kappa$, *then* $[\![[G/X]F]\!]_{\theta} = [\![F]\!]_{\theta[X \mapsto [\![G]\!]_{\theta}]}$ *for all* $\theta \in \mathsf{SAT}^{\Delta}.$

Proof. By induction on the structure of F.

Lemma 8 (Soundness of Equality). *If* $\Delta \vdash F = F' : \kappa$ *then* $[\![F]\!]_{\theta} = [\![F']\!]_{\theta}$ *for all* $\theta \in \mathsf{SAT}^{\Delta}.$

Proof. By induction on constructor equality, using the previous lemma for the first computation rule.

3.3   Interpretation of Terms

To complete our model, we define an interpretation $(\!|t|\!)_{\theta}$ of terms and then show $(\!|t|\!)_{\theta} \in [\![A]\!]_{\theta}$ for welltyped terms $\Delta \vdash t : A$ and sound valuations θ. For wellformed contexts Δ cxt a valuation is *sound*, $\theta \in [\![\Delta]\!]$, if $\theta \in \mathsf{SAT}^{\Delta}$ and $\theta(x) \in [\![A]\!]_{\theta}$ for each $(x{:}A) \in \Delta$. The term interpretation $(\!|t|\!)_{\theta}$ is simply the term t itself where all free variables x have been replaced by their value $\theta(x)$ in valuation θ. Note that theses values are strongly normalizing for sound valuations already; it remains to show that the full term $(\!|t|\!)_{\theta}$ is strongly normalizing for well-typed θ. This is a consequence of the following theorem.

Theorem 1 (Soundness of Typing). *If* $\Delta \vdash t : A$ *and* $\theta \in [\![\Delta]\!]$ *then* $(\!|t|\!)_{\theta} \in [\![A]\!]_{\theta}.$

Proof. By induction on $\Delta \vdash t : A$. Note that by Lemma 4 the context Δ and the type A are wellformed if the typing judgement is derivable. Since our term language is just pure lambda calculus, the proof is standard, for the rule of type equality use Lemma 8.

Corollary 2. *If* $\Delta \vdash t : A$, *the term* t *is strongly normalizing.*

Proof. By Theorem 1, choosing a valuation θ with $\theta(X) = \top^{\kappa}$ for all $X^{p\kappa} \in \Delta$ and $\theta(x) = x$ for all $(x : B) \in \Delta$. This valuation is sound since the type interpretation $[\![B]\!]_{\theta}$ is saturated, hence, contains x.

4   Primitive Recursion for Heterogeneous Datatypes

In this section, we propose a second way to equip System F^ω with fixed points of higher rank. Therein, we follow Mendler [Men87] who also—besides considering type equations in System F—gave an extension of F by least and greatest fixed points, together with elimination schemes which we refer to as Mendler (co)recursion. We carry Mendler's schemes to higher ranks and define a system $MRec^\omega$ as an extension of F^ω by least fixed points of type constructors, also called higher-order inductive types. In contrast to Fix^ω which possesses *equi*-recursive types, $MRec^\omega$ is in the style of *iso*-recursive type systems and has explicit introduction and elimination terms for inductive types. In analogy to Spławski and Urzyczyn [SU99] we conjecture that $MRec^\omega$ has no reduction preserving embedding into F^ω. However, it embeds into Fix^ω, as we will show in Section 5.

Our starting point is Curry-style System F^ω, enriched with unit type 1, binary products $A \times B$ and sums $A + B$ and the usual term constructors: $\langle\rangle$ for the inhabitant of the unit type, $\langle t_1, t_2 \rangle$, $\mathsf{fst}\, r$ and $\mathsf{snd}\, r$ for pairs and left and right projection, and $\mathsf{inl}\, t$, $\mathsf{inr}\, t$ and $\mathsf{case}\,(r, x. s, y. t)$ for left and right injection and case distinction. Note that there are no polarized kinds and no fixed-point constructors. An exposition of the exact rules for typing $\Gamma \vdash t : A$ and reduction $t \longrightarrow t'$ can be found in the appendix of Abel et al. [AMU03]. Since F^ω's notion of constructor equivalence is just plain β-equality, we even identify constructors with their β-normal form on the syntactic level.

4.1   Definition of System $MRec^\omega$

For every kind κ of F^ω, we add the constructor constant μ^κ of kind $(\kappa \to \kappa) \to \kappa$ to the system of constructors of F^ω, denoting least fixed-point formation. The term system of F^ω is extended by two families of constants: in^κ (fixed-point introduction) and $MRec^\kappa$ (fixed-point elimination) for every kind κ. In order to give their types, we need a notion of constructor containment: Every kind κ can uniquely be written in the form $\kappa_1 \to \ldots \kappa_n \to *$, in short $\boldsymbol{\kappa} \to *$. Define

$$F \subseteq^\kappa G := \forall \boldsymbol{X^\kappa}.\, F\boldsymbol{X} \to G\boldsymbol{X} : \kappa \to \kappa \to *,$$

for constructors $F, G : \kappa = \boldsymbol{\kappa} \to *$. The typing of the constants can now be given by $\mathsf{in}^\kappa : \forall F^{\kappa \to \kappa}.\, F(\mu^\kappa F) \subseteq^\kappa \mu^\kappa F$ and

$$MRec^\kappa : \forall F^{\kappa \to \kappa} \forall G^\kappa.\, (\forall X^\kappa.\, X \subseteq^\kappa \mu^\kappa F \to X \subseteq^\kappa G \to F X \subseteq^\kappa G) \to \mu^\kappa F \subseteq^\kappa G.$$

The notion $\longrightarrow$ of reduction for untyped terms is extended by the additional basic reduction rule of primitive recursion

$$MRec^\kappa\, s\, (\mathsf{in}^\kappa\, t) \longrightarrow_\beta s\, \mathsf{id}\, (MRec^\kappa\, s)\, t,$$

where $\mathsf{id} := \lambda x.x$ is the identity. Intuitively, subject reduction still holds because the type $\forall X^\kappa.\, X \subseteq^\kappa \mu^\kappa F \to X \subseteq^\kappa G \to F X \subseteq^\kappa G$ of the term s is instantiated with the fixed-point constructor $\mu^\kappa F$ itself. Therefore, the identity id qualifies as first argument to s. In general, the transformation from the blank type X back into the fixed-point, i.e., of type $X \subseteq^\kappa \mu^\kappa F$, which is the first formal argument of s, provides access to the predecessor of the recursion argument. This is the feature which distinguishes primitive recursion from iteration.

For $\kappa = *$, we have just restated Mendler's rules for recursive types [Men87]. At this point, let us remark that Mendler-style inductive types $\mu^\kappa F$—although not observed by Mendler—do not require positivity for F. This contrasts with the recursive types of Fix^ω. It also contrasts with formulations of primitive recursion in conventional style that have to rely on positivity or, less syntactically, on a monotonicity requirement such as that in [Mat01] for $\kappa = *$ or $* \rightarrow *$.

4.2   Example: Redecoration of Finite Triangular Matrices

As a non-trivial example of the use of MRec^ω for heterogeneous datatypes, we consider a redecoration operation for the diagonal elements of finite triangular matrices. In previous work with Uustalu, we have treated redecoration for *infinite* triangular matrices by higher-order coiteration [AMU03], and the finite ones by a computationally unsatisfactory encoding of recursion within iteration [AMU04].

Fix a type $E : *$ of matrix elements. The type $\mathsf{Tri}\,A$ of finite triangular matrices with diagonal elements in A and ordinary elements E can be obtained as follows, with $\kappa 1 := * \rightarrow *$:

$$\mathsf{TriF} := \lambda X^{\kappa 1} \lambda A^*.\ A \times (1 + X\,(E \times A)) : \kappa 1 \rightarrow \kappa 1$$
$$\mathsf{Tri} := \mu^{\kappa 1} \mathsf{TriF} \qquad\qquad\qquad\qquad : \kappa 1$$

We think of these triangles decomposed columnwise: The first column is a singleton of type A, the second a pair of type $E \times A$, the third a triple of type $E \times (E \times A)$, the fourth a quadruple of type $E \times (E \times (E \times A))$ etc. Hence, if some column has some type A' we obtain the type of the next column as $E \times A'$. By taking the left injection into the sum $1 + \ldots$, one can construct an element without further recurrence, the last column. We can visualize triangles like this:

$$
\begin{array}{c|c|c|c|c}
A & E & E & E & E \\ \cline{1-1}
 & A & E & E & E \\ \cline{2-2}
 & & A & E & E \\ \cline{3-3}
 & & & A & E \\ \cline{4-4}
 & & & & A
\end{array}
$$

The vertical lines hint at the decomposition scheme. In general, elements of type $\mathsf{Tri}\,A$ are constructed by means of

$$\mathsf{sg} := \lambda a.\ \mathsf{in}^{\kappa 1} \langle a, \mathsf{inl}\,\langle\rangle\rangle \quad : \forall A^*.\ A \rightarrow \mathsf{Tri}\,A$$
$$\mathsf{cons} := \lambda a \lambda t.\ \mathsf{in}^{\kappa 1} \langle a, \mathsf{inr}\,t\rangle : \forall A^*.\ A \rightarrow \mathsf{Tri}\,(E \times A) \rightarrow \mathsf{Tri}\,A$$

The function $\mathsf{top} : \forall A^*.\ \mathsf{Tri}\,A \rightarrow A = \mathsf{Tri} \subseteq^{\kappa 1} \lambda A^*.\ A$ that yields the topmost diagonal element, is defined as $\mathsf{top} := \mathsf{MRec}^{\kappa 1}(\lambda i \lambda top \lambda p.\ \mathsf{fst}\,p)$. As reduction behavior, we get

$$\mathsf{top}\,(\mathsf{sg}\,a) \longrightarrow^+ a$$
$$\mathsf{top}\,(\mathsf{cons}\,a\,t) \longrightarrow^+ a$$

If we remove the first column of a triangle $\mathsf{Tri}\,A$, we obtain a trapezium $\mathsf{Tri}\,(E \times A)$. We can get back a (smaller) triangle if we cut off the top row of the trapezium using the function $\mathsf{cut} : \forall A^*.\,\mathsf{Tri}\,(E \times A) \to \mathsf{Tri}\,A$. The exact definition of this function, which is like fcut in [AMU04, Example 34], has to be omitted due to lack of space.

Let $T\,A$ denote some sort of A-decorated (or A-labelled) trees. *Redecoration* [UV02] is an operation that takes an A-decorated tree $t : T\,A$ and a redecoration rule $f : T\,A \to B$ and returns a B-decorated tree $t' : T\,B$. For triangles, redecoration works as follows: In the triangle

$$A\ E\ E\ E\ E$$
$$A\ E\ E\ E$$
$$\underline{A}\ E\ E$$
$$A\ E$$
$$A$$

the underlined A (as an example) gets replaced by the B assigned by the redecoration rule to the sub triangle cut out by the horizontal line; similarly, every other A is replaced by a B.

For the definition of redecoration, we will need a means of lifting a redecoration rule on triangles to one on trapeziums.

$$\mathsf{lift} := \lambda f \lambda t.\, \langle \mathsf{fst}\,(\mathsf{top}\,t),\ f\,(\mathsf{cut}\,t) \rangle$$
$$:\ \forall A^* \forall B^*.\,(\mathsf{Tri}\,A \to B) \to \mathsf{Tri}\,(E \times A) \to E \times B$$

For a detailed explanation in which sense this is a lifting, see [AMU04]. Finally, we can define redecoration

$$\mathsf{redec} : \forall A^* \forall B^*.\,\mathsf{Tri}\,A \to (\mathsf{Tri}\,A \to B) \to \mathsf{Tri}\,B = \mathsf{Tri} \subseteq^{\kappa 1} G$$

with $G := \lambda A^* \forall B^*.\,(\mathsf{Tri}\,A \to B) \to \mathsf{Tri}\,B$. The definition makes essential use of primitive recursion in that it also uses the variable $i : X \subseteq^{\kappa 1} \mathsf{Tri}$ in the body of argument to $\mathsf{MRec}^{\kappa 1}$:

$$\mathsf{redec} := \mathsf{MRec}^{\kappa 1}\Big(\lambda i \lambda redec \lambda t \lambda f.\ \mathsf{case}\,(\mathsf{snd}\,t,$$
$$u.\,\mathsf{sg}\,(f\,(\mathsf{sg}\,(\mathsf{fst}\,t))),$$
$$r.\,\mathsf{cons}\,(f\,(\mathsf{cons}\,(\mathsf{fst}\,t)\,(i\,r)))\,(redec\,r\,(\mathsf{lift}\,f)))\Big)$$

Its reduction behavior is easy to calculate:

$$\mathsf{redec}\,(\mathsf{sg}\,a)\,f \quad \longrightarrow^+ \mathsf{sg}\,(f\,(\mathsf{sg}\,a))$$
$$\mathsf{redec}\,(\mathsf{cons}\,a\,r)\,f \longrightarrow^+ \mathsf{cons}\,(f\,(\mathsf{cons}\,a\,r))\,(\mathsf{redec}\,r\,(\mathsf{lift}\,f))$$

The reader is invited to compare this concise behaviour with the one obtained in [AMU04] within definitional extensions of system F^ω that therefore can only provide iteration schemes and no primitive recursion. Notice that the number of reduction steps does *not* depend on the terms a, r and f since these may just be variables. By a modification of the definition of redec above, it is easy to define a constant-time predecessor operation on triangles (a left inverse of $\mathsf{in}^{\kappa 1}$ for Tri even with respect to reductions of open terms): The access to r in the cons case of the reduction will be type-correct by using $(i\,r)$ instead of r, just as for redec.

5   Embedding of Mendler-Style Recursion into Fix$^\omega$

In this section, we prove—via an embedding into Fix$^\omega$—that Mendler recursion for higher ranks is strongly normalizing. The proof proceeds in two steps: First, we show that all constructions of MRec$^\omega$ can be defined in Fix$^\omega$ such that reduction is simulated. Then we map each welltyped term of MRec$^\omega$ onto a still welltyped term of Fix$^\omega$ of exactly the same shape (the translation is purely homomorphic). Thus, each infinite reduction sequence of MRec$^\omega$ would map onto an infinite sequence of Fix$^\omega$, which is a contradiction to the result of Section 3.

Products and sums can be defined in Fix$^\omega$ via the standard impredicative encoding (see Example 1). The interesting part is the definition of least fixed-points $\mu^\kappa F$ within Fix$^\omega$. We give their definition only for kinds carrying no polarity information, i.e., kinds of the form $\kappa = \mathtt{o}\kappa \to *$. This suffices because their purpose is just to serve as images in the translation of the least fixed points in the polarity-free system MRec$^\omega$. We define $\mu^\kappa := \lambda F^{\mathtt{o}(\mathtt{o}\kappa \to \kappa)}$. fix Φ_F with

$$\Phi_F := \lambda Y^{+\kappa} \lambda \boldsymbol{X}^{\mathtt{o}\kappa} \forall G^\kappa. \, (\forall X^\kappa. \, X \subseteq^\kappa Y \to X \subseteq^\kappa G \to F\,X \subseteq^\kappa G) \to G\,\boldsymbol{X}.$$

It is not hard to see that $F^{\mathtt{o}(\mathtt{o}\kappa \to \kappa)} \vdash \Phi_F : +\kappa \to \mathtt{o}\kappa \to *$, since the variable Y occurs twice to the left of an arrow in the body of the definition of Φ_F. Thus, for any $F : \mathtt{o}\kappa \to \kappa$ we have $\Phi_F : +\kappa \to \kappa$, and $\mu^\kappa : \mathtt{o}(\mathtt{o}\kappa \to \kappa) \to \kappa$ as required.

Once we have found a suitable representation of μ^κ in Fix$^\omega$, the definition of elimination and introduction falls into place:

$$\mathsf{MRec}^\kappa := \lambda s \lambda r. \, r\,s$$
$$\mathsf{in}^\kappa \quad := \lambda t \lambda s. \, s \; \mathsf{id} \; (\mathsf{MRec}^\kappa s) \, t$$

Note that the right-hand sides do not depend on κ. These definitions yield simulation of primitive recursion within Fix$^\omega$, as we can confirm by performing four β-reduction steps: $\mathsf{MRec}^\kappa s \, (\mathsf{in}^\kappa t) \longrightarrow^+ s \; \mathsf{id} \; (\mathsf{MRec}^\kappa s) \, t$.

Now, System MRec$^\omega$ can be translated into Fix$^\omega$ by replacing each arrow kind $\kappa \to \kappa'$ by $\mathtt{o}\kappa \to \kappa'$, and each annotated abstraction λX^κ by $\lambda X^{\mathtt{o}\kappa}$. All other syntactical constructions remain unchanged. Certainly, we can only map the constants in^κ and MRec^κ of MRec$^\omega$ onto their defined counterparts in Fix$^\omega$, if the types of source and target match. This can been seen by type-checking, for which the following chart might be an aid.

$$\Delta \; := \; F^{\mathtt{o}(\mathtt{o}\kappa \to \kappa)}, \; G^{\mathtt{o}\kappa}, \; X^{\mathtt{o}\kappa},$$
$$s : \forall X^\kappa. \, X \subseteq^\kappa \mu^\kappa F \to X \subseteq^\kappa G \to F\,X \subseteq^\kappa G,$$
$$r : \mu^\kappa F \, \boldsymbol{X} \; = \; \Phi_F(\mu^\kappa F) \, \boldsymbol{X},$$
$$t : F\,(\mu^\kappa F) \, \boldsymbol{X}$$

$$\Delta \; \vdash \; r\,s \; : \; G\,\boldsymbol{X}$$
$$\Delta \; \vdash \; \lambda s. \, s \; \mathsf{id} \; (\mathsf{MRec}^\kappa s) \, t \; : \; \Phi_F(\mu^\kappa F) \, \boldsymbol{X} \; = \; \mu^\kappa F \, \boldsymbol{X}$$

Theorem 2. *In System* MRec$^\omega$ *of Mendler recursion for arbitrary kinds there exists no infinite reduction sequence* $t_0 \longrightarrow t_1 \longrightarrow \dots$ *starting with a welltyped term* t_0.

Proof. By Corollary 2, using the abovementioned translation into Fix$^\omega$.

6   Conclusion and Future Work

We have presented two systems for total functions over higher-order and nested datatypes where the predecessor runs in constant time. The first system, Fix^ω, supports positive equi-recursive types of higher order. No primitive combinator for recursive functions is built in, but due to the strength of equi-recursive types in combination with impredicativity, customary recursion schemes can be defined. One instance is Mendler-style primitive recursion MRec, which for example can be used to define a redecoration algorithm for triangular matrices. We have shown that Mendler-style primitive recursion can be simulated in Fix^ω.

This simulation could have been extended to also account for coinductive type constructors, by defining Mendler-style corecursion for higher ranks in Fix^ω. A naturally corecursive program is substitution for the infinite version of de Bruijn terms coded as a nested datatype [AR99, BP99b]. Due to space restrictions we have to leave this direction to future work.

The systematic use of nested datatypes to represent datastructures with invariants is rather new [Hin98] (but also see [Oka96, Sections 10, 11] for earlier work). As an example, Hinze [Hin01] implemented Okasaki's functional version of red-black trees [Oka99] by help of a nested datatype to actually ensure the balancing properties of red-black trees by the type system. Most algorithms for nested datatypes published so far require just *iteration*, hence can be implemented in the framework of generalized folds [BP99a] or efficient folds [MGB04] or Mendler iteration [AMU04]. As more classical algorithms will find functional implementations using nested datatypes, we imagine many more examples requiring primitive recursion for higher-rank datatypes, and thus may infer termination of the respective algorithms.

References

[AMU03]   A. Abel, R. Matthes, and T. Uustalu. Generalized iteration and coiteration for higher-order nested datatypes. In A. Gordon, ed., *Proc. of FoSSaCS 2003*, vol. 2620 of *LNCS*, pp. 54–69. 2003.

[AMU04]   A. Abel, R. Matthes, and T. Uustalu. Iteration and coiteration schemes for higher-order and nested datatypes. *Theoretical Computer Science*, 2004. 79 pages, accepted for publication.

[AR99]   T. Altenkirch and B. Reus. Monadic presentations of lambda terms using generalized inductive types. In *Proceedings of CSL '99*, vol. 1683 of *LNCS*, pp. 453–468. 1999.

[BB85]   C. Böhm and A. Berarducci. Automatic synthesis of typed λ-programs on term algebras. *Theoretical Computer Science*, 39:135–154, 1985.

[BP99a]   R. Bird and R. Paterson. Generalised folds for nested datatypes. *Formal Aspects of Computing*, 11(2):200–222, 1999.

[BP99b]   R. S. Bird and R. Paterson. De Bruijn notation as a nested datatype. *Journal of Functional Programming*, 9(1):77–91, 1999.

[CP88]   T. Coquand and C. Paulin. Inductively defined types—preliminary version. In P. Martin-Löf and G. Mints, eds., *Proceedings of COLOG '88*, vol. 417 of *LNCS*, pp. 50–66. 1988.

[DC98]   D. Duggan and A. Compagnoni. Subtyping for object type constructors, 1998. Presented at FOOL 6.

[GHR93]   P. Giannini, F. Honsell, and S. Ronchi Della Rocca. Type inference: some results, some problems. *Fundamenta Informaticae*, 19(1-2):87 – 125, 1993.

[Gog95]   H. Goguen. Typed operational semantics. In M. Dezani-Ciancaglini and G. Plotkin, eds., *Proc. of TLCA '95*, vol. 902 of *LNCS*, pp. 186–200. 1995.

[Hin98]   R. Hinze. Numerical representations as higher-order nested datatypes. Tech. Rep. IAI-TR-98-12, Institut für Informatik III, Universität Bonn, 1998.

[Hin01]   R. Hinze. Manufacturing datatypes. *Journal of Functional Programming*, 11(5):493–524, 2001.

[Mat01]   R. Matthes. Monotone inductive and coinductive constructors of rank 2. In L. Fribourg, ed., *Proc. of CSL 2001*, vol. 2142 of *LNCS*, pp. 600–614. 2001.

[Men87]   N. P. Mendler. Recursive types and type constraints in second-order lambda calculus. In *Proceedings of LICS '87*, pp. 30–36. 1987.

[MGB04]   C. Martin, J. Gibbons, and I. Bayley. Disciplined, efficient, generalised folds for nested datatypes. *Formal Aspects of Computing*, 16(1):19–35, 2004.

[Oka96]   C. Okasaki. *Purely Functional Data Structures*. Ph.D. thesis, Carnegie Mellon University, 1996.

[Oka99]   C. Okasaki. Red-black trees in a functional setting. *Journal of Functional Programming*, 9(4):471–477, 1999.

[Ste98]   M. Steffen. *Polarized Higher-Order Subtyping*. Ph.D. thesis, Technische Fakultät, Universität Erlangen, 1998.

[SU99]   Z. Spławski and P. Urzyczyn. Type fixpoints: Iteration vs. recursion. In *Proceedings of ICFP'99*, pp. 102–113. SIGPLAN Notices, 1999.

[Urz96]   P. Urzyczyn. Positive recursive type assignment. *Fundamenta Informaticae*, 28(1–2):197–209, 1996.

[UV02]   T. Uustalu and V. Vene. The dual of substitution is redecoration. In K. Hammond and S. Curtis, eds., *Trends in Functional Programming 3*, pp. 99–110. Intellect, Bristol, Portland, OR, 2002.

[vBL+97] S. van Bakel, L. Liquori, S. Ronchi Della Rocca, and P. Urzyczyn. Comparing cubes of typed and type assignment systems. *Annals of Pure and Applied Logic*, 86(3):267–303, 1997.

[vRS95]   F. van Raamsdonk and P. Severi. On normalisation. Tech. Rep. CS-R9545, CWI, 1995.

[vRS+99] F. van Raamsdonk, P. Severi, M. H. Sørensen, and H. Xi. Perpetual reductions in lambda calculus. *Inf. and Comp.*, 149(2):173–225, 1999.

On the Building of Affine Retractions

Aleksy Schubert[*]

Institute of Informatics, Warsaw University,
ul. Banacha 2, 02-097 Warsaw, Poland
alx@mimuw.edu.pl

Abstract. A simple type σ is retractable to a simple type τ if there are
two terms $C : \sigma \to \tau$ and $D : \tau \to \sigma$ such that $D \circ C =_{\beta\eta} \lambda x.x$. The paper
presents a system which for given σ, τ derives affine retractability i.e. the
above relation with additional restriction that in C and D every bound
variable occurs at most once. A derivation in the system constructs
these terms. What is more, the complexity of building affine retractions is studied. The problem of affine retractability is NP-complete even
for the class of types over single type atom and having limited functional
order. A polynomial algorithm for types of orders less than 3 is also
presented.

1 Introduction

The notion of isomorphism, which renders the idea of identicality, appears very
frequently in many formal theories. In the simply typed lambda calculus it is defined as follows: two types σ, τ are isomorphic if there exists terms C of the type
$\sigma \to \tau$ and D of the type $\tau \to \sigma$ such that $D \circ C =_{\beta\eta} \lambda x.x$ and $C \circ D =_{\beta\eta} \lambda x.x$.
This notion has been studied since 1980s, see e.g. [BL85]. A complete and effec-
tive characterisation of the relation is given in [Cos95]. Moreover, the relation of
type isomorphism has already been successfully used in tools supporting search-
ing in software libraries e.g. [Cos95, Rit91].

The notion of retraction is a generalisation of isomorphism. In this case only
$D \circ C =_{\beta\eta} \lambda x.x$ is required in place of two abovementioned equalities. This
category theory based definition corresponds to a concept of coding — everything
that is encoded by means of C can be decoded back by D without any loss of
information.

The knowledge concerning type retractions is undeveloped even for very
simple formalisms, like the simply typed lambda calculus. There is a complete
and effective criterion of retractability in the calculus, but with respect to the
β-reduction [SU99]. There are also several sufficient conditions for $\beta\eta$-reduction
based retractions in [dPS92]. The last paper includes a complete characterisation of *affine retractions* in the lambda calculus with a single type atom

---

[*] This work was partly supported by KBN grant no 7 T11C 028 20, part of this work
was conducted while the author was a winner of the fellowship of The Foundation
for Polish Science (FNP).

J. Marcinkowski and A. Tarlecki (Eds.): CSL 2004, LNCS 3210, pp. 205–219, 2004.

where the word *affine* means that in terms C, D each bound variable is used at most once. A paper by Padovani [Pad01] contains an algorithm that decides full $\beta\eta$ retractability in the simply typed lambda calculus with types that have a single type atom. In a paper by Regnier and Urzyczyn [LR01], the authors give characterisations of $\beta\eta$ retractability in the lambda calculus with many type atoms, but these characterisations are either not effective or not complete. However, they give an effective and complete characterisation of affine retractability.

Type retractions (but in a richer type system) have already been used to prove that recursive polymorphic types cannot be encoded in the polymorphic lambda calculus $\lambda 2$ [SU99]. Type retractions may, in addition, turn out to be useful as a method to find more general operations in software libraries — similarly to the foregoing applications of type isomorphisms. By Curry-Howard isomorphism, retractions can also be used in automated provers as a vehicle supporting the reuse of already proved subproblems. Each of these applications requires an effective method to generate a link between the existing infrastructure (a software library, a library of proved lemmas) and the new requirement (a function to be found, a new formula to be proved). This link is provided by the terms C, D, mentioned in the definition of retraction. This paper presents a system for inferring *affine retractability* together with the accompanying terms C and D.

There are two reasons that justify the restriction to affine retractions. First, the type retraction problem can be defined in terms of the higher-order matching problem. This can be done using the following equation $\lambda x.X(Yx) = \lambda x.x$ where X, Y are unknown variables. The higher-order matching problem is known to be at least non-elementarily hard [Vor97], if not undecidable. Moreover, the construction by Padovani in [Pad01], which employs a special — known to be non-elementary — case of the higher-order matching problem suggests that the problem of finding retractions is highly intractable. Second, dealing with affine retractions is supported by the assumption that every transformation approved by a human must be fairly simple so that it can be understood. This kind of situation can occur in the afore-mentioned search in software libraries.

This paper contains a proof that the relation of affine retractability is polynomial for types of order at most 3. This provides a tight bound on the tractability of the problem as the problem is NP-complete already for types of order 4. The latter result holds for the class of types with unbounded number of atoms. If the number of type atoms is bounded by a number $k > 1$ the problem is NP-complete for types of order 5. If there is only one type atom the problem is NP-complete for types of order 7.

2   Basic Definitions

We assume that the reader is familiar with the simply typed lambda calculus. Thus we only sketch the basic definitions in order rather to settle the notation

Fig. 1. A schematic picture of $\sigma_1 \to \sigma_2 \to \cdots \to \sigma_n \to a$

than thoroughly introduce the notions. A more gentle introduction can be found e.g. in [Bar92].

The simply typed lambda calculus $\lambda_\to$ (in the Church fashion) has terms of the form MN or $\lambda x : \sigma.M$. The λ operator binds the variable x in the body of M so we deal with the notions of free and bound variables together with the α equivalence relation. We identify α-quivalent expressions. The types in $\lambda_\to$ are built using basic types from a nonempty set $\mathcal{B}$. The types are constructed by means of the $\to$ operator e.g. $\sigma \to \tau$. The types are assigned to terms by means of a type inference system. Rules of such a system can be found in [Bar92]. We sometimes use a notation $\Delta \to a$, where Δ is a set or a sequence of types $\{\delta_1, \ldots, \delta_n\}$, to denote $\delta_1 \to \cdots \to \delta_n \to a$. The types are categorised by an order. The order is defined as $\mathrm{ord}(a) = 1$ for $a \in \mathcal{B}$ and $\mathrm{ord}(\sigma \to \tau) = \max(\mathrm{ord}(\sigma) + 1, \mathrm{ord}(\tau))$. The set of types having the order k is denoted $\mathrm{T}^k_\to$. If $\sigma = \sigma_1 \to \cdots \to \sigma_n \to a$ or $\sigma = a$ then $head(\sigma) = a$. All the types $\sigma_1, \ldots, \sigma_n$ are called *arguments* of σ.

We also adopt a method of representing types on pictures which is more convenient in dealing with type retractions. The idea of the representation is presented on Fig. 1.

The simply typed lambda calculus is accompanied by a notion of evaluation. This is called β-reduction. This relation is generated by the basic rule $(\lambda x : \sigma.M)N \to_\beta M[x := N]$ where $[x := N]$ denotes the capture avoiding substitution. This relation is extended with the η-reduction reduction generated by $(\lambda x : \sigma.Mx) \to_\eta M$ where x does not occur freely in M. The combination of the two relations is denoted by $\to_{\beta\eta}$. We use also the reflexive transitive closure $\to^*_{\beta\eta}$ of $\to_{\beta\eta}$ as well as the least equivalence $=_{\beta\eta}$ containing $\to_{\beta\eta}$.

Definition 1. (retracts)
We say that a type σ is a *retract* of a type τ, and we write $\sigma \trianglelefteq_{\beta\eta} \tau$, iff there exists a pair of terms $C : \sigma \to \tau$ and $D : \tau \to \sigma$ such that $D \circ C =_{\beta\eta} \mathbf{I}_\sigma = \lambda x : \sigma.x$. We say that a type σ is an *affine retract* of a type τ, and we write $\sigma \trianglelefteq^1 \tau$ iff C, D are affine terms (i.e. terms with at most one occurrence of each bound variable). We usually omit subscript $\beta\eta$ and write $\trianglelefteq$ or $\trianglelefteq^1$.

The problem of finding affine retracts is defined as follows:

Definition 2. (problem of affine retractions)
Input: types σ, τ.
Question: is there a pair of affine terms C, D such that $C : \sigma \to \tau$ and $D : \tau \to \sigma$ such that $D \circ C =_{\beta\eta} \mathbf{I}_\sigma$?

(Ax)
$$a \trianglelefteq^1 a$$

(H)
$$\frac{\sigma \trianglelefteq^1 \sigma', \quad \tau \trianglelefteq^1 \tau'}{\sigma \to \tau \trianglelefteq^1 \sigma' \to \tau'}$$

(N)
$$\frac{\sigma \trianglelefteq^1 \tau}{\sigma \trianglelefteq^1 \Sigma \to \tau}$$

(D)
$$\frac{\Delta_1 \to a \trianglelefteq^1 \sigma_1, \cdots, \Delta_n \to a \trianglelefteq^1 \sigma_n}{\Delta_1 \cup \cdots \cup \Delta_n \to a \trianglelefteq^1 \{\Sigma_1 \to \sigma_1 \to a, \cdots, \Sigma_n \to \sigma_n \to a\} \to a}$$

Fig. 2. The syntax directed system

2.1   A System for Inference of Retraction Terms

We consider a system to infer inequalities $\trianglelefteq^1$. The system was proposed by Regnier and Urzyczyn (RU) and is presented at Fig. 2. The most important property of the system is the lack of any cut-like rule.

The system RU is a good starting point in design of our term system for deducing affine retractions. We modify the system RU since its original form has several notational conveniences which are inadequate in the context of generation of retracting terms. The term system presented on Fig. 3 infers sequents $\vdash C \bullet D : \sigma \trianglelefteq^1 \tau$ such that C, D correspond to terms certifying the retraction $\sigma \trianglelefteq^1 \tau$.

(Ax)
$$\vdash \mathbf{I}_a \bullet \mathbf{I}_a : a \trianglelefteq^1 a$$

(H)
$$\frac{\vdash T_1 \bullet T_2 : \sigma \trianglelefteq^1 \sigma', \quad \vdash T_1' \bullet T_2' : \tau \trianglelefteq^1 \tau'}{\vdash H_{\sigma,\tau,\sigma',\tau'} T_1' T_2 \bullet H_{\sigma',\tau',\sigma,\tau} T_2' T_1 : \sigma \to \tau \trianglelefteq^1 \sigma' \to \tau'}$$

(N)
$$\frac{\vdash T_1 \bullet T_2 : \sigma \trianglelefteq^1 \tau}{\vdash N_{\sigma,\tau,\Sigma} T_1 \bullet N'_{\sigma,\tau,\Sigma} T_2 : \sigma \trianglelefteq^1 \Sigma \to \tau}$$

(D)
$$\frac{\begin{array}{c}\vdash T_1 \bullet T_1' : \Sigma \to a \trianglelefteq^1 \tau', \\ \vdash T_2 \bullet T_2' : \sigma \trianglelefteq^1 \tau \\ head(\tau') = head(\sigma) = head(\tau) = a\end{array}}{\vdash DT_1 T_2 \bullet D' T_1' T_2' : \Sigma \to \sigma \trianglelefteq^1 (\Delta_1 \to \tau' \to \Delta_2 \to a) \to \tau}$$

(P)
$$\frac{\vdash T \bullet T' : \sigma_1 \to \cdots \to \sigma_n \to a \trianglelefteq^1 \tau_1 \to \cdots \to \tau_m \to a}{\vdash P_{\pi_2}^{\pi_1} T \bullet P_{\pi_1}^{\pi_2} T' : \sigma_{\pi_1(1)} \to \cdots \to \sigma_{\pi_1(n)} \to a \trianglelefteq^1 \atop \tau_{\pi_2(1)} \to \cdots \to \tau_{\pi_2(m)} \to a}$$

Fig. 3. The term system for linear retractions

The rule (D) in the system RU and in the term system are a little bit difficult to grasp. Figure 4 may help in understanding of them.

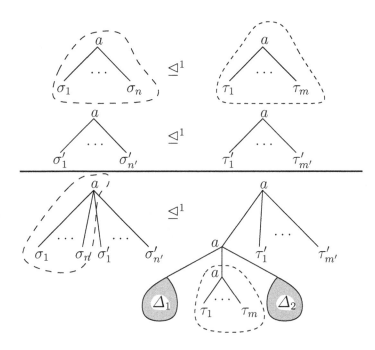

Fig. 4. The rule (D) from RU with emphasised deep inequality

Definition 3. (terms from the term system)
We define a transformation `lambdify(·)` for the retraction combinators used in Fig. 3

- `lambdify(`$\mathbf{I}_\sigma$`)` $= \lambda x : \sigma.x,$
- `lambdify(`$H_{\sigma,\tau,\sigma',\tau'}$`)` $=$
  $\lambda T_1 : \tau \to \tau'.\lambda T_2 : \sigma' \to \sigma.\lambda f : \sigma \to \tau.\lambda x : \sigma'.T_1(f(T_2x))$
- `lambdify(`$N_{\sigma,\tau,\Sigma}$`)` $= \lambda T : \sigma \to \tau.\lambda f : \sigma.\lambda x_1 : \sigma_1 \ldots x_n : \sigma_n.Tf$
  where $\Sigma = \{\sigma_1, \ldots, \sigma_n\}$
- `lambdify(`$N'_{\sigma,\tau,\Sigma}$`)` $= \lambda T : \tau \to \sigma.\lambda f : \Sigma \to \tau.T(fz_1 \cdots z_n)$
  where $\Sigma = \{\sigma_1, \ldots, \sigma_n\}$ and $z_i : \sigma_i$ for $i = 1, \ldots, n.$
- `lambdify(`D`)` $=$
  $\lambda T_1 : (\Sigma \to a) \to \tau'.\lambda T_2 : \sigma \to \tau.$
  $\lambda f : \Sigma \to \sigma.\lambda y : \Delta_1 \to \tau' \to \Delta_2 \to a.$
  $$\lambda z_1 : \tau_1 \ldots z_p : \tau_p. \ yc_1^{\delta_1} \cdots c_{q_1}^{\delta_{q_1}} \overline{D} c_{q_1+1}^{\delta_{q_1+1}} \cdots c_{q_2}^{\delta_{q_2}}$$
- `lambdify(`D'`)` $=$
  $$\lambda T'_1 : \tau' \to \Sigma \to a.\lambda T'_2 : \tau \to \sigma.$$
  $\lambda g : (\Delta_1 \to \tau' \to \Delta_2 \to a) \to \tau.\lambda x_1 : \rho_1, \ldots, x_r : \rho_r. \ T'_2\overline{D'}$

where

$$\tau = \tau_1 \to \cdots \to \tau_p \to a,$$
$$\Delta_1 = \{\delta_1, \ldots, \delta_{q_1}\}, \quad \Delta_2 = \{\delta_{q_1+1}, \ldots, \delta_{q_2}\}, \quad \Sigma = \{\rho_1, \ldots, \rho_r\}$$
$$\overline{D} = T_1 \lambda y_1 : \rho_1 \ldots y_r : \rho_r . T_2(f y_1 \cdots y_r) z_1 \cdots z_p,$$
$$\overline{D'} = g \lambda z_1 : \delta_1 \ldots z_{q_1} : \delta_{q_1} . \lambda y : \tau' . \lambda z_{q_1+1} : \delta_1 \ldots z_{q_2} : \delta_{q_2} . T_1' y x_1 \cdots x_r$$

with $c_1, \ldots, c_{q_2}$ being fresh variables of the types δ_k for $k = 1, \ldots, q_2$.
− lambdify$(P_{\pi_2}^{\pi_1}) =$

$$\lambda f : \sigma \to \tau.$$
$$\lambda x : \sigma_{\pi_1(1)} \to \cdots \to \sigma_{\pi_1(n)} \to a.$$
$$\lambda y_{\pi_2(1)} : \tau_{\pi_2(1)} \cdots y_{\pi_2(m)} : \tau_{\pi_2(m)}.$$
$$f(\lambda v_1 : \sigma_1 \ldots v_n : \sigma_n . x v_{\pi_1(1)} \cdots v_{\pi_1(n)}) y_1 \cdots y_m$$

where π_1, π_2 are permutations of the sets $\{1, \ldots, n\}$ and $\{1, \ldots, m\}$ respectively whereas $\sigma = \sigma_1 \to \cdots \to \sigma_n \to a$ and $\tau = \tau_1 \to \cdots \to \tau_m \to a$.

We omit type parametrisation of D, D' and $P_{\pi_2}^{\pi_1}$ as suitable notation is overwhelming, but we implicitly assume that these constants are annotated as follows

$$D, D' \text{ with } \Sigma \to \sigma, \text{ and } (\Delta_1 \to \tau' \to \Delta_2 \to a) \to \tau;$$
$$P_{\pi_2}^{\pi_1} \text{ with } \sigma_1 \to \cdots \to \sigma_n \to a, \text{ and } \tau_1 \to \cdots \to \tau_m \to a.$$

Proposition 4. (soundness)
Let

$$\frac{\vdash T_1 \bullet T_1' : \sigma_1 \trianglelefteq^1 \tau_1, \cdots, \vdash T_n \bullet T_n' : \sigma_n \trianglelefteq^1 \tau_n}{\vdash T \bullet T' : \sigma \trianglelefteq^1 \tau}$$

be a rule of the system on Fig. 3. If for all premises $\vdash T_i \bullet T_i' : \sigma_i \trianglelefteq^1 \tau_i$ the pair lambdify(T_i), lambdify(T_i') *certifies that σ_i is a retract of τ_i then the pair* lambdify(T), lambdify(T') *certifies that σ is a retract of τ.*

Proof. The proof is by cases according to the rules of the system from Fig. 3. □

Theorem 5. (completeness)
If there is a pair of terms T, T' which certifies that σ is a retract of τ then there is a pair of terms $\hat{T}, \hat{T}'$ over the signature defined in Definition 3 such that lambdify$(\hat{T})$, lambdify$(\hat{T}')$ *certifies that σ is a retract of τ.*

Proof. The proof is by induction with respect to the size of $\sigma \trianglelefteq^1 \tau$. □

2.2   Permutations of a Derivation

The term system allows us to perform several changes in particular form of a derivation. The possibilities are summarised in the following Prop. 6.

In order to make the formulation of the proposition easier we introduce a few conventions for referring to nodes in a derivation. For each rule on Fig. 3 the sequents above the rule line are called *premises* of the rule while the sequent below the line is called a *conclusion*. In the rule (H) the first sequent above the rule line is called a *left premise* of the rule whereas the second one is called a *right premise*. In the rule (D) the first sequent above the rule line is called the left premise and the second one is called the right premise.

Proposition 6. (permutations inside derivations) *Let D be a derivation for a retraction $\sigma \trianglelefteq^1 \tau$.*

1. *If D starts with two subsequent applications of the rule (P) followed by a derivation D' then there is a derivation D'' for $\sigma \trianglelefteq^1 \tau$ which starts with a single application of the rule (P) followed by the derivation D'.*
2. *If D starts with two subsequent applications of the rule (N) followed by a derivation D' then there is a derivation D'' for $\sigma \trianglelefteq^1 \tau$ which starts with a single application of the rule (N) followed by the derivation D'.*
3. *Let D start with an application of the rule (N) followed by a derivation D'. If D' starts with an application of the rule (P) with a derivation D'' applied to its premise then there is a derivation $\hat{D}$ for $\sigma \trianglelefteq^1 \tau$ which starts with an application of the rule (P) followed by an application of the rule (N) with D'' applied to its premise.*
4. *Let D start with an application of the rule (N) followed by a derivation D'. If D' starts with an application of one of the rules (H) or (D) with a derivation D_L applied to its left premise and D_R applied to its right premise then there is a derivation D'' for $\sigma \trianglelefteq^1 \tau$ which starts with an application of the rule (H) or (D) respectively with D_L applied to its left premise and D'_R applied to its right premise where D'_R starts with the rule (N) followed by the derivation D_R.*
5. *Let D start with an application of the rule (H) or (D) with derivations D_L and D_R applied to the left and to the right premise of the rule respectively. If D_R starts with an application of the rule (P) to the right premise of the rule which is further followed by a derivation D'_R then there is a derivation D'' for $\sigma \trianglelefteq^1 \tau$ which starts with an application of the rule (P) followed by an application of (H) or (D) respectively followed by D_L applied to the left premise and D'_R to the right premise.*
6. *Let D start with an application of the rule $(H)^1$ or $(D)^1$ with derivations D_L and D_R applied to the left and to the right premise of the rule. If D_R starts with an application of the rule $(H)^2$ or $(D)^2$ with derivations D_{RL} and D_{RR} applied to the left and to the right premise of the rule then there is a derivation D'' for $\sigma \trianglelefteq^1 \tau$ which starts with an application of the rule (P) followed by an application of the rule $(H)^2$ or $(D)^2$ respectively with the derivation D_{RL} applied to its left premise and derivation D'_R applied to its right premise where D'_R starts with an application of the rule $(H)^1$ or $(D)^1$ respectively followed by the derivation D_L applied to its left premise and D_{RR} to its right premise.*

Proof. We leave a routine case analysis for the reader.   □

Note that the rule (D) may be replaced by the rule (N) in case the former in has a sequent $a \trianglelefteq^1 a$ where $a \in \mathcal{B}$ as the left premise.

3   NP-Completeness

Theorem 7. (affine retractions are in NP)
There is a nondeterministic polynomial algorithm that solves the problem of affine retractions.

Proof. The problem of affine retractions immediately reduces to the problem of finding an affine solution to the following higher-order matching equation $X(Yx) = x$ where X, Y are unknowns of the types $\tau \to \sigma$ and $\sigma \to \tau$ respectively and x is a constant of the type σ. The problem of solving such equations has been proved NP-complete in [WD02].

A more direct argument could rely on the observation that in the system on Fig. 2 for each rule the different inequalities in the assumptions of the rule can be constructed from disjoint parts of the inequalities in the result and no part of the result is duplicated in the assumptions. Thus each derivation according to the system has only polynomially many uses of these rules. This means that it is enough to guess such a polynomial tree labelled with the rules and then check if the tree is a proper derivation.   □

3.1   NP-Hardness

We present here NP-hardness proofs for three languages of types. In the first one we deal with unbounded number of type atoms. In the second one the number of atoms is bounded by a number $k > 1$. In the third one there is only one type atom. In order to reuse a part of the NP-hardness construction we have to define a few notions which describe requirements on the language of types which enables the NP-hardness construction.

We deal with families of sets of types of the form $\mathcal{A} = \{A_i\}_{i \in \mathbb{N}}$ where $|A_i| = i$. Such families are called here for short *graded families*. We define $\sigma' = \sigma \to a$ for $\sigma \in \bigcup_{i \in \mathbb{N}} A_i$.

Definition 8. (properly isolated family)
We say that a graded family $\{A_i\}_{i \in \mathbb{N}}$ is *properly isolated* if for each i and for each $\sigma_\alpha, \sigma_{\alpha_1}, \ldots, \sigma_{\alpha_n}, \sigma_{\beta_1}, \ldots, \sigma_{\beta_m} \in A_i$ we have

1. if $\sigma_{\alpha_1} \trianglelefteq^1 \sigma_{\beta_1}$ then $\sigma_{\alpha_1} = \sigma_{\beta_1}$;
2. if $\sigma_\alpha \trianglelefteq^1 \sigma'_{\alpha_1} \to \cdots \to \sigma'_{\alpha_n} \to a$ then for some j we have $\sigma_\alpha \trianglelefteq^1 \sigma_{\alpha_j}$;
3. $\sigma_\alpha \ntrianglelefteq^1 \sigma_{\alpha_1} \to \cdots \to \sigma_{\alpha_n} \to a$;
4. if $\sigma_{\alpha_1} \to \cdots \to \sigma_{\alpha_n} \to a \trianglelefteq^1 \sigma_{\beta_1} \to \cdots \to \sigma_{\beta_m} \to a$ then $\{\alpha_1, \ldots, \alpha_n\} \subseteq \{\beta_1, \ldots, \beta_m\}$ where $\subseteq$ is the multiset inclusion;
5. if $\sigma_\alpha \trianglelefteq^1 (\sigma_{\alpha_1} \to \cdots \to \sigma_{\alpha_n} \to a) \to (\sigma_{\beta_1} \to \cdots \to \sigma_{\beta_m} \to a) \to a$ then $\sigma_\alpha \in \{\sigma_{\alpha_1}, \ldots, \sigma_{\alpha_n}, \sigma_{\beta_1}, \ldots, \sigma_{\beta_m}\}$.

The proof of NP-hardness relies on a reduction of the 3-SAT problem to the problem of affine retractions. First, we present a translation of 3-SAT problem instances into instances of the affine retractions problem. Then, we show that the instance of the former has a solution if and only if there is a pair of terms certifying retractability for the translation.

We fix a set X of denumerably many propositional variables. Let ϕ be an instance of 3-SAT. The formula has the form

$$\phi = \bigwedge_{i=1}^{n} \phi_i \tag{1}$$

where $\phi_i = y_{i,1} \vee y_{i,2} \vee y_{i,3}$ with $y_{i,j} = x$ or $y_{i,j} = \neg x$ for some $x \in X$. Let $\mathrm{FV}(\phi) = \{x_1, \ldots, x_m\}$ be the set of all variables occurring in ϕ and let $Y_0 = \{x_1, \ldots, x_m, \neg x_1, \ldots, \neg x_m\}$. The set Y_0 is called the set of literals in ϕ. For each literal $y = \neg x$ we define $\neg y$ to be x. As usual, a valuation $v : X \to \{0, 1\}$ allows us to define a value $v(\phi)$ of the formula ϕ. This is done by induction on the structure of ϕ according to the usual semantics of the connectives $\wedge, \vee,$ and $\neg$.

Suppose we have a properly isolated family $\{A_i\}_{i \in \mathbb{N}}$. Given the above-mentioned notation for subformulae of ϕ, we fix the notation for elements of A_{2m+n} as

$$A_{2m+n} = \{\sigma_{x_1}, \ldots, \sigma_{x_m}, \sigma_{\neg x_1}, \ldots, \sigma_{\neg x_m}, \sigma_{\phi_1}, \ldots, \sigma_{\phi_n}\}.$$

Let $B_y = \{\phi_i \mid \phi_i$ is a subformula of ϕ with $\phi_i = y_1 \vee y_2 \vee y_3,$
$y_1, y_2, y_3 \in Y_0,$ and y is one of $y_1, y_2, y_3\}$.

For each literal $y \in Y_0$, we define $\overline{\tau}_{B_y} = \sigma_{\neg y} \to \sigma_{\phi_{i_1}} \to \cdots \to \sigma_{\phi_{i_k}} \to a$ where $B_y = \{\phi_{i_1}, \ldots, \phi_{i_k}\}$. This allows us to define $\tau_{B_x} = \overline{\tau}_{B_x} \to \overline{\tau}_{B_{\neg x}} \to a$. Moreover, we define for $x \in \mathrm{FV}(\phi)$ a type $\tau_x = \sigma'_x \to \sigma'_{\neg x} \to a$.

This allows us to formulate the result of translating ϕ into an instance of the affine retractions problem σ_ϕ, τ_ϕ:

$$
\begin{aligned}
\sigma_\phi = \sigma_{x_1} &\to \cdots \to \sigma_{x_m} \to \\
\sigma_{\neg x_1} &\to \cdots \to \sigma_{\neg x_m} \to \\
\sigma_{\phi_1} &\to \cdots \to \sigma_{\phi_n} \to a
\end{aligned}
\qquad
\begin{aligned}
\tau_\phi = \tau_{x_1} &\to \cdots \to \tau_{x_m} \to \\
\tau_{B_{x_1}} &\to \cdots \to \tau_{B_{x_m}} \to a
\end{aligned}
\tag{2}
$$

Lemma 9. (a valuation induces a derivation)

Let $v : X \to \{0, 1\}$ be a valuation of propositional variables. If $v(\phi) = 1$ then there is a derivation for $\sigma_\phi \trianglelefteq^1 \tau_\phi$.

Proof. In order to prove the lemma we have to generalize the lemma so that it holds for wider range of τ_ϕ. We say that types σ_ϕ, τ'_ϕ are an acceptable coding of ϕ if σ_ϕ is as before and τ'_ϕ has the form $\tau'_\phi = \tau_{x_1} \to \cdots \to \tau_{x_m} \to \tau'_{B_{x_1}} \to \cdots \to \tau'_{B_{x_m}} \to a$ where $\tau'_{B_y} = \overline{\tau}'_{B_x} \to \overline{\tau}'_{B_{\neg x}} \to a$ and $\overline{\tau}'_{B_y} = \sigma_{\neg y} \to \sigma_{\phi_{i_1}} \to \cdots \to \sigma_{\phi_{i_k}} \to \rho_1 \to \cdots \rho_l \to a$ for $y \in Y_0$ with $\rho_1, \ldots, \rho_l$ being arbitrary types and $B_y = \{\phi_{i_1}, \ldots, \phi_{i_k}\}$. Moreover, we say that a coding is acceptable if in $\overline{\tau}'_{B_y}$ subtypes ρ_i are permuted with the types σ_α.

We produce a derivation for $\sigma_\phi \trianglelefteq^1 \tau'_\phi$ where σ_ϕ, τ'_ϕ are an acceptable coding of ϕ by induction on the number of variables in ϕ. Note that σ_ϕ, τ_ϕ are an acceptable coding for ϕ. First, we may permute $\sigma_\phi \trianglelefteq^1 \tau'_\phi$ using (P) so that we obtain

$$
\begin{aligned}
\sigma_\phi^1 = \sigma_{x_1} \to \sigma_\phi^2, & \quad \tau_\phi^{\prime 1} = \tau_{x_1} \to \tau_\phi^{\prime 2} \quad \text{in case } v(x_1) = 1 \\
\sigma_\phi^1 = \sigma_{\neg x_1} \to \sigma_\phi^2, & \quad \tau_\phi^{\prime 1} = \tau_{x_1} \to \tau_\phi^{\prime 2} \quad \text{in case } v(x_1) = 0
\end{aligned}
\tag{3}
$$

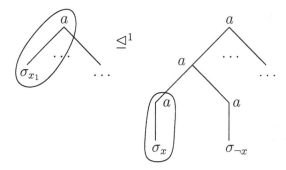

Fig. 5. The application of the rule (D) in order to remove the first variable in (3)

We apply the rule (D)^1 and obtain the left premise $\sigma'_{x_1} \trianglelefteq^1 \sigma'_{x_1}$ (or $\sigma'_{\neg x_1} \trianglelefteq^1 \sigma'_{\neg x_1}$) — see Fig. 5. This derivation can be done by the rule (H).

In order to construct a derivation for the right premise of (D)^1 we permute $\sigma^2_\phi \trianglelefteq^1 \tau'^2_\phi$ so that we obtain

$$\sigma_{\neg y} \to \sigma_{\phi_{i_1}} \to \cdots \to \sigma_{\phi_{i_r}} \to \sigma^3_\phi \trianglelefteq^1 \tau'_{B_{x_1}} \to \tau'^3_\phi \tag{4}$$

where y is either x_1 or $\neg x_1$ and $\{\phi_{i_1}, \ldots, \phi_{i_r}\} = B_y$. We apply the rule (D)^2 as on Fig. 6. The left premise of the rule can be derived by repeated application of the rule (H) to get rid of σ_α followed by an application of (N) to get rid of ρ_α.

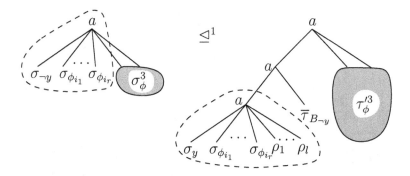

Fig. 6. The application of the rule (D) in order to decompose inequality (4)

The types $\sigma^3_\phi, \tau'^3_\phi$ are an acceptable coding of a formula ϕ' which is constructed from ϕ by erasing all the subformulae from B_y. Thus we obtain by induction a derivation for $\sigma^3_\phi \trianglelefteq^1 \tau'^3_\phi$. This derivation forms the lacking derivation for the right premise of the rule (D)^2.   □

Lemma 10. (a derivation induces a valuation)
If $\sigma_\phi \trianglelefteq^1 \tau_\phi$ has a derivation then there exists a valuation $v : X \to \{0, 1\}$ such that $v(\phi) = 1$.

Proof. In order to prove the lemma we have to generalize it to cover a wider class of terms σ_ϕ and τ_ϕ. We say that for a given formula ϕ and a partial function $v_0 : X \to \{0,1\}$ types σ_ϕ, τ_ϕ are an acceptable coding for ϕ, v if

$$
\begin{aligned}
\sigma_\phi &= \sigma_{x_1} \to \cdots \to \sigma_{x_m} \to & \tau_\phi &= \tau_{x_1} \to \cdots \to \tau_{x_m} \to \\
&\sigma_{\neg x_1} \to \cdots \to \sigma_{\neg x_m} \to & &\tau_{B_{x_1}} \to \cdots \to \tau_{B_{x_m}} \to a \\
&\sigma_{\phi_1} \to \cdots \to \sigma_{\phi_n} \to a
\end{aligned}
$$

where $\{x_1, \ldots, x_m\} \subseteq FV(\phi)\backslash Dom(v)$. We prove the following generalisation of our lemma:

Let σ_ϕ, τ_ϕ be an acceptable coding for a formula ϕ and a partial function v_0. If $\sigma_\phi \trianglelefteq^1 \tau_\phi$ has a derivation then there exists a valuation $v : X \to \{0,1\}$ which extends v_0 and for which $v(\phi) = 1$.

The proof is by induction on the number of variables in $FV(\phi)\backslash Dom(v_0)$. If there is a derivation for $\sigma_\phi \trianglelefteq^1 \tau_\phi$ then by Prop. 6 we may assume that it starts with the rule (P) followed by a certain number of the rules (H) and (D) applied to the subsequent right premises with a single (N) rule at the end followed by the axiom. Moreover, we assume that no rule (D) can be replaced by an equivalent rule (N). Derivations having this shape are called *well-formed derivations*.

For such derivations the first (P) rule may be followed by

1. the rule (H)^1 in which the left premise is $\sigma_{x_1} \trianglelefteq^1 \tau_\alpha$ for some α, or
2. the rule (D)^1 in which the left premise is

$$
\sigma_{x_1} \to \sigma_{\alpha_1} \to \cdots \to \sigma_{\alpha_n} \to a \trianglelefteq^1 \tau_{\beta_1} \to \cdots \to \tau_{\beta_m} \to a
$$

   for certain $\alpha_1, \ldots, \alpha_n, \beta_1, \ldots, \beta_m$, or
3. the rule (N)^1.

In case (1) we have two options either (a) $\tau_\alpha = \tau_{x_1}$, or (b) $\tau_\alpha = \tau_{B_{x_1}}$. This is guaranteed by the condition (5) on the properly isolated family.

In case (1.a) we have $\tau_\alpha = \tau_{x_1}$. We extend v_0 to v'_0 so that $v'_0(x_1) = 1$. In this situation we may assume that the right premise of the rule (H)^1 has a derivation which starts with the rule (P) followed by

i. the rule (H)^2 in which the left premise is $\sigma_{\neg x_1} \trianglelefteq^1 \tau_{\alpha'}$ or
ii. the rule (D)^2 in which the left premise is

$$
\sigma_{\neg x_1} \to \sigma_{\alpha_1} \to \cdots \to \sigma_{\alpha_n} \to a \trianglelefteq^1 \tau_{\beta_1} \to \cdots \to \tau_{\beta_m} \to a \quad \text{or} \quad (5)
$$

iii. the rule (N)^2.

For the situation (1.a.i) we observe that this is possible only if $\alpha' = x_j$ or $\alpha' = B_{x_j}$. The condition (2) on the properly isolated family implies that in the former case $\alpha' = x_1$. This is impossible, though, as τ_{x_1} is already used in the left premise of (H)^1. The condition (5) on the properly isolated family implies that in the latter case $B_{x_j} = B_{x_1}$. As we got rid of both σ_{x_1} and $\sigma_{\neg x_1}$ the right premise of $(H)^2$ is a formula which is an acceptable coding of ϕ and v_0' so by induction hypothesis v_0' extends into a valuation v such that $v(\phi) = 1$.

For the proof in the case (1.a.ii), we observe that the part (4) of Definition 8 implies that $\neg x_1 \in \{\beta_1, \ldots, \beta_m\}$. This is possible by an application the rule $(D)^2$ either to τ_{x_1} or to $\tau_{B_{x_1}}$. The former option is not possible since τ_{x_1} is already used in $(H)^1$. The latter option implies that the inequality (5) is in fact

$$\sigma_{\neg x_1} \to \sigma_{\alpha_1} \to \cdots \to \sigma_{\alpha_n} \to a \trianglelefteq^1 \sigma_{\neg x_1} \to \sigma_{\phi_{i_1}} \to \cdots \to \sigma_{\phi_{i_r}} \to a \text{ where}$$

$\{\phi_{i_1}, \ldots, \phi_{i_r}\} = B_{x_1}$. The condition (4) on the properly isolated family implies that $\{\alpha_1, \ldots, \alpha_n\} \subseteq \{\phi_{i_1}, \ldots, \phi_{i_r}\}$ (where the inclusion is taken as for multisets). This means that the right premise of $(D)^2$ — possibly after an application of the rule (P) — has the form

$$
\begin{array}{ll}
\sigma_{x_2} \to \cdots \to \sigma_{x_m} \to & \trianglelefteq^1 \quad \tau_{x_2} \to \cdots \to \tau_{x_m} \to \\
\sigma_{\neg x_2} \to \cdots \to \sigma_{\neg x_m} \to & \tau_{B_{x_2}} \to \cdots \to \tau_{B_{x_m}} \to a \\
\sigma_{\phi_{j_1}} \to \cdots \to \sigma_{\phi_{j_l}} \to a &
\end{array}
$$

The variables $x_2, \ldots, x_m$ belong to $FV(\phi) \setminus Dom(v_0')$ so this pair of types is an acceptable coding of ϕ. As this premise has its derivation, the induction hypothesis implies that v_0' can be extended to a valuation v such that $v(\phi) = 1$.

For the proof in the case (1.a.iii) we exploit the fact that the derivation is well-formed. This implies that the premise we consider has the form $a \trianglelefteq^1 \tau_1 \to \cdots \to \tau_r \to a$. This is impossible though, as each σ_x on the left hand side of the original inequality is accompanied by $\sigma_{\neg x}$.

In case (1.b) we have $\tau_\alpha = \tau_{B_{x_1}}$. We extend v_0 to v_0' so that $v_0'(x_1) = 0$. In this situation the right premise of the rule $(H)^1$ has a derivation which starts with the rule (P) followed by

i. the rule $(H)^2$ in which the left premise is $\sigma_{\neg x_1} \trianglelefteq^1 \tau_{\alpha'}$ or
ii. the rule $(D)^2$ in which the left premise is

$$\sigma_{\neg x_1} \to \sigma_{\alpha_1} \to \cdots \to \sigma_{\alpha_n} \to a \trianglelefteq^1 \tau_{\beta_1} \to \cdots \to \tau_{\beta_m} \to a \text{ or}$$

iii. the rule $(N)^2$.

The proof is similar to the one for the case (1.a), but we exchange the role of τ_{x_1} and $\tau_{B_{x_1}}$.

In case (2) the proof runs similarly as in the case (1.a.ii), but we have to put $v_0'(x_1) = 0$ and before stepping into the induction hypothesis we have to analyse the way $\sigma_{\neg x_1}$ is handled. This is done by an analysis similar to the one in the remaining parts of the case (1).

In case (3) we exploit the fact that the derivation is well-formed. This implies that the premise we consider has the form $a \trianglelefteq^1 \tau_1 \to \cdots \to \tau_r \to a$. This is impossible though, as each σ_x on the left hand side of the original inequality is accompanied by $\sigma_{\neg x}$. □

3.2   Properly Isolated Families

Definition 11. (families of types)
We now define three particular families of types.

- A family of types $\mathcal{A}^\infty = \{A_i^\infty\}_{i \in \mathbb{N}}$ is such that $A_n^\infty = \{a_1, \ldots, a_n\}$ where a_i for $i = 1, \ldots, n$ are distinct type atoms and each $a_i \neq a$ where a is the atom used in Sect. 3.1.
- A family of types $\mathcal{A}^2 = \{A_i^2\}_{i \in \mathbb{N}}$ is such that $A_n^2 = \{\sigma_1, \ldots, \sigma_n\}$ where σ_i for $i = 1, \ldots, n$ is defined as $\sigma_i = \underbrace{a \rightarrow \cdots \rightarrow a}_{n-i \text{ times}} \rightarrow \underbrace{b \rightarrow \cdots \rightarrow b}_{i+1 \text{ times}} \rightarrow b$ where $b \neq a$

  and b does not occur explicitly in the constructions from Sect. 3.1.
- A family of types $\mathcal{A}^1 = \{A_i^1\}_{i \in \mathbb{N}}$ is such that $A_n^1 = \{\sigma_1, \ldots, \sigma_n\}$ where σ_i for $i = 1, \ldots, n$ is defined as $\sigma_i = (\tilde{\sigma}_{2n-i+2} \rightarrow \tilde{\sigma}_{i+1} \rightarrow a) \rightarrow a$ where $\tilde{\sigma}_i = \underbrace{a \rightarrow \cdots \rightarrow a}_{i \text{ times}} \rightarrow a$.

Lemma 12. (isolation)
(A) The family $\mathcal{A}^\infty$ is a properly isolated family for the set of types built of infinitely many atoms. (B) The family $\mathcal{A}^2$ is a properly isolated family for the set of types built of k atomic types where $k \geq 2$. (C) The family $\mathcal{A}^1$ is a properly isolated family for the set of types built of a single atomic type.

Proof. The proof is by a routine case analysis.   □

Theorem 13. (NP-completeness)
The problem of affine retractions is NP-complete

1. *for the class of types of order 4 and having infinitely many type atoms,*
2. *for the class of types of order 5 and having k type atoms where $k \geq 2$,*
3. *for the class of types of order 7 and having a single type atom.*

Proof. It follows from Lemma 9 and 10 that the existence of a properly isolated family implies NP-completeness. From Lemma 12 we know that a properly isolated family does indeed exist in each case.   □

4   Polynomial Case

Here we introduce polynomial algorithms that decide affine retractability for types of order 1, 2 and 3. We present them as separate procedures instead of a single one. This is because we want to analyse separately the running time of the algorithms.

The algorithm for the order 1 is very obvious as we can only use the rule (Ax) in this case.

Definition 14. (algorithm for the order 1)
Let $\sigma = a$ and $\tau = b$ be the input types where a, b are atomic. If $a = b$ then the affine retractability holds else it does not hold.

The algorithms for orders 2 and 3 are based on the following observations.

Proposition 15. (the rule (D) in low orders)
If σ, τ have order strictly less than 4 then for each derivation of $\sigma \trianglelefteq^1 \tau$ there exists one which does not involve the rule (D).

Proof. As the right-hand side has the order at most 3 the rule (D) may give only a term of the order 1 on the right-hand side. Thus we may have the left-hand side of order 1. However, in such an application of (D) the right premise can also be obtained by the rule (N). □

The algorithm for the types in $T^2_\rightarrow$ (the order 2) requires an additional observation.

Definition 16. (weight of types)
For types of order 2 we introduce a weight function $\# : T^2_\rightarrow \to \mathbb{N}^\mathcal{B}$ defined as

$$\#(\sigma)(a) = |\{i \mid \sigma = \sigma_1 \to \cdots \to \sigma_n \to b, \text{ and } \sigma_i = a\}|.$$

We use the order $\leqslant$ on elements of $\mathbb{N}^\mathcal{B}$ defined as — $f \leqslant g$ iff for each $a \in \mathcal{B}$ we have $f(a) \leq g(a)$.

Proposition 17. (the weight and $\trianglelefteq^1$)
If $\sigma, \tau \in T^2_\rightarrow$ and $head(\sigma) = head(\tau)$ then $\sigma \trianglelefteq^1 \tau$ iff $\#(\sigma) \leqslant \#(\tau)$.

Proof. Induction on the size of σ. □

Definition 18. (algorithm for the order 2)
Let σ and τ be the input types. If $head(\sigma) \neq head(\tau)$ then the affine retractability does not hold. Otherwise, if $\#(\sigma) \leqslant \#(\tau)$ then $\sigma \trianglelefteq^1 \tau$.

Note that the above algorithm runs in time linear to the size of the input.
The correctness of the below defined algorithm for the order 3 is based on Prop. 15 and the correctness of the algorithm for the order 2.

Definition 19. (algorithm for the order 3)
Let σ and τ be the input types. If $head(\sigma) \neq head(\tau)$ then the affine retractability does not hold. In case $head(\sigma) = head(\tau)$ we construct a bipartite graph $G = (V_1 \cup V_2, E)$ in which

$$V_1 = \{\sigma_i \mid 1 \leq i \leq n \text{ and } \sigma = \sigma_1 \to \cdots \to \sigma_n \to a\}$$
$$V_2 = \{\tau_i \mid 1 \leq i \leq m \text{ and } \tau = \tau_1 \to \cdots \to \tau_m \to a\}$$
$$E = \{(\sigma_i, \tau_j) \mid \sigma_i \trianglelefteq^1 \tau_i, \sigma_i \in V_1, \tau_j \in V_2\}$$

Now, we find a perfect matching in G. If there is one then $\sigma \trianglelefteq^1 \tau$ holds otherwise it does not hold.

Note that the construction of the graph takes $O(n^2)$ where n is the size of the input and the size of G is $O(n^2)$. The running time for a perfect matching procedure is $O(m^3)$ where m is the size of an input graph (see e.g. Chap. 27 in [CLR90]). This makes the overall running time $O(n^6)$.

Discussion. The results in this section close the algorithmic gap for type languages with infinitely many type atoms. Formally, this is the kind of approach which is used in the currently used programming languages. Moreover, most of the currently used higher-oder functions have the order 3.

Although, one may want to restrict types so that they have at most k type atoms for some reasonably high fixed number e.g. 10. Unfortunately, the problem of affine retractability is NP-complete already for types with 2 atoms and having the order 5. The author concjectures, though, that this problem is polynomial for types of the order 4.

The algorithms presented here construct derivations for type retractions. The system on Fig. 3 can be used to compute retraction terms based on these derivations.

Acknowledgements. The author would like to thank Paweł Urzyczyn for pointing out the problem and for discussions on it. Thanks also go to Damian Niwiński for help in preparation of the paper.

References

[Bar92] Henk P. Barendregt, *Lambda calculi with types*, Handbook of Logic in Computer Science (S. Abramsky, D.M. Gabbay, and T.S.E. Maibaum, eds.), vol. 2, Oxford Science Publications, 1992, pp. 117–309.

[BL85] Kim Bruce and Giuseppe Longo, *Provable isomorphisms and domain equations in models of typed languages*, ACM Symposium on Theory of Computing (STOC'85), May 1985.

[CLR90] Thomas H. Cormen, Charles E. Leiserson, and Ronald L. Rivest, *Introduction to algorithms*, ch. 27, MIT Press, 1990.

[Cos95] Roberto Di Cosmo, *Isomorphisms of types: from lambda-calculus to information retrieval and language design*, Birkhauser, 1995.

[dPS92] Ugo de'Liguoro, Adolfo Piperno, and Richard Statman, *Retracts in Simply Typed $\lambda\beta\eta$-Calculus*, Proceedings of the Seventh Annual IEEE Symposium on Logic in Computer Science (Santa Cruz, CA), 1992, pp. 461–469.

[LR01] Paweł Urzyczyn Laurent Regnier, *Retractions of types with many atoms*, Tech. report, Institute of Informatics, Warsaw University, 2001.

[Pad01] Vincent Padovani, *Retracts in simple types*, Proc. of TLCA'2001, LNCS, no. 2044, Springer-Verlag, 2001, pp. 376–384.

[Rit91] Mikael Rittri, *Using types as search keys in function libraries*, Journal of Functional Programming **1** (1991), no. 1, 71–89.

[SU99] Zdzisław Spławski and Paweł Urzyczyn, *Type fixpoints: Iteration vs. recursion*, Proceedings of 4th International Conference on Functional Programming, ACM, 1999, pp. 102–113.

[Vor97] Sergei Vorobyov, *The "Hardest" Natural Decidable Theory*, Proceedings of the Twelfth Annual IEEE Symposium on Logic in Computer Science, 1997, pp. 294–305.

[WD02] Tomasz Wierzbicki and Dan Dougherty, *A Decidable Variant of Higher Order Matching*, Proc. 13th Conf. on Rewriting Techniques and Applications, RTA'02 (Sophie Tison, ed.), LNCS, no. 2378, Springer-Verlag, 2002, pp. 340–351.

Higher-Order Matching in the Linear λ-calculus with Pairing

Philippe de Groote and Sylvain Salvati

LORIA UMR n° 7503 – INRIA
Campus Scientifique, B.P. 239
54506 Vandœuvre lès Nancy Cedex – France
{degroote, salvati}@loria.fr

Abstract. We prove that higher-order matching in the linear λ-calculus with pairing is decidable. We also establish its NP-completeness under the assumption that the right-hand side of the equation to be solved is given in normal form.

1 Introduction

The decidability of higher-order matching (which consists in determining whether a simply typed λ-term is an instance of another one, modulo the conversion rules of the λ-calculus), has been intensively studied in the literature. In particular, second-order matching [12], third-order matching [5], and fourth-order matching [16] have been shown to be decidable (both modulo β and $\beta\eta$). On the other hand, it has been proved that, starting from the sixth order, higher-order matching modulo β is undecidable [14] (for $\beta\eta$, the problem is still open).

In two recent papers [9] and [10], we studied the decidability and the complexity of a quite restricted form of higher-order matching, namely, higher-order matching in the linear λ-calculus. This calculus corresponds, through the Curry-Howard isomorphism, to the implicative fragment of Girard's linear logic [7], and may be naturally extended by taking into account the other connectives of linear logic. We follow this line of research in the present paper by considering the linear λ-calculus with pairing, i.e., the calculus corresponding to the negative fragment of multiplicative additive linear logic.

The paper is organized as follows. Section 2 presents the necessary mathematical notions and notations that we use in the sequel. In section 3, we show that deciding whether a linear λ-term with pairs may be reduced to a given normal form may be done in polynomial time. Finally, section 4 shows that every term may be turned into another term that has the same behaviour with respect to reductions, and whose length is bounded in terms of the redices it contains and the size of its normal form. This technical result allows us to conclude that higher-order matching in the linear λ-calculus with pairing is decidable. We also obtain that the problem is NP-complete when the right member of the equation is given in normal form.

J. Marcinkowski and A. Tarlecki (Eds.): CSL 2004, LNCS 3210, pp. 220–234, 2004.

2   Mathematical Background

Definition 1. *Let $\mathscr{A}$ be a finite set of atomic types. The set $\mathscr{F}$ of types is defined according to the following grammar:*

$$\mathscr{F} \quad ::= \quad \mathscr{A} \mid (\mathscr{F} \multimap \mathscr{F}) \mid (\mathscr{F} \,\&\, \mathscr{F}).$$

Definition 2. *Let $(\Sigma_\alpha)_{\alpha \in \mathscr{F}}$ be a family of pairwise disjoint finite sets indexed by $\mathscr{F}$, whose almost every member is empty. Let $(\mathscr{X}_\alpha)_{\alpha \in \mathscr{F}}$ and $(\mathscr{Y}_\alpha)_{\alpha \in \mathscr{F}}$ be two families of pairwise disjoint countably infinite sets indexed by $\mathscr{F}$, such that $(\bigcup_{\alpha \in \mathscr{F}} \mathscr{X}_\alpha) \cap (\bigcup_{\alpha \in \mathscr{F}} \mathscr{Y}_\alpha) = \emptyset$. The set $\mathscr{T}$ of raw λ-terms is defined according to the following grammar:*

$$\mathscr{T} \quad ::= \quad \Sigma \mid \mathscr{X} \mid \mathscr{Y} \mid \lambda\mathscr{X}.\mathscr{T} \mid (\mathscr{T}\,\mathscr{T}) \mid \langle \mathscr{T}, \mathscr{T} \rangle \mid (\pi_1\mathscr{T}) \mid (\pi_2\mathscr{T}),$$

where $\Sigma = \bigcup_{\alpha \in \mathscr{F}} \Sigma_\alpha$, $\mathscr{X} = \bigcup_{\alpha \in \mathscr{F}} \mathscr{X}_\alpha$, and $\mathscr{Y} = \bigcup_{\alpha \in \mathscr{F}} \mathscr{Y}_\alpha$.

In the above definition, the elements of Σ correspond to constants, and the elements of $\mathscr{X}$ are the λ-variables. The elements of $\mathscr{Y}$ are called the *unknowns*, and will be denoted by uppercase bold letters ($\mathbf{X}, \mathbf{Y}, \mathbf{Z}, \ldots$).

The notions of free and bound occurrences of a λ-variable are defined as usual, and we write $FV(t)$ for the set of λ-variables that occur free in a λ-term t. A λ-term that does not contain any subterm of the form $\langle t, u \rangle$ is called a *purely applicative λ-term*. A λ-term that does not contain any unknown is called a *pure λ-term*.

The notion of linear λ-term is then defined as follows.

Definition 3. *The family $(\mathscr{T}_\alpha)_{\alpha \in \mathscr{F}}$ of sets of linear λ-terms is inductively defined as follows:*

1. *if $\mathrm{a} \in \Sigma_\alpha$ then $\mathrm{a} \in \mathscr{T}_\alpha$;*
2. *if $\mathbf{X} \in \mathscr{Y}_\alpha$ then $\mathbf{X} \in \mathscr{T}_\alpha$;*
3. *if $x \in \mathscr{X}_\alpha$ then $x \in \mathscr{T}_\alpha$;*
4. *if $x \in \mathscr{X}_\alpha$, $t \in \mathscr{T}_\beta$, and $x \in FV(t)$, then $\lambda x.t \in \mathscr{T}_{(\alpha \multimap \beta)}$;*
5. *if $t \in \mathscr{T}_{(\alpha \multimap \beta)}$, $u \in \mathscr{T}_\alpha$, and $FV(t) \cap FV(u) = \emptyset$, then $(t\,u) \in \mathscr{T}_\beta$;*
6. *if $t \in \mathscr{T}_\alpha$, $u \in \mathscr{T}_\beta$, and $FV(t) = FV(u)$ then $\langle t, u \rangle \in \mathscr{T}_{\alpha \& \beta}$;*
7. *if $t \in \mathscr{T}_{\alpha \& \beta}$ then $(\pi_1 t) \in \mathscr{T}_\alpha$;*
8. *if $t \in \mathscr{T}_{\alpha \& \beta}$ then $(\pi_2 t) \in \mathscr{T}_\beta$.*

The conditions on the free variables in clauses 4, 5, and 6 correspond to the linearity conditions. They constraint the way λ-variables may occur in a term.

We define $\mathscr{T}$ to be $\bigcup_{\alpha \in \mathscr{F}} \mathscr{T}_\alpha$ (which is a proper subset of the set of raw λ-terms). It is easy to prove that the sets $(\mathscr{T}_\alpha)_{\alpha \in \mathscr{F}}$ are pairwise disjoint. Consequently, we may define the type of a linear λ-term t to be the unique linear type α such that $t \in \mathscr{T}_\alpha$.

We let $t[x := u]$ denote the usual capture-avoiding substitution of a λ-variable by a λ-term, and $t[x_1 := u_1, \ldots, x_n := u_n]$ denote the usual notion of parallel substitution. If σ denotes such a parallel substitution $[x_1 := u_1, \ldots, x_n := u_n]$, we write $t.\sigma$ for $t[x_1:=u_1,\ldots,x_n:=u_n]$, $\sigma(x_i)$ for u_i, and we define $\mathrm{dom}(\sigma)$ to be the finite set of variables $\{x_1,\ldots,x_n\}$. We use the same notations to denote the substitutions of unknowns by λ-terms. The substitution the domain of which is empty is the identity and is noted Id.

We take for granted the usual notions of β reduction, left projection, and right projection:

$$(\lambda x.\,t)\,u \to t[x:=u], \qquad \pi_1\langle t,u\rangle \to t, \qquad \pi_2\langle t,u\rangle \to u.$$

The union of these three notions of reduction induces the relation of one step reduction $(\to)$, the relation of at most one step reduction $(\overset{=}{\to})$, the relations of n steps reduction $(\overset{n}{\to})$, and the relations of many steps reduction $(\overset{*}{\to})$. The equality between linear λ-terms $(=)$ is defined to be the reflexive, symmetric, transitive closure of the relation of reduction and we write $\equiv$ for syntatic equality. The linear λ-calculus with pairing is strongly normalizable, the equality $(=)$ is then decidable and every term has a unique normal form.

We now give a precise definition of the matching problem with which this paper is concerned.

Definition 4. *A matching problem (in the linear λ-calculus with pairing) consists of a pair of linear λ-terms (t,u) of the same type such that u does not contain any unknown.*

Such a problem admits a solution if and only if there exists a substitution $[\mathbf{X}_1:=t_1,\ldots,\mathbf{X}_n:=t_n]$ such that $t[\mathbf{X}_1:=t_1,\ldots,\mathbf{X}_n:=t_n] = u$

We end this section with two remarks about the previous definition:

1. In the substitution $[\mathbf{X}_1:=t_1,\ldots,\mathbf{X}_n:=t_n]$ we do not require $t_1,\ldots,t_n$ to be pure terms. This is mandatory. Consider, for instance, the following matching problem: $\pi_1\mathbf{X} = c$, where c is a constant of type α, and $\mathbf{X}$ an unknown of type α & β. This problem admits the solution $[\mathbf{X}:=\langle c,\mathbf{Y}\rangle]$, where $\mathbf{Y}$ is an unknown of type β. Now, if we would require the solution to be made of pure terms, we would face the problem of constructing a closed term of type β, which is undecidable.

2. In defining the notion of equality, we did not take into account η-reduction and surjective pairing. In fact, all the results we obtain in this paper also hold for this stronger notion of equality.

3   A Polynomially Bounded Reduction Strategy

One of the key properties in establishing the NP-completeness of higher-order matching in the linear λ-calculus [9] is that any linear λ-term (without pairs) may be reduced to its normal form in polynomial (actually, linear) time. This is a direct consequence of the fact that the length of the linear λ-terms strictly decreases under β-reduction.

This property does not hold in the presence of pairing. Indeed, in a linear λ-term of the form $\lambda x. \langle t, u \rangle$, there are at least two occurrences of x that are bound by the abstraction. Consequently, the length of a redex such as $(\lambda x. \langle t, u \rangle) v$ may be stricly less than the length of its contractum $\langle t, u \rangle [x:=v]$. In fact, it is even not the case (modulo $P \neq co\text{-}NP$) that a linear λ-term with pairs may be reduced to its normal form in polynomial time [15]. Consequently, in this section we establish the following weaker property: if $t \xrightarrow{*} u$ then there exists a reduction strategy $t \xrightarrow{n} u$ such that n is polynomially bounded with respect to the length of t and u.

In order to establish this property, we first define two notions of complexity.

Definition 5. *The complexity $\rho(\alpha)$ of a linear type α is defined to be the number of connectives it contains:*

1. *$\rho(a) = 0$, for a atomic;*
2. *$\rho(\alpha \multimap \beta) = \rho(\alpha) + \rho(\beta) + 1$;*
3. *$\rho(\alpha \,\&\, \beta) = \rho(\alpha) + \rho(\beta) + 1$.*

The complexity $\rho(t)$ of a linear λ-term t is defined to be the complexity of its type.

Definition 6. *The norm $\mu(t)$ of a linear λ-term t is inductively defined as follows:*

1. *$\mu(c) = 0$, for c being a constant, a λ-variable, or an unknown.*
2. *$\mu(\lambda x. t_1) = \mu(t_1)$*
3. *$\mu(t_1 t_2) = \begin{cases} \mu(t_1) + \mu(t_2) + \rho(t_1), & \text{if } t_1 t_2 \text{ is a redex.} \\ \mu(t_1) + \mu(t_2), & \text{otherwise.} \end{cases}$*
4. *$\mu(\langle t_1, t_2 \rangle) = \max(\mu(t_1), \mu(t_2))$*
5. *$\mu(\pi_1 t_1) = \begin{cases} \mu(t_1) + \rho(t_1), & \text{if } \pi_1 t_1 \text{ is a redex.} \\ \mu(t_1), & \text{otherwise.} \end{cases}$*
6. *$\mu(\pi_2 t_1) = \begin{cases} \mu(t_1) + \rho(t_1), & \text{if } \pi_2 t_1 \text{ is a redex.} \\ \mu(t_1), & \text{otherwise.} \end{cases}$*

The above norm does not strictly decrease when reducing a term. This is due to clause 4. Indeed, in case $t_1 \to t_2$ with $\mu(t_1) \leq \mu(u)$, we have that $\langle t_1, u \rangle \to \langle t_2, u \rangle$ while $\mu(\langle t_1, u \rangle) = \mu(\langle t_2, u \rangle)$. In fact, this is the only problematic case, and we will prove that the norm strictly decreases under reduction if there is no reduction step that takes place within a pair. To this end, we introduce the following notion of external reduction.

Definition 7. *The relation of external reduction ($\triangleright$) is defined by means of the following formal system.*

$$(\lambda x. t) u \triangleright t[x:=u] \qquad \pi_1 \langle t, u \rangle \triangleright t \qquad \pi_2 \langle t, u \rangle \triangleright u \qquad \frac{t \triangleright u}{\lambda x. t \triangleright \lambda x. u}$$

$$\frac{t \triangleright u}{t v \triangleright u v} \qquad \frac{t \triangleright u}{v t \triangleright v u} \qquad \frac{t \triangleright u}{\pi_1 t \triangleright \pi_1 u} \qquad \frac{t \triangleright u}{\pi_2 t \triangleright \pi_2 u}$$

We state two technical lemmas that will be useful in the sequel. Their proofs, which are not difficult, are left to the reader.

Lemma 1. *Let t and $\lambda x.\, u$ be two linear λ-terms such that $t \overset{*}{\to} \lambda x.\, u$. Then, there exists a linear λ-term $\lambda x.\, v$ such that $t \triangleright \lambda x.\, v \overset{*}{\to} \lambda x.\, u$.* □

Lemma 2. *Let t and $\langle u_1, u_2 \rangle$ be two linear λ-terms such that $t \overset{*}{\to} \langle u_1, u_2 \rangle$. Then, there exists a linear λ-term $\langle v_1, v_2 \rangle$ such that $t \triangleright \langle v_1, v_2 \rangle \overset{*}{\to} \langle u_1, u_2 \rangle$.* □

We are now in a position of proving that the norm of a term strictly decreases under external reduction. The keystone of the proof is the following substitution lemma.

Lemma 3. *Let $t \in \mathscr{T}$, and $u, x \in \mathscr{T}_\alpha$ be such that $x \in \mathrm{FV}(t)$. Then, we have that $\mu(t[x{:=}u]) \leq \mu(t) + \mu(u) + \rho(u)$.*

Proof. The proof proceeds by induction on the structure of t.

1. $t \equiv x$.
$$
\begin{aligned}
\mu(t[x{:=}u]) &= \mu(x[x{:=}u]) \\
&= \mu(u) \\
&\leq \mu(u) + \rho(u) \\
&= \mu(t) + \mu(u) + \rho(u)
\end{aligned}
$$

2. $t \equiv \lambda y.\, t_1$.
$$
\begin{aligned}
\mu(t[x{:=}u]) &= \mu(\lambda y.\, t_1[x{:=}u]) \\
&= \mu(t_1[x{:=}u]) \\
&\leq \mu(t_1) + \mu(u) + \rho(u) \\
&\qquad \text{(by induction hypothesis)} \\
&= \mu(t) + \mu(u) + \rho(u)
\end{aligned}
$$

3. $t \equiv t_1\, t_2$. We distinguish between two cases:
   (a) $x \in \mathrm{FV}(t_1)$. Because of the linearity of t, we have that $x \notin \mathrm{FV}(t_2)$. Consequently, $t[x{:=}u] = t_1[x{:=}u]\, t_2$. Then, there are three subcases:
   i. $t_1 \equiv x$ and $u \equiv \lambda y.\, u_1$.
$$
\begin{aligned}
\mu(t[x{:=}u]) &= \mu(x\, t_2[x{:=}u]) \\
&= \mu(u\, t_2) \\
&= \mu(t_2) + \mu(u) + \rho(u) \\
&= \mu(x) + \mu(t_2) + \mu(u) + \rho(u) \\
&= \mu(t) + \mu(u) + \rho(u)
\end{aligned}
$$

   ii. $t_1 \equiv \lambda y.\, t_{11}$.
$$
\begin{aligned}
\mu(t[x{:=}u]) &= \mu(t_1[x{:=}u]\, t_2) \\
&= \mu(t_1[x{:=}u]) + \mu(t_2) + \rho(t_1[x{:=}u]) \\
&= \mu(t_1[x{:=}u]) + \mu(t_2) + \rho(t_1) \\
&\qquad \text{(by stability of typing under substitution)} \\
&\leq \mu(t_1) + \mu(t_2) + \rho(t_1) + \mu(u) + \rho(u) \\
&\qquad \text{(by induction hypothesis)} \\
&= \mu(t) + \mu(u) + \rho(u)
\end{aligned}
$$

iii. Otherwise.

$$\begin{aligned}
\mu(t[x{:=}u]) &= \mu(t_1[x{:=}u]\, t_2) \\
&= \mu(t_1[x{:=}u]) + \mu(t_2) \\
&\leq \mu(t_1) + \mu(t_2) + \mu(u) + \rho(u) \\
&\qquad\qquad \text{(by induction hypothesis)} \\
&= \mu(t) + \mu(u) + \rho(u)
\end{aligned}$$

(b) $x \in FV(t_2)$. There are two subcases, which are similar to Subcases ii and iii of Case (a).

4. $t \equiv \langle t_1, t_2 \rangle$.

$$\begin{aligned}
\mu(t[x{:=}u]) &= \mu(\langle t_1[x{:=}u], t_2[x{:=}u] \rangle) \\
&= \max(\mu(t_1[x{:=}u]), \mu(t_2[x{:=}u])) \\
&\leq \max(\mu(t_1) + \mu(u) + \rho(u), \mu(t_2) + \mu(u) + \rho(u)) \\
&\qquad\qquad \text{(by induction hypothesis)} \\
&= \max(\mu(t_1), \mu(t_2)) + \mu(u) + \rho(u) \\
&= \mu(t) + \mu(u) + \rho(u)
\end{aligned}$$

5. $t \equiv \pi_1 t_1$. We distinguish between three cases:
   (a) $t_1 \equiv x$ and $u \equiv \langle u_1, u_2 \rangle$.

$$\begin{aligned}
\mu(t[x{:=}u]) &= \mu(\pi_1 u) \\
&= \mu(u) + \rho(u) \\
&= \mu(\pi_1 x) + \mu(u) + \rho(u) \\
&= \mu(t) + \mu(u) + \rho(u)
\end{aligned}$$

   (b) $t_1 \equiv \langle t_{11}, t_{12} \rangle$.

$$\begin{aligned}
\mu(t[x{:=}u]) &= \mu(\pi_1(t_1[x{:=}u])) \\
&= \mu(t_1[x{:=}u]) + \rho(t_1[x{:=}u]) \\
&= \mu(t_1[x{:=}u]) + \rho(t_1) \\
&\qquad\qquad \text{(by stability of typing under substitution)} \\
&\leq \mu(t_1) + \rho(t_1) + \mu(u) + \rho(u) \\
&\qquad\qquad \text{(by induction hypothesis)} \\
&= \mu(t) + \mu(u) + \rho(u)
\end{aligned}$$

   (c) Otherwise.

$$\begin{aligned}
\mu(t[x{:=}u]) &= \mu(\pi_1(t_1[x{:=}u])) \\
&= \mu(t_1[x{:=}u]) \\
&\leq \mu(t_1) + \mu(u) + \rho(u) \\
&\qquad\qquad \text{(by induction hypothesis)} \\
&= \mu(t) + \mu(u) + \rho(u)
\end{aligned}$$

6. $t \equiv \pi_2 t_1$. This case is similar to the previous one. $\square$

Proposition 1. *Let t and u be two linear λ-terms such that $t \vartriangleright u$. Then $\mu(u) < \mu(t)$.*

Proof. The proof proceeds by induction on the derivation of $t \vartriangleright u$. We only give the base cases, the induction steps are straightforward.

1. $t \equiv (\lambda x. t_1) \, t_2$ and $u \equiv t_1[x:=t_2]$.

$$\begin{aligned}
\mu(u) &= \mu(t_1[x:=t_2]) \\
&\leq \mu(t_1) + \mu(t_2) + \rho(t_2) \quad \text{(by Lemma 3)} \\
&< \mu(t_1) + \mu(t_2) + \rho(t_2) + \rho(t_1) + 1 \\
&= \mu(t_1) + \mu(t_2) + \rho(\lambda x. t_1) \\
&= \mu(t)
\end{aligned}$$

2. $t \equiv \pi_1 \langle t_1, t_2 \rangle$, and $u \equiv t_1$.

$$\begin{aligned}
\mu(u) &= \mu(t_1) \\
&\leq \max(\mu(t_1), \mu(t_2)) \\
&< \max(\mu(t_1), \mu(t_2)) + \rho(\langle t_1, t_2 \rangle) \\
&= \mu(\pi_1 \langle t_1, t_2 \rangle) \\
&= \mu(t)
\end{aligned}$$

3. $t \equiv \pi_2 \langle t_1, t_2 \rangle$, and $u \equiv t_2$. This case is similar to the previous one.   □

Corollary 1. *Let t and u be two linear λ-terms such that $t \overset{n}{\vartriangleright} u$. Then $n \leq \mu(t) - \mu(u)$.*

Proof. By iterating Proposition 1.   □

As we explained at the beginning of this section, we intend to establish that whenever $t \to u$, there exists a reduction strategy that is polynomially bounded by the size of both t and u. The idea is to use $\mu(t)$. This is not sufficient because it only works for external reduction. Now, a reduction step that takes place within one of the two components of a pair is useless if the residual of this component eventually disappears because of a subsequent projection. However, if there is no subsequent projection the residual of the pair will occur in u. These observations, which suggest that we must take into account the number of pair components that occur in u, motivate the next definition.

Definition 8. *The number of slices $\#(t)$ of a linear λ-term t is defined as follows:*

1. *if t is a purely applicative term then $\#(t) = 1$,*
2. *otherwise, the definition $\#(t)$ obeys the following equations:*
   (a) $\#(\lambda x. t_1) = \#(t_1)$
   (b) $\#(t_1 \, t_2) = \begin{cases} \#(t_1), & \text{if } t_2 \text{ is a purely applicative term,} \\ \#(t_2), & \text{if } t_1 \text{ is a purely applicative term,} \\ \#(t_1) + \#(t_2), & \text{otherwise.} \end{cases}$
   (c) $\#(\langle t_1, t_2 \rangle) = \#(t_1) + \#(t_2)$

(d) $\#(\pi_1 t_1) = \#(t_1)$
(e) $\#(\pi_2 t_1) = \#(t_1)$

We now state and prove the main proposition of this section.

Proposition 2. *Let t and u be two linear λ-terms such that $t \stackrel{*}{\to} u$. Then, there exists $n \in \mathbb{N}$ such that $t \stackrel{n}{\to} u$ and $n \leq \mu(t) \times \#(u)$*

Proof. The proof proceeds by induction on the subterm/reduction relation.

1. $t \equiv x$. We must have $u \equiv x$, and consequently $n = 0 = \mu(x)$.
2. $t \equiv \lambda x. t_1$. We must have $u \equiv \lambda x. u_1$, with $t_1 \stackrel{*}{\to} u_1$. Hence, the property holds by induction hypothesis.
3. $t \equiv t_1 t_2$. If $u \equiv u_1 u_2$, with $t_1 \stackrel{*}{\to} u_1$ and $t_2 \stackrel{*}{\to} u_2$, the induction is straightforward. Otherwise, there exist t'_{11} and t'_2 such that:

$$t_1 t_2 \stackrel{*}{\to} (\lambda x. t'_{11}) t'_2 \to t'_{11}[x:=t'_2] \stackrel{*}{\to} u,$$

where $t_1 \stackrel{*}{\to} \lambda x. t'_{11}$ and $t_2 \stackrel{*}{\to} t'_2$. Then, by Lemma 1, there exists t_{11} such that $t_1 \,\triangleright\, \lambda x. t_{11} \stackrel{*}{\to} \lambda x. t'_{11}$. Therefore there exists $n_1, n_2 \in \mathbb{N}$ such that $t_1 t_2 \stackrel{n_1}{\triangleright} (\lambda x. t_{11}) t_2 \,\triangleright\, t_{11}[x:=t_2] \stackrel{n_2}{\to} u$, because $t_{11}[x:=t_2] \stackrel{*}{\to} t'_{11}[x:=t'_2]$. Hence, by Corollary 1, we have:

$$n_1 + 1 \leq \mu(t_1 t_2) - \mu(t_{11}[x:=t_2]),$$

which implies $n_1 + 1 \leq (\mu(t_1 t_2) - \mu(t_{11}[x:=t_2])) \times \#(u)$, since $\#(u) > 0$. On the other hand, by induction hypothesis, we have $n_2 \leq \mu(t_{11}[x:=t_2])) \times \#(u)$. Consequently, we have that $n_1 + n_2 + 1 \leq \mu(t_1 t_2) \times \#(u)$. Then, we take $n = n_1 + n_2 + 1$.
4. $t \equiv \langle t_1, t_2 \rangle$. We must have that $u \equiv \langle u_1, u_2 \rangle$, with $t_1 \stackrel{*}{\to} u_1$ and $t_2 \stackrel{*}{\to} u_2$. Hence, by induction hypothesis, there exists $n_1, n_2 \in \mathbb{N}$ such that: $t_1 \stackrel{n_1}{\to} u_1$ with $n_1 \leq \mu(t_1) \times \#(u_1)$, and $t_2 \stackrel{n_2}{\to} u_1$ with $n_2 \leq \mu(t_2) \times \#(u_2)$. Consequently, we have that $\langle t_1, t_2 \rangle \stackrel{n_1}{\to} \langle u_1, t_2 \rangle \stackrel{n_2}{\to} \langle u_1, u_2 \rangle$. Then, we may take $n = n_1 + n_2$ because the following inequalities hold:

$$
\begin{aligned}
n_1 + n_2 &\leq \mu(t_1) \times \#(u_1) + \mu(t_2) \times \#(u_2) \\
&\leq \max(\mu(t_1), \mu(t_2)) \times \#(u_1) + \max(\mu(t_1), \mu(t_2)) \times \#(u_2) \\
&= \max(\mu(t_1), \mu(t_2)) \times (\#(u_1) + \#(u_2)) \\
&= \mu(\langle t_1, t_2 \rangle) \times \#(\langle u_1, u_2 \rangle)
\end{aligned}
$$

5. $t \equiv \pi_1 t_1$. If $u \equiv \pi_1 u_1$, with $t_1 \stackrel{*}{\to} u_1$, the induction is straightforward. Otherwise, there exist t'_{11} and t'_{12} such that $\pi_1 t_1 \stackrel{*}{\to} \pi_1 \langle t'_{11}, t'_{12} \rangle \to t'_{11} \stackrel{*}{\to} u$, where $t_1 \stackrel{*}{\to} \langle t'_{11}, t'_{12} \rangle$. Then, by Lemma 2, there exist t_{11} and t_{12} such that $t_1 \,\triangleright\, \langle t_{11}, t_{12} \rangle \stackrel{*}{\to} \langle t'_{11}, t'_{12} \rangle$. Consequently, there exists $n_1, n_2 \in \mathbb{N}$ such that: $\pi_1 t_1 \stackrel{n_1}{\triangleright} \pi_1 \langle t_{11}, t_{12} \rangle \,\triangleright\, t_{11} \stackrel{n_2}{\to} u$, because $t_{11} \stackrel{*}{\to} t'_{11}$. Then, by Corollary 1, we have $n_1 + 1 \leq \mu(\pi_1 t_1) - \mu(t_{11})$, which implies $n_1 + 1 \leq (\mu(\pi_1 t_1) - \mu(t_{11})) \times \#(u)$. By induction hypothesis, we also have that $n_2 \leq \mu(t_{11}) \times \#(u)$. Hence, we have that $n_1 + n_2 + 1 \leq \mu(\pi_1 t_1) \times \#(u)$, and we take $n = n_1 + n_2 + 1$.
6. $t \equiv \pi_1 t_2$. This case is similar to the previous one. $\square$

4   Decidability and NP-Completeness

In the presence of pairs, linear λ-terms may contain subterms which are bound to disappear during the reduction because of projections. Those subterms may be arbitrarily huge and contain many redices. The previous section showed how to cope with them in order not to reduce useless redices and to have a polynomial reduction. The purpose of this one is to prove that if $t \xrightarrow{n} u$ then there exists some t' obtained by deleting useless subterms from t and the size of which is polynomial with respect to n and the size of u. Together with the results of the previous section this property will help us to obtain decidability and complexity insights about the matching problem.

In order to model deletion in terms, we add a special constant ($\Diamond$) to the calculus. This constant may be used to replace any term of any type in the formation rules of Definition 3 (not taking into account the side condition on free variables in the case of the formation of a pair).

Within this new notion of term, we ditinguish those obtained by adding the following term formation rules to the formation rules of Definition 3:

$$\text{if } t \in \mathscr{T}_\alpha \text{ then } \langle t, \Diamond \rangle \in \mathscr{T}_{\alpha \& \beta} \text{ and } \langle \Diamond, t \rangle \in \mathscr{T}_{\beta \& \alpha}.$$

those terms are called *hollow terms*. A substitution is hollow if for all x (*resp.* $\mathbf{X}$) $\sigma(x)$ (*resp.* $\sigma(\mathbf{X})$) is hollow.

The fact that a certain term is obtained from another one by deleting one of its subterm induces a reflexive and transitive relation ($\sqsubseteq$) on terms defined by the following formal system:

$$\Diamond \sqsubseteq t \qquad \frac{t \sqsubseteq u}{\lambda x.\, t \sqsubseteq \lambda x.\, u} \qquad \frac{t_1 \sqsubseteq u_1 \quad t_2 \sqsubseteq u_2}{t_1\, t_2 \sqsubseteq u_1\, u_2}$$

$$\frac{t_1 \sqsubseteq u_1 \quad t_2 \sqsubseteq u_2}{\langle t_1, t_2 \rangle \sqsubseteq \langle u_1, u_2 \rangle} \qquad \frac{t \sqsubseteq u}{\pi_1 t \sqsubseteq \pi_1 u} \qquad \frac{t \sqsubseteq u}{\pi_2 t \sqsubseteq \pi_2 u}$$

This relation is naturally extended to the substitutions:

Definition 9. *Given two substitutions σ_1 and σ_2, we write $\sigma_1 \sqsubseteq \sigma_2$ if for all x (resp. for all $\mathbf{X}$) $\sigma_1(x) \sqsubseteq \sigma_2(x)$ (resp. $\sigma_1(\mathbf{X}) \sqsubseteq \sigma_2(\mathbf{X})$)*

Definition 10. *We define the length of term $|t|$ as follows:*

1. $|\Diamond| = 1$
2. $|h| = 1$ *if h is an atomic term*
3. $|\lambda x.t| = |t| + 1$
4. $|t_1 t_2| = |t_1| + |t_2|$
5. $|\langle t_1, t_2 \rangle| = |t_1| + |t_2|$
6. $|\pi_1(t)| = |t| + 1$ *and* $|\pi_2(t)| = |t| + 1$

Then we define the length of a substitution to be $|\sigma| = \sum_{x \in \mathrm{dom}(\sigma)} |\sigma(x)| + \sum_{\mathbf{X} \in \mathrm{dom}(\sigma)} |\sigma(\mathbf{X})|$

Lemma 4. *If t is a hollow term then there is a linear λ-term u such that $FV(u) = FV(t)$, $t \sqsubseteq u$, $|u| \leq |t|^2$.*

Proof. The proof is an induction on the structure of t. We only present the case where $t \equiv \langle \Diamond, t' \rangle$, the other ones being either similar to this one or straightforward.

If $t \equiv \langle \Diamond, t' \rangle$ the by induction hypothesis we have the existence of a linear term u' such that $FV(u') = FV(t')$, $t' \sqsubseteq u'$ and $|u'| \leq |t'|^2$. Let $FV(u') = \{x_1; \ldots; x_n\}$ and $\mathbf{X}$ be an unknown with a type so that $u \equiv \langle \mathbf{X}x_1 \ldots x_n, u' \rangle$ is a linear λ-term with the same type as t. Obviously we have $FV(u) = FV(t)$ and $t \sqsubseteq u$, it remains to show that $|u| \leq |t|$. As $FV(t') = \{x_1; \ldots; x_n\}$, $n \leq |t'|$ and :

$$|u| = |u'| + n + 1 \leq |t'|^2 + n + 1 \leq |t'|^2 + |t'| + 1 \leq (|t'| + 1)^2 \leq |t|^2$$

$\square$

Lemma 5. *If $u_1 \sqsubseteq u_2$ and $\sigma_1 \sqsubseteq \sigma_2$ then $u_1.\sigma_1 \sqsubseteq u_2.\sigma_2$.*

Proof. By induction on the structure of u_1 :

1. If $u_1 \equiv \Diamond$ then $u_1.\sigma_1 \equiv \Diamond$ and obviously $u_1.\sigma_1 \sqsubseteq u_2.\sigma_2$.
2. If $u_1 \equiv x$ then $u_2 \equiv x$ and $u_i.\sigma_i \equiv \sigma_i(x)$. As $\sigma_1 \sqsubseteq \sigma_2$, we have $\sigma_1(x) \sqsubseteq \sigma_2(x)$ and as a consequence we have $u_1.\sigma_1 \sqsubseteq u_2.\sigma_2$.
3. If $u_1 \equiv h$ where h is a constant, then $u_2 \equiv h$ and $u_i.\sigma_i \equiv h$. So $u_1.\sigma_1 \sqsubseteq u_2.\sigma_2$.
4. The other cases are direct consequences of the induction hypothesis. $\square$

With a specific strategy, the relation $\sqsubseteq$ can be preserved through reduction.

Lemma 6. *If $v_1 \sqsubseteq v_2$ and $v_1 \rightarrow w_1$ then there exists w_2 such that $v_2 \rightarrow w_2$ and $w_1 \sqsubseteq w_2$.*

Proof. By induction on the structure of v_1:

1. If $v_1 \equiv (\lambda x.t_1)t_2$ and $w_1 \equiv t_1[x := t_2]$ then $v_2 \equiv (\lambda x.t_1')t_2'$ where $t_i \sqsubseteq t_i'$. Then from lemma 5 if we let $w_2 \equiv t_1'[x := t_2']$ then $w_1 \sqsubseteq w_2$.
2. The other cases are straightforward. $\square$

Lemma 7. *If $v_1 \sqsubseteq v_2$ and $v_2 \rightarrow w_2$ then there exists w_1 such that $v_1 \overset{=}{\rightarrow} w_1$ and $w_1 \sqsubseteq w_2$.*

Proof. This lemma can be proved by induction on the structure of v_2 in a way similar to the previous one. The only difference appears in the case where $v_1 \equiv \Diamond$. In that case $w_1 \equiv \Diamond$ and obviously $w_1 \sqsubseteq w_2$. $\square$

As a consequence, under certain conditions, the relation $\sqsubseteq$ preserves equality between terms.

Lemma 8. *If v is the normal form of v_1, v does not contain any occurence of $\Diamond$ and $v_1 \sqsubseteq v_2$ then $v_1 = v_2$.*

Proof. The lemma can be proved by iterating Lemma 6 and remarking that as $\Diamond$ has no occurence in v if $v \sqsubseteq w$ then $v \equiv w$. $\square$

Lemma 9. *If $v_1 = v_3$ and $v_1 \sqsubseteq v_2 \sqsubseteq v_3$ then $v_1 = v_2$.*

Proof. If v is the common normal form of v_1 and v_3, then there exists n such that $v_1 \xrightarrow{n} v$. In order to prove the lemma we use an induction on n.

In case $n = 0$, then v_1 is in normal form. But v_3 is not necessarily in normal form. We are going to prove by induction on v_1 that the normal form of v_2 is v_1.

1. $v_1 \equiv \Diamond$: we then have to prove that if $v_3 = \Diamond$, the fact that $v_2 \sqsubseteq v_3$ implies that $v_2 = \Diamond$.
   We proceed by induction on the length of the reduction $v_3 \xrightarrow{p} \Diamond$. In case $p = 0$, $v_3 \equiv \Diamond$ and then $v_2 \sqsubseteq \Diamond$ so $v_2 \equiv \Diamond$. If $p > 0$ then $v_3 \to v_3' \xrightarrow{p-1} \Diamond$. By Lemma 7 there is v_2' such that $v_2 \xrightarrow{=} v_2'$ and $v_2' \sqsubseteq v_3'$. Then, by induction hypothesis, $v_2' = \Diamond$ and $v_2 = \Diamond$.
2. The other cases are simple consequences of induction.

Now if $n > 0$ then $v_1 \to v_1' \xrightarrow{n-1} v$. By Lemma 6 there exists v_2' such that $v_2 \to v_2'$ and $v_1' \sqsubseteq v_2'$. Still from Lemma 6 there exists v_3' such that $v_3 \to v_3'$ and $v_2' \sqsubseteq v_3'$. Thus we have $v_1' \sqsubseteq v_2' \sqsubseteq v_3'$ and $v_1' \xrightarrow{n-1} v$, the induction hypothesis gives that $v_2' = v_1'$ which allows us to conclude that $v_1 = v_2$.   □

When two terms are dominated (with respect to $\sqsubseteq$) by another one, they share a common syntactic structure but each of them can have specific subterms. The following lemma proves the existence of a term which possesses both their common and specific features.

Lemma 10. *If there are three hollow terms v_1, v_2 and v such that $v_1 \sqsubseteq v$ and $v_2 \sqsubseteq v$ then there is v_3 such that :*

1. *v_3 is a hollow term*
2. *$v_3 \sqsubseteq v$, $v_1 \sqsubseteq v_3$ and $v_2 \sqsubseteq v_3$*
3. *$|v_3| \le |v_1| + |v_2| - 1$*

Proof. We proceed by induction on the structure of v. The only interesting case consists in having $v \equiv \langle u_1, u_2 \rangle$, $v_1 \equiv \langle w_1, \Diamond \rangle$ and $v_2 \equiv \langle \Diamond, w_2 \rangle$ the other cases are straightforward. In that case, it suffices to take $v_3 \equiv \langle w_1, w_2 \rangle$ to respect the conditions of the lemma.   □

The next lemma is the generalisation of the previous one to the substitutions.

Lemma 11. *If σ, σ_1 and σ_2 are hollow substitutions, $\mathrm{dom}(\sigma) \ne \emptyset$, $\sigma_1 \sqsubseteq \sigma$ and $\sigma_2 \sqsubseteq \sigma$ then there is σ_3 such that:*

1. *$\sigma_3 \sqsubseteq \sigma$, $\sigma_1 \sqsubseteq \sigma_3$ and $\sigma_2 \sqsubseteq \sigma_3$.*
2. *$|\sigma_3| \le |\sigma_1| + |\sigma_2| - 1$.*

Proof. The proof of this lemma uses an induction on the size of $\mathrm{dom}(\sigma)$, the initial case is simply proved using the previous lemma and if $x \notin \mathrm{dom}(\sigma)$ we set $\sigma_3(x)$ to be equal to x.   □

If a term t is obtained from a term v by deleting some of its subterms (*i.e.* $t \sqsubseteq v$) then t and v still share a main global syntactic structure. In particular one can expect that if v is the result of a substitution σ applied to a term u then t is also *somehow* the result of applying a substitution to a certain term. The next lemma explicits precisely this fact.

Lemma 12. *If t and u are hollow terms, σ is a hollow substitution and $t \sqsubseteq u.\sigma$ then there exist u' and σ' such that:*

1. *u' and σ' are hollow*
2. *$u' \sqsubseteq u$ and $\sigma' \sqsubseteq \sigma$*
3. *$t \sqsubseteq u'.\sigma' \sqsubseteq u.\sigma$*
4. *$|u'| + |\sigma'| \leq |t| + 1$*

Proof. If $\text{dom}(\sigma) = \emptyset$ then $\sigma = Id$ and we just have to take $u' \equiv t$ and $\sigma' = Id$ to get all that is needed. The rest of the proof won't take this trivial case into account, and the condition $\text{dom}(\sigma) \neq \emptyset$ which will allow us to apply the Lemma 11 will be implicitly verified.

We prove this lemma using an induction on the structure of u:

1. In case $u \equiv x$ and $x \in \text{dom}(\sigma)$ we take $u' \equiv x$ and $\sigma' \equiv [x := t]$. Such u' and σ' verify the requiered properties.
2. In case $u \equiv \mathbf{X}$ and $\mathbf{X} \in \text{dom}(\sigma)$ then we also take $u' \equiv \mathbf{X}$ and $\sigma' = [\mathbf{X} := t]$.
3. In case $u \equiv h$ where h is an atomic term which is not in $\text{dom}(\sigma)$, we let $u' \equiv t$ and $\sigma' = Id$.
4. In case $u \equiv \langle u_1, u_2 \rangle$ then $t \equiv \langle t_1, t_2 \rangle$ so that $t_i \sqsubseteq u_i.\sigma$. The induction hypothesis implies the existence of two pairs u'_1, σ_1 and u'_2, σ_2 such that $t_i \sqsubseteq u'_i.\sigma_i \sqsubseteq u_i.\sigma$, $u'_i \sqsubseteq u_i$, $\sigma_i \sqsubseteq \sigma$ and $|u'_i| + |\sigma_i| \leq |t_i| + 1$. As $\sigma_1 \sqsubseteq \sigma$ and $\sigma_2 \sqsubseteq \sigma$, from Lemma 11, there exists σ' such that $\sigma' \sqsubseteq \sigma$, $\sigma_i \sqsubseteq \sigma'$ and $|\sigma'| \leq |\sigma_1| + |\sigma_2| - 1$. By Lemma 5, as $u'_i \sqsubseteq u_i$ and $\sigma_i \sqsubseteq \sigma' \sqsubseteq \sigma$ it comes that $t_i \sqsubseteq u'_i.\sigma_i \sqsubseteq u'_i.\sigma' \sqsubseteq u_i.\sigma$. We let $u' \equiv \langle u'_1, u'_2 \rangle$ and verify that $t \sqsubseteq u'.\sigma' \sqsubseteq u.\sigma$ and :

$$\begin{aligned} |u'| + |\sigma'| &\leq |u'_1| + |u'_2| + |\sigma_1| + |\sigma_2| - 1 \\ &\leq |t_1| + 1 + |t_2| + 1 - 1 \\ &\leq |t| + 1 \end{aligned}$$

5. the case where $u \equiv u_1 u_2$ can be solved in the same way as the previous one.
6. the other cases are straightforward. □

Corollary 2. *If t, t_1 and t_2 are hollow terms, $t = t_1[x := t_2]$ and $t \sqsubseteq t_1[x := t_2]$ then there exists t'_1 and t'_2 such that:*

1. *t'_1 and t'_2 are hollow terms*
2. *$t'_1 \sqsubseteq t_1$ and $t'_2 \sqsubseteq t_2$*
3. *$t = t'_1[x := t'_2]$*
4. *$|t'_1| + |t'_2| \leq |t| + 1$*

Proof. From the previous lemma we know that there exists t_1' and t_2' such that $t_1' \sqsubseteq t_1$, $t_2' \sqsubseteq t_2$, $t \sqsubseteq t_1'[x := t_2'] \sqsubseteq t_1[x := t_2]$ and $|t_1'| + |t_2'| \leq |t| + 1$. As $t = t_1[x := t_2]$ lemma 9 implies $t = t_1'[x := t_2']$.   □

We now establish the key lemma of this section. We get a bound on the size of the term t' obtained by deleting useless subterms of a term t.

Lemma 13. *If t, u and u' are hollow term, $t \to u$, $u' \sqsubseteq u$ and $u' = u$ then there is some t' such that:*

1. *t' is a hollow term*
2. *$t' \sqsubseteq t$*
3. *$t' = t$*
4. *$|t'| \leq |u'| + 2$*

Proof. We proceed by induction on the structure of t. We just present the cases where t is a redex and u is the result of the contraction of that redex, the other ones are direct consequences of the induction hypothesis:

1. If $t \equiv (\lambda x.t_1)t_2$ and $u \equiv t_1[x := t_2]$, from Corollary 2 there are two hollow terms t_1' and t_2' such that $t_i' \sqsubseteq t_i$, $t_1'[x := t_2'] = u' = u = t$ and $|t_1'| + |t_2'| \leq |u'| + 1$. Thus $(\lambda x.t_1')t_2'$ is a hollow term $(\lambda x.t_1')t_2' \sqsubseteq (\lambda x.t_1)t_2$, $(\lambda x.t_1')t_2' = t$ and $|(\lambda x.t_1')t_2'| = |t_1'| + |t_2'| + 1 \leq |u'| + 2$.
2. If $t \equiv \pi_1(\langle t_1, t_2 \rangle)$ *(resp. $t \equiv \pi_2(\langle t_1, t_2 \rangle)$)* and $u \equiv t_1$ *(resp. $u \equiv t_2$)* then we let $t' \equiv \pi_1(\langle u', \Diamond \rangle)$ *(resp. $t' \equiv \pi_2(\langle \Diamond, u' \rangle)$)* and we verify that t' fullfills the conditions of the lemma.   □

Lemma 14. *If t and u are hollow terms and $t \xrightarrow{n} u$ then there is some t' such that :*

1. *t' is a hollow term*
2. *$t' \sqsubseteq t$*
3. *$t' = t$*
4. *$|t'| \leq |u| + 2n$*

Proof. This result is obtained by iterating the previous lemma. The iteration can be initiated because $u \sqsubseteq u$ and $u = u$.   □

Proposition 3. *If t and u are hollow terms, σ is a hollow subsitution and $t.\sigma \xrightarrow{n} u$ then there exists σ' such that:*

1. *σ' is a hollow substitution*
2. *$\sigma' \sqsubseteq \sigma$*
3. *$t.\sigma' = u$*
4. *$|\sigma'| \leq |u| + 2n$*

Proof. From the previous lemma, we know that if $t.\sigma \xrightarrow{n} u$ then there exists a hollow term t' such that $t' \sqsubseteq t.\sigma$, $t' = u$ and $|t'| \leq |u| + 2n$. Lemma 12 leads to the existence of a hollow term t'' and a hollow subsitution σ' such that $t'' \sqsubseteq t$, $\sigma' \sqsubseteq \sigma$, $t' \sqsubseteq t''.\sigma' \sqsubseteq t.\sigma$ and $|t''| + |\sigma'| \leq |t'| + 1$. As a consequence $|\sigma'| \leq |u| + 2n$.

We now have to verify that $t.\sigma' = u$. Lemma 9 gives $t''.\sigma' = u$ because $t' = u$ and $t.\sigma = u$. As $t'' \sqsubseteq t$ and $\sigma' \sqsubseteq \sigma$ Lemma 5 gives $t''.\sigma' \sqsubseteq t.\sigma' \sqsubseteq t.\sigma$. But $t''.\sigma' = u$ and $t.\sigma = u$, then Lemma 9 leads to what we expected.   □

Theorem 1. *Given a matching problem in the linear λ-calculus with pairing (t, u), it is decidable whether it has a solution or not. Furthermore, if u is in normal form, this problem is NP-complete.*

Proof. Let v be the normal form of u. If the matching equation (t, u) admits a solution σ then, from Proposition 2, there exists n such that $t.\sigma \xrightarrow{n} v$ and $n \leq \mu(t.\sigma) \times \#(v)$. If we consider that the terms substituted to unknowns by σ are in normal form then the redices contained in $t.\sigma$ are those contained in t and those created by the substitution. Thus, if $\{\mathbf{X}_1, \ldots, \mathbf{X}_n\}$ is the multiset of unknowns that occure in t, we have:

$$\mu(t.\sigma) \leq \mu(t) + \sum_{i=1}^{n} \rho(\mathbf{X}_i)$$

From proposition 3 we know that there exists σ' such that $\sigma' \sqsubseteq \sigma$, $t.\sigma' = v$ and $|\sigma'| \leq |v| + 2n \leq |v| + 2\#(v)\mu(t) + \sum_{i=1}^{n} 2\#(v)\rho(\mathbf{X}_i)$. But σ' may substitute to some unknowns terms which contain some $\Diamond$. Lemma 4 gives us the existence of a substitution σ'' with the same domain as σ', which substitutes a linear λ-term to each unknown of its domain and such that $\sigma' \sqsubseteq \sigma''$ and:

$$|\sigma''| \leq |\sigma'|^2 \leq (|v| + 2\#(v)\mu(t) + \sum_{i=1}^{n} 2\#(v)\rho(\mathbf{X}_i))^2$$

As $\sigma' \sqsubseteq \sigma''$, Lemma 5 proves that $t.\sigma' \sqsubseteq t.\sigma''$. Finally, since it is a linear λ-term, v does not contain any $\Diamond$ and we have (Lemma 8) $t.\sigma'' = t.\sigma' = v$. Hence, if there is a solution to the equation then there is also a solution which is bounded. The problem is then decidable.

Furthermore, if u is in normal form then $|v| = |u|$ and the existence of a solution implies the existence of a polynomially bounded one. And since Proposition 2 entails, in that case, that verifying whether a substitution is a solution or not is polynomial, the problem is in NP if u is in normal form. And as it is an extension of linear λ-calculus which is NP-hard [9], matching in the linear λ-calculus with pairing is NP-complete when u is in normal form.   □

We have not found yet the precise complexity of matching in the linear λ-calculus with pairing in the case where the right part of the equation is not in normal form. We managed to prove that this problem was PSPACE-hard, but we did not find a PSPACE-algorithm which solves it. At worst, we still have the EXP-time algorithm which consists in normalizing the right part of the equation and then solving it.

References

1. H. Comon. Completion of rewrite systems with membership constraints. Part I: Deduction rules. *Journal of Symbolic Computation*, 25(4):397–419, 1998.
2. H. Comon. Completion of rewrite systems with membership constraints. Part II: Constraint solving. *Journal of Symbolic Computation*, 25(4):421–453, 1998.
3. S. A. Cook. The complexity of theorem proving procedures. *Proceedings of the 3rd annual ACM Symposium on Theory of Computing*, pages 151–158, 1971.
4. D. Dougherty and T. Wierzbicki. A decidable variant of higher order matching. In *Proc. 13th Conf. on Rewriting Techniques and Applications, RTA'02*, volume 2378, pages 340–351, 2002.
5. G. Dowek. Third order matching is decidable. *Annals of Pure and Applied Logic*, 69(2–3):135–155, 1994.
6. G. Dowek. Higher-order unification and matching. In A. Robinson and A. Voronkov (eds.), *Handbook of Automated Reasoning*, pp. 1009-1062, Elsevier, 2001.
7. J.Y. Girard Linear logic *Theoritical Computer Science*, 50:1–102, 1987.
8. W. D. Goldfarb. The undecidability of the second-order unification problem. *Theoretical Computer Science*, 13(2):225–230, 1981.
9. Ph. de Groote. Higher-order linear matching is NP-complete. *Lecture Notes in Computer Science*, 1833:127–140, 2000.
10. Ph. de Groote, S. Salvati. On the complexity of higher-order matching in the linear λ-calculus. *Lecture Notes in Computer Science*, 2706:234–245, 2003.
11. G. Huet. The undecidability of unification in third order logic. *Information and Control*, 22(3):257–267, 1973.
12. G. Huet. *Résolution d'équations dans les langages d'ordre $1, 2, \ldots, \omega$*. Thèse de Doctorat d'Etat, Université Paris 7, 1976.
13. J. Levy. Linear second-order unification. In H. Ganzinger, editor, *Rewriting Techniques and Applications, RTA'96*, volume 1103 of *Lecture Notes in Computer Science*, pages 332–346. Springer-Verlag, 1996.
14. R. Loader. Higher order β matching is undecidable. *Logic Journal of the IGPL*, 11(1): 51–68, 2002.
15. H. Mairson and K. Terui. On the Computational Complexity of Cut-Elimination in Linear Logic. *Theoretical Computer Science (Proceedings of ICTCS2003)*, LNCS 2841, Springer-Verlag, pp. 23–36 (2003)
16. V. Padovani. *Filtrage d'ordre supérieure*. Thèse de Doctorat, Université de Paris 7, 1996.
17. M. Schmidt-Schauß and J. Stuber. On the complexity of linear and stratified context matching problems. Rapport de recherche A01-R-411, LORIA, December 2001.

A Dependent Type Theory with Names and Binding

Ulrich Schöpp and Ian Stark

LFCS, School of Informatics, University of Edinburgh
JCMB, King's Buildings, Edinburgh EH9 3JZ

Abstract. We consider the problem of providing formal support for working with abstract syntax involving variable binders. Gabbay and Pitts have shown in their work on Fraenkel-Mostowski (FM) set theory how to address this through first-class names: in this paper we present a dependent type theory for programming and reasoning with such names. Our development is based on a categorical axiomatisation of names, with freshness as its central notion. An associated adjunction captures constructions known from FM theory: the freshness quantifier Ⅵ, name-binding, and unique choice of fresh names. The Schanuel topos — the category underlying FM set theory — is an instance of this axiomatisation. Working from the categorical structure, we define a dependent type theory which it models. This uses bunches to integrate the monoidal structure corresponding to freshness, from which we define novel multiplicative dependent products Π^* and sums Σ^*, as well as a propositions-as-types generalisation H of the freshness quantifier.

1 Introduction

The handling of variable binding in abstract syntax is a recognised challenge for machine-assisted reasoning about programming languages and logics. The problem is that a significant part of the formalisation effort may go into dealing with issues that are normally suppressed in informal practice: namely that one is working with α-equivalence classes of terms rather than raw terms.

Gabbay and Pitts have shown that FM set theory supports a notion of names that can make precise the informal practise of using concrete names for α-equivalence classes. They give a number of useful constructions: abstract syntax with binders can be encoded as an inductive data type, there is a useful syntax-independent notion of name-freshness, and a freshness quantifier simplifies reasoning with names.

The approach of Gabbay and Pitts has been studied in a number of other settings, among which are the first-order Nominal Logic [18], the higher-order logic FM-HOL [6] as well as the programming language FreshML [19]. Related [9] to FM theory, the Theory of Contexts [11] provides an axiomatisation of reasoning with names in dependent type theory. The ideas underlying FM have also proved useful in other areas such as Spatial Logic [2] or programming with semi-structured data with hidden labels [1]. These approaches typically focus either on programming with names, or reasoning about them. The Theory of Contexts, for example, supports reasoning with names, but does not admit functions that compare names or which (locally) choose fresh names.

J. Marcinkowski and A. Tarlecki (Eds.): CSL 2004, LNCS 3210, pp. 235–249, 2004.
© Springer-Verlag Berlin Heidelberg 2004

In this paper we take the first steps towards a dependent type theory incorporating FM concepts for both programming and reasoning with names. We introduce a dependent type theory, using as guidance the categorical structure of Schanuel topos, which is the category corresponding to FM set theory. In contrast to FM set theory, where swapping is the primitive notion for working with names, we take freshness as the central primitive of our type theory. This allows us to describe the constructions with names and binding in terms of universal constructions, and also avoids problems with extensional equality, which seems to be necessary for defining α-equivalence classes using swapping.

As the first contribution of the paper we introduce a *bunched* dependent type theory. Since freshness corresponds to a monoidal structure, bunches provide a natural way of integrating it into the type theory. Our bunched type theory may be seen as a generalisation of the $\alpha\lambda$-calculus of O'Hearn and Pym [17, 20]. The $\alpha\lambda$-calculus is a simple type theory corresponding to a category which is both cartesian closed and monoidal closed. Our type theory extends this situation, but only in the additive direction: we consider a category which is *locally* cartesian closed as well as monoidal closed. In this structure, we can model a dependent type theory with two function spaces $\Pi x{:}A.\ B$ and $\Pi^*x{:}C.\ D$. The first comes from the locally cartesian closed structure and consists of normal dependent functions. The second, which is subject to the restriction that C is closed, comes from the monoidal closed structure and may be thought of as consisting of functions which are only defined on arguments $x : C$ that contain just *fresh* names. In particular, with a type of names $\mathbf{N}$, we can use $\Pi^*n{:}\mathbf{N}.\ D$ to model α-equivalence classes, which corresponds to the well-known approach of modelling α-equivalence classes as 'fresh functions' [7, 4, 9, 5]. Another way of representing α-equivalence classes, as given in [7], is to consider them as pairs $n.x$ of a term x with a distinguished name n in such a way that the identity of n is hidden in the pair. This representation is also available in our type theory as fresh sum types Σ^*, dual to Π^*. The inhabitants of $\Sigma^*x{:}C.\ D$ may be thought of as pairs $M.N$ where $M : C$ and $N : D(M)$ and in which all the names in M have been hidden. To formulate Σ^*-types, we introduce a type $B^{*(M:A)}$, thought of as those elements of B which are free from all the names in the term $M : A$. These *freefrom* types are used to enforce that no use of a pair $M.N$ in $\Sigma^*x{:}C.\ D$ can reveal the hidden names.

As a second contribution of the paper, we give a new categorical axiomatisation of names and binding. The main feature of this axiomatisation is a propositions as types generalisation of the freshness quantifier of Gabbay and Pitts. To recall the freshness quantifier, consider quantifiers $\exists^*x{:}A.\ \varphi$ and $\forall^*x{:}A.\ \varphi$ expressing 'φ holds for some x containing only *fresh* names' and 'φ holds for any x containing only *fresh* names' respectively. The freshness quantifier V arises because, for the type of names $\mathbf{N}$, the propositions $\exists^*n{:}\mathbf{N}.\ \varphi$ and $\forall^*n{:}\mathbf{N}.\ \varphi$ are equivalent; and $\mathsf{V}\,n.\ \varphi$ is used to denote either of them. We have a propositions-as-types correspondence between $\exists^*$ and Σ^* as well as between $\forall^*$ and Π^*, so one may generalise the equivalence of $\exists^*n : \mathbf{N}.\ \varphi$ and $\forall^*n : \mathbf{N}.\ \varphi$ to an isomorphism between $\Sigma^*n{:}\mathbf{N}.\ D$ and $\Pi^*n{:}\mathbf{N}.\ D$.

This motivates our categorical axiomatisation of names. The central concept is freshness, giving rise to a certain 'fresh weakening' functor W. The types Σ^* and Π^* are left and right adjoints to W. Names are given by an object $\mathbf{N}$ having decidable equality.

Moreover, we require an isomorphism $\Sigma_{\mathbf{N}}^* \cong \Pi_{\mathbf{N}}^*$ generalising the freshness quantifier. We show that this structure includes not only the freshness quantifier, but also binding $(n.x)$ as in [7, 16] as well as unique choice of fresh names (new $n.M$) as in FreshML [19].

The semantics leads us to a type theory with names and binding. Based on the isomorphism $\Sigma_{\mathbf{N}}^* \cong \Pi_{\mathbf{N}}^*$, we introduce hidden-name types $\mathrm{H}n.D$ as a generalisation of the freshness quantifier. We may think of the elements of $\mathrm{H}n.D$ as elements of $\Sigma^*n{:}\mathbf{N}.D$, i.e. pairs with hidden names, but also as elements of $\Pi^*n{:}\mathbf{N}.D$, i.e. functions taking only fresh names. In analogy to the freshness quantifier, which has the rules from both $\exists^*$ and $\forall^*$, the rules for H are those from both Σ^* and Π^*. This dual view of hidden-name types turns out to be useful for working with abstract syntax: it allows us to use both HOAS-style constructions and FM-style constructions at the same time.

2   A Bunched Dependent Type Theory

In this section we introduce a first-order bunched dependent type theory and identify the categorical structure corresponding to it. The type theory has the following forms of sequents: $(\vdash \Gamma \text{ Bunch})$ — Γ is a bunch, or context; $(\Gamma \vdash A \text{ Type})$ — A is a type in context Γ; $(\Gamma \vdash M : A)$ — M is a term of type A in context Γ; as well as corresponding sequents for definitional equalities.

2.1   Bunches and Structural Rules

Bunches are built from the empty bunch $\diamond$ using two kinds of extension. First, the familiar additive context extension from dependent type theory, which takes a bunch Γ to the bunch $\Gamma, x : A$. Second, a multiplicative extension taking two bunches Γ and Δ to a new bunch $\Gamma * \Delta$. This extension is non-dependent in that no dependency is allowed across the $*$. The bunch $\Gamma * \Delta$ should be thought of as the context Γ, Δ with the restriction that the names occurring in Γ are disjoint from those in Δ. For example, if Lam is a type which encodes object-level λ-terms, then the bunch $(x{:}\mathsf{Lam}, y{:}\mathsf{Lam}) * (z{:}\mathsf{Lam})$ declares three terms x, y and z with the property that the names (representing the free variables of the encoded terms) in x and y are disjoint from those in z.

$$\frac{}{\vdash \diamond \text{ Bunch}} \qquad \frac{\Gamma \vdash A \text{ Type}}{\vdash \Gamma, x{:}A \text{ Bunch}} \; x \notin v(\Gamma) \qquad \frac{\vdash \Gamma \text{ Bunch} \quad \vdash \Delta \text{ Bunch}}{\vdash \Gamma * \Delta \text{ Bunch}} \; v(\Gamma) \cap v(\Delta) = \emptyset$$

In the side condition of these rules, we write $v(\Gamma)$ for the set of variables declared in Γ. We will frequently omit such side-conditions on the variable names, assuming tacitly that we encounter only bunches in which no variable is declared more than once.

We use the notation $\Gamma(\Delta)$ to indicate that Γ has a sub-bunch Δ, where sub-bunches are defined as follows: Δ is a sub-bunch of itself, and if Δ is a sub-bunch of Γ then it is also a subbunch of $(\Gamma, x : A)$, and $\Gamma * \Phi$, and $\Phi * \Gamma$. We write $\Gamma(\Phi)$ for the bunch which results from $\Gamma(\Delta)$ by replacing the (unique) occurrence of Δ in Γ with Φ.

Using this notation, we can formulate the structural rules:

$$\text{(Proj)} \; \frac{\Gamma \vdash A \text{ Type}}{\Gamma, x{:}A \vdash x : A} \; x \notin v(\Gamma)$$

$$(\text{Weak})\ \frac{\Gamma(\Delta) \vdash \mathcal{J} \quad \Delta \vdash A\ \text{Type}}{\Gamma(\Delta,\, x{:}A) \vdash \mathcal{J}}\ x \notin v(\Gamma, \Delta) \quad (\text{Subst})\ \frac{\Delta \vdash M : A \quad \Gamma(\Delta,\, x{:}A) \vdash \mathcal{J}}{\Gamma(\Delta)\,[M/x] \vdash \mathcal{J}\,[M/x]}$$

$$(\text{Unit})\ \frac{\Gamma(\Delta) \vdash \mathcal{J}}{\Gamma(\Delta * \Diamond) \vdash \mathcal{J}} \qquad (\text{Swap})\ \frac{\Gamma(\Delta * \Phi) \vdash \mathcal{J}}{\Gamma(\Phi * \Delta) \vdash \mathcal{J}} \qquad (\text{Assoc})\ \frac{\Gamma((\Delta * \Phi) * \Psi) \vdash \mathcal{J}}{\Gamma(\Delta * (\Phi * \Psi)) \vdash \mathcal{J}}$$

In these rules, we use $\mathcal{J}$ for an arbitrary judgement and double lines for bi-directional rules. We highlight the rule (Unit) which requires the empty bunch $\Diamond$ to be a unit for $*$, thus making $*$ affine. In particular, the multiplicative weakening rule

$$(*\text{-Weak})\ \frac{\Gamma(\Delta) \vdash \mathcal{J} \quad \vdash \Gamma(\Delta * \Phi)\ \text{Bunch}}{\Gamma(\Delta * \Phi) \vdash \mathcal{J}}$$

becomes admissible by using (Unit) together with (Weak).

Semantically, the bunches and structural rules can be modelled by a comprehension category [12] that in addition has an affine (i.e. the unit is isomorphic to the terminal object) symmetric monoidal structure $*$ in its base. We model the additive context-extension Γ, $x{:}A$ by the comprehension, and the multiplicative context-extension $\Gamma * \Delta$ by the monoidal product. To simplify the development, we make an additional assumption on the monoidal structure, given by the following definition [10].

Definition 1. *An* affine linear category *is a category* $\mathbb{B}$ *with finite products and an affine symmetric monoidal structure* $*$ *such that, for any two objects* A *and* B *of* $\mathbb{B}$, *the canonical map* $\langle \pi_1, \pi_2 \rangle : A * B \to A \times B$ *is a monomorphism.*

In most of the paper, we take a special comprehension category: the codomain fibration $cod : \mathbb{B}^{\to} \to \mathbb{B}$ for an affine linear category $\mathbb{B}$ having all pullbacks. Although technically the interpretation uses a corresponding split fibration to deal with well-known coherence issues [8], in the following we elide such details. We assume the reader to be familiar with the semantics of (first-order) dependent type theory, see e.g. [12, 22, 21].

2.2   Type Formers

In this section, we consider the types and terms, motivating them semantically. Starting from a codomain fibration $cod : \mathbb{B}^{\to} \to \mathbb{B}$ with an affine linear base $\mathbb{B}$, we step-by-step add more structure and introduce syntax based on it.

Type and Term Constants. Basic types and terms are given by constants. These can be formulated as usual. For example, a type constant T in context Γ may be introduced as $(\Gamma \vdash T(\boldsymbol{x})\ \text{Type})$, where $\boldsymbol{x}$ is the list of variables defined in Γ. That it is enough to annotate the constants just with the list of variables in Γ, ignoring any bunching structure, is a consequence of the assumption that the canonical map $A * B \rightarrowtail A \times B$ is a monomorphism.

Additive Types (Σ, Π). Types found in Martin-Löf type theory can also be formulated as usual. In this paper, we use dependent sums and products, but others such as identity types can be added without problem. To model Π-types in the codomain fibration, we assume $\mathbb{B}$ to be locally cartesian closed [21, 12].

Monoidal Product (*). We add types $A*B$ which internalise the context multiplication $\Gamma * \Delta$. The type $A*B$ may be thought of as containing all pairs $\langle M, N \rangle$ in $A \times B$ for which the sets of names underlying M and N are disjoint.

$$(*\text{-Ty}) \; \frac{\vdash A \text{ Type} \qquad \vdash B \text{ Type}}{\vdash A*B \text{ Type}} \qquad (*\text{-I}) \; \frac{\vdash A*B \text{ Type} \qquad \Gamma \vdash M : A \qquad \Delta \vdash N : B}{\Gamma * \Delta \vdash M*N : A*B}$$

$$(*\text{-E}) \; \frac{\Gamma(z:A*B) \vdash C \text{ Type} \quad \Delta \vdash M : A*B \quad \Gamma(x:A * y:B)\,[x*y/z] \vdash N : C\,[x*y/z]}{\Gamma(\Delta)\,[M/z] \vdash (\text{let } M \text{ be } x*y \text{ in } N) : C\,[M/z]}$$

Note that the type $A*B$ requires both A and B to be closed. This is because of substitution, as $(A*B)[\sigma]$ and $(A[\sigma]*B[\sigma])$ would not always have isomorphic interpretations.

Since the rule (*-Weak) is admissible, we can derive an inclusion $\imath_{A,B}$ of type $A*B \to A \times B$, given by the term $\imath_{A,B} =_{df} \lambda p : A*B. (\text{let } p \text{ be } x*y \text{ in } \langle x, y \rangle)$. Using this, we can state the equations for the monoidal product:

$$(*\text{-}\beta) \; \frac{\Gamma \vdash \text{let } M*N \text{ be } x*y \text{ in } R : C}{\Gamma \vdash (\text{let } M*N \text{ be } x*y \text{ in } R) = R\,[M/x]\,[N/y] : C}$$

$$(*\text{-}\eta) \; \frac{\Delta \vdash M : A*B \qquad \Gamma(z:A*B) \vdash N : C}{\Gamma(\Delta)[M/z] \vdash N\,[M/z] = \text{let } M \text{ be } x*y \text{ in } (N\,[x*y/z]) : C[M/z]}$$

$$(\text{Inject}) \; \frac{\Gamma \vdash M : A*B \qquad \Gamma \vdash N : A*B \qquad \Gamma \vdash \imath_{A,B}(M) = \imath_{A,B}(N) : A \times B}{\Gamma \vdash M = N : A*B}$$

Fresh Dependent Products (Π^*). We now make the further assumption on $\mathbb{B}$ that, for each object A in $\mathbb{B}$, the functor $- *A$ preserves pullbacks and has a right adjoint $A \multimap -$.

This gives rise the following situation. Let $gl(- * A)$ be the fibration defined by change-of-base as in the left square below. Let $W_A : \mathbb{B}^\to \to \mathbb{B}/(- * A)$ be the functor which maps an object $f : B \to G$ to $f * A : B * A \to G * A$. The assumption that $- * A$ preserves pullbacks amounts to saying that W_A is a *fibred* functor from *cod* to $gl(-*A)$. Moreover, it follows that W_A has a fibred right adjoint $\Pi_A^* : \mathbb{B}/(- * A) \to \mathbb{B}^\to$, see e.g. [14]. Explicitly, Π_A^* maps an object $g : C \to G * A$ to the the morphism $\Pi_A^* g$ as in the pullback on the right.

$$\begin{array}{ccc}
\mathbb{B}/(- * A) & \longrightarrow & \mathbb{B}^\to \\
{\scriptstyle gl(-*A)} \downarrow \quad {}^{\lrcorner} & & \downarrow {\scriptstyle cod} \\
\mathbb{B} & \underset{-*A}{\longrightarrow} & \mathbb{B}
\end{array} \qquad\qquad
\begin{array}{ccc}
\Pi_A^* C & \longrightarrow & A \multimap C \\
{\scriptstyle \Pi_A^* g} \downarrow \quad {}^{\lrcorner} & & \downarrow {\scriptstyle A \multimap g} \\
G & \underset{\eta}{\longrightarrow} & A \multimap (G * A)
\end{array}$$

Proposition 1. *For any object A of $\mathbb{B}$, the functor W_A as defined above has a fibred right adjoint Π_A^* if and only if $A*-$ preserves pullbacks and has a right adjoint $A \multimap -$.*

In this way, we can recast the monoidal closed structure in terms of a fibred adjunction, and introduce syntax for the fibred adjunction as follows.

$$(\Pi^*\text{-Ty}) \; \frac{\Gamma * x:A \vdash B \text{ Type}}{\Gamma \vdash \Pi^* x:A. B \text{ Type}}$$

$$(\Pi^*\text{-I}) \; \frac{\Gamma * x:A \vdash M : B}{\Gamma \vdash \lambda^* x:A. M : \Pi^* x:A. B} \qquad (\Pi^*\text{-E}) \; \frac{\Gamma \vdash M : \Pi^* x:A. B \qquad \Delta \vdash N : A}{\Gamma * \Delta \vdash M@N : B\,[N/x]}$$

$$(\Pi^*\text{-}\beta) \; \frac{\Gamma * x : A \vdash M : B \quad \Delta \vdash N : A}{\Gamma * \Delta \vdash (\lambda^* x : A.\, M)@N = M\,[N/x] : B\,[N/x]}$$

$$(\Pi^*\text{-}\eta) \; \frac{\Gamma \vdash M : \Pi^* x : A.\, B}{\Gamma \vdash \lambda^* x : A.\, (M@x) = M : \Pi^* x : A.\, B}$$

Notice that the fresh dependent product $\Pi^* x : A.\, B$ is only well-formed for closed types A, as bunching does not allow dependency across the $*$ in the bunch $\Gamma * x : A$.

The rules of Π^* derive from the adjoint correspondence

$$\frac{1_{G*A} = W_A(1_G) \to C \quad \text{in } \mathbb{B}/(G*A)}{1_G \to \Pi_A^*(C) \quad \text{in } \mathbb{B}/G},$$

since morphisms $1_G \to D$ in $\mathbb{B}/G$ correspond to terms in context G. Here, 1_G denotes the terminal object in $\mathbb{B}/G$. That Π_A^* is a *fibred* right adjoint means that substitution behaves as expected, that is we have $(\Pi^* x : A.\, B)[M/y] = \Pi^* x : A.\, (B[M/y])$ as well as $(\lambda^* x : A.\, N)[M/y] = \lambda^* x : A.\, (N[M/y])$.

Freefrom Types ($A^{*(N:B)}$). Having considered a fibred right adjoint Π_A^* to W_A, it is natural to ask for a fibred left adjoint Σ_A^* to W_A. To add syntax for such a left adjoint, we need to account for a one-to-one correspondence between maps $B \to W_A(C)$ in $\mathbb{B}/(G*A)$ and $\Sigma_A^*(B) \to C$ in $\mathbb{B}/G$. Hence, we need a syntactic equivalent for the map $B \to W_A(C)$, and so must introduce syntax for $W_A(C)$. Note that this is not necessary for Π^*, since there we only need the value of $W_A(1_G)$, which is 1_{G*A}.

We introduce types $B^{*(M:A)}$ as a syntax for working with $W_A(B)$. Intuitively, the type $B^{*(M:A)}$ comprises all those $p : B*A$ whose second component $\pi_2(p)$ is $M : A$. The functor W_A may then be understood as a 'fresh weakening' functor, taking the type $(\Gamma \vdash B \text{ Type})$ to $(\Gamma * x : A \vdash B^{*(x:A)} \text{ Type})$. Here, type A is necessarily closed, while B may in general depend on Γ. However, in the present paper we avoid the complexity of managing substitution in B by restricting to closed freefrom types:

$$(\text{F-Ty}) \; \frac{\vdash A \text{ Type} \quad \vdash B \text{ Type} \quad \Delta \vdash N : A}{\Delta \vdash B^{*(N:A)} \text{ Type}}$$

$$(\text{F-I}) \; \frac{\vdash A, B \text{ Type} \quad \Gamma \vdash M : B \quad \Delta \vdash N : A}{\Gamma * \Delta \vdash M^{*N} : B^{*(N:A)}}$$

$$(\text{F-E}) \; \frac{\Gamma(x : A, \, z : B^{*(x:A)}) \vdash C \text{ Type} \quad \Delta \vdash M : B^{*(N:A)} \quad \Gamma(y : B * x : A) \vdash R : C[y^{*x}/z]}{\Gamma(\Delta)[N/x][M/z] \vdash \text{let } M \text{ be } y^{*x} \text{ in } R : C[N/x][M/z]}$$

The equations[1], in which $\Gamma \vdash Q : B^{*(N:A)}$, are:

(β) let M^{*N} be y^{*x} in $R = R[N/x][M/y]$

(η) let Q be y^{*x} in $R[y^{*x}/z] = R[N/x][Q/z]$

Furthermore, we add a constant to 'join' two elements of freefrom types.

---

[1]   For brevity, from now on, we omit the contexts and typeability assumptions in the formulation of equations. Nevertheless, all equations are to be understood as equations-in-context, formulated under suitable typeability assumptions.

$$(\text{F-join}) \frac{\Gamma \vdash M : A^{*(R:C)} \qquad \Gamma \vdash N : B^{*(R:C)}}{\Gamma \vdash \text{join}_{A,B,C}(M,N) : (A \times B)^{*(R:C)}}$$

This constant is part of the syntax for W_A, arising from the fact that W_A is a *fibred functor*, equivalently that $- * A$ preserves pullbacks. It makes available the important property of freshness that if two objects x and y are fresh for some z then so is the pair $\langle x, y \rangle$. The behaviour of join is described by the equations

$$\text{let join}_{A,B,C}(M,N) \text{ be } y^{*x} \text{ in } (\pi_1 \, y)^{*x} = M,$$
$$\text{let join}_{A,B,C}(M,N) \text{ be } y^{*x} \text{ in } (\pi_2 \, y)^{*x} = N.$$

The semantic interpretation of (F-Ty) is given by the following diagram.

To see how this corresponds to W_A, recall that a closed type B in context Γ corresponds to the projection $\pi_B : \Gamma \times B \to \Gamma$. Using pullback-preservation of $- * A$, the following square is easily seen to be a pullback.

Since the bottom row of this diagram corresponds to the term $\Gamma * x : A \vdash x : A$, this means that $(\Gamma * x : A \vdash B^{*(x:A)} \text{ Type})$ receives an interpretation isomorphic to $\pi_B * A$, which, by definition, is just $W_A(\pi_B)$.

Fresh Dependent Sums (Σ^*). We now assume that W_A has a fibred left adjoint Σ_A^*. Using freefrom types as syntax for W_A, this gives rise to the following rules for Σ_A^*.

$$(\Sigma^*\text{-Ty}) \frac{\Gamma * x : A \vdash B \text{ Type}}{\Gamma \vdash \Sigma^* x : A. \, B \text{ Type}}$$

$$(\Sigma^*\text{-I}) \frac{x : A \vdash B \text{ Type} \qquad \Gamma \vdash M : A \qquad \Gamma \vdash N : B[M/x]}{\Gamma \vdash \text{bind}(M,N) : (\Sigma^* x : A. \, B)^{*(M:A)}}$$

$$(\Sigma^*\text{-E}) \frac{\Gamma \vdash M : \Sigma^* x : A. \, B \qquad (\Gamma * x : A), \, y : B \vdash N : C^{*(x:A)}}{\Gamma \vdash \text{let } M \text{ be } x.y \text{ in } N : C}$$

$$M.N =_{\text{df}} (\text{let } \text{bind}(M,N) \text{ be } u^{*m} \text{ in } u)$$

These rules are best explained using the intended model of names. The term bind(M,N) in (Σ^*-I) may be understood as the pair $\langle M, N \rangle$ with all the names in M made private, together with a proof that the names in M are indeed fresh for the pair. The abbreviation $M.N$ is a short-hand for the pair without the proof of freshness. The introduction rule (Σ^*-I) has a freefrom type in its conclusion because the constructor bind(N, M) comes from the unit $\eta : B \to W_A \Sigma_A^* B$ of the adjunction, whose codomain $W_A \Sigma_A^* B$ is the semantic equivalent of $(\Sigma^* x : A. \, B)^{*(x:A)}$. The elimination rule ($\Sigma^*$-E) formalises the intuition that an element M of type $\Sigma^* x : A. \, B$ is a pair with name-hiding. For this intuition to be valid, it should only be possible to use the components of the pair M in such a way that none of the hidden names is revealed. In $(\Sigma^*\text{-E})$ this is achieved using freefrom types: the term N has type $C^{*(x:A)}$, and such a term can be understood as an element of C whose value does not depend on the names in x.

The equations, in which $(\Gamma * x : A), y : B \vdash R : C^{*(x:A)}$ and $\Gamma, z : \Sigma^* x : A. B \vdash Q : D$, follow from the triangular identities for the adjunction $\Sigma_A^* \dashv W_A$.

(β) let bind(M, N) be z^{*u} in (let z be $x.y$ in $R)^{*u} = R[M/x][N/y]$

(η) let M be $x.y$ in (let bind(x, y) be z^{*x} in $Q^{*x}) = Q[M/z]$

We remark that the restriction on freefrom types that B must be closed in $B^{*(M:A)}$ makes the rules for Σ^* incomplete. For example, we have to restrict $(\Sigma^*\text{-I})$ so that B can only depend on x. More general rules are possible with unrestricted freefrom types.

2.3   Examples and Applications

As a simple example, we show that one can go from $\Pi x{:}A.\,B$ to $\Pi^* x{:}A.\,B$, as is the case in the affine $\alpha\lambda$-calculus.

$$
\dfrac{\dfrac{\dfrac{\dfrac{\vdots}{f{:}\Pi x{:}A.\,B \vdash f : \Pi x{:}A.\,B}\text{(Proj)}}{(f{:}\Pi x{:}A.\,B) * \Diamond \vdash f : \Pi x{:}A.\,B}\text{(Unit)}}{(f{:}\Pi x{:}A.\,B) * x{:}A \vdash f : \Pi x{:}A.\,B}\text{(Weak)} \qquad \dfrac{\dfrac{\dfrac{\dfrac{\vdots}{x{:}A \vdash x : A}\text{(Proj)}}{x{:}A * \Diamond \vdash x : A}\text{(Unit)}}{\Diamond * x{:}A \vdash x : A}\text{(Swap)}}{(f{:}\Pi x{:}A.\,B) * x{:}A \vdash x : A}\text{(Weak)}}{\dfrac{(f{:}\Pi x{:}A.\,B) * x{:}A \vdash f\,x : B}{f{:}\Pi x{:}A.\,B \vdash \lambda^* x{:}A.\,f\,x : \Pi^* x{:}A.\,B}\text{(Π^*-I)}}\text{(Π-E)}
$$

With type dependency and freefrom types, we can express freshness assumptions more precisely than with simply-typed bunches alone. For example, the freshness assertions in the context $x : A$, $y : A$, $u : A^{*(x:A)}$, $v : A^{*(\langle x,y\rangle:A\times A)}$ cannot be expressed with simply-typed bunches. On the other hand, the only way the freshness information in freefrom types $B^{*(M:A)}$ can ever be used is via bunches. We then have to ask the question if this is enough to derive useful statements involving freefrom types.

A useful set of rules for working with freefrom types appears in the type system of FreshML [19], which may be seen as a simply typed system with restricted freefrom types. Rules similar to those in FreshML are admissible in our system, thus allowing us to work with freefrom types in the style of FreshML. The main use of freshness in FreshML is for abstraction types (α-equivalence classes) and for the choice of fresh names (new $n.\,M$). Since we will see below that both constructions arise as instances of Π^* and Σ^*, we expect to have at our disposal at least the uses of names and binding as found in FreshML.

Furthermore, with dependent types we can also work with types that are not available in FreshML. For example, assume an inductive type L of lists of names. By structural recursion, we can define a function remove of type $\Pi n{:}\mathbf{N}.\,(\mathsf{L}\to\mathsf{L}^{*(n:\mathbf{N})})$ taking a name n and a list l to the list which results by removing n from l. As can be seen from the type, remove also provides a proof that n is fresh for the resulting list. Such freshness information is crucial for defining functions out of α-equivalence classes, to guarantee that the definition is independent of the choice of representative. An example of this, the function computing the free variables of a term, is given in Sec. 3.1 below.

2.4   Models

We summarise the structure required of a category $\mathbb{B}$ so that its codomain fibration models all of the syntax. The interpretation itself also requires this structure to be split, but due to space restrictions we omit the details of the interpretation.

Definition 2. *An affine linear category $\mathbb{B}$ is a* model of the bunched dependent type theory *if it is locally cartesian closed, and if, for each object A in $\mathbb{B}$, the functor W_A as defined above is a fibred functor from cod to $gl(-*A)$ having both fibred left and right adjoint $\Sigma_A^* \dashv W_A \dashv \Pi_A^*$.*

We have seen that the fibred adjunction $W_A \dashv \Pi_A^*$ can be formulated in terms of the monoidal structure. We know of no such non-fibred restatement for $\Sigma_A^* \dashv W_A$.

3   Names and Binding

In this section we consider how the bunched type theory can be used for working with names and binding. To this end, we consider a particular model of the type theory, the Schanuel topos $\mathbb{S}$, which is being widely used as a universe in which to work with names and binding. The Schanuel topos may be thought of as a category of sets involving names. For lack of space, we cannot present it in any detail; the reader is referred to e.g. [7] for its use for names and binding, and to e.g. [15, 13, 16] for categorical presentations. For the type theory we use the following categorical structure of $\mathbb{S}$.

Proposition 2. *The Schanuel topos $\mathbb{S}$ is a model of the bunched type theory having the following additional structure.*

1. *Finite coproducts which are stable under pullback.*
2. *An object $\mathbf{N}$ for which $[\delta, \iota] : \mathbf{N} + (\mathbf{N} * \mathbf{N}) \to (\mathbf{N} \times \mathbf{N})$ is an isomorphism. Here δ is the diagonal map and ι is the canonical monomorphism.*
3. *A vertical natural isomorphism $i : \Sigma_\mathbf{N}^* \to \Pi_\mathbf{N}^*$ such that the triangle below commutes.*

$$W_\mathbf{N}\Sigma_\mathbf{N}^* \xrightarrow{\quad W_\mathbf{N}(i) \quad} W_\mathbf{N}\Pi_\mathbf{N}^*$$
$$\underset{\eta}{\searrow} \quad \mathrm{Id} \quad \underset{\varepsilon}{\swarrow}$$

Here η is the unit of $\Sigma_\mathbf{N}^* \dashv W_\mathbf{N}$ and ε is the counit of $W_\mathbf{N} \dashv \Pi_\mathbf{N}^*$.
4. *For each object A and each monomorphism $m : B \rightarrowtail C$, the commuting square below is a pullback.*

$$
\begin{array}{ccc}
B * A & \xrightarrow{\pi_1} & B \\
{\scriptstyle m*A}\downarrow & \lrcorner & \downarrow{\scriptstyle m} \\
C * A & \xrightarrow{\pi_1} & C
\end{array}
$$

In the rest of this section we explain the structure in this proposition and how it can be integrated in the type theory. We argue informally towards the relation of the above structure to constructions in FM set theory.

As a model of the bunched type theory, $\mathbb{S}$ has both Σ^* and Π^* types. The fresh sums $\Sigma^*x{:}A.\,B$ may be constructed by taking certain equivalence classes of pairs $\langle M, N \rangle$ with $M : A$ and $N : B[M/x]$. Fresh products $\Pi^*x{:}A.\,B$ may be constructed as certain partial functions from A to B. This underpins the view of $\Sigma^*x{:}A.\,B$ and $\Pi^*x{:}A.\,B$ as non-standard sums and products. The difference from the standard sums and products is determined only by the names in A. For a type A that does not contain names, such as the natural numbers, the non-standard sums and products agree with the standard ones.

In Prop. 2.2 we ask for an object $\mathbf{N}$ of names with the property that any two names are either equal, i.e. a single element of $\mathbf{N}$, or they are fresh, i.e. an element of $\mathbf{N}*\mathbf{N}$. Thus, names have decidable equality, with two names being different precisely when they are fresh. This object of names plays the same role as the set of atoms $\mathbb{A}$ in FM set theory. We omit the rules for the type of names and its decidable equality, but remark that stable coproducts are used in the formulation of the term for deciding the equality.

Prop. 2.3 concerns the structure of the types $\Sigma^*n{:}\mathbf{N}.\,B$ and $\Pi^*n{:}\mathbf{N}.\,B$. Both types can be used for encoding of α-equivalence classes. An element $n.x$ of type $\Sigma^*n{:}\mathbf{N}.\,B$ is, by construction, an equivalence class and may be understood as the α-equivalence class of x with respect to n. This encoding of α-equivalence classes agrees with that of FM set theory. Indeed, for a closed type B, the construction of $\Sigma^*n{:}\mathbf{N}.\,B$ is (essentially) the same as that of the abstraction set $[\mathbb{A}]B$ of FM set theory. In the work on FM sets, it was also observed that α-equivalence classes may be constructed as partial functions from $\mathbf{N}$ to B. This construction is captured by the type $\Pi^*n{:}\mathbf{N}.\,B$. Therefore, $\Sigma^*n{:}\mathbf{N}.\,B$ and $\Pi^*n{:}\mathbf{N}.\,B$ are different encodings of the same α-equivalence classes, which means that the types should be isomorphic. This explains the isomorphism in Prop. 2.3. The isomorphism is useful for working with α-equivalence classes, as it allows us, for example, to form an α-equivalence class as a pair $n.x$ in $\Sigma^*n{:}\mathbf{N}.\,B$, and then to use it as a function in $\Pi^*n{:}\mathbf{N}.\,B$ to instantiate it at some other name $(n.M)@m$. We give further examples of this in Sec. 3.1, see also [7].

We integrate the isomorphism i in the type theory by means of hidden-name types $\mathrm{H}n.\,B$ which are isomorphic to both $\Sigma^*n{:}\mathbf{N}.\,B$ and $\Pi^*n{:}\mathbf{N}.\,B$. The rules for $\mathrm{H}n.\,B$ are those from both Σ^* and Π^*, giving H a self-dual nature.

$$(\text{H-Ty}) \ \frac{\Gamma * n{:}\mathbf{N} \vdash B \ \text{Type}}{\Gamma \vdash \mathrm{H}n.\,B \ \text{Type}}$$

$$(\text{H-I1}) \ \frac{\Gamma * n{:}\mathbf{N} \vdash M : B}{\Gamma \vdash \lambda^*_{\mathrm{H}}n.\,M : \mathrm{H}n.\,B} \qquad (\text{H-E1}) \ \frac{\Gamma \vdash M : \mathrm{H}n.\,B \qquad \Delta \vdash N : \mathbf{N}}{\Gamma * \Delta \vdash M@_{\mathrm{H}}N : B\,[N/n]}$$

$$(\text{H-I2}) \ \frac{n{:}\mathbf{N} \vdash B \ \text{Type} \qquad \Gamma \vdash M : \mathbf{N} \qquad \Gamma \vdash N : B[M/n]}{\Gamma \vdash \mathrm{bind}_{\mathrm{H}}(M, N) : (\mathrm{H}n.\,B)^{*(M:\mathbf{N})}}$$

$$(\text{H-E2}) \ \frac{\Gamma \vdash M : \mathrm{H}n.\,B \qquad (\Gamma * n{:}\mathbf{N}), y{:}B \vdash N : C^{*(n{:}\mathbf{N})}}{\Gamma \vdash \mathrm{let}\ M\ \mathrm{be}\ n._{\mathrm{H}}y\ \mathrm{in}\ N : C}$$

$$M._{\mathrm{H}}N =_{\mathrm{df}} (\mathrm{let}\ \mathrm{bind}_{\mathrm{H}}(M, N)\ \mathrm{be}\ u^{*m}\ \mathrm{in}\ u)$$

The type $\mathrm{H}n.\,B$ may be interpreted as either $\Sigma^*_{\mathbf{N}}B$ or $\Pi^*_{\mathbf{N}}B$. In the first case, the interpretation of $\lambda^*_{\mathrm{H}}n.\,M$ and $M@_{\mathrm{H}}N$ is given by $i^{-1}(\lambda^*n : \mathbf{N}.\,M)$ and $(i(M))@N$ respectively. With this interpretation, (β) and (η)-equations for H derive from those for Σ^* and Π^*. A further equation, which we omit, arises from the naturality of i.

$(\beta 1)$ $(\lambda_H^* n.\, M) @_H N = M\,[N/n]$

$(\eta 1)$ $\lambda_H^* n.\, (M @_H n) = M$ $n \notin \mathrm{FV}(M)$

$(\beta 2)$ let $\mathrm{bind}_H(M, N)$ be z^{*u} in (let z be $x.y$ in $R)^{*u} = R[M/x][N/y]$

$(\eta 2)$ let M be $x._H y$ in (let $\mathrm{bind}_H(x, y)$ be z^{*x} in $Q^{*x}) = Q[M/z]$

The commuting diagram in Prop. 2.3 provides two additional equations, which explain (to some extent) the interaction between the two roles of $Hn.\, B$ as $\Sigma^* n{:}\mathbf{N}.\, B$ and $\Pi^* n{:}\mathbf{N}.\, B$. The equations are formulated in context $\Gamma * n{:}\mathbf{N}$.

$(\beta 3)$ let $\mathrm{bind}_H(n, N)$ be x^{*m} in $x @_H m = N$

$(\eta 3)$ $\mathrm{bind}_H(n, \text{let } M \text{ be } x^{*m} \text{ in } x @_H m) = M$

From Prop. 2.4 it follows that hidden-name types are in propositions as types correspondence with the freshness quantifier V of Gabbay and Pitts. Consider the logic of subobjects of $\mathbb{S}$. From the fibred adjunction $\Sigma_A^* \dashv W_A \dashv \Pi_A^*$ we can derive a fibred adjunction $\exists_A^* \dashv W_A^S \dashv \forall_A^*$ on $\mathrm{Sub}(\mathbb{S})$, where W_A^S is the endofunctor on $\mathrm{Sub}(\mathbb{S})$ mapping a subobject $m : B \rightarrowtail C$ to $m * A : B * A \rightarrowtail C * A$ (note that $- * A$ preserves pullbacks, and so also monos). Prop. 2.4 then means that W_A^S is nothing but substitution along the projection $\pi_1 : (-) * A \to (-)$. Thus, the propositions as types analogues $\exists_A^*$ of Σ_A^* and $\forall_A^*$ of Π_A^* arise in terms of ordinary quantification along this projection. In the particular case where A is $\mathbf{N}$, it follows from $\Sigma_{\mathbf{N}}^* \cong \Pi_{\mathbf{N}}^*$ that $\exists_{\mathbf{N}}^* = \forall_{\mathbf{N}}^*$. We have thus shown that, along the projection $\pi_1 : (-) * \mathbf{N} \to \mathbf{N}$, the existential and the universal quantifier agree, and it may be seen [16] that this amounts the the freshness quantifier V, i.e. $\mathsf{V} = \exists_{\mathbf{N}}^* = \forall_{\mathbf{N}}^*$. As hidden-name types correspond to both $\exists_{\mathbf{N}}^*$ and $\forall_{\mathbf{N}}^*$, they thus correspond to V.

3.1   Examples and Applications

Unique Choice of Fresh Names.   For programming with names and binders, it is useful to have the ability to generate fresh names. In FreshML, one can write a term (new $n.\, M$), which is thought of as the unique value of M for an arbitrary freshly chosen name n. The existence of such a unique value can be guaranteed by a freshness condition on M. Using our notation, the introduction rule for new may be written as follows.

$$\frac{\Gamma * n{:}\mathbf{N} \vdash M : C^{*(n:\mathbf{N})}}{\Gamma \vdash \mathrm{new}\, n.\, M : C}$$

This is derivable in our system by means of the following derivation, in which we write 1 for the unit type with unique element $\diamond{:}1$.

$$\frac{\dfrac{\vdots}{\Gamma * n{:}\mathbf{N} \vdash \diamond : 1}}{\Gamma \vdash \lambda_H^* n.\, \diamond : Hn.\, 1} \quad \frac{\dfrac{\dfrac{\Gamma * n{:}\mathbf{N} \vdash M : C^{*(n:\mathbf{N})}}{(\Gamma * n{:}\mathbf{N}), u{:}1 \vdash M : C^{*(n:\mathbf{N})}}\;(\text{Weak})}{\Gamma, z{:}Hn.\, 1 \vdash \text{let } z \text{ be } n.u \text{ in } M : C}\;(\text{Weak}),(\text{H-E2})}{\Gamma \vdash \text{let } (\lambda_H^* n.\, \diamond) \text{ be } n.u \text{ in } M : C}\;(\text{Subst})$$

We use (new $n.\, M$) as an abbreviation for the term in the conclusion of this derivation.

In this way, we are using the fact that $Hn.\, 1$ is inhabited to obtain a supply of fresh names. This generalises the situation in FM set theory or the Theory of Contexts, where one uses the truth of the proposition ($\mathsf{V}\, n.\, \top$) as a supply of fresh names for reasoning.

Abstract Syntax with Variable Binding. A key application of names and binding is for working with abstract syntax involving variable binders. We encode abstract syntax as an inductive type, using hidden-name types $Hn.\,A$ for object-level binders. The duality of H offers two styles of working with abstract syntax: viewing H as Π^* allows us to work in the style of weak Higher Order Abstract Syntax (wHOAS) [3, 11], and viewing H as Σ^* supports the style of FM set theory. In the rest of this section, we give examples illustrating the advantages of both views as well as showing the benefits of mixing the two styles.

We take the syntax of the untyped λ-calculus as an example, encoding it as an inductive type Lam with three constructors: $\mathsf{var} : \mathbf{N} \to \mathsf{Lam}$, $\mathsf{app} : (\mathsf{Lam} \times \mathsf{Lam}) \to \mathsf{Lam}$ and $\mathsf{lam} : (Hn.\,\mathsf{Lam}) \to \mathsf{Lam}$. For example, the term $\lambda x.\,\lambda y.\,(x\ y)$ can be encoded as $\mathsf{lam}(\lambda_H^* x.\,\mathsf{lam}(\lambda_H^* y.\,\mathsf{app}(\mathsf{var}(x), \mathsf{var}(y))))$. In a context with two different names x and y, it may also be encoded as $\mathsf{lam}(x._H\mathsf{lam}(y._H\mathsf{app}(\mathsf{var}(x), \mathsf{var}(y))))$.

Semantically, Lam corresponds to an initial algebra, which lets us define functions by structural recursion. The following recursion principle follows from the initial algebra when $Hn.\,\mathsf{Lam}$ is viewed as $\Pi^* n{:}\mathbf{N}.\,\mathsf{Lam}$.

$$
\begin{array}{c}
x{:}\mathsf{Lam} \vdash A(x)\ \text{Type} \\
\Gamma \vdash f : \Pi n{:}\mathbf{N}.\,A(\mathsf{var}(n)) \\
\Gamma \vdash g : \Pi M, N{:}\mathsf{Lam}.\,A(M) \to A(N) \to A(\mathsf{app}(M, N)) \\
\Gamma \vdash h : \Pi M{:}(Hn.\,\mathsf{Lam}).\,(Hn.\,A(M@_H n)) \to A(\mathsf{lam}(M)) \\
\hline
\Gamma \vdash \mathsf{rec}(f, g, h) : \Pi M{:}\mathsf{Lam}.\,A(M)
\end{array}
$$

with equations (in which we write rec for $\mathsf{rec}(f, g, h)$)

$$
\mathsf{rec\ var}(n) = f\ n
$$
$$
\mathsf{rec\ app}(M, N) = g\ M\ N\ (\mathsf{rec}\ M)\ (\mathsf{rec}\ N)
$$
$$
\mathsf{rec\ lam}(M) = h\ M\ (\lambda_H^* n.\,(\mathsf{rec}\ (M@_H n))).
$$

For a closed type A, this structural recursion produces a unique function $\mathsf{Lam} \to A$ for given functions $f : \mathbf{N} \to A$, $g : \mathsf{Lam} \to \mathsf{Lam} \to A \to A \to A$ and $h : (Hn.\,\mathsf{Lam}) \to (Hn.\,A) \to A$. In FM set theory one has an apparently different recursion principle, where instead of h one is essentially given a function $k : Hn.\,\mathsf{Lam} \to A \to A^{*(n{:}\mathbf{N})}$. The above recursion principle is also applicable in this case, since from k we can define $h =_{\mathrm{df}} \lambda u : (Hn.\,\mathsf{Lam}).\,\lambda v : (Hn.\,A).\,\mathsf{new}\ n.\,((k@_H n)\ (u@_H n)\ (v@_H n))$. In this way, we get a second recursion operator $\mathsf{rec}'(f, g, k)$ with the following equation for the lam-case: $(\mathsf{rec}'(f, g, k)\ \mathsf{lam}(M)) = \mathsf{new}\ n.\,((k@_H n)\ (M@_H n)\ (\mathsf{rec}'(f, g, k)\ (M@_H n)))$.

As a first example of a recursively defined function, we define capture-avoiding substitution in the style of wHOAS and compare the definition to an FM-style encoding. Given $m{:}\mathbf{N}$ and $R{:}\mathsf{Lam}$, we can use rec to define $\mathsf{subst} : \mathsf{Lam} \to \mathsf{Lam}$ satisfying

$$
\mathsf{subst}(\mathsf{var}(n)) = \mathsf{ifeq}\ \langle m, n \rangle\ \text{then}\ n.\,R\ \text{else}\ n.\,\mathsf{var}(n)
$$
$$
\mathsf{subst}(\mathsf{app}(M, N)) = \mathsf{app}(\mathsf{subst}(M), \mathsf{subst}(N))
$$
$$
\mathsf{subst}(\mathsf{lam}(M)) = \mathsf{lam}(\lambda_H^* n.\,\mathsf{subst}(M@_H n)).
$$

This definition uses only the view of H as Π^* and is similar in spirit to wHOAS definitions. We can also define substitution in FM-style using rec'. For the lam-case, we then have $\mathsf{subst}(\mathsf{lam}(M)) = \mathsf{new}\ n.\ (\mathsf{let}\ \mathsf{bind}_H(n, \mathsf{subst}(M@_H n))\ \mathsf{be}\ w^{*n}\ \mathsf{in}\ (\mathsf{lam}(w))^{*n})$. However, this definition is more complex than the first one, since it involves a unique choice of fresh names via new. In the first definition we could do without the choice of a fresh name by using λ_H^* to 'rebind' the fresh name n.

As a second example, we define the function computing the free variables of a term. This example makes essential use of the view of H as Σ^*. We assume an inductive type L of lists of names, together with suitably defined functions $\mathsf{singleton} : \mathbf{N} \to \mathsf{L}$, $\mathsf{concat} : \mathsf{L} \to \mathsf{L} \to \mathsf{L}$, and $\mathsf{remove} : \Pi n{:}\mathbf{N}.\ (\mathsf{L} \to \mathsf{L}^{*(n:\mathbf{N})})$. Using rec, we can define $\mathsf{fv} : \mathsf{Lam} \to \mathsf{L}$ to satisfy the equations

$$\mathsf{fv}(\mathsf{var}(n)) = \mathsf{singleton}(n)$$
$$\mathsf{fv}(\mathsf{app}(M, N)) = \mathsf{concat}(\mathsf{fv}(M), \mathsf{fv}(N))$$
$$\mathsf{fv}(\mathsf{lam}(M)) = \mathsf{let}\ (\lambda_H^* n.\ \mathsf{fv}(M@n))\ \mathsf{be}\ n._H y\ \mathsf{in}\ (\mathsf{remove}\ n\ y)$$

This example demonstrates how let-terms can be used for 'pattern matching' elements of $Hn.\ A$. A similar pattern matching appears in FreshML. Moreover, the example shows that it is useful to mix the views of H as Π^* and Σ^*.

Note that, in the equation for lam, the subterm (remove n y) has type $\mathsf{L}^{*(n:\mathbf{N})}$, and that this freshness information is necessary for the let to be typeable. Intuitively, this is because the choice of representative $n.y$ must not affect the computation. Dependency in the type of remove is therefore essential for the pattern matching in the definition of fv. Without dependency we could write remove with type $\mathbf{N} \to \mathsf{L} \to \mathsf{L}$, but then fv as above would not be typeable. Indeed, this problem arises in FreshML, where fv cannot be defined using a remove function of this type (Nevertheless, fv can be defined in FreshML).

Again, we can use rec$'$ to give an alternative definition of fv so that it satisfies the equation $\mathsf{fv}(\mathsf{lam}(M)) = \mathsf{new}\ n.\ (\mathsf{remove}\ n\ (\mathsf{fv}(M@n)))$. Note that, by means of new, this encoding also uses the view of H as Σ^*, and this is in fact essential. The Theory of Contexts, for example, axiomatises a 'is not free in'-predicate rather than defining fv.

4 Discussion and Further Work

We have introduced a bunched dependent type theory that integrates FM concepts for working with names and binding.

One decision in the design of the bunches was to allow dependency for additive context extension but to forbid any dependency for multiplicative context extension. There are other possibilities for combining bunches and dependency. Pym [20, §15.15], for example, outlines a bunched dependent calculus allowing more dependency. The problem with using this for names and binding, which has lead us to the current design, is that it would require to generalise the monoidal product $*$ to a monoidal product on the slices of $\mathbb{S}$, and there seems to be no sensible way of doing this.

We stress that, although the examples in this paper concentrate on programming, reasoning with names and binding can also be accommodated in the type theory. Indeed, it is possible to define a higher-order logic over the dependent type theory [12, §11]. In addition to the usual logical connectives, this logic also features the multiplicative quantifiers $\exists^*$ and $\forall^*$, similar to $\forall_{\text{new}}$ and $\exists_{\text{new}}$ from **BI** [20], as well as the freshness quantifier N. This higher-order logic supports reasoning with names similar to the Theory of Contexts. For example, the Theory of Contexts has an 'extensionality' axiom, which may be expressed as $\Gamma \mid \exists^* n : \mathbf{N}. (M @_{\mathrm{H}} n =_A N @_{\mathrm{H}} n) \vdash (M =_{\mathrm{H}n. A} N)$, where M and N have type $\mathrm{H}n. A$ and $=_A$ denotes Leibniz equality. Making essential use of the equation $(\eta 3)$, this sequent is derivable in the logic. In another direction, one may also ask how the logic relates to Nominal Logic [18]. For this it is necessary to consider swapping, an essential ingredient of Nominal Logic that is absent from the type theory. We briefly discuss the possibilities of adding swapping below.

Another possibility for reasoning is to use dependent types to encode propositions as types. Alongside the usual encodings of $\forall$ as Π and $\exists$ as Σ, one can encode $\forall^*$ as Π^*, $\exists^*$ as Σ^*, and N as H. Although such an encoding is possible, the use of $\exists^*$ is very restricted, because the rules for Σ^* use types of the form $\varphi^{*(n:\mathbf{N})}$, and, at least in this paper, we allow such types only when φ is closed. Considering a higher-order logic is a way of side-stepping this problem, since, because of Prop. 2.4, we have an equivalence of $\varphi^{*(n:\mathbf{N})}$ and φ, so that freefrom types can be avoided altogether in the logic.

Although we have based our type theory on freshness rather than swapping, we nevertheless think that swapping can be useful in type theory. Swapping can be added to the type theory as a special kind of explicit substitution, as is done in [1, 23]. One application of swapping is to make available more information about the isomorphism $\Sigma_{\mathbf{N}}^* \cong \Pi_{\mathbf{N}}^*$ than is given by the commuting triangle in Prop. 2.3. The triangle only explains the instantiation of $n._{\mathrm{H}}x$ at n. With swapping, we can explain the instantiation of $n._{\mathrm{H}}x$ at names other than n by adding the equation $(n._{\mathrm{H}}x) @_{\mathrm{H}} m = (m\ n) \cdot M$. Furthermore, with swapping, we should get a logic close to Nominal Logic; see also [16].

Regarding the categorical semantics of the type theory, it is natural to ask how it compares to other categorical approaches to names and binding. Besides the Schanuel topos, two other categories used frequently [9, 4, 5, ...] for names and binding are $\mathrm{Set}^{\mathbb{V}}$, where $\mathbb{V}$ is the category of finite cardinals and all functions between them, and $\mathrm{Set}^{\mathbb{I}}$, where $\mathbb{I}$ is the category of finite cardinals and injections. However, neither category has all of the structure of Prop. 2. In $\mathrm{Set}^{\mathbb{V}}$ names do not have decidable equality, whereas $\mathrm{Set}^{\mathbb{I}}$ does not have a freshness quantifier and not all the canonical maps $A * B \rightarrow A \times B$ are monomorphic. In this light, Prop. 2 should be viewed as identifying the categorical structure underlying the work with names and binding, while for particular applications it may well be sufficient to have only some of this structure. Another example of such a substructure is Menni's axiomatisation of binders [16]. Nevertheless, there are categories other than the Schanuel topos having the structure of Prop. 2. One such category is a variation of the Schanuel topos in which the elements are allowed to contain countably many names rather than just finitely many, see [18, p.13]. There is also a realisability category having almost all of the structure of Prop. 2, the only restriction being that the type $\Sigma^* x : A. B$ can only be formed when A belongs to a certain restricted class of types (which includes all types with decidable equality). Moreover, this category models an impredicative universe, so that it should provide the basis for a bunched calculus of constructions.

There are many directions for further work. First, an immediate point requiring further work is the restriction that $B^{*(M:A)}$ can only be formed for closed B. Second, the proof theory of the bunched type theory needs further work. Also, variants such as a non-affine version of the type theory should be possible. Finally, algorithmic questions such as the decidability of type-checking should be considered.

Acknowledgements. We would like to thank Alex Simpson and John Power for interesting discussions on this work.

References

1. L. Cardelli, P. Gardner, and G. Ghelli. Manipulating trees with hidden labels. In *Proceedings of FOSSACS'03*, volume 2620 of *LNCS*. Springer, 2003.
2. L. Cardelli and A. Gordon. Logical properties of name restriction. In *Proceedings of TLCA'01*, volume 2044 of *LNCS*. Springer, 2001.
3. J. Despeyroux, A. Felty, and A. Hirschowitz. Higher-order abstract syntax in Coq. In *Proceedings of TLCA'95*, 1995.
4. M. Fiore, G. Plotkin, and D. Turi. Abstract syntax and variable binding. In *Proceedings of LICS99*, 1999.
5. M. Fiore and D. Turi. Semantics of name and value passing. In *Proceedings of LICS01*, 2001.
6. M. Gabbay. FM-HOL, a higher-order theory of names. In *Workshop on Thirty Five years of Automath*, 2002.
7. M. J. Gabbay and A. M. Pitts. A new approach to abstract syntax with variable binding. *Formal Aspects of Computing*, 13:341–363, 2002.
8. M. Hofmann. On the interpretation of type theory in locally cartesian closed categories. In *Proceedings of CSL94*, volume 933 of *LNCS*. Springer, 1994.
9. M. Hofmann. Semantical analysis of higher-order abstract syntax. In *Proceedings of LICS99*, 1999.
10. M. Hofmann. Safe recursion with higher types and BCK-algebra. *Annals of Pure and Applied Logic*, 104(1–3):113–166, 2000.
11. F. Honsell, M. Miculan, and I. Scagnetto. An axiomatic approach to metareasoning about nominal algebras in HOAS. In *Proceedings of ICALP01*, 2001.
12. B. Jacobs. *Categorical Logic and Type Theory*. Elsevier Science, 1999.
13. P.T. Johnstone. *Sketches of an Elephant: A Topos Theory Compendium*. Oxford University Press, 2002.
14. P. Lietz. A fibrational theory of geometric morphisms. Master's thesis, TU Darmstadt, May 1998.
15. S. MacLane and I. Moerdijk. *Sheaves in Geometry and Logic: A First Introduction to Topos Theory*. Springer-Verlag, 1992.
16. M. Menni. About И-quantifiers. *Applied Categorical Structures*, 11(5):421–445, 2003.
17. P. O'Hearn. On bunched typing. *Journal of Functional Programming*, 13(4):747–796, 2003.
18. A. M. Pitts. Nominal logic, a first order theory of names and binding. *Information and Computation*, 186:165–193, 2003.
19. A. M. Pitts and M. J. Gabbay. A metalanguage for programming with bound names modulo renaming. In *Proceedings of MPC2000*, volume 1837 of *LNCS*. Springer, 2000.
20. D. Pym. *The Semantics and Proof Theory of the Logic of Bunched Implications*. Kluwer Academic Publishers, 1999.
21. R.A.G. Seely. Locally cartesian closed categories and type theory. In *Math. Proc. Cambridge Philos. Soc.*, volume 95, pages 33–48, 1984.
22. P. Taylor. *Practical Foundations of Mathematics*. Cambridge University Press, 1999.
23. C. Urban, A. M. Pitts, and M. J. Gabbay. Nominal unification. In *Proccedings of CSL'03*, volume 2803 of *LNCS*. Springer, 2003.

Towards Mechanized Program Verification with Separation Logic[*]

Tjark Weber

Institut für Informatik, Technische Universität München
Boltzmannstr. 3, D-85748 Garching b. München, Germany
webertj@in.tum.de

Abstract. Using separation logic, this paper presents three Hoare logics (corresponding to different notions of correctness) for the simple While language extended with commands for heap access and modification. Properties of separating conjunction and separating implication are mechanically verified and used to prove soundness and relative completeness of all three Hoare logics. The whole development, including a formal proof of the Frame Rule, is carried out in the theorem prover Isabelle/HOL.

Keywords. Separation Logic, Formal Program Verification, Interactive Theorem Proving

1 Introduction

Since C. A. R. Hoare's seminal work in 1969 [9], extensions of his logic have been developed for a multitude of language constructs [1, 2], including recursive procedures, nondeterminism, and even object-oriented languages. Extending Hoare logic to pointer programs however is not without difficulties. Recently separation logic was proposed by O'Hearn, Reynolds et al. [15, 19, 16] to overcome the local reasoning problem that is raised by the treatment of record components as arrays [6, 5].

Machine support is indispensable for formal program verification. Manual proofs are error-prone, and the verification of medium-sized programs has become feasible only because systems like SVC [3] can automatically discharge many proof obligations. Separation logic, although its usability has been demonstrated in several case studies [18, 4], currently lacks such support. In this paper we show how separation logic can be embedded into the theorem prover Isabelle/HOL [14]. We thereby lay the foundations for the use of separation logic in a semi-automatic verification tool. Our work is based on a previous formalization of a simple imperative language [12] which however did not consider pointers or separation logic. The current focus is on fundamental semantic properties of the resulting Hoare logics.

This paper is organized as follows. In Section 2 we define the programming language, together with its operational and denotational semantics. Section 3

---

[*] Research supported by *Graduiertenkolleg Logik in der Informatik* (PhD Program Logic in Computer Science) of the Deutsche Forschungsgemeinschaft (DFG).

J. Marcinkowski and A. Tarlecki (Eds.): CSL 2004, LNCS 3210, pp. 250–264, 2004.

introduces separation logic. In Section 4 we present three Hoare logics for our language, all of which are proved to be sound and relative complete. Also the Frame Rule is adressed, and its soundness is proved for one of the Hoare logics. We discuss the mechanical verification of a simple pointer algorithm, in-place list reversal, in Section 5.

2   The Language

2.1   Semantic Domains

We use an unspecified type *var* of variables. Addresses are elements of a numerical type, namely naturals (*nat*), to permit address arithmetic. For simplicity, the same type is used for values. Thereby the value of a variable can immediately be used as an address, with no need for a conversion function (cf. [16]).

Stores map variables to values. Heaps are modelled as partial functions from addresses to values. Other possibilities would be to define heaps as subsets of *addr* × *val* (with functionality constraints), or as (*addr* × *val*) *list* (again with functionality constraints, and modulo order). However, our current definition is much easier to state and work with in Isabelle/HOL since it can make use of readily available function types and does not require subtyping. On the other hand it also permits infinite heaps. This seemingly minor difference will become important again in Section 4.4, when we consider the Frame Rule.

A program state is either a pair consisting of a store and a heap, or *None*. The latter value will be used in the semantics of the language to indicate that a memory error occurred during program execution. Arithmetic and boolean expressions are only modelled semantically: they are just functions on stores (and hence independent of the heap).

Most of Isabelle's syntax used in this paper is close to standard mathematical notation and should not require further explanation. Both $\implies$ and $\longrightarrow$ mean implication. $[\![\, P_1\, ; \, \ldots \, ; \, P_n \,]\!] \implies Q$ is an abbreviation for $P_1 \implies \ldots \implies P_n \implies Q$. We use $'a \Rightarrow 'b$ for the type of total functions from $'a$ to $'b$. Likewise, infix $\rightharpoonup$ is used to denote the type of partial functions. Other type constructors, e.g. *list*, are written postfix. Thus the abovementioned semantic domains can be formalized as follows:

types $addr\ =\ nat$
     $val\ \ =\ nat$
     $store\ =\ var \Rightarrow val$
     $heap\ \ =\ addr \rightharpoonup val$
     $state\ =\ (store \times heap)\ option$
     $aexp\ =\ store \Rightarrow val$
     $bexp\ =\ store \Rightarrow bool$

2.2   Syntax

We consider an extension of the simple While language [9, 12] with new commands for memory allocation (list, alloc), heap lookup, heap mutation, and memory deallocation (dispose).

Both list and alloc allocate memory on the heap. list can only be used when the number of addresses to be allocated is known beforehand, i.e. for allocation of fixed-size records. The list command takes a list of arithmetic expressions as its second argument. The number of consecutive addresses to be allocated is given by the length of the list; the allocated memory is then initialized with the values of the expressions in the list. alloc on the other hand is meant for dynamic allocation of arrays. Its second argument is a single arithmetic expression that specifies the number of consecutive addresses to be allocated. The allocated memory is initialized with arbitrary values.

The lookup command assigns the value of an (allocated) address to a variable, the heap mutation command modifies the value of the heap at a given address, and dispose finally deallocates a single address. The precise operational semantics is given in Section 2.4.

2.3   Basic Operations on Heaps

Before we can define the semantics of our language, we need to introduce some basic operations on heaps. We define four functions to retrieve the value of a heap at a specific address, remove an address from the domain of a heap, test whether a set of addresses is free in a heap, and update a set of consecutive addresses in a heap with specific values. To some extent these functions allow us to abstract from our particular implementation of heaps as partial functions.

$heap\text{-}lookup$ :: $heap \Rightarrow addr \Rightarrow val$
$heap\text{-}lookup$ h a $\equiv$ the $(h\ a)$
$heap\text{-}remove$ :: $heap \Rightarrow addr \Rightarrow heap$
$heap\text{-}remove$ h a $\equiv$ $h(a{:=}None)$
$heap\text{-}isfree$ :: $heap \Rightarrow addr \Rightarrow nat \Rightarrow bool$
$heap\text{-}isfree$ h a n $\equiv$ $set\ [a..a{+}n(] \cap dom\ h = \{\}$
$heap\text{-}update$ :: $heap \Rightarrow addr \Rightarrow (val\ list) \Rightarrow heap$
$heap\text{-}update$ h a vs $\equiv$ $h([a..a{+}length\ vs(][{\mapsto}]vs)$

Later we will also need notions of disjointness and union for heaps in order to define separating conjunction and separating implication. We say two heaps (or more generally, two partial functions) are *disjoint*, $\bowtie$, iff their domains are disjoint.

$f \bowtie g \equiv dom\ f \cap dom\ g = \{\}$

The *union* of heaps, $++$, is defined as one would expect, with the second heap having precedence over the first.

$f{+}{+}g \equiv \lambda x.\ case\ g\ x\ of\ None \Rightarrow f\ x\ |\ Some\ y \Rightarrow Some\ y$

We will only take the union of disjoint heaps however, and for those, $++$ is commutative:

Lemma. $f \bowtie g \Longrightarrow f ++ g = g ++ f$

2.4   Operational Semantics

The operational semantics of our language is defined via a (big-step) evaluation relation $\longrightarrow_c$. We write $\langle c,s \rangle \longrightarrow_c t$ for *execution of c, started in state s, may terminate in state t*. This evaluation relation is defined inductively.

$$\langle c, None \rangle \longrightarrow_c None$$
$$\langle skip, Some\ (s,h) \rangle \longrightarrow_c Some\ (s,h)$$
$$\langle x := a, Some\ (s,h) \rangle \longrightarrow_c Some\ (s[x \mapsto a\ s], h)$$
$$\langle c0, s \rangle \longrightarrow_c s'' \implies \langle c1, s'' \rangle \longrightarrow_c s' \implies \langle c0;\ c1,\ s \rangle \longrightarrow_c s'$$
$$b\ s \implies \langle c0, Some\ (s,h) \rangle \longrightarrow_c s' \implies \langle if\ b\ then\ c0\ else\ c1,\ Some\ (s,h) \rangle \longrightarrow_c s'$$
$$\neg b\ s \implies \langle c1, Some\ (s,h) \rangle \longrightarrow_c s' \implies \langle if\ b\ then\ c0\ else\ c1,\ Some\ (s,h) \rangle \longrightarrow_c s'$$
$$b\ s \implies \langle c, Some\ (s,h) \rangle \longrightarrow_c s'' \implies \langle while\ b\ do\ c,\ s'' \rangle \longrightarrow_c s'$$
$$\implies \langle while\ b\ do\ c,\ Some\ (s,h) \rangle \longrightarrow_c s'$$
$$\neg b\ s \implies \langle while\ b\ do\ c, Some\ (s,h) \rangle \longrightarrow_c Some\ (s,h)$$
$$[\![\ heap\text{-}isfree\ h\ a\ (length\ as);\ vs = map\ (\lambda e.\ e\ s)\ as\ ]\!]$$
$$\implies \langle x := list\ as,\ Some\ (s,h) \rangle \longrightarrow_c Some\ (s[x \mapsto a],\ heap\text{-}update\ h\ a\ vs)$$
$$(\forall\ a.\ \neg\ heap\text{-}isfree\ h\ a\ (length\ as)) \implies \langle x := list\ as,\ Some\ (s,h) \rangle \longrightarrow_c None$$
$$(heap\text{-}isfree\ h\ a\ (n\ s) \wedge (length\ vs = n\ s))$$
$$\implies \langle x := alloc\ n,\ Some\ (s,h) \rangle \longrightarrow_c Some\ (s[x \mapsto a],\ heap\text{-}update\ h\ a\ vs)$$
$$(\forall\ a.\ \neg\ heap\text{-}isfree\ h\ a\ (n\ s)) \implies \langle x := alloc\ n,\ Some\ (s,h) \rangle \longrightarrow_c None$$
$$a\ s \in dom\ h \implies \langle x := @a, Some\ (s,h) \rangle \longrightarrow_c Some\ (s[x \mapsto heap\text{-}lookup\ h\ (a\ s)], h)$$
$$a\ s \notin dom\ h \implies \langle x := @a, Some\ (s,h) \rangle \longrightarrow_c None$$
$$a\ s \in dom\ h \implies \langle @a := v, Some\ (s,h) \rangle \longrightarrow_c Some\ (s, heap\text{-}update\ h\ (a\ s)\ [v\ s])$$
$$a\ s \notin dom\ h \implies \langle @a := v, Some\ (s,h) \rangle \longrightarrow_c None$$
$$a\ s \in dom\ h \implies \langle dispose\ a, Some\ (s,h) \rangle \longrightarrow_c Some\ (s, heap\text{-}remove\ h\ (a\ s))$$
$$a\ s \notin dom\ h \implies \langle dispose\ a, Some\ (s,h) \rangle \longrightarrow_c None$$

The rules for skip, assignment, composition, if, and while are standard, and only shown for completeness. The rules for the pointer commands come in pairs, with one rule leading to a valid successor state, the other one to the error state *None*. Which rule can be applied depends on the current heap. Allocating memory in a heap that does not have enough free addresses will result in an error, as will the attempt to access, modify, or deallocate free addresses.

With the exception of the first rule, these rules are all syntax directed (i.e. applicable only to a specific command). The first rule is needed to ensure that programs "don't get stuck" when an error occurred. For the same reason it is important that we do not restrict the rule for sequential composition to valid states.

Nondeterminism is introduced by the rules for list and alloc. Both commands choose an arbitrary sequence of (consecutive) free addresses for the newly allocated memory. Furthermore, alloc initializes this memory with arbitrary values.

2.5   Denotational Semantics

In addition to the operational semantics, we also define the denotational semantics of commands. We will show that both semantics are equivalent, thus we could (in principle) do without a denotational semantics. However, we found that the denotational semantics, and in particular its fixed point characterization of while, is often easier to work with than the operational semantics. It enables us to prove semantic properties by induction on commands, rather than by induction on the evaluation relation. The denotational semantics of a command is given by a set of pairs of states.

types *com-den* $=$ *(state* $\times$ *state) set*

The following function Γ is used to define the semantics of the while command as a least fixed point. The O operator denotes relational composition.

$\Gamma :: bexp \Rightarrow com\text{-}den \Rightarrow (com\text{-}den \Rightarrow com\text{-}den)$
$\Gamma \; b \; cd \equiv (\lambda\varphi.$
$\qquad \{ \; (Some(s,h),t) \mid s \; h \; t. \; (Some(s,h),t) \in (\varphi \; O \; cd) \wedge b \; s \; \} \; \cup$
$\qquad \{ \; (Some(s,h),Some(s,h)) \mid s \; h. \; \neg b \; s \; \} \; \cup$
$\qquad \{ \; (None,None) \; \})$

The meaning function C, which maps each command to its denotational semantics, is now defined by primitive recursion.

$C \; \text{skip} = Id$
$C \; (x :== a) = \{ \; (Some(s,h),Some(s[x{\mapsto}a \; s],h)) \mid s \; h. \; True \; \} \; \cup$
$\qquad \{ \; (None,None) \; \}$
$C \; (c0;c1) = C(c1) \; O \; C(c0)$
$C \; (\text{if } b \text{ then } c1 \text{ else } c2) = \{ \; (Some(s,h),t) \mid s \; h \; t. \; (Some(s,h),t) \in C \; c1 \wedge b \; s \; \} \; \cup$
$\qquad \{ \; (Some(s,h),t) \mid s \; h \; t. \; (Some(s,h),t) \in C \; c2 \wedge \neg b \; s \; \} \; \cup$
$\qquad \{ \; (None,None) \; \}$
$C \; (\text{while } b \text{ do } c) = lfp \; (\Gamma \; b \; (C \; c))$
$C \; (x :== \text{list } as) = \{ \; (Some(s,h),Some(s[x{\mapsto}a],heap\text{-}update \; h \; a \; (map \; (\lambda e. \; e \; s) \; as))) $
$\qquad\qquad \mid s \; h \; a. \; heap\text{-}isfree \; h \; a \; (length \; as) \; \} \; \cup$
$\qquad \{ \; (Some(s,h),None) \mid s \; h. \; \forall \; a. \; \neg \; heap\text{-}isfree \; h \; a \; (length \; as) \; \} \; \cup$
$\qquad \{ \; (None,None) \; \}$
$C \; (x :== \text{alloc } n) = \{ \; (Some(s,h),Some(s[x{\mapsto}a],heap\text{-}update \; h \; a \; vs)) $
$\qquad\qquad \mid s \; h \; a \; vs. \; heap\text{-}isfree \; h \; a \; (n \; s) \wedge (length \; vs = n \; s) \; \} \; \cup$
$\qquad \{ \; (Some(s,h),None) \mid s \; h. \; \forall \; a. \; \neg \; heap\text{-}isfree \; h \; a \; (n \; s) \; \} \; \cup$
$\qquad \{ \; (None,None) \; \}$
$C \; (x :== @a) = \{ \; (Some(s,h),Some(s[x{\mapsto}heap\text{-}lookup \; h \; (a \; s)],h)) $
$\qquad\qquad \mid s \; h. \; a \; s \in dom \; h \; \} \; \cup$
$\qquad \{ \; (Some(s,h),None) \mid s \; h. \; a \; s \notin dom \; h \; \} \; \cup$
$\qquad \{ \; (None,None) \; \}$
$C \; (@a :== v) = \{ \; (Some(s,h),Some(s,heap\text{-}update \; h \; (a \; s) \; [v \; s])) $
$\qquad\qquad \mid s \; h. \; a \; s \in dom \; h \; \} \; \cup$
$\qquad \{ \; (Some(s,h),None) \mid s \; h. \; a \; s \notin dom \; h \; \} \; \cup$
$\qquad \{ \; (None,None) \; \}$
$C \; (\text{dispose } a) = \{ \; (Some(s,h),Some(s,heap\text{-}remove \; h \; (a \; s))) \mid s \; h. \; a \; s \in dom \; h \; \} \; \cup$
$\qquad \{ \; (Some(s,h),None) \mid s \; h. \; a \; s \notin dom \; h \; \} \; \cup$
$\qquad \{ \; (None,None) \; \}$

By induction on $\longrightarrow_c$, one can show that $\langle c,s \rangle \longrightarrow_c t$ implies $(s, t) \in C \; c$. The other direction, i.e. $(s, t) \in C \; c \Longrightarrow \langle c,s \rangle \longrightarrow_c t$, is shown by induction on c. For both directions, only the while case is not automatic (but still fairly simple). Taking these two results together, we obtain equivalence of denotational and operational semantics:

Theorem. $(s,t) \in C(c) = (\langle c,s \rangle \longrightarrow_c t)$

We will freely use this result in the following proofs whenever it is more convenient to reason using a particular semantics.

3   Assertions of Separation Logic

We only model the semantics of assertions, not their syntax. Assertions are predicates on stores and heaps:

types $assn = store \Rightarrow heap \Rightarrow bool$

This semantic approach (or *shallow embedding*) entails that any HOL term of the correct type can be used as an assertion, not just formulae of separation logic. If we had modelled assertions syntactically, we would have had to redefine most of HOL's logical connectives (including classical conjunction, implication, and first-order quantification), and the explicit definition of a formula's semantics would have introduced another layer of abstraction between separation logic and the lemmata and proof automation available in HOL. Our current definition on the other hand allows us to consider separation logic as an extension of higher-order logic, thereby giving us the features of HOL (almost) for free. The main drawback for our purposes is perhaps an esthetic one: when mixing classical and separating connectives, we have to use λ-abstractions to make their types compatible (cf. Section 3.1). A more detailed discussion of the respective strengths and weaknesses of shallow vs. *deep* embeddings is forthcoming [17].

Let us now introduce some abbreviations. *emp* asserts that the heap is empty (i.e. that no address is allocated), and $a \mapsto v$ is true of a heap iff a is the only allocated address, and it points to the value v.

$emp\ h \equiv dom\ h = \{\}$
$(a \mapsto v)\ h \equiv dom\ h = \{a\} \wedge heap\text{-}lookup\ h\ a = v$

Separation logic has two special connectives, separating conjunction ($\wedge_*$) and separating implication ($-_*$). $P \wedge_* Q$ states that the heap can be split into disjoint parts satisfying P and Q, respectively. $P -_* Q$ is true of a heap h iff Q holds for every extension of h with a disjoint part that satisfies P. These connectives are defined using quantification over heaps. The definitional approach allows us to *prove* their properties, rather than to introduce them as new axioms.

$sep\text{-}conj :: (heap \Rightarrow bool) \Rightarrow (heap \Rightarrow bool) \Rightarrow heap \Rightarrow bool$ (**infixl** $\wedge_*$)
$(P \wedge_* Q)\ h \equiv \exists h'\ h''.\ (h' \bowtie h'') \wedge (h' ++ h'' = h) \wedge P\ h' \wedge Q\ h''$

$sep\text{-}imp :: (heap \Rightarrow bool) \Rightarrow (heap \Rightarrow bool) \Rightarrow heap \Rightarrow bool$ (**infixr** $-_*$)
$(P -_* Q)\ h \equiv \forall h'.\ ((h' \bowtie h) \wedge P\ h') \longrightarrow Q\ (h ++ h')$

Although assertions of separation logic may depend on the store, they usually do so only in a completely homomorphic fashion (cf. [16]). Therefore this dependency can easily be eliminated from compound formulae, and it is sufficient to define separating conjunction and implication for predicates of type *heap* ⇒ *bool*. Further assertions denote that a heap contains exactly one allocated address a (written $a \mapsto -$), and that an address a points to a value v, where other addresses in the heap may be allocated as well ($a \hookrightarrow v$). Using address arithmetic (*Suc* is the successor function on naturals), we extend these notions to lists of values.

$$(a \mapsto -) \ h \equiv \exists v. \ (a \mapsto v) \ h$$
$$(a \hookrightarrow v) \equiv (a \mapsto v) \ \wedge * \ true$$
$$(a[\mapsto][]) \qquad = emp$$
$$(a[\mapsto](v \# vs)) = ((a \mapsto v) \ \wedge * \ ((Suc \ a)[\mapsto]vs))$$
$$(a[\hookrightarrow][]) \qquad = true$$
$$(a[\hookrightarrow](v \# vs)) = ((a \hookrightarrow v) \ \wedge * \ ((Suc \ a)[\hookrightarrow]vs))$$

3.1   Properties of Separating Conjunction and Separating Implication

We can relatively easily prove associativity and commutativity of $\wedge *$, identity of *emp* under $\wedge *$, and various distributive and semidistributive laws. Most of the proofs are automatic; sometimes however we need to manually instantiate the existential quantifiers obtained by unfolding the definition of $\wedge *$.

Lemma. $P \wedge * (Q \wedge * R) = (P \wedge * Q) \wedge * R$
Lemma. $P \wedge * Q = Q \wedge * P$
Lemma. $emp \wedge * P = P$
Lemma. $P \wedge * emp = P$
Lemma. $((\lambda h. \ P \ h \vee Q \ h) \wedge * R) \ h = (P \wedge * R) \ h \vee (Q \wedge * R) \ h$
Lemma. $((\lambda h. \ P \ h \wedge Q \ h) \wedge * R) \ h \longrightarrow ((P \wedge * R) \ h \wedge (Q \wedge * R) \ h)$
Lemma. $((\lambda h. \ \exists x. \ P \ x \ h) \wedge * Q) \ h = (\exists x. \ (P \ x \wedge * Q) \ h)$
Lemma. $((\lambda h. \ \forall x. \ P \ x \ h) \wedge * Q) \ h \longrightarrow (\forall x. \ (P \ x \wedge * Q) \ h)$
Lemma. $[\![\ \forall h. \ P \ h \longrightarrow P' \ h; \forall h. \ Q \ h \longrightarrow Q' \ h \ ]\!] \Longrightarrow (P \wedge * Q) \ h \longrightarrow (P' \wedge * Q') \ h$
Lemma. $[\![\ \forall h. \ (P \wedge * Q) \ h \longrightarrow R \ h \ ]\!] \Longrightarrow P \ h \longrightarrow (Q \ -* \ R) \ h$
Lemma. $[\![\ \forall h. \ P \ h \longrightarrow (Q \ -* \ R) \ h \ ]\!] \Longrightarrow (P \wedge * Q) \ h \longrightarrow R \ h$

Following Reynolds [16], we have also defined *pure, intuitionistic, strictly exact*, and *domain exact* assertions, and proved many of their properties. Our growing library of lemmata serves as a basis for verification proofs and increased proof automation.

4   Hoare Logics

4.1   Partial Correctness

In this subsection we present a Hoare logic for partial correctness. We say a Hoare triple $\{P\}c\{Q\}$ is *valid*, $\models_p$, iff every terminating execution of c that starts in a valid state (i.e. in a state of the form $Some\ (s,\ h)$) satisfying the precondition P ends up in a state that satisfies Q, *unless* a memory error occurs.

$$\models_p \{P\}c\{Q\} \equiv$$
$$\forall s\ h\ s'\ h'.\ (Some\ (s,h),\ Some\ (s',h')) \in C(c) \longrightarrow P\ s\ h \longrightarrow Q\ s'\ h'$$

Hence there are two ways in which a Hoare triple can be trivially valid: c, when executed in a state that satisfies the precondition, *i)* does not terminate at all, or *ii)* only terminates in the error state $None$.

Derivability, $\vdash_p$, of Hoare triples is defined inductively. The following set of Hoare rules is both sound and relative complete with respect to the notion of validity defined above.

$\vdash_p \{P\}\ \mathsf{skip}\ \{P\}$
$\vdash_p \{\lambda s\ h.\ P\ (s[x{\mapsto}(a\ s)])\ h\}\ x{:=}a\ \{P\}$
$[\![\ \vdash_p \{P\}c\{Q\};\ \vdash_p \{Q\}d\{R\}\ ]\!] \Longrightarrow \vdash_p \{P\}\ c;d\ \{R\}$
$[\![\ \vdash_p \{\lambda s\ h.\ P\ s\ h \wedge b\ s\}c\{Q\};\ \vdash_p \{\lambda s\ h.\ P\ s\ h \wedge \neg b\ s\}d\{Q\}\ ]\!] \Longrightarrow$
$\qquad \vdash_p \{P\}\ \mathsf{if}\ b\ \mathsf{then}\ c\ \mathsf{else}\ d\ \{Q\}$
$\vdash_p \{\lambda s\ h.\ P\ s\ h \wedge b\ s\}\ c\ \{P\} \Longrightarrow$
$\qquad \vdash_p \{P\}\ \mathsf{while}\ b\ \mathsf{do}\ c\ \{\lambda s\ h.\ P\ s\ h \wedge \neg b\ s\}$
$\vdash_p \{\lambda s\ h.\ (\forall\ a.\ ((a[{\mapsto}](map\ (\lambda e.\ e\ s)\ as))\ -\!\!*\ (P\ (s[x{\mapsto}a])))\ h)\}$
$\qquad x\ {:==}\ \mathsf{list}\ as\ \{P\}$
$\vdash_p \{\lambda s\ h.\ (\forall\ a\ vs.\ (length\ vs = n\ s) \longrightarrow ((a[{\mapsto}]vs)\ -\!\!*\ (P\ (s[x{\mapsto}a])))\ h)\}$
$\qquad x\ {:==}\ \mathsf{alloc}\ n\ \{P\}$
$\vdash_p \{\lambda s\ h.\ (a\ s \in dom\ h) \longrightarrow P\ (s[x{\mapsto}heap\text{-}lookup\ h\ (a\ s)])\ h\}\ x\ {:==}\ @a\ \{P\}$
$\vdash_p \{\lambda s\ h.\ (a\ s \in dom\ h) \longrightarrow P\ s\ (heap\text{-}update\ h\ (a\ s)\ [v\ s])\}\ @a\ {:==}\ v\ \{P\}$
$\vdash_p \{\lambda s\ h.\ (a\ s \in dom\ h) \longrightarrow P\ s\ (heap\text{-}remove\ h\ (a\ s))\}\ \mathsf{dispose}\ a\ \{P\}$
$[\![\ \forall s\ h.\ P'\ s\ h \longrightarrow P\ s\ h;\ \vdash_p \{P\}c\{Q\};\ \forall s\ h.\ Q\ s\ h \longrightarrow Q'\ s\ h\ ]\!] \Longrightarrow$
$\qquad \vdash_p \{P'\}c\{Q'\}$

Soundness is proved by a straightforward induction on $\vdash_p$. The only nontrivial case is the while rule; it requires fixed point induction.

Theorem. $\vdash_p \{P\}c\{Q\} \Longrightarrow \models_p \{P\}c\{Q\}$

To prove completeness, we employ the notion of *weakest (liberal) preconditions* [8].

$wp :: com \Rightarrow assn \Rightarrow assn$
$wp\ c\ Q \equiv \lambda s\ h.\ (\forall s'\ h'.\ (Some\ (s,h),\ Some(s',h')) \in C(c) \longrightarrow Q\ s'\ h')$

The key to the completeness proof is a lemma stating that Hoare triples of the form $\{wp\ c\ Q\}\ c\ \{Q\}$ are derivable. The lemma is proved by induction on c.

Lemma. $\forall Q.\ \vdash_p \{wp\ c\ Q\}\ c\ \{Q\}$

From this, relative completeness of the Hoare rules follows easily with the rule of consequence.

Theorem. $\models_p \{P\}c\{Q\} \Longrightarrow \vdash_p \{P\}c\{Q\}$

4.2   Tight Specifications

The Hoare logic from Section 4.1 does not guarantee the absence of memory errors. We now consider a slightly different Hoare logic for partial correctness, which perhaps better reflects the principle that "well-specified programs don't go wrong" [16]. In this logic, a Hoare triple $\{P\}c\{Q\}$ is *valid*, $\models_t$, iff every terminating execution of c that starts in a valid state satisfying P ends up in a valid state satisfying Q.

$$\models_t \{P\}c\{Q\} \equiv$$
$$\forall s\; h.\; ((P\; s\; h \longrightarrow (Some\; (s,h),\; None) \notin C(c))$$
$$\wedge\; (\forall s'\; h'.\; (Some\; (s,h),\; Some\; (s',h')) \in C(c) \longrightarrow P\; s\; h \longrightarrow Q\; s'\; h'))$$

Compared to the previous Hoare logic, we have added a safety constraint expressing that the error state *None* must be unreachable. Specifications are now *"tight"* in the sense that every address accessed by c must either be mentioned in the precondition, or allocated by c before it is used (in which case the precondition must ensure the existence of a free address).

Of course the preconditions in our Hoare rules must be modified to reflect this change in the definition of validity. The rules for skip, assignment, composition, if, and while, as well as the consequence rule, remain unchanged; therefore they are not shown below. The rules for list, alloc, lookup, mutate, and dispose however now have preconditions which consist of two parts: one guaranteeing the absence of an error, and the other one guaranteeing that the postcondition will hold in all reachable states.

$$\vdash_t \{\lambda s\; h.\; (\exists a.\; heap\text{-}isfree\; h\; a\; (length\; as)) \wedge (\forall a.\; ((a[\mapsto](map\; (\lambda e.\; e\; s)\; as))$$
$$-\!* (P\; (s[x\!\mapsto\!a]))) \; h)\} \; x :== list\; as\; \{P\}$$
$$\vdash_t \{\lambda s\; h.\; (\exists a.\; heap\text{-}isfree\; h\; a\; (n\; s)) \wedge (\forall a\; vs.\; (length\; vs = n\; s)$$
$$\longrightarrow ((a[\mapsto]vs) -\!* (P\; (s[x\!\mapsto\!a]))) \; h)\} \; x :== alloc\; n\; \{P\}$$
$$\vdash_t \{\lambda s\; h.\; (\exists v.\; ((a\; s)\!\hookrightarrow\!v)\; h \wedge P\; (s[x\!\mapsto\!v])\; h)\} \; x :== @a\; \{P\}$$
$$\vdash_t \{\lambda s\; h.\; ((a\; s)\!\mapsto\!- \wedge\!* (((a\; s)\!\mapsto\!(v\; s)) -\!* P\; s))\; h)\} \; @a :== v\; \{P\}$$
$$\vdash_t \{\lambda s\; h.\; ((a\; s)\!\mapsto\!- \wedge\!* P\; s)\; h)\} \; dispose\; a\; \{P\}$$

These rules are similar to the ones presented in [16], with the exception that for list and alloc, we need to assert the existence of available memory in the precondition. (In [16], free heap cells are guaranteed to exist because heaps are always finite.)

Using similar techniques as before – in particular, induction on $\vdash_t$ and a suitably modified notion of weakest liberal preconditions – we can prove soundness and relative completeness of this Hoare logic. Both properties are slightly more difficult to prove than for the logic in Section 4.1, since we do not just have to deal with the postcondition, but also with the safety constraint.

Theorem. $(\models_t \{P\}c\{Q\}) = (\vdash_t \{P\}c\{Q\})$

4.3   Total Correctness

So far we have only considered partial correctness, where a Hoare triple is valid iff every reachable state satisfies the postcondition. If we also want to take termination into account, we need to define a judgment $c \downarrow s$ that expresses guaranteed termination of c started in state s. The Hoare rules then differ from those for partial correctness only in the one place where nontermination can arise: the while rule. For the simple While language, the details have been carried out in [13]. Since the new pointer commands always terminate, the development would be almost identical for our extended language.

4.4 The Frame Rule

In Hoare logic for the simple While language, one can show that if $\models\{P\}c\{Q\}$, then $\models\{P \wedge R\}c\{Q \wedge R\}$, provided that no variables modified by c occur free in R. Under certain conditions (cf. the discussion in [19]), separation logic allows us to obtain a similar rule for our extended language:

$$\models\{P\}c\{Q\} \implies \models\{P \wedge *R\}c\{Q \wedge *R\} \quad,$$

with the same syntactic side condition on R. This *Frame Rule* is essential for modular verification, in particular in the presence of procedures. Unfortunately however, the Frame Rule does not hold in the two previously defined Hoare logics. As counterexamples consider

$$\models_p \{emp\} \text{ dispose } (\lambda s.\ 0) \{false\}$$
$$\neg(\models_p \{emp \wedge *true\} \text{ dispose } (\lambda s.\ 0) \{false \wedge *true\})$$

for the Hoare logic in Section 4.1, and

$$\models_t \{emp\} \ x{:}{=}{=}\text{alloc } (\lambda s.\ 1) \{true\}$$
$$\neg(\models_t \{emp \wedge *true\} \ x{:}{=}{=}\text{alloc } (\lambda s.\ 1) \{true \wedge *true\})$$

for the logic in Section 4.2. The reason why the Frame Rule does not hold in the second Hoare logic is that this logic, when used with potentially infinite heaps, does not validate *safety monotonicity* [19]. Safety monotonicity means that if executing c in a state with heap $h1$ is safe (i.e. cannot lead to *None*), then executing c in a state with an extended heap $h1 ++ h2$ (for $h1 \bowtie h2$) must be safe as well. This is in particular false for list and alloc, since there may not be enough free addresses left in the extended heap.

We could restore safety monotonicity by only considering finite heaps, as done in existing work on separation logic [16, 19]. Combined with an infinite contiguous address space, memory allocation will then always succeed. We note however that a slightly weaker property is sufficient to establish safety monotonicity: namely that heaps contain arbitrary long sequences of unallocated addresses. (Reynolds imposes an equivalent, but more complicated condition on the set of addresses in [16].) This motivates a Hoare logic where we only consider such *lacunary* heaps.

$$lacunary\ h \equiv \forall\, n.\ \exists\, a.\ heap\text{-}isfree\ h\ a\ n$$

Clearly every finite heap is lacunary, and every heap whose domain is contained in the domain of a lacunary heap is itself lacunary. Furthermore, lacunarity is invariant under execution of commands. This can be shown by induction on the evaluation relation $\longrightarrow_c$, with the rules for list, alloc, and dispose being the more interesting cases. Unlike finiteness however, lacunarity is not preserved under union of heaps.

Lemma. *finite* $(dom\ h) \Longrightarrow$ *lacunary* h
Lemma. $[\![\ dom\ h \subseteq dom\ h';\ lacunary\ h'\ ]\!] \Longrightarrow$ *lacunary* h
Lemma. $\langle c,Some\ (s,h)\rangle \longrightarrow_c Some\ (s',h') \Longrightarrow$ *lacunary* $h' = $ *lacunary* h

Based on the concept of lacunary heaps, we define yet another notion of validity, $\models_l$, for Hoare triples. The requirements are exactly the same as for $\models_t$ (i.e. the postcondition must hold in every reachable valid state, and the error state *None* must be unreachable), but for $\models_l$, they need to hold only if the initial heap is lacunary.

$$\models_l \{P\}c\{Q\} \equiv$$
$$\forall s\ h.\ lacunary\ h \longrightarrow ((P\ s\ h \longrightarrow (Some\ (s,h),\ None) \notin C(c))$$
$$\wedge\ (\forall s'\ h'.\ (Some\ (s,h),\ Some\ (s',h')) \in C(c) \longrightarrow P\ s\ h \longrightarrow Q\ s'\ h'))$$

A set of sound and relative complete Hoare rules is obtained by modifying the preconditions in the rules for skip, assignment, list, alloc, lookup, mutate, and dispose accordingly. The rules for list and alloc can then be simplified a little, since lacunarity already implies the existence of free addresses. The rules for composition, if, and while are the same as for $\vdash_t$. To prove completeness of the while rule, however, we need to strengthen the consequence rule.

$\vdash_l \{\lambda s\ h.\ lacunary\ h \longrightarrow P\ s\ h\}$ skip $\{P\}$
$\vdash_l \{\lambda s\ h.\ lacunary\ h \longrightarrow P\ (s[x\mapsto(a\ s)])\ h\}$ x:==a $\{P\}$
$\vdash_l \{\lambda s\ h.\ lacunary\ h \longrightarrow (\forall a.\ ((a[\mapsto](map\ (\lambda e.\ e\ s)\ as))$
$\qquad -\ast (\lambda hh.\ (P\ (s[x\mapsto a])\ hh)))\ h)\}$ x :== list as $\{P\}$
$\vdash_l \{\lambda s\ h.\ lacunary\ h \longrightarrow (\forall a\ vs.\ (length\ vs = n\ s)$
$\qquad \longrightarrow ((a[\mapsto]vs) -\ast (\lambda hh.\ (P\ (s[x\mapsto a])\ hh)))\ h)\}$ x :== alloc n $\{P\}$
$\vdash_l \{\lambda s\ h.\ lacunary\ h \longrightarrow (\exists v.\ ((a\ s)\hookrightarrow v)\ h \wedge P\ (s[x\mapsto v])\ h)\}$ x :== @a $\{P\}$
$\vdash_l \{\lambda s\ h.\ lacunary\ h \longrightarrow ((a\ s)\mapsto-\ \wedge\ast\ (((a\ s)\mapsto(v\ s)) -\ast P\ s))\ h\}$
$\qquad$ @a :== v $\{P\}$
$\vdash_l \{\lambda s\ h.\ lacunary\ h \longrightarrow ((a\ s)\mapsto-\ \wedge\ast P\ s)\ h\}$ dispose a $\{P\}$
$[\![\ \forall s\ h.\ lacunary\ h \longrightarrow P'\ s\ h \longrightarrow P\ s\ h;\ \vdash_l \{P\}c\{Q\};$
$\qquad \forall s\ h.\ lacunary\ h \longrightarrow Q\ s\ h \longrightarrow Q'\ s\ h\ ]\!] \Longrightarrow \vdash_l \{P'\}c\{Q'\}$

As usual, soundness is proved by induction on $\vdash_l$, and relative completeness is proved using (an adapted notion of) weakest liberal preconditions. The abovementioned properties of lacunary heaps are used in both directions of the proof.

Theorem. $(\models_l \{P\}c\{Q\}) = (\vdash_l \{P\}c\{Q\})$

4.5   Proving the Frame Rule

The proof of the Frame Rule presented in this subsection is largely based on [19]. Since we did not specify the syntax of assertions, our first step must be a semantic version of the Frame Rule's side condition. The set of variables that are *modified* by a command is defined as follows.

$$
\begin{array}{ll}
\mathit{ModifiedVars}\ \mathsf{skip} & = \{\} \\
\mathit{ModifiedVars}\ (x{:}{=}{=}a) & = \{x\} \\
\mathit{ModifiedVars}\ (c1;c2) & = \mathit{ModifiedVars}\ c1\ \cup\ \mathit{ModifiedVars}\ c2 \\
\mathit{ModifiedVars}\ (\mathsf{if}\ b\ \mathsf{then}\ c1\ \mathsf{else}\ c2) & = \mathit{ModifiedVars}\ c1\ \cup\ \mathit{ModifiedVars}\ c2 \\
\mathit{ModifiedVars}\ (\mathsf{while}\ b\ \mathsf{do}\ c) & = \mathit{ModifiedVars}\ c \\
\mathit{ModifiedVars}\ (x\ {:}{=}{=}\ \mathsf{list}\ as) & = \{x\} \\
\mathit{ModifiedVars}\ (x\ {:}{=}{=}\ \mathsf{alloc}\ n) & = \{x\} \\
\mathit{ModifiedVars}\ (x\ {:}{=}{=}\ @a) & = \{x\} \\
\mathit{ModifiedVars}\ (@a\ {:}{=}{=}\ v) & = \{\} \\
\mathit{ModifiedVars}\ (\mathsf{dispose}\ a) & = \{\}
\end{array}
$$

By induction on c, one can show that for $x \notin \mathit{ModifiedVars}\ c$, the value of x is invariant under execution of c.

Lemma. $\forall s\ h\ s'\ h'.\ (\mathit{Some}\ (s,h),\ \mathit{Some}\ (s',h')) \in C(c) \longrightarrow x \notin \mathit{ModifiedVars}(c)$
$\qquad \longrightarrow (s\ x = s'\ x)$

We say an assertion P is *independent* of a set of variables S, written $S \natural P$, iff P does not depend on the value of variables in S.

$$S \natural P \equiv \forall s\ s'.\ (\forall x.\ x \notin S \longrightarrow s\ x = s'\ x) \longrightarrow (P\ s = P\ s')$$

The key lemma is now proved by induction on c. It states that a memory error occuring in a lacunary heap can also occur in every subheap, and a valid execution either has a corresponding "restricted" execution in the subheap, or it corresponds to a memory error.

Lemma. $\forall s\ h1\ h2\ s'\ h'.\ h1{\bowtie}h2$
$\qquad \longrightarrow (\mathit{lacunary}\ (h1{+}{+}h2) \longrightarrow (\mathit{Some}\ (s,h1{+}{+}h2),\ \mathit{None}) \in C(c)$
$\qquad \longrightarrow (\mathit{Some}\ (s,h1),\ \mathit{None}) \in C(c))$
$\qquad \wedge\ ((\mathit{Some}\ (s,h1{+}{+}h2),\ \mathit{Some}(s',h')) \in C(c)$
$\qquad \longrightarrow (\mathit{Some}\ (s,h1),\ \mathit{None}) \in C(c)$
$\qquad \vee\ (\exists h1'.\ h1'{\bowtie}h2 \wedge h1'{+}{+}h2 = h' \wedge (\mathit{Some}\ (s,h1),\ \mathit{Some}(s',h1')) \in C(c)))$

Both safety monotonicity and the *frame property* [19] follow immediately.

Lemma. $[\![\ h1{\bowtie}h2\ ;\ \mathit{lacunary}\ (h1{+}{+}h2)\ ;\ (\mathit{Some}\ (s,h1),\ \mathit{None}) \notin C(c)\ ]\!]$
$\qquad \Longrightarrow (\mathit{Some}\ (s,h1{+}{+}h2),\ \mathit{None}) \notin C(c)$

Lemma. $[\![\ h1{\bowtie}h2\ ;\ (\mathit{Some}\ (s,h1),\ \mathit{None}) \notin C(c)\ ;$
$\qquad (\mathit{Some}\ (s,h1{+}{+}h2),\ \mathit{Some}(s',h')) \in C(c)\ ]\!]$
$\qquad \Longrightarrow \exists h1'.\ h1'{\bowtie}h2 \wedge h1'{+}{+}h2 = h' \wedge (\mathit{Some}\ (s,h1),\ \mathit{Some}(s',h1')) \in C(c)$

Finally we can prove the Frame Rule. Safety monotonicity is used to show that the error state is unreachable, and the frame property proves that every reachable state satisfies the postcondition.

Theorem. $[\![\ \models_l \{P\}c\{Q\};\ (\mathit{ModifiedVars}\ c)\natural R\ ]\!]$
$\qquad \Longrightarrow \models_l \{\lambda s\ h.\ (P\ s \wedge{*}\ R\ s)\ h\}c\{\lambda s\ h.\ (Q\ s \wedge{*}\ R\ s)\ h\}$

5 Example: In-place List Reversal

To evaluate the practical applicability of our framework, we verify an in-place list reversal algorithm. This relatively simple algorithm has been considered before [6, 5], also by Reynolds [16], who gave an (informal) correctness proof using separation logic, and by Mehta and Nipkow [11], who formally verified the algorithm in Isabelle/HOL, but without separation logic. The actual algorithm is shown below. i, j and k are variables: i contains a pointer to the current (initially, the first) list cell, j contains a pointer to the previous list cell (initially $null$), and k, which is initialized at the beginning of the loop body, contains a pointer to the next list cell. $null$ is just an abbreviation for 0, rather than a distinguished address. This resembles the treatment of the $NULL$ pointer in (e.g.) C [10].

$reverse\ i\ j\ k \equiv$
    $(j :== (\lambda s.\ null));$ $(*\ initially,\ there\ is\ no\ previous\ list\ cell\ *)$
    $\text{while}\ (\lambda s.\ s\ i \neq null)\ \text{do}$ $(*\ end\ of\ list\ reached?\ *)$
       $((((k :== @(\lambda s.\ Suc\ (s\ i)));$ $(*\ the\ next\ list\ cell\ *)$
       $(@(\lambda s.\ Suc\ (s\ i)) :== (\lambda s.\ s\ j)));$ $(*\ update\ pointer\ to\ next\ cell\ *)$
       $(j :== (\lambda s.\ s\ i)));$ $(*\ previous :== current\ *)$
       $(i :== (\lambda s.\ s\ k)))$ $(*\ current :== next\ *)$

The corresponding specification theorem states that if i, j, and k are distinct and i points to a list vs, then after execution of $reverse\ i\ j\ k$, j will point to the reversed list.

Theorem. $\models_t \{\ \lambda s\ h.\ heap\text{-}list\ vs\ (s\ i)\ h \wedge distinct\ [i,j,k]\ \}$
       $reverse\ i\ j\ k$
       $\{\ \lambda s\ h.\ heap\text{-}list\ (rev\ vs)\ (s\ j)\ h\ \}$

The predicate *heap-list* relates singly linked linear lists on the heap to Isabelle/HOL lists. *heap-list vs a h* is true iff the heap h contains a singly linked linear list whose cells contain the values vs, and whose first cell is at address a.

$heap\text{-}list\ []\ \ \ \ \ a\ h = ((a = null) \wedge emp\ h)$
$heap\text{-}list\ (v \# vs)\ a\ h = ((a \neq null) \wedge (\exists k.\ ((a[\mapsto][v,k]) \wedge * \ heap\text{-}list\ vs\ k)\ h))$

To prove the specification, we use soundness of $\vdash_t$ and apply appropriate Hoare rules until we are left with three verification conditions: namely that the precondition, after execution of $j :== (\lambda s.\ null)$, implies the loop invariant

$(\exists xs\ ys.\ (heap\text{-}list\ xs\ (s\ i)\ \wedge * \ heap\text{-}list\ ys\ (s\ j))\ h \wedge (rev\ vs) = (rev\ xs)@ys)$
$\wedge distinct\ [i,j,k]\ ,$

that the loop invariant is preserved during execution of the loop body, and finally that the loop invariant, together with $s\ i = null$, implies the postcondition.

Lemma. $heap\text{-}list\ vs\ a\ h \implies \exists xs\ ys.\ (heap\text{-}list\ ys\ null\ \wedge * \ heap\text{-}list\ xs\ a)\ h$
      $\wedge rev\ vs = rev\ xs\ @\ ys$
Lemma. $(heap\text{-}list\ ys\ j\ \wedge * \ heap\text{-}list\ (x\ \#\ xs)\ i)\ h \implies$
      $(heap\text{-}list\ xs\ (heap\text{-}lookup\ h\ (Suc\ i))\ \wedge * \ heap\text{-}list\ (x\ \#\ ys)\ i)$
          $(heap\text{-}update\ h\ (Suc\ i)\ [j])$
Lemma. $(heap\text{-}list\ xs\ null\ \wedge * \ heap\text{-}list\ ys\ a)\ h \implies heap\text{-}list\ (rev\ xs\ @\ ys)\ a\ h$

The first and last lemma are easily proved with the help of simple properties of *heap-list*. The proof of the second lemma is more difficult. Using the definition of separating conjunction, we obtain disjoint subheaps h' and h'' of h with *heap-list ys j h'* and *heap-list $(x\#xs)$ i h''*. The conclusion can then be shown by splitting *heap-update h (Suc i) [j]* into the two disjoint heaps $h''(i:=None,\ Suc\ i:=None)$ and $h'(i\mapsto x)(Suc\ i\mapsto j)$. Overall the separation logic proof is slightly less automatic than the proof in [11].

At the moment, the proof strategy employed here seems to be characteristic of formal program verification with separation logic. First Hoare rules are used to obtain a set of verification conditions; this step could easily be automated for programs with loop annotations. Some of the verification conditions can then be shown using simple algebraic properties (e.g. commutativity, associativity) of separating conjunction and implication and the involved predicates, while others presently require semantic arguments. Although it is known that separation logic is not finitely axiomatizable [7], we hope that further case studies will allow us to identify other useful laws of the separating connectives, so that the need for (usually involved) semantic arguments can be minimized.

6   Conclusions and Future Work

This work is a first step towards the use of separation logic in machine-assisted program verification. We have mechanically verified semantic properties of separation logic, and presented three different Hoare logics for pointer programs, all of which we proved sound and relative complete. The whole development, including a formal proof of the Frame Rule, was carried out in the semi-automatic theorem prover Isabelle/HOL.

From our experience, separation logic can be a useful tool to state program specifications in a short and elegant way. At this time, however, the advantage of concise specifications comes with a cost: verification proofs, when carried out at the level of detail that is required for mechanical verification, tend to become more intricate and less automatic. Further work is necessary to achieve a better integration of separation logic into the existing Isabelle/HOL framework, and to increase the degree of proof automation for the connectives of separation logic.

More immediate aims are a verification condition generator for an annotated version of the language, some syntactic sugar for the connectives of separation logic, and extensions to the programming language, e.g. recursive procedures and concurrency.

Acknowledgments. The author would like to thank Tobias Nipkow, Farhad Mehta and the anonymous referees for their valuable comments.

References

[1] Krzysztof R. Apt. Ten years of Hoare's logic: A survey – part I. *ACM Transactions on Programming Languages and Systems*, 3(4):431–483, October 1981.

[2] Krzysztof R. Apt. Ten years of Hoare's logic: A survey – part II: Nondeterminism. *Theoretical Computer Science*, 28:83–109, 1984.

[3] Clark W. Barrett, David L. Dill, and Jeremy R. Levitt. Validity checking for combinations of theories with equality. In Mandayam K. Srivas and Albert J. Camilleri, editors, *Formal Methods in Computer-Aided Design*, volume 1166 of *Lecture Notes in Computer Science*, pages 187–201. Springer, November 1996.

[4] Lars Birkedal, Noah Torp-Smith, and John C. Reynolds. Local reasoning about a copying garbage collector. In *Proceedings of the 31-st ACM SIGPLAN-SIGACT Symposium on Principles of Programming Languages (POPL)*, pages 220–231. ACM Press, January 2004.

[5] Richard Bornat. Proving pointer programs in Hoare logic. In *Mathematics of Program Construction*, pages 102–126, 2000.

[6] Rodney M. Burstall. Some techniques for proving correctness of programs which alter data structures. In Bernard Meltzer and Donald Michie, editors, *Machine Intelligence*, volume 7, pages 23–50. Edinburgh University Press, Edinburgh, Scotland, 1972.

[7] Cristiano Calcagno, Hongseok Yang, and Peter W. O'Hearn. Computability and complexity results for a spatial assertion language for data structures. In Ramesh Hariharan, Madhavan Mukund, and V. Vinay, editors, *FST TCS 2001: Foundations of Software Technology and Theoretical Computer Science*, volume 2245 of *Lecture Notes in Computer Science*, pages 108–119. Springer, 2001.

[8] E. Dijkstra. Guarded commands, non-determinacy and formal derivation of programs. *Communications of the ACM*, 18:453–457, 1975.

[9] C. A. R. Hoare. An axiomatic basis for computer programming. *Communications of the ACM*, 12(10):576–580, October 1969.

[10] Brian W. Kernighan and Dennis M. Ritchie. *The C Programming Language*. Prentice Hall, Inc., second edition, 1988.

[11] Farhad Mehta and Tobias Nipkow. Proving pointer programs in higher-order logic. In Franz Baader, editor, *Automated Deduction – CADE-19*, volume 2741 of *Lecture Notes in Artificial Intelligence*, pages 121–135. Springer, 2003.

[12] Tobias Nipkow. Winskel is (almost) right: Towards a mechanized semantics textbook. *Formal Aspects of Computing*, 10(2):171–186, 1998.

[13] Tobias Nipkow. Hoare logics in Isabelle/HOL. In H. Schwichtenberg and R. Steinbrüggen, editors, *Proof and System-Reliability*, pages 341–367. Kluwer, 2002.

[14] Tobias Nipkow, Lawrence C. Paulson, and Markus Wenzel. *Isabelle/HOL – A Proof Assistant for Higher-Order Logic*, volume 2283 of *Lecture Notes in Computer Science*. Springer, 2002.

[15] Peter W. O'Hearn, John Reynolds, and Hongseok Yang. Local reasoning about programs that alter data structures. In Laurent Fribourg, editor, *Computer Science Logic*, volume 2142 of *Lecture Notes in Computer Science*, pages 1–19. Springer, 2001.

[16] John C. Reynolds. Separation logic: A logic for shared mutable data structures. In *Proceedings of the 17th Annual IEEE Symposium on Logic in Computer Science (LICS'02)*, pages 55–74, 2002.

[17] Martin Wildmoser and Tobias Nipkow. Certifying machine code safety: shallow versus deep embedding. Accepted to the 17th International Conference on Theorem Proving in Higher Order Logics (TPHOLs 2004).

[18] Hongseok Yang. *Local Reasoning for Stateful Programs*. PhD thesis, University of Illinois, Urbana-Champaign, 2001.

[19] Hongseok Yang and Peter W. O'Hearn. A semantic basis for local reasoning. In *Foundations of Software Science and Computation Structure*, pages 402–416, 2002.

A Functional Scenario for Bytecode Verification of Resource Bounds*

Roberto M. Amadio, Solange Coupet-Grimal,
Silvano Dal Zilio, and Line Jakubiec

Laboratoire d'Informatique Fondamentale de Marseille (LIF),
CNRS and Université de Provence

Abstract. We consider a scenario where (functional) programs in pre-compiled form are exchanged among untrusted parties. Our contribution is a system of annotations for the code that can be verified at load time so as to ensure bounds on the time and space resources required for its execution, as well as to guarantee the usual integrity properties.

Specifically, we define a simple stack machine for a first-order functional language and show how to perform type, size, and termination verifications at the level of the bytecode of the machine. In particular, we show that a combination of size verification based on quasi-interpretations and of termination verification based on lexicographic path orders leads to an explicit bound on the space required for the execution.

1 Introduction

Research on mobile code has been a hot topic since the late 90's with many proposals building on the JAVA platform. Application scenarios include, for instance, programmable switches, network games, and applications for smart cards. A prevailing conclusion is that security issues are one of the fundamental problems that still have to be solved before mobile code can become a well-established and well-accepted technology. Initial proposals have focused on the integrity properties of the execution environment such as the absence of memory faults. In this paper, we consider an additional property of interest to guarantee the safety of a mobile code, that is, ensuring bounds on the (computational) resources needed for the execution of the code.

The interest of carrying on such analyses at bytecode level are now well understood [15, 16]. First, mobile code is shipped around in pre-compiled (or *bytecode*) form and needs to be analysed as such. Second, compilation is an error prone process and therefore it seems safer to perform static analyses at the level of the bytecode rather than at source level. In particular, we can reduce the size of the trusted code base and shift from the reliance on the correctness of the whole compilation chain to only the trust on the analyser.

Approach. The problem of bounding the usage made by programs of their resources has already attracted considerable attention. Automatic extraction of resource bounds

---

* This work was partly supported by ACI Sécurité Informatique, project CRISS.

has mainly focused on (first-order) functional languages starting from Cobham's characterisation [7] of polynomial time functions by bounded recursion on notation. Following work, see *e.g.*, [4, 8, 9, 11], has developed various inference techniques that allow for efficient analyses while capturing a sufficiently large range of practical algorithms.

We consider a rather standard first-order functional programming language with inductive types, pattern matching, and call-by value, that can be regarded as a fragment of various ML dialects. The language is also quite close to term rewriting systems (TRS) with constructor symbols. The language comes with three main varieties of static analyses: (i) a standard type analysis, (ii) an analysis of the size of the computed values based on the notion of *quasi-interpretation*, and (iii) an analysis that ensures termination; among the many available techniques we select here recursive path orderings.

The last two analyses, and in particular their combination, are instrumental to the prediction of the space and time required for the execution of a program as a function of the size of the input data. For instance, it is known [5] that a program admitting a polynomially bound quasi-interpretation and terminating by lexicographic path-ordering runs in polynomial space. This and other results can be regarded as generalisations and variations over Cobham's characterisation.

Contribution. The synthesis of termination orderings is a classical topic in term rewriting (see for instance [6]). The synthesis of quasi-interpretations — a concept introduced by Marion *et al.* [13] — is connected to the synthesis of polynomial interpretations for termination but it is generally easier because inequalities do not need to be strict and small degree polynomials are often enough [2]. We will not address synthesis issues in this paper. We suppose that the bytecode comes with annotations such as types and polynomial interpretations of function symbols and orders on function symbols.

We define a simple stack machine for a first-order functional language and show how to perform type, size, and termination verifications at the level of the bytecode of the machine. These verifications rely on certifiable annotations of the bytecode — we follow here the classical viewpoint that a program may originate from a malicious party and does not necessarily result from the compilation of a well-formed program.

Our main goal is to determine how these annotations have to be *formulated and verified* in order to entail size bounds and termination at *bytecode* level, *i.e.*, at the level of an assembler-like code produced by a compiler and executable on a simple stack machine. We carry on this program up to the point where it is possible to verify that a given bytecode will run in polynomial space thus providing a translation of the result mentioned above at byte code level. Beyond proving that a program "is in PSPACE" we extract a polynomial that bounds the size needed to run a program: given a function (identifier) f of arity n in a verified program, we obtain a polynomial $q(x_1, \ldots, x_n)$ such that for all values $v_1, \ldots, v_n$ of the appropriate types, the size needed for the evaluation of the call $f(v_1, \ldots, v_n)$ is bounded by $q(|v_1|, \ldots, |v_n|)$, where $|v|$ is the size of the value v.

A secondary goal of our work is of a pedagogical nature: present a minimal — the virtual machine includes only 6 instructions — but still relevant scenario in which problems connected to bytecode verification can be effectively discussed.

Our approach to resource bound certification follows distinctive design decisions. First, we allow the space needed for the execution of a program to vary depending on the size of its arguments. This is in contrast to most approaches that try to enforce a constant space bound. While this latter goal is reasonable for applications targeting embedded devices, we believe that it is not always relevant in the context of mobile code. Second, our method is applicable to a large class of algorithms and do not impose specific syntactical restrictions on programs. For example, we depart from works based on a linear usage of variables [8]. Given the specificities of our method, we may often ensure bounds on resources where other methods fail, but we may also give very rough estimate of the space needed, *e.g.* in cases where another method would have detected that memory operations may be achieved in-place. Hence, it may be interesting to couple our analysis with other methods for ensuring resource bounds.

Paper Organisation. The paper is organised as follows. Section 2 sketches a first-order functional language with simple types and call-by-value evaluation and recalls some basic facts about quasi-interpretations and termination. Section 3 describes a simple virtual machine comprising a minimal set of 6 instructions that suffice to compile the language described in the previous section. In Section 4, we define a type verification that guarantees that all values on the stack will be well typed. This verification assumes that constructors and function symbols in the bytecode are annotated with their type. In the following sections, we also assume that they are annotated with suitable functions to bound the size of the values on the stack (Section 6) and with an order to guarantee termination (Section 7). The size and termination verifications depend on a shape verification which is described in Section 5.

The presentation of each verification follows a common pattern: (i) definition of constraints on the bytecode and (ii) definition of a predicate which is invariant under machine reduction. The essential technical difficulty is in the structuring of the constraints and the invariants, the proofs are then routine inductive arguments. Additional technical details and omitted proofs can be found in a long version of this extended abstract [3].

2   A Functional Language

We consider a simple, typed, first-order functional language, with inductive types and pattern-matching. A program is composed of a list of mutually recursive type definitions followed by a list of mutually recursive first-order function definitions relying on pattern matching. Expressions and values in the language are built from a finite number of constructors, ranged over by $c, c_1, \ldots$. We use $f, f', \ldots$ to range over function identifiers and $x, x', \ldots$ for variables, and distinguish the following three syntactic categories:

$$
\begin{aligned}
v &::= c(v, \ldots, v) & \text{(values)} \\
p &::= x \mid c(p, \ldots, p) & \text{(patterns)} \\
e &::= x \mid c(e, \ldots, e) \mid f(e, \ldots, e) & \text{(expressions)}
\end{aligned}
$$

A function is defined by a sequence of pattern-matching *rules* of the form $f(p_1, \ldots, p_n) \Rightarrow e$, where e is an expression. We follow the usual hypothesis that the patterns $p_1, \ldots, p_n$ are linear and do not superpose. If e is an expression then $Var(e)$ is the set of variables occurring in it. The *size* of an expression $|e|$ is defined as 0 if e is a constant or a variable and $1 + \Sigma_{i \in 1..n} |e_i|$ if e is of the form $c(e_1, \ldots, e_n)$ or $f(e_1, \ldots, e_n)$.

Types. We use $t, t_1, \ldots$ to range over type identifiers. A type definition associates with each identifier the sequence of the types of its constructors, of the form c *of* $t_1 * \cdots * t_n$. For instance, we can define the type *bword* of binary words and the type *nat* of natural numbers in unary format:

$$bword = \mathsf{nil} \ | \ 0 \ of \ bword \ | \ 1 \ of \ bword \qquad\qquad nat = \mathsf{z} \ | \ \mathsf{s} \ of \ nat$$

In the following, we consider that constructors are declared with their functional type $(t_1, \ldots, t_n) \to t$. Similar types can be either assigned or inferred for the function symbols. We use the notation $f : (t_1, \ldots, t_n) \to t$ to refer to the type of f and $ar(f)$ for the arity of f. We use similar notations for constructors. The typing rules for the language are standard and are omitted — all the results given in this paper could be easily extended to a system with parametric polymorphism.

Evaluation. The following two rules define the standard call-by-value evaluation relation, where σ is a substitution from variables to values. In order to define the rule selected in the evaluation of a function call, we rely on the function *match* which returns the unique substitution (if any) defined on the variables in the patterns and matching the patterns against the vector of values. In particular, the condition $match((p_1, \ldots, p_n), (v_1, \ldots, v_n)) = \sigma$ imposes that $\sigma(p_i) = v_i$ for all $i \in 1..n$.

$$\frac{e_j \Downarrow v_j \quad j \in 1..n}{\mathsf{c}(e_1, \ldots, e_n) \Downarrow \mathsf{c}(v_1, \ldots, v_n)} \qquad \frac{f(p_1, \ldots, p_n) \Rightarrow e \ rule \quad e_j \Downarrow v_j \quad j \in 1..n \quad match((p_1, \ldots, p_n), (v_1, \ldots, v_n)) = \sigma \quad \sigma(e) \Downarrow v}{f(e_1, \ldots, e_n) \Downarrow v}$$

Example 1. The function *add* of type $(nat, nat) \to nat$, defined by the following two rules, computes the sum of two natural numbers.

$$add(\mathsf{z}, y) \Rightarrow y \qquad\qquad add(\mathsf{s}(x), y) \Rightarrow add(x, \mathsf{s}(y))$$

Quasi-Interpretations. Given a program, an *assignment* q associates with constructors $\mathsf{c}, \ldots$ and function symbols $f, \ldots$, functions $q_\mathsf{c}, q_f, \ldots$ over the non-negative reals $\mathbb{R}^+$ such that: (i) if c is a constant then q_c is the constant[1] 0, (ii) if c is a constructor with arity $n \geqslant 1$ then q_c is the function in $(\mathbb{R}^+)^n \to \mathbb{R}^+$ such that $q_\mathsf{c}(x_1, \ldots, x_n) = d + \Sigma_{i \in 1..n} x_i$, for some $d \geqslant 1$, and (iii) if f is a function (identifier) with arity n then $q_f : (\mathbb{R}^+)^n \to \mathbb{R}^+$ is monotonic and for all $i \in 1..n$ we have $q_f(x_1, \ldots, x_n) \geqslant x_i$. An assignment q is extended to all expressions as follows: $q_x = x$, $q_{\mathsf{c}(e_1, \ldots, e_n)} = q_\mathsf{c}(q_{e_1}, \ldots, q_{e_n})$, and $q_{f(e_1, \ldots, e_n)} = q_f(q_{e_1}, \ldots, q_{e_n})$.

Thus for every expression e we have a function expression q_e with variables in $Var(e)$. An assignment is a *quasi-interpretation* if for every rule $f(p_1, \ldots, p_n) \Rightarrow e$ in the program, the inequality $q_{f(p_1, \ldots, p_n)} \geqslant q_e$ holds over $\mathbb{R}^+$.

Example 2. With reference to Example 1, consider the assignment $q_\mathsf{s}(x) = 1 + x$ and $q_{add}(x, y) = x + y$. Since by definition $q_\mathsf{z} = 0$, we note that $q_v = |v|$ for all values v of

---

[1] We can choose any positive real constant for q_c, but this choice simplifies some of our proofs.

type nat. Moreover, it is easy to check that q is a quasi-interpretation as the inequalities $q_{add}(0, y) \geqslant y$ and $q_{add}(1 + x, y) \geqslant q_{add}(x, 1 + y)$ hold. □

Quasi-interpretations are designed so as to provide a bound on the size of the computed values as a function of the size of the input data. An interesting space for the synthesis of quasi-interpretations is the collection of max-plus polynomials [2], that is, functions equivalent to an expression of the form $\max_{i \in I}(\Sigma_{j \in 1..n} a_{i,j} x_j + a_i)$, with $a_{i,j} \in \mathbb{N}$ and $a_i \in \mathbb{Q}^+$, where $\mathbb{N}$ are the natural numbers and $\mathbb{Q}^+$ are the non-negative rationals. In this case, checking whether an assignment is a quasi-interpretation can be reduced to checking the satisfiability of a Presburger formula, and is therefore a decidable problem.

3   The Virtual Machine

We define a simple stack machine and a related set of bytecode instructions for the compilation and the evaluation of programs. We adopt the usual notation on words: ϵ is the empty sequence, $x \cdot x'$ is the concatenation of two sequences x, x'. We may also omit the concatenation operation $\cdot$ by simply writing $x\,x'$. Moreover, if x is a sequence then $|x|$ is its length and $x[i]$ its ith element counting from 1. We denote with $\boldsymbol{y}$ a vector $(y_1, \ldots, y_n)$ of elements. Then, $\boldsymbol{y}_i$ stands for the element y_i and $|\boldsymbol{y}|$ is the number n of elements in the vector. In the following, we will often manipulate *vectors of sequences* and use the notation $\boldsymbol{y}_i[k]$ to denote the kth element in the ith sequence of vector $\boldsymbol{y}$.

We suppose given a program with a set of *constructor names* and a disjoint set of *function names*. A function identifier f will also denote the sequence of instructions of the associated code. Then $f[i]$ stands for the ith instruction in the (compiled) code of f and $|f|$ for the number of instructions.

The virtual machine is built around a few components: (1) an *association list* between function identifiers and function codes; (2) a *configuration M*, which is a sequence of *frames* representing the memory of the machine; (3) a *bytecode interpreter* modelled as a reduction relation on configurations. In turn, a *frame* is a triple (f, pc, ℓ) composed of a function identifier, the value of the program counter (a natural number in $1..|f|$), and a *stack*. A *stack* is a sequence of values that serves both to store the parameters and the values computed during the execution. We work with a minimal set of instructions whose effect on the configuration is described in Table 1 and write $M \to M'$ if M reduces to M' by applying exactly one of the transformations.

The reduction $M \to M'$ is deterministic. The empty sequence of frames ϵ is a special state which cannot be accessed during a computation not raising an error, *i.e.*, not executing the instruction $\texttt{stop}$. A "good" execution starts with a configuration of the form $(f, 1, v_1 \cdots v_n)$, containing only one frame that corresponds to the evaluation of the expression $f(v_1, \ldots, v_n)$. The execution ends with a configuration of the form $(f, pc, \ell \cdot v_0)$ where $1 \leqslant pc \leqslant |f|$ and $f[pc] = \texttt{return}\ n$ (the integer n is the arity of f). In this case the result of the evaluation is v_0. By extension, we say that the configuration M is a result v_0, denoted $M \downarrow v_0$, if there exists a sequence ℓ such that $M \equiv (f, pc, \ell \cdot v_0)$ with $1 \leqslant pc \leqslant |f|$ and $f[pc] = \texttt{return}\ n$. All the other cases of blocked configuration, such that $M \not\to$, are considered as runtime errors.

Table 1. Bytecode Interpreter: $M \to M'$

$$\frac{f[pc] = \mathtt{load}\ i \qquad pc < |f| \qquad \ell[i] = v}{M \cdot (f, pc, \ell) \to M \cdot (f, pc+1, \ell \cdot v)} \qquad \frac{f[pc] = \mathtt{build}\ \mathsf{c}\,n \qquad pc < |f| \qquad \ell = \ell' \cdot v_1 \cdots v_n}{M \cdot (f, pc, \ell) \to M \cdot (f, pc+1, \ell' \cdot \mathsf{c}(v_1, \ldots, v_n))}$$

$$\frac{f[pc] = \mathtt{branch}\ \mathsf{c}\,j \qquad pc < |f| \qquad \ell = \ell' \cdot \mathsf{c}(v_1, \ldots, v_n)}{M \cdot (f, pc, \ell) \to M \cdot (f, pc+1, \ell' \cdot v_1 \cdots v_n)} \qquad \frac{f[pc] = \mathtt{branch}\ \mathsf{c}\,j \qquad 1 \leqslant j \leqslant |f| \qquad \ell = \ell' \cdot \mathsf{d}(\ldots) \quad \mathsf{c} \neq \mathsf{d}}{M \cdot (f, pc, \ell) \to M \cdot (f, j, \ell)}$$

$$\frac{f[pc] = \mathtt{call}\ g\ n \qquad pc < |f| \qquad \ell = \ell' \cdot v_1 \cdots v_n}{M \cdot (f, pc, \ell) \to M \cdot (f, pc, \ell) \cdot (g, 1, v_1 \cdots v_n)} \qquad \frac{f[pc] = \mathtt{stop}}{M \cdot (f, pc, \ell) \to \epsilon}$$

$$\frac{f[pc] = \mathtt{return}\ n \qquad \ell = \ell_0 \cdot v_0 \qquad \ell' = \ell'' \cdot v_1 \cdots v_n}{M \cdot (g, pc', \ell') \cdot (f, pc, \ell) \to M \cdot (g, pc'+1, \ell'' \cdot v_0)}$$

The language described in section 2 admits a direct compilation in our functional bytecode. Every function is compiled into a segment of instructions and linear pattern matching is compiled into a nesting of `branch` instructions. Finally, variables are replaced by offsets from the base of the stack frame.

Clearly, a realistic implementation should at least include a mechanism to execute efficiently tail recursive calls (when a `call` instruction is immediately followed by `return`) and a mechanism to share common sub-values in a configuration. For instance, using a stack of pointers to values allocated on a heap, it is possible to dispense with the copy performed by a `load` instructions. Our approach to size verification of the stack could be adapted to these possible enhancements of the virtual machine.

Preliminary Verifications. We define a minimal set of (syntactical) conditions on the shape of the code so as to avoid the simplest form of errors, *e.g.*, to guarantee that the program counter stays within the intended bounds.

A new frame may only originate from a `call` instruction, that is, for every pair of contiguous frames, $\cdots (f, pc, \ell)\,(g, pc', \ell') \cdots$, the instruction $f[pc]$ must be of the form `call` $g\ n$ and the stack ℓ must end with n values, say $v_1 \cdots v_n$, which are the parameters used in the call for g. We use the notation $arg(M, j)$ to refer to the vector of arguments with which the jth frame in M has been called: if $1 < j \leqslant m$ and $M \equiv (f_1, i_1, \ell_1) \cdots (f_m, i_m, \ell_m)$, we have $arg(M, j) = (v_1, \ldots, v_k)$ where $ar(f_j) = k$ and $\ell_{j-1} = \ell \cdot v_1 \cdots v_k$. (An alternative presentation of the reduction rules could be to carry these parameters explicitly as extra annotations on each frame.) By convention, we use $arg(M, 1)$ for the sequence of values used to initialise the execution of the machine.

We say that a function f is *well-formed* if the sequence of code of f terminates either with the `stop` or with the `return` instruction. Moreover, for every index $i \in 1..|f|$, we ask that: (1) if $f[i] = \mathtt{load}\ k$ then $k \geqslant 1$ and (2) if $f[i] = \mathtt{branch}\ \mathsf{c}\,j$ then $1 \leqslant j \leqslant |f|$. We assume that every function in the code is well-formed; the result of the compilation of functional programs clearly meets these well-formedness conditions. We say that a configuration $M \equiv (f_1, i_1, \ell_1) \cdots (f_m, i_m, \ell_m)$ is *well-formed* if for all $j \in 1..m$ we

Table 2. Well-Typed Instructions: $wt_i(f, \boldsymbol{T})$

---

case $f[i]$ of

load k	: $i < \|\boldsymbol{T}\|$, $\boldsymbol{T}_i[k] = t$ and $\boldsymbol{T}_{i+1} = \boldsymbol{T}_i \cdot t$
build $c\,n$	: let $c : (t_1, \ldots, t_n) \to t_0$ in $\exists T. i < \|\boldsymbol{T}\|, \boldsymbol{T}_i = T \cdot t_1 \cdots t_n$ and $\boldsymbol{T}_{i+1} = T \cdot t_0$
call $g\,n$	: let $g : (t_1, \ldots, t_n) \to t_0$ in $\exists T. i < \|\boldsymbol{T}\|, \boldsymbol{T}_i = T \cdot t_1 \cdots t_n$ and $\boldsymbol{T}_{i+1} = T \cdot t_0$
return n	: let $f : (t_1, \ldots, t_n) \to t_0$ in $\exists T. \boldsymbol{T}_i = T \cdot t_0$
stop	: true
branch $c\,j$	: let $c : (t_1, \ldots, t_n) \to t_0$ in
	$\quad \exists T. i < \|\boldsymbol{T}\|, \boldsymbol{T}_i = T \cdot t_0, \boldsymbol{T}_{i+1} = T \cdot t_1 \cdots t_n$ and $\boldsymbol{T}_j = \boldsymbol{T}_i$

---

have (1) the program counter i_j is in $1..\|f_j\|$; (2) the expression $arg(M, j)$ is defined; and (3) for all $j \in 1..m - 1$ we have $f_j[i_j] = $ call $f_{j+1}\, n_{j+1}$ — type verification will ensure, among other properties, that n_j is the arity of the function f_j. Well-formedness is preserved during execution (and the configuration ϵ is well-formed).

Proposition 1. *If M is a well-formed configuration and $M \to M'$ then the configuration M' is also well-formed.*

4   Type Verification

In this section, we define a simple type verification to ensure the well-formedness and well-typedness of the machine configurations during execution. This verification is very similar to the so called *bytecode verification* in the JAVA platform, see *e.g.* [1], and can be directly used as the basis of an algorithm for validating the bytecode before its execution by the interpreter. (A major difference is that we do not have to consider subroutines, access modifiers or object initialisation in our language.)

Type verification associates with every instruction (every step in the evaluation of a function code) an abstraction of the stack. In our case, an abstract stack is a sequence of types, or *type stack*, $T = t_1 \cdots t_n$, that should exactly match the types of the values present in the stack at the time of the execution. Accordingly, an *abstract execution* for a function f is a sequence $\boldsymbol{T}$ of type stacks such that $\|\boldsymbol{T}\| = \|f\|$.

To express that an abstract execution $\boldsymbol{T}$ is coherent with the instructions in f, we define the notion of *well-typed instruction* based on the auxiliary relation $wt_i(f, \boldsymbol{T})$, given below. Informally, we show that if $wt_i(f, \boldsymbol{T})$ and $\boldsymbol{T}_i = t_1 \cdots t_k$ then for every valid evaluation of f, the stack of values at the time of the execution of $f[i]$ is $\ell = v_1 \cdots v_k$ where v_j is a value of type t_j for every $j \in 1..k$. The definition of the relation $wt_i(f, \boldsymbol{T})$, where $\|f\| = \|\boldsymbol{T}\|$, is by case analysis on the instruction $f[i]$.

We define a well-typed function as a sequence of well-typed instructions. To verify a whole program, we simply need to verify every function separately.

Definition 1 (Well-Typed Function). *A sequence $\boldsymbol{T}$ is a valid abstract execution for the function f with signature $(t_1, \ldots, t_n) \to t_0$, denoted $wt(f, \boldsymbol{T})$, if and only if $\boldsymbol{T}_1 = t_1 \cdots t_n$ and $wt_i(f, \boldsymbol{T})$ for every $i \in 1..\|f\|$.*

We define the *flow graph* of function f as the directed graph $(\{1, \ldots, |f|\}, E_f)$ such that for all $i \in 1..|f| - 1$, the edge $(i, i + 1)$ is in E_f if $f[i]$ is a load, build, or call instruction and the edges $(i, i + 1)$ and (i, j) are in E_f if $f[i]$ is the instruction branch c j. If every node in the flow graph $(\{1, \ldots, |f|\}, E_f)$ is reachable from the node 1 then there is at most one abstract execution, T, such that $wt(f, T)$. Moreover, T can be effectively computed as the fixpoint of a function iterating the conditions given in Table 2, for example using Kildall's algorithm [10].

Example 3. We continue with our running example and display the type of each instruction in the (compiled) code of add. We also show the flow graph associated with the function that exhibits the two possible "execution paths" in the code of add.

```
1 :  nat nat          : load 1
2 :  nat nat nat      : branch s 7
3 :  nat nat nat      : load 2
4 :  nat nat nat nat  : build s 1
5 :  nat nat nat nat  : call add 2
6 :  nat nat nat      : return 2
7 :  nat nat nat      : load 2
8 :  nat nat nat nat  : return 2
```

In the following, we assume that every node in the flow graph is accessible. If T is "the" abstract execution of f, we say that f is a function (code) of type T. Next, we prove that the execution of verified programs never fails. As expected, we start by proving that type information is preserved during evaluation. This relies on the notions of well-typed frames and configurations. For instance, we say that a stack has type T, denoted $\ell : T$, if $T = t_1 \cdots t_n$ and $\ell = v_1 \cdots v_n$, where v_i is of type t_i for all $i \in 1..n$.

$$\frac{\mathsf{c} : (t_1, \ldots, t_n) \to t \qquad v_i : t_i \qquad i \in 1..n}{\mathsf{c}(v_1, \ldots, v_n) : t} \qquad \frac{v_i : t_i \qquad i \in 1..n}{v_1 \cdots v_n : t_1 \cdots t_n}$$

$$\frac{wt(f, T) \qquad \ell : T_i}{wt(f, i, \ell)} \qquad \frac{M \equiv (f_1, i_1, \ell_1) \ldots (f_m, i_m, \ell_m) \text{ well-formed} \qquad wt(f_j, i_j, \ell_j) \qquad j \in 1..m}{wt(M)}$$

Proposition 2 (Type Invariant). *Let M be a configuration. If $wt(M)$ and $M \to M'$ then $wt(M')$.*

We note that as a side result of the type verification, we obtain, for every instruction, the size of the stack at the time of its execution. The soundness of the type verification follows from a progress property.

Proposition 3 (Progress). *Assume M is a well-typed configuration. Then either $M \equiv \epsilon$, or M is a result, $M \downarrow v_0$, or M reduces, $\exists M' \, (M \to M')$.*

5   Shape Analysis

We define a shape analysis on the bytecode which appears to be original. Instead of computing the type of the values in the stack, we prove that we can also obtain partial

Table 3. Shape Constraints at Instruction i: $wsh_i(f, \sigma, E)$

---

case $f[i]$ of
    load k : $\sigma_{i+1} = \sigma_i$ and $E_{i+1} = E_i \cdot E_i[k]$
    build $c\,n$: $\sigma_{i+1} = \sigma_i$, $E_i = E \cdot e_1 \cdots e_n$ and $E_{i+1} = E \cdot c(e_1, \dots, e_n)$
    call $g\,n$: $\sigma_{i+1} = \sigma_i$, $E_i = E \cdot e_1 \cdots e_n$ and $E_{i+1} = E \cdot g(e_1, \dots, e_n)$
    branch $c\,j$: let $E_i = E \cdot p$ in
      if p is a variable x then
        let $\sigma' = [c(x_{i+1,h_i}, \dots, x_{i+1,h_{i+1}})/x]$ in
        $\sigma_j = \sigma_i$, $E_j = E_i$, $\sigma_{i+1} = \sigma' \circ \sigma_i$ and $E_{i+1} = \sigma'(E) \cdot x_{i+1,h_i} \cdots x_{i+1,h_{i+1}}$
      else if $p = c(e_1, \dots, e_n)$ then $\sigma_{i+1} = \sigma_i$ and $E_{i+1} = E \cdot e_1 \cdots e_n$
      else if $p = d(\dots)$ with $d \neq c$ then $\sigma_j = \sigma_i$ and $E_j = E_i$
                                      (where $h_{i+1} = h_i + ar(c) - 1$ and $h_j = h_i$)

---

information on their shape such as the identity of their top-most constructor. This verification is used in the following size and termination verifications (Sections 6 and 7). We suppose that the code of every function f in the program passes the type verification of Section 4 and that $wt(f, T)$ holds. We denote with h a vector of numbers such that h_i is the height of the stack for instruction i, that is $h_i = |T_i|$ for all $i \in 1..|f|$. Furthermore, for every instruction index i and position $k \in 1..h_i$ in the corresponding stack we assume a fresh variable $x_{i,k}$ ranging over expressions, that is terms built from variables, constructors and function symbols.

We show that under some restrictions on the form of the code, we can solve certain shape constraints and associate with every reachable instruction a substitution, σ_i, and to every position of the related stack an expression, $e_{i,j}$ (if f is well-typed and every node in its flow graph is reachable then the solution is unique). We can compare the shape analysis with the type verification of Section 4: we compute for each instruction a sequence of expressions, $E = e_1 \cdots e_n$, instead of a sequence of types $T = t_1 \cdots t_n$. The restrictions on the code are the following:

(1) the flow graph of the function is a tree rooted at instruction 1 whose leaves correspond to the instructions return or stop;
(2) every branch instruction is preceded only by load or branch instructions.

These conditions are satisfied by the bytecode obtained from the compilation of functional programs and entail that in every path from the root we cross a sequence of branch and load instructions, then a sequence of load, build, and call instructions, and finally either a stop or return instruction.

The shape constraints are displayed below. We note that applying a branch c j instruction to a stack whose head value is of the shape $d(\dots)$ with $d \neq c$ produces no effect which is fine since then the following instruction is not reachable (since the flow graph is a tree, we have $j \neq i + 1$). Hence the shape analysis may also be used to locate dead code. The definition of the relation $wsh_i(f, \sigma, E)$, where $|f| = |\sigma| = |E|$, is by case analysis on the instruction $f[i]$. There are no constraints on σ and E if $f[i]$ is a return or stop instruction.

The soundness of shape verification is obtained through the definition of a new predicate on configurations, *wsh*, which improves on the "well-typed" predicate introduced in the previous section.

Definition 2 (Well-Shaped Function). *A pair* (σ, E) *is a valid shape for the function* f *of type* $(t_1, \ldots, t_n) \to t_0$, *denoted* $wsh(f, \sigma, E)$, *if* σ_1 *is the identity substitution, id,* $E_1 = x_{1,1} \cdots x_{1,ar(f)}$, *and* $wsh_i(f, \sigma, E)$ *for all* $i \in 1..|f|$.

Assume we have a well-formed configuration M containing the frame (f, i, ℓ) in j^{th} position and that $arg(M, j) = (u_1, \ldots, u_k)$ are the parameters used to initialise this frame. The substitution σ_i relates the values $u_1, \ldots, u_k$ to the values occurring in ℓ. More precisely, $\sigma_i(x_{1,l})$ is a pattern with variables in $(x_{i,j})_{j \in 1..h_i}$ and there is at most one matching substitution ρ such that $\rho \circ \sigma_i(x_{1,l}) = u_l$ for all $l \in 1..k$. On the other hand, the expressions $e_{i,j}$ describe the values occurring in ℓ. If $e_{i,j}$ is a pattern, that is, if it does not contain a function symbol (which is always the case if the instruction $f[i]$ occurs before the first function call in the execution path), then $\ell[j] = \rho(e_{i,j})$.

For example, if we consider the shape constraints computed for the function *add* below, we have that for every frame (add, i, ℓ) originating from the parameters $(u_1 \, u_2)$, if $i = 5$ (at the point of the recursive call) then u_1 is of the form $s(u_3)$ and ℓ is the stack $(s(u_3) \, u_2 \, u_3 \, s(u_2))$.

$E_1 = x_{1,1} \, x_{1,2}$	: load 1	: $\sigma_1 = id$
$E_2 = x_{1,1} \, x_{1,2} \, x_{1,1}$	: branch s7	: $\sigma_2 = id$
$E_3 = s(x_{3,3}) \, x_{1,2} \, x_{3,3}$	: load 2	: $\sigma_3 = [s(x_{3.3})/x_{1,1}]$
$E_4 = s(x_{3,3}) \, x_{1,2} \, x_{3,3} \, x_{1,2}$	: build s1	: $\sigma_4 = [s(x_{3.3})/x_{1,1}]$
$E_5 = s(x_{3,3}) \, x_{1,2} \, x_{3,3} \, s(x_{1,2})$	: call *add* 2	: $\sigma_5 = [s(x_{3.3})/x_{1,1}]$
$E_6 = s(x_{3,3}) \, x_{1,2} \, add(x_{3,3}, s(x_{1,2}))$	: return 2	: $\sigma_6 = [s(x_{3.3})/x_{1,1}]$
$E_7 = x_{1,1} \, x_{1,2} \, x_{1,1}$	: load 2	: $\sigma_7 = id$
$E_8 = x_{1,1} \, x_{1,2} \, x_{1,1} \, x_{1,2}$	: return 2	: $\sigma_8 = id$

A configuration M is well-shaped if all the frames (f, i, ℓ) in M are well-shaped. This condition relies on the parameters used to initialise the frame.

$$\frac{wsh(f, \sigma, E) \qquad match((\sigma_i(x_{1,1}), \ldots, \sigma_i(x_{1,ar(f)})), u) = \rho}{wsh(f, u, i, \ell)} \quad \text{if } E_i[j] \text{ is a pattern then } \ell[j] = \rho(E_i[j])$$

$$\frac{M \equiv (f_1, i_1, \ell_1) \cdots (f_m, i_m, \ell_m) \qquad wt(M)}{wsh(M)} \quad u_j = arg(M, j) \qquad wsh(f_j, u_j, i_j, \ell_j) \qquad j \in 1..m$$

Assume the bytecode of the function f has passed the type and shape verifications. As for type verification, we prove that the shape predicate is invariant under reduction.

Proposition 4. *If* $wsh(M)$ *and* $M \to M'$ *then* $wsh(M')$.

The shape verification is particularly well-suited to the analysis of code obtained from the compilation of functional programs, but it may not scale well to optimised code, like the one obtained by the elimination of tail recursive calls. Nonetheless, we can easily define the size verification (see Section 6) without relying on the shape analysis and perform this verification on programs that do not meet the conditions given previously. Hence, we should not see the shape analysis as a required step of our method but rather as an elegant way to define simultaneously the core of our size and termination analyses.

6   Value Size Verification

We assume that we have synthesized suitable quasi-interpretations at the language level (before compilation) and that these informations are added to the bytecode. Hence, for every constructor c and function symbol f, the functions $q_c : (\mathbb{R}^+)^{ar(c)} \to \mathbb{R}^+$ and $q_f : (\mathbb{R}^+)^{ar(f)} \to \mathbb{R}^+$ are given.

We prove that we can check the validity of the quasi-interpretations at the bytecode level (and then prevent malicious code containing deceitful size annotations) and that we may infer a bound on the size of the frames on the stack.

We assume the bytecode passes the shape verification. Thus for every instruction index i in the segment of the function f, the sequence of expressions $\boldsymbol{E}_i$ and the substitution $\boldsymbol{\sigma}_i$ are determined. We also know $\boldsymbol{h}_i$, the height of the stack at instruction i, as computed during the type verification.

Definition 3. *We say that the size annotations for the function f are correct if the following condition holds for all $i \in 1..|f|$. Assume $\boldsymbol{E}_i = e_1 \cdots e_{h_i}$, then:*

$$\forall j \in 1..\boldsymbol{h}_i \qquad q_f(q_{\boldsymbol{\sigma}_i(x_{1,1})}, \ldots, q_{\boldsymbol{\sigma}_i(x_{1,ar(f)})}) \geqslant q_{e_j} \quad over \ \mathbb{R}^+ \qquad (1)$$

In the case of the (compiled) function add, for example, the correctness of the size annotations results from the validity of the inequality: $q_{add}(1 + x_{3,3}, x_{1,2}) \geqslant q_{add}(x_{3,3}, 1 + x_{1,2})$ (from (1) on the expressions obtained for instruction 6).

The complexity of verifying condition (1) depends on the choice of the quasi-interpretations space. This problem has the same complexity as verifying the correction of the quasi-interpretation at the level of the functional language, see Section 2. We also notice that the condition is quite redundant and can be optimised. Next, we show (Corollary 1) that the size of all the values occurring in a configuration during the evaluation of an expression $f(v_1, \ldots, v_n)$ are bounded by the quasi-interpretation of $f(v_1, \ldots, v_n)$. This follows from the definition of a new predicate $wsz(M)$ and a related invariant.

$$\frac{wsh(f, \boldsymbol{\sigma}, \boldsymbol{E}) \quad match((\boldsymbol{\sigma}_i(x_{1,1}), \ldots, \boldsymbol{\sigma}_i(x_{1,ar(f)})), u) = \rho}{\boldsymbol{E}_i = e_1 \cdots e_{h_i} \qquad \ell = v_1 \cdots v_{h_i} \qquad q_{\rho(e_j)} \geqslant q_{v_j} \qquad j \in 1..\boldsymbol{h}_i} {wsz(f, \boldsymbol{u}, i, \ell)}$$

$$\frac{M \equiv (f_1, i_1, \ell_1) \ldots (f_m, i_m, \ell_m) \quad wsh(M) \qquad \boldsymbol{u}_j = arg(M, j)}{wsz(f_j, \boldsymbol{u}_j, i_j, \ell_j) \qquad q_{f_k(\boldsymbol{u}_k)} \geqslant q_{f_{k+1}(\boldsymbol{u}_{k+1})} \qquad j \in 1..m \qquad k \in 1..m-1}{wsz(M)}$$

Assume the bytecode of the function f has passed the type and shape verifications. As for the type and shape verifications, we prove that the size predicate is invariant under reduction.

Proposition 5. *If $wsz(M)$ and $M \to M'$ then $wsz(M')$.*

Corollary 1. *Assume that all the functions in the program are well-sized. If the expression $f(v_1, \ldots, v_n)$ is well-typed and $(f, 1, v_1 \cdots v_n) \xrightarrow{*} M \cdot (g, i, \ell)$ then $|v| \leqslant q_{f(v_1, \ldots, v_n)}$ for all the values v occurring in ℓ.*

Proof. By definition, $wsz(f, 1, v_1 \cdots v_n)$. By proposition 5, it follows that $wsz(M \cdot (g, i, \ell))$. Let $\boldsymbol{u} = (u_1, \ldots, u_k)$ be the parameters used in the initialization of the top frame: $\boldsymbol{u} = arg(M \cdot (g, i, \ell), |M| + 1)$. Since the configuration is well-sized, we have $wsz(g, \boldsymbol{u}, i, \ell)$ and there is a substitution ρ such that: (c1) $q_{f(v_1, \ldots, v_n)} \geqslant q_{g(u_1, \ldots, u_k)}$, (c2) $\rho \circ \boldsymbol{\sigma}_i(x_{1,j}) = u_j$ for all $j \in 1..k$, and (c3) $wsh(g, \boldsymbol{\sigma}, E)$ and $\boldsymbol{E}_i = e_1 \cdots e_n$ and $q_{\rho(e_j)} \geqslant q_{\ell[j]}$ for $j \in 1..n$.

By definition, the size annotations in the bytecode are correct, which means that by the verification condition (1) we have: $q_{g(\boldsymbol{\sigma}_i(x_{1,1}), \ldots, \boldsymbol{\sigma}_i(x_{1,k}))} \geqslant q_{e_j}$ for all $j \in 1..n$. We conclude:

$$
\begin{aligned}
q_{f(v_1, \ldots, v_n)} &\geqslant q_{g(u_1, \ldots, u_k)} && \text{by (c1)} \\
&= q_{g(\rho \circ \boldsymbol{\sigma}_i(x_{1,1}), \ldots, \rho \circ \boldsymbol{\sigma}_i(x_{1,k}))} && \text{by (c2)} \\
&\geqslant q_{\rho(e_j)} && \text{by (1) and monotonicity} \\
&\geqslant q_v && \text{by (c3)} \\
&\geqslant |v| && q_v \text{ is a quasi-interpretation} \qquad \square
\end{aligned}
$$

We can use corollary 1 to give a rough estimate of the size of the frames occurring in a configuration. Assume that all the functions in the program are well-sized and consider a frame (g, i, ℓ) occurring in a configuration reached from the evaluation of the expression $f(v_1, \ldots, v_n)$.

From the type verification, we obtain a bound h_g on the length of the stack ℓ (for the function add in our examples we have $h_{add} = 4$). We may define the size of a frame as the sum of the size of the values in ℓ added to h_g — the quantity h_g makes allowance for the presence of constants stored in the stack and we neglect the space needed for storing the function identifier and the program counter. Hence the size of the frame (g, i, ℓ) is less than $h_g \cdot (q_{f(v_1, \ldots, v_n)} + 1)$. Likewise, if we define the size of a configuration M as the sum of the frames occurring in M, then we can bound the size of M by the expression $h_m \cdot (q_{f(v_1, \ldots, v_n)} + 1) \cdot l$, where l is the number of frames in M and h_m is the maximum of the h_g for all the functions g in the program.

In the next section, we use information obtained from a termination analysis to bound the number of frames that may appear in a reachable configuration. As a result, we obtain a bound on the maximal space needed for the evaluation of the expression $f(v_1, \ldots, v_n)$ (see Corollary 3). Moreover, if we can prove termination by lexicographic order, then this bound can be expressed as a polynomial expression on the values $|v_1|, \ldots, |v_n|$.

7   Termination Verification

In this section, we adapt recursive path orderings, a popular technique for checking termination (see, *e.g.*, [6]), to prove termination of the evaluation of the virtual machine. We suppose that the shape verification of the code succeeds. We assume given a preorder $\geqslant_\Sigma$ on the function symbols so that $f =_\Sigma g$ implies $ar(f) = ar(g)$. Recursive path ordering conditions force $f \geqslant_\Sigma g$ whenever f may call g, and $f =_\Sigma g$ whenever f

and g are mutually recursive. The pre-order $\geqslant_\Sigma$ is extended to the constructor symbols by assuming that a constructor is always smaller than a function symbol and that two distinct constructors are incomparable.

We recall that in the recursive path ordering one associates a *status* with each symbol specifying how its arguments have to be compared. It is required that if $f =_\Sigma g$ then f and g have the same status. Here we suppose that the status of every function symbol is lexicographic and that the status of every constructor symbol is the product. We denote with $>_l$ the induced path order. Note that on values $v >_l v'$ if and only if v embeds homomorphically v'. Hence, $v >_l v'$ implies $|v| > |v'|$.

The technical development resembles the one for the value size verification. First, we have to define when the termination annotations given with the bytecode are correct.

Definition 4. *We say that the termination annotations for the function f are correct if the following condition holds for all $i \in 1..|f|$. Assume $\boldsymbol{E}_i = e_1 \cdots e_{h_i}$, then:*

$$\forall j \in 1..\boldsymbol{h}_i \qquad f\big(\sigma_i(x_{1,1}), \ldots, \sigma_i(x_{1,ar(f)})\big) >_l e_j \qquad (2)$$

For the function add, the correctness of the termination annotations results primarily from the validity of the relation $add(\mathsf{s}(x_{3,3}), x_{1,2}) >_l add(x_{3,3}, \mathsf{s}(x_{1,2}))$ (for the lexicographic path ordering). Next, we introduce a predicate ter (for terminating) on well-shaped configurations M. As expected, the termination predicate is an invariant.

$$\frac{wsh(f, \sigma, \boldsymbol{E}) \qquad match\big((\sigma_i(x_{1,1}), \ldots, \sigma_i(x_{1,ar(f)}))\big), \boldsymbol{u}\big) = \rho}{\boldsymbol{E}_i = e_1 \cdots e_{h_i} \qquad \ell = v_1 \cdots v_{h_i} \qquad \rho(e_j) \geqslant_l v_j \qquad j \in 1..\boldsymbol{h}_i}{ter(f, \boldsymbol{u}, i, \ell)}$$

$$\frac{M \equiv (f_1, i_1, \ell_1) \ldots (f_m, i_m, \ell_m) \qquad wsh(M) \qquad \boldsymbol{u}_j = arg(M, j)}{ter(f_j, \boldsymbol{u}_j, i_j, \ell_j) f_k(\boldsymbol{u}_k) >_l f_{k+1}(\boldsymbol{u}_{k+1}) \qquad j \in 1..m \qquad k \in 1..m-1}{ter(M)}$$

Proposition 6. *If $ter(M)$ and $M \to M'$ then $ter(M')$.*

Corollary 2. *Assume that all the functions in the program have correct termination information (see Definition 4). Then the execution of a well-typed frame $(f, 1, v_1 \cdots v_n)$ terminates.*

Proof. We define a well-founded order on well-formed configurations that is compatible with the evaluation of the machine. If i is the index of an instruction in the code of f, let $acc(i)$ denotes the number of instructions reachable from i in the flow graph E_f. Since the flow graph is a tree, whenever we increment the counter or jump to another instruction this value decreases. Let $T = (T_\Sigma, >_l)$ be the collection of values with the lexicographic path order. It is well known that this is a well-founded order. Then consider $T \times \mathbb{N}$ with the lexicographic order from left to right. Again this is a well-founded order. Finally, consider $\mathcal{M}(T \times \mathbb{N})$ the finite multisets over $T \times \mathbb{N}$ with the induced well-founded order. We associate with a configuration $M \equiv (f_1, i_1, \ell_1) \cdots (f_m, i_m, \ell_m)$ the measure $\mu(M) = \{(f_1 arg(M, 1), acc(i_1)-1), \ldots, (f_{m-1} arg(M, m-1), acc(i_{m-1})-1), (f_m arg(M, m), acc(i_m))\}$.

Then, by case analysis, we check that all the reduction rules decrease this measure. This proof is by case analysis on the instruction $f_m[i_m]$. Assume $f_m[i_m] = \mathtt{call}\ g\ n$.

An element $(f(v), i)$ of the multiset is replaced by the two elements $(f(v), i - 1)$ and $(g(u), acc(1))$, where $f(v) >_l g(u)$ (by the invariant *ter*) so that, with respect to the lexicographic order: $(f(v), i) > (f(v), i - 1)$ and $(f(v), i) > (g(u), acc(1))$. In the other cases, an element $(f(v), i)$ is either removed or replaced by $(f(v), j)$ with $i > j$, as needed.   □

As observed in [5], termination by lexicographic order combined with a polynomial bound on the size of the values leads to polynomial space. We derive a similar result with a similar proof at bytecode level.

Corollary 3. *Suppose that the quasi-interpretations are bound by polynomials and that the value size and termination verifications of the bytecode succeeds. Then there exists a polynomial q such that every execution starting from a frame $(f, 1, v_1 \cdots v_n)$ (terminates and) runs in space bound by $q(|v_1|, \ldots, |v_n|)$.*

Proof. Note that if $f(v) >_l g(u)$ then either $f >_\Sigma g$ or $f =_\Sigma g$ and $v >_l u$. In a sequence $f_1(v_1) >_l \cdots >_l f_m(v_m)$, the first case can occur a constant number of times (the number of equivalence classes of function symbols with respect to $\geqslant_\Sigma$) thus it is enough to analyse the length of strictly decreasing sequences of tuples of values $(v_1, \ldots, v_k)$ lexicographically ordered where k is the largest arity of a function symbol. If b is a bound on the size of the values then since on values $v >_l v'$ implies $|v| > |v'|$ we derive that the sequence has length at most b^k. Since b is polynomial in the size of the arguments and the number of values on a frame is bound by a constant (via the stack height verification), a polynomial bound is easily derived.   □

From the type verification we obtain a bound h_m on the length of the stacks (for the function *add* in our examples we have $h_m = 4$). From the size verification we obtain a bound $q_m = q_{f(v_1,\ldots,v_n)}$ on the size of every value occurring in a stack (in our example $q_m = |v_1| + |v_2|$). Finally, the termination analysis provides a bound on the maximal number of frames. A crude analysis gives at most q_m^k frames, where k is the greatest arity among the functions occurring during the execution. Hence, the size needed for the execution of a correct program on the initial configuration $(f, 1, v_1 \cdots v_n)$ is bounded by the product $h_m \cdot (q_m + 1) \cdot q_m^k$. We may improve this bound using a finer analysis of the (proof of correctness of the) termination annotations. In the case of *add*, for example, we remark that the size of the first parameter decreases at every call — there could be at most $|v_1|$ frames in a reachable configuration — and therefore we may derive the stricter bound $4 \cdot (|v_1| + |v_2| + 1) \cdot |v_1|$ instead of $4 \cdot (|v_1| + |v_2| + 1) \cdot (|v_1| + |v_2|)^2$.

8   Conclusion and Related Work

The problem of bounding the size of the memory needed for executing a program has already attracted considerable attention but few works have addressed this problem at the level of the bytecode.

Most work in the literature on bytecode verification tends to guarantee the integrity of the execution environment. Work on resource bounds is carried on in the MRG project [17]. The main technical differences appear to be as follows: (i) they rely on a general proof carrying code approach while we are closer to a typed assembly language approach and (ii) their analyses focus on the size of the heap while we also consider the size of the stack and the termination of the execution. Another related work is due to Marion and Moyen [14] who perform a resource analysis of counter machines by reduction to a certain type of termination in Petri Nets. Their virtual machine is much more restricted than the one we study here as natural numbers is the only data type and the stack can only contain return addresses.

We have shown how to perform type, size, and termination verifications at the level of the bytecode running on a simple stack machine. We believe that the choice of a simple set of bytecode instructions has a pedagogical value: we can present a minimal but still relevant scenario in which problems connected to bytecode verification can be effectively discussed. We are in the process of formalising our virtual machine and the related invariants in the COQ proof assistant. We are also experimenting with the automatic synthesis of annotations at the source code level and with their verification at byte code level. Moreover, we plan to refine the predictions on the space needed for the execution of a program by referring to an optimised implementation of the virtual machine.

References

1. M. Abadi and R. Stata. A type system for Java bytecode subroutines. In *Proc. POPL*, 1998.
2. R. Amadio. Max-plus quasi-interpretations. In *Proc. TLCA*, Springer LNCS 2701, 2003.
3. R. Amadio, S. Coupet-Grimal, S. Dal Zilio and L. Jakubiec. A functional scenario for byte-code verification of resource bounds. Research Report LIF 17-2004.
4. S. Bellantoni and S. Cook. A new recursion-theoretic characterization of the poly-time functions. *Computational Complexity*, 2:97–110, 1992.
5. G. Bonfante, J.-Y. Marion and J.-Y. Moyen. On termination methods with space bound certifications. In *Proc. Perspectives of System Informatics*, Springer LNCS 2244, 2001.
6. F. Baader and T. Nipkow. *Term rewriting and all that*. Cambridge University Press, 1998.
7. A. Cobham. The intrinsic computational difficulty of functions. In *Proc. Logic, Methodology, and Philosophy of Science II*, North Holland, 1965.
8. M. Hofmann. The strength of non size-increasing computation. In *Proc. POPL*, 2002.
9. N. Jones. *Computability and complexity, from a programming perspective*. MIT-Press, 1997.
10. G. Kildall, A unified approach to global program optimization. In *Proc. POPL*, 1973.
11. D. Leivant. Predicative recurrence and computational complexity i: word recurrence and poly-time. *Feasible mathematics II, Clote and Remmel (eds.)*, Birkhäuser, 1994.
12. T. Lindholm and F. Yellin. *The Java virtual machine specification*. Addison-Wesley, 1999.
13. J.-Y. Marion. *Complexité implicite des calculs, de la théorie à la pratique*. Habilitation à diriger des recherches, Université de Nancy, 2000.
14. J.-Y. Marion, J.-Y. Moyen. *Termination and resource analysis of assembly programs by Petri Nets*. Technical Report, Université de Nancy, 2003.
15. G. Morriset, D. Walker, K. Crary and N. Glew. From system F to typed assembly language. In *ACM Transactions on Programming Languages and Systems*, 21(3):528-569, 1999.
16. G. Necula. Proof carrying code. In *Proc. POPL*, 1997.
17. D. Sannella. Mobile resource guarantee. IST-Global Computing research proposal, U. Edinburgh, 2001. http://www.dcs.ed.ac.uk/home/mrg/.

Proving Abstract Non-interference

Roberto Giacobazzi and Isabella Mastroeni

Dipartimento di Informatica
Università di Verona
Strada Le Grazie 15, I-37134 Verona, Italy
(roberto.giacobazzi@ | mastroeni@sci.)univr.it

Abstract. In this paper we introduce a compositional proof-system for certifying abstract non-interference in programming languages. Certifying abstract non-interference means proving that no unauthorized flow of information is observable by the attacker from confidential to public data. The properties of the computation that an attacker may observe are specified as an abstract domain. Assertions specify the secrecy of a program relatively to the given attacker and the proof-system specifies how these assertions can be composed in a syntax-directed a la Hoare deduction of secrecy. We prove that the proof-system is sound relatively to the standard semantics of an imperative programming language. This provides a sound proof-system for both certifying secrecy in language-based security and deriving attackers which do not violate secrecy inductively on program's syntax.

Keywords: Abstract interpretation, language-based security, abstract non-interference, verification

1   Introduction

Standard non-interference has been introduced by Goguen and Meseguer in [19] as a key feature to model information flows in security. The idea behind non-interference in security is that users have given access control privileges and higher privileges are required in order to access files containing confidential data. In this way, when authorized users accessing public data are non-interfering with those on private resources, no leakage of confidential information is possible by observing public input/output behavior of the system. On this pattern most security polices are specified in language-based security, where users are program components specified in some high-level programming language (see [26]). Most methods and techniques for checking *secure information flows* in software, ranging from standard data-flow/control-flow analysis techniques to type inference, are based on non-interference. All of these approaches are devoted to prove that a system as a whole, or parts of it, does not allow confidential data to flow towards public variables. Type-based approaches are designed in such a way that well-typed programs do not leak secrets. In a security-typed language, a type is inductively associated at compile-time with program statements in such a way that any statement showing a potential flow disclosing secrets is rejected [28, 30, 32]. Similarly, data-flow/control-flow analysis techniques are devoted to statically discover flows of secret data into public variables [6, 22, 23, 27]. The problem of weakening non-interference,

J. Marcinkowski and A. Tarlecki (Eds.): CSL 2004, LNCS 3210, pp. 280–294, 2004.

also known as refining security policies, has been recognized as a long standing major challenge in language-based security [26]. In standard non-interference, the attacker is able to fully analyze concrete computations. In this case, any conservative type/data-flow/control-flow analysis of information flows would discard all the programs which may provide any explicit or implicit concrete flows from confidential to public resources. Standard non-interference is therefore often too strict for any practical use in language-based security: most programs are rejected by static control/data flow analyzers or type checkers for non-interference. In order to adapt security policies to practical cases, it would be essential to know how much an attacker may learn from a program by (statically) analyzing its input/output behavior. This idea has recently lead to the definition of the notion of *abstract non-interference* [16]. Abstract non-interference captures a weaker form of non-interference, where non-interference is made parametric relatively to some abstract property of input/output behaviour. Consider the following program P written in a simple imperative language, where the **while**-statement iterates until x_1 is 0. Suppose x_1 : H is a secret variable and x_2 : L is a public variable:

$$\textbf{while } x_1 \textbf{ do } x_2 := x_2 * 2; \; x_1 := x_1 - 1 \textbf{ endw}$$

While in standard non-interference there is an implicit flow from x_1 to x_2, due to the **while**-statement, because x_2 changes depending on the initial value of x_1, this is not true for weaker abstractions of public data. In particular if the attacker can only observe the property of being power of 2 of public variables (x_2), since the operation cannot change its status of being or not a power of two. Then an attacker is unable to observe any interference due to the implicit flow. Abstract non-interference generalizes this idea to arbitrary abstractions of the semantics of a programming language. This provides both a characterization of the degree of secrecy of a program relatively to what an attacker can analyze about its input/output information flow and the possibility for certifying code relatively to some weaker form of non-interference.

The Problem

Abstract non-interference is based on the idea that the model of an attacker is an abstract interpretation of the semantics of the program. A program satisfies the abstract non-interference condition relatively to some given abstraction (attacker) if the abstraction obfuscates any possible interference between confidential and public data. In [16] the authors introduce a step-by-step weakening of Goguen and Meseguer's non-interference by specifying abstract non-interference as a property of the semantics of the program. The idea of modeling attackers as abstract domains provides advanced methods for deriving attackers by systematically transforming the corresponding abstract domains. An algebraic characterization of the most precise secure attacker, i.e., the most precise abstraction for which the program satisfies the abstract non-interference property, is given as a fixpoint domain construction. This abstraction, as well as any abstractions for which the program satisfies abstract non-interference, is both a model of an harmless attacker and a certificate for the security degree of the program. However the original definition of abstract non-interference is not specified inductively on program's syntax but rather it is derived as an abstraction of the concrete semantics of the whole program. This makes the use of abstract non-interference hard in automatic program certification mechanisms, such as in proof-carrying code architectures [24] and in type-based verification algorithms. The logical approach to secure information flow is not new. In [12] dynamic logic is used for characterizing secure information flows, deriving a theorem prover for checking programs. In [1] an axiomatic approach for checking secure information flows is provided. In particular the authors syntactically derive the secure information flows that may happen during the execution. Both these works don't characterize the power of the attacker.

Main Contribution

In this paper we introduce a compositional proof-system for certifying abstract non-interference in programming languages which means proving that the program satisfies an abstract non-interference constraint relatively to some given abstraction of its input/output. Abstractions are specified in the standard abstract interpretation [9] framework. Assertions in the proof-system have the form of Hoare triples: $(\eta)P(\rho)$ where P is a program fragment and η and ρ are abstractions of program's data. However the interpretation of abstract non-interference assertions is rather different from partial correctness assertions (see [3]): $(\eta)P(\rho)$ means that P is unable to disclose secrets if input and output values on public variables are approximated respectively in η and ρ. Hence, abstract non-interference assertions specify the secrecy of a program relatively to a given model of an attacker and the proof-system specifies how these assertions can be composed in a syntax-directed *a la* Hoare deduction of secrecy. We introduce two proof-systems for checking abstract non-interference. The first deals with a stronger notion of abstract non-interference called *narrow abstract non-interference* [16]. The advantage of narrow abstract non-interference is in the simplicity of the proof-system and in its natural derivation from the operational semantics of the language. This proof-system is necessary in order to derive a proof-system for most general abstract non-interference assertions. We prove that the proof-systems are sound relatively to the standard semantics of an imperative programming . Both proof-systems provide a deeper insight in abstract non-interference, by specifying how assertions concerning secrecy compose with each other. This is essential for any static semantics for secrecy devoted to derive certificates specifying the degree of secrecy of a program.

2   Basic Notions

If S and T are sets, then $\wp(S)$ denotes the powerset of S, $S \smallsetminus T$ denotes the set-difference between S and T, $S \subsetneq T$ denotes strict inclusion, and for a function $f : S \to T$ and $X \subseteq S$, $f(X) \stackrel{\text{def}}{=} \{f(x) \mid x \in X\}$. We will often denote $f(\{x\})$ as $f(x)$. $\langle P, \leq \rangle$ denotes a poset P with ordering relation $\leq$, while $\langle P, \leq, \vee, \wedge, \top, \bot \rangle$ denotes a complete lattice P, with ordering $\leq$, *lub* $\vee$, *glb* $\wedge$, greatest element (top) $\top$, and least element (bottom) $\bot$. Often, $\leq_P$ will be used to denote the underlying ordering of a poset P, and $\vee_P$, $\wedge_P$, $\top_P$ and $\bot_P$ denote the basic operations and elements if P is a complete lattice. $f : C \to A$ is (completely) additive if f preserves *lub*'s of all subsets of C (emptyset included). A proof-system $\mathcal{P}$ on a set of formulas $\varPhi$ is a finite set of axiom schemes and proof rules. A proof of φ in $\mathcal{P}$ is a finite sequence of formulas $\varphi_1, \ldots, \varphi_n$ such that $\varphi = \varphi_n$ and each φ_i is either an axiom in $\mathcal{P}$ or it can be obtained by applying a proof rule in $\mathcal{P}$. In this case φ is also called a *theorem* of $\mathcal{P}$, and denoted $\vdash_{\mathcal{P}} \varphi$.

2.1   Abstract Interpretation

Abstract domains can be equivalently formulated either in terms of Galois connections or closure operators [10]. An *upper closure operator* on a poset P is an operator $\rho : P \rightarrow P$ monotone, idempotent and extensive ($\forall x \in P.\ x \leq_P \rho(x)$). The set of all upper closure operators on P is denoted by $uco(P)$. Let $\langle C, \leq, \vee, \wedge, \top, \bot \rangle$ be a complete lattice. Closure operators are uniquely determined by the set of their fix-points $\rho(C)$. $\rho(C)$ is a complete sub-lattice of C iff ρ is additive. If C is a complete lattice then $uco(C)$ ordered point-wise is also a complete lattice, denoted by $\langle uco(C), \sqsubseteq, \sqcup, \sqcap, \lambda x.\top, \lambda x.x \rangle$, where for every $\rho, \eta \in uco(C)$, $\{\rho_i\}_{i \in I} \subseteq uco(C)$ and $x \in C$: $\rho \sqsubseteq \eta$ iff $\forall y \in C.\ \rho(y) \leq \eta(y)$ iff $\eta(C) \subseteq \rho(C)$; $(\sqcap_{i \in I} \rho_i)(x) = \wedge_{i \in I} \rho_i(x)$; and $(\sqcup_{i \in I} \rho_i)(x) = x \Leftrightarrow \forall i \in I.\ \rho_i(x) = x$. The *disjunctive completion* of a domain is the most abstract domain able to represent the concrete disjunction of its objects: $\Upsilon(\rho) = \sqcup\{\eta \in uco(C) | \eta \sqsubseteq \rho$ and η is additive$\}$. ρ is disjunctive iff $\Upsilon(\rho) = \rho$ (cf. [10, 18]). Closure operators and partitions are related concepts. A closure $\eta \in uco(\wp(X))$ induces a partition on X: $\{ [x]_\eta \mid x \in X \}$, where $[x]_\eta \overset{\text{def}}{=} \{ y \mid \eta(x) = \eta(y) \}$. The most concrete closure that induces the same partition of values as η is $\Pi(\eta) \overset{\text{def}}{=} \Upsilon(\{ [x]_\eta \mid x \in X \})$ (see Fig. 1). The idea is that $\Pi(\eta)$ is the

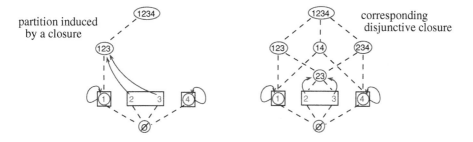

partition induced by a closure

corresponding disjunctive closure

Fig. 1. Example of partitions as disjunctive completion

most concrete closure such that for any $y \in \Pi(\eta(x))$: $\Pi(\eta(x)) = \Pi(\eta(y))$, while in general $\eta(y) \subseteq \eta(x)$.

2.2   The Deterministic Language

In the following we consider a simple imperative language, IMP [31] where programs are commands with the following syntax:

$$c ::= \mathbf{nil} \mid x := e \mid c; c \mid \mathbf{while}\ x\ \mathbf{do}\ c\ \mathbf{endw}$$

with e denoting expressions evaluated in the set of values $\mathbb{V}$ with standard operations, i.e., if $\mathbb{V} = \mathbb{Z}$ then e can be any arithmetical expression. As usual, $\mathbb{V}$ can be structured as a flatdomains, where bottom element, $\bot$, denotes the value of undefined variables. In the following we will denote by $Var(P)$ the set of variables of the program $P \in \text{IMP}$. Let's consider the well-known operational semantics of IMP [31]. The operational semantics naturally induces a transition relation on a set of states Σ, denoted $\rightarrow$, specifying the relation between a state and its possible successors. $\langle \Sigma, \rightarrow \rangle$ is a transition system. In our case, if $|Var(P)| = n$ then $\Sigma = \mathbb{V}^n$. We follow Cousot's construction [8, 11], defining semantics, at different levels of abstractions, as the abstract interpretation of the maximal trace semantics of a transition system associated with each well-formed program. In the following, Σ^+ and $\Sigma^\omega \stackrel{\text{def}}{=} \mathbb{N} \longrightarrow \Sigma$ denote respectively the set of finite nonempty and infinite sequences of symbols in Σ. Given a sequence $\sigma \in \Sigma^\infty \stackrel{\text{def}}{=} \Sigma^+ \cup \Sigma^\omega$, its length is denoted $|\sigma| \in \mathbb{N} \cup \{\omega\}$ and its i-th element is denoted σ_i. A non-empty finite (infinite) *trace* $\sigma \in \Sigma^\infty$ is a finite (infinite) sequence of program states where two consecutive elements are in the transition relation $\rightarrow$, i.e., for all $i < |\sigma|$: $\sigma_i \rightarrow \sigma_{i+1}$. The *maximal trace semantics* [11] of a transition system associated with a program P is $[\![P]\!]^\infty \stackrel{\text{def}}{=} [\![P]\!]^+ \cup [\![P]\!]^\omega$, where if $T \subseteq \Sigma$ is a set of final/blocking states then $[\![P]\!]^{\dot{n}} = \{\sigma \in \Sigma^+ | |\sigma| = n, \forall i \in [1, n) \cdot \sigma_{i-1} \rightarrow \sigma_i\}$, $[\![P]\!]^\omega = \{\sigma \in \Sigma^\omega | \forall i \in \mathbb{N} \cdot \sigma_i \rightarrow \sigma_{i+1}\}$, $[\![P]\!]^+ = \cup_{n>0} \{\sigma \in [\![P]\!]^{\dot{n}} | \sigma_{n-1} \in T\}$, and $[\![P]\!]^n = [\![P]\!]^{\dot{n}} \cap [\![P]\!]^+$. If $\sigma \in [\![P]\!]^+$, then $\sigma_\dashv$ and $\sigma_\vdash$ denote respectively the final and initial state of σ. The *denotational semantics* associates input/output functions with programs, by modeling non-termination by $\bot$. This semantics is derived in [8] by abstract interpretation from the maximal trace semantics with abstraction $\alpha^{\mathcal{D}}(X) \stackrel{\text{def}}{=} \lambda s \in \Sigma . \{\sigma_\dashv | \sigma \in X \cap \Sigma^+, s = \sigma_\vdash\} \cup \{\bot | \exists \sigma \in X \cap \Sigma^\omega, s = \sigma_\vdash\}$. Note that, since our programs are deterministic, $\alpha^{\mathcal{D}}(X)(s)$ is always a singleton. It is well known that we can associate with each program $P \in \text{IMP}$ a function $[\![P]\!]$ denoting its input/output relation, such that $[\![P]\!] \stackrel{\text{def}}{=} \alpha^{\mathcal{D}}([\![P]\!]^\infty)$.

3   Non-interference

Many security problems in language-based security are problems of interference. In order to keep some data confidential, a user might state a policy stipulating that no data visible to other users is affected by modifying confidential data. This policy allows programs to manipulate and modify private data, as long as visible outputs of those programs do not reveal information about the private data. A policy of this sort is a *non-interference* policy [19], also referred as *secrecy* [29]. Confidential data are considered *private*, labeled with H (high-level of secrecy), while all other data are public, labeled with L (low-level of secrecy) [14]. Secrecy is usually formulated saying that the *final* value of a low variable does not depend on the *initial* value of high-variables [29]. An attacker (or unauthorized user) is assumed to be allowed to view only information that is not secret. The usual method for showing that secrecy holds is to verify that the attacker cannot observe *any* difference between two executions that differ only in their secret input [29, 20]. In this case we say that the program has only *secure information flows* [22, 29, 14, 4, 5, 7, 13]. In order to model this situation we consider the denotational semantics $[\![P]\!]$ of the program P. We consider a typing function $t \in Var \longrightarrow \{H, L\}$, which associates with each variable in a program its security class. In the following, if $x \in Var(P)$ then we denote $x : t(x)$ the corresponding security typing. If $T \in \{H, L\}$, $v \in \mathbb{V}^n$, and $n = |\{x \in Var(P) | t(x) = T\}|$, we abuse notation by denoting $v \in \mathbb{V}^T$ the fact that v is a possible value for the vector of variables with security type T. Moreover, we assume that any state $s \in \Sigma$ can be seen as a pair $\langle h, l \rangle$ where $h \in \mathbb{V}^{\mathrm{H}}$ and $l \in \mathbb{V}^{\mathrm{L}}$ and we denote the projection on low values as $\langle h, l \rangle^{\mathrm{L}} = l$. In this case, *standard non-interference* can be formulated as follows.

$$\boxed{\begin{array}{c} \text{A program } P \text{ is } secure \text{ if} \\ \forall v \in \mathbb{V}^{\mathrm{L}}, \forall v_1, v_2 \in \mathbb{V}^{\mathrm{H}} . (\llbracket P \rrbracket(v_1, v))^{\mathrm{L}} = (\llbracket P \rrbracket(v_2, v))^{\mathrm{L}} \end{array}}$$

In [16] we introduced a weaker notion of non-interference modeling weaker information flows. The idea is that an attacker can observe only some properties of public concrete values. The observable properties are modeled as abstractions. As usual in abstract interpretation, a property is an *upper closure operator* on the concrete domain of computation [10]. It is clear that, any observation made on program input/output behaviour by abstract interpretation of its semantics strictly depends upon the chosen abstract domains. The *model of an attacker*, also called *attacker*, is therefore a pair of abstractions: $\langle \eta, \rho \rangle$, with $\eta, \rho \in uco(\wp(\mathbb{V}^{\mathrm{L}}))$, representing what an observer can see about respectively the input and output of a program. Given a program P, *narrow (abstract) non-interference*, denoted $[\eta]P(\rho)$, and *abstract non-interference*, denoted $(\eta)P(\rho)$, introduced in [16], represent a weakening of standard non-interference relatively to a given model of an attacker $\langle \eta, \rho \rangle$. In the following we will abuse notation by denoting with $\llbracket P \rrbracket$ also the additive lifting of $\llbracket P \rrbracket$ to sets of states. Moreover we will use the following simplified notation, $(\llbracket P \rrbracket(h_1, l_1))^{\mathrm{L}} = \llbracket P \rrbracket(h_1, l_1)^{\mathrm{L}}$.

Definition 1. Let $\eta, \rho \in uco(\wp(\mathbb{V}^{\mathrm{L}}))$. A program $P \in$ IMP is such that $[\eta]P(\rho)$ if $\forall h_1, h_2 \in \mathbb{V}^{\mathrm{H}}, \forall l_1, l_2 \in \mathbb{V}^{\mathrm{L}} . \eta(l_1) = \eta(l_2) \Rightarrow \rho(\llbracket P \rrbracket(h_1, l_1)^{\mathrm{L}}) = \rho(\llbracket P \rrbracket(h_2, l_2)^{\mathrm{L}})$. $P \in$ IMP is such that $(\eta)P(\rho)$ if $\forall h_1, h_2 \in \mathbb{V}^{\mathrm{H}}, \forall l \in \mathbb{V}^{\mathrm{L}} . \rho(\llbracket P \rrbracket(h_1, \eta(l))^{\mathrm{L}}) = \rho(\llbracket P \rrbracket(h_2, \eta(l))^{\mathrm{L}})$.

The difference between abstract and narrow non-interference lies upon what the attacker may observe of the input property η. Due to the possible presence of *deceptive flows* in narrow non-interference (see [16]), abstract non-interference represents a weaker notion.

Proposition 1. $[\mathtt{id}]P(\mathtt{id}) \Rightarrow (\eta)P(\rho)$ $[\eta]P(\rho) \Rightarrow (\eta)P(\rho)$

Example 1. Let $Sign = \{\mathbb{Z}, +, -, \varnothing\}$ and $Par = \{\mathbb{Z}, 2\mathbb{Z}, 2\mathbb{Z} + 1, \varnothing\}$, and consider the program $P \stackrel{\text{def}}{=} l := 2 * l * h^2$ with security typing $t = \langle h : \mathrm{H}, l : \mathrm{L} \rangle$ and $\mathbb{V} = \mathbb{Z}$. Note that $Par(-2) = Par(4) = 2\mathbb{Z}$, but $Sign(\llbracket P \rrbracket(h, -2)^{\mathrm{L}}) = 0 - \neq 0 + = Sign(\llbracket P \rrbracket(h, 4)^{\mathrm{L}})$. Namely $\not\models [Par]P(Sign)$ due to a deceptive flow generated by a change of low inputs having the same property in $Sign$.

In [16] two methods for deriving the most concrete output observation for a program, given the input one, for both narrow and abstract non-interference are provided. In particular the idea is that of collecting in the same abstract object all the elements that, if distinguished, would generate a visible flow. This most concrete output observation that is not able to get information from the program P observing η in input for narrow and abstract non-interference is respectively denoted $[\eta]\llbracket P \rrbracket(\mathtt{id})$ and $(\eta)\llbracket P \rrbracket(\mathtt{id})$.

Theorem 1 ([16]). $[\eta][\![P]\!](\mathrm{id}) \sqsubseteq \rho \;\Leftrightarrow\; [\eta]P(\rho)$ *and* $(\eta)[\![P]\!](\mathrm{id}) \sqsubseteq \rho \;\Leftrightarrow\; (\eta)P(\rho)$.

In the following whenever ρ is such that $(\eta)[\![P]\!](\mathrm{id}) \sqsubseteq \rho$ we will write $\models (\eta)P(\rho)$. The same holds for the narrow non-interference.

The main limitation on the use of either $[\eta][\![P]\!](\mathrm{id})$ or $(\eta)[\![P]\!](\mathrm{id})$ for checking abstract non-interference is due to their dependence from the final result of the concrete semantics of the program itself. This makes the construction of $(\eta)[\![P]\!](\mathrm{id})$ and $[\eta][\![P]\!](\mathrm{id})$ a hard task for large programs. In particular, no evidence is made in [16] on how these abstract domains can be derived inductively on program's syntax. This problem is solved in the next section, where a proof-system is introduced for both narrow and abstract non-interference.

4   Axiomatic Abstract Non-interference

In this section we introduce a proof-system for certifying abstract non-interference of programs. We assume a set Φ of basic formulas which can be freely generated from some given set of predicates on $\mathbb{V}^{\mathrm{L}}$ with the basic connectives $\wedge$, $\vee$ and $\neg$. An abstract domain $\rho \in uco(\wp(\mathbb{V}^{\mathrm{L}}))$ can therefore be represented as a $\wedge$-closed set of formulas in Φ. $\rho = \curlyvee(\rho)$, i.e., ρ is disjunctive, iff it is closed under $\vee$ [18]. Note also that $\rho = \Pi(\rho)$ iff ρ is closed both by $\vee$ and $\neg$ (cf. [17, 25]). The semantics of a set of formulas is the corresponding abstract domain. The interpretation of $\sqcap$ and $\sqcup$ are therefore straightforward. As in most programming languages, IMP allows both explicit (through assignment) and implicit (through conditionals) flows [13]. The source of implicit flows in IMP is the **while**-statement.

In order to certify secrecy when implicit flows may occur, we need to model the properties that are invariant during the executions of programs. Intuitively an abstraction is invariant for a program fragment P, written $\{\rho\}_{\mathrm{L}} P \{\rho\}_{\mathrm{L}}$, when by observing the property ρ of public input of P, we are not able to observe any differences in the ρ property of the public output. In other words $\{\rho\}_{\mathrm{L}} P \{\rho\}_{\mathrm{L}}$ means that P is observably equivalent to **nil** as regards as the property ρ. This information is essential in order to certify the lack of implicit flows relatively to an abstraction. These invariant abstractions are obtained with an *a la* Hoare proof-system, where assertions are invariant properties of the form $\{\rho\}_{\mathrm{L}} P \{\rho\}_{\mathrm{L}}$, with $\rho \in uco(\wp(\mathbb{V}^{\mathrm{L}}))$. Invariants of expressions are parametric on a public variable: $\models \{\rho\} \langle e, x \rangle \{\rho\}$ iff $\forall l \in \mathbb{V}^{\mathrm{L}}, \forall h \in \mathbb{V}^{\mathrm{H}} . \rho(\mathcal{E}[\![e]\!](h,l)) = \rho(l_{|_x})$, where for any expression e, $\mathcal{E}[\![e]\!] : \Sigma \longrightarrow \mathbb{V}$ is the standard semantics of expressions and $l_{|_x}$ denotes the value that in $l \in \mathbb{V}^{\mathrm{L}}$ is assigned to x. The intuition is that e does not change the property ρ of the value of x. The public invariants of programs are defined as $\models \{\rho\}_{\mathrm{L}} P \{\rho\}_{\mathrm{L}}$ iff $\forall l \in \mathbb{V}^{\mathrm{L}}, \forall h \in \mathbb{V}^{\mathrm{H}} . \rho([\![P]\!](h,l)^{\mathrm{L}}) = \rho(l)$. Public invariants for programs can be derived by induction on the syntax of IMP by using the proof-system $\mathcal{I} = \{\mathbf{I1}, \ldots, \mathbf{I8}\}$ as defined in Table 1. Rule **I1** says that the property $\top$, which is unable to distinguish any value, is invariant for any program. **I2** says that any property is invariant for the program **nil**. The same holds if the program is an assignment to high variables (**I3**), since by definition invariants are constraints on low variables only. In **I4** if a property is invariant for the evaluation of an expression as regards as the input value of a low variable x, then it is invariant for the assignment to x. Consider for example

Table 1. Derivation of public invariants of programs

$$
\mathbf{I1:}\ \{\top\}_L\, c\, \{\top\}_L \qquad \mathbf{I2:}\ \{\rho\}_L\ \mathbf{nil}\ \{\rho\}_L \qquad \mathbf{I3:}\ \dfrac{x:H}{\{\rho\}_L\ x := e\ \{\rho\}_L} \qquad \mathbf{I4:}\ \dfrac{\{\rho\}\ \langle e,x\rangle\ \{\rho\},\ x:L}{\{\rho\}_L\ x := e\ \{\rho\}_L}
$$

$$
\mathbf{I5:}\ \dfrac{\{\rho\}_L\, c_1\, \{\rho\}_L,\ \{\rho\}_L\, c_2\, \{\rho\}_L}{\{\rho\}_L\ c_1;c_2\ \{\rho\}_L} \qquad \mathbf{I6:}\ \dfrac{\{\rho\}_L\, c\, \{\rho\}_L}{\{\rho\}_L\ \mathbf{while}\ x\ \mathbf{do}\ c\ \mathbf{endw}\ \{\rho\}_L} \qquad \mathbf{I7:}\ \dfrac{\{\rho'\}_L\, c\, \{\rho'\}_L,\ \rho'\sqsubseteq\rho}{\{\rho\}_L\, c\, \{\rho\}_L}
$$

Table 2. Axiomatic narrow (abstract) non-interference

$$
\mathbf{N0:}\ \dfrac{[\eta]\,[\![c]\!]\,(\mathrm{id})\sqsubseteq\rho}{[\eta]c(\rho)} \qquad \mathbf{N1:}\ [\eta]c(\top) \qquad \mathbf{N2:}\ \dfrac{\Pi(\eta)\sqsubseteq\Pi(\rho)}{[\eta]\mathbf{nil}(\rho)} \qquad \mathbf{N3:}\ \dfrac{[\eta]e(\rho),\ [\Pi(\eta)\sqsubseteq\Pi(\rho)],\ x:L}{[\eta]x := e(\rho)}
$$

$$
\mathbf{N4:}\ \dfrac{x:H,\ \Pi(\eta)\sqsubseteq\Pi(\rho)}{[\eta]x := e(\rho)} \qquad \mathbf{N5:}\ \dfrac{[\eta]c_1(\rho),\ [\rho]c_2(\beta)}{[\eta]c_1;c_2(\beta)} \qquad \mathbf{N6:}\ \dfrac{\{\rho\}_L\, c\, \{\rho\}_L}{[\rho]\mathbf{while}\ x\ \mathbf{do}\ c\ \mathbf{endw}(\rho)}
$$

$$
\mathbf{N7:}\ \dfrac{[\eta']c(\rho'),\ \eta\sqsubseteq\eta',\ \rho'\sqsubseteq\rho}{[\eta]c(\rho)} \qquad \mathbf{N8:}\ \dfrac{\forall i\in I\,.\,[\eta]c(\rho_i)}{[\eta]c(\bigsqcup_{i\in I}\rho_i)} \qquad \mathbf{N9:}\ \dfrac{\forall i\in I\,.\,[\eta]c(\rho_i)}{[\eta]c(\bigsqcap_{i\in I}\rho_i)}
$$

the expression $l + 2$, then the property *Sign* (which abstracts on the sign of an integer variable) is not invariant, since if we consider the input value $l = -1$, then we have that $Sign(l + 2) = Sign(1) = + \neq Sign(l) = -$. On the other hand, we have that *Par* (which abstract on the parity of an integer variable) is invariant for this expression as regards as the variable l, since the operation $l + 2$ doesn't change the parity of the value assigned to l. At this point if the statement is $l := l + 2$, then we have that $\{Par\}_L\ l := l + 2\ \{Par\}_L$. Note that, in order to apply this rule, if $\mathbb{V}^L = \mathbb{V}_1 \times \ldots \times \mathbb{V}_n$, then $\rho \in uco(\mathbb{V}^L)$ is such that $\rho(\langle x_1,\ldots,x_n\rangle) = \langle \rho(x_1),\ldots,\rho(x_n)\rangle$. Rule **I5** says that the invariants distribute on the sequential composition. **I6** states that, given a **while**-statement, if a property is invariant for the body, then the same property is invariant for the whole statement. This rule holds since the only modifications of variables made by the **while**, are made by its body. Weakening (**I7**) says that any more abstract property of an invariant is still invariant. A derivation in the proof-system of public invariants in Table 1 is denoted $\vdash_{\mathcal{I}}$.

Theorem 2. *Let $P \in$ IMP and $\rho \in uco(\mathbb{V}^L)$. If $\vdash_{\mathcal{I}} \{\rho\}_L\, P\, \{\rho\}_L$ then $\models \{\rho\}_L\, P\, \{\rho\}_L$.*

We can now introduce a proof-system for narrow abstract non-interference. This is specified as in Table 2. Rule **N0** derives from Th. 1. It states that given a program c and an input observation η we can derive the most concrete output observation that makes the program secret. This corresponds to finding the strongest post-condition (viz. the most concrete abstract domain) for the program c with precondition η such that narrow abstract non-interference holds. This is a "semantic rule", because it involves the construction of the abstract domain $[\eta]\,[\![c]\!]\,(\mathrm{id})$, which is equivalent to compute the concrete semantics of the command c [16]. However, this rule allows us to include in the narrow abstract non-interference proofs, also assertions which can be systematically derived as an abstract domain transformation as shown in [16]. Rules **N1** says that if the output observation is the property $\top$, then the input can be any property. **N2** says that **nil** is secret for any possible attacker such that the partition induced by input observation is more concrete than the output one. This condition is necessary since in this case abstract non-interference corresponds to saying $\forall l_1, l_2 . \eta(l_1) = \eta(l_2) \Rightarrow \rho(l_1) = \rho(l_2)$ which holds iff $\Pi(\eta) \sqsubseteq \Pi(\rho)$. Rule **N3** considers a notion of secrecy extended to expressions, i.e., $\models [\eta]e(\rho)$ iff $\forall l_1, l_2 \in \mathbb{V}^L . \eta(l_1) = \eta(l_2)$ we have $\forall h_1, h_2 \in \mathbb{V}^H . \rho(\mathcal{E}[\![e]\!](h_1, l_1)) = \rho(\mathcal{E}[\![e]\!](h_2, l_2))$. Being the variable public, the secrecy of the expression distributes on the assignment when the partition induced by the input observation is more concrete than the output one. This condition on the induced partitions is necessary only when there are public variables for which the assignment behaves as **nil** (see **N2**). Rule **N4** says that an assignment to a high variable is always secret when the partition induced by input observation is more concrete than the output one since an assignment to private variables behaves as **nil** for the public variables. Indeed note that if, for example, we have the statement $h := h + 1$, then clearly $\rho([\![h := h + 1]\!](h, l_1)^L) = \rho(l_1)$ and $\rho([\![h := h + 1]\!](h, l_2)^L) = \rho(l_2)$. This means that also in this case narrow non-interference corresponds to saying $\eta(l_1) = \eta(l_2) \Rightarrow \rho(l_1) = \rho(l_2)$. Both **N3** and **N4** consider closures on tuples that are tuples of abstractions, as in **I4**. Rule **N5** shows how we can compose the attackers in presence of sequential composition of programs. In particular two programs c_1 and c_2 compose secretly when c_1 is secret for the output observation which is the input one that makes c_2 secret. **N6** controls the **while**-statement. In particular the condition $\{\rho\}_L c \{\rho\}_L$ states that the program c is not acting on the property ρ of the public data, namely ρ is invariant in the execution of c, in the sense that the property ρ of public data is not changed by the execution of c. If this happens then the behaviour of c observed from ρ is the same as the program **nil**, and therefore the fact that the **while** is executed or not is not distinguishable from an observer. We apply this rule also when the guard is a low variable, because narrow non-interference may observe also deceptive flows. **N7** is the consequence rule, which states that we can concretize the input observation and we can abstract the output one, as observed in [16]. Finally **N8** and **N9** says that both the least upper bound and the greatest lower bound of output observations making a program secret, still make the program secret. We denote by $\mathcal{N} = \mathcal{I} \cup \{\mathbf{N0}, \ldots, \mathbf{N9}\}$ the proof-system for narrow abstract non-interference and by $\mathcal{N}_0 = \mathcal{I} \cup \{\mathbf{N1}, \ldots, \mathbf{N9}\}$ the same proof system without the semantic rule **N0**. Next result specifies that the proof-system, without the rule **N0**, is sound.

Theorem 3. *Let $P \in$ IMP and $\eta, \rho \in uco(\mathbb{V}^L)$. If $\vdash_{\mathcal{N}_0} [\eta]P(\rho)$ then $\models [\eta]P(\rho)$.*

Example 2. Consider the closure *Par* which observes parity, and the program:

$$P \stackrel{\text{def}}{=} l := 2 * h; \ \textbf{while } h \textbf{ do } l := l + 2; \ h := h - 1 \textbf{ endw}$$

with security typing: $t = \langle h : \text{H}, l : \text{L} \rangle$ and $\mathbb{V}^H = \mathbb{V}^L = \mathbb{Z}$. We have $[\top]2 * h(\rho_1)$ where ρ_1 is the closure which is not able to distinguish even numbers, i.e., $\rho_1 = \Upsilon(\{2\mathbb{Z}\} \cup \{ \{n\} \mid n \text{ odd} \})$. Therefore, by **N3**, we obtain $[\top]l := 2 * h(\rho_1)$ (note that since there is only one low variable we ignore the condition $\Pi(\eta) \sqsubseteq \Pi(\rho)$). Consider now the **while**-statement. We note that the operation $l + 2$ leaves unchanged the parity of l, this means that if the input is even the the output is even, and similarly if it is odd. Namely for each n such that $Par(n) = Par(l)$ then $Par(\mathcal{E}[\![l + 2]\!](h, n)) = Par(n + 2) = Par(n) = Par(l)$. Therefore $\{Par\} \langle l + 2, l \rangle \{Par\}$ which implies

$$\frac{\{Par\} \langle l + 2, l \rangle \{Par\}}{\{Par\}_{\text{L}}\, l := l + 2 \, \{Par\}_{\text{L}}} \qquad \frac{h : \text{H}}{\{Par\}_{\text{L}}\, h := h + 1 \, \{Par\}_{\text{L}}}$$

Therefore, by **I5**, we have that $\{Par\}_{\text{L}}\, l := l + 2; \; h := h - 1 \, \{Par\}_{\text{L}}$. Now we can apply rule **N6** obtaining

$$\frac{\{Par\}_{\text{L}}\, l := l + 2; \; h := h - 1 \, \{Par\}_{\text{L}}}{[Par]\textbf{while } h \textbf{ do } l := l + 2; \; h := h - 1 \textbf{ endw}(Par)}$$

Finally note that $\rho_1 \sqsubseteq Par$ hence by **N7** we have also that $[\top]l := 2 * h(Par)$, therefore we can apply rule **N5** and we obtain that $[\top]P(Par)$.

Unfortunately the system $\mathcal{N}_0$ is not complete, and in particular **N5** is the rule that introduces incompleteness.

Example 3. Consider the property Par observing parity, and the following program P with security typing: $t = \langle h : \text{H}, l : \text{L} \rangle$ and $\mathbb{V}^{\text{H}} = \mathbb{V}^{\text{L}} = \mathbb{Z}$.

$$P \stackrel{\text{def}}{=} l := 4 * h^2 + 4; \; \textbf{while } h \textbf{ do } l := l \bmod 4; \; h := 0 \textbf{ endw}$$

Let us denote $c \stackrel{\text{def}}{=} \textbf{while } h \textbf{ do } l := l \bmod 4; \; h := 0 \textbf{ endw}$. We can prove that we have $\models [\top]l := 4h^2 + 4(\rho_1)$, where ρ_1 is defined in Example 2, and $\models [\top]P(\rho_1)$. But we can show that we have $\not\models [\rho_1]c(\rho_1)$ since $\rho_1([\![c]\!](0, 5)^{\text{L}}) = 5 \neq \rho_1([\![c]\!](1, 5)^{\text{L}}) = 1$. This means that $\not\vdash_{\mathcal{N} \smallsetminus \{\textbf{N1}\}} [\top]P(\rho_1)$.

It is clear that rule **N0** makes the proof system complete. This is a straight consequence of Theorem 1.

Corollary 1. *The proof-system $\mathcal{N}$ is complete.*

We now introduce in Table 3 a proof-system for abstract non-interference, i.e., modeling how $(\eta)P(\rho)$ assertions compose inductively on program's syntax. The rules **A0** and **A1** in Table 3 are similar to the ones in Table 2. The rule **A2** differs from **N2** since abstract non-interference avoids deceptive flows. In rule **A3** we consider the generalization of the notion of abstract non-interference to expressions as we made for the narrow one. Moreover, as in **N3**, we consider here only abstractions of tuples that are tuples of abstractions. **A4** is straightforward. In order to understand the difference from **N4** consider the example used for explaining **N4**, i.e., $h := h + 1$ then in abstract non-interference we compute the following sets: $\rho([\![h := h + 1]\!](h, \eta(l_1))^{\text{L}}) = \rho(\eta(l_1))$ and $\rho([\![h := h + 1]\!](h, \eta(l_2))^{\text{L}}) = \rho(\eta(l_2))$, which are clearly the same when $\eta(l_1) = \eta(l_2)$. The major difference between narrow and abstract non-interference is in rules **A5**. In this case we need to consider a narrow assertion for c_2 involving disjunctive domains. This is due to the fact that by definition abstract non-interference checks input properties on singletons while the output of the abstract non-interference assertion for c_1

Table 3. Axiomatic abstract non-interference

A0: $\dfrac{(\eta)[\![c]\!](\mathrm{id}) \sqsubseteq \rho}{(\eta)c(\rho)}$ **A1:** $(\eta)c(\top)$ **A2:** $(\eta)\mathrm{nil}(\rho)$ **A3:** $\dfrac{(\eta)e(\rho),\ x:\mathrm{L}}{(\eta)x := e(\rho)}$	

A4: $\dfrac{x:\mathrm{H}}{(\eta)x := e(\rho)}$ **A5:** $\dfrac{(\eta)c_1(\curlyvee(\rho)),\ [\rho]c_2(\curlyvee(\beta))}{(\eta)c_1;c_2(\curlyvee(\beta))}$ **A6:** $\dfrac{\{\rho\}_\mathrm{L}\ c\ \{\rho\}_\mathrm{L},\ x:\mathrm{H}}{(\rho)\mathbf{while}\ x\ \mathbf{do}\ c\ \mathbf{endw}(\rho)}$

A7: $\dfrac{(\eta)c(\rho),\ x:\mathrm{L}}{(\eta)\mathbf{while}\ x\ \mathbf{do}\ c\ \mathbf{endw}(\rho)}$ **A8:** $\dfrac{(\eta)c(\rho'),\ \rho'\sqsubseteq\rho}{(\eta)c(\rho)}$ **A9:** $\dfrac{\forall i \in I.(\eta)c(\rho_i)}{(\eta)c(\bigsqcup_{i\in I}\rho_i)}$ **A10:** $\dfrac{\forall i \in I.(\eta)c(\rho_i)}{(\eta)c(\bigsqcap_{i\in I}\rho_i)}$

deals with properties of sets of values. In order to cope with this 'type mismatch", we need to strengthen the natural counterpart of rule **N5** for abstract non-interference. Next example shows that by considering abstract non-interference for c_2 is not sufficient to achieve soundness.

Example 4. Consider Par and the program P in Example 3. We can prove that we have $(\top)l := 4h^2 + 4(Par)$ and $(Par)c(\rho)$, where $\rho \overset{\mathrm{def}}{=} Par \cup \{0\}$. But we can show that $\not\models (\top)l := 4h^2 + 4;\ c(\rho)$ since $\rho([\![l := 4h^2 + 4;\ c]\!](0, \mathbb{Z})^\mathrm{L}) = \rho(4) = 2\mathbb{Z}$ while we have $\rho([\![l := 4h^2 + 4;\ c]\!](1, \mathbb{Z})^\mathrm{L}) = \rho(0) = \{0\}$, namely they are different.

Moreover note that **A5** requires that for both c_1 and c_2 the output closures are additive maps, i.e., disjunctive abstract domains, as shown in the following example.

Example 5. Consider the following program P with security typing: $t = \langle h : \mathrm{H}, l : \mathrm{L}\rangle$ and $\mathbb{V}^\mathrm{H} = \mathbb{V}^\mathrm{L} = \mathbb{Z}$

$$P \overset{\mathrm{def}}{=} c_1; c_2 = l := (h \bmod 2)(2l \bmod 4) + (1 - (h \bmod 2))(l \bmod 2 + 1);$$
$$l := (l \bmod 2) * 4h + (1 - (l \bmod 2)) * (4h + 1)$$

Consider $\rho = \{\mathbb{Z}, 4\mathbb{Z}, 4\mathbb{Z} + 1, 4\mathbb{Z} + 2, 4\mathbb{Z} + 3, \varnothing\}$ (not additive), then $(\top)c_1(\rho)$ since $\forall h \in 2\mathbb{Z}\ \rho([\![c_1]\!](h, \mathbb{Z})^\mathrm{L}) = \rho(\{1, 2\}) = \mathbb{Z}$, and $\forall h \in 2\mathbb{Z} + 1$ we have $\rho([\![c_1]\!](h, \mathbb{Z})^\mathrm{L}) = \rho(\{0, 2\}) = \mathbb{Z}$. On the other hand it is simple to show that $[\rho]c_2(\rho)$ since this statement leaves unchanged the abstraction of l. But if we consider the composition then we have that $\not\models (\top)P(\rho)$ because if $h \in 2\mathbb{Z}$ then $\rho([\![P]\!](h, \mathbb{Z})^\mathrm{L}) = \rho(\{4h, 4h + 1\}) = \mathbb{Z}$ while if $h \in 2\mathbb{Z} + 1$ then $\rho([\![P]\!](h, \mathbb{Z})^\mathrm{L}) = \rho(\{4h + 1\}) = 4\mathbb{Z} + 1$. Note that the first statement is not secret if we consider the disjunctive completion of ρ in output.

Rule **A6** is equal to **N6**, since also **A5** requires narrow non-interference. Rule **A7** is straightforward from the definition of abstract non-interference and was absent in narrow one for the presence of deceptive flows. The last three rules (**A8**, **A9** and **A10**) change since in abstract non-interference we cannot concretize the input observation. The proof-system in Table 3 is denoted $\mathcal{A} = \mathcal{N} \cup \{\mathbf{A0}, \ldots, \mathbf{A10}\}$ and the proof system without the semantic rules **A0** is denoted as $\mathcal{A}_0 = \mathcal{N}_0 \cup \{\mathbf{A1}, \ldots, \mathbf{A10}\}$. The following theorem proves the soundness of the proof-system $\mathcal{A}_0$ with respect to the standard semantics of IMP.

Theorem 4. *Let $P \in$ IMP be a program and $\eta, \rho \in uco(\mathbb{V}^L)$. If $\vdash_{\mathcal{A}_0} (\eta)P(\rho)$ then $\models (\eta)P(\rho)$.*

Example 6. Consider Par and the program P in Example 3. We can prove that $(\top)l := 4h^2 + 4(Par)$ and $[Par]c(Par)$ therefore $(\top)l := 4h^2 + 4; c(Par)$. Indeed if we consider $l_1 = 4$ and $l_2 = 8$ then clearly $\top(4) = \top(8) = \top$ but $Par([\![l := 4h^2 + 4; c]\!](0, \top)^L) = Par([\![c]\!](0, 4h^2 + 4)^L) = Par(4h^2 + 4) = 2\mathbb{Z}$, while $Par([\![l := 4h^2 + 4; c]\!](1, \top)^L) = Par([\![c]\!](1, 4h^2 + 4)^L) = Par(0) = 2\mathbb{Z}$, namely they are the same.

Example 7. Consider the program fragment

$$P \stackrel{\text{def}}{=} l := 2^h; \textbf{ while } h \textbf{ do } l := 2 * l; \ h := h - 1 \textbf{endw}$$

with security typing: $t = \langle h : \text{H}, l : \text{L} \rangle$ and $\mathbb{V}^H = \mathbb{V}^L = \mathbb{N}$. First of all we note that $(\top)2^h(\rho_1)$, where $\rho_1 \stackrel{\text{def}}{=} \curlyvee(\{\{2\}^{\mathbb{N}}\} \cup \{n \mid n \notin \{2\}^{\mathbb{N}}\})$. This means that we can apply **A3**, obtaining $(\top)l := 2^h(\rho_1)$. Consider now the **while**-statement that we denote by c and the closure $\rho_2 \stackrel{\text{def}}{=} \curlyvee(\{n\{2\}^{\mathbb{N}} \mid n \in \mathbb{N} \text{ odd }\})$. We note that $\{\rho_2\} \langle 2 * l, l \rangle \{\rho_2\}$ and therefore, by **I4** we have $\{\rho_2\}_L \ l := 2 * l \ \{\rho_2\}_L$. On the other hand, by **I3** we have $\{\rho_2\}_L \ h := h - 1 \{\rho_2\}_L$, and therefore by **I5** we obtain $\{\rho_2\}_L \ l := 2 * l; \ h := h - 1 \{\rho_2\}_L$. Now we can apply **A6** obtaining $[\rho_2]$**while** h **do** $l := 2 * l; \ h := h - 1$ **endw**(ρ_2) and therefore we use **A5** obtaining $(\top)P(\rho_2)$.

The following example shows that the proof-system $\mathcal{A}_0$ for abstract non interference in Table 3 is not complete.

Example 8. Consider the closure $\rho \stackrel{\text{def}}{=} \{\mathbb{Z}, 2\mathbb{Z}, 4\mathbb{Z}, \varnothing\}$ and consider the program

$$P \stackrel{\text{def}}{=} \textbf{while } h \textbf{ do } l := (l \bmod 4) * (l \div 4); \ h := 0 \textbf{ endw}$$

with security typing: $t = \langle h : \text{H}, l : \text{L} \rangle$ and $\mathbb{V}^H = \mathbb{V}^L = \mathbb{Z}$. Note that $(\rho)P(\rho)$ since, for example, $\rho([\![P]\!](1, 2\mathbb{Z})^L) = 2\mathbb{Z} = \rho([\![P]\!](0, 2\mathbb{Z})^L)$. But we have that $\not\models \{\rho\}_L P \{\rho\}_L$ since $\rho([\![P]\!](1, 2)^L) = \rho(0) = 4\mathbb{Z} \neq \rho(2) = 2\mathbb{Z}$.

The example above shows that **A6** is not complete, but it is not the only incomplete rule. In particular, by the same argument used in Example 3 for **N5**, **A5** is also incomplete. Even **A7** is incomplete as shown in the following example.

Example 9. Consider $\rho \stackrel{\text{def}}{=} \{\mathbb{Z}, \{0\}, 2\mathbb{Z}_0, 2\mathbb{Z} + 1, \varnothing\}$, where $2\mathbb{Z}_0 \stackrel{\text{def}}{=} 2\mathbb{Z} \smallsetminus \{0\}$, and

$$P \stackrel{\text{def}}{=} \textbf{while } l_1 \textbf{ do } l_2 := \texttt{iszero}(l_1) * h^2; \ l_1 := 0 \textbf{ endw}$$

with security typing: $t = \langle h : \text{H}, l_1, l_2 : \text{L} \rangle$ and $\texttt{iszero}(x) = 1$ if $x = 0$ and $\texttt{iszero}(x) = 0$ otherwise. Then we show that $\not\models (\rho)l_2 := \texttt{iszero}(l_1) * h^2; \ l_1 := 0(\rho)$ since, if we take the low input $\langle 0, 2\mathbb{Z}_0 \rangle$ then we have $\rho([\![l_2 := \texttt{iszero}(l_1) * h^2; \ l_1 := 0]\!](1, \langle 0, 2\mathbb{Z}_0 \rangle)^L) = \rho(\langle 0, 1 \rangle) = \langle 0, 2\mathbb{Z} + 1 \rangle \neq \langle 0, 2\mathbb{Z}_0 \rangle = \rho([\![l_2 := \texttt{iszero}(l_1) * h^2; \ l_1 := 0]\!](2, \langle 0, 2\mathbb{Z}_0 \rangle)^L)$. But it is worth noting that $(\rho)P(\rho)$ since for example $\rho([\![P]\!](h, \langle 0, 2\mathbb{Z}_0 \rangle)^L) = \langle 0, 2\mathbb{Z}_0 \rangle$ and $\rho([\![P]\!](h, \langle 2\mathbb{Z}_0, 2\mathbb{Z}_0 \rangle)^L) = \langle 0, 0 \rangle$.

All the other rules are complete. As above, for the proof-system for narrow abstract non-interference $\mathcal{N}$, also for abstract non-interference, the semantic rule **A0** makes $\mathcal{A}$ complete. This is a straight consequence of Th. 1.

Corollary 2. *The proof-system $\mathcal{A}$ is complete.*

Next result specifies a relation between derivations in the narrow and abstract non-interference proof systems. This result is in accordance with the expected relation between narrow and abstract non-interference, the first being stronger.

Theorem 5. *Let $P \in$ IMP be a program and $\eta, \rho \in uco(\mathbb{V}^{\mathrm{L}})$. If $\vdash_{\mathcal{N}_0} [\eta]P(\rho)$ then $\vdash_{\mathcal{A}_0} (\eta)P(\rho)$.*

Next example shows that $\mathcal{A}$ is strictly weaker than $\mathcal{N}$. We show that if $\models [\eta]P(\rho)$ and $\vdash_{\mathcal{A}_0} (\eta)P(\rho)$, the fact that $[\eta]P(\rho) \Rightarrow (\eta)P(\rho)$ does not imply that $\vdash_{\mathcal{N}_0} [\eta]P(\rho)$.

Example 10. Consider the property Par and the program: $P \stackrel{\text{def}}{=} h := h+1; \ l := 2*h$, with security typing: $t = \langle h : \mathrm{H}, l : \mathrm{L} \rangle$ and $\mathbb{V}^{\mathrm{H}} = \mathbb{V}^{\mathrm{L}} = \mathbb{Z}$. Note that $[Sign]P(Par)$ since $\forall l \in \mathbb{V}^{\mathrm{L}}$, $h \in \mathbb{V}^{\mathrm{H}}$ we have $Par([\![P]\!](h,l)^{\mathrm{L}}) = Par(2*h) = 2\mathbb{Z}$. This means also that $\models (Sign)P(Par)$. But note that $\not\models [Sign]h := h + 1(Par)$ since $Sign(2) = Sign(3) = \mathbb{Z}^+$ and $Par([\![h := h+1]\!](h,2)^{\mathrm{L}}) = Par(2) = 2\mathbb{Z} \neq Par([\![h := h+1]\!](h,3)^{\mathrm{L}}) = Par(3) = 2\mathbb{Z} + 1$. This means that $\not\vdash_{\mathcal{N}_0} [Sign]P(Par)$. On the other hand we have that $\vdash_{\mathcal{A}_0} (Sign)h := h + 1(Par)$ and $\vdash_{\mathcal{N}_0} [Par]l := 2 * h(Par)$, therefore we can use **A5** since Par is disjunctive, and therefore we infer $(Sign)P(Par)$.

5   Discussion

We have introduced a sound proof-system for both narrow and abstract non-interference. The advantage of a proof-system for abstract non-interference is that checking abstract non-interference can be easily mechanized. Both $\mathcal{N}$ and $\mathcal{A}$ can benefit of standard abstract interpretation methods for generating basic certificates for simple program fragments (rules **N0** and **A0**). The other rules allow us to combine certificates from program fragments in a proof-theoretic derivation of harmless models of attackers, certifying program secrecy. The interest in this technology is mostly related with its use in *a la* proof carrying code (PCC) verification of abstract non-interference, when mobile code is allowed. In this case in a PCC architecture, the code producer may create an abstract non-interference certificate that attests to the fact that the code secrecy cannot be violated by the corresponding model of the attacker. Then the code consumer may validate the certificate to check that the foreign code is secure for the corresponding model of attacker. The implementation of this technology requires an appropriate choice of a logic for specifying abstractions and an adequate logical framework where the logic can be manipulated. We believe that predicate abstraction [15, 21] is a fairly simple and easily mechanizable way for reasoning about abstract domains. More appropriate logics can be designed following the ideas in [2], even though a mechanizable logic for reasoning about abstractions is currently a major challenge in this field and deserves further investigations. The language we used is quite simple. Even though abstract non-interference made secrecy a purely semantics problem, any extension of IMP and its semantics with for example probabilistic choice, non terminating computations, and concurrency, may require a redesign of the proof-systems for narrow and abstract non-interference. It would be particularly interesting to extend IMP with concurrency. The main interest in this extension deals both with the chance to reduce protocol verification to non-interference problems and with the possibility of modeling active attackers as abstract interpretations in language-based security. The models of attackers developed in abstract non-interference are indeed passive [16]. Active attackers would be particularly relevant in order to extend abstract non-interference as a language-based tool for protocol validation.

Acknowledgments. This work is supported by the Italian-MIUR Projects *COFIN02-CoVer: Constraint-based verification of reactive systems* and *FIRB: Abstract interpretation and model checking for the verification of embedded systems*. We also thank the anonymous referees for their helpful comments to the previous versions of this paper.

References

1. G. R. Andrews and R. P. Reitman. An axiomatic approach to information flow in programs. *ACM Trans. Program. Lang. Syst.*, 2(1):56–76, 1980.
2. A. Appel. Foundational proof-carrying code. In *Proc. of the 16th IEEE Symp. on Logic in Computer Science (LICS '01)*, pages 247–258. IEEE Computer Society Press, Los Alamitos, Calif., 2001.
3. K. Apt and E.-R. Olderog. *Verification of sequential and concurrent programs.* Springer-Verlag, Berlin, 1997.
4. D. E. Bell and E. Burke. A software validation technique for certification: The methodology. Technical Report MTR-2932, MITRE Corp. Badford, MA, 1974.
5. D. E. Bell and L. J. LaPadula. Secure computer systems: Mathematical foundations and model. Technical Report M74-244, MITRE Corp. Badford, MA, 1973.
6. D. Clark, C. Hankin, and S. Hunt. Information flow for algol-like languages. *Computer Languages*, 28(1):3–28, 2002.
7. E. S. Cohen. Information transmission in computational systems. *ACM SIGOPS Operating System Review*, 11(5):133–139, 1977.
8. P. Cousot. Constructive design of a hierarchy of semantics of a transition system by abstract interpretation. *Theor. Comput. Sci.*, 277(1-2):47,103, 2002.
9. P. Cousot and R. Cousot. Abstract interpretation: A unified lattice model for static analysis of programs by construction or approximation of fixpoints. In *Proc. of Conf. Record of the 4th ACM Symp. on Principles of Programming Languages (POPL '77)*, pages 238–252. ACM Press, New York, 1977.
10. P. Cousot and R. Cousot. Systematic design of program analysis frameworks. In *Proc. of Conf. Record of the 6th ACM Symp. on Principles of Programming Languages (POPL '79)*, pages 269–282. ACM Press, New York, 1979.
11. P. Cousot and R. Cousot. Inductive definitions, semantics and abstract interpretation. In *Proc. of Conf. Record of the 19th ACM Symp. on Principles of Programming Languages (POPL '92)*, pages 83–94. ACM Press, New York, 1992.
12. A. Darvas, R. Hähnle, and D. Sands. A theorem proving approach to analysis of secure information flow. In R. Gorrieri, editor, *Workshop on Issues in the Theory of Security, WITS.* IFIP WG 1.7, ACM SIGPLAN and GI FoMSESS, 2003.
13. D. E. Denning. A lattice model of secure information flow. *Communications of the ACM*, 19(5):236–242, 1976.
14. D. E. Denning and P. Denning. Certification of programs for secure information flow. *Communications of the ACM*, 20(7):504–513, 1977.

15. C. Flanagan and S. Qadeer. Pedicate abstraction for software verification. In *Proc. of Conf. Record of the 29th ACM Symp. on Principles of Programming Languages (POPL '02)*, pages 191–202. ACM Press, New York, 2002.

16. R. Giacobazzi and I. Mastroeni. Abstract non-interference: Parameterizing non-interference by abstract interpretation. In *Proc. of the 31st Annual ACM SIGPLAN-SIGACT Symposium on Principles of Programming Languages (POPL '04)*, pages 186–197. ACM-Press, NY, 2004.

17. R. Giacobazzi and E. Quintarelli. Incompleteness, counterexamples and refinements in abstract model-checking. In P. Cousot, editor, *Proc. of The 8th Internat. Static Analysis Symp. (SAS'01)*, volume 2126 of *Lecture Notes in Computer Science*, pages 356–373. Springer-Verlag, 2001.

18. R. Giacobazzi and F. Ranzato. Optimal domains for disjunctive abstract interpretation. *Sci. Comput. Program*, 32(1-3):177–210, 1998.

19. J. A. Goguen and J. Meseguer. Security policies and security models. In *Proc. IEEE Symp. on Security and Privacy*, pages 11–20. IEEE Computer Society Press, 1982.

20. J. A. Goguen and J. Meseguer. Unwinding and inference control. In *Proc. IEEE Symp. on Security and Privacy*, pages 75–86. IEEE Computer Society Press, 1984.

21. S. Graf and H. Saïdi. Construction of abstract state graphs with PVS. In *Proc. of the 9th Internat. Conf. on Computer Aided Verification (CAV '97)*, volume 1254 of *Lecture Notes in Computer Science*, pages 72–83. Springer-Verlag, Berlin, 1997.

22. R. Joshi and K. R. M. Leino. A semantic approach to secure information flow. *Science of Computer Programming*, 37:113–138, 2000.

23. P. Laud. Semantics and program analysis of computationally secure information flow. In *In Programming Languages and Systems, 10th European Symp. On Programming, ESOP*, volume 2028 of *Lecture Notes in Computer Science*, pages 77–91. Springer-Verlag, 2001.

24. G. Necula. Proof-carrying code. In *Proc. of Conf. Record of the 24th ACM Symp. on Principles of Programming Languages (POPL '97)*, pages 106–119. ACM Press, New York, 1997.

25. F. Ranzato and F. Tapparo. Making abstract model checking strongly preserving. In M. Hermenegildo and G. Puebla, editors, *Proc. of The 9th Internat. Static Analysis Symp. (SAS'02)*, volume 2477 of *Lecture Notes in Computer Science*, pages 411–427. Springer-Verlag, 2002.

26. A. Sabelfeld and A.C. Myers. Language-based information-flow security. *IEEE J. on selected ares in communications*, 21(1):5–19, 2003.

27. A. Sabelfeld and D. Sands. A PER model of secure information flow in sequential programs. *Higher-Order and Symbolic Computation*, 14(1):59–91, 2001.

28. C. Skalka and S. Smith. Static enforcement of security with types. In *ICFP'00*, pages 254–267. ACM Press, New York, 2000.

29. D. Volpano. Safety versus secrecy. In *Proc. of the 6th Static Analysis Symp. (SAS'99)*, volume 1694 of *Lecture Notes in Computer Science*, pages 303–311. Springer-Verlag, 1999.

30. D. Volpano, G. Smith, and C. Irvine. A sound type system for secure flow analysis. *Journal of Computer Security*, 4(2,3):167–187, 1996.

31. G. Winskel. *The formal semantics of programming languages: an introduction*. MIT press, 1993.

32. M. Zanotti. Security typings by abstract interpretation. In M. Hermenegildo and H. Puebla, editors, *Proc. of The 9th Internat. Static Analysis Symp. (SAS'02)*, volume 2477 of *Lecture Notes in Computer Science*, pages 360–375. Springer-Verlag, 2002.

Intuitionistic LTL and a New Characterization of Safety and Liveness

Patrick Maier

Max-Planck-Institut für Informatik
Stuhlsatzenhausweg 85, 66123 Saarbrücken, Germany
maier@mpi-sb.mpg.de

Abstract. Classical linear-time temporal logic (LTL) is capable of specifying of and reasoning about infinite behaviors only. While this is appropriate for specifying non-terminating reactive systems, there are situations (e. g., assume-guarantee reasoning, run-time verification) when it is desirable to be able to reason about finite and infinite behaviors. We propose an interpretation of the operators of LTL on finite and infinite behaviors, which defines an intuitionistic temporal logic (ILTL). We compare the expressive power of LTL and ILTL. We demonstrate that ILTL is suitable for assume-guarantee reasoning and for expressing properties that relate finite and infinite behaviors. In particular, ILTL admits an elegant logical characterization of safety and liveness properties.

1 Introduction

Linear-time temporal logic (LTL) [18] is a convenient specification language for reactive systems. The underlying computational model is that of an infinite behavior, i. e., a non-terminating sequence of interactions between the system and its environment, which makes LTL a specification language for infinite behaviors only. In theory, this is not a problem because every reactive system with finite (and infinite) behaviors can be transformed into one which exhibits only infinite behaviors. In practice, however, it is sometimes essential to reason about finite and infinite behaviors simultaneously and, perhaps, to distinguish finite from infinite behaviors. For example, in run-time verification one needs to relate observed (real) finite behaviors to specified (ideal) infinite behaviors in order to determine whether the observations violate the specification or not. Or, in modular verification, one has to check that a component satisfies an assume-guarantee specification, which amounts to checking that the component keeps satisfying the guarantee at least as long an arbitrary environment satisfies the assumption. Here again, assumption and guarantee are specified as sets of infinite behaviors whereas it is natural to view the component as a prefix-closed set of finite (and possibly infinite) behaviors.

There are various suggestions as how to extend LTL to finite behaviors. For instance, [12] extends the logic with weak and strong next operators whose interpretations differ at the end of finite behaviors. Likewise, [7] interprets LTL formulas by weak and strong semantics, which also differ on finite behaviors. In

J. Marcinkowski and A. Tarlecki (Eds.): CSL 2004, LNCS 3210, pp. 295–309, 2004.

contrast, we propose a semantics for LTL that treats finite and infinite behaviors uniformly. Inspired from the above view of reactive systems as prefix-closed sets of finite and infinite behaviors, our semantics is based on prefix-closed sets. This gives rise to a Heyting algebra of prefix-closed sets rather than a Boolean algebra (because the complement of a prefix-closed set need not be prefix-closed), so we end up with ILTL, an intuitionistic variant of LTL. The idea of using the Heyting algebra of prefix-closed sets of behaviors as the semantic basis for an intuitionistic logic can also be found in [3], [2] and [13]. However, the interpretation of the temporal operators of LTL in this Heyting algebra seems novel to this paper. Departing from the semantic approach to temporal logic, [6] studies a fragment of ILTL, namely the one generated by the temporal next-operator, using proof-theoretic methods.

In temporal verification, the classification of safety and liveness properties, informally introduced by Lamport [11] and made precise by Alpern and Schneider [4], plays an important role because many (deductive) verification methods are applicable only to safety or liveness properties. Still, these methods are universal thanks to the decomposition theorem [4] (and its effective version for ω-regular properties [5]) stating that every linear-time temporal property can be expressed as a conjunction of a safety and a liveness property. Clearly, a similar classification of safety and liveness properties and a decomposition theorem for our intuitionistic logic ILTL would be desirable. We present a novel abstract classification of safety and liveness properties in a Heyting algebra, which is immediately applicable to all intuitionistic linear-time temporal logics including ILTL, and we prove a decomposition theorem. As the classification only uses the operators of the Heyting algebras, we obtain a simple logical characterization of safety and liveness and an effective decomposition theorem for free.

Over the years, there has been a body of work about safety and liveness. In the direction of generalizing the topology-based results of Alpern and Schneider, [9] proves a decomposition theorem for disjunctively complete Boolean algebras, which [16] generalizes to modular complemented lattices. Our results subsume [9] because every Boolean algebra is a Heyting algebra. However, a modular complemented lattice need not be a Heyting algebra, and vice versa, so [16] is neither subsumed nor does it subsume our results. Beyond linear-time, [15] proposes a classification of safety and liveness for branching time. Concerning effective reasoning with safety and liveness properties, [12] gives syntactic characterizations of safety and liveness properties in LTL with past operators; [19] does the same without using past operators. Interestingly, in the introduction to [17], Plotkin and Stirling shortly put forward some ideas about an intuitionistic linear-time temporal logic and a corresponding classification of safety and liveness properties. We consider it likely that their ideas give rise to the same classification of safety and liveness as ours.

Plan. Section 2 introduces some notation. Section 3 defines the intuitionistic temporal logic ILTL, compares it to its classical companion LTL and illustrates the use ILTL as a semantic basis for assume-guarantee specifications. Section 4 introduces intuitionistic safety and liveness and compares these notions to the classical ones proposed by Alpern and Schneider [4], and Section 5 presents a more abstract algebraic view on intuitionistic safety and liveness. Section 6 concludes. Proofs which have been omitted here due to lack of space can be found in [14].

2   Preliminaries

Behaviors. We fix a non-empty set AP of atomic propositions. By Σ, we denote the power set of AP. Given $p \in AP$, we abbreviate the set of sets containing p by Σ_p, i.e., $\Sigma_p = \{a \in \Sigma \mid p \in a\}$. By Σ^∞, we denote the set of non-empty words over the alphabet Σ. Words can be of finite or infinite length, so Σ^∞ is partitioned into Σ^+ and Σ^ω, the sets of finite and infinite words, respectively. Here in the context of discrete linear-time, a behavior is just a word in Σ^∞.

Power Set Lattice of Behaviors. By $\boldsymbol{\mathcal{P}(\Sigma^\infty)} = \langle \mathcal{P}(\Sigma^\infty), \cap, \cup \rangle$, we denote the power set lattice of Σ^∞, ordered by $\subseteq$. Frequently, we will refer to the elements of this lattice as languages or properties.

We call a function $C : \mathcal{P}(\Sigma^\infty) \to \mathcal{P}(\Sigma^\infty)$ a closure operator on Σ^∞ if C is inflationary, idempotent and monotone, i.e., for all $L, L' \subseteq \Sigma^\infty$, $L \subseteq C(L)$ and $C(C(L)) = C(L)$ and $L \subseteq L'$ implies $C(L) \subseteq C(L')$. We call C a topological closure operator on Σ^∞ if C is a closure operator which distributes over finite joins, i.e., $C(\emptyset) = \emptyset$ and for all $L_1, L_2 \subseteq \Sigma^\infty$, $C(L_1 \cup L_2) = C(L_1) \cup C(L_2)$.

Boolean Algebra of Sets of Infinite Behaviors. We define inf : $\mathcal{P}(\Sigma^\infty) \to \mathcal{P}(\Sigma^\infty)$ as mapping a language L to $\inf(L) = L \cap \Sigma^\omega$, the set of infinite behaviors in L. Note that inf is an endomorphism of the complete lattice $\boldsymbol{\mathcal{P}(\Sigma^\infty)}$, in particular inf preserves infinite joins and meets. By INF, we denote the range of inf, i.e., $INF = \{\inf(L) \mid L \subseteq \Sigma^\infty\} = \mathcal{P}(\Sigma^\omega)$. Due to inf being an endomorphism, INF induces a complete sublattice of $\boldsymbol{\mathcal{P}(\Sigma^\infty)}$, which turns out to be a complete lattice of sets. In fact, $\boldsymbol{INF} = \langle INF, \cap, \cup, -, \Sigma^\omega, \emptyset \rangle$ is a complete Boolean algebra, where the unary operator $-$ denotes complementation, i.e., $-L = \{w \in \Sigma^\omega \mid w \notin L\}$.

Heyting Algebra of Prefix-Closed Sets of Behaviors. Let $\preceq$ be the prefix order on Σ^∞, and let $\mathrm{pref}(w) = \{u \in \Sigma^\infty \mid u \preceq w\}$ denote the set of all prefixes of a behavior $w \in \Sigma^\infty$. Thus, pref : $\Sigma^\infty \to \mathcal{P}(\Sigma^\infty)$ is a function from behaviors to languages. We extend the domain of pref to languages in the usual way, i.e., we define pref : $\mathcal{P}(\Sigma^\infty) \to \mathcal{P}(\Sigma^\infty)$ by $\mathrm{pref}(L) = \bigcup_{w \in L} \mathrm{pref}(w)$. Note that pref is a closure operator on Σ^∞, which is why we call a language in the range of pref prefix-closed. Moreover, pref preserves infinite joins, yet in general, it does not preserve meets, not even finite ones. By $PREF$, we denote the range of pref, i.e., $PREF = \{\mathrm{pref}(L) \mid L \subseteq \Sigma^\infty\}$ is the set of prefix-closed languages. Despite pref not preserving all meets, $PREF$ induces a complete sublattice of $\boldsymbol{\mathcal{P}(\Sigma^\infty)}$, which turns out to be a complete lattice of sets. In fact, $\boldsymbol{PREF} = \langle PREF, \cap, \cup, \Rightarrow, \Sigma^\infty, \emptyset \rangle$ is a complete Heyting algebra, i.e., for all languages $L_1, L_2 \in PREF$ there is a greatest language $L \in PREF$, namely $L = \{w \in \Sigma^\infty \mid \mathrm{pref}(w) \cap L_1 \subseteq L_2\}$, such that $L_1 \cap L \subseteq L_2$. We call L the relative pseudo-complement of L_1 and L_2 and denote it by $L_1 \Rightarrow L_2$.

3   Linear-Time Temporal Logics

The set of formulas *Form* of the linear-time temporal logics considered in this paper is defined by the following grammar, where p ranges over the atomic propositions AP, and φ and ψ range over *Form*.

$$Form ::= \top \mid \bot \mid p \mid \varphi \wedge \psi \mid \varphi \vee \psi \mid \varphi \rightarrow \psi \mid \neg\varphi \mid \mathbf{X}\varphi \mid \mathbf{F}\varphi \mid \mathbf{G}\varphi \mid \varphi \,\mathbf{U}\, \psi \mid \varphi \,\mathbf{W}\, \psi$$

For $\varphi, \psi \in Form$, we treat $\varphi \leftrightarrow \psi$ as a shorthand for $(\varphi \rightarrow \psi) \wedge (\psi \rightarrow \varphi)$. To save on parenthesis, we adopt the convention that the unary operators $\neg$ (negation), $\mathbf{X}$ (next), $\mathbf{F}$ (eventually) and $\mathbf{G}$ (always) have the highest binding power, followed by the binary operators $\mathbf{U}$ (until) and $\mathbf{W}$ (weak until). The remaining binary operators follow with binding power decreasing in the usual order from $\wedge$ (conjunction) to $\vee$ (disjunction) to $\rightarrow$ (implication) to $\leftrightarrow$ (equivalence).

We say that a formula is in negation normal form (NNF) if it does not contain implication nor equivalence and negation is applied only to atomic propositions.

3.1   Classical Semantics

By interpreting formulas over the Boolean algebra $\mathbf{INF}$, we provide a semantical definition of the classical linear-time temporal logic LTL[1], where the classical interpretation function $\mathrm{Mod}_c : Form \rightarrow INF$ is defined recursively in figure 1. This definition makes use of the monotone functions next_c and $\mathrm{untilnext}[L_1, L_2]_c$ (with parameters $L_1, L_2 \in INF$) on INF, which map a language L to $\mathrm{next}_c(L) = \Sigma L$ and $\mathrm{untilnext}[L_1, L_2]_c(L) = L_2 \cup (L_1 \cap \mathrm{next}_c(L))$, respectively.

Given sets of formulas Φ and Ψ, we say that Φ classically entails Ψ, denoted by $\Phi \models_c \Psi$, if $\bigcap_{\varphi \in \Phi} \mathrm{Mod}_c(\varphi) \subseteq \bigcap_{\psi \in \Psi} \mathrm{Mod}_c(\psi)$. If Φ is a singleton set $\{\varphi\}$, we may omit set braces and write $\varphi \models_c \Psi$ in place of $\{\varphi\} \models_c \Psi$; similarly for $\Psi = \{\psi\}$. If Φ is the empty set, we may write $\models_c \Psi$ in place of $\emptyset \models_c \Psi$. We call ψ a classical tautology if $\models_c \psi$.

3.2   Intuitionistic Semantics

Similar to the classical logic LTL above, we define an intuitionistic variant called $ILTL$ by interpreting formulas over the Heyting algebra $\mathbf{PREF}$, where the intuitionistic interpretation function $\mathrm{Mod}_i : Form \rightarrow PREF$ is defined recursively in figure 2. This definition uses the monotone functions next_i and $\mathrm{untilnext}[L_1, L_2]_i$ (with parameters $L_1, L_2 \in PREF$) on $PREF$, which map a language $L \in PREF$ to $\mathrm{next}_i(L) = \Sigma \cup \Sigma L$ and $\mathrm{untilnext}[L_1, L_2]_i(L) = L_2 \cup (L_1 \cap \mathrm{next}_i(L))$, respectively.

---

[1] Although presented differently, this semantics agrees with the standard semantical definition of LTL, cf. [18] or [8].

$$\mathrm{Mod_c}(\top) = \Sigma^\omega \qquad\qquad\qquad \mathrm{Mod_c}(\bot) = \emptyset$$
$$\mathrm{Mod_c}(\varphi \wedge \psi) = \mathrm{Mod_c}(\varphi) \cap \mathrm{Mod_c}(\psi) \qquad \mathrm{Mod_c}(\neg\varphi) = -\mathrm{Mod_c}(\varphi)$$
$$\mathrm{Mod_c}(\varphi \vee \psi) = \mathrm{Mod_c}(\varphi) \cup \mathrm{Mod_c}(\psi) \qquad \mathrm{Mod_c}(\varphi \rightarrow \psi) = \mathrm{Mod_c}(\neg\varphi \vee \psi)$$
$$\mathrm{Mod_c}(p) = \Sigma_p \Sigma^\omega = \{w \in \Sigma^\omega \mid \exists a \in \Sigma_p \exists u \in \Sigma^\omega : w = au\}$$
$$\mathrm{Mod_c}(\mathbf{X}\varphi) = \mathrm{next_c}(\mathrm{Mod_c}(\varphi))$$
$$\mathrm{Mod_c}(\varphi \, \mathbf{U} \, \psi) = \bigcup_{n<\omega} \mathrm{untilnext}[\mathrm{Mod_c}(\varphi), \mathrm{Mod_c}(\psi)]_c^n(\emptyset)$$
$$\mathrm{Mod_c}(\varphi \, \mathbf{W} \, \psi) = \bigcap_{n<\omega} \mathrm{untilnext}[\mathrm{Mod_c}(\varphi), \mathrm{Mod_c}(\psi)]_c^n(\Sigma^\omega)$$
$$\mathrm{Mod_c}(\mathbf{F}\varphi) = \bigcup_{n<\omega} \mathrm{next}_c^n(\mathrm{Mod_c}(\varphi)) = \mathrm{Mod_c}(\top \, \mathbf{U} \, \varphi)$$
$$\mathrm{Mod_c}(\mathbf{G}\varphi) = \bigcap_{n<\omega} \mathrm{next}_c^n(\mathrm{Mod_c}(\varphi)) = \mathrm{Mod_c}(\varphi \, \mathbf{W} \, \bot)$$

Fig. 1. Classical interpretation of formulas

Given sets of formulas Φ and Ψ, we say that Φ intuitionistically entails Ψ, denoted by $\Phi \models_i \Psi$, if $\bigcap_{\varphi \in \Phi} \mathrm{Mod_i}(\varphi) \subseteq \bigcap_{\psi \in \Psi} \mathrm{Mod_i}(\psi)$. As in the classical case, we may omit set braces around single formulas, and we may omit the empty set on the left-hand side. We call ψ an intuitionistic tautology if $\models_i \psi$.

Proposition 1. *For all formulas φ and ψ, $\varphi \models_i \psi$ if and only if $\models_i \varphi \rightarrow \psi$.*

In summary, the definition of the intuitionistic semantics is largely analogous to the definition of the classical semantics, except for the intuitionistic interpretation of implication and negation and a slight difference in the treatment of the next operator. Note that these differences are forced by the carrier *PREF* of the Heyting algebra, as the classical interpretations do not result in prefix-closed sets.

3.3 Expressive Power

Comparing the expressive power of *LTL* and *ILTL* amounts to comparing the sets of behaviors that can be specified by formulas in these logics. Unfortunately, *LTL*

$$\mathrm{Mod_i}(\top) = \Sigma^\infty \qquad\qquad\qquad \mathrm{Mod_i}(\bot) = \emptyset$$
$$\mathrm{Mod_i}(\varphi \wedge \psi) = \mathrm{Mod_i}(\varphi) \cap \mathrm{Mod_i}(\psi) \qquad \mathrm{Mod_i}(\varphi \rightarrow \psi) = \mathrm{Mod_i}(\varphi) \Rightarrow \mathrm{Mod_i}(\psi)$$
$$\mathrm{Mod_i}(\varphi \vee \psi) = \mathrm{Mod_i}(\varphi) \cup \mathrm{Mod_i}(\psi) \qquad \mathrm{Mod_i}(\neg\varphi) = \mathrm{Mod_i}(\varphi \rightarrow \bot)$$
$$\mathrm{Mod_i}(p) = \Sigma_p \cup \Sigma_p \Sigma^\infty = \{w \in \Sigma^\infty \mid \exists a \in \Sigma_p \exists u \in \Sigma^\infty : w = a \text{ or } w = au\}$$
$$\mathrm{Mod_i}(\mathbf{X}\varphi) = \mathrm{next_i}(\mathrm{Mod_i}(\varphi))$$
$$\mathrm{Mod_i}(\varphi \, \mathbf{U} \, \psi) = \bigcup_{n<\omega} \mathrm{untilnext}[\mathrm{Mod_i}(\varphi), \mathrm{Mod_i}(\psi)]_i^n(\emptyset)$$
$$\mathrm{Mod_i}(\varphi \, \mathbf{W} \, \psi) = \bigcap_{n<\omega} \mathrm{untilnext}[\mathrm{Mod_i}(\varphi), \mathrm{Mod_i}(\psi)]_i^n(\Sigma^\infty)$$
$$\mathrm{Mod_i}(\mathbf{F}\varphi) = \bigcup_{n<\omega} \mathrm{next}_i^n(\mathrm{Mod_i}(\varphi)) = \mathrm{Mod_i}(\top \, \mathbf{U} \, \varphi)$$
$$\mathrm{Mod_i}(\mathbf{G}\varphi) = \bigcap_{n<\omega} \mathrm{next}_i^n(\mathrm{Mod_i}(\varphi)) = \mathrm{Mod_i}(\varphi \, \mathbf{W} \, \bot)$$

Fig. 2. Intuitionistic interpretation of formulas

and *ILTL* interpret formulas over the two different algebras **INF** and **PREF**, so we cannot directly compare their interpretations. However, using the defining mappings inf : $\mathcal{P}(\Sigma^\infty) \rightarrow$ *INF* and pref : $\mathcal{P}(\Sigma^\infty) \rightarrow$ *PREF* of these algebras, we can map the carrier of each algebra to (a subset of) the carrier of the other and thus compare.

Expressive Power in **INF**. First, we compare *LTL* and *ILTL* in the Boolean algebra of sets of infinite behaviors **INF**, i.e., we restrict the intuitionistic semantics to infinite words via inf. The proposition below relates the semantics for formulas in negation normal form. From this proposition follows that intuitionistic entailment of formulas in NNF implies classical entailment and that in **INF**, *ILTL* is at least as expressive as *LTL*.

Proposition 2. *If φ is a formula in NNF then* $\mathrm{Mod}_c(\varphi) = \inf(\mathrm{Mod}_i(\varphi))$.

Corollary 3. *Let Φ and Ψ be sets of formulas in NNF. If $\Phi \models_i \Psi$ then $\Phi \models_c \Psi$.*

Corollary 4. *In* **INF***, ILTL is at least as expressive as LTL.*

It is unknown whether the converse of Corollary 4 is also true, i.e., whether for all formulas ψ there exist formulas φ such that $\inf(\mathrm{Mod}_i(\psi)) = \mathrm{Mod}_c(\varphi)$. We conjecture that this is the case. However, this seems difficult to prove since in intuitionistic logics, we cannot use equivalence transformations to normal forms like NNF.

Expressive Power in **PREF**. Now, we compare *LTL* and *ILTL* in the Heyting algebra of prefix-closed sets of behaviors **PREF**, i.e., we extend the classical semantics into prefix-closed sets via pref. The proposition below shows that the two logics cannot be equally expressive in **PREF**.

Proposition 5. *There is no formula φ with* $\mathrm{pref}(\mathrm{Mod}_c(\varphi)) = \Sigma = \mathrm{Mod}_i(\mathbf{X}\bot)$.

This implies that either the two logics are incomparable or *ILTL* is strictly more expressive than *LTL*, but it is not known which case holds true. We conjecture that *ILTL* is more expressive than *LTL*, yet proving this, i.e., proving that for all formulas φ there exist formulas ψ such that $\mathrm{pref}(\mathrm{Mod}_c(\varphi)) = \mathrm{Mod}_i(\psi)$, might require a lemma similar to Proposition 2. However, such a lemma seems difficult to obtain. In particular, the proof of Proposition 2 cannot be directly adapted since it exploits the fact that inf distributes over intersections, which pref does not do.

3.4   Application: Assume-Guarantee Specifications

Modular verification naturally demands for so-called assume-guarantee specifications (A-G specs), which are pairs of formulas in some temporal logic. Informally, a component of a system satisfies an A-G spec $\varphi \xrightarrow{+} \psi$ if the component satisfies the guarantee ψ at least as long as its environment (including the other components) meets the assumption φ. Once A-G specs are available for all components, properties of the global system may be deduced from the composition (i. e., conjunction) of these A-G specs instead of the (potentially large) parallel composition of all components. Due to possibly circular dependencies between assumptions and guarantees, composing A-G specs in a sound way requires non-trivial composition rules, see for instance [1], [10] or [13].

In the Heyting algebra of prefix-closed sets of finite behaviors, [3] demonstrates that under a suitable notion of concurrency (shared variables and interleaving execution) an A-G spec $\varphi \xrightarrow{+} \psi$ corresponds to an intuitionistic implication $\varphi \rightarrow \psi$, which gave rise to composition rules based on conjunction of intuitionistic implication. Later, Abadi and Merz [2] found a more general interpretation of the operator $\xrightarrow{+}$, which again can be reduced to intuitionistic implication. Here, we present their interpretation of $\xrightarrow{+}$ in **PREF**, the Heyting algebra of prefix-closed sets of finite and infinite behaviors. For $\varphi, \psi \in Form$, the semantics of $\varphi \xrightarrow{+} \psi$ is defined by

$$\mathrm{Mod}_i(\varphi \xrightarrow{+} \psi)$$
$$= \{w \in \Sigma^\infty \mid \forall v \in \mathrm{pref}(w) : \mathrm{pref!}(v) \subseteq \mathrm{Mod}_i(\varphi) \text{ implies } v \in \mathrm{Mod}_i(\psi)\} \,,$$

where $\mathrm{pref!} : \Sigma^\infty \rightarrow PREF$ maps behaviors to their sets of proper prefixes, i. e., $\mathrm{pref!}(v) = \mathrm{pref}(v) \setminus \{v\}$. By well-founded induction on the prefix order, [2] proves that for all $\varphi, \psi \in Form$, $\mathrm{Mod}_i(\varphi \xrightarrow{+} \psi) = \mathrm{Mod}_i((\psi \rightarrow \varphi) \rightarrow \psi)$, hence in **PREF**, A-G specs are merely short hands for intuitionistic implication. This fact is exploited in [2] to develop concise soundness proofs of various proof rules for conjoining circularly dependent A-G specs.

A general observation about such composition rules for A-G specs is that they essentially only admit circular dependencies on safety properties. In classical linear-time temporal logics, this can be achieved by decomposing properties into their safety and liveness parts — which is always possible thanks to the decomposition theorems in [4] and [5] — and disallowing circular dependencies on the liveness parts. Therefore, it is natural to ask for similar decomposition theorems for intuitionistic temporal logics.

4   Safety and Liveness

In this section, we introduce notions of safety and liveness for the intuitionistic temporal logic *ILTL* and compare them to the corresponding notions for *LTL* as proposed by Alpern and Schneider [4]. Actually, Alpern and Schneider did not define safety and liveness for *LTL* but for the Boolean algebra **INF** of sets of infinite behaviors, over which *LTL* formulas are interpreted. Consequently, we define safety and liveness for the Heyting algebra **PREF** of prefix-closed sets of finite and infinite behaviors.

4.1   Safety and Liveness in Classical Logics

We start by reviewing the standard notions of safety and liveness for classical linear-time temporal logics as introduced in [4]. There, safety and liveness are defined in terms of a topology on Σ^ω — in fact, the Cantor topology on Σ^ω if Σ is finite — which is induced by the topological closure operator C_c on Σ^ω with $C_c(L) = \{w \in \Sigma^\omega \mid \mathrm{pref}(w) \cap \Sigma^+ \subseteq \mathrm{pref}(L)\}$ for all $L \subseteq \Sigma^\omega$. We call $L \in INF$ a *classical safety property* if L is closed, i.e., $C_c(L) = L$, and a *classical liveness property* if L is dense, i.e., $C_c(L) = \Sigma^\omega$.

As closed sets of a topological space, classical safety properties are closed under finitary disjunction and infinitary conjunction. And as dense sets, classical liveness properties are closed under infinitary disjunction and under implication.

Proposition 6. *Let* $w \in \Sigma^\omega$, *let* $L_1, L_2 \in INF$ *and let* $\mathcal{L} \subseteq INF$.

1. Σ^ω *is a classical safety property.*
2. $\emptyset$ *is a classical safety property.*
3. $\{w\}$ *is a classical safety property.*
4. *If* L_1 *and* L_2 *are classical safety properties then so is* $L_1 \cup L_2$.
5. *If all* $L \in \mathcal{L}$ *are classical safety properties then so is* $\bigcap_{L \in \mathcal{L}} L$.

Proposition 7. *Let* $L_1, L_2 \in INF$ *and let* $\mathcal{L} \subseteq INF$.

1. Σ^ω *is a classical liveness property.*
2. *If some* $L_0 \in \mathcal{L}$ *is a classical liveness property then so is* $\bigcup_{L \in \mathcal{L}} L$.
3. *If* L_2 *is a classical liveness property then so is* $L_1 \Rightarrow L_2 = -L_1 \cup L_2$.

It is instructive to see which logical operations do not preserve classical safety or liveness properties. In the following examples, let p and q be atomic propositions.

- Neither safety nor liveness properties are closed under negation. For instance, $\mathrm{Mod}_c(\mathbf{G}p)$ is a safety property but $\mathrm{Mod}_c(\neg\mathbf{G}p) = \mathrm{Mod}_c(\mathbf{F}\neg p)$ is a liveness property.
- Safety properties are not closed under implication. E.g., $\mathrm{Mod}_c(\mathbf{G}p)$ and $\mathrm{Mod}_c(\mathbf{G}q)$ are safety properties but $\mathrm{Mod}_c(\mathbf{G}p \to \mathbf{G}q) = \mathrm{Mod}_c(\mathbf{F}\neg p \vee \mathbf{G}q)$ is a liveness property.
- Safety properties cannot be closed under infinitary disjunction. Otherwise, every $L \in INF$ would be a safety property because $L = \bigcup_{w \in L}\{w\}$.
- Liveness properties are not closed under intersection. E.g., $\mathrm{Mod}_c(\mathbf{GF}p)$ and $\mathrm{Mod}_c(\mathbf{FG}\neg p)$ are both liveness properties but $\mathrm{Mod}_c(\mathbf{GF}p \wedge \mathbf{FG}\neg p) = \mathrm{Mod}_c(\mathbf{GF}p \wedge \neg\mathbf{GF}p)$ is not.

The (trivial) property Σ^ω is the only one which is both a safety and liveness property, but there are many properties which are neither. For instance, $L = \mathrm{Mod}_c(p \mathbf{U} q)$ is such a property because $C_c(L) = \mathrm{Mod}_c(p \mathbf{U} q \vee \mathbf{G}p) \neq L$ and $C_c(L) \neq \Sigma^\omega$. However, [4] at least proves that all properties in classical linear-time temporal logics can be decomposed into their safety and liveness parts.

Proposition 8. *Every* $L \in INF$ *is the conjunction of a classical safety and a classical liveness property. More precisely,* $L = C_c(L) \cap (-C_c(L) \cup L)$.

4.2   Safety and Liveness in Intuitionistic Logics

To transfer the notions of safety and liveness to the Heyting algebra $\boldsymbol{PREF}$, we generalize the closure operator $C_c : \mathcal{P}(\Sigma^\omega) \to \mathcal{P}(\Sigma^\omega)$ to $C_i : \mathcal{P}(\Sigma^\infty) \to \mathcal{P}(\Sigma^\infty)$ by defining $C_i(L) = \{w \in \Sigma^\infty \mid \mathrm{pref}(w) \cap \Sigma^+ \subseteq \mathrm{pref}(L)\}$. It turns out that C_i is a topological closure operator on Σ^∞ and hence induces a topology — in fact, it induces the Scott topology on Σ^∞ (ordered by the prefix order) if Σ is countable. Thus, we can reuse the topological definitions of safety and liveness and call $L \in PREF$ an *intuitionistic safety property* if $C_i(L) = L$ and an *intuitionistic liveness property* if $C_i(L) = \Sigma^\infty$.

Note that C_i is algebraically definable in $\boldsymbol{PREF}$ because for all $L \in PREF$, $C_i(L) = \{w \in \Sigma^\infty \mid \mathrm{pref}(w) \cap \Sigma^+ \subseteq L\} = \Sigma^+ \Rightarrow L$. Therefore, L is an intuitionistic safety property iff $\Sigma^+ \Rightarrow L = L$ iff $\Sigma^+ \Rightarrow L \subseteq L$, and L is an intuitionistic liveness property iff $\Sigma^+ \Rightarrow L = \Sigma^\infty$ iff $\Sigma^+ \subseteq L$ iff $\Sigma^+ \cup L = L$. For comprehending these algebraic definitions, the following intuition might help. Safety and liveness properties differ fundamentally in the way they constrain finite and infinite behaviors. If a safety property is refuted then it can always be refuted by a finite behavior, whereas a liveness property can never be refuted by a finite behavior. So one could say that a safety property L essentially only constrains finite behaviors in the sense that whenever all finite prefixes of an infinite behavior w satisfy L (i. e., $w \in \Sigma^+ \Rightarrow L$) then w satisfies L. Likewise, a liveness property L essentially only constrains infinite behaviors in the sense that all finite behaviors satisfy L.

Intuitionistic safety and liveness properties are closed under essentially the same logical operations as their classical counterparts. Additionally, intuitionistic safety properties are closed under (intuitionistic) implication and negation, and intuitionistic liveness properties are closed under infinitary conjunction.

Proposition 9. *Let $w \in \Sigma^\infty$, let $L, L_1, L_2 \in PREF$ and let $\mathcal{L} \subseteq PREF$.*

1. Σ^∞ *is an intuitionistic safety property.*
2. $\emptyset$ *is an intuitionistic safety property.*
3. $\mathrm{pref}(w)$ *is an intuitionistic safety property.*
4. *If L_1 and L_2 are intuitionistic safety properties then so is $L_1 \cup L_2$.*
5. *If all $L \in \mathcal{L}$ are intuitionistic safety properties then so is $\bigcap_{L \in \mathcal{L}} L$.*
6. *If L_2 is an intuitionistic safety property then so is $L_1 \Rightarrow L_2$.*
7. *If L is an intuitionistic safety property then so is $-L = L \Rightarrow \emptyset$.*

Proof. Claims 2 and 3 follow from the definition of safety because $\Sigma^+ \Rightarrow \emptyset = \emptyset$ and $\Sigma^+ \Rightarrow \mathrm{pref}(w) = \{v \in \Sigma^\infty \mid \mathrm{pref}(v) \cap \Sigma^+ \subseteq \mathrm{pref}(w)\} = \mathrm{pref}(w)$. All other claims follow from Propositions 15 and 17 and Corollary 16, see next section.   □

Proposition 10. *Let $L_1, L_2 \in PREF$ and let $\mathcal{L} \subseteq PREF$.*

1. Σ^∞ *is an intuitionistic liveness property.*
2. *If some $L_0 \in \mathcal{L}$ is an intuitionistic liveness property then so is $\bigcup_{L \in \mathcal{L}} L$.*
3. *If L_2 is a intuitionistic liveness property then so is $L_1 \Rightarrow L_2$.*
4. *If all $L \in \mathcal{L}$ are intuitionistic liveness properties then so is $\bigcap_{L \in \mathcal{L}} L$.*

Proof. Follows from Proposition 18, see next section.   □

We notice that intuitionistic safety properties are not closed under infinitary disjunction, for the same reason as in the classical case. And intuitionistic liveness properties are not closed under (intuitionistic) negation. E. g., $\mathrm{Mod}_i(\mathbf{F}p)$ is a liveness property but $\mathrm{Mod}_i(\neg\mathbf{F}p) = \mathrm{Mod}_i(\bot)$ is not.

Similar to the classical case, Σ^∞ is the only property which is both an intuitionistic safety and liveness property, cf. Proposition 20. Again, there are many properties which are neither; this follows from Proposition 13 below. Yet, there is also the following decomposition theorem.

Proposition 11. *Every $L \in PREF$ is the conjunction of an intuitionistic safety and an intuitionistic liveness property. More precisely, $L = (\Sigma^+ \Rightarrow L) \cap (\Sigma^+ \cup L)$.*

Proof. Follows from Proposition 19, see next section.   □

So far, our approach to safety and liveness was purely semantical, relying only on the operators of the Heyting algebra **PREF** and the constant Σ^+. However, these operators correspond to the intuitionistic connectives of *ILTL*, and Σ^+ is expressible in *ILTL*, namely $\Sigma^+ = \mathrm{Mod}_i(\mathbf{F}\bot)$. Immediately, this gives us a simple logical characterization of intuitionistic safety and liveness and a logical formulation of the decomposition theorem.

Corollary 12. *Let φ be a formula.*

1. *φ is an intuitionistic safety property if and only if $\models_i (\mathbf{F}\bot \to \varphi) \to \varphi$.*
2. *φ is an intuitionistic liveness property if and only if $\models_i \mathbf{F}\bot \to \varphi$.*
3. *$\models_i \varphi \leftrightarrow (\mathbf{F}\bot \to \varphi) \wedge (\mathbf{F}\bot \vee \varphi)$.*

4.3   Classical Versus Intuitionistic Safety and Liveness

In Section 3, the mappings inf : $\mathcal{P}(\Sigma^\infty) \to INF$ and pref : $\mathcal{P}(\Sigma^\infty) \to PREF$ were used to compare the expressive power of the logics *LTL* and *ILTL*. Now, we will use the same mappings to investigate the relationship between the classical notions of safety and liveness and their intuitionistic counterparts.

It turns out that the intuitionistic notions of safety and liveness subsume the classical ones because every classical safety resp. liveness property is mapped to a corresponding intuitionistic property via pref. However, only the classical notion of safety subsumes the intuitionistic one in the sense that every intuitionistic safety property is mapped to a corresponding classical property via inf. For liveness this is not the case. For instance, Σ^+ is an intuitionistic liveness property to which no corresponding classical property exists, in particular $\inf(\Sigma^+) = \emptyset$ is not a classical liveness property.

Proposition 13. *Let $L \in INF$.*

1. *L is a classical safety property iff $\mathrm{pref}(L)$ is an intuitionistic one.*
2. *L is a classical liveness property iff $\mathrm{pref}(L)$ is an intuitionistic one.*

Proposition 14. *Let $L \in PREF$.*

1. *If L is an intuitionistic safety property then $\inf(L)$ is a classical one.*
2. *If $\inf(L)$ is a classical liveness property then L is an intuitionistic one.*

Note that the statements of Proposition 14 cannot be reversed. To see this let $L = \mathrm{Mod}_i(\mathbf{F}\bot \vee \mathbf{G}p)$, where p is an atomic proposition. Then $\inf(L) = \mathrm{Mod}_c(\mathbf{G}p)$. Thus, $\inf(L)$ is a classical safety property but $\varSigma^+ \Rightarrow L = \varSigma^\infty \neq L$, so L is not an intuitionistic safety property. However, $\varSigma^+ \subseteq L$, so L is an intuitionistic liveness property but $\inf(L)$ is not a classical one.

5   Algebraic Characterization of Safety and Liveness

In this section, we develop notions of safety and liveness and prove a decomposition theorem for arbitrary Heyting algebras. Thus, we provide abstract algebraic proofs for the claims of the previous section about safety and liveness in the concrete Heyting algebra of prefix-closed sets of behaviors $\boldsymbol{PREF}$.

Let $\boldsymbol{H} = \langle H, \sqcap, \sqcup, \Rightarrow, \top, \bot \rangle$ be a Heyting algebra. We denote the order relation on this algebra by $\sqsubseteq$. Recall that $\langle H, \sqcap, \sqcup \rangle$ is a distributive lattice with $\top$ and $\bot$ and for all $x, y, z \in H$, $z \sqsubseteq x \Rightarrow y$ if and only if $x \sqcap z \sqsubseteq y$. This equivalence can be seen as the definition of $x \Rightarrow y$, the pseudo-complement of x relative to y. For $x \in H$, we denote by $-x$ the pseudo-complement of x, which is defined as $-x = x \Rightarrow \bot$. Note that if the law of excluded middle holds in $\boldsymbol{H}$ (i.e., if $x \sqcup -x = \top$ for all $x \in H$) then $x \Rightarrow y = -x \sqcup y$.

By $\mathcal{J}(\boldsymbol{H})$, we denote the join-irreducible elements in $\boldsymbol{H}$, where $j \in H$ is join-irreducible iff $j \neq \bot$ and for all $x, y \in H$, $j = x \sqcup y$ implies $j = x$ or $j = y$. Note that for $j \in \mathcal{J}(\boldsymbol{H})$ and $x, y \in H$, $j \sqsubseteq x \sqcup y$ implies $j \sqsubseteq x$ or $j \sqsubseteq y$ because $\boldsymbol{H}$ is distributive. We call a subset $S \subseteq H$ join-dense in $\boldsymbol{H}$ iff for every $x \in H$ there exists $T \subseteq S$ such that $x = \bigsqcup T$. We call a subset $S \subseteq H$ a forest iff for each $x \in S$, the set $T = \{y \in S \mid y \sqsubseteq x\}$ induces a linear suborder of $\boldsymbol{H}$, i.e., for all $u, v \in T$, $u \sqsubseteq v$ or $v \sqsubseteq u$.

Throughout this section, we fix an arbitrary element $a \in H$, relative to which we will define safety and liveness. In $\boldsymbol{H}$, this a plays the role of $\varSigma^+$ in $\boldsymbol{PREF}$, i.e., it separates the 'finite' from the 'infinite' behaviors. Remarkably, the closure properties (except for closure under negation) and the decomposition theorem below hold independent of the choice of a. Thus in $\boldsymbol{PREF}$, we may well choose non-standard separating elements, for instance $\varSigma$, to define interesting non-standard notions of safety and liveness.

5.1   Safe Elements

We define the function $\mathrm{safe}_a : H \to H$ by $\mathrm{safe}_a(x) = a \Rightarrow x$. The function safe_a is a closure operator, hence we call safe_a the *safety closure*. We call an element $x \in H$ *a-safe* if x is a fixpoint of this closure, i.e., $\mathrm{safe}_a(x) = x$.

We investigate whether safe elements are closed under the operations of the Heyting algebra and hence under the corresponding intuitionistic connectives. It turns out that safe elements are closed under implication and conjunction, even

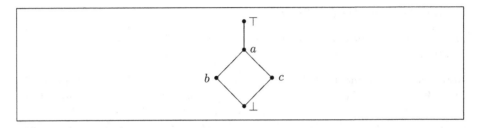

Fig. 3. A Heyting algebra where the a-safe elements are not closed under join

under infinitary conjunction. Whether safe elements are closed under negation depends on $\bot$ being safe.

Proposition 15. *Let $x, y \in H$, and let $S \subseteq H$ such that $\bigsqcap S$ exists.*

1. $\top$ *is a-safe.*
2. *If y is a-safe then $x \Rightarrow y$ is a-safe.*
3. *If all $s \in S$ are a-safe then $\bigsqcap S$ is a-safe.*

Proof. Assume that y and all $s \in S$ are a-safe.

1. $\mathrm{safe}_a(\top) = a \Rightarrow \top = \top$.
2. $\mathrm{safe}_a(x \Rightarrow y) = a \Rightarrow (x \Rightarrow y) = (a \sqcap x) \Rightarrow y = (x \sqcap a) \Rightarrow y = x \Rightarrow (a \Rightarrow y) = x \Rightarrow y$, where the last equality holds because y is a-safe.
3. $\mathrm{safe}_a(\bigsqcap S) = a \Rightarrow \bigsqcap S = \bigsqcap_{s \in S}(a \Rightarrow s) = \bigsqcap_{s \in S} s = \bigsqcap S$, where the second equality holds because $\Rightarrow$ completely distributes over meets on the right-hand side, and the third equality holds because all s are a-safe. $\qquad \square$

Corollary 16. *The following statements are equivalent:*

1. *For all $x \in H$, if x is a-safe then $-x$ is a-safe.*
2. $\bot$ *is a-safe.*

In general, safe elements are not closed under disjunction. For instance, in the Heyting algebra in figure 3, b and c are a-safe because $a \Rightarrow b = b$ and $a \Rightarrow c = c$, but $a \Rightarrow (b \sqcup c) = a \Rightarrow a = \top$, so $b \sqcup c$ is not a-safe. Yet, if the Heyting algebra $\boldsymbol{H}$ satisfies a natural condition, namely that the join-irreducible elements form a join-dense forest, then safe elements are closed under finite disjunction.

Proposition 17. *Let $\mathcal{J}(\boldsymbol{H})$ be a forest, which is join-dense in $\boldsymbol{H}$. Let $x, y \in H$. If x and y are a-safe then $x \sqcup y$ is a-safe.*

Proof. Omitted due to lack of space; see [14] instead. $\qquad \square$

Note that in the Heyting algebra in figure 3, safe elements fail to be closed under disjunction because the join-irreducibles b, c and $\top$ do not form a forest. However, in the Heyting algebra $\boldsymbol{PREF}$ of prefix-closed sets of behaviors, the join-irreducibles are the prefix-closures of single behaviors, i.e., $\mathcal{J}(\boldsymbol{PREF}) = \{\mathrm{pref}(w) \mid w \in \Sigma^\infty\}$. Obviously, $\mathcal{J}(\boldsymbol{PREF})$ forms a forest, which is join-dense in $\boldsymbol{PREF}$. Hence, safety properties in $\boldsymbol{PREF}$ are closed under finite disjunction.

5.2   Live Elements

We define the function $\text{live}_a : H \rightarrow H$ by $\text{live}_a(x) = a \sqcup x$. The function live_a is a closure operator, hence we call live_a the *liveness closure*. We call an element $x \in H$ *a-live* if x is a fixpoint of this closure, i. e., $\text{live}_a(x) = x$.

Similar to the case for safe elements, we investigate whether live elements are closed under the operations of the Heyting algebra and hence under the corresponding intuitionistic connectives. It turns out that live elements are closed under implication and under finitary and infinitary conjunction and disjunction.

Proposition 18. *Let $x, y \in H$, and let $S, T \subseteq H$ such that $\bigsqcap S$ and $\bigsqcup T$ exist.*

1. $\top$ *is a-live.*
2. *If y is a-live then $x \Rightarrow y$ is a-live.*
3. *If all $s \in S$ are a-live then $\bigsqcap S$ is a-live.*
4. *If some $t_0 \in T$ is a-live then $\bigsqcup T$ is a-live.*

Proof. Assume that y and all $s \in S$ are a-live, and let $t_0 \in T$ be a-live.

1. $\text{live}_a(\top) = a \sqcup \top = \top$.
2. As $\text{live}_a(y) = a \sqcup y = y$, we have $a \sqsubseteq y = y \sqcap (x \Rightarrow y) \sqsubseteq x \Rightarrow y$. Hence $\text{live}_a(x \Rightarrow y) = a \sqcup (x \Rightarrow y) = x \Rightarrow y$.
3. As $\text{live}_a(s) = a \sqcup s = s$ for all $s \in S$, we have $a \sqsubseteq s$ for all $s \in S$, so $a \sqsubseteq \bigsqcap S$. Hence $\text{live}_a(\bigsqcap S) = a \sqcup \bigsqcap S = \bigsqcap S$.
4. $\text{live}_a(\bigsqcup T) = a \sqcup \bigsqcup T = a \sqcup t_0 \sqcup \bigsqcup T = t_0 \sqcup \bigsqcup T = \bigsqcup T$, where the third equality holds because t_0 is a-live.   □

5.3   Decomposition Theorem

With the above notions of safety and liveness, just simple reasoning with the laws of Heyting algebras proves that every element of the algebra can be decomposed into a conjunction of a safe and a live part.

Proposition 19. *Every $x \in H$ is the meet of an a-safe and an a-live element. More precisely, $x = \text{safe}_a(x) \sqcap \text{live}_a(x)$.*

Proof. $\text{safe}_a(x) \sqcap \text{live}_a(x) = (a \Rightarrow x) \sqcap (a \sqcup x) = ((a \Rightarrow x) \sqcap a) \sqcup ((a \Rightarrow x) \sqcap x) = (a \sqcap x) \sqcup x = x$, where the third equality holds due to the cancellation laws for the relative pseudo-complement in Heyting algebras, which say that $y \sqcap (y \Rightarrow z) = y \sqcap z$ and $(y \Rightarrow z) \sqcap z = z$ for all $y, z \in H$.   □

The above decomposition might be trivial, for instance in the case that x is both safe and live. However, the following proposition shows that this cannot happen for non-trivial x because safe and live elements are separated.

Proposition 20. *No non-trivial element in H is both a-safe and a-live. More precisely, if $x \in H$ is a-safe and a-live then $x = \top$.*

Proof. Let $x \in H$ be a-safe and a-live. Then $x = \text{safe}_a(x) = \text{safe}_a(\text{live}_a(x)) = a \Rightarrow (a \sqcup x) = \top$, where the last equality holds because $y \Rightarrow z = \top$ for all $y, z \in H$ with $y \sqsubseteq z$. $\qquad \square$

Whether there are elements which are neither safe nor live (so that the above decomposition is really non-trivial) depends on the Heyting algebra. For example, all elements in figure 3 are a-safe ($\bot$, b, c, $\top$) or a-live (a, $\top$). However as shown in the previous section, in the Heyting algebra **PREF** of prefix-closed sets of behaviors, there are elements which are neither Σ^+-safe nor Σ^+-live.

Finally, we note that when the Heyting algebra H happens to be a Boolean algebra, the definition of the liveness closure can be reduced to the safety closure, as it is the case in most decomposition theorems, see for instance [4] or [16].

Proposition 21. *If the law of excluded middle holds in H then for all $x \in H$,* $\text{live}_a(x) = \text{safe}_a(x) \Rightarrow x.$

Proof. $\text{safe}_a(x) \Rightarrow x = x \sqcup -\text{safe}_a(x) = x \sqcup -(a \Rightarrow x) = x \sqcup -(-a \sqcup x) = x \sqcup (a \sqcap -x) = (x \sqcup a) \sqcap (x \sqcup -x) = (x \sqcup a) \sqcap \top = x \sqcup a = live_a(x).$ $\qquad \square$

6   Conclusion

We have presented *ILTL*, an intuitionistic variant of the linear-time temporal logic *LTL*, which is capable of specifying sets of finite and infinite behaviors simultaneously. The intuitionistic nature of *ILTL* comes in handy when doing assume-guarantee reasoning, because special temporal operators that have been introduced to reason about assume-guarantee specifications are definable via the intuitionistic implication. Furthermore, we have given an abstract algebraic definition of notions of safety and liveness suitable for intuitionistic temporal logics. These intuitionistic notions are similar to the classical ones, yet they are more compatible with the logical connectives; in particular, intuitionistic liveness properties are closed under conjunction. The logic *ILTL* admits an elegant logical characterization of intuitionistic safety and liveness. It remains to be investigated whether our abstract algebraic definition of safety and liveness also applies to other intuitionistic temporal logics, e. g., to intuitionistic variants of CTL.

There are a still number of unresolved questions concerning the logic *ILTL*. The exact expressive power should be determined, one should give an axiomatization, and one should address decidability and complexity of the satisfiability and model checking problems. Whether *ILTL* can be considered a useful specification language depends on the answers to these questions.

Acknowledgment. The author thanks Viorica Sofronie-Stokkermans for the many intensive discussions that greatly helped to develop and clarify the ideas of this paper. Thanks to an anonymous reviewer for directing our attention to the connection between intuitionistic safety/liveness and the Scott topology on strings.

References

1. Martín Abadi and Leslie Lamport. Conjoining specifications. *ACM Transactions on Programming Languages and Systems*, 17(3):507–534, 1995.
2. Martín Abadi and Stephan Merz. An abstract account of composition. In *20th International Symposium on Mathematical Foundations of Computer Science (MFCS)*, LNCS 969, pages 499–508. Springer, 1995.
3. Martín Abadi and Gordon D. Plotkin. A logical view of composition. *Theoretical Computer Science*, 114:3–30, 1993.
4. Bowen Alpern and Fred B. Schneider. Defining liveness. *Information Processing Letters*, 21(4):181–185, 1985.
5. Bowen Alpern and Fred B. Schneider. Recognizing safety and liveness. *Distributed Computing*, 2(3):117–126, 1987.
6. Rowan Davies. A temporal-logic approach to binding-time analysis. In *Proceedings of the 11th IEEE Symposium on Logic in Computer Science (LICS)*, pages 184–195. IEEE Computer Society, 1996.
7. Cindy Eisner, Dana Fisman, John Havlicek, Yoad Lustig, Anthony McIsaac, and David Van Campenhout. Reasoning with temporal logic on truncated paths. In *15th International Conference on Computer Aided Verification (CAV)*, LNCS 2725, pages 27–39. Springer, 2003.
8. E. Allen Emerson. Temporal and modal logic. In Jan van Leeuwen, editor, *Handbook of Theoretical Computer Science*, volume B, pages 995–1072. Elsevier, 1990.
9. H. Peter Gumm. Another glance at the Alpern-Schneider characterization of safety and liveness in concurrent executions. *Information Processing Letters*, 47(6):291–294, 1993.
10. Bengt Jonsson and Yih-Kuen Tsay. Assumption/guarantee specifications in linear-time temporal logic. *Theoretical Computer Science*, 167:47–72, 1996.
11. Leslie Lamport. Proving the correctness of multiprocess programs. *IEEE Transactions on Software Engineering*, 3(2):125–143, 1977.
12. Orna Lichtenstein, Amir Pnueli, and Lenore Zuck. The glory of the past. In *Logic of Programs*, LNCS 193, pages 196–218. Springer, 1985.
13. Patrick Maier. *A Lattice-Theoretic Framework For Circular Assume-Guarantee Reasoning*. PhD thesis, Universität des Saarlandes, Saarbrücken, July 2003.
14. Patrick Maier. Intuitionistic LTL and a new characterization of safety and liveness. Technical Report MPI-I-2004-2-002, Max-Planck-Institut für Informatik, 2004.
15. Panagiotis Manolios and Richard Trefler. Safety and liveness in branching time. In *Proceedings of the 16th IEEE Symposium on Logic in Computer Science (LICS)*, pages 366–374. IEEE Computer Society, 2001.
16. Panagiotis Manolios and Richard Trefler. A lattice-theoretic characterization of safety and liveness. In *Proceedings of the 22nd ACM Symposium on Principles of Distributed Computing (PODC)*, pages 325–333. ACM Press, 2003.
17. Gordon Plotkin and Colin Stirling. A framework for intuitionistic modal logics. In *Proceedings of the 1st Conference on Theoretical Aspects of Reasoning about Knowledge (TARK)*, pages 399–406. Morgan Kaufmann, 1986.
18. Amir Pnueli. The temporal semantics of concurrent programs. *Theoretical Computer Science*, 13:45–60, 1981.
19. A. Prasad Sistla. Safety, liveness and fairness in temporal logic. *Formal Aspects of Computing*, 6:495–511, 1994.

Moving in a Crumbling Network:
The Balanced Case

Philipp Rohde

RWTH Aachen, Informatik VII
rohde@informatik.rwth-aachen.de

Abstract. In this paper we continue the study of 'sabotage modal logic' SML which was suggested by van Benthem. In this logic one describes the progression along edges of a transition graph in alternation with moves of a saboteur who can delete edges. A drawback of the known results on SML is the asymmetry of the two modalities of 'moving' and 'deleting': Movements are local, whereas there is a global choice for edge deletion. To balance the situation and to obtain a more realistic model for traffic and network problems, we require that also the sabotage moves (edge deletions) are subject to a locality condition. We show that the new logic, called path sabotage logic PSL, already has the same complexities as SML (model checking, satisfiability) and that it lacks the finite model property. The main effort is finding a pruned form of SML-models that can be enforced within PSL and giving appropriate reductions from SML to PSL.

Keywords: modal logics, dynamic logics, model checking

1 Introduction

In the 'classical' framework of model checking one considers movements of agents within a system, but the underlying structure is assumed to be static. So in many formalisms only properties of unchanged systems are expressible. This motivates a more general approach where dynamic changes of the underlying structure are relevant. For example, consider a computer network where connections may break down. Some natural questions arise for such a system: Is it possible – regardless of the removed connections – to interchange information between two designated servers? Another task of this kind arises for navigation systems: Is it possible to find a way between cities within a traffic network where connections are canceled, e.g., because of roadworks or traffic jams?

To specify problems of this nature, van Benthem considered 'sabotage modal logics' which are modal logics over changing models (cf. [1]). He introduced a cross-model modality referring to submodels from which objects have been removed. SML consists of standard modal logic equipped with a 'edge-deleting' modality and is capable of expressing elementary changes of transition systems itself. One could express problems related to this situation by first order specifications, but then one has to put up with the high complexity of FO. So SML seems to be a moderate strengthening of modal logic for this kind of problems.

J. Marcinkowski and A. Tarlecki (Eds.): CSL 2004, LNCS 3210, pp. 310–324, 2004.

But in [3] and [4] we showed that the new operator already strengthens modal logic in such a way that all the nice algorithmic and model-theoretic properties of modal logic get lost. In fact, from the viewpoint of complexity, SML much more resembles FO than modal logic: Uniform model checking for SML is PSPACE-complete and the satisfiability problem is undecidable. But after all, an advantage of SML over FO is a linear formula and a polynomial program complexity of model checking.

A drawback of SML is the asymmetry of the two modalities of 'moving' and 'deleting': Movements are local, whereas the choice for edge deletion is global. So SML seems to be an appropriate specification for dynamic problems like the traffic problem mentioned above: The canceling of connections is global and (almost) independent of a movement within the system. But for other dynamic tasks SML fails to be a realistic model, especially if the 'saboteur' also has to move within the system using the same connections as the 'runner'. For example, a computer virus needs to use the same internet connections before it reaches the target that it wants to block. In this paper we introduce the path sabotage logic PSL to balance the situation: We require that the saboteur moves within the system such that exactly those edges are deleted that were taken along his path. Hence also the sabotage moves are subject to a locality condition. We show that PSL already has the same complexities as SML and that PSL also fails to have the finite model property.

In Sect. 2 we repeat the definition of SML and introduce the logic PSL. In Sect. 3 we show that model checking for PSL is PSPACE-complete and that PSL has an effective formula and program complexity. To reduce the satisfiability problem for SML to the same problem for PSL we need a kind of normal form for SML-models (relative to a given SML-formula), namely pruned models. In Sect. 4 we introduce this notion and show that every SML-model can be transformed into a pruned form. In Sect. 5 we show how to enforce within PSL that a model of a given SML-formula contains a pruned submodel together with some additional properties that we need for the reduction of the satisfiability problem.

I would like to thank Christof Löding for several comments and Benedikt Löwe who had the idea of the path sabotage logic.

2   Preliminaries

In this section we repeat the definition of the sabotage modal logic SML with a global 'edge-deleting' modality and introduce the balanced version of SML with a 'deleting by moving' modality which we call path sabotage logic PSL. We interpret both logics over edge-labeled transition systems. For that let Prop be a finite set of unary predicate symbols. A transition system $\mathcal{T}$ is a tuple (S, Σ, R, L) with a set of states S, a finite alphabet Σ, a ternary transition relation $R \subseteq S \times \Sigma \times S$ and a labeling function $L : S \to 2^{\text{Prop}}$.

Let $p \in \text{Prop}$ and $a \in \Sigma$. Formulae of the *sabotage modal logic* SML are inductively defined by the grammar

$$\varphi ::= \top \mid p \mid \neg\varphi \mid \varphi \vee \varphi \mid \Diamond_a\varphi \mid \Theta_a\varphi.$$

As usual, $\bot$ is an abbreviation for $\neg\top$. The dual modalities are defined by $\Box_a\varphi :=$
$\neg\Diamond_a\neg\varphi$ and $\boxminus_a\varphi := \neg\Diamond_a\neg\varphi$.

Let $\mathcal{T} = (S, \Sigma, R, L)$ be a transition system. For a set $E \subseteq R$ we define the
transition system $\mathcal{T} \setminus E := (S, \Sigma, R \setminus E, L)$. The semantics of SML relative to a
current position $s \in S$ are inductively defined by

$(\mathcal{T}, s) \models \top$ is true,

$(\mathcal{T}, s) \models p$ iff $p \in L(s)$,

$(\mathcal{T}, s) \models \neg\varphi$ iff not $(\mathcal{T}, s) \models \varphi$,

$(\mathcal{T}, s) \models \varphi \vee \psi$ iff $(\mathcal{T}, s) \models \varphi$ or $(\mathcal{T}, s) \models \psi$,

$(\mathcal{T}, s) \models \Diamond_a\varphi$ iff there is $s' \in S$ with $(s, a, s') \in R$ and $(\mathcal{T}, s') \models \varphi$,

$(\mathcal{T}, s) \models \Diamond_a\varphi$ iff there is $(t, a, t') \in R$ with $(\mathcal{T} \setminus \{(t, a, t')\}, s) \models \varphi$.

The sabotage modality $\Diamond$ has the global power to delete transitions somewhere in the system whereas the standard modality $\Diamond$ only allows of moving
locally. To balance the situation we introduce a new sabotage modality $\Diamond$ such
that deletion is combined with a movement that is independent of the one according to the standard modalities. Hence a current position in the system becomes
a pair of states. The syntax of the *path sabotage logic* PSL is defined in the same
way, but using the modality $\Diamond_a$ instead of $\Diamond_a$, for $a \in \Sigma$. The dual modality $\Box_a$
is defined analogously.

The semantics of PSL relative to a current position $[s, t]$ for $s, t \in S$ are
inductively defined by

$(\mathcal{T}, s, t) \models \top$ is true,

$(\mathcal{T}, s, t) \models p$ iff $p \in L(s)$,

$(\mathcal{T}, s, t) \models \neg\varphi$ iff not $(\mathcal{T}, s, t) \models \varphi$,

$(\mathcal{T}, s, t) \models \varphi \vee \psi$ iff $(\mathcal{T}, s, t) \models \varphi$ or $(\mathcal{T}, s, t) \models \psi$,

$(\mathcal{T}, s, t) \models \Diamond_a\varphi$ iff there is $s' \in S$ with $(s, a, s') \in R$ and $(\mathcal{T}, s', t) \models \varphi$,

$(\mathcal{T}, s, t) \models \Diamond_a\varphi$ iff there is $t' \in S$ with $(t, a, t') \in R$ and

         $(\mathcal{T} \setminus \{(t, a, t')\}, s, t') \models \varphi$.

Note that propositions can only be checked on paths built up by standard
modalities.

A measure for the complexity of an SML-formula φ is the number of nested
sabotage modalities. We call this the *sabotage depth* $\mathrm{sd}(\varphi)$ of φ and define inductively

$$\mathrm{sd}(\top) := \mathrm{sd}(p) := 0, \qquad\qquad \mathrm{sd}(\varphi_1 \vee \varphi_2) := \max\{\mathrm{sd}(\varphi_1), \mathrm{sd}(\varphi_2)\},$$
$$\mathrm{sd}(\neg\psi) := \mathrm{sd}(\Diamond_a\psi) := \mathrm{sd}(\psi), \qquad \mathrm{sd}(\Diamond_a\psi) := \mathrm{sd}(\psi) + 1.$$

The number of nested path sabotage operators of a PSL-formula φ is called
path sabotage depth $\mathrm{pd}(\varphi)$ and is defined analogously.

For a fixed $a \in \Sigma$, the number $\mathrm{sd}_a(\varphi)$ of nested modalities $\Diamond_a$ is defined in
the same way, but using $\mathrm{sd}_a(\Diamond_a\psi) := \mathrm{sd}_a(\psi) + 1$ and $\mathrm{sd}_a(\Diamond_b\psi) := \mathrm{sd}_a(\psi)$ for

$b \neq a$. In the next section we will see that the path sabotage depth pd(φ) of a formula φ is the main factor in the complexity of the model checking problem for PSL. But first we repeat some known results on the logic SML. The combined complexity of SML model checking, i.e., the complexity measured in terms of the size of the formula and the size of the structure, was already settled in [3]. The formula and program complexity of SML model checking was determined in [4].

Theorem 1. *1. Combined complexity: Model checking for* SML *is* PSPACE-*complete.*

 2. Formula complexity: Model checking for SML *with a fixed transition system can be solved in linear time in the size of the formula.*

 3. Program complexity: Model checking for a fixed SML*-formula can be solved in polynomial time in the size of the transition system.* □

In [4] it was also shown that, in contrast to modal logic where each satisfiable formula has a finite model, this property does not hold for SML.

Theorem 2. *There is an* SML*-formula that has only infinite models.* □

Further it was proven that the satisfiability problem for SML is undecidable. To be more precise:

Theorem 3. *The problems of deciding whether a given* SML*-formula has a model (Satisfiability), has a finite model (Finite Satisfiability), or is satisfiable, but only has infinite models (Infinity Axiom) are undecidable.* □

3   Model Checking for PSL

In this section we show that model checking for PSL is also PSPACE-complete. For membership we give a translation of PSL into first order logic. The completeness is shown by a reduction of the SML model checking problem to the one for PSL. In the rest of the section we show that PSL has an effective formula and program complexity. We do that by translating the model checking problem for PSL into the one for standard modal logic. Some proofs are slight modifications of the ones for SML that are presented in [3] and [4], so we omit the details.

By heavy use of variables one can translate PSL into first order logic. Since FO model checking is in PSPACE we obtain:

Lemma 4. *For every* PSL*-formula φ there is an effectively constructible* FO*-formula $\hat{\varphi}(x, y)$ such that for every transition system $\mathcal{T}$ and states s, t of $\mathcal{T}$ one has:*

$$(\mathcal{T}, s, t) \models \varphi \iff \mathcal{T} \models \hat{\varphi}[s, t].$$

The size of $\hat{\varphi}(x, y)$ is polynomial in the size of φ. In particular, model checking for PSL *is in* PSPACE. □

Next we give a reduction of SML model checking to PSL model checking. For an alphabet Σ and $m \geq 1$ let $\Sigma_m := \Sigma \,\dot{\cup}\, \{1, \ldots, m\}$ (w.l.o.g. we assume that $i \notin \Sigma$ for every $1 \leq i \leq m$). For a transition system $\mathcal{T} = (S, \Sigma, R, L)$ we define the transition system $\mathcal{T}_m := (S, \Sigma_m, R_m, L)$, where

$$R_m := R \,\dot{\cup}\, \{(s, i, s') \mid s, s' \in S \wedge 1 \leq i \leq m\}.$$

For $m = 0$ let $\Sigma_0 = \Sigma$ and $\mathcal{T}_0 = \mathcal{T}$. For a given SML-formula φ over Σ let the PSL-formula $\varphi^{\#}$ over $\Sigma_{\mathrm{sd}(\varphi)}$ be inductively defined as follows: $(\top)^{\#} := \top$, $(p)^{\#} := p$ and the operator $\#$ is homeomorphic for $\vee, \neg$ and $\Diamond_a$. For $\varphi = \Diamond_a \psi$ and $i = \mathrm{sd}(\varphi)$ let $\varphi^{\#} := \Diamond_i \Diamond_a \psi^{\#}$. Note that $|R_m| = |R| + m \cdot |S|^2$ and that $|\varphi^{\#}|$ is polynomial in $|\varphi|$.

Lemma 5. *For every* SML*-formula* φ, *transition system* $\mathcal{T}$, *and* $s, t \in \mathcal{T}$ *it holds*

$$(\mathcal{T}, s) \models \varphi \iff (\mathcal{T}_{\mathrm{sd}(\varphi)}, s, t) \models \varphi^{\#}.$$

Proof. By induction on the structure of φ. Let $m := \mathrm{sd}(\varphi)$. Since the standard modalities in φ do not speak about the symbols $1, \ldots, m$, the only interesting case is for $\varphi = \Diamond_a \psi$. Let $\mathcal{T} = (S, \Sigma, R, L)$ with $(\mathcal{T}, s) \models \varphi$. Then there is $(u, a, u') \in R$ such that $(\mathcal{T} \setminus \{(u, a, u')\}, s) \models \psi$. Since $\mathrm{sd}(\psi) = m - 1$, it holds $((\mathcal{T} \setminus \{(u, a, u')\})_{m-1}, s, u') \models \psi^{\#}$ by induction. Clearly we have

$$\mathcal{T}_n \setminus \{(v, a, v')\} = (\mathcal{T} \setminus \{(v, a, v')\})_n \tag{1}$$

for any transition $(v, a, v') \in R$ and $n \in \mathbb{N}$. Hence $(\mathcal{T}_{m-1} \setminus \{(u, a, u')\}, s, u') \models \psi^{\#}$ and therefore $(\mathcal{T}_{m-1}, s, u) \models \Diamond_a \psi^{\#}$. Since the symbol m does not occur in $\psi^{\#}$, we can arbitrarily add m-transitions to the model without affecting the truth of $\psi^{\#}$. So we also have $(\mathcal{T}_m \setminus \{(t, m, u)\}, s, u) \models \Diamond_a \psi^{\#}$. Since (t, m, u) is a transition in $\mathcal{T}_m$ we get $(\mathcal{T}_m, s, t) \models \Diamond_m \Diamond_a \psi^{\#}$.

For the converse let $\varphi^{\#} = \Diamond_m \Diamond_a \psi^{\#}$ with $(\mathcal{T}_m, s, t) \models \varphi^{\#}$. Then there are $u, u' \in S$ with $(u, a, u') \in R$ such that

$$(\mathcal{T}_m \setminus \{(t, m, u), (u, a, u')\}, s, u') \models \psi^{\#}.$$

Since the symbol m does not occur in $\psi^{\#}$ and by (1) it holds

$$((\mathcal{T} \setminus \{(u, a, u')\})_{m-1}, s, u') \models \psi^{\#}.$$

By induction we have $(\mathcal{T} \setminus \{(u, a, u')\}, s) \models \psi$, hence $(\mathcal{T}, s) \models \Diamond_a \psi$. $\qquad \square$

Corollary 6. *The model checking problem for* PSL *is* PSPACE*-complete.*

Proof. By Lemma 4, PSL model checking is in PSPACE. As noted above the size of $\varphi^{\#}$ is polynomial in $|\varphi|$ and the size of $\mathcal{T}_{\mathrm{sd}(\varphi)}$ is polynomial in $|\mathcal{T}|$ and $|\varphi|$. By the previous lemma we have a polynomial time reduction of the PSPACE-hard SML model checking to PSL model checking. $\qquad \square$

In the rest of the section we give a reduction of PSL model checking to the one for standard modal logic. For a transition system $\mathcal{T} = (S, \Sigma, R, L)$ we define the transition system $\mathcal{T}^{\diamond} := (S^{\diamond}, \Sigma^{\diamond}, R^{\diamond}, L^{\diamond})$ that encodes all possible ways of sabotaging $\mathcal{T}$:

$$S^{\diamond} := S \times S \times 2^{R}, \quad \Sigma^{\diamond} := \Sigma \mathbin{\dot{\cup}} \{\bar{a} \mid a \in \Sigma\},$$
$$R^{\diamond} := \{((s, t, E), a, (s', t, E)) \mid (s, a, s') \in R \setminus E\} \cup$$
$$\{((s, t, E), \bar{a}, (s, t', E')) \mid (t, a, t') \in R \setminus E \wedge E' = E \cup \{(t, a, t')\}\},$$
$$L^{\diamond}(s, t, E) := L(s) \text{ for each } s, t \in S \text{ and } E \subseteq R.$$

Over this system one can simulate the sabotage operator $\Diamond_a$ by using an $\bar{a}$-transition, i.e., by the modal operator $\Diamond_{\bar{a}}$. This motivates the following inductive definition of the ML-formula $\varphi^{\diamond}$ for a given PSL-formula φ: $(\top)^{\diamond} := \top$, $(p)^{\diamond} := p$ and the operator $\diamond$ is homeomorphic for $\vee, \neg$ and $\Diamond_a$. For $\varphi = \Diamond_a \psi$ let $\varphi^{\diamond} := \Diamond_{\bar{a}} \psi^{\diamond}$.

Recall that $\mathrm{pd}(\varphi)$ denotes the depth of nested path sabotage operators of a PSL-formula φ (cf. Sect. 2). If $\mathrm{pd}(\varphi)$ is small then we do not need the complete transition system $\mathcal{T}^{\diamond}$ to evaluate $\varphi^{\diamond}$. So, for $n \in \mathbb{N}$, we define $\mathcal{T}_n^{\diamond}$ to be the transition system $\mathcal{T}^{\diamond}$ restricted to the states (s, t, E) with $|E| \leq n$. Note that $\mathcal{T}_n^{\diamond} = \mathcal{T}^{\diamond}$ for $n \geq |R|$. The proof of the following lemma is a slight modification of the one for SML presented in [4].

Lemma 7. *For every* PSL*-formula* φ, *transition system* $\mathcal{T}$, *and* $s, t \in \mathcal{T}$ *it holds*

$$(\mathcal{T}, s, t) \models \varphi \iff (\mathcal{T}_{\mathrm{pd}(\varphi)}^{\diamond}, (s, t, \emptyset)) \models \varphi^{\diamond}. \qquad \Box$$

This reduction can be used to determine the formula complexity and the program complexity of PSL model checking:

Corollary 8. *1. Formula complexity: Model checking for* PSL *with a fixed transition system can be solved in linear time in the size of the formula.*
2. Program complexity: Model checking for PSL *with a fixed formula can be solved in polynomial time in the size of the transition system.*

Proof. It is well known that the model checking problem for modal logic over transition systems can be solved in time $\mathcal{O}(|\psi| \cdot |\mathcal{T}|)$, where $|\psi|$ is the size of the given ML-formula ψ and $|\mathcal{T}|$ is the size of the given transition system $\mathcal{T}$ (cf. [2]). Hence, by Lemma 7, we can solve the model checking problem for a PSL-formula φ and $\mathcal{T}$ in time $\mathcal{O}(|\varphi^{\diamond}| \cdot |\mathcal{T}_{\mathrm{pd}(\varphi)}^{\diamond}|)$. From the definition of $\varphi^{\diamond}$ we get $|\varphi^{\diamond}| = |\varphi|$.

1. For a fixed transition system $\mathcal{T}$ we can estimate the size of $\mathcal{T}_{\mathrm{pd}(\varphi)}^{\diamond}$ by $|\mathcal{T}_{\mathrm{pd}(\varphi)}^{\diamond}| \in \mathcal{O}(|\mathcal{T}|^2 \cdot 2^{|\mathcal{T}|})$. Hence the formula complexity is in $\mathcal{O}(|\varphi|)$.

2. Since the number of subsets $E \subseteq R$ with $|E| \leq \mathrm{pd}(\varphi)$ is in $\mathcal{O}(|\mathcal{T}|^{\mathrm{pd}(\varphi)})$ we get $|\mathcal{T}_{\mathrm{pd}(\varphi)}^{\diamond}| \in \mathcal{O}(|\mathcal{T}|^{\mathrm{pd}(\varphi)+2})$. So the model checking complexity with a fixed PSL-formula φ is polynomial in $|\mathcal{T}|$. $\qquad \Box$

4   Pruned SML-Models

In the last section we gave a reduction of the model checking problem for SML to the one for PSL. For a reduction of the satisfiability problem we need a more sophisticated approach. In this section we show that each model of a given SML-formula φ can be pruned such that it consists only of those states that are reachable from the initial state by the standard modalities in φ together with a bounded number of additional states (we call it a *pruned model relative to* φ). We define the pruned form of a model in two steps. In the next section we show how to enforce within PSL that a model of a given SML-formula φ contains a pruned submodel (relative to φ) where each two states are connected by i-transitions for $1 \leq i \leq \mathrm{sd}(\varphi)$ and such that one cannot escape the pruned submodel by using the modalities of φ. Then we can use the same argument as before to translate SML-modalities into PSL-modalities.

Let φ be an SML-formula over Σ. We define inductively the set of path labels $P_\varphi \subseteq \Sigma^*$ corresponding to the standard modalities in φ:

$$
P_\varphi := \begin{cases}
\{\varepsilon\} & \text{if } \varphi = \top \text{ or } \varphi = p, \\
P_{\varphi_1} \cup P_{\varphi_2} & \text{if } \varphi = \varphi_1 \vee \varphi_2, \\
P_\psi & \text{if } \varphi = \neg\psi \text{ or } \varphi = \Diamond_a\psi, \\
\{\varepsilon\} \cup \{a \cdot \pi \mid \pi \in P_\psi\} & \text{if } \varphi = \Diamond_a\psi.
\end{cases}
$$

For $\mathcal{T} = (S, \Sigma, R, L)$ and $s \in \mathcal{T}$ let $\mathcal{T}_{\varphi,s} := (S_{\varphi,s}, \Sigma, R_{\varphi,s}, L|_{S_{\varphi,s}})$ be the transition system restricted to paths in P_φ starting in s:

$$
S_{\varphi,s} := \{t \mid t \in S \text{ and there is a } \pi\text{-path from } s \text{ to } t \text{ in } \mathcal{T} \text{ for some } \pi \in P_\varphi\},
$$
$$
R_{\varphi,s} := \{(t, a, t') \mid (t, a, t') \in R \text{ and there is a } \pi\text{-path from } s \text{ to } t \text{ in } \mathcal{T}
$$
$$
\text{for some } \pi \in P_\varphi, \text{ such that } \pi \cdot a \in P_\varphi\}.
$$

Note that, if $(\mathcal{T}, s) \models \varphi$, then $\mathcal{T}_{\varphi,s}$ does not need to be a model of φ. There may be 'dummy' transitions in $\mathcal{T}$ that have to be deleted to satisfy φ, but which are not reachable by the standard modalities of φ.

Example 9. Consider the formula $\varphi := \Diamond_a\top \wedge \Diamond_a\Box_a\bot \wedge \Diamond_a\Diamond_a\top$. The following transition system $(\mathcal{T}, s)$ is a model of φ:

$$
s \xrightarrow{\ a\ } s' \xrightarrow{\ a\ } s''
$$

Since $P_\varphi = \{\varepsilon, a\}$ the transition system $\mathcal{T}_{\varphi,s}$ consists only of the states s, s' and the transition (s, a, s'). Since we cannot delete two different a-transitions, it fails to be a model of φ.

But in fact, the exact position of a 'dummy' transition in $\mathcal{T}$ is irrelevant, hence we can equip $\mathcal{T}_{\varphi,s}$ with these transitions in a canonical way. Further we can bound the number of these transitions: One only needs $\mathrm{sd}_a(\varphi)$ many additional a-transitions for each $a \in \Sigma$, where $\mathrm{sd}_a(\varphi)$ is the depth of nested $\Diamond_a$ in φ (cf. Sect. 2). We show this in the rest of the section.

For two sets R and R' of transitions with $R' \subseteq R$ and $a \in \Sigma$ let $\mathrm{diff}_a(R, R') :=$ $|R \setminus R' \cap S \times \{a\} \times S| \in \mathbb{N} \cup \{\infty\}$. Let $\kappa^a_{\varphi,s} \in \mathbb{N}$ be the minimum of $\mathrm{sd}_a(\varphi)$ and the number of a-transitions in $\mathcal{T}$ that are not present in $\mathcal{T}_{\varphi,s}$:

$$\kappa^a_{\varphi,s} := \min\{\mathrm{sd}_a(\varphi), \mathrm{diff}_a(R, R_{\varphi,s})\}.$$

The *pruned form* $\mathcal{T}^*_{\varphi,s}$ of an SML-model $\mathcal{T}$ relative to φ and s is defined by $\mathcal{T}^*_{\varphi,s} := (S^*_{\varphi,s}, \Sigma, R^*_{\varphi,s}, L_{\varphi,s})$, where

$$S^*_{\varphi,s} := S_{\varphi,s} \dot\cup \{s^a_i \mid a \in \Sigma \wedge 1 \le i \le \kappa^a_{\varphi,s}\},$$
$$R^*_{\varphi,s} := R_{\varphi,s} \dot\cup \{(s^a_i, a, s) \mid a \in \Sigma \wedge 1 \le i \le \kappa^a_{\varphi,s}\}.$$

Example 10. For the formula φ and the model $\mathcal{T}$ of Example 9 the transition system $\mathcal{T}^*_{\varphi,s}$ is

$$s^a_1 \xrightarrow{\ a\ } s \xrightarrow{\ a\ } s'$$

Theorem 11. *For every SML-formula φ, transition system $\mathcal{T}$, and $s \in \mathcal{T}$ it holds*

$$(\mathcal{T}, s) \models \varphi \iff (\mathcal{T}^*_{\varphi,s}, s) \models \varphi.$$

Proof. By induction on the structure of φ. For the atomic cases $\varphi = \top$ and $\varphi = p$ we have $P_\varphi = \{\varepsilon\}$ and $\kappa^a_{\varphi,s} = 0$ for every $a \in \Sigma$. Hence $S^*_{\varphi,s} = \{s\}$ and $(\mathcal{T}, s)$ is a model of φ iff $(\mathcal{T}^*_{\varphi,s}, s)$ is a model of φ. By induction and the fact that $\mathcal{T}^*_{\psi,s} \cong \mathcal{T}^*_{\neg\psi,s}$ the case $\varphi = \neg\psi$ is also clear.

Claim 1. For $\varphi = \psi \vee \chi$ it holds $(\mathcal{T}^*_{\varphi,s})^*_{\psi,s} \cong \mathcal{T}^*_{\psi,s}$ and $(\mathcal{T}^*_{\varphi,s})^*_{\chi,s} \cong \mathcal{T}^*_{\chi,s}$.

Proof (of Claim). By symmetry it is enough to show the first statement. Since $P_\psi \subseteq P_\varphi$ it is easy to see that $S_{\psi,s} \subseteq S_{\varphi,s}$ and $R_{\psi,s} \subseteq R_{\varphi,s}$. Since the additional states s^a_i in $\mathcal{T}^*_{\varphi,s}$ do not belong to $S_{\varphi,s}$ it follows $(\mathcal{T}^*_{\varphi,s})_{\psi,s} \cong \mathcal{T}_{\psi,s}$. Hence it suffices to show that the same number of additional states s^a_i is added to both models, for each $a \in \Sigma$. For that let $a \in \Sigma$ and λ^a be the number of states s^a_i in $(\mathcal{T}^*_{\varphi,s})^*_{\psi,s}$:

$$\lambda^a := \min\{\mathrm{sd}_a(\psi), \mathrm{diff}_a(R^*_{\varphi,s}, R_{\psi,s})\}.$$

Case 1: $\mathrm{sd}_a(\varphi) \le \mathrm{diff}_a(R, R_{\varphi,s})$. Since $(s^a_i, a, s) \in R^*_{\varphi,s} \setminus R_{\varphi,s} \subseteq R^*_{\varphi,s} \setminus R_{\psi,s}$ for every $1 \le i \le \kappa^a_{\varphi,s}$, we have

$$\mathrm{diff}_a(R^*_{\varphi,s}, R_{\psi,s}) \ge \kappa^a_{\varphi,s} = \mathrm{sd}_a(\varphi) \ge \mathrm{sd}_a(\psi),$$

hence $\lambda^a = \mathrm{sd}_a(\psi)$. On the other hand, since $R_{\psi,s} \subseteq R_{\varphi,s}$ we have

$$\mathrm{sd}_a(\psi) \le \mathrm{sd}_a(\varphi) \le \mathrm{diff}_a(R, R_{\varphi,s}) \le \mathrm{diff}_a(R, R_{\psi,s}),$$

hence $\kappa^a_{\psi,s} = \mathrm{sd}_a(\psi)$, i.e., $\kappa^a_{\psi,s} = \lambda^a$.

Case 2: $\mathrm{sd}_a(\varphi) > \mathrm{diff}_a(R, R_{\varphi,s})$. Then there is exactly one a-transition in $R^*_{\varphi,s}$ for each a-transition in R and vice versa. Since $R_{\psi,s} \subseteq R$ and $R_{\psi,s} \subseteq R^*_{\varphi,s}$ we therefore get $\mathrm{diff}_a(R^*_{\varphi,s}, R_{\psi,s}) = \mathrm{diff}_a(R, R_{\psi,s})$ and hence

$$\kappa^a_{\psi,s} = \min\{\mathrm{sd}_a(\psi), \mathrm{diff}_a(R, R_{\psi,s})\} = \lambda^a.$$

In both cases the same number of states s^a_i together with transitions (s^a_i, a, s) is added for each $a \in \Sigma$. Hence we get $(\mathcal{T}^*_{\varphi,s})^*_{\psi,s} \cong \mathcal{T}^*_{\psi,s}$.   □

Now we are ready to show the induction step for $\varphi = \psi \vee \chi$. We have

$$
\begin{aligned}
&(\mathcal{T}, s) \models \varphi \\
\Longleftrightarrow\ &(\mathcal{T}, s) \models \psi \text{ or } (\mathcal{T}, s) \models \chi \\
\Longleftrightarrow\ &(\mathcal{T}^*_{\psi,s}, s) \models \psi \text{ or } (\mathcal{T}^*_{\chi,s}, s) \models \chi && \text{by induction} \\
\Longleftrightarrow\ &((\mathcal{T}^*_{\varphi,s})^*_{\psi,s}, s) \models \psi \text{ or } ((\mathcal{T}^*_{\varphi,s})^*_{\chi,s}, s) \models \chi && \text{by Claim 1} \\
\Longleftrightarrow\ &(\mathcal{T}^*_{\varphi,s}, s) \models \psi \text{ or } (\mathcal{T}^*_{\varphi,s}, s) \models \chi && \text{by induction} \\
\Longleftrightarrow\ &(\mathcal{T}^*_{\varphi,s}, s) \models \varphi.
\end{aligned}
$$

Claim 2. For $\varphi = \Diamond_a \psi$ and $t \in S$ with $(s, a, t) \in R$ it holds $(\mathcal{T}^*_{\varphi,s})^*_{\psi,t} \cong \mathcal{T}^*_{\psi,t}$.

Proof (of Claim). If there is a π-path from t to v in $\mathcal{T}$ for some $\pi \in P_\psi$, then there is an $a \cdot \pi$-path from s to v and $a \cdot \pi \in P_\varphi$ by definition of P_φ. Hence $S_{\psi,t} \subseteq S_{\varphi,s}$. Analogously we have $R_{\psi,t} \subseteq R_{\varphi,s}$. So $(\mathcal{T}^*_{\varphi,s})_{\psi,t} \cong \mathcal{T}_{\psi,t}$ and it suffices to show that the same number of additional states s^b_i for $b \in \Sigma$ is added to both models. Using the fact that $\mathrm{sd}_b(\varphi) = \mathrm{sd}_b(\psi)$ for every $b \in \Sigma$, the proof is almost the same as for the previous claim (using $R_{\psi,t}$ and $\kappa^b_{\psi,t}$ instead of $R_{\psi,s}$ and $\kappa^a_{\psi,s}$). $\qquad \square$

Let $\varphi = \Diamond_a \psi$. Since $\varepsilon \in P_\psi$ we have $a \in P_\varphi$. By definition of $\mathcal{T}^*_{\varphi,s}$ there is no a-transition from s to some s^b_i, $b \in \Sigma$. Hence

$$
\begin{aligned}
t \in S \wedge (s, a, t) \in R &\Longleftrightarrow t \in S_{\varphi,s} \wedge (s, a, t) \in R_{\varphi,s} \\
&\Longleftrightarrow t \in S^*_{\varphi,s} \wedge (s, a, t) \in R^*_{\varphi,s}.
\end{aligned}
\tag{2}
$$

Therefore it holds

$$
\begin{aligned}
&(\mathcal{T}, s) \models \varphi \\
\Longleftrightarrow\ &\exists t \in S : (s, a, t) \in R \wedge (\mathcal{T}, t) \models \psi \\
\Longleftrightarrow\ &\exists t \in S : (s, a, t) \in R \wedge (\mathcal{T}^*_{\psi,t}, t) \models \psi && \text{by induction} \\
\Longleftrightarrow\ &\exists t \in S : (s, a, t) \in R \wedge ((\mathcal{T}^*_{\varphi,s})^*_{\psi,t}, t) \models \psi && \text{by Claim 2} \\
\Longleftrightarrow\ &\exists t \in S : (s, a, t) \in R \wedge (\mathcal{T}^*_{\varphi,s}, t) \models \psi && \text{by induction} \\
\Longleftrightarrow\ &\exists t \in S^*_{\varphi,s} : (s, a, t) \in R^*_{\varphi,s} \wedge (\mathcal{T}^*_{\varphi,s}, t) \models \psi && \text{by (2)} \\
\Longleftrightarrow\ &(\mathcal{T}^*_{\varphi,s}, s) \models \varphi.
\end{aligned}
$$

Claim 3. Let $\varphi = \Diamond_a \psi$.

1. For every $t, t' \in S^*_{\varphi,s}$ with $(t, a, t') \in R^*_{\varphi,s}$ there are $u, u' \in S$ with $(u, a, u') \in R$ such that $(\mathcal{T}^*_{\varphi,s} \setminus \{(t, a, t')\})^*_{\psi,s} \cong (\mathcal{T} \setminus \{(u, a, u')\})^*_{\psi,s}$.
2. For every $u, u' \in S$ with $(u, a, u') \in R$ there are $t, t' \in S^*_{\varphi,s}$ with $(t, a, t') \in R^*_{\varphi,s}$ such that $(\mathcal{T}^*_{\varphi,s} \setminus \{(t, a, t')\})^*_{\psi,s} \cong (\mathcal{T} \setminus \{(u, a, u')\})^*_{\psi,s}$.

Proof (of Claim). By definition it holds $P_\varphi = P_\psi$ and therefore $S_{\varphi,s} = S_{\psi,s}$ and $R_{\varphi,s} = R_{\psi,s}$.

1. Case I: If $(t, a, t') \in R_{\varphi,s}$ then also $(t, a, t') \in R$ and we set $u := t$ and $u' := t'$. First we show

$$(\mathcal{T}_{\varphi,s}^* \setminus \{(t, a, t')\})_{\psi,s} \cong (\mathcal{T} \setminus \{(u, a, u')\})_{\psi,s}. \qquad (3)$$

Let S_1 and S_2 be the state sets of the left hand side, resp., right hand side and let R_1, R_2 be the corresponding transition relations. It suffices to show $S_1 = S_2$ and $R_1 = R_2$. It holds $v \in S_1$ iff there is a π-path from s to v in $\mathcal{T}_{\varphi,s}^* \setminus \{(t, a, t')\}$ for some $\pi \in P_\psi$, i.e., there is a sequence $\rho = (v_0, a_0, v_1), \ldots, (v_{n-1}, a_{n-1}, v_n)$ with $v_0 = s$, $v_n = v$, $(v_i, a_i, v_{i+1}) \in R_{\varphi,s}^* \setminus \{(t, a, t')\}$ for every $i < n$ and $a_0 \cdots a_{n-1} \in P_\psi$. Since none of the additional states s_i^b, $b \in \Sigma$ has an incoming transition it holds $v_i \in S_{\varphi,s}$ for every $i \leq n$ and $(v_i, a_i, v_{i+1}) \in R_{\varphi,s} \setminus \{(t, a, t')\}$ for every $i < n$. By definition we also have $v_i \in S$ for every $i \leq n$ and $(v_i, a_i, v_{i+1}) \in R \setminus \{(t, a, t')\}$ for every $i < n$. Hence ρ is also a π-path from s to v in $\mathcal{T} \setminus \{(t, a, t')\}$ and therefore $v \in S_2$.

On the other hand, let $v \in S_2$ and ρ be a π-path from s to v in $\mathcal{T} \setminus \{(t, a, t')\}$ for $\pi \in P_\psi$ as above. Then $\rho[0, i]$ is a $\pi[0, i]$-path from s to v_i with $\pi[0, i] \in P_\varphi$ for every $i < n$. Hence $v_i \in S_{\varphi,s}$ for every $i \leq n$, $(v_i, a_i, v_{i+1}) \in R_{\varphi,s} \setminus \{(t, a, t')\}$ for every $i < n$ and ρ is a π-path from s to v in $\mathcal{T}_{\varphi,s} \setminus \{(t, a, t')\} \subseteq \mathcal{T}_{\varphi,s}^* \setminus \{(t, a, t')\}$, i.e., $v \in S_1$ and therefore $S_1 = S_2$. $R_1 = R_2$ is shown analogously.

Next we show that the same number of additional states s_i^b is added to both models in (3), for every $b \in \Sigma$. If $\mathrm{sd}_b(\varphi) > \mathrm{diff}_b(R, R_{\varphi,s})$ then the set of b-transitions in $R_{\varphi,s}^*$ has the same cardinality as the set of b-transitions in R. Since $R_1 = R_2$ we therefore get

$$\mathrm{diff}_b(R_{\varphi,s}^* \setminus \{(t, a, t')\}, R_1) = \mathrm{diff}_b(R \setminus \{(t, a, t')\}, R_2).$$

If $\mathrm{sd}_b(\varphi) \leq \mathrm{diff}_b(R, R_{\varphi,s})$ then the number of additional states s_i^b in $\mathcal{T}_{\varphi,s}^*$ is equal to $\mathrm{sd}_b(\varphi)$ and there are just as many b-transitions in $R_{\varphi,s}^* \setminus R_{\varphi,s}$. Since $R_1 \subseteq R_{\varphi,s}$, $\mathrm{sd}_a(\varphi) = \mathrm{sd}_a(\psi) + 1$ and $\mathrm{sd}_b(\varphi) = \mathrm{sd}_b(\psi)$ for $b \neq a$ it holds

$$\mathrm{diff}_b(R \setminus \{(t, a, t')\}, R_2) \geq \mathrm{diff}_b(R_{\varphi,s}^* \setminus \{(t, a, t')\}, R_1) \geq \mathrm{sd}_b(\varphi) \geq \mathrm{sd}_b(\psi).$$

Therefore the number of additional states s_i^b in both models is equal to $\mathrm{sd}_b(\psi)$.

Case II: If $(t, a, t') = (s_i^a, a, s)$ for some $1 \leq i \leq \kappa_{\varphi,s}^a$ then by definition, there are $u, u' \in R$ with $(u, a, u') \in R \setminus R_{\varphi,s}$. With the notation as before it is easy to see that $S_1 = S_2 = S_{\varphi,s}$ and $R_1 = R_2 = R_{\varphi,s}$, hence (3) is also true for this case. If $b \neq a$ then $\min\{\mathrm{sd}_b(\psi), \mathrm{diff}_b(R_{\varphi,s}^* \setminus \{(s_i^a, a, s)\}, R_1)\} = \min\{\mathrm{sd}_b(\varphi), \mathrm{diff}_b(R_{\varphi,s}^*, R_{\varphi,s})\} = \min\{\mathrm{sd}_b(\varphi), \kappa_{\varphi,s}^b\} = \kappa_{\varphi,s}^b$. On the other hand, $\min\{\mathrm{sd}_b(\psi), \mathrm{diff}_b(R \setminus \{(u, a, u')\}, R_2)\} = \min\{\mathrm{sd}_b(\varphi), \mathrm{diff}_b(R, R_{\varphi,s})\} = \kappa_{\varphi,s}^b$.

For $b = a$ it holds $\min\{\mathrm{sd}_a(\psi), \mathrm{diff}_a(R_{\varphi,s}^* \setminus \{(s_i^a, a, s)\}, R_1)\} = \min\{\mathrm{sd}_a(\varphi) - 1, \kappa_{\varphi,s}^a - 1\} = \kappa_{\varphi,s}^a - 1$ and $\min\{\mathrm{sd}_a(\psi), \mathrm{diff}_a(R \setminus \{(u, a, u')\}, R_2)\} = \min\{\mathrm{sd}_a(\varphi), \mathrm{diff}_a(R, R_{\varphi,s})\} - 1 = \kappa_{\varphi,s}^a - 1$ (note that $(u, a, u') \notin R_2$). So in both cases the number of additional states s_i^b is the same.

2. If $(u, a, u') \in R_{\varphi,s} \subseteq R$ then we set $t := u$ and $t' := u'$ and the proof is exactly the same as for Case I above. Now let $(u, a, u') \in R \setminus R_{\varphi,s}$. Since $\mathrm{sd}_a(\varphi) \geq 1$ we have $\kappa_{\varphi,s}^a \geq 1$ and there is $s_1^a \in S_{\varphi,s}^* \setminus S_{\varphi,s}$ with $(s_1^a, a, s) \in R_{\varphi,s}^*$. Then we set $t := s_1^a$ and $t' := s$ and repeat the proof of Case II above. $\square$

By using Claim 3 we are able to prove the last induction step. For that let $\varphi = \Diamond_a \psi$. Then

$$(\mathcal{T}, s) \models \varphi$$
$$\iff \exists u, u' \in S : (u, a, u') \in R \wedge (\mathcal{T} \setminus \{(u, a, u')\}, s) \models \psi$$
$$\iff \exists u, u' \in S : (u, a, u') \in R \wedge ((\mathcal{T} \setminus \{(u, a, u')\})^*_{\psi, s}, s) \models \psi \qquad \text{by ind.}$$
$$\iff \exists t, t' \in S^*_{\varphi, s} : (t, a, t') \in R^*_{\varphi, s} \wedge ((\mathcal{T}^*_{\varphi, s} \setminus \{(t, a, t')\})^*_{\psi, s}, s) \models \psi \qquad \text{by Cl. 3}$$
$$\iff \exists t, t' \in S^*_{\varphi, s} : (t, a, t') \in R^*_{\varphi, s} \wedge (\mathcal{T}^*_{\varphi, s} \setminus \{(t, a, t')\}, s) \models \psi \qquad \text{by ind.}$$
$$\iff (\mathcal{T}^*_{\varphi, s}, s) \models \varphi.$$

This concludes the proof of the theorem. □

5   Finite Model Property and Satisfiability for PSL

In this section we present five PSL-formulae (α_i, $\beta^a_{k,i}, \gamma_i, \delta_i$ and ζ_i). Together they ensure that a model of an SML-formula φ contains a pruned submodel (relative to φ) such that each two states of the submodel are connected by i-transitions for $1 \leq i \leq \text{sd}(\varphi)$. Further one cannot escape the submodel either by using the standard modalities or by using the sabotage modalities of φ. For technical reasons we additionally use the symbol 0 as a kind of anchor: Deletion of 0-transitions allow us to mark and identify states. Then we are ready to show the main results of the paper: PSL lacks the finite model property and the satisfiability problem for PSL is undecidable.

Let φ be an SML-formula over Σ and let P_φ be as in the last section. We assume that $\Sigma \cap \{0, \ldots, \text{sd}(\varphi)\} = \emptyset$. For a transition system $\mathcal{T} = (S, \Sigma', R, L)$ with $\Sigma \subseteq \Sigma'$ and $s \in S$ let $S_{\varphi, s} \subseteq S$ be defined as before. For a language $A \subseteq \Sigma^*$ the modal operator $\Diamond_A$ is defined by

$$\Diamond_A \psi := \bigvee_{a_1 \cdots a_n \in A} \Diamond_{a_1} \cdots \Diamond_{a_n} \psi.$$

The operator $\Box_A$ is defined analogously. Note that $\Diamond_\emptyset = \bot$ and $\Diamond_{\{\varepsilon\}} \psi = \psi$. In the sequel let $\Sigma_m := \Sigma \stackrel{.}{\cup} \{0, \ldots, m\}$ and $\mathcal{T} = (S, \Sigma_m, R, L)$ be a transition system over Σ_m. The PSL-formula α_i over Σ_m is defined by

$$\alpha_i := \Diamond_0 \top \wedge \Diamond_0 \Box_0 \bot \wedge \Box_{P_\varphi} (\Diamond_0 \top \wedge \Diamond_i \Diamond_0 \Box_0 \bot).$$

Lemma 12. *If* $(\mathcal{T}, s, t) \models \alpha_i$, *then* $s = t$ *and for every* $u \in S_{\varphi, s}$ *there is* $(s, i, u) \in R$ *and* u *has exactly one 0-successor. In particular,* $(s, i, s) \in R$.

Proof. It is easy to see that the first two terms imply $s = t$. If the current position is $[s, s]$, then the last term says that for every $u \in S_{\varphi, s}$ it holds: u has a 0-successor (by $\Diamond_0 \top$) and there is a sabotage path $(s, i, v), (v, 0, w)$ such that u has no 0-successor anymore. Hence it must be $u = v$ and there is only one 0-successor of u. □

For $a \in \Sigma$ the PSL-formula $\beta_{k,i}^a$ over Σ_m is inductively defined by

$$\beta_{0,i}^a := \boxdot_i(\boxdot_a \bot \vee \Diamond_0 \Diamond_{P_\varphi} \boxdot_0 \bot),$$
$$\beta_{k+1,i}^a := \Diamond_i(\Diamond_0 \boxdot_{P_\varphi} \Diamond_0 \top \wedge \Diamond_a \top \wedge \boxdot_{\Sigma \setminus \{a\}} \bot \wedge \boxdot_a(\Diamond_0 \boxdot_0 \bot \wedge \beta_{k,i}^a)).$$

Lemma 13. *If* $(\mathcal{T}, s, t) \models \alpha_i \wedge \beta_{k,i}^a$ *for some* $k \in \mathbb{N}$, *then there are pairwise different* $s_1^a, \ldots, s_k^a \in S$ *such that for every* $1 \leq j \leq k$:

1. $s_j^a \in S \setminus S_{\varphi,s}$, *it has a 0-successor and there is* $(s, i, s_j^a) \in R$,
2. *there is* $(s_j^a, a, s) \in R$ *and if* $(s_j^a, a, v) \in R$ *for some* $v \in S$, *then* $v = s$,
3. s_j^a *has no b-successor for* $b \in \Sigma$, $b \neq a$.

On the other hand, if there is $v \in S \setminus S_{\varphi,s}$ *with* $(s, i, v) \in R$ *and* $(v, a, s) \in R$, *then* $v = s_j^a$ *for some* $1 \leq j \leq k$. *In particular,* $(\mathcal{T}, s, t) \not\models \beta_{l,i}^a$ *for any* $l \neq k$.

Proof. By induction on k. By the previous lemma α_i implies $s = t$, so the current position is $[s, s]$. For $k = 0$ assume that there is $v \in S$ with $(s, i, v) \in R$ and $(v, a, s) \in R$. If (s, i, v) is removed and the current position becomes $[s, v]$ then, since v has an a-successor, the second disjunct of $\beta_{0,i}^a$ must be true. This means that there is an outgoing 0-transition of v and, if it is removed, there is a π-path from s to some $u \in S$ for $\pi \in P_\varphi$ such that u has no 0-successor. But by α_i every such u has a 0-successor in the initial model, hence it must be $u = v$ and therefore $v \in S_{\varphi,s}$.

For the induction step we assume that the statement holds for k. If the current position is $[s, s]$ then the first conjunct of $\beta_{k+1,i}^a$ implies that there are $u, v \in S$ and a sabotage path $(s, i, u), (u, 0, v)$ such that every $w \in S_{\varphi,s}$ still has a 0-successor. Hence $u \notin S_{\varphi,s}$. If the current position is $[s, u]$, the second and third conjunct say that u has an a-successor and no b-successor for $b \neq a$. The last term forces that for every a-successor v of u, if the current position is $[s, v]$, then there is $(v, 0, w) \in R$ for some $w \in S$ and, if this transition is removed, s has no 0-successor anymore. But by α_i state s has an initial 0-successor, therefore it must be $v = s$. The current position becomes $[s, s]$ again and by induction $\beta_{k,i}^a$ implies the existence of $s_1^a, \ldots, s_k^a$ with the stated properties. Since the transition (s, i, u) was removed we have $u \neq s_j^a$ for every $1 \leq j \leq k$. Hence we can set $s_{k+1}^a = u$.

Assume that there is $v \in S \setminus S_{\varphi,s}$ and there are $(s, i, v) \in R$ and $(v, a, s) \in R$. If $u \neq v$ then both transitions were not deleted until the current position becomes $[s, s]$ again. By induction, $\beta_{k,i}^a$ implies $v = s_j^a$ for some $1 \leq j \leq k$. $\square$

Let γ_i be the following PSL-formula over Σ_m:

$$\gamma_i := \boxdot_i(\Diamond_0 \Diamond_{P_\varphi} \boxdot_0 \bot \vee \Diamond_\Sigma \Diamond_0 \boxdot_0 \bot) \wedge \boxdot_{P_\varphi} \boxdot_\Sigma(\Diamond_0 \top \wedge \Diamond_{P_\varphi} \Diamond_0 \boxdot_0 \bot).$$

Lemma 14. *Let* $k_a \in \mathbb{N}$ *for* $a \in \Sigma$ *such that*

$$(\mathcal{T}, s, t) \models \alpha_i \wedge \bigwedge_{a \in \Sigma} \beta_{k_a,i}^a \wedge \gamma_i.$$

Then for every i-successor v *of* s, *either* $v \in S_{\varphi,s}$ *or* $v = s_j^a$ *for some* $a \in \Sigma$ *and* $1 \leq j \leq k_a$ *as given in Lemma 13. Further, every Σ-successor of a state in $S_{\varphi,s}$ also belongs to $S_{\varphi,s}$.*

Proof. By Lemma 12 the current position is $[s, s]$ and every state $u \in S_{\varphi,s}$ has exactly one 0-successor. By removing (s, i, v) one reaches position $[s, v]$. The first disjunct in the first brackets of γ_i is satisfied if and only if $v \in S_{\varphi,s}$. If $v \in S \setminus S_{\varphi,s}$ then, by the second disjunct, there is a sabotage path $(v, a, w), (w, 0, w')$ for some $a \in \Sigma$ such that s has no 0-successor anymore. Hence $w = s$ and there is $(v, a, s) \in R$. By Lemma 13 we have $v = s_j^a$ for some $1 \le j \le k_a$.

Now let $u \in S_{\varphi,s}$ and $v \in S$ with $(u, a, v) \in R$ for some $a \in \Sigma$. By the second conjunct of γ_i, v has a 0-successor and for some $\pi \in P_\varphi$, there is a sabotage π-path from s to some $w \in S_{\varphi,s}$ such that, if the path is extended to some 0-successor of w, then v has no 0-successor anymore. Hence $v = w$ and v belongs to $S_{\varphi,s}$.   □

Let δ_i be the following PSL-formula over Σ_m

$$\delta_i := \Box_i \Box_i (\Diamond_0 \top \wedge \Diamond_i \Diamond_0 \Box_0 \bot) \wedge \Box_i \Box_i (\Diamond_0 \top \wedge \Diamond_i \Diamond_0 \Box_0 \bot).$$

Lemma 15. *If* $(\mathcal{T}, s, s) \models \delta_i$, *then:*

1. *If* $(s, i, u) \in R$ *and* $(s, i, v) \in R$ *for* $u \ne v \in S$, *then also* $(u, i, v) \in R$.
2. *If* $(s, i, u) \in R$ *and* $(u, i, v) \in R$ *for* $u, v \in S$, *then also* $(s, i, v) \in R$.

Proof. 1. Let $u, v \in S$, $u \ne v$ with $(s, i, u) \in R$ and $(s, i, v) \in R$. By the first conjunct of δ_i, starting from position $[s, s]$ and removing the transition (s, i, u) the current position becomes $[s, u]$. Since (s, i, v) is still available we can reach position $[v, u]$. Then v has a 0-successor and there is a sabotage path $(u, i, w), (w, 0, w')$ such that v has no 0-successor anymore. Hence $v = w$, i.e., there is $(u, i, v) \in R$.

2. Let $u, v \in S$ with $(s, i, u) \in R$ and $(u, i, v) \in R$. By the second conjunct we can reach position $[v, s]$ from the initial position $[s, s]$ and v has a 0-successor. Further there is a sabotage path $(s, i, w), (w, 0, w')$ such that v has no 0-successor anymore. Hence $w = v$, i.e., there is $(s, i, v) \in R$.   □

Let ζ_i be the following PSL-formula over Σ_m:

$$\zeta_i := \Box_i \left(\Diamond_0 \top \wedge \left(\Diamond_0 \Box_0 \bot \vee \Diamond_i (\Diamond_0 \Box_0 \bot \wedge \Diamond_i \Diamond_0 \Box_0 \bot) \right) \right).$$

Lemma 16. *If* $(\mathcal{T}, s, s) \models \zeta_i$, *then for every* $u \in S$ *with* $(s, i, u) \in R$ *there is* $(u, i, u) \in R$ *and* u *has exactly one 0-successor.*

Proof. Let the initial position be $[s, s]$ and let $u \in S$ with $(s, i, u) \in R$. If the position becomes $[u, s]$, then u has a 0-successor by the first conjunct of ζ_i. The first disjunct is true if and only if $u = s$ and s has a single 0-successor. In this case we have $(s, i, s) \in R$ by the assumption. If $u \ne s$, then the second disjunct must be satisfied. To satisfy $\Diamond_0 \Box_0 \bot$ state u can only have a single 0-successor and one has to remove the transition (s, i, u) such that the current position becomes $[u, u]$. But then one has to use (and remove) an i-transition leading back to u to satisfy the last term, i.e., there must be $(u, i, u) \in R$.   □

Now let $m := \operatorname{sd}(\varphi)$ and let $\varphi^{\#}$ be the PSL-formula as defined in Sect. 3. Let $\varphi^{\dagger}$ be the following PSL-formula over Σ_m:

$$\varphi^{\dagger} := \bigwedge_{i=1}^{m} \left(\alpha_i \wedge \gamma_i \wedge \delta_i \wedge \zeta_i \wedge \Box_i \bigwedge_{n=1}^{m} \Diamond_n \Diamond_0 \Box_0 \bot \right) \wedge \bigwedge_{a \in \Sigma} \bigvee_{k=0}^{\operatorname{sd}_a(\varphi)} \bigwedge_{i=1}^{m} \beta_{k,i}^{a} \wedge \varphi^{\#}.$$

The additional term ensures together with ζ_i that, if $(s, i, u) \in R$ for some $u \in S$ and $1 \le i \le m$ then there is also $(s, j, u) \in R$ for every $1 \le j \le m$ with $j \ne i$. In particular, the additional states s_j^a due to $\beta_{k,i}^a$ are identical with the ones given by $\beta_{k,n}^a$ for $n \ne i$.

For $\mathcal{T} = (S, \Sigma, R, L)$ and $s \in S$ let $\mathcal{T}_{\varphi,s}^* = (S_{\varphi,s}^*, \Sigma, R_{\varphi,s}^*, L_{\varphi,s}^*)$ be defined as in Sect. 4. The transition system $\mathcal{T}_{\varphi,s}^{\dagger}$ is defined by $\mathcal{T}_{\varphi,s}^{\dagger} := (S_{\varphi,s}^*, \Sigma_m, R^{\dagger}, L_{\varphi,s}^*)$ where

$$R^{\dagger} := R_{\varphi,s}^* \,\dot{\cup}\, \{(u, 0, u) \mid u \in S_{\varphi,s}^*\} \,\dot{\cup}\, \{(u, i, v) \mid u, v \in S_{\varphi,s}^* \wedge 1 \le i \le m\}.$$

Theorem 17. *Let φ be an SML-formula over Σ. Then φ is satisfiable iff $\varphi^{\dagger}$ is satisfiable, and φ has a finite model iff $\varphi^{\dagger}$ has a finite model.*

Proof. Let $\mathcal{T} = (S, \Sigma, R, L)$ and $s \in S$ with $(\mathcal{T}, s) \models \varphi$. By Theorem 11 it holds $(\mathcal{T}_{\varphi,s}^*, s) \models \varphi$. Since the symbol 0 does not occur in $\varphi^{\#}$ the same argument as for Lemma 5 shows that $(\mathcal{T}_{\varphi,s}^*, s, t) \models \varphi^{\#}$ for any $t \in S$. On the other hand, it is easy to check that $(\mathcal{T}_{\varphi,s}^{\dagger}, s, s)$ satisfies α_i, γ_i, δ_i and ζ_i for every $1 \le i \le m$ and that for any $a \in \Sigma$, there is exactly one k with $0 \le k \le \operatorname{sd}_a(\varphi)$ (namely $\kappa_{\varphi,s}^a$), such that $\beta_{k,i}^a$ is true for every $1 \le i \le m$. Hence $(\mathcal{T}_{\varphi,s}^{\dagger}, s, s) \models \varphi^{\dagger}$, i.e., $\varphi^{\dagger}$ is satisfiable and if $\mathcal{T}$ is a finite model of φ, then $\mathcal{T}_{\varphi,s}^{\dagger}$ is a finite model of $\varphi^{\dagger}$.

For the converse let $\mathcal{T} = (S, \Sigma_m, R, L)$ and $s, t \in S$ such that $(\mathcal{T}, s, t) \models \varphi^{\dagger}$. By Lemma 12 it holds $s = t$. By Lemma 13 there is exactly one k_a for every $a \in \Sigma$ with $0 \le k_a \le \operatorname{sd}_a(\varphi)$ such that $\beta_{k_a,i}^a$ is satisfied. Let $S_{\varphi,s} \subseteq S$ be as before and let $S' \subseteq S$ be defined by

$$S' := S_{\varphi,s} \,\dot{\cup}\, \{s_j^a \mid a \in \Sigma \wedge 1 \le j \le k_a\},$$

where the s_j^a's in $S \setminus S_{\varphi,s}$ are according to Lemma 13. Note that we have $(s, i, s_j^a) \in R$ for every $1 \le i \le m$ by the additional term in $\varphi^{\dagger}$. Each s_j^a has a single outgoing Σ-transition which is labeled by a and leads to s. By Lemmata 12, 13, 16, there is $(s, i, u) \in R$ for every $u \in S'$ and u has exactly one 0-successor. Since only the existence of 0-successors is used in all subformulae, but none of these transitions is actually traversed, we can assume that all 0-transitions occur as loops, i.e., there is $(u, 0, u) \in R$ for every $u \in S'$ and $(u, 0, v) \notin R$ for $u, v \in S', u \ne v$. Further, there is $(u, i, v) \in R$ for every $u, v \in S'$ (by Lemma 15, if $u \ne v$ and by Lemma 16, if $u = v$). Let $u \in S'$ and $v \in S$ with $(u, i, v) \in R$. Since there is $(s, i, u) \in R$, there is also $(s, i, v) \in R$ by Lemma 15. By Lemma 14 it follows $v \in S'$, i.e., one cannot escape S' by using i-transitions. On the other hand, again by Lemma 14, one cannot escape $S_{\varphi,s}$ by using Σ-transitions. In other words, using any modality in $\varphi^{\dagger}$ – either a standard or a sabotage one –

one stays in S'. It is easy to see that we can therefore restrict T to the states in S', i.e., for the transition system $T' := (S', \Sigma_m, R \cap S' \times \Sigma_m \times S', L|_{S'})$ it also holds $(T', s, s) \models \varphi^\dagger$. In particular, (T', s, s) is a model of $\varphi^\#$ with i-transitions between any two states. Let T'' be the restriction of T' to the alphabet Σ. Since the symbol 0 does not occur in $\varphi^\#$ and by the same argument as for Lemma 5 we get $(T'', s) \models \varphi$, i.e., φ is satisfiable. Further, if T is a finite model of $\varphi^\dagger$, then T'' is a finite model of φ. □

Now we are ready to transfer the results on SML to PSL. By using the reduction $\varphi \mapsto \varphi^\dagger$ together with Theorem 2 and Theorem 3 we get

Corollary 18. *The logic* PSL *does not have the finite model property. The decision problems* Satisfiability, Finite Satisfiability, *and* Infinity Axiom *for* PSL *are undecidable.* □

6 Conclusion

We have considered the path sabotage logic PSL which is a balanced version of SML. Both logics are extensions of modal logic that are capable of describing elementary changes of structures. We have shown that the model checking complexity for the logic PSL with a localized sabotage modality is as hard as for SML that has a global 'edge-deleting' modality. Also the satisfiability problem stays undecidable. In fact, from the viewpoint of complexity, both logics much more resemble first-order logic than modal logic, except for a linear formula and a polynomial program complexity.

There are other restrictions to the global power of the sabotage operator, for example the localized version of SML where only those edges can be deleted that start at the current position within the system. Interpreting the modalities as movements of the agents 'runner' and 'saboteur' in a crumbling network, this localized sabotage logic corresponds to the situation that the saboteur can only block adjacent nodes and that the runner gives the saboteur a 'pickaback' while moving in the network. An argument (to be presented elsewhere) which resembles the proofs above shows that the complexities stay the same: Uniform model checking is PSPACE-complete and the satisfiability problem is undecidable.

References

1. J. v. Benthem. An essay on sabotage and obstruction. In D. Hutter and S. Werner, editors, *Festschrift in Honour of Prof. Jörg Siekmann*, LNAI. Springer, 2002.
2. R. Fagin, J. Y. Halpern, Y. Moses, and M. Y. Vardi. *Reasoning about Knowledge*. MIT Press, 1995.
3. Ch. Löding, Ph. Rohde. Solving the sabotage game is PSPACE-hard. In Proceedings of MFCS 2003. Vol. 2747 of LNCS, Springer (2003), pp. 531–540
4. Ch. Löding, Ph. Rohde. Model checking and satisfiability for sabotage modal logic. In Proceedings of FSTTCS 2003. Vol. 2914 of LNCS, Springer (2003), pp. 302–313

Parameterized Model Checking of Ring-Based Message Passing Systems*

E. Allen Emerson and Vineet Kahlon

Department of Computer Sciences,
The University of Texas at Austin,
Austin, TX 78712, USA

Abstract. The *Parameterized Model Checking Problem (PMCP)* is to decide whether a temporal property holds for a uniform family of systems, U^n, comprised of finite, but arbitrarily many, copies of a *template* process U. Unfortunately, it is undecidable in general [3]. In this paper, we consider the PMCP for systems comprised of processes arranged in a ring that communicate by passing messages via tokens whose values can be updated at most a bounded number of times. Correctness properties are expressed using the stuttering-insensitive linear time logic LTL\X. For bidirectional rings we show how to reduce reasoning about rings with an arbitrary number of processes to rings with up to a certain finite *cutoff* number of processes. This immediately yields decidability of the PMCP at hand. We go on to show that for unidirectional rings small cutoffs can be achieved, making the decision procedure provably efficient. As example applications, we consider protocols for the leader election problem.

1   Introduction

The *Parameterized Model Checking Problem (PMCP)* is to decide whether a temporal property holds for a uniform family of systems U^n comprised of finite, but arbitrarily many, copies of a *template* process U. Unfortunately, PMCP is undecidable because a system of size n can simulate a Turing machine for n steps [3]. The Halting problem for Turing Machines can then be easily formulated as a PMCP for reachability of the halting state, viz., EF*halt*. This argument can be refined even in the case where the parameterized system is a unidirectional ring [17]. It follows from a result by Shannon [16] that the undecidability result holds even when the head (token circulating in the ring) can have only two possible states [16]. An essential part of the undecidability proof of the latter is that the message token changes value an arbitrary number of times.

We show in this paper that if there is a bound b on the number of times the token changes value during a run of the system, then the PMCP is decidable. This boundedness assumption can be justified by the fact that protocols for a number of ring based applications have the property that the value of each message bearing token can be changed

---

* Research supported in part by NSF grants CCR-020-5483 and CCR-009-8141, and SRC contract 2002-TJ-1026; {emerson, kahlon}@cs.utexas.edu

J. Marcinkowski and A. Tarlecki (Eds.): CSL 2004, LNCS 3210, pp. 325–339, 2004.

only a bounded number of times in any run of the protocol. For instance, in standard protocols for the leader election problem [15], every token makes at most one value change during any run of each of the protocols.

We express correctness properties using the stuttering-insensitive linear temporal logic LTL\X. The basic assertions are of the form Ah, or Eh, where formula h is built using F "sometimes", G "always", U "until" but without X "next-time"; and A "for all futures" and E "for some future" are the usual path quantifiers. Use of stuttering-insensitive logics is natural when model checking parameterized systems as the next-time operator X gives us the ability to count, often leading to undecidability of the PMCP [10].

In the case of unidirectional (or certain restricted bidirectional) rings, we argue that arbitrarily "large" systems of size n can be imitated up to stuttering by a small system of a certain cutoff size c, where $c = O(b)$. Thus to solve PMCP, checking correctness over all sizes n, it is necessary and sufficient to check all sizes m up to c. In the context of rings, this style of "cutoff" argument has been used in [11], where it was shown how to reduce reasoning about properties expressed using the branching time temporal logic CTL*\X from a system with an arbitrary number of processes to systems with up to a small cutoff number of processes. However, the results were established only for unidirectional rings where the token could not carry values, viz., processes could not exchange messages among themselves, resulting in a framework with limited modeling power. For example, it is not clear how standard protocols for the Leader Election problem (see, for example, [15]) that require tokens to change values, viz., messages to be exchanged, can be encoded in this framework. Our unidirectional ring framework has a broader modeling power but with an efficiently decidable PMCP.

The case of bidirectional rings is more involved. Here we find it convenient to exploit the viewpoint that a ring of many (n) similar processes is tantamount to a Turing machine on a circular tape (CTM for short) with n tape cells. To see this, we note that a token in a ring can be viewed as the head of the CTM, with the value of the token representing the control state of the head. Cell i of the circular tape corresponds to P_i, the ith process in the ring, with the tape symbol in cell i representing the local state of P_i. This, in effect, reduces the PMCP for bidirectional rings in which the token makes only a bounded number of value changes to the study of the PMCP for CTMs in which the head only makes a bounded number of state changes. To analyze the behavior of CTMs we in turn study (Linear Tape) Turing Machines with bounded state changes to the head. For an arbitrary Turing machine, the associated PMCP again amounts to the halting problem and is undecidable. However, we demonstrate that for a Turing machine that can make at most a bounded number b of state changes, the halting problem is decidable, and, hence for the associated ring system where token values change at most b times, the PMCP is decidable. The latter result is established by induction on b. The base case $b = 1$ represents a Turing machine with a single (non-halting) state.

The rest of the paper is organized as follows. The unidirectional (or restricted bidirectional) ring model is introduced and the related cutoff results shown in section 2 while the cutoff results for bidirectional rings are given in section 3. Applications are handled in section 4 and we conclude with some remarks in section 5.

2   Unidirectional Rings

Communication in computer networks is usually carried out via message passing using packets or value-bearing tokens in which the sender puts the data and the address of the intended receiver. However, apart from data transfer, tokens also play a crucial role in the *implementation* of network protocols. In a typical network protocol, a process sends out a token *owned* by it to gather information about other processes in the network. In leader election [15], for example, a process sends out a token bearing its identifier to find out whether there exists another process with an identifier of greater value. In this role tokens play a passive role in that they do not cause any state change in processes other than the ones owning it but are used merely for information gathering. A key reason for this might be that most protocols are data independent. In this section, we propose a simple framework to model such protocols and show how to reduce reasoning about linear time properties for such a system with an arbitrary number of processes to one with a few.

The Process Framework. We consider systems comprised of processes arranged in the form of a ring communicating using multiple message-bearing tokens, each of whose value can be modified at most a bounded number of times (see remark 2.2), say b. All tokens move in the same direction, say clockwise. In a ring $\mathcal{R}$ comprised of the n processes $P_0, ..., P_{n-1}$ listed in clockwise order of occurrence around the ring, the ith process, P_i, is given by a tuple of the form $(Q_i, \Sigma_i, T_i, R_i, i_i)$, where Q_i is the finite set of states of P_i, Σ_i the set of labels of P_i, T_i the set of tokens owned by P_i, R_i its transition relation and i_i the initial state. Let $\mathsf{T} = \cup_i T_i$. Each token in T can take on values from the set V. In any global state of the ring, a token is in the *possession* of exactly one process. A process may, however, possess multiple tokens.

Transitions of P_i can be classified as either *internal* or *token dependent*. An internal transition of P_i is of the general form $a \xrightarrow{l} b$, and can always be fired irrespective of the current global state of the system. A token dependent transition of process P_i, on the other hand, is of the general form $tr : a \xrightarrow{l:g \to A} b$, where $g : V \to \{true, false\}$ is a boolean valued function and action A is either the expression *skip* or of the form $\mathsf{t} := v$. Token dependent transition tr can be fired only if P possesses a token t from the subset $T_{tr} \subseteq T$ with a value that enables guard g. We then say that transition tr *involves* token t. After tr is fired, process P_i transits to local state b and t is passed on to the clockwise neighbor. If action A is the expression *skip*, then the token is simply passed on with its value unchanged, else if A is the expression $\mathsf{t} := v$, then the token is passed with its value updated to v.

For each token dependent transition tr of process P_i, the set T_{tr} is either T_i or $\mathsf{T} \setminus T_i$. If $T_{tr} = T_i$, viz., tr involves tokens owned by it, then tr is termed an *endogenous* token dependent transition, else if $T_{tr} = \mathsf{T} \setminus T_i$, viz., tr involves tokens not owned by it, then tr is termed an *exogenous* token dependent transition. Exogenous transitions of process P_i can be thought of as constituting the communication layer of P_i responsible for handling tokens owned by other processes but causing no change in the local state of P_i. On executing an exogenous transition involving token t, it is passed on to the clockwise neighbor with a possible change in the value of t but without changing the local state of P_i. Thus every exogenous transition is of the general form $a \xrightarrow{l:g \to A} a$. We assume that the action A of exogenous transitions is oblivious of the current local state of P_i and depends only on the value of the token. Thus if $a \xrightarrow{l:g \to A} a$ is a exogenous transition, then for each $b \in Q_i$, there exists an exogenous transition in R_i of the form $b \xrightarrow{l:g \to A} b$. To prevent a process from indefinitely taking possession of a token not owned by it, we assume that from any local state of P_i, for any possible value of $t \notin T_i$, there always exists an exogenous transition of P_i that is enabled. We use $\mathcal{R} = (S^n, \Sigma, \mathsf{T}, R^n, \mathsf{i}^n)$, to denote the ring comprised of the n processes $P_0, ..., P_{n-1}$ executing asynchronously with interleaving semantics and is defined in the usual way.

Reduction Result. We show a one way reduction for properties of the form $\mathsf{A}h(i,j)$, where $h(i,j)$ is a LTL\X formula with atomic propositions over the local states of processes P_i and P_j, from a ring $\mathcal{R}$ of arbitrary size comprised of possibly distinct non-isomorphic processes to a ring of size at most $\mathsf{b}(|T_i| + |T_j|)$, where b is the bound on the number of times the value of each token of $\mathcal{R}$ can be modified. We assume that each process P_i of $\mathcal{R}$ is *deterministic*, viz., for every local state a of P_i, the following conditions hold

(i) for every possible value of a token not owned by P_i, there is a unique exogenous transition of P_i from a that is enabled, and

(ii) there is either an internal transition or for every possible value of token $t \in T_i$, a unique endogenous transition that is enabled from a, but not both.

Using the fact that exogenous transitions are state oblivious and the deterministic nature of processes it can be shown that $P(t)$ is independent of the global computation $\mathcal{R}$ executes. Thus we have the following.

Lemma 2.0. $P(t)$ *is well-defined.*

For set T of tokens, we let $P(T)$ denote $\bigcup_{t \in T} P(t)$. Let P_i and P_j be processes belonging to ring $\mathcal{R}$. We let $\mathcal{R}(i,j)$ denote the ring comprised of the processes $\{P_i, P_j\} \cup P(T_{P_i}) \cup P(T_{P_j})$ occurring in the same relative clockwise order as along $\mathcal{R}$.

Proposition 2.1 (Reduction Result). *Let $\mathcal{R}$ be a ring with processes P_i and P_j. Then $\mathcal{R} \models \mathsf{E}h(i,j)$ implies that $\mathcal{R}(i,j) \models \mathsf{E}h(i,j)$, where $h(i,j)$ is a LTL\X formula over the local states of P_i and P_j.*

Proof Idea. Given a computation x of $\mathcal{R}$, we construct a computation y of $R(i,j)$ such that $x[i,j]$, viz., x projected onto processes P_i and P_j, is a stuttering of $y[i,j]$. $\quad\square$

Remark 2.2 (Boundedness). In general, the number of value changes for tokens of a given ring might not be bounded and hence the above result may not yield any reduction. However, for special cases we can deduce from merely a static analysis of the syntax of the processes that each token undergoes only a bounded number of value changes. One such useful case results by treating each token t as essentially a counter with an integer value which decreases each time t is updated. This gives rise to a ring model where we have integer-valued tokens such that for each local state a of a process the token dependent transitions from a involving t are of the form $a \xrightarrow{l:t>c \to t:=d} b$, where $c > d$, and $a \xrightarrow{l':t \leq c \to skip} f$. Thus token t can be thought of as a counter that is set initially and each time a token dependent transition modifies the value of t there is a decrease in its value. Once t is modified by a transition of the form $tr : a \xrightarrow{l:t>c \to t:=d} b$ of process P_k, then the next token dependent transition to modify the value of t is the next transition of the form $tr' : a' \xrightarrow{l':t>c' \to t_k:=d'} b$, where $c' < d$, that is encountered as we traverse the ring in a clockwise direction from process P_k. This transition is either an exogenous transition of a process other than P_i (which is the same for each local state of the process) or an endogenous transition from the current local state of P_i. We call such a pair of transitions 'adjacent'. Thus the maximum number of times the value of t can be updated is the maximal length of a sequence of adjacent transitions. For the LCR protocol (section 4), the maximum length of such a sequence for each token is 1.

Extensions. The results also hold for the following two extensions of our model.

(a) **Adding FIFO Queues.** Queues may be necessary to ensure that tokens sent to a process are handled in the order received. This guarantees weak and strong fairness requirements are met for the verification of liveness properties.

(b) **Restricted Bidirectional Tokens.** The model can also be generalized by allowing restricted bidirectional tokens where instead of always moving in a fixed clockwise direction we can allow a token to be able to change direction when it is assigned a new value. For a fixed value, however, the token always moves in the same direction.

3   Bidirectional Rings

We present a generalization of the unidirectional ring model proposed in [11] by allowing (a) bidirectional rings, and (b) the token to carry values. We consider systems comprised of finite, but arbitrarily many, copies of a single process template P arranged in the form of a ring executing concurrently, viz., with interleaving semantics. We only consider the case where processes communicate using a *solitary* token t that is allowed to carry values. Template process P has two types of transitions: (1) *token dependent* that require P to possess t in order to fire, and (2) *internal* that can be fired irrespective of whether P possesses t or not. In addition, P uses transitions labeled with the *receive* action to take possession of t from its counterclockwise neighbor and transitions labeled with *send* actions to relinquish possession of t to its clockwise neighbor. In any computation, the system is allowed to change the value of the token at most a bounded number of times, say b. Allowing an unbounded number of value changes to the token could, in general, make a family of such systems Turing-powerful [17] and hence the corresponding PMCP undecidable.

Formally, process P is defined to be a labeled transition system given by the tuple $(S \times (V \cup \{\perp\}), \Sigma, R, (i, \perp))$, where

- V is the finite set of values that t can take, with $\perp \notin V$.
- $S \times (V \cup \{\perp\})$ is the set of states of P with pair $(a, v) \in S \times (V \cup \{\perp\})$ indicating that P is in local state a; and v is the value of t in case P possesses t, else $v = \perp$.
- $(i, \perp)$ is the initial state of P.
- Σ, the set of actions, is the disjoint union of the set of "internal" actions Σ_i, the set of "token dependent" actions Σ_{td} and the set of token transfer actions $\bigcup_{v \in V} \{snd_v, rcv_v\}$.

- R, the transition relation, is the set of all transitions $(a, v) \xrightarrow{l} (b, v', D)$, with $D \in$ Dir $= \{\text{counterclockwise, clockwise, undefined}\}$, where
  - $l \in \Sigma_i$ implies that $v = v'$ and $D = $ undefined.
  - $l \in \Sigma_{td}$ implies that $v = v'$, $v \in V$ and $D = $ undefined.
  - $l = rcv_u$ implies that $v = \perp$, $v' = u$ and $D = $ undefined.
  - $l = snd_u$ implies that $v \in V$, $v' = \perp$ and $D \in \{\text{clockwise, counterclockwise}\}$.
  - if $a = $ i and $v = \perp$, then $l = rcv_u$, for some $u \in V$, viz., the only possible initial action is a receive. We also assume that along any path of P send and receive actions alternate.

In this paper, for simplicity we consider only bidirectional rings where the processes are deterministic. A bidirectional ring system comprised of n copies of a process template P is denoted by P^n and is represented as $(P_0, ..., P_{n-1})$ to emphasize the fact that process P_{i+1}[1] has P_i as its counterclockwise and P_{i+2} its clockwise neighbor. Analogously, $(s_0, ..., s_{n-1})$, where for each i, $s_i \in S_P$, represents a global 'cyclic' state of P^n, with process P_i in local state s_i. We assume that the send and receive actions of two neighboring processes synchronize when transferring a token.

The (Single Index) PMCP for Bidirectional Rings. To decide whether for all n, $P^{n+l}, (xa^n) \models h(\text{m})$, where (xa^n), the initial configuration of P^{n+l}, is such that x is a fixed sequence of local states of P of length $l \geq$ m and $h(\text{m})$ is a LTL\X formula over the local states of process P_m. We assume that initially the token is in the possession of process P_0.

Linear Tape and Circular Tape Turing Machines. A (linear tape) Turing Machine M is defined to be a tuple of the form $M = (Q, \Sigma \cup \{\sqcup, \Gamma\}, \delta, q_0)$ where,

- Q is the set of states of M
- $q_0 \in Q$ is the initial state of M
- $\Sigma \cup \{\sqcup, \Gamma\}$ is the set of tape symbols with '$\sqcup$' being the blank symbol and 'Γ' the left end tape marker such that $\Sigma \cap \{\sqcup, \Gamma\} = \emptyset$.
- $\delta \subseteq Q \times \Sigma \cup \{\sqcup, \Gamma\} \times Q \times \Sigma \times \{L, R\}$ is the transition relation. Since Γ is the left-end tape marker, we assume that if $(p, \Gamma, q, b, D) \in \delta$, then $b = \Gamma$ and $D = R$, i.e., cell 0, containing Γ, always reflects back the head to the right.

In this paper, for Turing Machines with linear tapes, the cell containing Γ will be referred to as cell 0 while the ith cell to its right is referred to as cell i.

Analogously we define a Circular Tape Turing machine (CTM), $M = (Q, \Sigma, \delta, q_0)$ on the tape cells $0, ..., m$ where, transition relation $\delta \subseteq Q \times \Sigma \times Q \times \Sigma \times \{L, R\}$, has the property that on a right move from cell m the head ends up at cell 0 and on a left move from cell 0 the head ends up at cell m. Note that in this case because of the circular topology of the system, the left and right directions are not well defined but we interpret them as the clockwise and counterclockwise directions, respectively.

Modeling Bidirectional Token Rings as Circular Tape Turing Machines. Consider the sequence of transitions of $a_0 \xrightarrow{snd_u} a_1 \xrightarrow{l_1} ... \xrightarrow{l_{k-1}} a_k \xrightarrow{rcv_v} a_{k+1}$ of process P where

---

[1] Here '+' denotes addition modulo n.

for each $i \in [1 : k - 1]$, l_i is an internal or a token dependent transition. Note that since we consider only deterministic systems, it is clear that after firing the send transition $a_0 \xrightarrow{snd_u} a_1$, a process has to execute all the actions $l_1, ..., l_{k-1}, rcv_v$ in the order listed to receive the token again. Thus we can, in effect, replace the firing of the above sequence of transitions with the firing of just one receive transition $a_1 \xrightarrow{l_1...l_{k-1}rcv_v} a_{k+1}$. A similar observation holds for all internal and token dependent transitions sandwiched between a receive and a send transition in which case we can replace all these transitions with a single send transition. Thus, it suffices to consider processes P where each transition of P is either a send or a receive transition with send and receive transitions alternating along any path in the transition diagram of P.

Using this assumption, we can now readily see that the ring $P^n = (P_0...P_{n-1})$ with token t comprised of n copies of process template $P = (S \times (V \cup \{\bot\}), \Sigma, R, (i, \bot))$ can be looked upon as the CTM, $C^n = (V, S, \delta, i)$ with one head and tape cells $0, ..., n - 1$. Here cell i corresponds to process P_i with the local state of P_i being looked upon as the tape symbol in cell i. The token t can be thought of as the head of the CTM with the value of t being the state of the head. Transition $(p, a) \rightarrow (q, c, D) \in \delta$ iff for some $b \in S$, both the transitions $(a, \bot) \xrightarrow{rcv_p} (b, p)$ and $(b, p) \xrightarrow{snd_q} (c, \bot, D)$ are in R.

Thus the PMCP defined before can now be reformulated as follows: To decide whether for all n, $C^{n+l}, (xa^n) \models h(m)$, where x is a sequence of tape symbols of S of length l and $h(m)$ is a LTL\X formula over the tape alphabet of cell m, with $m \leq l$. We assume that for each n, in the initial cyclic configurations (xa^n), the head is placed at cell 0.

3.1   Linear Tape Turing Machines

We begin by showing that the behavior of a given deterministic one state Turing Machine M can be deduced from an analysis of the structure of the transition diagram of the control state of M.

Let $M = (Q, \Sigma \cup \{\Gamma, \sqcup\}, \delta, q_0)$ be a given deterministic Linear Tape Turing Machine. We assume that M has just one control state, say q, and that the head of M is initially placed at cell 0 with the rest of the tape cells each containing the empty symbol '$\sqcup$'.

We define the *transition graph* of M as the directed graph $G = (V, E)$, where $V = \Sigma \cup \{\sqcup\}$ and $E = \{(a, b)|a, b \in \Sigma \cup \{\sqcup\}, \delta(q, a) = (q, b, D), with D \in \{L, R\}\}$. Since we are considering a Turing machine with a solitary control state, in any configuration, the direction in which the head of M moves depends only on the symbol it is currently reading. Thus each tape symbol in $\Sigma \cup \{\Gamma, \sqcup\}$ can be characterized as either a *left-symbol* or a *right-symbol* depending on whether the head moves left or right upon reading it. Given symbol $a \in \Sigma \cup \{\sqcup\}$, let G_a denote the subgraph of G induced by the set of symbols reachable from a in G. We say that symbol $a \in \Sigma$ is *writable* iff M starting at cell 0 on the empty input, with each non-zero cell containing $\sqcup$, writes a in some tape cell in finitely many moves. Symbol $a \in \Sigma \cup \{\sqcup\}$ is *readable* iff M, starting on the empty input, reaches a configuration in finitely many steps in which the head is positioned at a cell containing a.

Since M is a deterministic Turing Machine, each node of G has out-degree at most one. To start with, each non-zero tape cell contains $\sqcup$ and so for a symbol to be writable it has to be reachable from $\sqcup$ in G. We may therefore assume, without loss of generality, that all symbols are reachable from $\sqcup$ in G. Thus G is either a simple path starting at $\sqcup$ or a 'lollipop' of the form $a_0 \rightarrow \ldots \rightarrow a_k \rightarrow \ldots \rightarrow a_d \rightarrow a_k$, where $a_0 = \sqcup$. We begin by considering the case where G is the simple path $a_0 \rightarrow \ldots \rightarrow a_k$ starting at $\sqcup$. Later we show how to reduce the analysis for the case where G is a lollipop to this case.

Definitions and Notation. Let $a, b \in \Sigma \cup \{\sqcup\}$ be such that there is a path from a to b in G. We define the *depth* of b with respect to a, denoted by $d(b, a)$, to be the number of states, not including b, in the unique path from a to b. Analogously, the *left-depth* (*right-depth*) of symbol b with respect to a, denoted by $d_L(b, a)$ ($d_R(b, a)$), are defined to be the number of left (right) symbols, not including b, along the path from a to b in G. We abbreviate $d(a, \sqcup)$ as $d(a)$ and refer to it simply as the *depth* of a. Similarly, $d_L(a, \sqcup)$ ($d_R(a, \sqcup)$) is abbreviated by $d_L(a)$ ($d_R(a)$) and called the *left-depth* (*right-depth*) of a. We write $a < b$ to mean $d(a) < d(b)$.

The content of the ith tape cell after the nth move of M is denoted by $t(i, n)$. For $j \geq 1$, we call the portion of the tape comprised of cells numbered greater than or equal to j, the *interval* starting with j and denote it as $I(j)$. For each interval $I(j)$, we define the *traversal number* of $I(j)$ after move n of M, denoted by $trav(j, n)$, as the ordered pair (k, l), where k is the number of times the head moved from cell $j - 1$ to j, viz., *entered* interval $I(j)$, among the first n moves of M, and l is the number of moves made by the head from a cell inside the interval, viz., the cells $j, j + 1, \ldots$, after it entered the interval for the kth (last) time.

Key Results. The analysis of the behavior of a single state deterministic Turing machine rests on the following two facts:

1. If in the initial configuration of M, the head is placed at cell 0 and each cell of the tape contains the empty symbol $\sqcup$, then after finitely many steps of M the contents of the tape form a non-increasing (depth wise) sequence of tape symbols.

2. Symbol $a \in \Sigma$ is readable iff for each $b \leq a$, $d_R(b) \leq d_L(b)$.

Proposition 3.1 (Monotonicity Result). *For $i \geq 1$, we have $t(i, n) \geq t(i + 1, n)$. Furthermore, if after n moves the head is positioned at cell $h < i$ and $t(i, n) \neq \sqcup$, then $t(i, n) > t(i + 1, n)$.*

An immediate consequence is the following.

Corollary 3.2 *For $i < j$, we have $t(i, n) \geq t(j, n)$.*

Using the above results, we next show that a necessary and sufficient condition for a tape symbol a to be readable is that for all $b \leq a$, we have $d_R(b) \leq d_L(b)$.

Proposition 3.3 *If $a \in \Sigma$ is readable then for all $b \leq a$, $d_R(b) \leq d_L(b)$.*

Proposition 3.4 *Let $a_i \in \Sigma$. If for all $j \leq i$, $d_R(a_j) \leq d_L(a_j)$, then a_i is readable.*

Predicting the Behavior of Linear Tape Turing Machines. Let a_j be written in cell k in move m_{j_k} and in cell $k + 1$ in move $m_{j_{k+1}}$. Consider the configuration of the tape between moves m_{j_k} and $m_{j_{k+1}}$. All the cells from 0 to k have a_j written in them.

Since cell $k + 1$ gets written by a_j in finitely many steps, only finitely many, say $k + l$, cells of the tape are visited in $m_{j_{k+1}}$ moves. Then $t(j, n) = \sqcup$ for all $j \geq k + l + 1$. Consider now the configuration of the tape after execution of step $m_{j_{k+1}} - 1$. Since in the very next step a_j is written in cell $k + 1$, the head is currently at cell $k + 1$. Then using proposition 3.1, we have that $a_j = t(k, m_{j_k}) > t(k+1, m_{j_k}) \geq t(k+2, m_{j_k}) > t(k+3, m_{j_k}) > \dots > t(k+l, m_{j_k})$. Therefore it follows that $l \leq k+1 = |G|$. Thus we see that configuration of the tape forms a non-increasing sequence of length $k + l$ with the remaining cells containing the blank symbol. We consider two cases.

(a) Simple Paths. First assume that G is the simple path $a_0 \rightarrow \dots \rightarrow a_k$. There are two sub-cases to consider:

Assume first that a_k is readable. By definition of readability, there is a reachable configuration **c** of M wherein the head after, say n moves, is at tape cell $i \geq 1$ containing a_k. Since the tape configuration forms a non-increasing sequence, it follows that if a_k is readable, then it will be read first in cell 1. Clearly, after reading a_k, the head cannot make any more moves and so M deadlocks in cell 1. Thus by the above comment, in this case only $k + 1$ tape cells were visited during the computation before M deadlocks in cell 1 and the visited tape cells contain a non-increasing sequence of non-empty tape symbols of length at most $k + 1 = |G|$.

Next assume that a_k is not readable. In this case M cannot deadlock, for otherwise symbol a_k would be readable. Let a_j be the symbol of least depth, j, such that $d_R(a_j) > d_L(a_j)$. Clearly, a_{j-1} is a right symbol and $d_R(a_{j-1}) = d_L(a_{j-1})$ and so $d_R(a_j) = d_L(a_j) + 1$. Then using propositions 3.3 and 3.4, we have that all symbols less than or equal to a_{j-1} are readable but a_j is not. Thus a_j is writable but a_{j+1} is not. Since by corollary 3.2, for all $n, i \geq 1$, we have that $t(i, n) \geq t(i + 1, n)$, we have, using the same argument as in the previous case, that the first cell into which a_j is written is cell 1. Since a_j is a right symbol, after writing a_j the head move to the right to cell 2 and then never visits cell 1 again, for otherwise a_{j+1} would be writable. Thus from the above comments we have that when a_{j+1} is written into cell k all cells $0, \dots, k$ contain a_{j+1}, all cells $k + l + 1, \dots$ contains the blank symbols and cell $k + 1, \dots, k + l$ form a non-increasing sequence with $l \leq j + 1$. Thus we can liken the computation to a *wave front* that moves along the tape from left to right such that to the right of the front all cells have the symbol $\sqcup$ while to the left all cell have the symbol a_j. Thus, in this case the computation is unbounded, viz., every cell of the tape is visited at least once. We say that the computation *diverges*.

(b) Lollipops. We now consider the case when G is a lollipop, say $L = a_0 \rightarrow \dots \rightarrow a_k \rightarrow \dots \rightarrow a_d \rightarrow a_k$. Note that in this case the machine never deadlocks because no matter what symbol the head is currently reading, there is always a move it can make. Let L_ω be the 'unrolling' $\{a'_i\}_{i=0}^\infty = a_0 \dots a_k (a_{k+1} \dots a_d a_k)^\omega$ of L. From the discussion in the previous section, it follows that we all we need to do is decide whether there exists an i such that $d_R(a'_i) > d_L(a'_i)$ and, if yes, find the least such i. Let C_L and C_R denote the number of left and right symbols, respectively, in the cycle $a_k \dots a_d$ and $l_C = d - k + 1$ denote the length of the cycle. Then we can show that if for some i, $d_R(a'_i) > d_L(a'_i)$ then there exists such an $i \in [0 : k(d - k + 2)]$ and hence such an i can be determined efficiently in time $O(|G|^2 log(|G|))$.

Using the result for the case when G is a simple path we see that if there exists an i such that $d_R(a_i') > d_L(a_i')$, then a front develops writing a_j' to its left, where j is the least i with the above mentioned property. If no such i exists then no front develops and thus cell 1 is visited infinitely often during the computation. In this case if the cycle of the lollipop contains a right symbol then the computation of M on the empty string is unbounded. On the other hand, if all symbols in the cycle are left symbols then since a_k is readable and all symbols appearing after a_k in the lollipop are left symbols, so after reading a_k the head shuttles between cells 0 and 1 without visiting any other cell thereafter with tape symbols being written repeatedly in following cyclic fashion $a_{k+1} \rightarrow \ldots \rightarrow a_d \rightarrow a_k$ in cell 1.

The above discussion can be summed up as follows.

Proposition 3.5 (Behavior Lemma). *Let M be a given linear tape Turing machine with only one control state. Then one of the following holds.*

– *the head of M eventually deadlocks in cell 1*
– *the head of M diverges*
– *the head of M eventually shuttles between cells 0 and 1 indefinitely.*

Furthermore, if G is the transition graph of M, then the behavior of M can be decided in time $O(|G|^2 log|G|)$.

3.2   The PMCP for Bidirectional Rings

We now show how the results for linear tape Turing machines with a solitary state can be leveraged to give decision procedures for the PMCP for bidirectional rings. The connection between Turing machines and rings is established via the *Ring Traversal Lemma* using the notion of *crossing numbers* discussed below. The PMCP for rings can equivalently be formulated as follows: given a LTL\X formula $h(m)$ with atomic propositions over the local states of process P_m, where m $\in [0 : l - 1]$, does there exist n such that $C^{l+n}, (xa^n) \models Eh(m)$? We assume that in each of the initial cyclic configurations (xa^n), the head is placed at cell 0.

Notation. We refer to the counterclockwise and clockwise directions along the circular tape of C^{n+l} as *right* and *left* directions, respectively. We assume that tape cells $0, \ldots, n + l - 1$ of C^{n+l} are arranged in a counterclockwise direction in the order listed. For any *interval*, viz., a finite set of adjacent cells along the circular tape, when traversing the cells of the interval in the counterclockwise direction, the cell encountered first is called the *left end* of the interval whereas the cell encountered last is called the *right end* of the interval. For C^{n+l}, cells $0, \ldots, l - 1$ containing the input sequence x is designated as interval X while the set of remaining cells, each containing the tape symbol a, is designated the *outer ring*. As for Turing machines with linear tapes, we let G denote the transition graph of C^{n+l} and G_a the subgraph of G induced by the set of all symbols reachable from a in G.

Strategy. We begin by outlining our strategy. For a ring of size $n+l$, starting at the initial cyclic configuration (xa^n), we construct a transition diagram $G_X(n)$ on the configurations of interval X, where each configuration is given by the contents of the tape cells constituting X along with the cell number of X on which the head is currently placed. If from a configuration **c** of X, the head moves outside interval X, then if the head does not re-enter X, then **c** has no successor in $G_X(n)$, else the successor is the configuration that results when the head re-enters X. In the second case, the transition that results is called an *external* transition of $G_X(n)$. All transitions of $G_X(n)$ that are not external are called *internal* and correspond to movements of the head within interval X. Since M is a deterministic Turing machine, $G_X(n)$ is either a simple path or a lollipop. Note that since we are interested in the 'behaviour' of cell m belonging to interval X, there exists n such that $M, (xa^n) \models Eh(m)$ iff there exists n such that $G_X(n) \models Eh(m)$, where in both cases the formula $h(m)$ is interpreted over the tape alphabets in cell m. We show the existence of a cutoff $c \geq l$ such that for all $j \geq c$, the transition diagram $G_X(j)$ is the same as $G_X(c)$. This reduces the PMCP to determining whether there exists $i \in [l : c]$ such that $M, (xa^i) \models Eh(m)$, i.e., model checking at most c finite state systems, which is clearly decidable. We point out that we do not actually construct $G_X(n)$ but merely use it to prove our cutoff result. Towards that end, however, we need to elucidate the structure of $G_X(n)$. Note that the internal transitions of $G_X(n)$ are easy to figure out as they correspond to movements of the head within interval X. But for the external transition, the key question that needs to be answered is that in case the head leaves interval X whether it re-enters X again and, if yes, then the direction from which it re-enters X and the configurations of both interval X and the outer ring on re-entry in relation to the configuration of the outer ring on the last exit. We address this issue next.

Ring Traversals. Let (xa^*) denote the set of cyclic configurations wherein all cells other than the one containing sequence x contain the tape symbol a. We now show that if M starts at the configuration (xa^*) with the head positioned inside interval X, then the above result says that if the head exits X for the kth time, then it cannot shuttle in the outer ring forever, but (a) it either re-enters X, or (b) it deadlocks outside X, and in both cases when that happens the configuration of the ring is of the form $(yx'zb^*)$ where $|x'| = |x|$ and y and z constitute the 'out-growth' of the sequence in interval X during the kth 'excursion' of the head outside X. The ring traversal lemma given below allows us to quantify the length of this outgrowth. The key idea is that starting from a cyclic configuration of the form $(yxza^n)$ with interval X containing the sequence x, if the head exits X on the right, then the head may re-enter X on the right thus completing an external transition or deadlock outside X without diverging in the outer ring. The interesting case occurs when the head diverges in the outer ring, say from the right end of interval Z (containing z) in the counterclockwise direction. Because of the circular nature of the tape, the head enters interval Y (containing sequence y) from the left end. There are three possibilities now. The head may in finitely many moves either (1) re-enter X from the left without diverging again in the outer ring again, thus completing the external transition, or (2) deadlock without re-entering interval X and without diverging again in the outer ring, or (3) diverge in the outer ring again, this time in the clockwise direction. In this fashion, we see that the head may keep on diverging back and forth in the outer ring till it either re-enters X from either the right or the left end, or it deadlocks without re-entering X. This is formalized in the ring traversal lemma, the statement of which requires the notion of crossing numbers defined next.

Crossing Numbers. Let y be a given finite string of tape symbols and let interval Y comprised of cells $1, ..., n$ of a linear tape, contain y. Let cell $n + 1$ contain Δ, where δ, the transition relation for M has the property that $\delta(q, \Delta) = (q, \Delta, L)$. Thus Δ merely 'reflects' back the head to the left into Y.

Then the *left-right crossing number* of y, denoted by $C_{LR}(y)$, is intended to capture the number of moves made by the head on the left end of interval Y, viz., from cell 1 to 0, after the head enters interval Y at the left end and before it exits Y at the right end for the first time. Formally, $C_{LR}(y)$ is defined as follows. Starting at cell 0 (containing Γ), if the head ever exits interval Y on the right, viz., makes a right move from cell n to $n + 1$, then we define $C_{LR}(y)$ as the number of moves made by the head from cell 1 to 0 before it exits Y to the right for the first time. If the head never exits Y on the right, there are three possible cases (1) the head either deadlocks in Y in which case $C_{LR}(Y)$ is defined to the number of moves made by the head from cell 1 to cell 0, viz., at the *left end* of interval X, before it deadlocks, or (2) the interval Y is exited to the left an unbounded number of times in which case we define $C_{LR}(y)$ as ∞, or (3) after finitely steps the head keeps on shuttling in Y without exiting Y on either side thereafter. In that case, we define $C_{LR}(y)$ as $\bot$.

In general, for $D_1, D_2 \in \{L, R\}$, we may define $C_{D_1 D_2}(y)$, to capture the number of moves made by the head on the D_2'th end of interval Y, where $D_2' \in \{L, R\} \setminus \{D_2\}$, viz., the opposite end from which the head is supposed to exit Y, after the head enters interval Y at the D_1th end.

Proposition 4.1 (Ring Traversal Lemma). *Starting at the initial configuration (xa^n) of C^{l+n}, suppose that when the head exits interval X for the kth time, the ring configuration is of the form $(yx'zb^*)$, where x' is the content of X. If n is greater than the maximum of the minimum of $C_{LL}(z)|G|$ and $C_{LR}(y)|G|$, and the minimum of $C_{RL}(z)|G|$ and $C_{RR}(y)|G|$, viz., the ring is of sufficiently large size, then one of the following holds.*

1. *the head deadlocks before entering interval X again.*
2. *the head re-enters interval X after finitely many steps.*

In both cases, the resulting configuration is of one of the two forms: $(y''y'x''z'c^*)$ or $(y'x''z'z''c^*)$, where $|x''| = |x|$, $|y'| = |y|$, $|z'| = |z|$ and $|y''|, |z''|$ are less than or equal to the minimum of $C_{RL}(z)|G|$ and $C_{RR}(y)|G|$ or the minimum of $C_{LR}(y)|G|$ and $C_{LL}(z)|G|$ accordingly as the head exits X to the left or to the right.

A crucial consequence is that the behavior of the head (as far as interval X is concerned) after exiting X for the kth time is the same for all n greater than a *threshold* value, viz., the minimum of $C_{RL}(z)|G|$ and $C_{RR}(y)|G|$ or the minimum of $C_{LR}(y)|G|$ and $C_{LL}(z)|G|$, the only difference being the number of cells in the outer ring containing the symbol c. This observation gives us the cutoff which we derive next.

Generating the Cutoff. Using the above result, we next show the existence of cutoff $c \geq l$ such that for all $j \geq c$, the transition graphs $G_X(j)$ is the same as $G_X(c)$. Let (xa^n) be the initial tape configuration with the head at cell 0. Recall that G_a is the subgraph of G induced by the set of all tape symbols reachable from a in G. Here we consider only one case where G_a is a simple path with the other being handled in a similar fashion.

Let G_a be the simple path $a_0 \rightarrow ... \rightarrow a_d$. In this case, we have that the head can make at most d + 1 moves from any cell of the ring without deadlocking. Then, from the definition of crossing numbers, it follows that $C_{DD'}(w) \leq d + 1$, for any sequence w of tape symbols and any $D, D' \in \{L, R\}$. Hence from the Ring Traversal Lemma 4.1, it follows that after the head exits X in configuration $(yx'zb^*)$ the length of the newly added intervals Y' and Z' containing respectively y'' and z'' is at most $(d + 1)^2$. Since the head can exit interval X at most d + 1 times (without deadlocking), at most d + 1 external transitions can be fired in $G_X(n)$ for any n. Then, using proposition 4.1 repeatedly, we have that in all exits and re-entries of X, the total length of the newly added intervals is at most $(d + 1)^3$. Thus in this case, for each $j \geq c = l + (d + 1)^3$, $G_X(j)$ is the same as $G_X(c)$ and so the value of the cutoff is $c = l + (d + 1)^3$.

Multiple but Bounded Number of States. Using the fact that M is deterministic, we can reduce the analysis of the case where $b \geq 1$ changes are allowed to the control state to the repeated application of the case with one control case. Starting from the initial configuration (xa^*) in state q_0, the first step is to decide whether a state change occurs to the head and if, yes, then the resulting configuration c_0 after the move in which the change occurs. If a state change does occurs then we repeat the above step but starting in c_0 as the initial configuration. But this is just an instance of the original problem but with one lesser state change allowed. Thus to study the behavior of M we need to carry out this procedure at most b times.

Proposition 4.2 (Decidability Result). *The PMCP for LTL\X properties is decidable for bidirectional rings with a token that is allowed to change value a bounded number of times.*

4 Applications

The framework(s) presented in this paper are broad enough to model a variety of ring based applications. Our framework can model the Leader Election Problem and Token Ring LANs, neither of which could be handled by [11]. Examples that require bidirectional rings include bidirectional variants of all applications considered in [11]. However for lack of space, we consider only the Leader Election Problem.

Leader Election Protocols. In local area token networks, a single token circulates around the ring giving it owner the sole right to initiate communication. If the token is lost, then the *Leader Election Problem* [15], is to elect a new unique leader to act as the new owner of the regenerated token.

The LCR Leader Election Protocol. The Le Lann, Chang and Roberts (LCR) protocol assumes that each process P_i in a given unidirectional ring has an integer $id_i > 0$ as a unique identifier not necessarily in increasing or decreasing order around the ring. The protocol works as follows: Each process P_i sends token t_i with its identifier value id_i around the ring. We model this by letting P_i own t_i. When a process receives a token, it compares the value of the token to its own identifier. If the value is greater than its

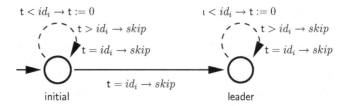

Fig. 1. The LCR Protocol

identifier, it passes the token unchanged. If the value is less than its own it changes its value to 0. If the value is equal to its own, the process declares itself leader. The transition diagram for process P_i is shown in figure 1. Then from the discussion in section 2, it follows that $P(t_i)$ is the first (in case there exists one) process occurring along the ring in the clockwise direction with identifier greater than id_i.

We need to verify that for any arbitrarily large ring it is never the case that two distinct processes P_i and P_j declare themselves leaders, viz., $f = \mathsf{EF}(leader_i \wedge leader_j)$ is not satisfied. Since $|P(t_i)|, |P(t_j)| \leq 1$, $\mathcal{R}(i,j)$ has at most 4 processes. Then, we see using proposition 2.1, that we can reduce the reasoning of the leader election protocols for *all* rings containing processes P_i and P_j irrespective of their size to 6 canonical ring systems each with at most 4 processes. Since i and j were arbitrarily chosen, we have that since the LCR protocol is correct for the 6 canonical systems it is correct for any arbitrary ring.

5   Concluding Remarks

The generally undecidable PMCP has received a good deal of attention in the literature. A number of interesting proposals have been put forth, and successfully applied to certain examples (e.g., [1, 2, 5, 6, 14, 18]). However a lot of these methods suffer from the following drawbacks: much human ingenuity may be required to develop, e.g., network invariants; the method may not terminate; the complexity may be intractably high; and the underlying abstraction may only be conservative rather than exact.

However for frameworks that handle specialized application domains decision procedures can be given that are both sound and complete, fully automatic and in some cases efficient ([4, 7, 8, 11, 12])). In this paper, we have considered the PMCP for LTL\X properties for parameterized families of rings wherein processes communicate using message passing via tokens. Previous work, to the best of our knowledge, has only considered unidirectional rings with a solitary token that could not carry any values [11] and so messages could not be exchanged between processes. Such systems have limited expressive power and cannot model, for instance, standard solutions for the leader election problem. We have extended the known envelope of decidability of the PMCP for ring systems to bidirectional token rings wherein the token can carry messages but only a bounded number of value changes to the token are permitted. Our reduction technique involves showing how to reduce reasoning about a ring with an arbitrary number of processes to a ring with up to a cutoff number of processes. In this paper, the reduction results were established for a bidirectional ring with a single token. A possible direction for future research is to study bidirectional rings with multiple tokens.

We have also identified a broad unidirectional ring framework which allows multiple tokens with each token being allowed a bounded number of value changes. For this framework, we have shown that small cutoffs can indeed be obtained making our technique truly efficient. For bidirectional rings, our methods are *exact*, viz., both sound and complete, and fully automated and for unidirectional rings provably efficient. Moreover the use of cutoffs has the added advantage that the reduced system is a replica of the original system but with a fewer number of processes. This is beneficial for several reasons. First it gives us a clean reduction as there is no need, e.g., to construct an abstract graph which may have a complex, non-obvious structure very different from the original system. Secondly, it caters to automatic error trace recovery.

References

1. P. Abdulla, A. Boujjani, B. Jonsson and M. Nilsson. Handling global conditions in parameterized systems verification. CAV 1999.
2. P. Abdulla and B. Jonsson. On the existence of network invariants for verifying parameterized systems. In *Correct System Design - Recent Insights and Advances*, 1710, LNCS, pp 180-197, 1999.
3. K. Apt and D. Kozen. Limits for automatic verification of finite-state concurrent systems. *Information Processing Letters*, 15, pages 307-309, 1986.
4. T. Arons, A. Pnueli, S. Ruah, J, Xu and L. Zuck. Parameterized Verification with Automatically Computed Inductive Assertions. CAV 2001, LNCS 2102, 2001.
5. M.C. Browne, E.M. Clarke and O. Grumberg. Reasoning about Networks with Many Identical Finite State Processes. *Information and Control*, 81(1), pages 13-31, April 1989.
6. E.M. Clarke, O. Grumberg and S. Jha. Verifying Parameterized Networks using Abstraction and Regular Languages. CONCUR 95. LNCS 962, pages 395-407, Springer-Verlag, 1995.
7. E.A. Emerson and V. Kahlon. Reducing Model Checking of the Many to the Few. CADE-17. LNCS , Springer-Verlag, 2000.
8. E.A. Emerson and V. Kahlon. Model Checking Large-Scale and Parameterized Resource Allocation Systems. TACAS, 2002.
9. E.A. Emerson and V. Kahlon. Rapid Parameterized Model Checking of Snoopy Cache Coherence Protocols. TACAS, 2003.
10. E.A. Emerson and V. Kahlon. Model Checking Guarded Protocols. LICS, 2003.
11. E.A. Emerson and K.S. Namjoshi. Reasoning about Rings. POPL. pages 85-94, 1995.
12. E.A. Emerson and K.S. Namjoshi. Automatic Verification of Parameterized Synchronous Systems. CAV. LNCS, Springer-Verlag, 1996.
13. S.M. German and A.P. Sistla. Reasoning about Systems with Many Processes. *J. ACM*,39(3), July 1992.
14. R.P. Khurshan and L. McMillan. A Structural Induction Theorem for Processes. PODC. pages 239-247, 1989.
15. N. Lynch. *Distributed Algorithms*, Morgan-Kaufmann, 1996.
16. C.E. Shannon, A Universal Turing Machine with Two Internal States. Automata Studies. Princeton, NJ: Princeton University Press, pp. 157-165, 1956.
17. I. Suzuki. Proving properties of a ring of finite state systems. *IPL*, 28, pages 213-314,1988.
18. P. Wolper and V. Lovinfosse. Verifying Properties of Large Sets of Processes with Network Invariants. In J. Sifakis(ed) *Automatic Verification Methods for Finite State Systems*, Springer-Verlag, LNCS 407, 1989.

A Third-Order Bounded Arithmetic Theory for PSPACE

Alan Skelley*

Department of Computer Science, University of Toronto
10 King's College Road, Toronto, ON M5S 3G4 Canada
skelley@acm.org

Abstract. We present a novel third-order theory W_1^1 of bounded arithmetic suitable for reasoning about PSPACE functions. This theory has the advantages of avoiding the smash function symbol and is otherwise much simpler than previous PSPACE theories. As an example we outline a proof in W_1^1 that from any configuration in the game of Hex, at least one player has a winning strategy. We then exhibit a translation of theorems of W_1^1 into families of propositional tautologies with polynomial-size proofs in BPLK (a recent propositional proof system for PSPACE and an alternative to G). This translation is clearer and more natural in several respects than the analogous ones for previous PSPACE theories.

Keywords: Bounded arithmetic, propositional proof complexity, PSPACE, quantified propositional calculus

1 Introduction

Theories of bounded arithmetic such as S_2^i and T_2^i of Buss [1] are interesting for their close ties to computational complexity. For example, the S_2 hierarchy collapses if and only if S_2 proves that the polynomial-time hierarchy collapses [3, 25, 16]. An important property of a theory is the computational complexity of functions that can be defined in it, and theories are known that correspond in this way to many natural complexity classes; see for example [7], [2], [13], [6].

Another important feature of theories of bounded arithmetic is that theorems can often be translated into families of tautologies with polynomial-sized proofs in a related propositional proof system. For example, propositional translations of theorems of Cook's equational theory of polynomial-time functions, PV, have polynomial-sized extended Frege proofs [12].

1.1 Our Results and Related Work

In his thesis [1], Buss introduced the first-order S_2 hierarchy but he also gave second-order theories U_2^1 and V_2^1 whose Σ_1^B-definable functions are exactly the classes PSPACE and EXPTIME, respectively. The ability to reason about the

* Research supported by Canadian Natural Sciences and Engineering Research Council grant PGSB-208264-2000

J. Marcinkowski and A. Tarlecki (Eds.): CSL 2004, LNCS 3210, pp. 340–354, 2004.

exponentially-large second-order objects gives the theory greatly increased power; for example, V_2^1 is otherwise identical to T_2, whose Σ_1^b-definable functions are from the polynomial hierarchy.

Now, Razborov [19] and Takeuti [24] independently showed a general method (the RSUV isomorphism) by which a first-order theory could be shown equivalent to a second-order theory: for example, the Σ_1^b-definable number functions of S_2^1 are the same as the Σ_1^B-definable *string* functions of V_1^1. Zambella [25] then gave a very elegant presentation of a second-order hierarchy $\{V^i\}$ equivalent to $\{S_2^i\}$. This second-order "viewpoint" has been adopted by other authors [8, 9] and has the advantages of greatly reducing the number of axioms required due to the absence of '#' (the smash function symbol) from the language and also simplifying the bootstrapping of the theories.

In this paper we introduce a new third-order theory called W_1^1 designed to exploit both the above uses of a higher order in bounded arithmetic: Firstly to simplify the language, presentation and bootstrapping and secondly to reason about exponentially large objects. We show that the Σ_1^B-definable string functions of this theory are exactly those computable in polynomial space (PSPACE). Our witnessing theorem is much simpler than the analogous one for U_2^1 since it completely eliminates the complicated witnessing formulas of [1] and also uses a simpler comprehension scheme that does not necessitate adding comprehension rules to the sequent calculus.

We also discuss a recent propositional proof system, BPLK [22], corresponding to PSPACE and give a translation of theorems of W_1^1 into families of propositional tautologies with polynomial-size proofs in this new system. This translation is very much simpler than the analogous one for U_2^1 and G, the quantified propositional calculus that is the only previously studied propositional proof system for PSPACE. This latter translation is from [17] and lacks many technical details that we suspect would be very tricky if written out in full.

2   A Third-Order Language

We consider a three-sorted ("third-order") predicate calculus with free and bound variables of the first sort named $a, b, c, ...$ and $x, y, z, ...$, respectively, and free and bound variables of the second sort named $A, B, C, ...$ and $X, Y, Z, ...$, and likewise of the third sort named $\mathcal{A}, \mathcal{B}, \mathcal{C}, ...$ and $\mathcal{X}, \mathcal{Y}, \mathcal{Z}, ...$ The first sort is intended to represent natural numbers; the second, finite sets of natural numbers; and the third, finite sets of finite sets. The language $\mathcal{L}_A^3$ consists of the following set of non-logical symbols: $\mathcal{L}_A^3 = \{0, 1, +, \cdot, | \cdot |_2, \in_2, \in_3, \leq, =\}$, the same as the set $\mathcal{L}_A^2$ for V^1 but with the addition of the third-order membership predicate $A \in_3 \mathcal{B}$. Note in particular the absence of the smash function symbol. The expression $|X|_2$ is intended to represent the largest element of the set X. Such sets are interchangeable with finite binary strings under the following mapping, as in [9]: The set X represents the string with length $|X|_2 - 1$ whose ith bit is 1 exactly when $i \in_2 X$. This map is a bijection with the exception that the string corresponding to the empty set would be undefined, so we define it to be the empty string. Third-order objects can then be thought of as sets of strings.

Number terms are defined identically as in V^1, in particular not including any reference to third-order variables. Formulas additionally may have third-order variables and quantifiers. The hierarchy $\Sigma_i^{\mathcal{B}}$ of classes of formulas in this language is analogous to Σ_i^B and Σ_i^b for second- and first-order formulas: $\Sigma_i^{\mathcal{B}}$ consists of those formulas with arbitrarily many bounded first- and second-order quantifiers, and exactly i alternations of third-order quantifiers, the outer-most being **restricted**, i.e. equivalent to an existential quantifier. We shall be concerned only with $i \in \{0, 1\}$. Now, **strict** $\Sigma_1^{\mathcal{B}}$-formulas are those consisting of a single existential third-order quantifier followed by a formula with no third-order quantifiers; we shall be mainly concerned with a slightly more inclusive class of formulas called $\forall^2 \Sigma_1^{\mathcal{B}}$, consisting of a single bounded universal second-order quantifier followed by a **strict** $\Sigma_1^{\mathcal{B}}$-formula. Restricting several schemes in our theory to this class will be justified in section 4 by the fact that an appropriate replacement scheme will be provable in our theory.

Note that third-order quantifiers are not bounded, and in fact there does not seem to be any way to bound them since terms cannot reference third-order variables. Fortunately, in the appropriate fragment of the theory we shall be concerned with, these variables will always be implicitly bounded.

3   The Theory W_1^1

W_1^1 is a theory over the above-defined third-order language. The axioms of W_1^1 are B1-B12 and L1,L2 of [8] (open axioms defining the function and predicate symbols in the language), (strict) $\forall^2 \Sigma_1^{\mathcal{B}}$-IND, and the following two comprehension schemes $\Sigma_0^{\mathcal{B}}$-2COMP:

$$(\exists Y \leq t(\overline{x}, \overline{X}))(\forall z \leq s(\overline{x}, \overline{X}))[\phi(\overline{x}, \overline{X}, \overline{\mathcal{X}}, z) \leftrightarrow Y(z)]$$

and $\Sigma_0^{\mathcal{B}}$-3COMP:

$$(\exists \mathcal{Y})(\forall Z \leq s(\overline{x}, \overline{X}))[\phi(\overline{x}, \overline{X}, \overline{\mathcal{X}}, Z) \leftrightarrow \mathcal{Y}(Z)],$$

where in each case $\phi \in \Sigma_0^{\mathcal{B}}$ subject to the restriction that neither Y nor $\mathcal{Y}$, as appropriate, occurs free in ϕ. $\mathcal{Y}(Z)$ abbreviates $X \in_3 \mathcal{Y}$, and similarly for $Y(z)$.

4 $\Sigma_1^{\mathcal{B}}$-Replacement Schemes

In this section we shall show that W_1^1 proves various replacement schemes, allowing third-order existential quantifiers to be moved past lower-order quantifiers.

First, though, it is convenient to note that adding to W_1^1 function symbols for its number- and string-valued $\Sigma_1^{\mathcal{B}}$-definable functions results in a conservative extension. The proof of the present claim is analogous to that for first-order bounded arithmetic theories in section 2.3 of [1]. In that proof, a given Σ_1^b-formula in the augmented language is shown to be equivalent to a constructed Σ_1^b-formula in the original language, and preserves strictness of the quantifier syntax.

$W_1^1 \supset V (= \bigcup V^i)$ since all the axioms of the latter theory are in the former. W_1^1 can therefore Σ_∞^B-define all number- and string-valued functions of number and string arguments from the polynomial-time hierarchy. By the remarks in the previous paragraph, we can add symbols for these functions to W_1^1 and obtain a conservative extension. In particular, pairing functions such as $\langle x, y \rangle$, $\langle X, Y \rangle$ and $\langle X, y \rangle$ may be added. For a third-order variable $\mathcal{X}$ define $\mathcal{X}^{[x]}(X) \equiv \mathcal{X}(\langle x, X \rangle)$ and $\mathcal{X}^{[X]}(Y) \equiv \mathcal{X}(\langle X, Y \rangle)$, which make $\mathcal{X}$ into an array, with rows indexed by number or strings respectively, each row of which is a third-order object. Let '$\frown$' represent string concatenation and $\dot{-}$ represent limited subtraction. With this in mind, we can state the Σ_1^B replacement schemes:

Definition 1 (Σ_1^B Replacement Schemes). Σ_1^B-1REPL is:

$$\forall x \le y \exists \mathcal{X} \phi(x, y, \mathcal{X}) \leftrightarrow \exists \mathcal{X} \forall x \le y \phi(x, y, \mathcal{X}^{[x]})$$

and Σ_1^B-2REPL is:

$$\forall X \le y \exists \mathcal{X} \phi(X, y, \mathcal{X}) \leftrightarrow \exists \mathcal{X} \forall X \le y \phi(X, y, \mathcal{X}^{[X]}),$$

where in each case ϕ is a (general) Σ_1^B-formula that may have other free variables than those indicated.

Theorem 2. The Σ_1^B replacement schemes are theorems of W_1^1.

Proof (Proof Sketch). Although the Σ_1^B-1REPL scheme has a simpler proof, it can also be proved in the same way as the Σ_1^B-2REPL scheme, so we sketch only a proof of the latter.

$\leftarrow$: This direction of the equivalence, namely that for $\phi(X, y, \mathcal{X}) \in \Sigma_1^B$

$$W_1^1 \vdash \exists \mathcal{X} \forall X \le y \phi(X, y, \mathcal{X}^{[X]}) \supset \forall X \le y \exists \mathcal{X} \phi(X, y, \mathcal{X})$$

is immediate.

$\rightarrow$: The existence of a proof in W_1^1 of this direction of the equivalence is itself proved by structural induction on ϕ. The base case of the induction is when ϕ is Σ_0^B. let ψ be $\forall X \le y \exists \mathcal{X} \phi(X, y, \mathcal{X})$. Let $\theta(c)$ be the formula

$$\forall X \le (y \dot{-} c) \exists \mathcal{X} \forall Y \le c \phi(X \frown Y, y, \mathcal{X}^{[Y]}).$$

$\theta(0)$ is a simple logical consequence of ψ, and $W_1^1 \vdash \psi \wedge \theta(c) \supset \theta(c+1)$ by use of Σ_0^B-3COMP to combine two third-order objects (coding the two arrays of third-order objects for all strings of length smaller than y starting with $X \frown 0$ and $X \frown 1$ respectively) into one third-order object coding the array for all strings of length smaller than y starting with X. Thus $W_1^1 \vdash \psi \supset \theta(y)$ by $\forall^2 \Sigma_1^B$-IND, and clearly $W_1^1 \vdash \theta(y) \supset \exists \mathcal{X} \forall X \le y \phi(X, y, \mathcal{X}^{[X]})$. This induction, incidentally, is the only place where $\forall^2 \Sigma_1^B$-IND, rather than strict Σ_1^B-IND, seems to be necessary.

The induction step ($\phi \notin \Sigma_0^B$) is proved by putting the formula in prenex form and then applying the induction hypothesis several times to manipulate the quantifiers. We omit the details.   □

The following is an immediate, useful corollary:

Corollary 3. Let $\phi \in \Sigma_1^B$. Then there exists $\psi \in$ strict Σ_1^B such that $W_1^1 \vdash \phi \leftrightarrow \psi$.

5   Definability in W_1^1

We know that W_1^1 can Σ_0^B-define all functions (of string variables) from the polynomial-time hierarchy. In fact, W_1^1 can Σ_1^B-define all string functions computable in polynomial space:

Theorem 4. *Let $f \in PSPACE$ be of polynomial growth rate. Then there is a strict Σ_1^B-formula ϕ such that*

1. $W_1^1 \vdash \forall X \exists Y \phi(X, Y)$
2. $W_1^1 \vdash \forall X \forall Y \forall Z (\phi(X, Y) \wedge \phi(X, Z) \supset Y =_2 Z$ *($Y =_2 Z$ may be defined as* $(|Y|_2 = |Z|_2 \wedge \forall x \leq |Y|_2 (Y(x) \leftrightarrow Z(x)))))$
3. *For all strings X, $\phi(X, f(X))$ is true.*

Proof (Proof Sketch). The proof is by induction on the logarithm of the length (number of steps) of the PSPACE computation that for any initial configuration there is a unique ending configuration. In the induction step two computations of length 2^i are pieced together using Σ_0^B-3COMP.   □

5.1   Strategies in Hex

As an example, consider the game of Hex, which has recently achieved some notoriety in the form of propositional tautologies due to Buss [5], Urquhart and others. These tautologies state that a finished game of Hex has a winner and are generally provable in Frege, resolution or weaker systems, depending on the formulation. The winner can be found in logarithmic space by solving a related graph reachability problem. A related problem is to determine which player has the winning strategy from a given configuration, which is PSPACE complete [20]. A Hex configuration is easily coded as a string, compared to which a strategy is an exponential-sized object coding a map from partially filled boards to moves. Thus there is a Σ_1^B formula Strategy1(X) stating that there exists a strategy such that for any (game sized) sequence of moves by player 2, when player 1 responds according to his strategy then he is the winner. There is similarly a Σ_1^B-formula Strategy2(X), and as expected,

Theorem 5.
$$W_1^1 \vdash \forall X [Strategy1(X) \vee Strategy2(X)].$$

Proof (Proof Sketch). Given a configuration X, we can define **continuations** of X as those configurations reachable from X by play. This is a simple matter of counting the numbers of added pieces of each colour, and checking that no existing pieces have been changed or removed. Then it is proved by induction on the number of remaining moves in the game that from any continuation of X, some player has a winning strategy. The base case is a reformulation of the above tautologies and is thus easily provable in W_1^1. In the induction step, from a given position Y the induction hypothesis gives a winning strategy for some player from each possible next position. If the current player can reach a winning position with a move then the strategy for the current position is amended to apply to the configuration Y by adding that move. Otherwise, a strategy for the other player is the merger of his strategies for all possible next positions.   □

5.2   A Witnessing Theorem for W_1^1

To prove the converse of Theorem 4, namely that functions provably total in W_1^1 are in PSPACE, we shall use a Buss-style witnessing argument, which requires that we define an equivalent sequent calculus formulation $LK^3 - W_1^1$ of W_1^1. We omit this for brevity, but it is essentially LK with the addition of second- and third-order quantifier introduction rules (replacing only free variables by quantifiers) plus the following $\forall^2 \Sigma_1^{\mathcal{B}}$-IND rule:

$$\frac{\Gamma, \phi(b) \longrightarrow \phi(b+1), \Delta}{\Gamma, \phi(0) \longrightarrow \phi(t), \Delta},$$

where b appears only as indicated and $\phi \in \forall^2 \Sigma_1^{\mathcal{B}}$. As initial sequents we allow all substitution instances of the axioms (other than induction) of W_1^1. Note that all rules of $LK^3 - W_1^1$ are valid in W_1^1, and furthermore, $LK^3 - W_1^1$ proves the induction and comprehension schemes of W_1^1. Formally, $LK^3 - W_1^1$ also adopts the usual conventions concerning free and bound variables, as in [4].

The standard definition of an anchored cut in LK^3 is extended for $LK^3 - W_1^1$ by allowing cuts on the descendents of principal formulas of the $\forall^2 \Sigma_1^{\mathcal{B}}$-IND rule, in addition to cuts on descendents of formulas in non-logical axioms. The anchored completeness theorem for LK^3 can then be extended to $LK^3 - W_1^1$ in the usual way to cope with the induction rules, as detailed in [23].

With this in mind, we can now state the witnessing theorem we wish to prove, followed by several definitions:

Theorem 6. *Suppose $W_1^1 \vdash \exists Y \phi(X, Y)$, for $\phi(X, Y) \in \Sigma_1^{\mathcal{B}}$ with all free variables displayed. Then there exists a function $f \in PSPACE$ of polynomial growth rate such that for every string X, $\phi(X, f(X))$ is true.*

Definition 7. *Let $\psi \equiv \forall X \leq t \exists \mathcal{X} \phi(X, \mathcal{X}) \in \forall^2 \Sigma_1^{\mathcal{B}}$, with other free variables not shown. Consider an assignment to the free variables of ψ. Then the string relation $\mathcal{A}(A, B)$ **satisfies** ψ (with respect to the assignment to the free variables of ψ) iff for every string A of no more than t bits, $\phi(A, \{B\}(\mathcal{A}(A, B)))$ is true in the standard model, where $\{B\}(\mathcal{A}(A, B))$ denotes the unary string predicate (with argument B) obtained by fixing the first argument of relation $\mathcal{A}$ to A.*

Definition 8. *Let S be the sequent $\Gamma \longrightarrow \Delta$ such that $\Gamma \bigcup \Delta \subset \forall^2 \Sigma_1^{\mathcal{B}}$, i.e.*

$$\Gamma = \{\forall A_i \leq s_i \exists \mathcal{A}_i \gamma_i(A_i, \mathcal{A}_i, \overline{B}, \overline{B}, \overline{b})\} \qquad and$$

$$\Delta = \{\forall C_i \leq t_i \exists \mathcal{C}_i \delta_i(C_i, \mathcal{C}_i, \overline{B}, \overline{B}, \overline{b})\},$$

with $\{\gamma_i\} \bigcup \{\delta_i\} \subset \Sigma_0^{\mathcal{B}}$. (Leading quantifiers are written for simplicity but may be absent.)

Then **PSPACE Oracle Witnessing Operators (POWOs)** *for S are operators, or type-2 predicates. For each formula from Δ*

$$\forall C_i \le t_i \exists \mathcal{C}_i \delta_i(C_i, \mathcal{C}_i, \overline{\mathcal{B}}, \overline{B}, \overline{b})$$

that is not $\Sigma_0^{\mathcal{B}}$ (and may or may not have the leading string quantifier as pictured), the POWO f_i is a predicate with arguments $\{\overline{\mathcal{B}}, \overline{B}, \overline{b}\}$ (for the free variables of the sequent), $\{\mathcal{A}_j(A_j, X)\}$ (for the string relations satisfying the formulas in the antecedent) and finally $\{C_i, X\}$, making f_i into a two-place string relation when the other arguments are fixed. The f_i must have the property that for any assignment to the free variables $\overline{\mathcal{B}}, \overline{B}, \overline{b}$ of S and string relations $\{\mathcal{A}_j(A_j, X)\}$, if each formula γ_j on the left is satisfied by the corresponding $\mathcal{A}_j$, then some δ_i on the right is satisfied by the string relation $\{C_i, X\}f_i$, obtained by fixing all but the last two arguments to the operator f_i.

Furthermore, each f_i is computable by an oracle Turing machine in space (including on the query tapes) polynomial in the lengths of its string and number inputs.

Now the theorem will follow from the following lemma:

Lemma 9. *Suppose $LK^3 - W_1^1 \vdash \Gamma \longrightarrow \Delta$, where $\Gamma \bigcup \Delta \subset \forall^2 \Sigma_1^{\mathcal{B}}$. Then there exist PSPACE Oracle Witnessing Operators for $\Gamma \longrightarrow \Delta$.*

Proof (Proof of Theorem 6 from Lemma 9). Suppose $W_1^1 \vdash \exists Y \phi(X, Y)$, for $\phi(X, Y) \in \Sigma_1^{\mathcal{B}}$ with all free variables displayed. By Parikh's theorem, $W_1^1 \vdash \exists Y \le t(|X|_2)\phi(X, Y)$, for some term t. By Corollary 3, $W_1^1 \vdash \phi(X, Y) \leftrightarrow \exists \mathcal{Y} \psi(X, Y, \mathcal{Y})$, for some $\psi \in \Sigma_0^{\mathcal{B}}$. Also, $W_1^1 \vdash \exists Y \le t(|X|_2)\exists \mathcal{Y}\psi(X, Y, \mathcal{Y}) \leftrightarrow \exists \mathcal{Y} \exists Y \le t(|X|_q)\psi(X, Y, \mathcal{Y})$. Applying the lemma to the sequent $\longrightarrow \exists \mathcal{Y} \exists Y \le t(|X|_2)\psi(X, Y, \mathcal{Y})$, we obtain a PSPACE (in $|X|$) predicate for $\mathcal{Y}$ satisfying that sequent, and so for particular X the string Y can be obtained in PSPACE by evaluating ψ, with access to the predicate $\mathcal{Y}$, on each string of length $\le t(|X|_2)$ in turn. It is easy to see that the computed string Y satisfies $\phi(X, Y)$ (for the same fixed X). □

All that remains is to prove the lemma:

Proof (Proof of Lemma 9). Suppose $LK^3 - W_1^1 \vdash \Gamma \longrightarrow \Delta$, where $\Gamma \bigcup \Delta \subset \forall^2 \Sigma_1^{\mathcal{B}}$, and consider an anchored proof π of this sequent. Since both the endsequent of π and every non-logical axiom of $LK^3 - W_1^1$ is $\forall^2 \Sigma_1^{\mathcal{B}}$, and since the induction rule is limited to this same class of formulas, every formula in π is $\forall^2 \Sigma_1^{\mathcal{B}}$.

We now show by induction on the number of sequents in π that POWOs exist for $\Gamma \longrightarrow \Delta$.

Base Case: The base case is that $\Gamma \longrightarrow \Delta$ is either an initial sequent of LK^3 or an instance of an axiom. The only such sequents requiring POWOs are those with a third-order quantifier in the succedent, namely an instance

$$\longrightarrow (\exists \mathcal{Y})(\forall Z \le s(\overline{B}, \overline{b}))[\phi(\overline{\mathcal{B}}, \overline{B}, \overline{b}, Z) \leftrightarrow \mathcal{Y}(Z)]$$

of $\Sigma_0^{\mathcal{B}}$-3COMP, where $\phi \in \Sigma_0^{\mathcal{B}}$, subject to the restriction that $\mathcal{Y}$ does not occur free in ϕ. The only POWO required for this sequent is computed by the predicate

$$f(\overline{\mathcal{B}}, \overline{B}, \overline{b}, \overline{\mathcal{A}}, Z) \leftrightarrow |Z|_2 \le s(\overline{B}, \overline{b}) \wedge \phi(\overline{\mathcal{B}}, \overline{B}, \overline{b}, Z),$$

which is in some level of the polynomial-time hierarchy, and thus certainly in PSPACE.

Induction Step: The induction step has several cases depending on which rule has been used to derive $\Gamma \longrightarrow \Delta$.

1.-8. Weakening; Contraction; Exchange, introduction of $\neg$, $\vee$ on the right and $\wedge$ on the left; Introduction of $\vee$ on the left and $\wedge$ on the right; First- or second-order $\forall$: **left** and $\exists$: **right**; First- or second-order $\forall$: **right** and $\exists$: **left**; Third-order $\exists$: **left**; and Third-order $\exists$: **right**: These cases are all easy and are omitted for brevity.

9. The **cut** rule:

The inference is

$$\frac{\Gamma \longrightarrow \phi, \Delta \qquad \Gamma, \phi \longrightarrow \Delta}{\Gamma \longrightarrow \Delta}.$$

A POWO for the conclusion proceeds in two phases: First, it evaluates its formula using the POWO from the left hypothesis, and if that POWO satisfies the formula, it emulates it. Otherwise, it emulates the POWO from the right hypothesis, and uses the POWO for ϕ from the left hypothesis to supply a value for the oracle argument. The whole procedure uses at most the sum of the space requirements of the two POWOs from the hypotheses. If any free variables are eliminated, then as before a dummy argument of the correct type is supplied to the POWOs.

10. $\forall^2 \Sigma_1^B$-IND:

The inference is:

$$\frac{\Gamma, \phi(b) \longrightarrow \phi(b+1), \Delta}{\Gamma, \phi(0) \longrightarrow \phi(t), \Delta}.$$

The POWOs for the conclusion will iterate the construction from the previous case, as the current instance of the induction rule could be simulated by t instances of the cut rule, along with some weakenings.

More precisely, let f_ϕ be the POWO for the instance of ϕ in the succedent of the hypothesis. Let ψ be any formula in the succedent of the hypothesis (including ϕ) and f_ψ its POWO. We construct a POWO f'_ψ for ψ in the conclusion in stages:

$f^0_\psi(X, Y) \leftrightarrow f_\psi(X, Y)$.

$f^k_\psi(X, Y) \leftrightarrow (\psi(f^{k-1}_\psi) \wedge f^{k-1}_\psi(X, Y)) \vee (\neg\psi(f^{k-1}_\psi) \wedge f^{k-1}_\psi(f_\phi, X, Y))$.

f^1_ψ checks if f_ψ satisfies ψ and if so, simulates f_ψ. If not, f^1_ψ computes $f_\psi(f_\phi)$, that is to say, uses f_ϕ to answer queries to the oracle argument corresponding to ϕ.

f^k_ψ checks if f^{k-1}_ψ satisfies ψ and if so, simulates f^{k-1}_ψ. If not, f^k_ψ computes $f^{k-1}_\psi(f_\phi)$.

f'_ψ, then, evaluates t and computes f^t_ψ. Computing f^t_ψ requires t times the space required to compute f_ϕ plus the space requirements of f_ψ, and so only increases the space usage of POWOs by a polynomial factor.

$\square$

6   The Propositional System BPLK

In this section we review the sequent system BPLK [22], which is basically PK (i.e., the propositional fragment of LK) enhanced with the reasoning power of Boolean programs, defined below. These (Boolean programs) were introduced in [10] and are a way of specifying Boolean functions. They are something like a generalization of the technique of using new atoms to replace part of a Boolean formula, which idea is the basis of extended Frege systems. The following definition is from that paper:

Definition 10 (Cook-Soltys). *A Boolean Program P is specified by a finite sequence $\{f_1, ..., f_m\}$ of function symbols, where each symbol f_i has an associated arity k_i, and an associated defining equation*

$$f_i(\overline{p_i}) := A_i$$

where $\overline{p_i}$ is a list $p_1, ..., p_{k_i}$ of variables and A_i is a formula all of whose variables are among $\overline{p_i}$ and all of whose function symbols are among $f_1, ..., f_{i-1}$. In this context the definition of a formula is expanded to allow the use of function symbols as connectives.

The semantics are as for propositional formulas, except that when evaluating an application $f_i(\overline{\phi})$ of a function symbol, the value is defined, using the defining equation, to be $A_i(\overline{\phi})$. There is no free/bound distinction between variables in the language of Boolean programs.

An interesting property of Boolean programs from [10] that demonstrates their comparability to quantified Boolean formulas is that evaluating them is PSPACE-complete.

Definition 11 (BPLK). *The system BPLK is like the propositional system PK, but with the following changes:*

1. *In addition to sequents, a proof also includes a Boolean program that defines functions. Whenever we refer to a BPLK-proof, we shall always explicitly write it as the pair $\langle \pi, P \rangle$ of the proof (sequents) and the Boolean program defining the function symbols occurring in the sequents.*
2. *Formulas in sequents are formulas in the context of Boolean programs, as defined earlier.*
3. *If the Boolean program contains a definition of the form $f(\overline{p}) := A(\overline{p})$, the new LK rules*

$$f : \textbf{left} \quad \frac{A(\overline{\phi}), \Gamma \longrightarrow \Delta}{f(\overline{\phi}), \Gamma \longrightarrow \Delta} \qquad and \qquad f : \textbf{right} \quad \frac{\Gamma \longrightarrow \Delta, A(\overline{\phi})}{\Gamma \longrightarrow \Delta, f(\overline{\phi})}$$

   may be used, where $\overline{\phi}$ are precisely as many formulas as $\overline{p}$ are variables.
4. *(**Substitution Rule**) The new inference rule **subst***

$$\frac{\Delta(q, \overline{p}) \longrightarrow \Gamma(q, \overline{p})}{\Delta(\phi, \overline{p}) \longrightarrow \Gamma(\phi, \overline{p})}$$

   may be used, where all occurrences of q have been substituted for.

Simultaneous substitutions can be simulated with several applications of **subst**. The following is the main result of [22]:

Theorem 12. *BPLK and G are polynomially equivalent for proofs of propositional tautologies.*

7   Translation into BPLK

In this section we define a translation $||\cdot||$ of Σ_∞^B formulas in the language $\mathcal{L}_A^3$ of W_1^1 (i.e. with no third-order quantifiers) into families of propositional sequents in the language of Boolean programs. Our main result is

Theorem 13. *If $\phi(A) \in \Sigma_\infty^B$ and if $W_1^1 \vdash \phi(A)$ then BPLK has polynomial-sized proofs of the translations $||\phi||$; furthermore, these proofs are definable in S_2^1 and V^1 (or any theory defining polytime functions).*

This will follow directly from lemma 17 below. The definability of the proofs follows from the fact that they can actually be constructed in polynomial time.

First, we can extend the definitions of a Boolean Program and of a BPLK proof as follows:

Definition 14. *A **Boolean Semiprogram** is like a Boolean program, except we allow some function symbols used in the program to be undefined ("free").*

Definition 15. *A **BPLK-Sequence** is the same as a BPLK proof except that the requirement that all function symbols occurring in the sequence be defined by the accompanying Boolean program is dropped. Furthermore, the accompanying Boolean program is instead a Boolean semiprogram. Any undefined function symbol appearing in the sequence or the semiprogram is called "free".*

The following translation is defined for the larger class Σ_0^B (including free third-order variables) and is necessary for the main lemma in the proof:

Definition 16. *Let $\phi(\mathcal{A}_1, ..., \mathcal{A}_j, A_1, ..., A_k)$ be Σ_0^B in the language $\mathcal{L}_A^3$. For every $m_1, ..., m_k$ (lengths of the string objects) we construct a Boolean semiprogram $P_\phi^{m_1,...,m_k}$ and a formula $||\phi||^{m_1,...,m_k}$ in the language of Boolean programs, with the atoms $\bar{\bar{p}} = (\bar{p}_i, i = 1, ..., k)$, where each $\bar{p}_i = (p_{i,0}, ..., p_{i,m_k})$. By induction on the structure of ϕ:*

- *If ϕ is an atomic formula $s = t$, $t \leq s$ or $t \in_2 T_i$ then s and t are first-order terms with no free first-order variables and refer only to the length of strings, which is known. They can be evaluated and $||\phi||^{m_1,...,m_k}$ is a constant.*
- *The cases where ϕ is formed with a propositional connective are trivial and we omit the details.*
- *If ϕ is the atomic formula $A_i \in_3 \mathcal{A}_j$ then $||\phi||^{m_i} := g_{\mathcal{A}_j}(p_{i,0}, ..., p_{i,m_i})$. $P_\phi^{m_i} := \emptyset$. The intention is that $g_{\mathcal{A}_j}$ be a free function symbol and we shall be careful not to add a definition for any function symbol of this form to our Boolean semiprograms. Furthermore, this is the only case in the construction where a free function symbol is produced.*

- If ϕ is $\exists x \leq t\psi(x)$ then $||\phi||^{m_1,...,m_k} := \bigvee_{n \leq t} ||\psi(n)||^{m_1,...,m_k}$ ($\phi(n)$ is

  $\phi(x)[s/x]$ where s is a constant term of value n, say $\overbrace{1 + ... + 1}^{n}$). $P_\phi^{m_1,...,m_k} := P_\psi^{m_1,...,m_k}$.
- If ϕ is $\forall x \leq t\psi(x)$ then $||\phi||^{m_1,...,m_k} := \bigwedge_{n \leq t} ||\psi(n)||^{m_1,...,m_k}$. $P_\phi^{m_1,...,m_k} := P_\psi^{m_1,...,m_k}$.
- If ϕ is $\exists X \leq t\psi(X)$ then $||\phi||^{m_1,...,m_k} := f_\phi(\overline{p})$ and $P_\phi^{m_1,...,m_k}$ is as follows:

$$f_{\phi,0}^l(\overline{p}, q_0, ..., q_l) := ||\psi||^{m_1,...,m_k,l}$$

for each $l \leq \underline{t}$.

$$f_{\phi,i}^l(\overline{p}, q_i, ..., q_l) := f_{\phi,i-1}^l(\overline{p}, 0, q_i, ..., q_l) \vee f_{\phi,i-1}^l(\overline{p}, 1, q_i., , , .q_l)$$

for each $l \leq \underline{t}$ and $i \leq l + 1$.

$$f_\phi(\overline{p}) := \bigvee_{l \leq \underline{t}} f_{\phi,l+1}^l(\overline{p}).$$

- The case where ϕ is $\forall X \leq t\psi(X)$ is symmetric to the previous one.

It is clear that for fixed ϕ, the size of $||\phi||^{m_1,...,m_k}$ is polynomial in $m_1., , , .m_k$. Whenever we talk of BPLK proofs or BPLK-sequences involving translations of this form, we shall insist that the associated Boolean (semi-)program extend the (semi-)program resulting from the translation.

The following lemma is the main lemma of the proof. In the previous section, since it is not possible to translate a general Σ_1^B formula into the language of BPLK, we defined POWOs and used them to witness a sequent containing third-order quantifiers. Similarly, in the lemma below we shall translate sequents with third-order quantifiers as if those third-order variables were free, and then show that BPLK can prove the existence of a function symbol witnessing the sequent in much the same way. This aspect of the statement of the lemma is greatly simplified compared to the analogous lemma in [17], where to talk about an arbitrary witness to the antecedent of the sequent, the authors stated the lemma with arbitrary formulas of the appropriate class substituted for the third-order variables.

Since formulas in the proof are not all guaranteed to be **strict** Σ_1^B, due to the slightly more complicated induction scheme in W_1^1, the translations used in the lemma are actually translations of the equivalent form given by the replacement theorem (i.e. with the third-order quantifier moved to the front).

Lemma 17. *Let* $LK^3 - W_1^1 \vdash \Gamma \longrightarrow \Delta$ *where* $\Gamma \bigcup \Delta \subset \forall^2 \Sigma_1^B$, *i.e.*

$$\Gamma = \{\forall A_i \leq s_i \exists \mathcal{A}_i \gamma_i(A_i, \mathcal{A}_i, \overline{B}, \overline{B}, \overline{b})\} \qquad and$$

$$\Delta = \{\forall C_i \leq t_i \exists \mathcal{C}_i \delta_i(C_i, \mathcal{C}_i, \overline{B}, \overline{B}, \overline{b})\},$$

with $\{\gamma_i\} \bigcup \{\delta_i\} \subset \Sigma_0^B$, *and although we write for simplicity the initial string and third-order quantifiers for each formula, in fact for some of the formulas either the initial string quantifier or both initial quantifiers may be absent.*

Then for each $m_1, ..., m_k$ and $n_1, ..., n_l$ there are function symbols $h_i^{\overline{m},\overline{n}}$ and BPLK-sequences with endsequents

$$..., ||\forall A_i \gamma_i(A_i, \mathcal{A}_i^{[A_i]}, \overline{\mathcal{B}}, \overline{B}, \underline{n})||^{m_1,...,m_k}, ...$$
$$\longrightarrow ..., ||\forall C_i \delta_i(C_i, \mathcal{C}_i^{[C_i]}, \overline{\mathcal{B}}, \overline{B}, \underline{n})||^{m_1,...,m_k} [h_i^{\overline{m},\overline{n}} / g_{\mathcal{C}_i}], ...$$

where $h_i^{\overline{m},\overline{n}}$ are called witnessing function symbols and are not free, but may be defined in terms of free function symbols (in particular, $g_{\mathcal{A}_i}$). Furthermore, these sequences have size polynomial in $m_1, ..., m_k$ and $n_1, ..., n_l$.

The notation $...[h_i^{\overline{m},\overline{n}}/g_{\mathcal{C}_i}]$ in the succedent means that one should first perform the translation, and then substitute function symbol h_i for the free symbol $g_{\mathcal{C}_i}$ in the result.

Proof. We show the existence of the desired BPLK-sequence by induction on the number of sequents in the W_1^1 proof, in a manner very similar to the witnessing theorem of the previous section. The witnessing function symbols of the present lemma are analogous to POWOs.

Base Case: This is trivial for initial sequents and the witnessing function symbol, if required, is defined to be the constant false predicate. For translations of axioms B1-B12, L1, L2 and instances of $\Sigma_0^{\mathcal{B}}$-2COMP, it follows from the analogous result for V_1^1 and Extended Frege. For translations of instances of $\Sigma_0^{\mathcal{B}}$-3COMP, the witnessing function symbol has defining formula identical to the comprehension formula, and then the translation of the instance is proved using the introduction rule for this symbol followed by repeated substitutions and $\wedge :$ **right** inferences.

Induction Step: There are cases depending on the final inference of the W_1^1 proof:

1.-5. Weakening, Exchange, introduction of $\neg$, $\vee$ on the right and $\wedge$ on the left; Contraction, introduction of $\vee$ on the left and $\wedge$ on the right; introduction of first-, second- and third- order quantifiers:
These cases are all straightforward and are omitted.

6. Cut, Induction:
The cut rule is handled by defining new witnessing function symbols for the conclusion by cases, using the witnessing function symbol for the cut formula. For induction this procedure is iterated as many times as the value of the induction bound.

For example, if the cut formula is $\forall C_i \leq t_i \exists \mathcal{C}_i \delta_i(C_i, \mathcal{C}_i)$, then a new witnessing function symbol h_j for $\forall C_j \leq t_j \exists \mathcal{C}_j \delta_j(C_j, \mathcal{C}_j)$ would be defined as follows, where h_j' is the witnessing function symbol for the hypothesis with the cut formula on the right, and h_j'' that for the hypothesis with the cut formula on the left:

$$h_j := (||\delta_j(C_j, \mathcal{C}_j^{[C_j]})||[h_j'/g_{c_j}] \wedge h_j) \vee h_j''(h_i).$$

$\square$

7.1   Consistency and Polynomial Simulation

Now Cook [12] and later others [15], [17], [14], etc. showed that some bounded arithmetic theories can prove the consistency of related propositional proof systems, and furthermore that any proof system whose consistency can be proved in the theory can be polynomially simulated by the related proof system. For completeness we mention the analogous results for W_1^1 and BPLK.

Let $BPTAUT(X)$ be a formula stating that the string X codes a tautological propositional formula in the language of Boolean programs, as follows: "for any assignment to the free variables, there exists a transcript of the exponential-length computation of the Boolean function symbol terms occurring in the formula such that the resulting truth-values satisfy the formula". Clearly a $\Sigma_1^{\mathcal{B}}$ formula will suffice. Let $Prf_{BPLK}(X, Y)$ be a $\Sigma_0^{\mathcal{B}}$ formula stating that X codes a BPLK-proof of the formula coded by Y. Then

Theorem 18.

$$W_1^1 \vdash \forall X, Y [Prf_{BPLK}(X, Y) \supset BPTAUT(Y)]$$

The formula in the theorem is called RFN(BPLK).

Proof (Proof Sketch). By induction on the length of the proof, similar to the witnessing theorem, a transcript is constructed, for each assignment, of evaluating the formula at that assignment.   □

The next thing to show would be that if P is a proof system whose consistency can be proved in W_1^1, i.e. $W_1^1 \vdash RFN(P)$, then BPLK polynomially simulates P. For U_2^1, what is known is actually the weaker statement that if U_2^1 proves $i - RFN(P)$, which is the consistency of P for Σ_i^q formulas, then G polynomially simulates P for proofs of those formulas. An analogous statement is almost certainly true of W_1^1 and BPLK simply because BPLK polynomially simulates G, and because W_1^1 and U_2^1 are most likely related by an RSUV-style isomorphism. Of more interest is the statement for RFN(P), but this formula is likely not $\Sigma_0^{\mathcal{B}}$ for interesting proof systems (G or even BPLK, for instance), and so the usual techniques do not seem to apply due to the expressibility of formulas in the language of BPLK. A proof system with more expressive formulas, however, would be a candidate for this kind of statement. See the open problems for details.

8   Open Problems

Several future directions are indicated. First of all, one motivation for the definition W_1^1 was to simplify the axioms as much as possible, yet we were unable to limit induction to strict $\Sigma_1^{\mathcal{B}}$ formulas. One problem, then, is to prove the replacement theorems of W_1^1 with this more restricted induction. There does not seem to be any good reason why this should not be possible. On the other hand, Cook and Thapen [11] have recently used KPT-like witnessing theorems to show independence of certain replacement schemes from various theories of bounded arithmetic, and their techniques may apply in this case.

Next, there are some unresolved technical issues regarding BPLK: The most pressing is to eliminate the substitution rule (analogously to in G) but at least the current proof of BPLK's p-equivalence to G seems to rely on this rule in an essential way. See [21] for details.

Another idea is to extend W_1^1 to obtain theories for higher complexity classes. For example, by analogy to V_2^1, extending the induction in W_1^1 to full induction on the strings should yield a theory for EXPTIME, but this would be inelegant to state (although a more natural formulation may exist). Nevertheless, it should be possible to obtain theories for each level of the exponential-time hierarchy in this way, and with more work, for the linear-exponential-time hierarchy and others.

Finally, the idea of having free function symbols in a BPLK proof seems quite general and suggests a direction for even stronger proof systems obtained by allowing function symbol quantifiers in a new kind of BPLK proof. Indeed, this would seem to be a modern version of the *Protothetic* of Stanisław Leśniewski [18] and would hopefully match the stronger theories envisaged in the previous paragraph.

9   Acknowledgment

Many thanks to Stephen Cook for countless helpful discussions on this topic. Thanks also to the reviewers for several important comments.

References

1. S. Buss. *Bounded Arithmetic*. Bibliopolis, Naples, 1986.
2. Samuel Buss, Jan Krajíček, and Gaisi Takeuti. On provably total functions in bounded arithmetic theories R_3^i, U_2^i and V_2^i. In Peter Clote and Jan Krajíček, editors, *Arithmetic, proof theory and computational complexity*, pages 116–61. Oxford University Press, Oxford, 1993.
3. Samuel R. Buss. Relating the bounded arithmetic and polynomial time hierarchies. *Annals of Pure and Applied Logic*, 75(1–2):67–77, 12 September 1995.
4. Samuel R. Buss, editor. *Handbook of Proof Theory*. Elsevier Science B. V., Amsterdam, 1998.
5. Samuel R. Buss. Polynomial-size frege and resolution proofs of st-connectivity and hex tautologies. Typewritten manuscript, 2003.
6. Mario Chiari and Jan Krajíček. Witnessing functions in bounded arithmetic and search problems. *The Journal of Symbolic Logic*, 63(3):1095–1115, September 1998.
7. P. Clote and G. Takeuti. Bounded arithmetic for NC, ALogTIME, L and NL. *Annals of Pure and Applied Logic*, 56(1–3):73–117, 29 April 1992.
8. S. Cook and A. Kolokolova. A second-order system for polytime reasoning using Grädel's theorem. In *16th Annual IEEE Symposium on Logic in Computer Science (LICS '01)*, pages 177–186, Washington - Brussels - Tokyo, June 2001. IEEE.
9. S. A. Cook. CSC 2429S: Proof Complexity and Bounded Arithmetic. Course notes, URL: "http://www.cs.toronto.edu/~sacook/csc2429h", Winter 2002.
10. Stephen Cook and Michael Soltys. Boolean programs and quantified propositional proof systems. *Bulletin of the Section of Logic*, 28(3), 1999.

11. Stephen Cook and Neil Thapen. The strength of replacement in weak arithmetic. In *LICS04*, 2004. To appear.

12. Stephen A. Cook. Feasibly constructive proofs and the propositional calculus (preliminary version). In *Conference Record of Seventh Annual ACM Symposium on Theory of Computing*, pages 83–97, Albuquerque, New Mexico, 5–7 May 1975.

13. Stephen A. Cook. Relating the provable collapse of P to NC^1 and the power of logical theories. *DIMACS Series in Discrete Math. and Theoretical Computer Science*, 39, 1998.

14. Jan Krajíček. On Frege and Extended Frege proof systems. In *P. Clote, J. Remmel (eds.): Feasible Mathematics II*, pages 284–319. Birkhäuser, Boston, 1995.

15. Jan Krajíček and Pavel Pudlák. Quantified propositional calculi and fragments of bounded arithmetic. *Zeitschr. f. Mathematikal Logik u. Grundlagen d. Mathematik*, 36:29–46, 1990.

16. Jan Krajíček, Pavel Pudlák, and Gaisi Takeuti. Bounded arithmetic and the polynomial hierarchy. *Annals of Pure and Applied Logic*, 52(1–2):143–153, 1991.

17. Jan Krajíček and Gaisi Takeuti. On bounded Σ_1^1 polynomial induction. In S. R. Buss and P. J. Scott, editors, *FEASMATH: Feasible Mathematics: A Mathematical Sciences Institute Workshop*, pages 259–80. Birkhauser, 1990.

18. Stanisław Leśniewski. Grundzüge eines neunen Systems der Grundlagen der Mathematik. *Fundamenta Mathematicae*, 14:1–81, 1929.

19. Alexander A. Razborov. An equivalence between second order bounded domain bounded arithmetic and furst order bounded arithmetic. In Peter Clote and Jan Krajíček, editors, *Arithmetic, proof theory and computational complexity*, pages 247–77. Oxford University Press, Oxford, 1993.

20. Stefan Reisch. Hex ist PSPACE-vollständig. *Acta Informatica*, 15:167–191, 1981.

21. Alan Skelley. Relating the PSPACE reasoning power of Boolean programs and quantified Boolean formulas. Master's thesis, University of Toronto, 2000. Available from ECCC in the 'theses' section.

22. Alan Skelley. Propositional PSPACE reasoning with Boolean programs vs.quantified Boolean formulas. In *ICALP04*, 2004. To appear.

23. Michael Soltys. A model-theoretic proof of the completeness of LK proofs. Manuscript, available on author's web page, 1999.

24. Gaisi Takeuti. RSUV isomorphism. In Peter Clote and Jan Krajíček, editors, *Arithmetic, proof theory and computational complexity*, pages 364–86. Oxford University Press, Oxford, 1993.

25. D. Zambella. Notes on polynomially bounded arithmetic. *The Journal of Symbolic Logic*, 61(3):942–966, 1996.

Provably Total Primitive Recursive Functions: Theories with Induction

Andrés Cordón-Franco, Alejandro Fernández-Margarit, and
F. Félix Lara-Martín

Dpto. Ciencias de la Computación e Inteligencia Artificial.
Facultad de Matemáticas. Universidad de Sevilla
C/ Tarfia, s/n. Sevilla, 41012, (Spain)
{acordon,fflara}@us.es

Abstract. A natural example of a function algebra is $\mathcal{R}(\mathbf{T})$, the class of provably total computable functions (p.t.c.f.) of a theory $\mathbf{T}$ in the language of first order Arithmetic. In this paper a simple characterization of that kind of function algebras is obtained. This provides a useful tool for studying the class of primitive recursive functions in $\mathcal{R}(\mathbf{T})$. We prove that this is the class of p.t.c.f. of the theory axiomatized by the induction scheme restricted to (parameter free) $\Delta_1(\mathbf{T})$–formulas (i.e. Σ_1–formulas which are equivalent in $\mathbf{T}$ to Π_1–formulas).

Moreover, if $\mathbf{T}$ is a sound theory and proves that exponentiation is a total function, we characterize the class of primitive recursive functions in $\mathcal{R}(\mathbf{T})$ as a function algebra described in terms of bounded recursion (and composition). Extensions of this result are related to open problems on complexity classes. We also discuss an application to the problem on the equivalence between (parameter free) Σ_1–collection and (uniform) Δ_1–induction schemes in Arithmetic.

The proofs lean upon axiomatization and conservativeness properties of the scheme of $\Delta_1(\mathbf{T})$–induction and its parameter free version.

1 Introduction

A function algebra is a family of functions that can be described as the smallest class of functions that contains some initial functions and is closed under certain operators. Classical examples of function algebras include the class of primitive recursive functions, $\mathcal{PR}$, classes $\mathcal{E}^n$, $(n \geq 1)$, in the Grzegorczyk hierarchy and the class of Kalmár elementary functions, $\mathcal{E}$ (see [6, 13]). Another important example is given by $\mathcal{R}(\mathbf{T})$, the class of provably total computable functions (p.t.c.f.) of a theory $\mathbf{T}$ in the language of first order Arithmetic. The class $\mathcal{R}(\mathbf{T})$ can be used to obtain independence results for $\mathbf{T}$ and to separate it from other theories. On the other hand, if a function algebra, $\mathcal{C}$, is the class of p.t.c.f. of a theory, $\mathbf{T}$, then proof–theoretic and model–theoretic properties of $\mathbf{T}$ can be used to establish results on $\mathcal{C}$. This increases the methods available in the study of function algebras by adding to them techniques from Proof Theory and Model Theory. As surveyed in [6], function algebras provide machine–independent characterizations of many complexity classes and offer an alternative view of important

J. Marcinkowski and A. Tarlecki (Eds.): CSL 2004, LNCS 3210, pp. 355–369, 2004.
© Springer-Verlag Berlin Heidelberg 2004

open problems in Complexity Theory. In this way, classes of p.t.c.f. constitute a link among Complexity Theory, Proof Theory and Model Theory that has been exploited in the work on Bounded Arithmetic (see [12]).

In this paper we present a new example of the fruitful interactions among fragments of Arithmetic, function algebras and computational complexity. Given a function algebra, $\mathcal{C}$, we introduce the algebra $\mathcal{E}^{\mathcal{C}}$ defined as the smallest class containing the basic functions (zero, successor and projections) and closed under composition and $\mathcal{C}$–bounded recursion. We study the relationship between $\mathcal{C}$ and $\mathcal{E}^{\mathcal{C}}$ when $\mathcal{C}$ is the class of p.t.c.f. of a theory $\mathbf{T}$. If $\mathcal{C} = \mathcal{R}(\mathbf{T})$ then

- (Theorem 4) $\mathcal{C} \cap \mathcal{PR} \subseteq \mathcal{E}^{\mathcal{C}}$. Moreover, if $\mathcal{C}$ is closed under bounded minimization, $\mathcal{E}^{\mathcal{C}}$ is the closure of $\mathcal{C} \cap \mathcal{PR}$ under composition and bounded recursion.
- (Theorem 5) Assume that $\mathcal{C}$ is closed under bounded minimization. Then $\mathcal{C} \cap \mathcal{PR} = \mathcal{E}^{\mathcal{C}}$ if and only if there exists a theory $\mathbf{T}'$ such that $\mathcal{E}^{\mathcal{C}} = \mathcal{R}(\mathbf{T}')$.

For the proof of these results the concept of a Δ_0–generated function algebra is introduced. A function algebra, $\mathcal{C}$, is Δ_0–generated if (it contains Grzegorcyk's class $\mathcal{M}^2$ and) each function in $\mathcal{C}$ can be obtained as a composition of two functions in $\mathcal{C}$ with Δ_0-definable graph. We prove (see Theorem 1) that a function algebra is Δ_0–generated if and only if it is the class $\mathcal{R}(\mathbf{T})$ for some theory $\mathbf{T}$ (extending $\mathbf{I}\Delta_0$).

If $\mathcal{C} \subseteq \mathcal{PR}$ is closed under bounded minimization, then Theorem 5 states that $\mathcal{C} = \mathcal{E}^{\mathcal{C}}$ if and only if $\mathcal{E}^{\mathcal{C}}$ is Δ_0–generated. This fact has interesting applications to complexity classes as $\mathcal{F}\text{PH}$ (computable functions in the Polynomial Time Hierarchy, that is, $\bigcup_{i=1}^{\infty} \square_i^p$ in S. Buss' terminology, see [10]) and $\mathcal{F}\text{LTH}$ (computable functions in the Linear Time Hierarchy, see [6]). Both classes are contained in $\mathcal{PR}$ and are Δ_0–generated and closed under bounded minimization:

- $\mathcal{F}\text{LTH} = \mathcal{M}^2 = \mathcal{R}(\mathbf{I}\Delta_0)$ (see [6, 16]), and
- $\mathcal{F}\text{PH} = \mathcal{R}(\mathbf{I}\Delta_0 + \Omega_1)$ (see [10]), where $\mathbf{I}\Delta_0 + \Omega_1$ is the theory introduced by A. Wilkie and J. Paris in [17].

But $\mathcal{E}^{\mathcal{F}\text{LTH}} = \mathcal{E}^2 = \mathcal{F}\text{LINSPACE}$ (R.W. Ritchie, see [6]) and $\mathcal{E}^{\mathcal{F}\text{PH}} = \mathcal{F}\text{PSPACE}$ (D.B. Thompson, see [6]). Therefore, by Theorem 5 it follows that:

1. If $\mathcal{F}\text{PSPACE}$ is Δ_0–generated then $\mathcal{F}\text{PSPACE} = \mathcal{F}\text{PH}$.
2. If $\mathcal{F}\text{LINSPACE}$ is Δ_0–generated then $\mathcal{F}\text{LINSPACE} = \mathcal{F}\text{LTH}$. Or, equivalently, $\mathcal{E}^2 = \mathcal{M}^2$ if and only if $\mathcal{E}^2$ is Δ_0–generated.

These facts suggest that a deeper knowledge of structural properties of Δ_0–generated function algebras (specially, construction of non Δ_0–generated function algebras) could be relevant in the study of complexity classes. They also raise a natural question: if $\mathcal{C} = \mathcal{R}(\mathbf{T})$ and $\mathcal{E}^{\mathcal{C}}$ is Δ_0–generated, is there a natural theory $\mathbf{T}'$ such that $\mathcal{R}(\mathbf{T}') = \mathcal{E}^{\mathcal{C}}$? We obtain an answer to this question from the study of induction schemes for Δ_1–formulas. Let $\mathbf{I}\Delta_1(\mathbf{T})^-$ be the theory axiomatized by induction scheme restricted to parameter free $\Delta_1(\mathbf{T})$–formulas.

- (Theorem 2) $\mathcal{R}(\mathbf{I}\Delta_1(\mathbf{T})^-) = \mathcal{C} \cap \mathcal{PR}$.

So, from Theorem 5 we get that, if $\mathcal{C}$ is closed under bounded minimization and $\mathcal{E}^{\mathcal{C}}$ is Δ_0–generated, then $\mathcal{E}^{\mathcal{C}} = \mathcal{R}(\mathbf{I}\Delta_1(\mathbf{T})^-)$.

Next step is to find conditions ensuring $\mathcal{E}^{\mathcal{C}}$ is Δ_0–generated. Classes $\mathcal{E}^{\mathcal{C}}$ are a generalization of Grzegorcyzk's classes $\mathcal{E}^n$ and it is well–known that if exponential function is in $\mathcal{E}^n$, then bounded recursion can be reduced to bounded minimization (see [6, 13]). But bounded minimization has a straightforward formulation in the language of first order Arithmetic and as a consequence (for $n \geq 3$) $\mathcal{E}^n$ is Δ_0–generated. The key ingredients in the proof of this fact are exponential function (which allows for coding of sequences of arbitrary length) and Σ_1–collection principle (as a suitable formulation of the combinatorial principles involved). These arguments lead to a natural condition for $\mathcal{E}^{\mathcal{C}}$ to be a Δ_0–generated function algebra and relate the study of $\mathcal{E}^{\mathcal{C}}$ to the problem on the equivalence between the schemes of Σ_1–collection and Δ_1–induction in Arithmetic (see [7]). In [3, 5], L. Beklemishev obtains Π_2–axiomatized theories that are not closed under Σ_1–collection rule or Δ_1–induction rule. He proposes classes of p.t.c.f. as a tool to separate the fragments $\mathbf{I}\Delta_1$ and $\mathbf{B}\Sigma_1$. Recently (see [15]) T. Slaman has proved that $\mathbf{I}\Delta_1 + \mathbf{exp}$ is equivalent to $\mathbf{B}\Sigma_1 + \mathbf{exp}$ (where $\mathbf{exp}$ is a Π_2–sentence expressing that exponentiation defines a total function with Δ_0 definable graph (see [10])). So, Beklemishev's approach must fail. Nevertheless, as we shall show, classes of p.t.c.f. could be used to obtain positive results on fragments of Arithmetic. Motivated by Beklemishev's work in [3, 4, 5], we study the classes of p.t.c.f. of the theories $\mathbf{I}\Delta_1(\mathbf{T})$ and $\mathbf{L}\Delta_1(\mathbf{T})$ introduced in [9], and their relationship with the uniform counterpart of Slaman's result.

Theorem 5 holds for $\mathcal{C}$ closed under bounded minimization. We prove that if Theorem 5 also holds under the (apparently weaker) following hypothesis:

(IC) $\mathcal{C} = \mathcal{R}(\mathbf{T})$ and $\mathbf{T}$ extends $\mathbf{I}\Delta_1(\mathbf{T})$,

then a (weak) uniform counterpart of Slaman's result can be obtained, namely, theories $\mathbf{B}\Sigma_1^- + \mathbf{exp}$ and $\mathbf{UI}\Delta_1 + \mathbf{exp}$ are equivalent, modulo Π_1–true sentences (see Theorem 7). Last equivalence can be also obtained from Slaman's theorem and Σ_3–conservativeness between $\mathbf{I}\Delta_1 + \mathbf{exp}$ and $\mathbf{UI}\Delta_1 + \mathbf{exp}$ (see corollary 6 in [5]). However, we present an independent approach stressing the role of function algebras via classes of p.t.c.f.

Our main tools for the proofs are axiomatization and conservativeness results for $\mathbf{I}\Delta_{n+1}(\mathbf{T})$ and Herbrand analyses, essentially along the lines presented by W. Sieg in [14]; however, we work in a model–theoretic framework, following J. Avigad's work in [1].

2   Fragments of Arithmetic and Function Algebras

Through this paper we deal with classes of p.t.c.f. of a number of theories. We are mainly interested in characterizations of these classes as function algebras. So, first of all, we introduce the theories and classes of functions we are concerned with. These theories are axiomatized by axiom schemes expressing classical principles in Arithmetic as induction, minimization and collection.

Let $\mathcal{L} = \{0, 1, <, +, \cdot\}$ be the language of first order Arithmetic. The induction and minimization axioms for a formula $\varphi(x, \vec{v})$ with respect to x are, respectively,

$$\mathbf{I}_{\varphi,x}(\vec{v}) \equiv \varphi(0, \vec{v}) \wedge \forall x \, [\varphi(x, \vec{v}) \rightarrow \varphi(x + 1, \vec{v})] \rightarrow \forall x \, \varphi(x, \vec{v}),$$
$$\mathbf{L}_{\varphi,x}(\vec{v}) \equiv \exists x \, \varphi(x, \vec{v}) \rightarrow \exists x \, (\varphi(x, \vec{v}) \wedge \forall z < x \, \neg\varphi(z, \vec{v})).$$

The collection axiom for a formula $\varphi(x, y, \vec{v})$ with respect to x, y is

$$\mathbf{B}_{\varphi,x,y}(z, \vec{v}) \equiv \forall x \leq z \, \exists y \, \varphi(x, y, \vec{v}) \rightarrow \exists u \, \forall x \leq z \, \exists y \leq u \, \varphi(x, y, \vec{v}).$$

As usual, we write $\mathbf{I}_\varphi$ instead of $\mathbf{I}_{\varphi,x}$ and similarly we use $\mathbf{L}_\varphi$ and $\mathbf{B}_\varphi$.

All theories considered in this paper are extensions of $\mathbf{P}^-$ a finite set of Π_1 formulas whose models are the nonnegative part of a discretely ordered commutative ring (see [11]). Other theories are defined by restricting the schemes just introduced to formulas in the classes Σ_n or Π_n in the Arithmetical Hierarchy. If Γ is a class of formulas of $\mathcal{L}$, then $\mathbf{I}\Gamma = \mathbf{P}^- + \{\mathbf{I}_\varphi : \varphi \in \Gamma\}$. The theory $\mathbf{L}\Gamma$ is similarly defined using $\mathbf{L}_\varphi$ instead of $\mathbf{I}_\varphi$. For collection, $\mathbf{B}\Gamma = \mathbf{I}\Delta_0 + \{\mathbf{B}_\varphi : \varphi \in \Gamma\}$, where Δ_0 denotes the class of bounded formulas of $\mathcal{L}$ (see [10, 11]).

Induction schemes for Δ_{n+1}–formulas will be also considered, $\mathbf{I}\Delta_{n+1}$ is the theory given by:

$$\mathbf{P}^- + \{\forall x \, (\varphi(x, \vec{v}) \leftrightarrow \psi(x, \vec{v})) \rightarrow \mathbf{I}_\varphi(\vec{v}) : \varphi(x, \vec{v}) \in \Sigma_{n+1}, \psi(x, \vec{v}) \in \Pi_{n+1}\}.$$

If parameters, $\vec{v}$, are not allowed, then we obtain the theory $\mathbf{I}\Delta_{n+1}^-$. The uniform version of induction scheme, $\mathbf{UI}\Delta_{n+1}$, was introduced by R. Kaye. It is defined by considering the scheme $\forall \vec{v} \forall x \, (\varphi(x, \vec{v}) \leftrightarrow \psi(x, \vec{v})) \rightarrow \forall \vec{v} \, \mathbf{I}_\varphi(\vec{v})$. This theory is also studied by Beklemishev in [5], where it is denoted by $sI\Delta_1$.

Definition. Let $\mathbf{T}$ be a theory in the language $\mathcal{L}$. We say that $f : \omega^k \rightarrow \omega$ is a provably total computable function of $\mathbf{T}$ if there exists a formula $\varphi(\vec{x}, y) \in \Sigma_1$ such that

1. $\mathbf{T} \vdash \forall \vec{x} \, \exists! y \, \varphi(\vec{x}, y)$.
2. For all $a_1, \ldots, a_k, b \in \omega$, $\quad f(\vec{a}) = b \Longleftrightarrow \mathcal{N} \models \varphi(\vec{a}, b)$.

Where $\mathcal{N}$ denotes the standard model of Arithmetic whose universe is the set of natural numbers, ω. In such a case, we say that $\varphi(\vec{x}, y)$ defines f in $\mathbf{T}$.

This definition is sensitive to changes in the language of the theory. If $\mathbf{T}$ is a theory in a language $\mathcal{L}'$ extending $\mathcal{L}$, then $\mathcal{R}(\mathbf{T})$ will denote the class obtained by considering $\Sigma_1(\mathcal{L}')$–formulas instead of Σ_1–formulas.

The class $\mathcal{R}(\mathbf{T})$ has turned out to be a natural object, its closure properties (under certain operators) reflecting axiom schemes (or inference rules) provable in $\mathbf{T}$. Thus, closure under primitive recursion corresponds to Σ_1–induction and bounded minimization to Σ_1–collection (see [2]). In particular, $\mathcal{R}(\mathbf{I}\Sigma_1) = \mathcal{PR}$ and $\mathcal{R}(\mathbf{I}\Delta_0 + \mathbf{exp}) = \mathcal{E}$.

It is easy to check that if $\Phi \subseteq \mathbf{Th}_{\Pi_1}(\mathcal{N})$ and $\mathbf{T}$ is a sound theory (that is, $\mathcal{N} \models \mathbf{T}$) then $\mathcal{R}(\mathbf{T}) = \mathcal{R}(\mathbf{T} + \Phi)$ (see [14]). The class $\mathcal{R}(\mathbf{T})$ is determined by $\mathbf{Th}_{\Pi_2}(\mathbf{T})$ (the set of Π_2–sentences provable in $\mathbf{T}$). The converse also holds modulo Π_1–true sentences.

Proposition 1. *Let* $\mathbf{T}_1$ *and* $\mathbf{T}_2$ *be* Π_2*–axiomatized sound extensions of* $\mathbf{I}\Delta_0$*. The following conditions are equivalent:*

1. $\mathcal{R}(\mathbf{T}_1) = \mathcal{R}(\mathbf{T}_2)$.
2. $\mathbf{T}_1 + \mathbf{Th}_{\Pi_1}(\mathcal{N}) \Longleftrightarrow \mathbf{T}_2 + \mathbf{Th}_{\Pi_1}(\mathcal{N})$.

Proof. We only prove $(1) \Longrightarrow (2)$. By symmetry, it is enough to show that $\mathbf{T}_1 + \mathbf{Th}_{\Pi_1}(\mathcal{N}) \Longrightarrow \mathbf{T}_2$. Let $\theta(x,y)$ be a Δ_0–formula such that $\mathbf{T}_2 \vdash \forall x \exists y\, \theta(x,y)$. Let $\theta'(x,y)$ be the formula $\theta(x,y) \wedge \forall z < y\, \neg\theta(x,z)$. Since $\mathbf{T}_2 \Longrightarrow \mathbf{I}\Delta_0$, $\mathbf{T}_2 \vdash \forall x \exists! y\, \theta'(x,y)$. Let f be the computable function defined by θ' in $\mathcal{N}$. Then, $f \in \mathcal{R}(\mathbf{T}_2)$; so, by (1), $f \in \mathcal{R}(\mathbf{T}_1)$. Hence, there is $\varphi(x,y) \in \Sigma_1$ defining f in $\mathbf{T}_1$. Thus, $\mathcal{N} \models \varphi(x,y) \leftrightarrow \theta'(x,y)$. In particular, $\mathcal{N} \models \forall x,y\,(\varphi(x,y) \to \theta'(x,y))$; so, since this last formula is a Π_1 sentence, $\mathbf{T}_1 + \mathbf{Th}_{\Pi_1}(\mathcal{N}) \vdash \forall x \exists y\, \theta(x,y)$. $\quad\square$

Functions with a Δ_0–definable graph will play a prominent role throughout this work. Let us introduce the following notation.

We denote by Δ_0^0 the class of sets Δ_0 definable in the standard model. The graph of a function f is denoted by $Gr(f) = \{(\vec{a}, b) \in \omega^{k+1} : f(\vec{a}) = b\}$. If $\mathcal{C}$ is a class of functions, then $\mathcal{C}_*$ denotes the class of subsets of ω^k whose characteristic functions are in $\mathcal{C}$. Finally, given $f, g : \omega^k \to \omega$, we write $f \leq g$ to mean that for each $\vec{a} \in \omega^k$, $f(\vec{a}) \leq g(\vec{a})$.

One of the aims of this work is to obtain descriptions of $\mathcal{R}(\mathbf{T})$ as a function algebra generated by means of some operators from a small set of basic functions. The considered classes of basic functions will always contain the set

$$\mathcal{B} = \{S, O\} \cup \{\Pi_i^n : 1 \leq i \leq n\}$$

where $S, O : \omega \to \omega$ are given by $S(a) = a + 1$ and $O(a) = 0$, and $\Pi_i^n : \omega^n \to \omega$, by $\Pi_i^n(a_1, \ldots, a_n) = a_i$. As operators, beside *composition*, we consider:

Bounded minimization, $\mu_\leq$: If $g : \omega^{m+1} \to \omega$, then $f = \mu_\leq(g)$ is the function $f : \omega^{m+1} \to \omega$ defined by

$$f(a_1, \ldots, a_m, b) = \begin{cases} \min(\{z : g(\vec{a}, z) = 0\}), & \text{if } \exists z \leq b\,(g(\vec{a}, z) = 0); \\ 0, & \text{otherwise.} \end{cases}$$

Bounded recursion, **BR**: A function $f : \omega^{n+1} \to \omega$ is defined from $g : \omega^n \to \omega$, $h : \omega^{n+2} \to \omega$ and $C : \omega^{n+1} \to \omega$ by bounded recursion, if $f \leq C$ and

$$f(\vec{x}, 0) = g(\vec{x}); \quad f(\vec{x}, y+1) = h(\vec{x}, y, f(\vec{x}, y)).$$

In this case we write, $f = \mathbf{BR}_C(g, h)$ and we shall say that f is defined by C*–bounded recursion* from g and h.

Let $\mathcal{F}$ be a class containing $\mathcal{B}$. In this paper, $\mathbf{C}(\mathcal{F})$ will denote the closure of $\mathcal{F}$ under composition and $\mathbf{E}(\mathcal{F})$ the closure of $\mathcal{F}$ under composition and bounded recursion. We also consider the following slight (but crucial) modification of closure under bounded recursion.

Definition. $\mathcal{E}^{\mathcal{F}}$ is the smallest class of functions containing $\mathcal{B}$ and closed under composition and $\mathcal{F}$–bounded recursion; that is, closed under C–bounded recursion for every $C \in \mathcal{F}$.

Let us observe that $\mathcal{E}^{\mathcal{F}} \subseteq \mathcal{PR}$ and if $\mathcal{F} \subseteq \mathcal{E}^{\mathcal{F}}$, then $\mathbf{C}(\mathcal{F}) \subseteq \mathcal{E}^{\mathcal{F}} \subseteq \mathbf{E}(\mathcal{F})$.

Grzegorczyk's classes, $\mathcal{E}^n$, can be described in the form $\mathcal{E}^{\mathcal{F}}$. For instance, let $\mathcal{P}_0$, $\mathcal{P}_1$ and $\mathcal{P}_2$ be, respectively, the classes of functions $\mathbf{C}(\mathcal{B})$, $\mathbf{C}(\mathcal{B} \cup \{+\})$ and $\mathbf{C}(\mathcal{B} \cup \{+, \times\})$, then, for $j = 0, 1, 2$, it holds that $\mathcal{E}^{\mathcal{P}_j} = \mathcal{E}^j$ (see [13]).

The basic function algebra in this paper will be Grzegorczyk's class $\mathcal{M}^2$: the closure of $\mathcal{B} \cup \{+, \times\}$ under composition and $\mu_\leq$ (see [6, 13]). As we shall see (Proposition 3), $\mathcal{M}^2$ is the class $\mathcal{R}(\mathbf{I}\Delta_0)$ and, therefore, all function algebras considered in this paper contain it. This motivates the following definition.

Definition. An F–algebra is a family, $\mathcal{C}$, of computable functions containing $\mathcal{B}$ and closed under composition. We shall say that $\mathcal{C}$ is rudimentary if $\mathcal{M}^2 \subseteq \mathcal{C}$.

A pairing function is available in $\mathcal{M}^2$. Let $J : \omega^2 \to \omega$ be Cantor's function:

$$J(a, b) = \frac{(a+b)(a+b+1)}{2} + a.$$

Its lateral inverses K, L are given by $K(a) = (\mu z)_{\leq a}(\exists y \leq a\,(J(z, y) = a))$ and $L(a) = (\mu z)_{\leq a}(\exists x \leq a\,(J(x, z) = a))$. Then $J, K, L \in \mathcal{M}^2$. We shall write $\langle x, y \rangle = J(x, y)$ and $(z)_0 = K(z)$, $(z)_1 = L(z)$.

Basic properties of $\mathcal{M}^2$ are summed up in next proposition (see [6] or [13]).

Proposition 2.   1. $\Delta_0^0 = \mathcal{M}_*^2$.
2. For each $f : \omega^k \to \omega$, the following conditions are equivalent:
   (a) $f \in \mathcal{M}^2$.
   (b) $Gr(f) \in \Delta_0^0$ and there exists a term t of $\mathcal{L}$ such that $f \leq t$.

As a consequence a characterization of $\mathcal{R}(\mathbf{I}\Delta_0)$ can be obtained. A proof–theoretic proof of this result was obtained by G. Takeuti (see [16]).

Proposition 3. $\mathcal{M}^2 = \mathcal{R}(\mathbf{I}\Delta_0)$.

Hence, for every extension, $\mathbf{T}$, of $\mathbf{I}\Delta_0$ the class $\mathcal{R}(\mathbf{T})$ is a rudimentary F–algebra. Now we introduce a necessary and sufficient condition under which a rudimentary F–algebra is the class of p.t.c.f. of some theory. The following results seem to be folklore and have appeared more or less explicitly in the literature (see proposition 4.1 in [2] and previous remarks in that paper). However, Theorem 1 below does not seem to be known. As it was remarked in the Introduction, it provides interesting insights on open problems in Complexity Theory.

Lemma 1. Let $\mathcal{C}$ be a rudimentary F–algebra and $f : \omega^k \to \omega$ such that $Gr(f) \in \Delta_0^0$. If there exists $g \in \mathcal{C}$ such that $f \leq g$, then $f \in \mathcal{C}$.

Proof. Let $h : \omega^{k+1} \to \omega$ given by $h(\vec{a}, b) = (\mu z)_{\leq b}[f(\vec{a}) = z]$. Since $Gr(f) \in \Delta_0^0$, then $Gr(h) \in \Delta_0^0$. By Proposition 2–(2), $h \in \mathcal{M}^2 \subseteq \mathcal{C}$. Let $g \in \mathcal{C}$ such that $f \leq g$. Then $f(\vec{a}) = h(\vec{a}, g(\vec{a}))$. Since $\mathcal{C}$ is closed under composition, $f \in \mathcal{C}$.   □

Lemma 2. Let $\mathbf{T}$ be an extension of $\mathbf{I}\Delta_0$. Then for each $f \in \mathcal{R}(\mathbf{T})$ there exists $g \in \mathcal{R}(\mathbf{T})$ such that $Gr(g) \in \Delta_0^0$ and $f = K \circ g$.

Proof. For each $f \in \mathcal{R}(\mathbf{T})$ and $\varphi(\vec{x}, y, z) \in \Delta_0$ such that $\exists z \, \varphi(\vec{x}, y, z)$ defines f in $\mathbf{T}$, let $\psi(\vec{x}, v) \in \Delta_0$ be the formula

$$\exists y, z \leq v \, [\langle y, z \rangle = v \wedge \varphi(\vec{x}, y, z) \wedge \forall z' < z \, \neg \varphi(\vec{x}, y, z')].$$

Then $\mathbf{T} \vdash \forall \vec{x} \, \exists! y \, \psi(\vec{x}, y)$. Let g be the function defined in $\mathcal{N}$ by $\psi(\vec{x}, v)$. Then $g \in \mathcal{R}(\mathbf{T})$, $Gr(g) \in \Delta_0^0$ and $f(\vec{a}) = K(g(\vec{a}))$. $\qquad \square$

The above lemma motivates the following definition.

Definition. Let $\mathcal{C}$ be a rudimentary F–algebra. We say that $\mathcal{C}$ is a Δ_0–generated F–algebra if for each $f \in \mathcal{C}$ there exist $g_1, g_2 \in \mathcal{C}_0$ such that $f = g_1 \circ g_2$.

The following result shows that Δ_0–generated F–algebras correspond to classes of p.t.c.f. of extensions of $\mathbf{I}\Delta_0$.

Theorem 1. *The following conditions are equivalent:*

1. *$\mathcal{C}$ is a Δ_0–generated F–algebra.*
2. *There exists a (sound) $\mathcal{L}$–theory, $\mathbf{T}$, extending $\mathbf{I}\Delta_0$ such that $\mathcal{R}(\mathbf{T}) = \mathcal{C}$.*

Proof. $(2) \Longrightarrow (1)$: It follows from Proposition 3 and Lemma 2.

$(1) \Longrightarrow (2)$: For each $f \in \mathcal{C}_0 = \{h \in \mathcal{C} : Gr(h) \in \Delta_0^0\}$, let $\theta_f(x, y)$ be a Δ_0–formula defining f in $\mathcal{N}$. Let $\Gamma = \{\forall x \, \exists y \, \theta_f(x, y) : f \in \mathcal{C}_0, f : \omega \to \omega\}$. Next claim is a slight generalization of proposition 4.2 in [2] and it can also be proved along the lines sketched there.

Claim: $\mathcal{R}(\mathbf{I}\Delta_0 + \Gamma) = \mathbf{C}(\mathcal{M}^2 \cup \mathcal{C}_0)$.

Thus, $\mathcal{R}(\mathbf{I}\Delta_0 + \Gamma) = \mathbf{C}(\mathcal{M}^2 \cup \mathcal{C}_0) = \mathcal{C}$, last equality since $\mathcal{C}$ is Δ_0– generated. $\qquad \square$

3   Axiomatizing $\Delta_{n+1}(\mathbf{T})$–Induction

The aim of this section is to characterize the class of primitive recursive functions in $\mathcal{R}(\mathbf{T})$, where $\mathbf{T}$ is an extension of $\mathbf{I}\Delta_0$, as the class of p.t.c.f. of a suitable theory. To this end, we consider the class of $\Delta_{n+1}(\mathbf{T})$–formulas:

$$\Delta_{n+1}(\mathbf{T}) = \{\varphi(x, \vec{v}) \in \Sigma_{n+1} : \text{ there exists } \psi(x, \vec{v}) \in \Pi_{n+1}, \mathbf{T} \vdash \varphi \leftrightarrow \psi\}.$$

When the schemes of induction and minimization are restricted to these classes of formulas we obtain the theories $\mathbf{I}\Delta_{n+1}(\mathbf{T})$ and $\mathbf{L}\Delta_{n+1}(\mathbf{T})$ introduced in [9]. There the following version of the collection scheme is also considered

$$\mathbf{B}^* \Delta_{n+1}(\mathbf{T}) = \mathbf{I}\Delta_0 + \{\mathbf{B}_{\varphi, x, y}(z, \vec{v}) : \varphi \in \Pi_n, \exists y \, \varphi(x, y, \vec{v}) \in \Delta_{n+1}(\mathbf{T})\}.$$

Let us state here some basic properties of these theories, for details and proofs see [9]. If $\varphi \in \Sigma_{n+1}$ and $\psi \in \Pi_{n+1}$ then $\varphi \leftrightarrow \psi$ is a Π_{n+2}–formula. Therefore,

- If $\mathbf{Th}_{\Pi_{n+2}}(\mathbf{T}) = \mathbf{Th}_{\Pi_{n+2}}(\mathbf{T}')$ then $\mathbf{I}\Delta_{n+1}(\mathbf{T}) \Longleftrightarrow \mathbf{I}\Delta_{n+1}(\mathbf{T}')$.

A similar result holds for minimization and collection. The following basic properties will be used without explicit mention.

- $\mathbf{L}\Delta_{n+1}(\mathbf{T}) \Longrightarrow \mathbf{I}\Delta_{n+1}(\mathbf{T})$ and $\mathbf{B}^*\Delta_{n+1}(\mathbf{T}) \Longrightarrow \mathbf{I}\Sigma_n$.
- If $\mathbf{T}$ is an extension of $\mathbf{I}\Sigma_n$ then $\mathbf{L}\Delta_{n+1}(\mathbf{T}) \Longrightarrow \mathbf{B}^*\Delta_{n+1}(\mathbf{T})$.

As noticed in [9], last property follows by an argument that mimics the proof of Gandy's Theorem, $\mathbf{L}\Delta_1 \Longrightarrow \mathbf{B}\Sigma_1$, given in [10], lemma I.2.17. A variation of that argument, considering also lemma I.2.16 in [10], gives us that

Lemma 3. $\mathbf{Th}_{\Pi_{n+2}}(\mathbf{T}) + \mathbf{L}\Delta_{n+1}(\mathbf{T}) \Longleftrightarrow \mathbf{Th}_{\Pi_{n+2}}(\mathbf{T}) + \mathbf{B}^*\Delta_{n+1}(\mathbf{T})$.

The following notion, introduced in [9], has turned out to be useful for the study of $\Delta_{n+1}(\mathbf{T})$–induction.

Definition. We say that $\mathbf{T}$ has Δ_{n+1}–induction if $\mathbf{T} \Longrightarrow \mathbf{I}\Delta_{n+1}(\mathbf{T})$.

Theories $\mathbf{I}\Delta_{n+1}(\mathbf{T})$ and $\mathbf{L}\Delta_{n+1}(\mathbf{T})$ are Π_{n+3}–axiomatizable. But adding to them $\mathbf{Th}_{\Pi_{n+2}}(\mathbf{T})$, their quantifier complexity is reduced to Π_{n+2}.

Lemma 4. $\mathbf{Th}_{\Pi_{n+2}}(\mathbf{T}) + \mathbf{I}\Delta_{n+1}(\mathbf{T})$ and $\mathbf{Th}_{\Pi_{n+2}}(\mathbf{T}) + \mathbf{L}\Delta_{n+1}(\mathbf{T})$ are Π_{n+2}–axiomatizable.

In this section we shall obtain a useful axiomatization of $\mathbf{I}\Delta_{n+1}(\mathbf{T})$ in terms of $\mathbf{I}\Sigma_{n+1}$ and $\mathbf{Th}_{\Pi_{n+2}}(\mathbf{T})$. To this end we introduce the disjunction of two theories, which corresponds to intersection between classes of p.t.c.f.

If $\mathbf{T}_1$ and $\mathbf{T}_2$ are theories in the language $\mathcal{L}$, then $\mathbf{T}_1 \vee \mathbf{T}_2$ denotes the theory axiomatized by the set of formulas $\{\varphi_1 \vee \varphi_2 : \varphi_1 \in \mathbf{T}_1 \text{ and } \varphi_2 \in \mathbf{T}_2\}$.

Lemma 5. If $\mathbf{T}_1, \mathbf{T}_2 \Longrightarrow \mathbf{I}\Delta_0$ then $\mathcal{R}(\mathbf{T}_1 \vee \mathbf{T}_2) = \mathcal{R}(\mathbf{T}_1) \cap \mathcal{R}(\mathbf{T}_2)$.

Proof. Since $\mathbf{T}_1, \mathbf{T}_2 \Longrightarrow \mathbf{T}_1 \vee \mathbf{T}_2$, it holds that $\mathcal{R}(\mathbf{T}_1 \vee \mathbf{T}_2) \subseteq \mathcal{R}(\mathbf{T}_1) \cap \mathcal{R}(\mathbf{T}_2)$.

Conversely, if $f \in \mathcal{R}(\mathbf{T}_1) \cap \mathcal{R}(\mathbf{T}_2)$, then there exist $\psi_1(\vec{x}, y, z)$, $\psi_2(\vec{x}, y, z) \in \Delta_0$ such that $\exists z\, \psi_i(\vec{x}, y, z)$ defines f in $\mathbf{T}_i$. Let $\theta_0(\vec{x}, u) \in \Delta_0$ the formula $(\psi_1(\vec{x}, (u)_0, (u)_1) \vee \psi_2(\vec{x}, (u)_0, (u)_1))$ and let $\theta(\vec{x}, y)$ be the formula

$$\exists z\, [\theta_0(\vec{x}, \langle y, z \rangle) \wedge \forall w < \langle y, z \rangle\, \neg\theta_0(\vec{x}, w)].$$

Then $\theta(\vec{x}, y)$ defines f in $\mathbf{T}_1 \vee \mathbf{T}_2$. So, $f \in \mathcal{R}(\mathbf{T}_1 \vee \mathbf{T}_2)$. $\square$

Next proposition is theorem 2.2 in [8]. Now it can be rephrased as follows.

Proposition 4. If $\mathfrak{A} \not\models \mathbf{Th}_{\Pi_{n+2}}(\mathbf{T})$ and $\mathfrak{A} \models \mathbf{I}\Delta_{n+1}(\mathbf{T})$, then $\mathfrak{A} \models \mathbf{I}\Sigma_{n+1}$. Hence,

$$\mathbf{I}\Delta_{n+1}(\mathbf{T}) \Longrightarrow \mathbf{I}\Sigma_{n+1} \vee \mathbf{Th}_{\Pi_{n+2}}(\mathbf{T}).$$

Moreover, if $\mathbf{T}$ has Δ_{n+1}–induction then $\mathbf{I}\Delta_{n+1}(\mathbf{T}) \Longleftrightarrow \mathbf{I}\Sigma_{n+1} \vee \mathbf{Th}_{\Pi_{n+2}}(\mathbf{T})$.

From this proposition, using Lemma 5, a first result on $\mathcal{R}(\mathbf{T}) \cap \mathcal{PR}$ is obtained for theories with Δ_1–induction.

Corollary 1.   *1.* $\mathcal{PR} \cap \mathcal{R}(\mathbf{T}) \subseteq \mathcal{R}(\mathbf{I}\Delta_1(\mathbf{T}))$.
2. If $\mathbf{T}$ *has* Δ_1*-induction then* $\mathcal{PR} \cap \mathcal{R}(\mathbf{T}) = \mathcal{R}(\mathbf{I}\Delta_1(\mathbf{T}))$.

By Lemma 5, $\mathcal{R}(\mathbf{T}) \cap \mathcal{PR}$ is always a Δ_0-generated F–algebra. To get a natural theory $\mathbf{T}'$ such that $\mathcal{R}(\mathbf{T}') = \mathcal{R}(\mathbf{T}) \cap \mathcal{PR}$ without assuming that $\mathbf{T}$ has Δ_1-induction, we consider parameter free versions of $\mathbf{I}\Delta_{n+1}(\mathbf{T})$ and $\mathbf{I}\Sigma_{n+1}$, denoted by $\mathbf{I}\Delta_{n+1}(\mathbf{T})^-$ and $\mathbf{I}\Sigma_{n+1}^-$, respectively.

Theorem 2. *For each sound theory,* $\mathbf{T}$*,* $\mathcal{R}(\mathbf{I}\Delta_1(\mathbf{T})^-) = \mathcal{R}(\mathbf{T}) \cap \mathcal{PR}$.

Proof. By a similar argument to that of theorem 2.2 in [8], it is shown that:

$$\mathbf{I}\Delta_{n+1}(\mathbf{T})^- \Longrightarrow \mathbf{I}\Sigma_{n+1}^- \vee \mathbf{Th}_{\Pi_{n+2}}(\mathbf{T}).$$

Since $\mathcal{PR} = \mathcal{R}(\mathbf{I}\Sigma_1^-)$, by Lemma 5, $\mathcal{PR} \cap \mathcal{R}(\mathbf{T}) = \mathcal{R}(\mathbf{I}\Sigma_1^- \vee \mathbf{T}) \subseteq \mathcal{R}(\mathbf{I}\Delta_1(\mathbf{T})^-)$.
    Let us prove $\mathcal{R}(\mathbf{I}\Delta_1(\mathbf{T})^-) \subseteq \mathcal{R}(\mathbf{T}) \cap \mathcal{PR}$. Obviously, $\mathcal{R}(\mathbf{I}\Delta_1(\mathbf{T})^-) \subseteq \mathcal{PR}$. Now, let us observe that $\mathbf{I}\Delta_1^-$ is Σ_2-axiomatizable and, therefore,

$$\mathbf{T} + \mathbf{Th}_{\Pi_1}(\mathcal{N}) \Longrightarrow \mathbf{T} + \mathbf{I}\Delta_1^- \Longrightarrow \mathbf{T} + \mathbf{I}\Delta_1(\mathbf{T})^- \Longrightarrow \mathbf{I}\Delta_1(\mathbf{T})^-.$$

So, by Proposition 1, $\mathcal{R}(\mathbf{I}\Delta_1(\mathbf{T})^-) \subseteq \mathcal{R}(\mathbf{T} + \mathbf{Th}_{\Pi_1}(\mathcal{N})) = \mathcal{R}(\mathbf{T})$.   □

4 $\mathcal{C}$–Bounded Recursive Arithmetic: $\mathcal{C}$–BRA

In this section we characterize $\mathcal{R}(\mathbf{T}) \cap \mathcal{PR}$ in terms of bounded recursion. Our main tool will be a version of the well–known system **PRA** (Primitive Recursive Arithmetic). Our analysis of this theory follows the lines sketched in [1].

Definition. Let $\mathcal{C}$ be a rudimentary F–algebra. The theory $\mathcal{C}$–**BRA**, $\mathcal{C}$–Bounded Recursive Arithmetic, is given by:

– Language: $\mathcal{L}_{pr}^{\mathcal{C}} = \bigcup_{i \in \omega} \mathbf{L}_i$, where
  • $\mathbf{L}_0 = \mathcal{L}$ plus a function symbol B_f for each basic function, $f \in \mathcal{B}$.
  • $\mathbf{L}_{j+1} = \mathbf{L}_j$ plus a function symbol, $\mathbf{f}_t$ for each term of $\mathbf{L}_j$, and a function symbol $\mathbf{f}_{t_1,t_2,g}$ for each function $g \in \mathcal{C}$ and terms $t_1(\vec{x})$, $t_2(\vec{x}, y, z)$ of $\mathbf{L}_j$ such that the function defined from t_1 and t_2 by primitive recursion is bounded by g, i. e., $h \leq g$, where $h : \omega^{n+1} \to \omega$ is the function given by

$$h(\vec{x}, 0) = t_1(\vec{x}), \qquad h(\vec{x}, y+1) = t_2(\vec{x}, y, h(\vec{x}, y)).$$

– Axioms:
  (1) $\mathbf{P}^-$.
  (2) $B_S(x) = x + 1$, $B_{\Pi_i^n}(x_1, \ldots, x_n) = x_i$, $B_O(x) = 0$.
  (3) $\mathbf{f}_t(\vec{x}) = t(\vec{x})$.
  (4) $\mathbf{f}_{t_1,t_2,g}(\vec{x}, 0) = t_1(\vec{x})$, $\mathbf{f}_{t_1,t_2,g}(\vec{x}, y+1) = t_2(\vec{x}, y, \mathbf{f}_{t_1,t_2,g}(\vec{x}, y))$.
  (5) Open Induction: The induction scheme for open formulas of $\mathcal{L}_{pr}^{\mathcal{C}}$.

It is routine to check that $\mathcal{C}$–**BRA** satisfies the following properties stated for **PRA** in [1].

Lemma 6. *It holds that:*

1. *In* C–**BRA** *the class of open formulas is closed under bounded quantification.*
2. *In* C–**BRA** *every* Δ_0–*formula is equivalent to an open formula.*
3. C–**BRA** *supports definition by cases.*
4. C–**BRA** *is universally axiomatizable.*

Lemma 7. $\mathcal{R}(C$–**BRA**$) = \mathcal{E}^C$.

Proof. Obviously $\mathcal{E}^C \subseteq \mathcal{R}(C$–**BRA**$)$. Since C–**BRA** is a universal theory and supports definition by cases, the result follows from Herbrand's theorem.   □

Next we investigate the relations between $\mathcal{PR} \cap \mathcal{R}(\mathbf{T})$ and $\mathcal{E}^C$. The key ingredient is Theorem 3 below stating a conservation result between C–**BRA** and $\mathbf{L}\Delta_1(\mathbf{T})$. In the proof of that theorem we use the model–theoretic framework developed by Avigad in [1]. Let us recall some definitions and results from that paper that will be used in what follows.

Definition. We say that a structure $\mathfrak{A}$ is $\exists_2$–closed (or Herbrand saturated, in Avigad's terminology) if for each $\varphi(\vec{x}) \in \exists_2$ and $\vec{a} \in \mathfrak{A}$ such that $\mathfrak{B} \models \varphi(\vec{a})$, for some $\mathfrak{B}$, $\mathfrak{A} \prec_{\forall_1} \mathfrak{B}$, it holds $\mathfrak{A} \models \varphi(\vec{a})$.

As it is proved in [1], every universal theory has a $\exists_2$–closed model. For these models the following results hold (see [1], theorems 3.3 and 3.4):

Proposition 5. *Let* $\mathfrak{A}$ *be an* $\exists_2$-*closed model and* $\theta(\vec{x}, y, \vec{z})$ *an open formula such that* $\mathfrak{A} \models \forall \vec{x} \exists y \, \theta(\vec{x}, y, \vec{a})$. *Then there exist a universal formula* $\psi(\vec{z}, \vec{w})$ *and terms* $t_1, \ldots, t_k$ *such that* $\mathfrak{A} \models \exists \vec{w} \, \psi(\vec{a}, \vec{w})$ *and*

$$\models \psi(\vec{z}, \vec{w}) \rightarrow \theta(\vec{x}, t_1(\vec{x}, \vec{z}, \vec{w}), \vec{z}) \vee \cdots \vee \theta(\vec{x}, t_k(\vec{x}, \vec{z}, \vec{w}), \vec{z}).$$

Proposition 6. *Let* $\mathbf{T}_2$ *be a universal theory and let* $\mathbf{T}_1$ *be a theory in the language of* $\mathbf{T}_2$. *If every* $\exists_2$-*closed model of* $\mathbf{T}_2$ *is a model of* $\mathbf{T}_1$, *then every* $\forall_2$-*theorem of* $\mathbf{T}_1$ *is a theorem of* $\mathbf{T}_2$.

Last proposition is used in [1] to obtain new proofs of a number of conservation results. In what follows we use it to prove our main conservation result. First of all, we show that $\exists_2$–closed models of C–**BRA** satisfy Σ_1–collection.

Lemma 8. *Let* $\mathfrak{A} \models C$–**BRA** *be an* $\exists_2$-*closed model.*

1. *In* $\mathfrak{A}$ *each* Δ_1-*formula is equivalent to an open formula.*
2. $\mathfrak{A} \models \mathbf{B}\Sigma_1$.

Proof. **(1)** Let $\varphi(x, y, v), \psi(x, y, v) \in \Delta_0$ and $a \in \mathfrak{A}$ such that

$$\mathfrak{A} \models \exists y \, \varphi(x, y, a) \leftrightarrow \forall y \, \psi(x, y, a).$$

Let $\theta(x, y, v)$ be the formula $\varphi(x, y, v) \vee \neg\psi(x, y, v)$. Then $\mathfrak{A} \models \forall x \exists y \, \theta(x, y, a)$. By Lemma 6, there exist φ_0 and θ_0 quantifier–free formulas such that

$$C\text{–}\mathbf{BRA} \vdash (\varphi_0 \leftrightarrow \varphi) \wedge (\theta_0 \leftrightarrow \theta).$$

Then $\mathfrak{A} \models \forall x \, \exists y \, \theta_0(x, y, a)$ and, by Proposition 5 (recall that $\mathcal{C}\text{–}\mathbf{BRA}$ supports definition by cases), there exist $b \in \mathfrak{A}$ and a term $t(x, v, w)$ such that $\mathfrak{A} \models \forall x \, \theta_0(x, t(x, a, b), a)$. As a consequence, $\mathfrak{A} \models \exists y \, \varphi(x, y, a) \leftrightarrow \varphi_0(x, t(x, a, b), a)$.

(2) By Lemma 6, the class of open formulas is closed in $\mathcal{C}\text{–}\mathbf{BRA}$ under bounded quantification. Since $\mathcal{C}\text{–}\mathbf{BRA}$ proves open induction, a standard argument (see lemma I.2.12 in [10]) shows that minimization scheme for open formulas holds in $\mathcal{C}\text{–}\mathbf{BRA}$. So, by (1), $\mathfrak{A} \models \mathbf{L}\Delta_1$. But $\mathbf{L}\Delta_1 \iff \mathbf{B}\Sigma_1$ (see [10] lemmas I.2.16, I.2.17), hence $\mathfrak{A} \models \mathbf{B}\Sigma_1$.   □

Theorem 3. *Let* $\mathbf{T}$ *be a (sound)* Π_2*–axiomatized extension of* $\mathbf{I}\Delta_0$ *and* $\mathcal{C} = \mathcal{R}(\mathbf{T})$. *For each* Π_2*–sentence* θ, *if* $\mathbf{L}\Delta_1(\mathbf{T}) \vdash \theta$ *then* $\mathcal{C}\text{–}\mathbf{BRA} \vdash \theta$.

Proof. Since $\mathcal{C}\text{–}\mathbf{BRA}$ is a universal theory, following [1], we prove that every $\exists_2$–closed model of $\mathcal{C}\text{–}\mathbf{BRA}$, $\mathfrak{A}$, is a model of $\mathbf{L}\Delta_1(\mathbf{T})$. Then the result follows by Proposition 6. In a first step we prove $\mathfrak{A} \models \mathbf{I}\Delta_1(\mathbf{T})$.

Let $\varphi(x, y, \vec{v}), \psi(x, y, \vec{v}) \in \Delta_0$ such that $\mathbf{T} \vdash \exists y \, \varphi(x, y, \vec{v}) \leftrightarrow \forall y \, \psi(x, y, \vec{v})$. We may assume that $\mathbf{T} \vdash \varphi(x, y_1, \vec{v}) \wedge \varphi(x, y_2, \vec{v}) \to y_1 = y_2$ (if not, we take as φ the formula $\varphi(x, y, \vec{v}) \wedge \forall z < y \, \neg\varphi(x, z, \vec{v})$). Let $\theta(x, y, \vec{v}) \in \Delta_0$ the formula $\varphi(x, y, \vec{v}) \vee \neg\psi(x, y, \vec{v})$. Then,

$$\mathbf{T} \vdash \forall x \, \exists y \, (\theta(x, y, \vec{v}) \wedge \forall z < y \, \neg\theta(x, z, \vec{v})).$$

Since $\mathbf{T}$ is a sound theory, the formula $\theta(x, y, \vec{v}) \wedge \forall z < y \, \neg\theta(x, z, \vec{v})$ defines a p.t.c.f. of $\mathbf{T}$, say f. Then $\mathcal{N} \models \forall y \, (\varphi(x, y, \vec{v}) \to y \leq f(x, \vec{v}))$. Now, we continue the proof as in [1], theorem 4.1.

Let $\mathfrak{A}$ be an $\exists_2$–closed model of $\mathcal{C}\text{–}\mathbf{BRA}$ and φ_0 an open formula equivalent to φ in $\mathcal{C}\text{–}\mathbf{BRA}$. Let us see that $\mathfrak{A} \models \mathbf{I}_{\exists y \, \varphi_0}$. Assume that, for some $\vec{a} \in \mathfrak{A}$,

$$\mathfrak{A} \models \exists y \, \varphi_0(0, y, \vec{a}) \wedge \forall x \, (\exists y \, \varphi_0(x, y, \vec{a}) \to \exists y \, \varphi_0(x+1, y, \vec{a})).$$

Then, as in [1], since $\mathcal{C}\text{–}\mathbf{BRA}$ supports definition by cases, by Proposition 5, there exist $\vec{b}, c \in \mathfrak{A}$ and a function symbol $\mathbf{g}(x, y, \vec{v}, \vec{w})$ such that

$$\mathfrak{A} \models \varphi_0(0, c, \vec{a}) \wedge \forall x, y \, (\varphi_0(x, y, \vec{a}) \to \varphi_0(x+1, \mathbf{g}(x, y, \vec{a}, \vec{b}), \vec{a})).$$

Let us denote by g the function defined by $\mathbf{g}$ in $\mathcal{N}$. Let $h_0 : \omega^{n+2} \to \omega$ be defined in $\mathcal{N}$ by

$$h_0(x, y, z, \vec{v}, \vec{w}) = \begin{cases} g(x, z, \vec{v}, \vec{w}), & \text{if } \varphi_0(x+1, g(x, z, \vec{v}, \vec{w}), \vec{v}); \\ 0, & \text{otherwise.} \end{cases}$$

Then $h_0 \in \mathcal{E}^{\mathcal{C}}$. Let f_0 be the function defined by primitive recursion as follows:

$$f_0(0, y, \vec{v}, \vec{w}) = y, \qquad f_0(x+1, y, \vec{v}, \vec{w}) = h_0(x, y, f_0(x, y, \vec{v}, \vec{w}), \vec{v}, \vec{w}).$$

Then $f_0(x, y, \vec{v}, \vec{w}) \leq f'(x, y, \vec{v}, \vec{w}) = f(x+1, \vec{v}) + y$. So, $f_0 \in \mathcal{E}^{\mathcal{C}}$, since it is defined by f'–bounded recursion and $f' \in \mathcal{C}$. Let $\mathbf{f}_0$ be the function symbol corresponding to f_0. Then $\mathfrak{A}$ satisfies that

$$\varphi_0(0, \mathbf{f}_0(0, c, \vec{a}, \vec{b}), \vec{a}) \wedge \forall x \, (\varphi_0(x, \mathbf{f}_0(x, c, \vec{a}, \vec{b}), \vec{a}) \to \varphi_0(x+1, \mathbf{f}_0(x+1, c, \vec{a}, \vec{b}), \vec{a})).$$

Since $\mathfrak{A}$ satisfies open induction, $\mathfrak{A} \models \forall x \, \varphi_0(x, \mathbf{f}_0(x, c, \vec{a}, \vec{b}), \vec{a})$. Hence, it follows that $\mathfrak{A} \models \forall x \, \exists y \, \varphi(x, y, \vec{a})$. So, $\mathfrak{A} \models \mathbf{I}\Delta_1(\mathbf{T})$.

Let us see that $\mathfrak{A} \models \mathbf{L}\Delta_1(\mathbf{T})$. We distinguish two cases:

1. If $\mathfrak{A} \not\models \mathbf{T}$ then, since $\mathbf{T}$ is Π_2–axiomatized, by Proposition 4, $\mathfrak{A} \models \mathbf{I}\Sigma_1$.
2. If $\mathfrak{A} \models \mathbf{T}$, then, by Lemma 8, $\mathfrak{A} \models \mathbf{T} + \mathbf{B}\Sigma_1$; hence, $\mathfrak{A} \models \mathbf{L}\Delta_1(\mathbf{T})$.   □

As a consequence, we get some results relating $\mathcal{E}^{\mathcal{C}}$ and $\mathcal{C} \cap \mathcal{PR}$. The notion of Δ_0–generativeness provides a necessary and sufficient condition for $\mathcal{E}^{\mathcal{C}} = \mathcal{C} \cap \mathcal{PR}$.

Theorem 4. *Let $\mathcal{C}$ be a Δ_0–generated F–algebra. Then*

1. $\mathcal{PR} \cap \mathcal{C}$ *is Δ_0–generated.*
2. $\mathcal{PR} \cap \mathcal{C} \subseteq \mathcal{E}^{\mathcal{C}} = \mathcal{E}^{\mathcal{PR} \cap \mathcal{C}} \subseteq \mathbf{E}(\mathcal{PR} \cap \mathcal{C})$.
3. *If $\mathcal{C}$ is closed under bounded minimization, then $\mathcal{E}^{\mathcal{C}} = \mathbf{E}(\mathcal{PR} \cap \mathcal{C})$.*

Proof. By Theorem 1, there is a sound extension of $\mathbf{I}\Delta_0$, $\mathbf{T}$, such that $\mathcal{R}(\mathbf{T}) = \mathcal{C}$.

(1) Since $\mathcal{C} \cap \mathcal{PR} = \mathcal{R}(\mathbf{T} \vee \mathbf{I}\Sigma_1)$, by Theorem 1, $\mathcal{C} \cap \mathcal{PR}$ is Δ_0–generated.

(2) By Corollary 1, $\mathcal{PR} \cap \mathcal{C} \subseteq \mathcal{R}(\mathbf{I}\Delta_1(\mathbf{T}))$. Moreover, by Theorem 3 and Lemma 7, $\mathcal{R}(\mathbf{I}\Delta_1(\mathbf{T})) \subseteq \mathcal{E}^{\mathcal{C}}$; so, $\mathcal{PR} \cap \mathcal{C} \subseteq \mathcal{E}^{\mathcal{C}}$. It is trivial that $\mathcal{E}^{\mathcal{C} \cap \mathcal{PR}} \subseteq \mathbf{E}(\mathcal{C} \cap \mathcal{PR})$.

Now, let us see that $\mathcal{E}^{\mathcal{C}} = \mathcal{E}^{\mathcal{C} \cap \mathcal{PR}}$. It is enough to prove that $\mathcal{E}^{\mathcal{C}} \subseteq \mathcal{E}^{\mathcal{C} \cap \mathcal{PR}}$. We proceed by induction on the definition of $f \in \mathcal{E}^{\mathcal{C}}$. The critical step is the definition by $\mathcal{C}$–bounded recursion. But, let us observe that if $f \in \mathcal{E}^{\mathcal{C}}$ is defined by $\mathcal{C}$–bounded recursion, then f is bounded by a function $g_1 \in \mathcal{C}$ and by a function $g_2 \in \mathcal{PR}$ (in fact, $f \in \mathcal{PR}$). We prove that in this case f is bounded by a function $h \in \mathcal{C} \cap \mathcal{PR}$.

Let $\psi_1(\vec{x}, y, z), \psi_2(\vec{x}, y, z) \in \Delta_0$ such that $\exists z \, \psi_1(\vec{x}, y, z)$ and $\exists z \, \psi_2(\vec{x}, y, z)$ define g_1 and g_2 in $\mathbf{T}$ and $\mathbf{I}\Sigma_1$, respectively. Let $\theta_0(\vec{x}, u) \in \Delta_0$ be the formula

$$\begin{cases} (\psi_1(\vec{x}, (u)_0, (u)_1) \vee \psi_2(\vec{x}, (u)_0, (u)_1)) \wedge \\ \forall v < u \, (\neg\psi_1(\vec{x}, (v)_0, (v)_1) \wedge \neg\psi_2(\vec{x}, (v)_0, (v)_1)). \end{cases}$$

Then $\exists z \, \theta_0(\vec{x}, \langle y, z \rangle)$ defines in $\mathbf{T} \vee \mathbf{I}\Sigma_1$ a function h such that for all $\vec{a} \in \omega$, $h(\vec{a}) = g_1(\vec{a})$ or $h(\vec{a}) = g_2(\vec{a})$. So, $h \in \mathcal{R}(\mathbf{T} \vee \mathbf{I}\Sigma_1) = \mathcal{C} \cap \mathcal{PR}$ and, since $f \leq g_1$ and $f \leq g_2$, we have $f \leq h$.

(3) Since $\mathcal{C}$ is closed under $\mu_{\leq}$, each function in $\mathcal{C}$ is bounded by a nondecreasing function also in $\mathcal{C}$. By induction on the construction of $f \in \mathcal{E}^{\mathcal{PR} \cap \mathcal{C}}$, it is proved that each element of $\mathcal{E}^{\mathcal{C}} \, (= \mathcal{E}^{\mathcal{PR} \cap \mathcal{C}})$ is bounded by an element of $\mathcal{PR} \cap \mathcal{C}$. So, $\mathcal{E}^{\mathcal{C} \cap \mathcal{PR}}$ is closed under bounded recursion. Hence, **(3)** follows from part **(2)**.   □

Theorem 5. *Let $\mathcal{C}$ be a Δ_0–generated F–algebra. If $\mathcal{C}$ is closed under bounded minimization, then the following conditions are equivalent:*

1. $\mathcal{E}^{\mathcal{C}}$ *is Δ_0–generated. Or, by Theorem 4, $\mathcal{E}^{\mathcal{C} \cap \mathcal{PR}}$ is Δ_0–generated.*
2. $\mathcal{E}^{\mathcal{C}} = \mathcal{C} \cap \mathcal{PR}$.

Proof. **(2)** $\Longrightarrow$ **(1)**: By Theorem 4, $\mathcal{C} \cap \mathcal{PR}$ is Δ_0–generated; so, **(1)** holds.

(1) $\Longrightarrow$ **(2)**: Let $\mathcal{F}_0$ be the class of all functions in $\mathcal{E}^{\mathcal{C}}$ having a Δ_0–definable graph. Then $\mathcal{E}^{\mathcal{C}} = \mathbf{C}(\mathcal{F}_0)$. By the proof of part (3) of Theorem 4, each function in $\mathcal{E}^{\mathcal{C}}$ is bounded by a function in $\mathcal{C}$. Since $\mathcal{E}^{\mathcal{C}}$ is a rudimentary F–algebra, by Lemma 1, $\mathcal{F}_0 \subseteq \mathcal{C}$ and, as a consequence, $\mathcal{E}^{\mathcal{C}} = \mathbf{C}(\mathcal{F}_0) \subseteq \mathcal{C} \cap \mathcal{PR}$. On the other hand, by Theorem 4, $\mathcal{C} \cap \mathcal{PR} \subseteq \mathcal{E}^{\mathcal{C}}$. This proves **(2)**. □

Corollary 2. *The following statements are equivalent:*

1. $\mathcal{E}^2$ *is a* Δ_0*–generated F–algebra.*
2. $\mathcal{M}^2 = \mathcal{E}^2.$
3. *There exists an extension of* $\mathbf{I}\Delta_0$, $\mathbf{T}$, *such that* $\mathcal{E}^2 = \mathcal{R}(\mathbf{T})$.

Now we give a characterization of $\mathcal{R}(\mathbf{I}\Delta_1(\mathbf{T}))$ in terms of $\mathcal{C}$–bounded recursion.

Theorem 6. *Let* $\mathbf{T}$ *be a sound extension of* $\mathbf{I}\Delta_0 + \mathbf{exp}$ *and* $\mathcal{C} = \mathcal{R}(\mathbf{T})$. *If* $\mathcal{C}$ *is closed under bounded minimization, then*

$$\mathcal{R}(\mathbf{I}\Delta_1(\mathbf{T})) = \mathcal{R}(\mathbf{L}\Delta_1(\mathbf{T})) = \mathcal{E}^{\mathcal{C}} = \mathcal{C} \cap \mathcal{PR}.$$

Proof. Without loss of generality we can assume that $\mathbf{T}$ is Π_2 axiomatized.

First we prove the result for $\mathbf{T}$ satisfying $\mathbf{I}\Sigma_1 \Longrightarrow \mathbf{T}$. Then $\mathcal{C} = \mathcal{C} \cap \mathcal{PR}$. By Proposition 4, $\mathbf{L}\Delta_1(\mathbf{T}) \Longrightarrow \mathbf{I}\Delta_1(\mathbf{T}) \Longrightarrow \mathbf{T}$; hence, $\mathbf{T}+\mathbf{L}\Delta_1(\mathbf{T}) \Longleftrightarrow \mathbf{L}\Delta_1(\mathbf{T})$. By Lemma 3, $\mathbf{T}+\mathbf{L}\Delta_1(\mathbf{T}) \Longleftrightarrow \mathbf{T}+\mathbf{B}^*\Delta_1(\mathbf{T})$, so $\mathcal{R}(\mathbf{L}\Delta_1(\mathbf{T})) = \mathcal{R}(\mathbf{T}+\mathbf{B}^*\Delta_1(\mathbf{T}))$. In [9] (see remark 2.8), it is proved that $\mathbf{T} + \mathbf{B}^*\Delta_1(\mathbf{T}) \Longleftrightarrow [\mathbf{T}, \Sigma_1\text{–CR}]$ (the closure of $\mathbf{T}$ under unnested applications of Σ_1–collection rule). Therefore, by corollary 5.6 in [2], since $\mathbf{T} \vdash \mathbf{exp}$ we get

$$\mathcal{R}(\mathbf{L}\Delta_1(\mathbf{T})) = \mathcal{R}([\mathbf{T}, \Sigma_1\text{–CR}]) = \mathbf{E}(\mathcal{C}).$$

By Theorem 4–(3), $\mathcal{E}^{\mathcal{C}} = \mathbf{E}(\mathcal{C} \cap \mathcal{PR})$; so, $\mathcal{R}(\mathbf{L}\Delta_1(\mathbf{T})) = \mathcal{E}^{\mathcal{C}}$, since $\mathcal{C} = \mathcal{C} \cap \mathcal{PR}$. As a consequence, $\mathcal{E}^{\mathcal{C}}$ is Δ_0–generated and, by Theorem 5, $\mathcal{E}^{\mathcal{C}} = \mathcal{C}$. Now the result follows from the chain of inclusions below:

$$\mathcal{R}(\mathbf{L}\Delta_1(\mathbf{T})) = \mathcal{E}^{\mathcal{C}} = \mathcal{C} \subseteq \mathcal{R}(\mathbf{I}\Delta_1(\mathbf{T})) \subseteq \mathcal{R}(\mathbf{L}\Delta_1(\mathbf{T})).$$

Let us prove the general case. Let $\mathbf{T}^I$ be the theory $\mathbf{I}\Sigma_1 \vee \mathbf{T}$. By Lemma 5, $\mathcal{R}(\mathbf{T}^I) = \mathcal{R}(\mathbf{T}) \cap \mathcal{R}(\mathbf{I}\Sigma_1)$. Since $\mathbf{I}\Sigma_1 \Longrightarrow \mathbf{T}^I \Longrightarrow \mathbf{I}\Delta_0 + \mathbf{exp}$, by previous case $\mathcal{R}(\mathbf{L}\Delta_1(\mathbf{T}^I)) = \mathcal{E}^{\mathcal{R}(\mathbf{T}^I)} = \mathcal{R}(\mathbf{T}^I)$, and by Theorem 4, $\mathcal{E}^{\mathcal{R}(\mathbf{T}^I)} = \mathcal{E}^{\mathcal{C} \cap \mathcal{PR}} = \mathcal{E}^{\mathcal{C}}$. So, $\mathcal{E}^{\mathcal{C}}$ is Δ_0–generated and, by Theorem 5, $\mathcal{E}^{\mathcal{C}} = \mathcal{C} \cap \mathcal{PR}$. By Theorem 3,

$$\mathcal{R}(\mathbf{L}\Delta_1(\mathbf{T})) \subseteq \mathcal{E}^{\mathcal{C}} = \mathcal{C} \cap \mathcal{PR} \subseteq \mathcal{R}(\mathbf{I}\Delta_1(\mathbf{T})) \subseteq \mathcal{R}(\mathbf{L}\Delta_1(\mathbf{T})).$$

This concludes the proof of the theorem. □

The hypothesis on $\mathcal{C}$ in Theorem 6, namely, $\mathcal{C}$ is closed under $\mu_\leq$, is equivalent to the existence of a theory $\mathbf{T}'$ such that $\mathcal{C} = \mathcal{R}(\mathbf{T}')$ and $\mathbf{T}'$ extends $\mathbf{L}\Delta_1(\mathbf{T}')$.

Below we discuss if this hypothesis can be weakened. This is related to the problem on the equivalence between $\mathbf{UI}\Delta_1$ and $\mathbf{B}\Sigma_1^-$. Here, $\mathbf{B}\Sigma_1^-$ denotes parameter free Σ_1–collection (see [9] for a deeper background on these theories).

First of all, let us observe that the hypothesis cannot be omitted. In [3] Beklemishev obtains $f \in \mathcal{E}^4$ ($\subseteq \mathcal{PR}$) such that $\mathcal{C} = \mathbf{C}(\mathcal{E}^3 \cup \{f\})$ is not closed under bounded recursion. Let $\mathbf{T}$ be the theory given by $\mathbf{I}\Delta_0 + \exp +$ "f is total". Then $\mathcal{R}(\mathbf{T}) = \mathcal{C}$ and, since $\mathbf{I}\Sigma_1 \Longrightarrow \mathbf{T}$, as in the first part of the proof of Theorem 6, $\mathcal{R}(\mathbf{L}\Delta_1(\mathbf{T})) = \mathbf{E}(\mathcal{C})$. So, $\mathcal{C} = \mathcal{C} \cap \mathcal{PR} \neq \mathcal{R}(\mathbf{L}\Delta_1(\mathbf{T}))$.

A suitable hypothesis on $\mathcal{C}$ to be used in Theorems 5 and 6 instead of the closure under bounded minimization is the following one:

(IC) There exists a theory $\mathbf{T}$ such that $\mathcal{C} = \mathcal{R}(\mathbf{T})$ and $\mathbf{T}$ has Δ_1–induction.

Observe that if Theorem 5 holds under hypothesis **(IC)**, so does Theorem 6. Next lemma allows us to avoid using Theorem 4–(3) in the proof of Theorem 6.

Lemma 9. *Let $\mathbf{T}$ be a sound Π_2–axiomatized extension of $\mathbf{I}\Delta_0 + \exp$. Let $\mathcal{C} = \mathcal{R}(\mathbf{T})$. Then $\mathbf{E}(\mathcal{C}) = \mathbf{C}(\mathcal{C} \cup \mathcal{E}^{\mathcal{C}})$.*

Proof. By the first part of the proof of Theorem 6, $\mathbf{E}(\mathcal{C}) = \mathcal{R}(\mathbf{T} + \mathbf{L}\Delta_1(\mathbf{T}))$. Let $\mathbf{T}_{\mathcal{C}}$ be the theory obtained by extending $\mathcal{C}$–$\mathbf{BRA}$ as follows:

For each formula $\varphi(\vec{x}, y) \in \Delta_0$ such that $\mathbf{T} \vdash \forall \vec{x} \exists y\, \varphi(\vec{x}, y)$, we add a new symbol function $\mathbf{f}_\varphi$ and take as an axiom the formula

$$\mathbf{f}_\varphi(\vec{x}) = y \leftrightarrow \varphi(\vec{x}, y) \wedge \forall z < y\, \neg\varphi(\vec{x}, z).$$

Since each Δ_0–formula is equivalent in $\mathcal{C}$–$\mathbf{BRA}$ to an open formula, $\mathbf{T}_{\mathcal{C}}$ is a universal theory and supports definition by cases. Then, by a standard Herbrand analysis we get that $\mathcal{R}(\mathbf{T}_{\mathcal{C}}) = \mathbf{C}(\mathcal{C} \cup \mathcal{E}^{\mathcal{C}})$.

Moreover, every $\exists_2$–closed model of $\mathbf{T}_{\mathcal{C}}$ is a model of $\mathbf{T} + \mathbf{B}\Sigma_1$, since it is a $\exists_2$–closed model of $\mathcal{C}$–$\mathbf{BRA}$ (Lemma 8). So, as in Theorem 3, we get that for each formula $\theta \in \Pi_2$,

$$\mathbf{T} + \mathbf{B}\Sigma_1 \vdash \theta \Longrightarrow \mathbf{T}_{\mathcal{C}} \vdash \theta.$$

Since $\mathbf{T} \vdash \exp$, then $\mathbf{E}(\mathcal{C}) = \mathcal{R}(\mathbf{T} + \mathbf{B}\Sigma_1) \subseteq \mathcal{R}(\mathbf{T}_{\mathcal{C}}) = \mathbf{C}(\mathcal{C} \cup \mathcal{E}^{\mathcal{C}}) \subseteq \mathbf{E}(\mathcal{C})$. □

We conclude studying the equivalence between $\mathbf{UI}\Delta_1$ and $\mathbf{B}\Sigma_1^-$.

Proposition 7. *Assume that Theorem 5 holds under hypothesis (IC). Let $\mathbf{T}$ be a Π_2–axiomatized extension of $\mathbf{Th}_{\Pi_1}(\mathcal{N}) + \exp$. If $\mathbf{T}$ has Δ_1–induction then $\mathbf{T}$ extends $\mathbf{L}\Delta_1(\mathbf{T})$.*

Proof. As noticed in the proof of Lemma 9, $\mathcal{R}(\mathbf{T} + \mathbf{L}\Delta_1(\mathbf{T})) = \mathbf{E}(\mathcal{C})$. Moreover, by Lemma 9, $\mathbf{E}(\mathcal{C}) = \mathbf{C}(\mathcal{C} \cup \mathcal{E}^{\mathcal{C}})$ and by Theorem 6, $\mathcal{E}^{\mathcal{C}} = \mathcal{R}(\mathbf{I}\Delta_1(\mathbf{T}))$. So,

$$\mathbf{E}(\mathcal{C}) = \mathbf{C}(\mathcal{C} \cup \mathcal{E}^{\mathcal{C}}) \subseteq \mathcal{R}(\mathbf{T} + \mathbf{I}\Delta_1(\mathbf{T})) \subseteq \mathcal{R}(\mathbf{T} + \mathbf{L}\Delta_1(\mathbf{T})) = \mathbf{E}(\mathcal{C}).$$

Therefore, $\mathcal{R}(\mathbf{T}) = \mathcal{R}(\mathbf{T} + \mathbf{L}\Delta_1(\mathbf{T}))$, since $\mathbf{T}$ has Δ_1–induction. By Lemma 4, $\mathbf{T} + \mathbf{L}\Delta_1(\mathbf{T})$ is Π_2 axiomatizable. Hence, by Proposition 1, $\mathbf{T} \Longleftrightarrow \mathbf{T} + \mathbf{L}\Delta_1(\mathbf{T})$ (recall that $\mathbf{T}$ extends $\mathbf{Th}_{\Pi_1}(\mathcal{N})$). In particular, $\mathbf{T} \Longrightarrow \mathbf{L}\Delta_1(\mathbf{T})$. □

Theorem 7. *Assume that Theorem 5 holds under hypothesis* **(IC)**. *Then*

$$\mathbf{B}\Sigma_1^- + \mathbf{Th}_{\Pi_1}(\mathcal{N}) + \mathbf{exp} \quad \Longleftrightarrow \quad \mathbf{UI}\Delta_1 + \mathbf{Th}_{\Pi_1}(\mathcal{N}) + \mathbf{exp}.$$

Proof. It is known that $\mathbf{B}\Sigma_1^- \Longrightarrow \mathbf{UI}\Delta_1$; so, we only prove that

$$\mathbf{UI}\Delta_1 + \mathbf{Th}_{\Pi_1}(\mathcal{N}) + \mathbf{exp} \Longrightarrow \mathbf{B}\Sigma_1^- + \mathbf{Th}_{\Pi_1}(\mathcal{N}) + \mathbf{exp}.$$

Let $\mathfrak{A} \models \mathbf{UI}\Delta_1 + \mathbf{Th}_{\Pi_1}(\mathcal{N}) + \mathbf{exp}$ and $\mathbf{T} = \mathbf{Th}_{\Pi_2}(\mathfrak{A})$. Then it is easy to check that $\mathbf{T}$ has Δ_1–induction and extends $\mathbf{Th}_{\Pi_1}(\mathcal{N}) + \mathbf{exp}$. By Proposition 7, $\mathbf{T}$ extends $\mathbf{L}\Delta_1(\mathbf{T})$. As a consequence, $\mathbf{T}$ extends $\mathbf{B}^*\Delta_1(\mathbf{T})$, since $\mathbf{L}\Delta_1(\mathbf{T}) \Longrightarrow \mathbf{B}^*\Delta_1(\mathbf{T})$. From this it follows that $\mathfrak{A} \models \mathbf{B}\Sigma_1^-$ (see remark 2.5.3 in [9]). □

References

1. Avigad, J. Saturated models of universal theories. *Annals of Pure and Applied Logic*, 118:219–234, 2002.
2. Beklemishev, L. D. Induction rules, reflection principles, and provably recursive functions. *Annals of Pure and Applied Logic*, 85:193–242, 1997.
3. Beklemishev, L. D. A proof-theoretic analysis of collection. *Archive for Mathematical Logic*, 37:275–296, 1998.
4. Beklemishev, L. D. Open Least Element Principle and Bounded Query Computation. In *Computer Science Logic, CSL'99*. LNCS 1683, pages 389–404. Springer–Verlag, 1999.
5. Beklemishev, L. D. On the induction schema for decidable predicates. *The Journal of Symbolic Logic*, 68(1):17–34, 2003.
6. Clote, P. Computation Models and Function Algebras. In *Handbook of Computability Theory*, pages 589–681. North-Holland, 1999.
7. Clote, P.; Krajíček, J. Open Problems. In *Arithmetic, Proof Theory and Computational Complexity*, pages 1–19. Clarendon Press, Oxford 1993.
8. Cordón Franco, A.; Fernández Margarit, A.; Lara Martín, F. F. On the quantifier complexity of $\Delta_{n+1}(\mathbf{T})$–induction. *Archive for Mathematical Logic*, 43:371-398, 2004.
9. Fernández Margarit, A.; Lara Martín, F. F. Induction, minimization and collection for $\Delta_{n+1}(\mathbf{T})$–formulas. *Archive for Mathematical Logic*, 43:505-541, 2004.
10. Hájek, P.; Pudlak, P. *Metamathematics of First-Order Arithmetic*. Springer-Verlag, 1993.
11. Kaye, R. *Models of Peano Arithmetic*. Clarendon Press, Oxford, 1991.
12. Krajíček, J. *Bounded arithmetic, Propositional logic, and Complexity theory*. Cambridge University Press, Cambridge, 1995.
13. Rose, H. E. *Subrecursion: Functions and hierarchies*. Clarendon Press, Oxford, 1984.
14. Sieg, W. Herbrand analyses. *Archive for Mathematical Logic*, 30:409–441, 1991.
15. Slaman, T. Σ_n–bounding and Δ_n–induction. Proceedings of the American Mathematical Society, 132(8):2449–2456, 2004.
16. Takeuti, G. Grzegorcyk's hierarchy and $Iep\Sigma_1$. *The Journal of Symbolic Logic*, 59(4):1274–1284, 1994.
17. Wilkie, A.; Paris, J.B. On the scheme of induction for bounded arithmetic formulas. *Annals of Pure and Applied Logic* 35:261–302, 1987.

Logical Characterizations of PSPACE

David Richerby[*]

University of Cambridge Computer Laboratory,
William Gates Building,
J.J. Thomson Avenue,
Cambridge, CB3 0FD, UK
David.Richerby@cl.cam.ac.uk

Abstract. We present two, quite different, logical characterizations of the computational complexity class **PSPACE** on unordered, finite relational structures. The first of these, the closure of second-order logic under the formation of partial fixed points is well-known in the folklore but does not seem to be in the literature. The second, the closure of first-order logic under taking partial fixed points and under an operator for nondeterministic choice, is novel. We also present syntactic normal forms for the two logics and compare the second with other choice-based fixed-point logics found in the literature.

Keywords. Finite model theory, descriptive complexity, choice operators, partial fixed points.

1 Introduction

Fixed-point extensions of first-order logic have played a central rôle in descriptive complexity theory: on ordered structures, FO(IFP) captures **P** [Imm86, Var82], FO(PFP) captures **PSPACE** [AV89] and extensions of first-order logic with transitive closure operators, which can be seen as fragments of FO(IFP), capture **L** and **NL** [Imm87]. All of these characterizations depend on the structures being ordered: without the ordering, none of these logics can define even simple counting queries.

The requirement for an ordering is unsatisfactory as it allows queries such as, 'there is an edge from the first vertex of the graph to the second,' which is as much a property of the ordering as it is of the graph. Restricting to formulae that give the same result for any ordering does not help as the 'order-invariant' fragment of each of the logics mentioned above is undecidable.

One alternative to requiring ordered structures is to move to second-order logic, SO, where the ordering relation can be quantified into existence. This does not seem to help with capturing **P** or **PSPACE**, as even the existential fragment of SO already captures **NP** [Fag74] and SO itself only captures the polynomial hierarchy [Sto76]. Here, we show that the combination of second-order quantification and partial fixed points captures **PSPACE**. This characterization appears in the folklore but there does not seem to be a proof in the literature.

---

[*] Research supported by EPSRC grant GR/S06721.

J. Marcinkowski and A. Tarlecki (Eds.): CSL 2004, LNCS 3210, pp. 370–384, 2004.

Another alternative is to use a choice operator. Several such mechanisms have appeared in the literature — see, for example, [BG00, DR03, GH98]. Of these, the most general is Blass and Gurevich's δ operator, which forms terms $\delta x\,\varphi$ with the semantics, 'choose an x that satisfies φ.' Choice and ordering are closely related: given an ordering, choice can be simulated by taking the least element with a property; given choice and iteration, an ordering can be built up by choosing an element to be first and iterating to extend the ordering until it includes every element. Formulae with choice are nondeterministic: the meaning of a formula may depend on which choices are taken during its evaluation, though the set of possible answers depends only on the formula and the structure on which it is evaluated. We investigate $\mathrm{FO}(\mathrm{PFP}, \delta)$, the combination of partial fixed-point logic with choice, to obtain our second characterization of **PSPACE** on unordered structures. We also provide normal forms for this logic and for second-order logic with partial fixed-points.

All structures in this paper are finite and all vocabularies are finite and purely relational, though constant symbols are omitted for notational convenience only. We write $|\mathfrak{A}|$ for the universe of structure $\mathfrak{A}$ and $\|\mathfrak{A}\|$ for the size of $|\mathfrak{A}|$.

2 Partial Fixed Points

In this section, we give a brief introduction to partial fixed-point logics, which were originally introduced by Abiteboul and Vianu [AV89]; see also [EF99]. Partial fixed points are of particular interest in descriptive complexity.

Theorem 1 ([AV89]). FO(PFP) *captures* **PSPACE** *on the class of ordered, finite relational structures.*

Let S be a set and let f be a map $\mathcal{P}(S) \to \mathcal{P}(S)$. A fixed point of f is some $s \subseteq S$ such that $f(s) = s$. Consider the sequence of stages $\langle s^i \rangle_{i \geqslant 0}$, where $s^0 = \emptyset$ and $s^{i+1} = f(s^i)$. If this sequence eventually becomes stationary, i.e., if there is an n such that $s^i = s^n$ for all $i > n$, set $\mathbf{pfp}\,f = s^n$. Clearly, there is no reason to suppose that a general map f will have a fixed point at all, let alone one that can be reached by an iteration like this. If the sequence does not eventually become stationary, set $\mathbf{pfp}\,f = \emptyset$. We shall refer to $\mathbf{pfp}\,f$ as the *partial fixed-point* of f, even though there is no guarantee that it really is a fixed point.

Let $\mathcal{L}$ be a logic and let φ be an $\mathcal{L}$-formula with a distinguished r-ary relation symbol X, in which we consider the variables $x_1 \ldots x_r$ to be free. On a structure $\mathfrak{A}$, we may associate with φ a map $f_\varphi^{\mathfrak{A}} : \mathcal{P}(|\mathfrak{A}|^r) \to \mathcal{P}(|\mathfrak{A}|^r)$ with $f_\varphi^{\mathfrak{A}}(X) = \{\,\bar{a} : (\mathfrak{A}, \bar{a}) \vDash \varphi\,\}$. $\mathcal{L}(\mathrm{PFP})$ is the closure of $\mathcal{L}$ under the rule that, if φ is a formula, X is an r-ary relation symbol and $\bar{t}$ is an r-tuple of terms, $\Phi \equiv (\mathbf{pfp}_{X,\,x_1 \ldots x_r} \varphi)(\bar{t})$ is a formula. Within the expression $\mathbf{pfp}_{X,\,\bar{x}} \varphi$, the variables in $\bar{x}$ are bound. If $\bar{a}$ is an interpretation for the free variables in Φ, we write $(\mathfrak{A}, \bar{a}) \vDash \Phi$ if, and only if, $\langle\, t_1(\bar{a}) \ldots t_r(\bar{a})\,\rangle \in \mathbf{pfp}\,f_\varphi^{\mathfrak{A}}$, where $t_i(\bar{a})$ denotes the value of the term t_i under given interpretation for the free variables.

Theorem 2. *Each formula of* FO(PFP) *is equivalent to a formula of the form* $\exists y\,(\mathbf{pfp}_{X,\,\bar{x}} \varphi)(y \ldots y)$, *where* $\varphi \in$ FO.

We can also define simultaneous partial fixed points. For $i \in [1, k]$, let f_i be a map $S^{r_1} \times \cdots \times S^{r_k} \to S^{r_i}$ and define the sequence of stages

$$\bar{R}^0 = \langle \emptyset, \ldots, \emptyset \rangle$$
$$\bar{R}^{i+1} = \langle f_1(\bar{R}^i), \ldots, f_k(\bar{R}^i) \rangle.$$

If there is an n such that $\bar{R}^i = \bar{R}^n$ for all $i > n$, we set $\mathbf{pfp}\,(f_1, \ldots, f_k) = R_1^n$; otherwise, it is $\emptyset$ as before. We denote by $\mathcal{L}$(S-PFP) the closure of the logic $\mathcal{L}$ under simultaneous partial fixed points. Allowing such simultaneous definitions does not increase the power of reasonable logics, i.e., logics that are regular in the sense of Ebbinghaus [Ebb85]. In particular, this result applies to SO(PFP) and we shall see in Section 5 that it also goes through for partial fixed points of first-order formulae with choice. Proof is by coding a tuple of relations into a single relation of increased arity; this relation will reach a fixed point if, and only if, all of its component relations do.

Theorem 3. *Let $\mathcal{L}$ be a regular logic. $\mathcal{L}$(PFP) = $\mathcal{L}$(S-PFP). Further, any simultaneous induction $\Phi \equiv \mathbf{pfp}_{X_1, \ldots, X_n, \bar{x}}(\varphi_1, \ldots, \varphi_n)(\bar{t})$ is equivalent to a simple induction of the form $\widehat{\Phi} \equiv \exists y\,(\mathbf{pfp}_{X, \bar{u}\bar{x}}\,\varphi)(y \ldots y)$.*

3 Second-Order Logic and Partial Fixed Points

We now consider SO(PFP), the closure of second-order logic under the partial fixed-point operator. The following result appears as an exercise in [EF99] and is well-known in the folklore but there seems to be no proof in the literature.

Theorem 4. *SO(PFP) captures **PSPACE** on the class of unordered finite relational structures.*

Proof. (**PSPACE** $\leqslant$ SO(PFP)) By Theorem 1, FO(PFP) captures **PSPACE** on ordered structures. It follows that any **PSPACE** property of unordered structures is defined by a sentence of the form $\exists R\,(\theta \wedge \varphi)$, where R is a new binary relation symbol, θ asserts that R is interpreted by a linear order and φ is a formula of FO(PFP).

(SO(PFP) $\leqslant$ **PSPACE**) By induction on the structure of formulae. FO $<$ **L** and the case of the partial fixed-point operator is standard (see, e.g., [EF99]). The only remaining case is $\exists R\,\varphi$. Assuming inductively that we have a polynomial-space algorithm for φ, we can determine whether $\mathfrak{A} \vDash \exists R\,\varphi$ by checking in turn for each interpretation X of R, whether $(\mathfrak{A}, X) \vDash \varphi$. Interpretations for R can be written down in at most $\|\mathfrak{A}\|^{\mathrm{ar}(R)}$ bits. $\square$

Corollary 5. *Let $\Phi \in$ SO(PFP)$[\sigma]$. There is a formula $\widehat{\Phi} \equiv \exists R\,\exists y\,(\mathbf{pfp}_{X, \bar{x}}\,\varphi)$ $(y \ldots y)$, where R is a new binary relation symbol and φ is first-order such that, on all $\mathfrak{A} \in$ STRUC$[\sigma]$, $\widehat{\Phi}^{\mathfrak{A}} = \Phi^{\mathfrak{A}}$.*

We cannot drop the requirement that R be at least binary in Corollary 5: the result fails for any monadic existential second-order quantifier prefix.

Theorem 6. *No formula of* SO(PFP) *of the form* $\exists R_1 \ldots \exists R_n \exists y\, (\mathbf{pfp}_{X,\bar{x}}\, \varphi)$ $(y \ldots y)$, *with* $\varphi \in$ FO *and the* R_i *unary, defines the class of finite, even sets.*

Proof. We use Fagin's game for the existential fragment of monadic second-order logic [Fag75]. Fix n and k and let $N = k2^n$. Let $\mathfrak{A} = [1, N]$ and $\mathfrak{B} = [1, N+1]$. Player I chooses $R_1^{\mathfrak{A}}, \ldots, R_n^{\mathfrak{A}} \subseteq |\mathfrak{A}|$. Let $c(a, \mathfrak{A}) = \{\, i : a \in R_i^{\mathfrak{A}} \,\}$. By the pigeon-hole principle, there must be at least one colour $\kappa \subseteq [1, n]$ such that $\|\{\, a : c(a, \mathfrak{A}) = \kappa \,\}\| \geqslant k$. Player II sets the $R_i^{\mathfrak{B}}$ so that, for $j \in [1, N]$, $c(j, \mathfrak{B}) = c(j, \mathfrak{A})$ and $c(N+1, \mathfrak{B}) = \kappa$.

Player II now wins the k-pebble Ehrenfeucht-Fraïssé game on $(\mathfrak{A}, R_1^{\mathfrak{A}}, \ldots, R_n^{\mathfrak{A}})$ and $(\mathfrak{B}, R_1^{\mathfrak{B}}, \ldots, R_n^{\mathfrak{B}})$. The only difference between the structures is that the latter has an extra element but Player I can never expose this difference as both structures have at least k nodes of that colour. It follows that no formula $\exists R_1 \ldots \exists R_n\, \varphi$, where $\varphi \in \mathcal{L}_{\infty\omega}^{\omega}$ (the finite-variable fragment of infinitary first-order logic), defines the class of sets of even cardinality. The result follows as any formula of FO(PFP) is equivalent to one of $\mathcal{L}_{\infty\omega}^{\omega}$ [KV92]. □

An alternative characterization of **PSPACE** by an extension of second-order logic is given by Harel and Peleg [HP84]. Second-order formulae formulae with free relation variables can be seen as defining 'super-relations', i.e., relations on relations (the relation sets of the next section are a special case of this). Harel and Peleg show that **PSPACE** is captured by second-order logic equipped with an operator for forming transitive closures of super-relations.

4   Nondeterministic Choice

We now consider first-order logic with choice. Blass and Gurevich introduce two choice operators, δ and δ', in [BG00], forming terms $\delta x\, \varphi$ with the effect of choosing an x that satisfies φ. The choices made by Hilbert's better-known ε choice operator [HB39] (also discussed by Blass and Gurevich) are deterministic, defined by a choice function f such that choosing from a set S will always give element $f(S)$; in contrast, δ and δ' are *nondeterministic*: choosing twice from the same set is not guaranteed to give the same result.

Definition 7. *Let* σ *be a vocabulary. The formulae of* FO(δ)[σ] *are the least set containing* FO[σ] *and closed under the rule that, if* φ *is a formula and* x *is a variable,* $\delta x\, \varphi$ *is a term.*

Choosing an x that satisfies φ invites the question of what to do when there is no such x in some structure $\mathfrak{A}$. In this case, $\delta x\, \varphi$ evaluates to some 'default element' of $|\mathfrak{A}|$. To define the semantics of formulae on ordinary structures, which do not have default elements, we borrow the behaviour of the δ' operator in [BG00], namely that choosing from the empty set is the same as choosing from the whole universe, while retaining the unbiased choice of δ. This operator is mentioned in [BG00] but not given a name: for simplicity of notation, we call it δ and will not again refer to the operator that requires default elements.

Blass and Gurevich give a three-valued semantics to the logic. Following our treatment of logics with choice in [DR03], we give a different semantics, in which formulae denote sets of relations, each the result of taking different choices. The semantics of terms and formulae are defined by mutual recursion; a term denotes a set of functions from the values of variables to the value of the term.

Definition 8. *The semantics of terms is given by:*

$$\llbracket x_i \rrbracket^{\mathfrak{A}} = \{\, \lambda \bar{x}.\, x_i \,\}$$

$$\llbracket \delta x_i\, \varphi \rrbracket^{\mathfrak{A}} = \bigcup_{R \in \llbracket \varphi \rrbracket^{\mathfrak{A}}} \{\, \lambda \bar{x}.\, a : a \in \pi_i(\bar{x}, R) \,\},$$

where $\pi_i(\bar{x}, R) = \{\, a : \bar{x}[a/x_i] \in R \,\}$, if this set is non-empty, and $|\mathfrak{A}|$, otherwise. The semantics of formulae is given by:

$$\llbracket t_1 = t_2 \rrbracket^{\mathfrak{A}} = \{\, \{\, \bar{a} : f_1(\bar{a}) = f_2(\bar{a}) \,\} : f_i \in \llbracket t_i \rrbracket^{\mathfrak{A}} \,\}$$

$$\llbracket R(t_1 \ldots t_n) \rrbracket^{\mathfrak{A}} = \{\, \{\, \bar{a} : \langle\, f_1(\bar{a}) \ldots f_n(\bar{a}) \,\rangle \in R^{\mathfrak{A}} \,\} : f_i \in \llbracket t_i \rrbracket^{\mathfrak{A}} \,\}$$

$$\llbracket \neg \varphi \rrbracket^{\mathfrak{A}} = \{\, |\mathfrak{A}|^k \setminus S : S \in \llbracket \varphi \rrbracket^{\mathfrak{A}} \,\}$$

$$\llbracket \varphi \wedge \psi \rrbracket^{\mathfrak{A}} = \{\, S_1 \cap S_2 : S_1 \in \llbracket \varphi \rrbracket^{\mathfrak{A}}, S_2 \in \llbracket \psi \rrbracket^{\mathfrak{A}} \,\}$$

$$\llbracket \exists x_i\, \varphi \rrbracket^{\mathfrak{A}} = \{\, \{\, \bar{a} : \bar{a}[a/a_i] \in S \text{ for some } a \,\} : S \in \llbracket \varphi \rrbracket^{\mathfrak{A}} \,\}$$

Our semantics is subtly different from that of Blass and Gurevich, in which a term denotes a function from the values of variables to the set of possible values of the term and a formula denotes a 'nondeterministic relation'. An ordinary relation can be seen as a function from tuples to truth values, with $R = \{\, \bar{a} : f(\bar{a}) = \text{true} \,\}$; a nondeterministic relation can be seen as a function from tuples to non-empty sets of truth values. Intuitively, if $f(\bar{a}) = \{\, \text{true} \,\}$, $\bar{a} \in R$ for all evaluations; if $f(\bar{a}) = \{\, \text{true, false} \,\}$, $\bar{a} \in R$ for some evaluations and not for others; and, if $f(\bar{a}) = \{\, \text{false} \,\}$, $\bar{a} \notin R$ for all evaluations.

The difference in semantics of terms (set of functions versus set-valued function) is insignificant and is for notational convenience only. The change from a three-valued semantics to a semantics based on sets of relations gives much greater flexibility: in particular, the three-valued semantics can be recovered from it by associating with a set of relations R the nondeterministic relation corresponding the function f given by true $\in f(\bar{a}) \Leftrightarrow \bar{a} \in \bigcup R$ and false $\in f(\bar{a}) \Leftrightarrow \bar{a} \notin \bigcap R$. We showed in [DR03] a close correspondence between the relation-set semantics of fixed-point logics with choice and the computation paths of nondeterministic Turing machines, which, we feel, makes our semantics better suited to descriptive complexity. Further, note that the semantics of **ifp** in FO(IFP, δ) is defined in [BG00] by using a relation set as an intermediate stage, from which a three-valued relation is produced.

Definition 9. $\varphi \in \text{FO}(\delta)$ *is FO-approximable if there are first-order formulae* $\varphi_\cup$ *and* $\varphi_\cap$ *such that*

$$\varphi_\cup^{\mathfrak{A}} = \bigcup \llbracket \varphi \rrbracket^{\mathfrak{A}} \qquad \text{and} \qquad \varphi_\cap^{\mathfrak{A}} = \bigcap \llbracket \varphi \rrbracket^{\mathfrak{A}}.$$

FO-approximability corresponds to the FO-definability of ranges in [BG00]. The following lemma and its two corollaries appear there, for Blass and Gurevich's semantics; though our presentation and proofs are rather different as we are working in a relational, rather than functional, setting. Lemma 13 is new. All results in this section carry through to $\mathcal{L}^\omega_{\infty\omega}(\delta)$, the extension of $\mathcal{L}^\omega_{\infty\omega}$ by δ-terms.

Lemma 10. *Every $\varphi \in$ FO(δ) is FO-approximable.*

Proof. By induction on the structure of φ.

$$\big(R(t_1 \dots t_n)\big)_\cup \equiv \exists y_1 \dots y_n \big(\bigwedge_{i \in [1,n]} \theta_i \wedge R(\bar{y})\big)$$

$$\big(R(t_1 \dots t_n)\big)_\cap \equiv \forall y_1 \dots y_n \big((\bigwedge_{i \in [1,n]} \theta_i) \to R(\bar{y})\big),$$

where the y_i are variables that do not occur in any of the t_i and $\theta_i \equiv \big(\varphi_i[y_i/x] \vee \forall x \, \neg\varphi_i\big)_\cup$ if $t_i \equiv \delta x \, \varphi_i$, and $\theta_i \equiv y_i = v$ if $t_i \equiv v$.

$$(\neg\varphi)_\cup \equiv \neg\,\varphi_\cap \qquad\qquad (\neg\varphi)_\cap \equiv \neg\,\varphi_\cup$$
$$(\varphi \wedge \psi)_\cup \equiv \varphi_\cup \wedge \psi_\cup \qquad (\varphi \wedge \psi)_\cap \equiv \varphi_\cap \wedge \psi_\cap$$
$$(\exists x \, \varphi)_\cup \equiv \exists x \, \varphi_\cup \qquad\quad (\exists x \, \varphi)_\cap \equiv \exists x \, \varphi_\cap \qquad\qquad \square$$

A related choice operator is Abiteboul and Vianu's witness operator, **W** (see, e.g., [ASV90]). **W** is a much more powerful choice operator than δ: it can be used to simulate existential second-order quantification and, hence, capture **NP** without the need for fixed-points.

Corollary 11. *Every formula $\varphi \in$ FO(δ) is equivalent to one in which every δ-term is of the form $\delta y \, \psi$, where ψ is first-order.*

Proof. It is immediate from Definition 8 that $[\![\delta y \, \psi]\!]^{\mathfrak{A}} = [\![\delta y \, \psi_\cup]\!]^{\mathfrak{A}}$. $\square$

We call a formula *deterministic* if, on every structure, it defines a single relation, i.e., $\|[\![\varphi]\!]^{\mathfrak{A}}\| = 1$ for every finite structure $\mathfrak{A}$ of appropriate vocabulary. Clearly, if φ is deterministic, $[\![\varphi]\!]^{\mathfrak{A}} = \{\varphi^{\mathfrak{A}}_\cup\} = \{\varphi^{\mathfrak{A}}_\cap\}$.

Corollary 12. *Any deterministic formula $\varphi \in$ FO(δ) is equivalent to a first-order formula.*

This is in contrast with FO(ε), where the corresponding notion to determinism is ε-*invariance*: a formula is ε-invariant if defines the same query irrespective of which choice function is used. Otto shows that there are formulae of FO(ε) that are ε-invariant over the class of finite structures but not equivalent to any first-order formula [Ott00], though any formula that is ε-invariant on the class of arbitrary structures is equivalent to some first-order formula [BG00].

While the deterministic fragment of FO(δ) has the same expressive power as ordinary first-order logic, it is not a decidable fragment. On the other hand, the syntax of first-order logic is clearly decidable so we can consider FO as a recursive syntax for deterministic FO(δ).

Lemma 13. *The set of formulae in* $\mathrm{FO}(\delta)[\sigma]$ *that are deterministic is undecidable if* σ *contains an at-least-binary relation symbol.*

Proof. Let ψ be a sentence of $\mathrm{FO}[\sigma]$ and let $\varphi(x) \equiv \neg\psi \wedge x = \delta y\,(y = y)$. φ is deterministic if, and only if, ψ is valid. However, by Trakhtenbrot's Theorem [Tra50], the set of valid first-order formulae over a vocabulary containing a relation symbol of arity at least two is undecidable. $\qquad\square$

5   Partial Fixed Points with Choice

We now turn our attention to the combination of $\mathrm{FO}(\delta)$ with partial fixed-points. To do this, we must redefine the partial fixed-point operator to take nondeterministic formulae as arguments.

Definition 14. *Let* $\mathfrak{A}$ *be a structure and* $F^{\mathfrak{A}} : \mathcal{P}(|\mathfrak{A}|^r) \to \mathcal{P}(\mathcal{P}(|\mathfrak{A}|^r))\backslash\{\,\emptyset\,\}$, *i.e., a map from* r-*ary relations over* $|\mathfrak{A}|$ *to non-empty sets of* r-*ary relations over* $|\mathfrak{A}|$. *Let* $T_F^{\mathfrak{A}}$ *be the least tree satisfying:*

- *the root is labelled* $\emptyset$;
- *a node labelled* S *has a child labelled* S' *for each* $S' \in F^{\mathfrak{A}}(S)$.

   Let $P = \langle\, P^i \,\rangle_{i\geqslant 0}$ *be a maximal path in* $T_F^{\mathfrak{A}}$ *and let* $\Lambda P = P^n$ *if* n *is minimal such that* $P^i = P^n$ *for all* $i \geqslant n$ *and* $\Lambda P = \emptyset$ *if there is no such* n. $\mathbf{pfp}\,F^{\mathfrak{A}} = \{\,\Lambda P : P$ *labels a maximal path in* $T_F^{\mathfrak{A}}\,\}$.

   Each path in the tree can be seen as a sequence of choices taken in evaluating the argument: this viewpoint is useful when informally describing the semantics of a formula.

Example 15. Let X be a binary relation symbol and F be the map defined by

$$\varphi(X, xy) \equiv X(xy) \vee \big[\big(\exists u\,\neg X(uu)\big) \wedge \big(y = \delta u\,\neg X(uu)\big) \wedge \big(x = y \vee X(xx)\big)\big].$$

   When evaluated on a structure $\mathfrak{A}$, $\mathbf{pfp}\,F$ defines the set of linear orders of $|\mathfrak{A}|$. Consider a single path in $T_F^{\mathfrak{A}}$: at each stage, if there are elements that do not yet appear in the ordering, one is nondeterministically chosen and becomes the new greatest element. The fixed point is reached when all elements are ordered.

   Although the $\mathbf{pfp}$ operator is defined in terms of infinite paths in infinite trees, $\mathbf{pfp}\,F$ is determined by a finite portion of the tree, as shown by the following lemma.

Lemma 16. *Let* $F^{\mathfrak{A}}$ *be a map as in Definition 14, let* P *be a maximal path in* $T_F^{\mathfrak{A}}$ *with nodes labelled* $\langle\, R^i \,\rangle_{i\geqslant 0}$ *and let* $n = 2^{\|\mathfrak{A}\|^r}$. *Either* $R^m = R^{m+1} \in \mathbf{pfp}\,F^{\mathfrak{A}}$ *for some* $m < n$ *or* $\emptyset \in \mathbf{pfp}\,F^{\mathfrak{A}}$.

Proof. Suppose $R^m = R^{m+1}$ for some $m < n$. $T_F^{\mathfrak{A}}$ must contain a path Q labelled $\langle\, R^0, \ldots, R^m, R^m, R^m, \ldots \,\rangle$, with $\Lambda Q = R^m$. (Of course, we may have $R^m = \emptyset$.)
   Conversely, suppose $R^m \neq R^{m+1}$ for all $m < n$. By the pigeon-hole principle, $R^i = R^j$ for some $i < j \leqslant n+1$. It follows that $T_F^{\mathfrak{A}}$ contains a path Q' labelled $\langle\, R^0, \ldots, R^j, R^{i+1}, \ldots, R^j, R^{i+1}, \ldots \,\rangle$, with $\Lambda Q' = \emptyset$. $\qquad\square$

Definition 17. *The formulae of* $\mathrm{FO}(\mathrm{PFP}, \delta)[\sigma]$ *are the least set that contains* $\mathrm{FO}(\delta)[\sigma]$ *and is closed under the rule for forming fixed-point formulae in Section 2. The semantics is that of* $\mathrm{FO}(\delta)$ *from Definition 8, plus the rule*

$$[\![(\mathbf{pfp}_{X, \bar{y}}\, \varphi)(t_1 \ldots t_r)]\!]^{\mathfrak{A}} =$$

$$\{\, \{\, \bar{a} : \langle\, f_1(\bar{a}) \ldots f_r(\bar{a})\, \rangle \in R \,\} : R \in \mathbf{pfp}\, F_{\varphi}^{\mathfrak{A}},\, f_i \in [\![t_i]\!]^{\mathfrak{A}}\, \}.$$

We do not describe $\mathrm{FO}(\mathrm{PFP}, \delta)$ as a logic: in our view, a logic consists of both a set of formulae and a satisfaction relation that connects formulae with the structures in which they are true. For *deterministic* formulae, i.e., those that define a single relation, the satisfaction relation is obvious: we may write $(\mathfrak{A}, \bar{a}) \vDash \varphi$ if, and only if, $\bar{a} \in R$, where $[\![\varphi]\!]^{\mathfrak{A}} = \{ R \}$. In general, though, formulae are not restricted to defining singleton sets of relations and there are many possible definitions of satisfaction for such formulae. For the time being, we shall consider formulae just in terms of defining nondeterministic queries and defer the consideration of satisfaction relations until Section 5.2.

In common with $\mathrm{FO}(\mathrm{PFP})$, we can allow parameterized definitions without increasing the expressive power of the logic. Let $\Phi \equiv (\mathbf{pfp}_{X, \bar{x}}\, \varphi)(\bar{t})$, where the free variables of φ are among $\bar{x}\bar{z}$. $[\![\Phi]\!]^{\mathfrak{A}} = \{\, \{\, \bar{b}\bar{a} : \bar{b} \in R \,\} : R \in [\![\Phi]\!]^{(\mathfrak{A}, \bar{a})}\, \}$.

Theorem 18. *Every application of the* **pfp** *operator is equivalent to one in which every free variable of the operand is bound.*

Proof. Let $\Phi \equiv (\mathbf{pfp}_{X, \bar{x}}\, \varphi)(\bar{t})$, where φ has free variables $\bar{x}\bar{z}$ and none of the variables in $\bar{z}$ appears bound in φ. For any $\mathfrak{A}$ of appropriate vocabulary, $[\![\Phi]\!]^{\mathfrak{A}} = [\![(\mathbf{pfp}_{Z, \bar{x}\bar{z}}\, \varphi^*)(\bar{t}\bar{z})]\!]^{\mathfrak{A}}$, where Z is a new relation symbol of arity $|\bar{x}\bar{z}|$ and φ^* is the result of replacing every subformula $X(\bar{u})$ in φ with $Z(\bar{u}\bar{z})$. □

We can also define simultaneous fixed points, much as in the deterministic case. Again, this does not increase expressive power.

Definition 19. *Let* $\mathfrak{A}$ *be a structure and, for* $i \in [1, n]$, *let* $F_i^{\mathfrak{A}}$ *be a map with domain* $\mathcal{P}(|\mathfrak{A}|^{r_1}) \times \cdots \times \mathcal{P}(|\mathfrak{A}|^{r_n})$ *and range* $\mathcal{P}(\mathcal{P}(|\mathfrak{A}|^{r_i})) \setminus \{\emptyset\}$. *Let* $T_{F_1, \ldots, F_n}^{\mathfrak{A}}$ *be the least tree satisfying:*

- *the root is labelled with the n-tuple* $\langle \emptyset, \ldots, \emptyset \rangle$;
- *a node labelled* $\bar{R} = \langle R_1, \ldots, R_n \rangle$ *has a child labelled* $\bar{S}$ *for each* $\bar{S} \in F_1^{\mathfrak{A}}(\bar{R}) \times \cdots \times F_n^{\mathfrak{A}}(\bar{R})$.

Let $P = \langle \bar{R} \rangle^{i \geqslant 0}$ *be a maximal path in* $T_{F_1, \ldots, F_n}^{\mathfrak{A}}$ *and let* $\Lambda P = R_1^n$ *if* n *is minimal such that* $\bar{R}^i = \bar{R}^n$ *for all* $i \geqslant n$ *and* $\Lambda P = \emptyset$ *if there is no such* n. $\mathbf{pfp}\,(F_1^{\mathfrak{A}}, \ldots, F_n^{\mathfrak{A}}) = \{\, \Lambda P : P$ *is a maximal path in* $T_{F_1, \ldots, F_n}^{\mathfrak{A}}\, \}.$

Theorem 20. *Let* $\Phi \in \mathrm{FO}(\mathrm{PFP}, \delta)$. *There is a formula* $\widehat{\Phi}$ *using no simultaneous inductions, such that* $[\![\widehat{\Phi}]\!]^{\mathfrak{A}} = [\![\Phi]\!]^{\mathfrak{A}}$ *for all* $\mathfrak{A}$ *of appropriate vocabulary. Further, we may take* $\widehat{\Phi} \equiv \exists y\, (\mathbf{pfp}_{X, \bar{x}}\, \varphi)(y \ldots y)$.

The proof is omitted as it is largely standard: the simultaneously-defined relations are coded into a single relation of greater arity. We do not include simultaneous definitions in the formal definition of the syntax and, when we write simultaneous inductions, they should be treated as an abbreviation for the equivalent simple induction.

A second consequence of the theorem is that we may, from this point, assume that the t_i in any formula $(\mathbf{pfp}_{X,\bar{x}}\,\varphi)(\bar{t})$ are variables and not δ-terms.

5.1   Normal Forms

Say that a formula of $\mathrm{FO}(\mathrm{PFP},\delta)$ is *totally defined* if, for all of its subformulae $(\mathbf{pfp}_{X,\bar{x}}\,\varphi)(\bar{y})$, every path in $T_\varphi^{\mathfrak{A}}$ leads to a fixed point, for every structure $\mathfrak{A}$ of appropriate vocabulary. That is, for every path $P = \langle R^i \rangle_{i\geqslant 0}$ in $T_\varphi^{\mathfrak{A}}$, there is an n such that $R^i = R^n$ for all $i \geqslant n$.

Theorem 21. *Each formula of* $\mathrm{FO}(\mathrm{PFP},\delta)$ *is equivalent to one that is totally defined.*

Proof. Let $\Phi \equiv (\mathbf{pfp}_{X,\bar{x}}\,\varphi)(\bar{y})$, with $\mathrm{ar}(X) = r$. Let $P = \langle R^i \rangle_{i\geqslant 0}$ be a path in $T_\varphi^{\mathfrak{A}}$. By Lemma 16, if $R^i = R^{i+1}$ for some $i \leqslant 2^{\|\mathfrak{A}\|^r}$ then $T_\varphi^{\mathfrak{A}}$ contains a path P' with $\Lambda P' = R^i$; if not, the tree contains a path P' with $\Lambda P' = \emptyset$.

We simulate Φ on a structure $\mathfrak{A}$ with a totally-defined formula $\widehat{\Phi}$, which simulates Φ for $n = 2^{\|\mathfrak{A}\|^r}$ stages. If, during that time, two successive stages have the same value, every subsequent stage of the simulation will also have that value; if not, then all stages after the nth will have value $\emptyset$. A new r-ary relation Y will be used to remember the value of X from the previous stage.

To perform the simulation, we first build up a linear order of $|\mathfrak{A}|$ in a new binary relation $\leqslant$. We shall abuse notation slightly and use the symbol $\leqslant$ to stand for the lexicographic ordering that $\leqslant$ induces on r-tuples as this induced ordering is FO-definable from $\leqslant$. Using this linear order, we can treat an r-ary relation C as an $\|\mathfrak{A}\|^r$-bit binary number: let $o(\bar{a}) = \|\{\,\bar{b} : \bar{b} < \bar{a}\,\}\|$ and associate C with the number $\sum_{\bar{c}\in C} 2^{o(\bar{c})}$. Given a relation C, the relation C' such that $n(C') = n(C) + 1 \pmod{2^{\|\mathfrak{A}\|^r}}$ is defined by the first-order formula

$$\theta_{\mathrm{succ}}(\bar{x}) \equiv C(\bar{x}) \leftrightarrow \exists\bar{y}\,\bigl(\bar{y} < \bar{x} \wedge \neg C(\bar{y})\bigr).$$

That is, the least unset bit becomes set, all bits below that become unset and all remaining bits are unaffected. (A similar technique for cycling through all possible relations with a partial fixed point is used in [Daw98].)

$$\widehat{\Phi} \equiv \mathbf{pfp}_{X,Y,\leqslant,C,\bar{x}}(\varphi_X, X(\bar{y}), \varphi_\leqslant, \varphi_C)(\bar{y}).$$

We have seen how to construct linear orders in Example 15.

$$\varphi_X(\bar{x}) \equiv \bigl(\forall y\, y \leqslant y\bigr) \wedge \bigl(X = Y \neq \emptyset\ ?\ X(\bar{x}) : \exists\bar{y}\,\neg C(\bar{y}) \wedge \varphi(\bar{x})\bigr),$$

where $\alpha\ ?\ \beta : \gamma$ abbreviates $(\alpha \wedge \beta) \vee (\neg\alpha \wedge \gamma)$. The first conjunct ensures that the linear order has been constructed before the fixed point is simulated. Once that has been done, if the simulation has reached a non-empty fixed point X, all subsequent stages will maintain that fixed point; otherwise, the next stage is simulated, as long as the counter relation has not yet filled up. If the counter relation is full, the simulation is complete — if no non-empty fixed point has been reached, no tuples satisfy φ_X, as required.

The counter relation C is defined by:

$$\varphi_C(\bar{x}) \equiv \left(\forall y\, y \leqslant y\right) \wedge \left(\forall \bar{y}\, C(\bar{y})\,?\,x_1 = x_1 : \theta_{\mathrm{succ}}(\bar{x})\right).$$

Once the linear order has been constructed, this formula causes $n(C)$ to be incremented once per stage until its maximal value is reached.   □

Our first normal form shows that, in common with most fixed-point logics, every formula is equivalent to (in the sense of defining the same nondeterministic query as) a formula containing just one instance of the fixed-point operator. Call a tuple *diagonal* if all its elements are equal.

Theorem 22. *Let $\Phi \in \mathrm{FO}(\mathrm{PFP}, \delta)[\sigma]$. There is a $\widehat{\Phi} \equiv \exists y\,(\mathbf{pfp}_{X,\bar{x}}\,\varphi)(y \ldots y)$, where $\varphi \in \mathrm{FO}(\delta)$, such that $[\![\widehat{\Phi}]\!]^{\mathfrak{A}} = [\![\Phi]\!]^{\mathfrak{A}}$ for all $\mathfrak{A} \in \mathrm{STRUC}[\sigma]$.*

Proof (Sketch). By induction on the structure of Φ. All relevant subformulae Φ_i will inductively be assumed to be of the form $\Phi_i \equiv \exists y\,(\mathbf{pfp}_{X_i,\bar{x}}\,\varphi_i)(y \ldots y)$. Further, by Theorem 21, we may assume that Φ_i is totally defined. We give the formula $\widehat{\Phi}$ as a simultaneous induction; this is equivalent to a formula of the required form by Theorem 20.

$(\Phi \equiv R(t_1 \ldots t_n))$ We may assume that each t_i is of the form $\delta x\, \Phi_i$, where Φ_i has the properties given in the previous paragraph: the term v is equivalent to $\delta x\,(\mathbf{pfp}_{X_i}\, x = v)$, where X_i is a new nullary relation symbol. (The fixed-point expression evaluates to true if, and only if, x and v are equal so the meaning of the term is 'choose an x that is equal to v'.) We may further assume that each Φ_i takes x as a parameter so, by Theorems 18 and 20, $\Phi_i \equiv \exists y\,(\mathbf{pfp}_{X_i,\bar{x}}\,\varphi_i)(y \ldots yx)$, which we may assume to be totally defined by the previous theorem.

$\widehat{\Phi}$ simulates the Φ_i until each has reached its fixed point, simulates the evaluation of the terms t_i and asks whether the resulting tuple $\bar{a}$ appears in R. This is straightforward with a simultaneous induction; we omit the details.

$(\Phi \equiv \neg\Phi_1)$ Let A be a new nullary relation symbol.

$$\widehat{\Phi} \equiv \mathbf{pfp}_{A,X_1,\bar{x}}(\neg\exists y\, X_1(y \ldots y), \varphi_1).$$

Since Φ_1 is totally defined, X_1 will eventually reach a fixed point. If that fixed point contains no diagonal tuples, A (and, hence, $\widehat{\Phi}$) will be true at all future stages; otherwise, it will be false at all future stages, as required. Since A and X_1 have both reached fixed points, the simultaneous induction has reached a fixed point and we are done. The cases of conjunction and existential quantification are similar.

$(\Phi \equiv (\mathbf{pfp}_{X, \bar{x}} \Phi_1(\bar{u}))$ To simulate a single stage of Φ, we simulate Φ_1 until it reaches a fixed point. We then set X to be the set of interpretations for $\bar{x}$ (the variables in which are treated as parameters to Φ_1) that result in the presence of a diagonal tuple in the fixed point of Φ_1. At the next stage of the simulation, X_1 is reset to be empty and the simulation is repeated until X reaches a fixed point, which we may assume it does by Theorem 21. $\widehat{\Phi}$ will be true if, and only if, this fixed point includes the tuple $\bar{u}$. We omit the details.   □

The second normal form, Theorem 23, shows that, in addition to only requiring one occurrence of the **pfp** operator in any formula, it suffices to make only one choice at each stage of the construction of the fixed point. For the proof, we code relations $R_1^{r_1}, \dots, R_n^{r_n}$ into a single relation R of arity $n + 1 + r$, where $r = \max \{ r_1, \dots, r_n \}$, such that

$$\langle a_1 \dots a_{r_i} \rangle \in R_i \iff \langle b_1 \dots b_{n+1} a_1 \dots a_r \rangle \in R \text{ for some } a_{r_i+1} \dots a_r$$
$$\text{and } b_1 = \cdots = b_i = a_1 \neq b_{i+1} = \cdots = b_{n+1}.$$

Each tuple in R is prefixed by a label whose equality type indicates the coded relation to which it belongs. We write $\widetilde{R}_i(\bar{x})$ for the first-order formula that says that the tuple $\bar{x}$ is an element of the relation R_i coded in R and $L_i(\bar{x})$ for the formula that says that $\bar{x}$ has the correct label to code an element of R_i. This requires structures to have at least two elements but the result applies to all structures as there are only finitely many single-element structures of any given vocabulary, which can be treated a special cases.

Theorem 23. *Let $\Phi \in \mathrm{FO}(\mathrm{PFP}, \delta)[\sigma]$. There is a formula $\widehat{\Phi}$ of the form $\widehat{\Phi} \equiv \exists y \left(\mathbf{pfp}_{X, \bar{x}} \exists d \left((d = \delta x\, \xi) \wedge \zeta \right) \right)(y \dots y)$, where $\xi, \zeta \in \mathrm{FO}$, such that $[\![\widehat{\Phi}]\!]^{\mathfrak{A}} = [\![\Phi]\!]^{\mathfrak{A}}$ for all $\mathfrak{A} \in \mathrm{STRUC}[\sigma]$.*

Proof. By the previous theorem and Theorem 18, we may assume that $\Phi \equiv \exists y \, (\mathbf{pfp}_{X, \bar{x}} \varphi)(y \dots y)$, where $\varphi \in \mathrm{FO}(\delta)$. We may further assume Φ is totally defined and that every δ-term in φ occurs in a formula $v = \delta x\, \theta$ as any formula $R(t_1 \dots t_r)$ is equivalent to $\exists v_1 \dots v_r \left(R(\bar{v}) \wedge v_1 = t_1 \wedge \dots \wedge v_r = t_r \right)$.

Define $T(\varphi)$, the *syntax tree* of φ recursively on the structure of the formula, as follows, writing $-$ for the empty tree and $[L, \ell, r]$ for a node labelled L with left and right subtrees ℓ and r, respectively.

$$T(R(x_{i_1} \dots x_{i_r})) = [R(x_{i_1} \dots x_{i_r}), -, -]$$
$$T(x_i = x_j) = [x_i = x_j, -, -]$$
$$T(x_i = \delta x_j\, \varphi) = [x_i = \delta x_j, T(\varphi), -]$$
$$T(\neg\varphi) = [\neg, T(\varphi), -]$$
$$T(\varphi \wedge \psi) = [\wedge, T(\varphi), T(\psi)]$$
$$T(\exists x_i\, \varphi) = [\exists x_i, T(\varphi), -].$$

Let $N_1, \dots, N_s$ be a full enumeration of the nodes of $T(\varphi)$ such that, for $i < j$, N_i is never in a subtree rooted at N_j. We use this enumeration of the nodes of $T(\varphi)$ to evaluate φ over several stages of a fixed point. Each of these stages will require at most one choice to be made.

Let D and $D_1, \ldots, D_s$ be new nullary relation symbols and $R_1 \ldots R_s$ be new k-ary relation symbols, where $k = \text{ar}(X)$. Let $\bar{u}$ be a $(2s + 2)$-ary tuple of new variables and Y be a new relation symbol of arity $2s + 2 + k$. Y will store the standard coding of X, D, $D_1, \ldots, D_s$ and $R_1, \ldots, R_s$.

Simulation of one stage of Φ will take $s+1$ stages. Initially, all the D_i are false. The formulae represented by $N_s, \ldots, N_1$ are evaluated in turn, with the relations they define stored in $X_s, \ldots, X_1$. As each X_i is evaluated, the corresponding D_i is set to true. D will be set to true once the fixed point has been reached in order to ensure that the simulation also reaches a fixed point. The fixed point has been reached when X_1 has been evaluated and contains exactly the same tuples as X (X_1 holds the evaluation of φ). At this point, if X contains a diagonal tuple, φ_A adds all diagonal tuples to Y, so $\widehat{\Phi}$ will be true, as required.

To get the desired normal form, we must perform the coding 'by hand' rather than by appealing to Theorem 20. $\widehat{\Phi} \equiv \exists y \, (\mathbf{pfp}_{Y, \bar{u}\bar{x}} \, \varphi^*)(y \ldots y)$, where

$$\varphi^*(\bar{u}\bar{x}) \equiv \exists d \left[(d = \delta x \, \psi) \wedge \left(\varphi_A \vee \varphi_D \vee \varphi_X \vee \bigvee_{i \in [1,s]} (\theta_i \wedge \varphi_i) \right) \right]$$

and θ_i says that $D = $ false and that $D_j = $ false for $j \leqslant i$ and true for $j > i$, i.e., that the simulation is not yet over and N_i is the next node to be evaluated.

$$\varphi_D(\bar{u}\bar{x}) \equiv L_D(\bar{u}\bar{x}) \wedge \widetilde{D}_1 \wedge \widetilde{X} = \widetilde{X}_1$$

$$\varphi_A(\bar{u}\bar{x}) \equiv \left(\widetilde{D} \wedge \theta_{\text{diag}}(\bar{u}\bar{x}) \wedge Y(\bar{u}\bar{x}) \right) \vee \left(\exists \bar{u}\bar{x} \, \varphi_D(\bar{u}\bar{x}) \wedge \theta_{\text{diag}}(\bar{u}\bar{x}) \wedge \exists y \, \widetilde{X}(y \ldots y) \right)$$

$$\varphi_X(\bar{u}\bar{x}) \equiv \widetilde{D}_1 \wedge L_X(\bar{u}\bar{x}) \wedge \widetilde{X}_1(\bar{x})$$

where $\theta_{\text{diag}}(\bar{u}\bar{x})$ asserts that its argument is diagonal. φ_D sets D to true when the fixed point has been reached, i.e., when $D_1 = $ true and $X_1 = X$. The first disjunct of φ_A maintains any diagonal tuples in Y once $D = $ true; the second adds diagonal tuples to Y if the fixed point has been reached (i.e., D will be true at the next iteration) and X contains diagonal tuples. If $D_1 = $ true (i.e., the evaluation of φ is complete), φ_X copies X_1 to X.

To understand the remaining subformulae, it is helpful to fix in one's mind an interpretation for $\bar{u}$ and $\bar{x}$.

$$\psi(\bar{u}\bar{x}x) \equiv \bigvee_{i \in I} \left(\theta_i \wedge \widetilde{X}_\ell(\bar{x}[x/x_n]) \right),$$

where $I = \{ i : N_i = [x_m = \delta x_n, N_\ell, -] \}$. If the node of $T(\varphi)$ currently being evaluated is a choice node, ψ defines the choice set, from which d is chosen.

$$\varphi_i(\bar{u}\bar{x}) \equiv L_{D_i}(\bar{u}\bar{x}) \vee \left(L_{X_i}(\bar{u}\bar{x}) \wedge \chi_i(\bar{x}) \right),$$

where $\chi_i(\bar{x})$ depends on N_i. The difficult case is choice nodes: if $N_i = [x_m = \delta x_n, N_\ell, -]$, $\chi_i(\bar{x}) \equiv x_m = d \wedge \exists x_m \, \widetilde{X}_\ell(\bar{x})$. The other cases are reasonably obvious: for example, if $N_i = [\neg, N_\ell, -]$, $\chi_i(\bar{x}) \equiv \neg \widetilde{X}_\ell(\bar{x})$. $\qquad \square$

Every formula being equivalent to one with a single occurrence of the fixed-point operator, with only one choice taken per stage suggests a link with the fixed-point operators introduced by Gire and Hoang [GH98]. These operators perform inflationary (rather than partial) inductions about two formulae: one defines a 'choice set', from which a single tuple is chosen; the other uses this chosen tuple to define the tuples to be added to the relation under construction. In [DR03], we show that their logic IFP$_c$, which we denote C-IFP, has the same expressive power as FO(IFP, δ). From the previous theorem, we can see that a partial fixed-point logic in the style of C-IFP would define the same queries as the apparently more general FO(PFP, δ). We can also use the iteration provided by a fixed-point operator to simulate the more powerful **W** operator with repeated δ choices so FO(PFP, δ) and FO(PFP, **W**) define the same queries.

5.2   Expressive Power

So far, we have considered FO(PFP, δ) only in terms of defining nondeterministic queries: we have not called it a logic as we have not presented a satisfaction relation. We first deal with the deterministic case. Recall that a formula φ is deterministic if, on all structures $\mathfrak{A}$, it defines a single relation, i.e., $\|[\![\varphi]\!]^{\mathfrak{A}}\| = 1$. In this case, satisfaction is easily defined to coincide with satisfaction of ordinary deterministic logics. Denote by FO(PFP, δ)$_{\text{det}}$ the FO(PFP, δ) formulae that are deterministic on the class of all finite structures of appropriate vocabulary and write $(\mathfrak{A}, \bar{a}) \vDash_{\text{det}} \varphi$ if, and only if, $\bar{a} \in R$, where $[\![\varphi]\!]^{\mathfrak{A}} = \{ R \}$.

Theorem 24. FO(PFP, δ)$_{\text{det}}$ *captures* **PSPACE** *on the class of unordered finite structures.*

Proof. (FO(PFP, δ)$_{\text{det}} \leqslant$ **PSPACE**) Consider evaluating a formula φ on a structure $\mathfrak{A}$. We may assume that φ is totally defined and in the normal form of Theorem 22. Since φ is deterministic, we need evaluate only one path through $\mathcal{T}_{\varphi}^{\mathfrak{A}}$ and, by Lemma 16, it suffices to evaluate that path to depth $N = 2^{\|\mathfrak{A}\|^r}$, where r is the arity of the relation constructed by the fixed-point operator. Counting out the stages can be done with a $\|\mathfrak{A}\|^r$-bit binary counter, evaluating each stage requires choosing a fixed number of elements of $|\mathfrak{A}|$ and evaluating a first-order formula, which can be done in space $\mathcal{O}(\log |\mathfrak{A}|)$. The total space requirement is, therefore, polynomial in $\|\mathfrak{A}\|$.

(**PSPACE** $\leqslant$ FO(PFP, δ)$_{\text{det}}$) Any **PSPACE** query can be defined by a deterministic FO(PFP, δ) formula by first defining a linear order and then simulating an order-invariant FO(PFP) sentence on the resulting ordered structure. $\square$

Notice that we do not describe FO(PFP, δ)$_{\text{det}}$ as a logic as it does not have recursive syntax. The following theorem is similar to Theorem 28 of [DR03], though the restriction to vocabularies containing an at-least-binary relation symbol is lifted. The same techniques can be applied to remove the restriction there.

Theorem 25. *The set of deterministic* FO(PFP, δ) *formulae is not decidable, even over the empty vocabulary.*

Proof. The case where the vocabulary contains an at-least binary relation symbol is covered by Lemma 13. For the empty vocabulary, let A and E be new relation symbols of arity 0 and 2, respectively, and consider the sentence

$$\varphi \equiv \mathbf{pfp}_{A,E,xy}(\psi, \varphi_E),$$

where ψ is some first-order sentence in vocabulary $\langle E \rangle$ and

$$\varphi_E(xy) \equiv E(xy) \vee \big(x = \delta u\,(u = u) \wedge x = \delta v\,(v = v) \wedge y = \delta w\,(w = w)\big)$$

At each stage of evaluating φ on a structure $\mathfrak{A}$, φ_E chooses three elements u, v and w. If $u = v$, the pair uw is added to E; otherwise E remains unchanged. If $(\mathfrak{A}, E) \vDash \psi$, A is set to true; otherwise, it remains false. Along every path in $T_\varphi^{\mathfrak{A}}$, E reaches some fixed point and, for any $S \subseteq |\mathfrak{A}|^2$, there is a path in the tree where E reaches the fixed point S. true $\in \llbracket \varphi \rrbracket^{\mathfrak{A}}$ if, and only if $(\mathfrak{A}, E) \vDash \psi$ for some E and false $\in \llbracket \varphi \rrbracket^{\mathfrak{A}}$ if, and only if, $(\mathfrak{A}, E) \nvDash \psi$ for some E. φ is deterministic if, and only if, ψ or its complement is valid, so the undecidability of the set of deterministic $\mathrm{FO}(\mathrm{PFP}, \delta)$ formulae follows from Trakhtenbrot's Theorem [Tra50]. $\qquad\square$

We now consider the satisfaction of nondeterministic formulae.

Definition 26. *The formulae of* $\mathrm{FO}(\mathrm{PFP}, \delta)_\exists$ *are those of* $\mathrm{FO}(\mathrm{PFP}, \delta)$ *and we write* $(\mathfrak{A}, \bar{a}) \vDash_\exists \varphi$ *if, and only if,* $\bar{a} \in R$ *for some* $R \in \llbracket \varphi \rrbracket^{\mathfrak{A}}$.

Equivalently, $(\mathfrak{A}, \bar{a}) \vDash_\exists \varphi$ if, and only if, $\bar{a} \in \bigcup \llbracket \varphi \rrbracket^{\mathfrak{A}}$. $\mathrm{FO}(\mathrm{PFP}, \delta)_\exists$ is our second characterization of **PSPACE** on unordered finite structures.

Theorem 27. $\mathrm{FO}(\mathrm{PFP}, \delta)_\exists$ *captures* **PSPACE**.

Proof. (**PSPACE** $\leqslant \mathrm{FO}(\mathrm{PFP}, \delta)_\exists$) Let $\Phi \in \mathrm{SO}(\mathrm{PFP})$; by Corollary 5, we may assume $\Phi \equiv \exists E\, \varphi$, where E is a new binary relation symbol and $\varphi \in \mathrm{FO}(\mathrm{PFP})$. Let A be a new nullary relation symbol and let $\widehat{\Phi} \equiv \mathbf{pfp}_{A,E,xy}(\varphi, \varphi_E)$, where φ_E is as in the proof of the previous theorem.

For every $R \subseteq |\mathfrak{A}|^2$, there is a path in $T_{\widehat{\Phi}}^{\mathfrak{A}}$ labelled $\langle A^i, E^i \rangle_{i \geqslant 0}$ such that, for all sufficiently large i, $E^i = R$ and A^i is true if and only if $(\mathfrak{A}, R) \vDash \varphi$. Therefore, $\mathfrak{A} \vDash \widehat{\Phi}$ if, and only if, if there is some interpretation of E that satisfies φ, if, and only if, $\mathfrak{A} \vDash \widehat{\Phi}$, as required.

($\mathrm{FO}(\mathrm{PFP}, \delta)_\exists \leqslant$ **PSPACE**) $\Phi \in \mathrm{FO}(\mathrm{PFP}, \delta)_\exists$ can be evaluated using a nondeterministic Turing machine using polynomial space bounds. The fixed-point operator can be evaluated using standard techniques (see, e.g., [EF99]) and δ-terms can be evaluated using the nondeterminism of the machine. By Savitch's Theorem [Sav70], **NPSPACE** = **PSPACE** and we are done. $\qquad\square$

We could also consider the analogous $\mathrm{FO}(\mathrm{PFP}, \delta)_\forall$. It is easy to see that this has the same expressive power as the existential semantics, since nondeterministic space classes are closed under complementation [Imm88, Sze88].

An interesting feature of fixed-point logics with choice is that, in contrast to the conventional fixed-point logics, they do not embed into infinitary logics. Both $\mathrm{FO}(\mathrm{PFP}, \delta)$ and $\mathrm{FO}(\mathrm{IFP}, \delta)$ can express the parity query on pure sets but, since $\mathcal{L}_{\infty\omega}^\omega(\delta) = \mathcal{L}_{\infty\omega}^\omega$ and $\mathcal{L}_{\infty\omega}^\omega$ has a 0-1 law [KV92], there is no formula of $\mathcal{L}_{\infty\omega}^\omega(\delta)$ that expresses this query.

References

[ASV90] Serge Abiteboul, Eric Simon and Victor Vianu. Non-deterministic languages to express deterministic transformations. In *Proceedings of 9th ACM Symposium on Principles of Database Systems*, pp. 218–229. ACM, 1990.

[AV89] Serge Abiteboul and Victor Vianu. Fixpoint extensions of first-order and datalog-like languages. In *Proceedings of 4th IEEE Annual Symposium on Logic in Computer Science*, pp. 71–79. IEEE Computer Society, 1989.

[BG00] Andreas Blass and Yuri Gurevich. The logic of choice. *Journal of Symbolic Logic*, 65(3):1264–1310, 2000.

[Daw98] Anuj Dawar. A restricted second order logic for finite structures. *Information and Computation*, 143(2):154–174, 1998.

[DR03] Anuj Dawar and David M. Richerby. Fixed-point logics with nondeterministic choice. *Journal of Logic and Computation*, 13(4):503–530, 2003.

[Ebb85] Heinz-Dieter Ebbinghaus. Extended logics: The general framework. In John Barwise and Solomon Feferman, editors, *Model-Theoretic Logics*, Perspectives in Mathematical Logic, pp. 25–76. Springer-Verlag, 1985.

[EF99] Heinz-Dieter Ebbinghaus and Jörg Flum. *Finite Model Theory*. Springer-Verlag, 2nd edition, 1999.

[Fag74] Ronald Fagin. Generalized first-order spectra and polynomial-time recognizable sets. In Richard Karp, editor, *Complexity of Computation*, volume 7 of *SIAM-AMS Proceedings*, pp. 43–73. SIAM-AMS, 1974.

[Fag75] Ronald Fagin. Monadic generalized spectra. *Zeitschrift für Mathematische Logik und Grundlagen der Mathematik*, 21:89–96, 1975.

[GH98] F. Gire and H.K. Hoang. An extension of fixpoint logic with a symmetry-based choice construct. *Information and Computation*, 144(1):40–65, 1998.

[HB39] David Hilbert and Paul Bernays. *Grundlagen der Mathematik*, volume 2. Springer-Verlag, 1939. (In German.)

[HP84] David Harel and David Peleg. On static logics, dynamic logics, and complexity classes. *Information and Control*, 60(1–3):86–102, 1984.

[Imm86] Neil Immerman. Relational queries computable in polynomial time. *Information and Control*, 68(1–3):86–104, 1986.

[Imm87] Neil Immerman. Logics that capture complexity classes. *SIAM Journal of Computing*, 16(4):236–244, 1987.

[Imm88] Neil Immerman. Nondeterministic space is closed under complementation. *SIAM Journal of Computing*, 17(5):935–938, 1988.

[KV92] Phokion G. Kolaitis and Moshe Y. Vardi. Infinitary logics and 0-1 laws. *Information and Computation*, 98(2):258–294, 1992.

[Ott00] Martin Otto. Epsilon-logic is more expressive than first-order logic over finite structures. *Journal of Symbolic Logic*, 65(4):1749–1757, 2000.

[Sav70] Walter J. Savitch. Relationships between nondeterministic and deterministic tape complexities. *J. Computer and System Sciences*, 4(2):177–192, 1970.

[Sto76] Larry J. Stockmeyer. The polynomial-time hierarchy. *Theoretical Computer Science*, 3(1):1–22, 1976.

[Sze88] Róbert Szelepcsényi. The method of forced enumeration for nondeterministic automata. *Acta Informatica*, 26(3):279–284, 1988.

[Tra50] Boris A. Trakhtenbrot. Impossibility of an algorithm for the decision problem in finite classes. *Doklady Akademii Nauk SSSR*, 70:569–572, 1950. (In Russian; English translation: AMS Translations, Series 2, 23:1–5, 1963.)

[Var82] Moshe Y. Vardi. Complexity of relational query languages. In *Proceedings of 14th ACM Symposium on Theory of Computing*, pp. 137–146. ACM, 1982.

The Logic of the
Partial λ-Calculus with Equality

Lutz Schröder[*]

BISS, Department of Computer Science, University of Bremen

Abstract. We investigate the logical aspects of the partial λ-calculus with equality, exploiting an equivalence between partial λ-theories and partial cartesian closed categories (pcccs) established here. The partial λ-calculus with equality provides a full-blown intuitionistic higher order logic, which in a precise sense turns out to be almost the logic of toposes, the distinctive feature of the latter being unique choice. We give a linguistic proof of the generalization of the fundamental theorem of toposes to pcccs with equality; type theoretically, one thus obtains that the partial λ-calculus with equality encompasses a Martin-Löf-style dependent type theory. This work forms part of the semantical foundations for the higher order algebraic specification language HASCASL.

Introduction

Partial functions play an important role in modern algebraic specification, serving to model both non-termination and irregular termination; specification languages featuring partial functions include RSL [8], SPECTRUM [3], and CASL [2, 15]. The natural generalization of the simply typed λ-calculus to partial functions is the partial λ-calculus [13, 14, 18], which forms the basis for the recently introduced wide-spectrum language HASCASL [23, 25]. HASCASL offers a setting for both specification and implementation of higher order functional programs; moreover, it has served as a background formalism for the development of monad-generic computational logics [22, 24]. A central role in all this is played by the fact that the partial λ-calculus *with equality* induces a full intuitionistic higher order logic, corresponding to HASCASL's internal logic [23]. Here, we investigate the character and expressivity of this logic more closely.

The central tool for this investigation is an equivalence between partial λ-theories with equality and *partial cartesian closed categories (pcccs)* with equality proved here. One associates to each pccc a partial λ-theory, its internal language, and conversely to each partial λ-theory a classifying category; the two constructions are essentially mutually inverse. Thus, one can freely move back and forth between logical and categorical formulations and arguments.

It turns out that in a hierarchy of categorical notions comprising in ascending order of strength locally cartesian closed categories, quasitoposes, and toposes,

[*] Research supported by the DFG project HasCASL (KR 1191/7-1).

pcccs with equality fit between locally cartesian closed categories and quasito-poses. In terms of logic, locally cartesian closed categories correspond to Martin-Löf style dependent type theory [27], and toposes to intuitionistic type theory with power types [11] (the precise logical counterpart of quasitoposes is, to our knowledge, open). In particular, this means that the partial λ-calculus with equality encodes a dependent type theory; more precisely, one even has a partial version of dependent product types, which categorically relates to a novel notion of locally partial cartesian closed category. Moreover, we show that topos logic is characterized within the partial λ-calculus by the axiom of unique choice; differently put, topos logic can be recovered from the partial λ-calculus with equality by giving up the distinction between functions and functional relations.

Related work includes [13], where a semantics for the partial λ-calculus in left exact pcccs is given, as well as [18], where a classifying category construction for the pure partial λ-calculus is described, using however different categorical notions. A fuller exposition of some of the results presented here can be found in [19].

1   The Partial λ-Calculus

The partial λ-calculus [13, 14, 18] is a typed higher-order formalism that explicitly handles partial functions. It is formally similar to the simply typed λ-calculus, the crucial difference being that function types are thought of as types of partial functions. This is reflected both in the semantics [13] and in the deductive system, which has to keep track of definedness of terms.

We now give a brief definition of the syntax and deduction system of the partial λ-calculus, with one modification in comparison to [13]: there are various types of equations between partial terms, two of the more common being *existential* equations, to be read 'both sides are defined and equal, and *strong* equations, to be read 'one side is defined iff the other is, and then the two sides are equal'. While the presentation in [13] is based on strong equations, we focus mainly on existential equations, since these are slightly better suited for our categorical treatment, and give a correspondingly adapted deduction system; the expressivity of both types of equations is the same [13]. We will, moreover, for the purposes of this paper mostly be interested in theories that possess an equality predicate, which has the effect of transforming a simple λ-calculus into a full-blown higher order logic.

A *partial λ-theory* consists of *axioms* over a *signature*. A signature is given by sets of basic *sort* and *operation* symbols, where the latter (thought of as representing partial maps) consist of their *name* and *profile*, written in the form $f : \bar{s} \rightharpoonup t$. Here, t is a *type* and $\bar{s} = (s_1, \ldots, s_n)$ is a *multi-type*, i.e. a list of types (the bar notation is used to indicate lists of items throughout). Types are freely generated from the basic sorts by closing them under the formation of *partial function types* written

$$\bar{s} \rightharpoonup t,$$

with $\bar{s}$ and t as above (one cannot resort to currying for multi-argument partial functions [13]). Following [13], we assume application operators $(\bar{s} \multimap t)\bar{s} \rightharpoonup t$ in the signature, so that application does not require extra typing or deduction rules. We use the notation $\bar{s} \multimap \bar{t}$ to denote the multi-type with components $\bar{s} \multimap t_i$ (which is *not* the same as a function type into the product of the t_i).

Given a signature, typed *terms* and *multi-terms*, i.e. lists $\bar{\alpha} = (\alpha_1, \ldots, \alpha_n)$ of terms, in a context $\Gamma = (\bar{x} : \bar{s}) = (x_1 : s_1, \ldots, x_n : s_n)$ of distinct variables x_i with assigned types s_i are formed according to the typing rules

$$\frac{x : s \text{ in } \Gamma}{\Gamma \triangleright x : s} \qquad \frac{\begin{array}{c}\Gamma \triangleright \bar{\alpha} : \bar{t} \\ f : \bar{t} \rightharpoonup u\end{array}}{\Gamma \triangleright f(\bar{\alpha}) : u} \qquad \frac{\Gamma, \bar{y} : \bar{t} \triangleright \alpha : u}{\Gamma \triangleright \lambda \bar{y} : \bar{t}. \alpha : \bar{t} \multimap u},$$

where the judgement $\Gamma \triangleright \bar{\alpha} : \bar{t}$ is read '(multi-)term $\bar{\alpha}$ has (multi-)type $\bar{t}$ in context Γ'; here, typing judgements for multi-terms are just collections of typing judgements for the constituent terms. The higher order application operator is denoted by juxtaposition, while term formation using operators from the signature is written, as above, with brackets. Where convenient, terms will be regarded as singleton multi-terms, similarly for types. For convenience, we regard the empty multi-type and the empty multi-term also as a type and a term denoted by 1 and by $*$, respectively.

An *existential equation* between two terms in context Γ is written $\Gamma \triangleright \alpha \stackrel{e}{=} \beta$ or $\alpha \stackrel{e}{=} \beta$, to be understood as indicated above. Equations between multi-terms are regarded as sets of equations between terms; the union of such sets is denoted by $\wedge$, and the empty set of equations by $\top$. Equations of the form $\alpha \stackrel{e}{=} \alpha$ just state that α is defined; they are abbreviated as def α (and e.g. def(α, β) codes the same set of equations as def $\alpha \wedge$ def β). An *existentially conditioned equation (ece)* [5] in context Γ is a sentence of the form $\Gamma \triangleright$ def $\bar{\alpha} \Rightarrow \psi$, where $\bar{\alpha}$ is a multi-term and ψ is an existential equation in context Γ. The axioms of a partial λ-theory are given as eces.

The deduction system for the partial λ-calculus is shown in Figure 1. Deduction takes place over a fixed context Γ and in a theory with the set $\mathcal{A}$ of axioms. We write $\Gamma \triangleright$ def $\bar{\alpha} \vdash \phi$ if an equation ϕ can be deduced from def $\bar{\alpha}$ in context Γ by means of these rules; in this case, $\Gamma \triangleright$ def $\bar{\alpha} \Rightarrow \phi$ is a *theorem*. Subderivations are also denoted in the form $\Delta \triangleright$ def $\bar{\alpha} \vdash \phi$, where the context Δ and the assumption def $\bar{\alpha}$ are to be understood as extending the ambient context and assumptions. Strong equations $\Delta \triangleright \alpha \stackrel{s}{=} \beta$ are used as abbreviations for subderivations $\Delta \triangleright$ def $\alpha \vdash$ def β and $\Delta \triangleright$ def $\beta \vdash \alpha \stackrel{e}{=} \beta$.

The usual forms of the β- and η-rules can be derived by means of the substitution rule. Rule (ξ) implies that all λ-terms are defined.

A *morphism* between two signatures is a pair of maps for sorts and operators, respectively, that is compatible with operator profiles. A *translation* between partial λ-theories is a signature morphism which transforms axioms into theorems. Partial λ-theories and translations form a category **pλTh**.

$$
\text{(var)} \ \frac{x : s \text{ in } \Gamma}{\text{def } x} \quad \text{(st)} \ \frac{\text{def } f(\bar{\alpha})}{\text{def } \bar{\alpha}} \quad \text{(unit)} \ \frac{x : 1 \text{ in } \Gamma}{x \stackrel{e}{=} *} \quad \text{(sym)} \ \frac{\alpha \stackrel{e}{=} \beta}{\beta \stackrel{e}{=} \alpha} \quad \text{(tr)} \ \frac{\begin{array}{c} \alpha \stackrel{e}{=} \beta \\ \beta \stackrel{e}{=} \gamma \end{array}}{\alpha \stackrel{e}{=} \gamma}
$$

$$
\text{(cg)} \ \frac{\begin{array}{c} \bar{\alpha} \stackrel{e}{=} \bar{\beta} : \bar{t} \\ f : \bar{t} \rightharpoonup u \end{array}}{f(\bar{\alpha}) \stackrel{s}{=} f(\bar{\beta})} \quad \text{(ax)} \ \frac{\begin{array}{c} (\bar{y} : \bar{t} \rhd \text{def } \bar{\alpha} \Rightarrow \phi) \in \mathcal{A} \\ \bar{y} : \bar{t} \text{ in } \Gamma \\ \text{def } \bar{\alpha} \end{array}}{\phi} \quad \text{(sub)} \ \frac{\begin{array}{c} \bar{y} : \bar{t} \rhd \text{def } \bar{\alpha} \vdash \phi \\ \text{def}(\bar{\beta}, \bar{\alpha}[\bar{\beta}/\bar{y}]) \end{array}}{\phi[\bar{\beta}/\bar{y}]}
$$

$$
\text{(}\eta\text{)} \ \frac{x : \bar{t} \multimap u \text{ in } \Gamma}{(\lambda \bar{y} : \bar{t}. x \, \bar{y}) \stackrel{e}{=} x} \quad \text{(}\beta\text{)} \ \frac{\bar{y} : \bar{t} \text{ in } \Gamma}{(\lambda \bar{y} : \bar{t}. \alpha) \, \bar{y} \stackrel{s}{=} \alpha} \quad \text{(}\xi\text{)} \ \frac{\Delta \rhd \alpha \stackrel{s}{=} \beta}{(\lambda \Delta. \alpha) \stackrel{e}{=} \lambda \Delta. \beta}
$$

Fig. 1. Deduction rules for existential equality in context Γ

We introduce an important shorthand notation: for (multi-)terms $\bar{\alpha}, \bar{\beta}$, we have *conditioned terms* [5, 14]

$$
\bar{\alpha} \upharpoonright \bar{\beta} := (\lambda \bar{x}, \bar{y}. \bar{x})(\bar{\alpha}, \bar{\beta})
$$

denoting $\bar{\alpha}$ with its domain restricted to the common domain of $\bar{\alpha}$ and $\bar{\beta}$. A first use of conditioned terms is λ-abstraction of multi-terms: $\lambda \bar{y} : \bar{t}. \bar{\alpha}$ denotes the multi-term with components $\lambda \bar{y} : \bar{t}. \alpha_i \upharpoonright \bar{\alpha}$.

The partial λ-calculus automatically comes with a rudimentary logic: we can regard $\Omega = 1 \multimap 1$ as a type of truth values, $s \multimap 1$ as the type of predicates on s, and (partial) terms $\phi : 1$ as formulas. For such ϕ, we will shortly write ϕ in place of def ϕ, and we can turn definedness assertions into formulas by observing that for any term α, def α is, in this notation, equivalent to $(\lambda x. *) \alpha$. In the way of connectives, however, one generally does not have more than conjunction and truth, expressed e.g. via $p \wedge q = (\lambda x, y : 1. *)(p, q)$ and $\top = *$.

The picture changes completely in the presence of an equality predicate:

Definition 1. A partial λ-theory *has equality* if there exists, for each type s, a (defined) closed term $eq_s : ss \multimap 1$ (i.e. a binary predicate on s) such that

$$
x, y : s \rhd eq_s(x, y) \Rightarrow x \stackrel{e}{=} y \quad \text{and} \quad x : s \rhd eq_s(x, x)
$$

(See also [6, 13].) Note that the axioms for eq_s are eces. When there is no danger of confusion, we shall write $\alpha \stackrel{e}{=} \beta$ in place of $eq(\alpha, \beta)$. Equality gives rise to a full-fledged intuitionistic logic, along much the same lines as in [6, 11]: letting p and q range over (partial) terms of type 1, we can put

$$p \Rightarrow q := ((\lambda. p) \stackrel{e}{=} \lambda. p \wedge q),$$
$$\forall \bar{y} : \bar{t}. p := ((\lambda \bar{y} : \bar{t}. p) \stackrel{e}{=} \lambda \bar{y} : \bar{t}. \top),$$
$$\bot := \forall a : \Omega. a*,$$
$$\neg p := p \Rightarrow \bot,$$
$$p \vee q := \forall a : \Omega. ((p \Rightarrow a*) \wedge (q \Rightarrow a*)) \Rightarrow a*, \text{ and}$$
$$\exists \bar{y} : \bar{t}. p := \forall a : \Omega. (\forall \bar{y} : \bar{t}. p \Rightarrow a*) \Rightarrow a*,$$

where we omit unused variables of type 1 from λ-abstractions (note that all right hand sides are partial terms of type 1). The usual deduction rules of intuitionistic higher order logic are obtained as lemmas. The main topic of this paper is the closer investigation of this logic.

2   Partial Cartesian Closed Categories

We now give a brief outline of the categorical setting for the semantics, and indeed the syntax, of the partial λ-calculus.

Given a category whose morphisms are thought of as total functions, partial functions $A \rightharpoonup B$ correspond to *partial morphisms*, i.e. spans (m, f) of the form

$$\begin{array}{ccc} \bullet & \xrightarrow{f} & B \\ {\scriptstyle m}\downarrow & & \\ A & & \end{array},$$

where m is a monomorphism of a restricted class $\mathcal{M}$ representing the domain of definition, taken modulo isomorphism in the obvious sense. The composite of (m, f) and a partial morphism (n, g) from B to C is defined as $(mf^{-1}(n), gf*)$, where

$$\begin{array}{ccc} \bullet & \xrightarrow{f*} & \bullet \\ {\scriptstyle f^{-1}(n)}\downarrow & & \downarrow{\scriptstyle n} \\ \bullet & \xrightarrow{f} & B \end{array}$$

is a pullback. In order for this to be possible, we have to require a few closure properties of $\mathcal{M}$:

Definition 2. A class of monomorphisms in a category $\mathbf{C}$ is called a *dominion* [18] if it contains all identities and is closed under composition and pullbacks, i.e. pullbacks of $\mathcal{M}$-morphisms (along arbitrary morphisms) exist and are in $\mathcal{M}$. A *dominional category* is a pair $(\mathbf{C}, \mathcal{M})$, where $\mathcal{M}$ is a dominion on $\mathbf{C}$; an *admissible subobject* is an element of $\mathcal{M}$. A functor between dominional categories is called *dominional* if it preserves admissible subobjects and their pullbacks. An equivalence functor between dominional categories is called a *dominional equivalence* if it preserves and reflects admissible subobjects; it is then automatically a dominional functor.

A dominion $\mathcal{M}$ is closed under intersections. If m is a monomorphism and $mg \in \mathcal{M}$, then $g \in \mathcal{M}$. In particular, $\mathcal{M}$ contains all isomorphisms. For a dominional category $(\mathbf{C}, \mathcal{M})$, the partial morphisms form a category $\mathbf{P}(\mathbf{C}, \mathcal{M})$, which contains $\mathbf{C}$ as a (non-full) subcategory [17, 18].

As usual, we call a category (functor, subcategory) *cartesian* if it has (preserves, is closed under) finite products; the terminal object is denoted by 1. In a cartesian dominional category $(\mathbf{C}, \mathcal{M})$, $\mathcal{M}$ is closed under products (but not under pairing). Cartesian dominional categories are equivalent to first order partial equational theories [20].

Definition 3. A cartesian dominional category $(\mathbf{C}, \mathcal{M})$ *has equality* if $\mathcal{M}$ contains all diagonals $A \to A \times A$.

If $(\mathbf{C}, \mathcal{M})$ has equality, then $\mathbf{C}$ has equalizers (hence is finitely complete, shortly: left exact or *lex*), and $\mathcal{M}$ contains all regular monomorphisms and is closed under pairing.

The semantics of the partial λ-calculus has been given in terms of a class of dominional categories called partial cartesian closed categories (pcccs) [13]. The crucial feature of a pccc is that it admits the interpretation of partial function types as *partial function spaces* $A \rightharpoonup B$, which are defined by the property that partial morphisms from $C \times A$ to B are in bijective correspondence with total morphisms $C \to (A \rightharpoonup B)$. More formally,

Definition 4. A cartesian dominional category $(\mathbf{C}, \mathcal{M})$ is called a *partial cartesian closed category (pccc)* if the composite functor

$$\mathbf{C} \xrightarrow{-\times A} \mathbf{C} \hookrightarrow \mathbf{P}(\mathbf{C}, \mathcal{M})$$

has a right adjoint for each object A in $\mathbf{C}$.

(This definition is weaker than the one given in [13] in that we do not require left exactness.) Every lex pccc is cartesian closed [6].

Partial function spaces $A \rightharpoonup B$ in a pccc come with a co-universal partial *evaluation morphism* ev from $(A \rightharpoonup B) \times A$ to B. Explicitly, every partial morphism f from $C \times A$ to B factors uniquely as $ev \circ (\hat{f} \times A)$ in $\mathbf{P}(\mathbf{C}, \mathcal{M})$ by a total morphism $\hat{f} : C \to (A \rightharpoonup B)$ called its *abstraction*.

For a pccc $(\mathbf{C}, \mathcal{M})$, the embedding $\mathbf{C} \hookrightarrow \mathbf{P}(\mathbf{C}, \mathcal{M})$ is left adjoint, being isomorphic to $\_ \times 1$. Spelling this out yields that $\mathcal{M}$-partial morphisms in $(\mathbf{C}, \mathcal{M})$ are representable [1, 18], with the partial morphisms into A represented by $1 \rightharpoonup A$. In particular, $\Omega = 1 \rightharpoonup 1$ classifies $\mathcal{M}$-subobjects. By consequence, *every map in $\mathcal{M}$ is a regular monomorphism.*

Of particular interest is the case that the pccc $(\mathbf{C}, \mathcal{M})$ has equality. In this case, $\mathcal{M} = \text{RegMono}(\mathbf{C})$, so that we can *omit the mention of $\mathcal{M}$*. In fact, pcccs with equality can be succinctly characterized as cartesian closed categories with representable regular partial morphisms in which regular monos are stable under composition; a further characterization will be given in Section 5. In particular, every quasi-topos is a pccc with equality (but not conversely [1]). A typical example of a pccc without equality is the category of cpos and continuous functions with Scott open sets [26] as admissible subobjects.

Definition 5. A cartesian dominional functor between two pcccs is called *partial cartesian closed (pcc)* if it preserves partial function spaces. This defines the category **PCCC** of pcccs.

Remark 6. There is a large number of axiomatizations of categories where partial morphisms are directly treated as arrows. Essentially, these axiomatizations characterize full subcategories of $\mathbf{P}(\mathbf{C}, \mathcal{M})$ for some cartesian dominional category $(\mathbf{C}, \mathcal{M})$ or, in the higher order case equivalently, of Kleisli categories arising from representations of partial morphisms, i.e. from the adjunction between $(\mathbf{C}, \mathcal{M})$ and $\mathbf{P}(\mathbf{C}, \mathcal{M})$ for some $(\mathbf{C}, \mathcal{M})$ [4, 6, 17]. In these approaches, categories of the form $\mathbf{P}(\mathbf{C}, \mathcal{M})$ are typically distinguished by a splitting condition for subfunctions of the identity which ensures that domains of partial functions are actually objects. For the purposes of this paper, as well as the (logically posterior) paper on Henkin models [21], it appears to be more convenient to work directly with the underlying dominional categories. In particular, this makes the relation of pcccs with toposes and locally cartesian closed categories more immediate; moreover, certain categorical techniques such as in particular the use of representable functors (which plays a crucial role in [21]) are more directly available.

3   The Internal Language and Its Interpretation

We now establish an equivalence between partial λ-theories with equality and pcccs with equality, proceeding as follows: we associate to each pccc $(\mathbf{C}, \mathcal{M})$ an *internal language* $\mathsf{L}(\mathbf{C}, \mathcal{M})$, thus obtaining a functor

$$\mathsf{L} : \mathbf{PCCC} \to \mathbf{p\lambda Th}.$$

In this process, we will introduce an interpretation of $\mathsf{L}(\mathbf{C}, \mathcal{M})$ in $(\mathbf{C}, \mathcal{M})$, for which we prove a soundness theorem. We will then construct classifying categories for partial λ-theories with equality, i.e. free objects w.r.t. the functor L. It will turn out that every pccc with equality is equivalent to the classifying category of its internal language, and that the internal logic of the classifying category of a partial λ-theory is a conservative extension, so that pcccs with equality are essentially the same as partial λ-theories with equality.

To begin, we associate a signature Σ to a pccc $(\mathbf{C}, \mathcal{M})$. The sorts in Σ are the objects of $\mathbf{C}$. An interpretation $[\![\_]\!]$ in $\mathbf{C}$ for types and multi-types is defined recursively in the obvious way using products and partial function spaces. The operators of profile $\bar{s} \to t$ in Σ are the partial morphisms from $[\![\bar{s}]\!]$ to $[\![t]\!]$ in $(\mathbf{C}, \mathcal{M})$, with evaluation morphisms (cf. Section 2) as application operators.

The interpretation $[\![\_]\!]$ is then extended to contexts, terms, multi-terms, and definedness conditions: for a context $\Gamma = (\bar{x} : \bar{s})$, $[\![\Gamma]\!] = [\![\bar{s}]\!]$. Given a term or multi-term $\Gamma \triangleright \bar{\alpha} : \bar{t}$, we define a partial morphism denoted

$$[\![\Gamma]\!] \longleftarrow\!\!\!\supset [\![\Gamma.\,\mathrm{def}\,\bar{\alpha}]\!] \xrightarrow{\;[\![\Gamma.\,\bar{\alpha}]\!]\;} [\![\bar{t}]\!]$$

by recursion over the term structure: variables are interpreted as (total) product projections, and operator application as composition of partial morphisms. Multi-terms are modelled by intersecting the domains of the components and tupling the resulting restrictions. Finally, $[\![\Gamma.\,\lambda\bar{y}:\bar{u}.\,\beta]\!]$ is defined as the abstraction of $[\![\Gamma,\bar{y}:\bar{u}.\,\beta]\!]$ (cf. Section 2). We will denote any existing domain-codomain restrictions of $[\![\Gamma.\,\bar{\alpha}]\!]$ to subobjects of $[\![\Gamma]\!]$ and $[\![\bar{t}]\!]$, respectively, by $[\![\Gamma.\,\bar{\alpha}]\!]$ as well.

This interpretation leads to a notion of satisfaction in $\mathbf{C}$:

Definition 7. An ece $\Gamma \rhd \mathrm{def}\,\bar{\alpha} \Rightarrow \beta_1 \overset{e}{=} \beta_2$ in Σ holds in $(\mathbf{C}, \mathcal{M})$ if $[\![\Gamma.\,\mathrm{def}\,\bar{\alpha}]\!]$ is contained in $[\![\Gamma.\,\mathrm{def}(\beta_1,\beta_2)]\!]$ and the restrictions of the $[\![\Gamma.\,\beta_i]\!]$ to $[\![\Gamma.\,\mathrm{def}\,\bar{\alpha}]\!]$ coincide.

The definition of $\mathsf{L}(\mathbf{C}, \mathcal{M})$ is completed by taking the eces that hold in $(\mathbf{C}, \mathcal{M})$ as the axioms of $\mathsf{L}(\mathbf{C}, \mathcal{M})$. The theory $\mathsf{L}(\mathbf{C}, \mathcal{M})$ has equality iff $(\mathbf{C}, \mathcal{M})$ does.

The deduction system of Figure 1 is sound for this interpretation:

Theorem 8 (Soundness). *All theorems of* $\mathsf{L}(\mathbf{C}, \mathcal{M})$ *hold in* $(\mathbf{C}, \mathcal{M})$.

The proof hinges on the following lemma:

Lemma 9 (Substitution). *Let* $\Gamma \rhd \bar{\alpha}:\bar{t}$ *and* $\Delta \rhd \bar{\beta}:\bar{u}$ *be multi-terms in* $\mathsf{L}(\mathbf{C}, \mathcal{M})$, *where* $\Delta = (\bar{y}:\bar{t})$. *Then* $[\![\Gamma.\,\bar{\alpha}]\!]$ *has a restriction* $[\![\Gamma.\,\mathrm{def}(\bar{\beta}[\bar{\alpha}/\bar{y}],\bar{\alpha})]\!] \rightarrow [\![\Delta.\,\mathrm{def}\,\bar{\beta}]\!]$. *In the arising diagram*

$$
\begin{array}{ccc}
 & [\![\Gamma.\,\mathrm{def}(\bar{\beta}[\bar{\alpha}/\bar{y}],\bar{\alpha})]\!] & \hookrightarrow & [\![\Gamma.\,\mathrm{def}\,\bar{\alpha}]\!] \\[1ex]
{\scriptstyle[\![\Gamma.\,\bar{\beta}[\bar{\alpha}/\bar{y}]]\!]}\nearrow & \downarrow {\scriptstyle[\![\Gamma.\,\bar{\alpha}]\!]} & & \downarrow {\scriptstyle[\![\Gamma.\,\bar{\alpha}]\!]} \\[1ex]
[\![\bar{u}]\!] \longleftarrow & [\![\Delta.\,\mathrm{def}\,\bar{\beta}]\!] & \hookrightarrow & [\![\Delta]\!] \\
\phantom{[\![\bar{u}]\!]}{\scriptstyle[\![\Delta.\,\bar{\beta}]\!]} & & &
\end{array}
$$

the triangle commutes and the square is a pullback.

4   Classifying Categories

Thanks to the internal logic, the effort required for the construction of a classifying category for a partial λ-theory with equality is essentially no greater than in the first order case as carried out in [20]: in that case, objects of the classifying category are pairs $(\Gamma.\,\phi)$ consisting of a context Γ and a definedness assertion ϕ in that context. This construction can be copied literally for partial λ-theories with equality; the point is that partial function spaces $(\Gamma.\,\phi) \rightharpoonup (\Delta.\,\psi)$ may safely be regarded as subobjects of the partial function space $\Gamma \rightharpoonup \Delta$, so that no additional objects are required to obtain partial cartesian closedness.

In general, given a partial λ-theory $\mathcal{T}$, we construct a *syntactic category* $\mathsf{Sy}(\mathcal{T})$ as follows: the objects are definedness-assertions-in-context $(\Gamma.\,\phi)$ as indicated above (i.e. $\phi \equiv \mathrm{def}\,\bar{\beta}$ for some multiterm $\bar{\beta}$). Morphisms $(\Gamma.\,\phi) \rightarrow (\Delta.\,\psi)$, where $\Gamma = (\bar{x}:\bar{s})$ and $\Delta = (\bar{y}:\bar{t})$, are multi-terms $\Gamma \rhd \bar{\alpha}:\bar{t}$ such that

$$\Gamma \rhd \phi \vdash \psi[\bar{\alpha}/\bar{y}] \wedge \mathrm{def}\,\bar{\alpha},$$

taken modulo equality deducible from ϕ. The identity on $(\Gamma.\phi)$ is $\bar{x}$; composition is substitution. A subobject is admissible (regular) in $\mathsf{Sy}(\mathcal{T})$ iff it has a representative of the form

$$\bar{x} : (\Gamma.\phi) \hookrightarrow (\Gamma.\psi).$$

It is shown as in the first order case [20] that $\mathsf{Sy}(\mathcal{T})$ is a cartesian dominional category, with products given by concatenation of contexts and conjunction of definedness assertions, $1 = (())$, and the pullback of $(\Gamma.\phi) \hookrightarrow (\Gamma.\psi)$ along a morphism $\bar{\alpha} : (\Delta.\chi) \to (\Gamma.\psi)$ being $(\Delta.\chi \wedge \phi[\bar{\alpha}/\bar{x}])$.

Theorem 10. *If $\mathcal{T}$ has equality, then $\mathsf{Sy}(\mathcal{T})$ is a pccc.*

Proof. Recall the logic defined using equality as described in Section 1. Given objects $(\Gamma.\phi)$ and $(\Delta.\psi)$ as above, and $\bar{z} : \bar{s} \dashrightarrow \bar{t}$ (cf. Section 1), let $dc(\bar{z},\bar{x})$ abbreviate the formula

$$(\mathrm{def}\; z_1\,\bar{x} \Rightarrow \mathrm{def}\; z_2\,\bar{x}) \wedge \ldots \wedge (\mathrm{def}\; z_m\,\bar{x} \Rightarrow \mathrm{def}\; z_1\,\bar{x})$$

ensuring that all components of $\bar{z}$ have the same domain of definition, and write $\bar{z}(\bar{x})$ for $(z_1(\bar{x}),\ldots,z_m(\bar{x}))$. Then the object

$$\left(\bar{z} : \bar{s} \dashrightarrow \bar{t}.\; \forall\bar{x}.\,dc(\bar{z},\bar{x}) \wedge \left(\mathrm{def}\; \bar{z}\,\bar{x} \Rightarrow (\phi \wedge \psi[\bar{z}\,\bar{x}/\bar{y}])\right)\right)$$

is the partial function space $(\Gamma.\phi) \dashrightarrow (\Delta.\psi)$, with $\bar{z}(\bar{x})$ as evaluation map. $\square$

A translation $\sigma : \mathcal{T}_1 \to \mathcal{T}_2$ between partial λ-theories induces a functor

$$\mathsf{Sy}(\sigma) : \mathsf{Sy}(\mathcal{T}_1) \to \mathsf{Sy}(\mathcal{T}_2)$$

which, in the case with equality, preserves the pccc structure since this structure has the syntactical description given above.

We will now show that $\mathsf{Sy}(\mathcal{T})$ is, for $\mathcal{T}$ with equality, free over $\mathcal{T}$ w.r.t. L. The corresponding unit is the translation

$$\eta_{\mathcal{T}} : \mathcal{T} \to \mathsf{L}(\mathsf{Sy}(\mathcal{T}))$$

that maps a sort s to $(x : s)$ and an operator $f : \bar{s} \to t$ to the operator in $\mathsf{L}(\mathsf{Sy}(\mathcal{T}))$ given by the partial morphism $(\bar{x} : \bar{s}) \longleftarrow\!\!\!\supset (\bar{x} : \bar{s}.\,\mathrm{def}\, f(\bar{x})) \xrightarrow{f(\bar{x})} (x : t)$. By the soundness theorem (for $\mathsf{L}(\mathsf{Sy}(\mathcal{T}))$), η *is a conservative extension.*

Given a pccc $(\mathbf{C},\mathcal{M})$, the co-unit

$$E_{(\mathbf{C},\mathcal{M})} : \mathsf{Sy}(\mathsf{L}(\mathbf{C},\mathcal{M})) \to (\mathbf{C},\mathcal{M})$$

of the adjunction maps an object $(\Gamma.\phi)$ in $\mathsf{Sy}(\mathsf{L}(\mathbf{C}))$ to $[\![\Gamma.\phi]\!]$ and a morphism $\bar{\alpha} : (\Gamma.\phi) \to (\Delta.\psi)$ to the composite

$$[\![\Gamma.\phi]\!] \hookrightarrow [\![\Gamma.\psi[\bar{\alpha}/\bar{y}] \wedge \mathrm{def}\, \bar{\alpha}]\!] \xrightarrow{[\![\Gamma.\bar{\alpha}]\!]} [\![\Delta.\psi]\!]$$

of the inclusion provided by soundness theorem and the restriction of $[\![\Gamma.\bar{\alpha}]\!]$ according to the substitution lemma. Using the soundness theorem and the substitution lemma, it is shown that this defines a dominional *equivalence*. In particular, $\mathsf{Sy}(\mathsf{L}(\mathbf{C},\mathcal{M}))$ is a pccc, and $E_{(\mathbf{C},\mathcal{M})}$ is a pcc functor.

This is all we need in order to prove

Theorem 11. *If $\mathcal{T}$ has equality, then $\mathsf{Sy}(\mathcal{T})$ is the free pccc over $\mathcal{T}$ in the sense that any translation $\sigma : \mathcal{T} \to \mathsf{L}(\mathbf{C},\mathcal{M})$, where $(\mathbf{C},\mathcal{M})$ is a pccc, factors essentially uniquely as $\mathsf{L}(\sigma^{\#})\eta_{\mathcal{T}}$, where $\sigma^{\#} : \mathsf{Sy}(\mathcal{T}) \to (\mathbf{C},\mathcal{M})$.*

Here, 'essentially' means that $\sigma^{\#}$ is unique up to a unique natural isomorphism. Thus, $\mathsf{Sy}(\mathcal{T})$ is determined up to equivalence by this property.

Proof. The uniqueness statement is clear. To prove existence, just note that $\sigma^{\#} := E_{(\mathbf{C},\mathcal{M})} \circ \mathsf{Sy}(\sigma)$ has the required properties. $\qquad\qquad\square$

This theorem justifies calling $\mathsf{Sy}(\mathcal{T})$ the *classifying category* of $\mathcal{T}$, denoted from now on by $\mathsf{Cl}(\mathcal{T})$. (For partial λ-theories without equality, $\mathsf{Sy}(\mathcal{T})$ and $\mathsf{Cl}(\mathcal{T})$ will in general be different.) Since $E_{(\mathbf{C},\mathcal{M})}$ is an equivalence, the category of pcccs with equality is essentially (i.e. up to 2-dimensional equivalence) the Kleisli category of the (2-)adjunction $\mathsf{Cl} \dashv \mathsf{L}$. The objects of this category are the partial λ-theories with equality; the morphisms from $\mathcal{T}_1$ to $\mathcal{T}_2$ are the translations from $\mathcal{T}_1$ to $\mathsf{L}(\mathsf{Cl}(\mathcal{T}_2))$. These morphisms are naturally generalized translations: sorts are mapped to 'types', i.e. domains of multi-terms, and symbols are mapped to multi-terms (all partial morphisms in $\mathsf{Sy}(\mathcal{T}_2)$ may be written in the form $(\Gamma.\phi) \xleftarrow{\quad} (\Gamma.\,\mathsf{def}\,\bar{\alpha}) \xrightarrow{\ \bar{\alpha}\ } \bullet$); moreover, two morphisms of this kind are identified if they map all symbols to strongly equal multi-terms. We have established that

> pcccs with equality are equivalent to partial λ-theories with equality.

Remark 12. Without equality, the construction of the classifying category becomes more complex, since it is no longer possible to define partial function spaces $(\Gamma.\phi) \multimap (\Delta.\psi)$ as subspaces of $\Gamma \multimap \Delta$. (The obvious idea of using the Yoneda extension is probably not the right one, for reasons laid out in [19].) In [19], this problem is solved by moving to an extended theory with equality and a dominance [18]; the classifying category of the original theory is then obtained as a subcategory of the classifying category of the extended theory, the latter being constructed as above. This establishes an equivalence between partial λ-theories and pcccs.

5   Unique Choice

We now proceed with the investigation of the higher order logic induced by a partial λ-theory with equality, exploiting the equivalence result proved above. In particular, we make use of the fact that every pccc $\mathbf{C}$ with equality is equivalent to $\mathsf{Cl}(\mathsf{L}(\mathbf{C}))$; in fact, in the following we shall not distinguish between these two categories at all.

It is at first sight slightly puzzling that pcccs with equality are equivalent to intuitionistic HOL, although the latter is more commonly associated with toposes; see e.g. [11], where in fact toposes are constructed from type theories that can be translated into the partial λ-calculus with equality. We stress that pcccs with equality are substantially weaker than even quasitoposes [1], which in turn are way more general than toposes — e.g., there are many non-trivial quasitoposes in topology [28], while the only topos which is at the same time a topological category over **Set** is **Set** itself. It turns out that the crucial point here is unique choice.

For the remainder of this section, let **C** be a pccc with equality. It can be shown that a morphism in **C** is a monomorphism (epimorphism) iff the obvious internal formula expressing injectivity (surjectivity) holds in **C**. In particular, given a morphism $f : A \rightarrow B$, its factorization through the subobject

$$(b : B. \exists a : A. f(a) \stackrel{e}{=} b)$$

of B is an (Epi, Regular Mono)-factorization, i.e. (Epi, Regular Mono) is a factorization structure on **C** (thus, **C** automatically satisfies condition 19.1.1. in [28]). Hence, all extremal monomorphisms in **C** are regular; thus, we obtain a further characterization of pcccs with equality as *cartesian closed categories with representable extremal partial morphisms*. In particular, quasitoposes are precisely the finitely cocomplete pcccs with equality.

For each object A in **C**, we have a type

$$Sg(A) := (x : A \multimap 1. \exists! a : A. x\, a)$$

of singleton subsets (predicates) of A. A morphism $sg_A : A \rightarrow Sg(A)$ is given by the term $(a : A) \triangleright \lambda b : A. b \stackrel{e}{=} a$.

Proposition 13. *Let A be an object in* **C**. *Then the following are equivalent:*

(i) *sg_A is an isomorphism.*
(ii) *Every partial functional relation with codomain A is a partial function; i.e. in context $R : BA \multimap 1$, with B a further* **C**-*object, the formula*

$$(\forall x : B, y, z : A. R(x, y) \wedge R(x, z) \Rightarrow y \stackrel{e}{=} z) \implies$$
$$(\exists f : B \multimap A. \forall x : B, y : A. f\, x \stackrel{e}{=} y \Leftrightarrow R(x, y))$$

holds.
(iii) *Every monomorphism with domain A is extremal.*

An object that satisfies the equivalent conditions in the above proposition is called *coarse*, following [16] and [28], where Conditions (i) and (iii) are used. Condition (ii) is often referred to as *unique choice* (although it is usually formulated in terms of total functions). An inverse f of sg_A can be regarded as a partial morphism from $A \multimap 1$ to A. Thus, we can define the *unique description operator* by

$$\iota a : A. \phi := f(\lambda a : A. \phi)$$

for a formula ϕ in context $a : A$ — i.e. $\iota a : A. \phi$ is the unique element of A satisfying ϕ, if such an element indeed exists uniquely, and is otherwise undefined.

An immediate consequence of the above proposition is that we can axiomatize toposes in the partial λ-calculus:

Theorem 14. *The classifying category of a partial λ-theory $\mathcal{T}$ with equality is a topos iff $\mathcal{T}$ implies the unique choice axiom for all types.*

Thus, the initial question of what class of categories intuitionistic HOL really corresponds to may the resolved as follows:

> *Pcccs with equality 'are' intuitionistic HOL,*
> *toposes are intuitionistic HOL with unique description.*

In the construction of the classifying topos given in [11], unique choice is implicit in that morphisms *are* functional relations; in the same way, one can construct a topos from **C** (i.e. from a partial λ-theory with equality). An alternative way of obtaining an equivalent topos is the following observation.

Theorem 15. *The subcategory $\mathbf{Ind}(\mathbf{C})$ of coarse objects is a bireflective subtopos of $\mathbf{C}$ (i.e. the reflective arrows are bimorphisms), with reflective arrows sg_A. Moreover, $\mathbf{Ind}(\mathbf{C})$ is a topos coreflection of $\mathbf{C}$ in the sense that every pcc functor $\mathbf{E} \to \mathbf{C}$, with $\mathbf{E}$ a topos, factors through $\mathbf{Ind}(\mathbf{C})$.*

(The first clause of this theorem slightly generalizes results of [16, 28].) In particular, $\mathbf{Ind}(\mathbf{C})$ contains Ω and all objects of the form $Sg(A)$, since for these objects, the unique choice function can actually be written as a term. $\mathbf{Ind}(\mathbf{C})$ is equivalent to the topos of functional relations over $\mathbf{C}$ because sg_A becomes an isomorphism when regarded as a functional relation.

An important consequence of Theorem 14 is that results obtained using the interplay of partial λ-theories and pcccs apply also to toposes. This includes in particular the equivalence result for Henkin models of pcccs [21], which in its thus obtained specialized form states that, given a topos $\mathbf{E}$, models (logical morphisms) of $\mathbf{E}$ in toposes are essentially equivalent to Henkin models of $\mathbf{E}$, i.e. lex functors $\mathbf{E} \to \mathbf{Set}$.

Remark 16. A further point regarding the relationship of these results to [11] that requires clarification is the following. In loc. cit., it is claimed (correctly) that the extension of a type theory to the internal language of its classifying topos, constructed as the topos of functional relations, is conservative. The type theory used in loc. cit. can be regarded as a sublanguage of the partial λ-calculus with internal equality; the fact that the latter does not prove unique choice, which however holds in all toposes, appears at first sight to contradict the mentioned conservativity result. However, this is resolved by noting that the type theory of [11] (like most other versions of topos logic including the original Mitchell-Bénabou language [12]) in fact cannot *express* unique choice, since it does not have actual function types. In other words, the logic of pcccs with internal equality differs from topos logic in that it takes functions rather than subsets as the primitive notion and then distinguishes between 'maps' (functional relations) and 'morphisms' (functions).

6 Dependent Types

An important aspect of toposes is that they admit a Martin-Löf style dependent type theory; categorically, this means that every topos $\mathbf{E}$ is *locally cartesian closed* [9, 27], i.e. every slice $\mathbf{E}/A$ is cartesian closed — this is the non-trivial part of the *fundamental theorem of toposes* [10]. It is known that this theorem holds also for quasitoposes [16, 28], and a proof that the statement generalizes to pcccs with equality can be extracted from [28]. This implies that the partial λ-calculus with equality already includes dependent type theory, in particular has dependent product types. We now give a simple linguistic proof of the fundamental theorem; moreover, we present a novel notion of locally partial cartesian closed category.

The intuition behind the correspondence between local cartesian closedness and dependent types is the following. Let $\mathbf{C}$ be locally cartesian closed. A type C depending on a variable $y : B$ is regarded as a *bundle*, i.e. a morphism $g : C \to B$, with $C(y)$ being the fibre of g over y. Dependent sum types are then defined simply by composition: if the type $B = B(x)$ depends on a variable $x : A$, i.e. is a bundle $f : B \to A$, then $\sum y : B(x). C(y) = \sum_f g$ is just the composite $fg : C \to A$. Dependent product types, on the other hand, arise by exponentiation in the slice category. The point here is that local cartesian closedness is equivalent to the existence of right adjoints Π_f for all pullback functors $f^* : \mathbf{C}/A \to \mathbf{C}/B$, $f : B \to A$ [7]. Intuitively, for types $B(x)$, $D(x)$ depending on $x : A$, i.e. bundles $f : B \to A$, $h : D \to A$, the fibre over $x : A$ of the function space $f \to h$ in $\mathbf{C}/A$ is the function space $B(x) \to D(x)$. For $g : C \to B$ and $f : B \to A$ as above, the fibre over $x : A$ of $\Pi y : B. C(y) = \Pi_f g$ is the subspace of sections of g in the fibre of $f \to fg$; this fibre may be thought of as $B(x) \to \sum y : B(x). C(y)$.

The mentioned characterization of local cartesian closedness can be generalized to the partial setting: for an object A in a dominional category $(\mathbf{C}, \mathcal{M})$, the $\mathcal{M}$-carried morphisms form a dominion on $\mathbf{C}/A$, also denoted $\mathcal{M}$. For $f : B \to A$, $h : D \to A$, the morphisms $f \to h$ in $\mathbf{P}(\mathbf{C}/A, \mathcal{M})$ are *lax* commutative triangles, i.e. partial morphisms $k : B \rightharpoonup D$ such that $hk = f$ holds *on the domain of k*.

Theorem and Definition 17. *Let $(\mathbf{C}, \mathcal{M})$ be a lex dominional category. Then $(\mathbf{C}/A, \mathcal{M})$ is a pccc for each object A iff the pullback functor*

$$f^* : \mathbf{C}/A \to \mathbf{C}/B \hookrightarrow \mathbf{P}(\mathbf{C}/B, \mathcal{M})$$

has a right adjoint Π_f^p for each morphism $f : B \to A$. In this case, $(\mathbf{C}, \mathcal{M})$ is called locally partial cartesian closed.

The elements of $\Pi_f^p g$ are *partial* sections of g, i.e. partial functions h such that $gh = id$ on the domain of h. In particular, partial function spaces $A \rightharpoonup B$ in $(\mathbf{C}, \mathcal{M})$ may be recovered as $\Pi^p x : A. B$.

In this terminology, the fundamental theorem reads as follows.

Theorem 18. *Every pccc with equality is locally partial cartesian closed.*

As mentioned above, a categorical proof can essentially be found in [28]. The equivalence result of Section 4 allows a simple and transparent linguistic proof:

Proof. Let $\mathbf{C}$ be a pccc with equality. The partial function space $f \rightharpoonup g$ for two objects $f : B \to A$, $g : C \to A$ of $\mathbf{C}/A$, expressed in $\mathsf{Cl}(\mathsf{L}(\mathbf{C}))$, is

$$(x : A, y : B \rightharpoonup C. \ \forall z : B. \ \mathrm{def} \, y \, z \Rightarrow (f(z) \stackrel{e}{=} x \wedge g(y \, z) \stackrel{e}{=} x)) \qquad \square$$

Conclusions and Future Work

We have identified the logic of the partial λ-calculus with equality as the internal logic of partial cartesian closed categories (pcccs) with equality. Building on this result, we have clarified the relationship of this logic with various other higher order logics that have categorical counterparts. In particular, a partial λ-theory with equality is the internal language of a topos iff it satisfies the unique choice axiom, and the partial λ-calculus with equality encodes a dependent type theory with partial dependent products — a known generalization of the fundamental theorem of toposes to pcccs with equality [16, 28], for which we give a transparent linguistic proof. An open problem that remains is to find, somewhere between topos logic and the partial λ-calculus, the precise internal logic of quasitoposes.

This work forms part of the semantical foundations of HASCASL. The equivalence of pcccs and partial λ-theories is needed to prove the equivalence between the semantics of the partial λ-calculus in pcccs on the one hand and a Henkin-style set-theoretic model theory on the other hand [21]. The relevance of unique choice, universally or for certain types, has become apparent e.g. in [22]. The implications of the fact that dependent types are encodable in the partial λ-calculus w.r.t. the specification methodology of HASCASL are under investigation.

Acknowledgements

The author wishes to thank Till Mossakowski and Christoph Lüth for useful comments and discussions.

References

1. J. Adámek, H. Herrlich, and G. E. Strecker, *Abstract and concrete categories*, Wiley Interscience, 1990.
2. M. Bidoit and P. D. Mosses, CASL *user manual*, LNCS, vol. 2900, Springer, 2004.
3. M. Broy, C. Facchi, R. Grosu, R. Hettler, H. Hussmann, D. Nazareth, F. Regensburger, and K. Stølen, *The requirement and design specification language SPECTRUM*, Tech. report, Technical University of Munich, 1993.
4. A. Bucalo, C. Führmann, and A. Simpson, *Equational lifting monads*, Category Theory and Computer Science, ENTCS, vol. 29, 1999.
5. P. Burmeister, *A model theoretic oriented approach to partial algebras*, Akademie-Verlag, 1986.

6. P.-L. Curien and A. Obtułowicz, *Partiality, cartesian closedness and toposes*, Information and Computation **80** (1989), 50–95.
7. P. Freyd, *Aspects of topoi*, Bull. Austral. Math. Soc. **7** (1972), 1–76.
8. C. George, P. Haff, K. Havelund, A. Haxthausen, R. Milne, C. Nielson, S. Prehn, and K. Wagner, *The Raise Specification Language*, Prentice Hall, 1992.
9. M. Hofmann, *On the interpretation of type theory in locally cartesian closed categories*, Computer Science Logic, LNCS, vol. 933, 1995, pp. 427–441.
10. P. Johnstone, *Sketches of an elephant: a topos theory compendium*, vol. 2, Clarendon, 2002.
11. J. Lambek and P. Scott, *Introduction to higher order categorical logic*, Cambridge University Press, 1986.
12. W. Mitchell, *Boolean topoi and the theory of sets*, J. Pure Appl. Algebra **2** (1972), 261–274.
13. E. Moggi, *Categories of partial morphisms and the λ_p-calculus*, Category Theory and Computer Programming, LNCS, vol. 240, Springer, 1986, pp. 242–251.
14. _____, *The partial lambda calculus*, Ph.D. thesis, University of Edinburgh, 1988.
15. P. D. Mosses (ed.), CASL *reference manual*, LNCS, vol. 2960, Springer, 2004.
16. J. Penon, *Quasi-topos*, C. R. Acad. Sci., Paris, Sér. A **276** (1973), 237–240.
17. E. Robinson and G. Rosolini, *Categories of partial maps*, Inform. and Comput. **79** (1988), 95–130.
18. G. Rosolini, *Continuity and effectiveness in topoi*, Ph.D. thesis, Merton College, Oxford, 1986.
19. L. Schröder, *The* HASCASL *prologue: categorical syntax and semantics of the partial λ-calculus*, available as **http://www.informatik.uni-bremen.de/~lschrode/hascasl/plam.ps**
20. _____, *Classifying categories for partial equational logic*, Category Theory and Computer Science, ENTCS, vol. 69, 2002.
21. _____, *Henkin models of the partial λ-calculus*, Computer Science Logic, LNCS, vol. 2803, Springer, 2003, pp. 498–512.
22. L. Schröder and T. Mossakowski, *Monad-independent dynamic logic in* HASCASL, J. Logic Comput., to appear. Earlier version in Recent Developments in Algebraic Development Techniques (WADT 02), LNCS, vol. 2755, Springer, 2003, 425–441.
23. _____, HASCASL*: Towards integrated specification and development of functional programs*, Algebraic Methodology And Software Technology, LNCS, vol. 2422, Springer, 2002, pp. 99–116.
24. _____, *Monad-independent Hoare logic in* HASCASL, Fundamental Aspects of Software Engineering (M. Pezze, ed.), LNCS, vol. 2621, 2003, pp. 261–277.
25. L. Schröder, T. Mossakowski, and C. Maeder, HASCASL – *Integrated functional specification and programming. Language summary*, available at **http://www.informatik.uni-bremen.de/agbkb/forschung/formal_methods/CoFI/HasCASL**
26. D. Scott, *Continuous lattices*, Toposes, Algebraic Geometry and Logic, LNM, vol. 274, Springer, 1972, pp. 97–136.
27. R. A. G. Seely, *Locally cartesian closed categories and type theory*, Math. Proc. Cambridge Philos. Soc. **95** (1984), 33–48.
28. O. Wyler, *Lecture notes on topoi and quasitopoi*, World Scientific, 1991.

Complete Lax Logical Relations for Cryptographic Lambda-Calculi*

Jean Goubault-Larrecq[1], Sławomir Lasota[2,**], David Nowak[1], and Yu Zhang[1,***]

[1] LSV/CNRS & INRIA Futurs & ENS Cachan, Cachan, France
{goubault,nowak,zhang}@lsv.ens-cachan.fr
[2] Institute of Informatics, Warsaw University, Warszawa, Poland
sl@mimuw.edu.pl

Abstract. Security properties are profitably expressed using notions of contextual equivalence, and logical relations are a powerful proof technique to establish contextual equivalence in typed lambda calculi, see e.g. Sumii and Pierce's logical relation for a cryptographic lambda-calculus. We clarify Sumii and Pierce's approach, showing that the right tool is prelogical relations, or lax logical relations in general: relations should be lax at encryption types, notably. To explore the difficult aspect of fresh name creation, we use Moggi's monadic lambda-calculus with constants for cryptographic primitives, and Stark's name creation monad. We define logical relations which are lax at encryption and function types but strict (non-lax) at various other types, and show that they are sound and complete for contextual equivalence at all types.

Keywords: Logical relations, Monads, Cryptographic lambda-calculus, Subscone

1   Introduction

There are nowadays many existing models for cryptographic protocol verification. The most well-known are perhaps the Dolev-Yao model (after [7], see [6] for a survey) and the spi-calculus of [1]. A lesser known model was introduced by Sumii and Pierce [18], the *cryptographic lambda-calculus*. This has certain advantages; notably, higher-order behaviors are naturally taken into account, which is ignored in other models (although, at the moment, higher order is not perceived as a needed feature in cryptographic protocols). Better, second-order terms naturally encode asymmetric encryption. It may also be appealing to consider that proving security properties in the cryptographic lambda-calculus can be achieved through the use of well-crafted *logical relations*, a tool that has been used many times with considerable success in the λ-calculus: see [12, Chapter 8],

* Partially supported by the RNTL project Prouvé, the ACI Sécurité Informatique Rossignol, the ACI jeunes chercheurs "Sécurité informatique, protocoles cryptographiques et détection d'intrusions", and the ACI Cryptologie "PSI-Robuste".
** Partially supported by the KBN grant 7 T11C 002 21 and by the Research Training Network *Games*. Part of this work was performed during the author's stay at LSV.
*** PhD student under an MENRT grant on ACI Cryptologie funding, École Doctorale Sciences Pratiques (Cachan).

J. Marcinkowski and A. Tarlecki (Eds.): CSL 2004, LNCS 3210, pp. 400–414, 2004.

for numerous examples. Sumii and Pierce [18] in particular define three logical relations that can be used to establish contextual equivalence, hence prove security properties, but completeness remains open.

Our contributions are twofold: first, we clarify the import of Sumii and Pierce as far as the behavior of logical relations on encryption types is concerned, and simplify it to the point that we reduce it to prelogical relations [10] and more generally to lax logical relations [16]; while standard recourses to the latter were usually required because of arrow types, here we require the logical relations to be lax at *encryption types*. Second, we prove various completeness results: two terms are contextually equivalent if and only if they are related by some lax logical relation. This holds at all types, not just first-order types as in previous works. An added bonus of using lax logical relations is that they extend directly to more complex models of encryption, where cryptographic primitives may obey algebraic laws. Proofs omitted in the sequel are to be found in the full version of this paper, available as a technical report [9].

Outline. We survey related work in Section 2. We focus on the approach of Sumii and Pierce, in which they define several rather complex logical relations as sound criteria of contextual equivalence. We take a new look at this approach in Section 3 and Section 4, and gradually deconstruct their work to the point where we show the power of prelogical relations in action. This is shown in the absence of fresh name creation, for added clarity. We tackle the difficult issue of names in Section 5, using Moggi's elegant computational λ-calculus framework with Stark's name creation monad.

2   Related Work

Logical relations have often been used to prove various properties of typed lambda calculi. We are interested here in using logical relations or variants thereof as sound criteria for establishing *contextual equivalence* of two programs. This is instrumental in defining security properties. As noticed in [1, 18], a datum M of type τ is *secret* in some term $t(M)$ of type τ' if and only if no intruder can say anything about M just by looking at $t(M)$, i.e., if and only if $t(M) \approx_{\tau'} t(M')$ for any two M and M', where $\approx_{\tau'}$ denotes contextual equivalence at type τ'. We are using λ-calculus notions here, following [18], but the idea of using contextual equivalence to define security properties was pioneered by Abadi and Gordon [1], where both secrecy and authentication are investigated.

We shall define precisely what we mean by contextual equivalence in a calculus without names (Section 3.2), then with names (Section 5.3). Both notions are standard, the latter being inspired by [15], only adapted to Moggi's computational λ-calculus [14]. In [15] and some other places, this kind of equivalence, which states that two values (or terms) a and a' are equivalent provided every context of type bool must give identical results on a and on a', is called observational equivalence. We stress that this should not be confused with observational equivalence as it is defined for data refinement [12], where *models* are related, not *values* in the same model as here.

The main point in passing from contextual equivalence to logical relations is to avoid the universal quantification over contexts in the former. But there are two kinds of technical difficulties one must face in defining logical relations for cryptographic λ-

calculi. The first, and hardest one, is *fresh name creation*. The second is dealing with encryption and decryption. We shall see that the latter has an elegant solution in terms of *prelogical* relations [10], which we believe is both simpler and more general than Sumii and Pierce's proposal [18]; this is described in Section 3, although we ignore fresh name creation there, for clarity.

Dealing with fresh name creation is harder. The work of Sumii and Pierce [18] is inspired in this respect by Pitts and Stark [15], who proposed a λ-calculus devoted to the study of fresh name creation, the *nu-calculus*. They define a so-called operational logical relation to establish observational equivalence of nu-calculus expressions. They prove that this logical relation is complete up to first-order types.

In [8], Goubault-Larrecq, Lasota and Nowak define a Kripke logical relation for the dynamic name creation monad, which is extended by Zhang and Nowak in [19] so that it coincides with Pitts and Stark's operational logical relation up to first-order types. We continue this work here, relying on the elegance of Moggi's computational λ-calculus [14] to describe side effects, and in particular name creation, using Stark's insights [17].

Further comparisons will be made in the course of this paper, especially with bisimulations for spi-calculus [1, 4, 5]. This continues the observations pioneered in [8], where notions of logical relations for various monads were shown to be proper extensions of known notions of bisimulations. The precise relation with hedged and framed bisimulation [5] remains to be stated precisely.

3   Deconstructing Sumii and Pierce's Approach

The starting point of this paper was the realization that the rather complex family of logical relations proposed by Sumii and Pierce [18] could be simplified in such a way that it could be described as merely *one* way of building logical relations that have all desired properties. It turned out that the only property we really need to be able to deal with encryption and decryption primitives is that the logical relations should relate the encryption function with itself, and the decryption function with itself.

3.1   The Toy Cryptographic λ-Calculus

To show the idea in action, let us use a minimal extension of the simply-typed λ-calculus with encryption and decryption, and call it the *toy cryptographic λ-calculus*. We shall show how the idea works on this calculus, which is just a fragment of Sumii and Pierce's [18] cryptographic λ-calculus. The main thing that is missing here is nonce creation, i.e., fresh name creation.

For this moment, we restrict the types to:

$$\tau ::= b \mid \tau_1 \rightarrow \tau_2 \mid \text{key}[\tau] \mid \text{bits}[\tau]$$

where b ranges over a set Σ of so-called *base types*, e.g., integers, booleans, etc. Sumii and Pierce's calculus in addition has cartesian product and coproduct types. $\text{key}[\tau]$ is the type of (symmetric) keys that can be used to encrypt values of type τ, $\text{bits}[\tau]$ is the type of *ciphertexts* obtained by encrypting some value of type τ—necessarily with a key of type $\text{key}[\tau]$. There is no special type for nonces, which are being thought as objects of type $\text{key}[\tau]$ for some τ.

The terms of the toy cryptographic λ-calculus are given by the grammar:

$$t, u, v, \ldots ::= \quad x \mid \lambda x \cdot t \mid tu \mid \{t\}_u \mid \texttt{let } \{x\}_t = u \texttt{ in } v_1 \texttt{ else } v_2$$

where x ranges over a countable set of variables, $\{t\}_u$ denotes the ciphertext obtained by encrypting t with key u (t is called the *plaintext*), and $\texttt{let } \{x\}_t = u \texttt{ in } v_1 \texttt{ else } v_2$ is meant to evaluate u, attempt to decrypt it using key t, then proceed to evaluate v_1 with plaintext stored in x if decryption succeeded, or evaluate v_2 if decryption failed. Definitions of free and bound variables and α-renaming are standard, hence omitted; x is bound in $\lambda x \cdot t$, with scope t, as well as in $\texttt{let } \{x\}_t = u \texttt{ in } v_1 \texttt{ else } v_2$, with scope v_1.

Typing is as one would expect. *Judgments* are of the form $\Gamma \vdash t : \tau$, where Γ is a *context*, i.e., a finite mapping from variables to types. If Γ maps x_1 to $\tau_1, \ldots, x_n$ to τ_n, we write it $x_1 : \tau_1, \ldots, x_n : \tau_n$. Typing rules for encryption and decryption are

$$\frac{\Gamma \vdash t : \tau \qquad \Gamma \vdash u : \texttt{key}[\tau]}{\Gamma \vdash \{t\}_u : \texttt{bits}[\tau]} \ (Enc)$$

$$\frac{\Gamma \vdash t : \texttt{key}[\tau] \qquad \Gamma \vdash u : \texttt{bits}[\tau] \qquad \Gamma, x : \tau \vdash v_1 : \tau' \qquad \Gamma \vdash v_2 : \tau'}{\Gamma \vdash \texttt{let } \{x\}_t = u \texttt{ in } v_1 \texttt{ else } v_2 : \tau'} \ (Dec)$$

A simple denotational semantics for the typed toy cryptographic calculus is as follows. Let $[\![\_]\!]$ be any function mapping types τ to sets so that $[\![\tau_1 \to \tau_2]\!]$ is the set $[\![\tau_1]\!] \to [\![\tau_2]\!]$ of all functions from $[\![\tau_1]\!]$ to $[\![\tau_2]\!]$, for all types τ_1 and τ_2. Let $[\![b]\!]$ be arbitrary for every base type b, $[\![\texttt{key}[\tau]]\!]$ be arbitrary. For every $V \in [\![\tau]\!]$, $K \in [\![\texttt{key}[\tau]]\!]$, write $E(V, K)$ the pair (V, K), to suggest that this really denotes the encryption of V with key K. (That ciphertexts are just modeled as pairs is exactly as in modern versions of the Dolev-Yao model [7], or in the spi-calculus [1].) Then, let $[\![\texttt{bits}[\tau]]\!]$ be the set of all pairs $E(V, K)$, $V \in [\![\tau]\!]$, $K \in [\![\texttt{key}[\tau]]\!]$.

For any set A, let $A_\perp$ be the disjoint sum of A with $\{\perp\}$, where $\perp$ is an element outside A, and let ι be the canonical injection of A into $A_\perp$. While we have defined $E(V, K)$ as the pair (V, K), we define the inverse decryption function from $[\![\texttt{bits}[\tau]]\!] \times [\![\texttt{key}[\tau]]\!]$ to $[\![\tau]\!]_\perp$ by letting $D(V', K')$ be $\iota(V)$ if V' is of the form (V, K) with $K = K'$, and $\perp$ otherwise. We then describe the value $[\![t]\!] \rho$ of the term t in the environment ρ by structural induction on t,

$$[\![\Gamma, x : \tau \vdash x : \tau]\!] \rho = \rho(x)$$
$$[\![\Gamma \vdash \lambda x \cdot t : \tau_1 \to \tau_2]\!] \rho = (V \in [\![\tau_1]\!] \mapsto [\![\Gamma, x : \tau_1 \vdash t : \tau_2]\!] \rho[x := V])$$
$$[\![\Gamma \vdash tu : \tau_2]\!] \rho = [\![\Gamma \vdash t : \tau_1 \to \tau_2]\!] \rho([\![\Gamma \vdash u : \tau_1]\!] \rho)$$
$$[\![\Gamma \vdash \{t\}_u : \texttt{bits}[\tau]]\!] \rho = E([\![\Gamma \vdash t : \tau]\!] \rho, [\![\Gamma \vdash u : \texttt{key}[\tau]]\!] \rho)$$
$$[\![\texttt{let } \{x\}_t = u \texttt{ in } v_1 \texttt{ else } v_2]\!] \rho = \begin{cases} [\![v_1]\!] \rho[x := V_1] & \text{if } V = \iota(V_1) \\ [\![v_2]\!] \rho & \text{if } V = \perp \end{cases}$$
$$\text{where } V = D([\![u]\!] \rho, [\![t]\!] \rho)$$

More formally, for any context Γ, a Γ-*environment* ρ is a map such that, for every $x : \tau$ in Γ, $\rho(x)$ is an element of $[\![\tau]\!]$. Write $\rho[x := V]$ the environment mapping x to V and every other variable y to $\rho(y)$. Write $[x := V]$ the environment mapping just x to

V. We write $(V \in A \mapsto f(V))$ the (set-theoretic) function mapping V in A to $f(V)$ to distinguish it from the (syntactic) λ-abstraction $\lambda x \cdot f(x)$. In $[\![\Gamma \vdash tu : \tau_2]\!] \rho$, we assume that the premises of the last rule of the implicit typing derivation are $\Gamma \vdash t : \tau_1 \to \tau_2$ and $\Gamma \vdash u : \tau_1$. We write $[\![t]\!]$ instead of $[\![t]\!] \rho$ when the environment ρ is irrelevant, e.g., an empty environment.

3.2   What Are Logical Relations for Encryption?

We first fix a subset Obs of Σ, of so-called *observation types*. Typically, Obs will contain just the type bool of Booleans, one of the base types. We say that $a, a' \in [\![\tau]\!]$ are *contextually equivalent*, and we write $a \approx_\tau a'$, in the set-theoretic model above if and only if, whatever the term $\mathcal{C}$ such that $x : \tau \vdash \mathcal{C} : o$ is derivable ($o \in$ Obs), $[\![\mathcal{C}]\!] [x := a] = [\![\mathcal{C}]\!] [x := a']$.

In the λ-calculus setting, a (binary) *logical relation* is a family $(\mathcal{R}_\tau)_{\tau \text{ type}}$ of binary relations $\mathcal{R}_\tau$, one for each type τ, on $[\![\tau]\!]$, such that:

(Log) $\forall f, f' \in [\![\tau_1 \to \tau_2]\!], f\ \mathcal{R}_{\tau_1 \to \tau_2}\ f' \Leftrightarrow (\forall a\ \mathcal{R}_{\tau_1}\ a',\ f(a)\ \mathcal{R}_{\tau_2}\ f'(a'))$.

Here we write $a\ \mathcal{R}\ a'$ to say that a and a' are related by the binary relation $\mathcal{R}$. In other words, logical relations relate exactly those functions that map related arguments to related results. This is the standard definition of logical relations in the λ-calculus [12]. Note that there is no constraint on base types. In the typed λ-calculus, i.e., without encryption and decryption, the condition above forces $(\mathcal{R}_\tau)_{\tau \text{ type}}$ to be uniquely determined, by induction on types, from the relations $\mathcal{R}_b$, $b \in \Sigma$. More importantly, it entails the so-called *basic lemma*. To state it, first say that two Γ-environments ρ, ρ' are *related* by the logical relation, in notation $\rho\ \mathcal{R}_\Gamma\ \rho'$, if and only if $\rho(x)\ \mathcal{R}_\tau\ \rho'(x)$ for every $x : \tau$ in Γ. The basic lemma states that if $\Gamma \vdash t_0 : \tau$ is derivable, and ρ, ρ' are two related Γ-environments, then $[\![t_0]\!] \rho\ \mathcal{R}_\tau\ [\![t_0]\!] \rho'$. This is a simple induction on (the typing derivation of) t_0.

We are interested in the basic lemma because, as observed e.g. in [18], this implies that for any logical relation that coincides with equality on observation types, any two terms with logically related values are contextually equivalent.

In the toy cryptographic λ-calculus, we have left the definition of $\mathcal{R}_{\text{key}[\tau]}$ and $\mathcal{R}_{\text{bits}[\tau]}$ open. Here are conditions under which the basic lemma holds in the toy cryptographic λ-calculus. For any type τ, let $\mathcal{R}_{\tau \text{ option}}$ be the binary relation on $[\![\tau]\!]_\perp$ defined by $V\ \mathcal{R}_{\tau \text{ option}}\ V'$ if and only if $V = V' = \perp$, or $V = \iota(V_1)$, $V' = \iota(V_1')$ for some V_1, V_1', and $V_1\ \mathcal{R}_\tau\ V_1'$.

Lemma 1. *Assume that:*
1. for every $V\ \mathcal{R}_\tau\ V'$ and $K\ \mathcal{R}_{\text{key}[\tau]}\ K'$, $\mathsf{E}(V, K)\ \mathcal{R}_{\text{bits}[\tau]}\ \mathsf{E}(V', K')$;
2. for every $V\ \mathcal{R}_{\text{bits}[\tau]}\ V'$ and $K\ \mathcal{R}_{\text{key}[\tau]}\ K'$, $\mathsf{D}(V, K)\ \mathcal{R}_{\tau \text{ option}}\ \mathsf{D}(V', K')$.
Then the basic lemma holds: if $\Gamma \vdash t_0 : \tau$ is derivable, and ρ, ρ' are two related Γ-environments, then $[\![t_0]\!] \rho\ \mathcal{R}_\tau\ [\![t_0]\!] \rho'$.

Before we proceed, let us remark that we do not need *any* property of E or D in the proof of this lemma. The property that $\mathsf{D}(\mathsf{E}(V, K), K) = \iota(V)$ is only needed to show that let $\{x\}_t = \{u\}_t$ in v_1 else v_2 and $v_1[u/x]$ have the same semantics, which we do not care about here. The property that $E(V, K)$ is the pair (V, K), or that E is even injective, is just never needed. This means that Lemma 1 also holds if we use encryption primitives that obey algebraic laws.

There is a kind of converse to Lemma 1. Assume that we have an additional type former τ option, with constructors SOME $: \tau \to \tau$ option and NONE $: \tau$ option. Assume their semantics is given by $[\![\tau \text{ option}]\!] = [\![\tau]\!]_{\perp}$, $[\![\text{SOME } t]\!] = \iota([\![t]\!])$, $[\![\text{NONE}]\!] = \perp$. Finally, assume that $\mathcal{R}_{\tau \text{ option}}$ is defined as above. Then we may define an encryption primitive $enc = \lambda v \cdot \lambda k \cdot \{v\}_k$ and a decryption primitive in the toy cryptographic lambda-calculus by $dec = \lambda v \cdot \lambda k \cdot \text{let } \{x\}_k = v \text{ in SOME } x \text{ else NONE}$. If the basic lemma holds, then we must have $[\![enc]\!] \ \mathcal{R}_{\tau \to \text{key}[\tau] \to \text{bits}[\tau]} \ [\![enc]\!]$ and $[\![dec]\!] \ \mathcal{R}_{\text{bits}[\tau] \to \text{key}[\tau] \to \tau \text{ option}} \ [\![dec]\!]$. These are just conditions 1. and 2.

Call cryptographic logical relation any logical relation for which the basic lemma holds. Conditions 1. and 2. can therefore be rephrased as the following motto: a cryptographic logical relation should relate encryption with itself, and decryption with itself.

3.3   Existence of Logical Relations for Encryption

How can we *build* a cryptographic logical relation inductively on types? We first need to address the question of *existence* of logical relations satisfying the basic lemma.

Let us fix a type τ, and assume that we have already constructed $\mathcal{R}_\tau$ and $\mathcal{R}_{\text{key}[\tau]}$. Let $\mathcal{R}_{\text{bits}[\tau]}^{\perp}$ be the smallest relation on $[\![\text{bits}[\tau]]\!]$ satisfying condition 1., i.e., such that $E(V, K) \ \mathcal{R}_{\text{bits}[\tau]}^{\perp} \ E(V', K)$ for all $V \ \mathcal{R}_\tau \ V'$ and $K \ \mathcal{R}_{\text{key}[\tau]} \ K'$. Let $\mathcal{R}_{\text{bits}[\tau]}^{\top}$ be the largest relation on $[\![\text{bits}[\tau]]\!]$ satisfying condition 2., i.e., such that whenever $V \ \mathcal{R}_{\text{bits}[\tau]}^{\top} \ V'$, then $D(V, K) \ \mathcal{R}_{\tau \text{ option}} \ D(V', K')$ for every $K \ \mathcal{R}_{\text{key}[\tau]} \ K'$. These two relations clearly exist. Conditions 1. and 2. state that we should choose $\mathcal{R}_{\text{bits}[\tau]}$ so that $\mathcal{R}_{\text{bits}[\tau]}^{\perp} \subseteq \mathcal{R}_{\text{bits}[\tau]} \subseteq \mathcal{R}_{\text{bits}[\tau]}^{\top}$. This exists if and only if $\mathcal{R}_{\text{bits}[\tau]}^{\perp} \subseteq \mathcal{R}_{\text{bits}[\tau]}^{\top}$.

In turn, $\mathcal{R}_{\text{bits}[\tau]}^{\perp} \subseteq \mathcal{R}_{\text{bits}[\tau]}^{\top}$ is equivalent to: for every $V \ \mathcal{R}_\tau \ V'$ and $K \ \mathcal{R}_{\text{key}[\tau]} \ K'$, for every $K_1 \ \mathcal{R}_{\text{key}[\tau]} \ K_1'$, $D(E(V, K), K_1) \ \mathcal{R}_{\tau \text{ option}} \ D(E(V', K'), K_1') \ (*)$. Let therefore $V \ \mathcal{R}_\tau \ V'$, and fix $K \ \mathcal{R}_{\text{key}[\tau]} \ K'$. By choosing $K_1 = K$, $(*)$ becomes $\iota(V) \ \mathcal{R}_{\tau \text{ option}} \ D(E(V', K'), K_1')$, which is equivalent to $K' = K_1'$ and $V \ \mathcal{R}_\tau \ V'$. Similarly by choosing $K' = K_1'$, we get $K = K_1$ and $V \ \mathcal{R}_\tau \ V'$. In other words, as soon as $\mathcal{R}_\tau$ is not empty, $\mathcal{R}_{\text{key}[\tau]}$ must be a *partial bijection* on $[\![\text{key}[\tau]]\!]$, i.e., the graph of a bijection between two subsets of $[\![\text{key}[\tau]]\!]$.

Proposition 1. *Let $\mathcal{R}_b^0$ be given binary relations on $[\![b]\!]$ for every base type b. Let $\mathcal{R}_{\text{key}[\tau]}^0$ be any partial bijection on $[\![\text{key}[\tau]]\!]$ for every type τ. There exists a cryptographic logical relation $(\mathcal{R}_\tau)_{\tau \text{ type}}$ such that $\mathcal{R}_b = \mathcal{R}_b^0$ for every base type b, and such that $\mathcal{R}_{\text{key}[\tau]} = \mathcal{R}_{\text{key}[\tau]}^0$ for every type τ. We may define $\mathcal{R}_{\text{bits}[\tau]}$, for any type τ, as any relation such that $\mathcal{R}_{\text{bits}[\tau]}^{\perp} \subseteq \mathcal{R}_{\text{bits}[\tau]} \subseteq \mathcal{R}_{\text{bits}[\tau]}^{\top}$.*

Proposition 1 shows that cryptographic logical relations exist that coincide with given relations on base types. But contrarily to logical relations in the λ-calculus, they are far from being uniquely determined: we have considerable freedom as to the choice of the relations at key and bits types.

To define $\mathcal{R}_{\text{key}[\tau]}$, notably, we may use the intuition that some keys are observable by an intruder, and some others are not. Letting fr_τ be the set of observable keys, define

$\mathcal{R}_{\text{key}[\tau]}$ as relating the key K with itself provided $K \in fr_\tau$, and not relating any non-observable key with any key. This is clearly a partial bijection, in fact the identity on the subset fr_τ of $[\![\text{key}[\tau]]\!]$. This is a popular choice: fr_τ is what Abadi and Gordon [2] call a *frame*, up to the fact that frames are defined there as sets of names, not of keys.

To define $\mathcal{R}_{\text{bits}[\tau]}$, we may choose any relation sandwiched between $\mathcal{R}_{\text{bits}[\tau]}^\perp$ and $\mathcal{R}_{\text{bits}[\tau]}^\top$. For every $V_0, V_0' \in [\![\text{bits}[\tau]]\!]$, $V_0 \mathcal{R}_{\text{bits}[\tau]}^\perp V_0'$ if and only if V_0 is of the form $\mathsf{E}(V, K)$, V_0' is of the form $\mathsf{E}(V', K')$, $V \mathcal{R}_\tau V'$ and $K = K' \in fr_\tau$. In other words, V_0 and V_0' are related by $\mathcal{R}_{\text{bits}[\tau]}^\perp$ if and only if they are encryptions of related plaintexts by a unique key that the intruder may observe. On the other hand, $V_0 \mathcal{R}_{\text{bits}[\tau]}^\top V_0'$ if and only if $V_0 = \mathsf{E}(V, K)$ and $V_0' = \mathsf{E}(V', K')$ with either $V \mathcal{R}_\tau V'$ and $K = K' \in fr_\tau$, or $K, K' \notin fr_\tau$ (whatever V, V').

So, $\mathcal{R}_{\text{bits}[\tau]}$ is completely characterized by the datum of fr_τ, plus a function ψ_τ mapping pairs of keys K, K' in $[\![\text{key}[\tau]]\!] \setminus fr_\tau$ to a binary relation $\psi_\tau(K, K')$ on $[\![\tau]\!]$: if $\mathcal{R}_{\text{bits}[\tau]}$ is given, then let $\psi_\tau(K, K')$ be defined as relating V with V' if and only if $\mathsf{E}(V, K) \mathcal{R}_{\text{bits}[\tau]} \mathsf{E}(V', K')$; on the other hand, given ψ_τ, the relation $\mathcal{R}_{\text{bits}[\tau]}$ that relates $\mathsf{E}(V, K)$ with $\mathsf{E}(V', K')$ if and only if $V \mathcal{R}_\tau V'$ and $K = K' \in fr_\tau$, or $K, K' \notin fr_\tau$ and $V \psi_\tau(K, K') V'$, is such that $\mathcal{R}_{\text{bits}[\tau]}^\perp \subseteq \mathcal{R}_{\text{bits}[\tau]} \subseteq \mathcal{R}_{\text{bits}[\tau]}^\top$.

Given parameters fr and ψ, we then get the following definition of a *unique* cryptographic logical relation by induction on types, so that it coincides with given relations on base types:

Proposition 2. *Let fr_τ be some subset of $[\![\text{key}[\tau]]\!]$, for each type τ, and ψ_τ be any function from $([\![\text{key}[\tau]]\!] \setminus fr_\tau)^2$ to the set $\mathbb{P}([\![\tau]\!] \times [\![\tau]\!])$ of binary relations on $[\![\tau]\!]$. For any family $\mathcal{R}_b^0$ of binary relations on $[\![b]\!]$, b a base type, let $(\mathcal{R}_\tau^{fr,\psi})_{\tau \text{ type}}$ be the family of relations defined by:*

- $\mathcal{R}_b^{fr,\psi} = \mathcal{R}_b^0$ *for each base type b;*
- *for every $f, f' \in [\![\tau_1 \to \tau_2]\!]$, $f \mathcal{R}_{\tau_1 \to \tau_2}^{fr,\psi} f'$ if and only if for every $a \mathcal{R}_{\tau_1}^{fr,\psi} a'$, $f(a) \mathcal{R}_{\tau_2}^{fr,\psi} f'(a')$;*
- *for every $K, K' \in [\![\text{key}[\tau]]\!]$, $K \mathcal{R}_{\text{key}[\tau]}^{fr,\psi} K'$ if and only if $K = K' \in fr_\tau$;*
- *for every $V, V' \in [\![\tau]\!]$, for every $K, K' \in [\![\text{key}[\tau]]\!]$, $\mathsf{E}(V, K) \mathcal{R}_{\text{bits}[\tau]}^{fr,\psi} \mathsf{E}(V', K')$ if and only if $V \mathcal{R}_\tau^{fr,\psi} V'$ and $K = K' \in fr_\tau$, or $K, K' \notin fr_\tau$ and $V \psi_\tau(K, K') V'$.*

Whatever the choices of fr_τ and ψ_τ, $(\mathcal{R}_\tau^{fr,\psi})_{\tau \text{ type}}$ is a cryptographic logical relation.

Clearly, Proposition 2 generalizes to the case where fr_τ and ψ_τ are not given *a priori*, but defined using the relations $\mathcal{R}_{\tau'}^{fr,\psi}$ for (not necessarily strict) subtypes τ' of τ. That is, when not just $\mathcal{R}_\tau^{fr,\psi}$ but also fr_τ and ψ_τ are defined by mutual induction on types.

It is interesting, too, to relate the definition of $\mathcal{R}_\tau^{fr,\psi}$ to selected parts of the notion of framed bisimulation [2]. Slightly adapting [2] again, call a *theory* (on type $\text{bits}[\tau]$) any *finite* binary relation th_τ on $[\![\text{bits}[\tau]]\!]$. By finite, we mean that it should be finite as a set of pairs of values. A frame-theory pair (fr_τ, th_τ) is *consistent* if and only if th_τ is a partial bijection, and $\mathsf{E}(V, K) th_\tau \mathsf{E}(V', K')$ implies $K \notin fr_\tau$ and $K' \notin fr_\tau$. Any consistent frame-theory pair determines a ψ_τ function by $V \psi_\tau(K, K') V'$ if and only if $\mathsf{E}(V, K) th_\tau \mathsf{E}(V', K')$. It follows that frame-theory pairs, as explained here, are special cases of pairs of a frame fr_τ and a function ψ_τ.

4 A Uniform Cryptographic λ-Calculus, and Prelogical Relations

Reflecting on the developments above, we see that it would be more natural to use, instead of the toy cryptographic λ-calculus, a simply-typed λ-calculus with two constants enc and dec, with respective semantics given by E and D. While we are at it, it is clear from the way we define $\mathcal{R}^0_{\text{key}[\tau]}$ in Proposition 2 that the type key$[\tau]$ behaves more like a base type than a type constructed from another type. It is therefore relevant to change the algebra of types to something like:

$$\tau ::= b \mid \tau_1 \to \tau_2 \mid \text{bits}[\tau] \mid \text{key} \mid \tau \text{ option} \mid \ldots$$

where b ranges over Σ, Σ now contains a collection of *key types* $\text{key}_1, \ldots, \text{key}_n$ (wlog., we shall use just one, which we write key), and the τ option type is used to give a typing to dec : bits$[\tau] \to$ key $\to \tau$ option; enc is assumed to have type $\tau \to$ key $\to$ bits$[\tau]$. The final ellipsis is meant to indicate that there may be other type formers (products, etc.): we do not wish to be too specific here.

The language we get is just the simply-typed λ-calculus with constants... up to the fact that we need option types τ option. The constants to consider here are at least dec, enc, SOME : $\tau \to \tau$ option, NONE : τ option, and case : τ option $\to (\tau \to \tau') \to \tau' \to \tau'$. (The case constant implements the elimination principle for τ option; we write case s of SOME $x \Rightarrow t \mid$ NONE $\Rightarrow t'$ instead of case $s(\lambda x \cdot t)t'$, and leave the semantics of case as an exercise to the reader.)

The fact that the constants dec, enc, are required to have their denotations, D and E, related to themselves is reminiscent of *prelogical relations* [10]. These can be defined in a variety of ways. Following [10, Definition 3.1, Proposition 3.3], a *prelogical relation* is any family $(\mathcal{R}_\tau)_{\tau \text{ type}}$ of relations such that:

1. for every $f, f' \in [\![\tau_1 \to \tau_2]\!]$, if $f \mathcal{R}_{\tau_1 \to \tau_2} f'$ and $a \mathcal{R}_{\tau_1} a'$ then $f(a) \mathcal{R}_{\tau_2} f'(a')$;
2. K $\mathcal{R}_{\tau_1 \to \tau_2 \to \tau_1}$ K, where K is the function mapping $x \in [\![\tau_1]\!]$, $y \in [\![\tau_2]\!]$ to x;
3. S $\quad \mathcal{R}_{(\tau_1 \to \tau_2 \to \tau_3) \to (\tau_1 \to \tau_2) \to \tau_1 \to \tau_3}$ S, where S is the function mapping $x \in [\![\tau_1 \to \tau_2 \to \tau_3]\!]$, $y \in [\![\tau_1 \to \tau_2]\!]$, $z \in [\![\tau_1]\!]$ to $x(z)(y(z))$;
4. and for every constant $a : \tau$, $[\![a]\!] \mathcal{R}_\tau [\![a]\!]$.

where $[\![a]\!]$ denotes $[\![a]\!]\rho$ for any environment ρ. Condition 1. is just one half of (**Log**). The basic lemma for prelogical relations [10, Lemma 4.1] is stronger than for logical relations: prelogical relations are *exactly* those families of relations indexed by types such that the basic lemma holds.

Note that the use of prelogical relations also requires us to relate the semantics of SOME with itself, that of NONE with itself, and that of case with itself.

Then, we may observe that prelogical relations are not just sound for contextual equivalence, they are *complete*, at all types, even higher-order. Recall that a value $a \in [\![\tau]\!]$ is *definable* if and only if there exists a (necessarily closed) term t such that $\vdash t : \tau$ is derivable, and $a = [\![t]\!]$. The main point in our completeness argument is that there is a lax logical relation built by considering the trace of $\approx_\tau$ on definable elements. The relation is necessarily a partial equality on observation types $o \in \text{Obs}$.

Theorem 3 (Completeness). *Prelogical relations are complete for contextual equivalence in the λ-calculus, in the strong sense that there is a prelogical relation $(\mathcal{R}_\tau)_{\tau \text{ type}}$ such that for every t_1, t_2 s.t. $\vdash t_1 : \tau, \vdash t'_2 : \tau$, $[\![t_1]\!] \approx_\tau [\![t_2]\!]$ if and only if $[\![t_1]\!] \mathcal{R}_\tau [\![t_2]\!]$.*

The argument before Proposition 2 applies here without further ado: every prelogical relation must be a partial bijection at the key type, and conversely, any prelogical relation that is the equality on $fr \subseteq [\![\mathtt{key}]\!]$ at the key type satisfies the basic lemma, hence can be used to establish contextual equivalence. Specializing the prelogical relation $(\mathcal{R}_\tau)_{\tau \text{ type}}$ of Theorem 3 (its proof is in the full version [9]), we get that $\mathcal{R}_{\mathtt{key}}$ is exactly equality on the set $fr = \{[\![t]\!] \mid \vdash t : \mathtt{key}\}$ of definable keys.

Similarly, we may define the binary relation $\psi_\tau(K, K')$, for every $K, K' \in [\![\mathtt{key}]\!] \setminus fr$, (i.e., for all non-definable keys) by $V \psi_\tau(K, K') V'$ if and only if $\mathsf{E}(V, K)\mathcal{R}_{\mathtt{bits}[\tau]}\mathsf{E}(V', K')$, i.e., if and only if $\mathsf{E}(V, K)$ and $\mathsf{E}(V', K')$ are definable at type $\mathtt{bits}[\tau]$, and $\mathsf{E}(V, K) \approx_{\mathtt{bits}[\tau]} \mathsf{E}(V', K')$.

From this, we infer immediately the following combination of the analogue of Proposition 2 (soundness) with Theorem 3 (completeness):

Proposition 4. *There is a prelogical relation* $(\mathcal{R}^{fr,\psi}{}_\tau)_{\tau \text{ type}}$, *parameterized by fr and ψ, which is:*

- strict *at the key type: i.e., for every $K, K' \in [\![\mathtt{key}]\!]$, $K \; \mathcal{R}^{fr,\psi}_{\mathtt{key}} \; K'$ if and only if $K = K' \in fr$;*
- strict *at $\mathtt{bits}[\tau]$ types: i.e., for every $V, V' \in [\![\tau]\!]$, for every $K, K' \in [\![\mathtt{key}]\!]$, $\mathsf{E}(V, K) \; \mathcal{R}^{fr,\psi}_{\mathtt{bits}[\tau]} \; \mathsf{E}(V', K')$ if and only if $V \; \mathcal{R}^{fr,\psi}_\tau \; V'$ and $K = K' \in fr$, or $K, K' \notin fr$ and $V \psi_\tau(K, K') V'$;*
- *and such that, for some fr and ψ, for every closed terms t, t' of type τ, $[\![t]\!] \approx_S \tau [\![t']\!]$ if and only if $[\![t]\!] \; \mathcal{R}^{fr,\psi}_\tau \; [\![t']\!]$.*

The idea of being *strict* at some type τ is, in all cases, that the (pre)logical relation at type τ should be defined uniquely as a function of the (pre)logical relations at all immediate subterms of τ. The prelogical relation of Proposition 4 is strict at option types, too, provided there is a closed term of type τ or $[\![\tau]\!]$ has no junk.

While the point in prelogical relations in [10] is mainly of being not strict at arrow types, the point here is to argue that it is meaningful either not to be strict at $\mathtt{bits}[\tau]$ types, as in Section 3.2 (in the sense that $\mathcal{R}_{\mathtt{bits}[\tau]}$ was not determined uniquely from $\mathcal{R}_\tau$), or equivalently to be strict at $\mathtt{bits}[\tau]$, given parameters fr and τ. We believe that just saying that we do not require strictness at $\mathtt{bits}[\tau]$, thus omitting the fr and τ parameters, leads to some simplification.

5   Name Creation and Lax Logical Relations

No decent calculus for cryptographic protocols can dispense with fresh name creation. This is most easily done by following Stark [17], who defined a categorical semantics for a calculus with fresh name creation based on Moggi's monadic λ-calculus [14]. We just take his language, adding all needed constants as in Section 4.

5.1   The Moggi-Stark Calculus

The *Moggi-Stark calculus* is obtained by adding a new type former T (the *monad*), to the types of the λ-calculus of Section 4, so that $T\tau$ is a type as soon as τ is:

$$\tau ::= \quad b \mid \tau_1 \to \tau_2 \mid \mathtt{bits}[\tau] \mid \mathtt{key} \mid \tau \text{ option} \mid T\tau \mid \ldots$$

(We continue to leave the definition of our calculi open, as shown with the ellipsis ..., to facilitate the addition of new types and constants, if needed.) Following Stark, we also require the existence of a new base type $\nu \in \Sigma$ of *names*. (This will take the place of the type key of keys, which we shall equate with names.) The λ-calculus of Section 4 is enriched with constructs val t and let $x \Leftarrow t$ in u (not to be confused with the let construct of Section 3.1), with typing rules as following, and two constants new : $T\nu$ (fresh name creation) and $\doteq: \nu \rightarrow \nu \rightarrow$ bool (equality of names).

$$\frac{\Gamma \vdash t : \tau}{\Gamma \vdash \text{val } t : T\tau} \text{ (val)} \qquad \frac{\Gamma \vdash t : T\tau \quad \Gamma, x : \tau \vdash u : T\tau'}{\Gamma \vdash \text{let } x \Leftarrow t \text{ in } u : T\tau'} \text{ (let)}$$

In Stark's semantics (notations are ours here), given any finite set s (of names), $[\![t]\!] \, s\rho$ is the value of t in environment ρ assuming that all previously created names are in s. This allows one to describe the creation of fresh names as returning any name outside s. This is most elegantly described by letting the values of terms be taken in the presheaf category $\mathbf{Set}^{\mathcal{I}}$ [17], where $\mathcal{I}$ is the category whose objects are finite sets and whose morphisms $s \xrightarrow{i} s'$ are injections. Given any type τ, $[\![\tau]\!] \, s$ is intuitively the set of all values of type τ in a world where all created names are in s. Since $[\![\tau]\!]$ is a functor, for every injection $s \xrightarrow{i} s'$ there is a conversion $[\![\tau]\!] \, i$ that sends any value a of $[\![\tau]\!] \, s$ to one in $[\![\tau]\!] \, s'$, intuitively by renaming the names in a using i. By extension, if Γ is any context $x_1 : \tau_1, \ldots, x_n : \tau_n$, let $[\![\Gamma]\!]$ be $[\![\tau_1]\!] \times \ldots \times [\![\tau_n]\!]$, using the products in $\mathbf{Set}^{\mathcal{I}}$—i.e., products at each world s. Then, as usual in categorical semantics [11], given any term t such that $\Gamma \vdash t : \tau$ is derivable, $[\![t]\!]$ is a morphism from $[\![\Gamma]\!]$ to $[\![\tau]\!]$. This means that $[\![t]\!]$ is a natural transformation from $[\![\Gamma]\!]$ to $[\![\tau]\!]$, in particular that, for every finite set s, $[\![t]\!] \, s$ maps any Γ, s-*environment* ρ (a map sending each x_i such that $x_i : \tau_i$ is in Γ to some element of $[\![\tau_i]\!] \, s$) to some value $[\![t]\!] \, s\rho$ in $[\![\tau]\!] \, s$; and all this is natural in s, i.e., compatible with renaming of names.

Interestingly, $T\tau$, the type of computations that result in a value of type τ, possibly creating fresh names during the course of computation, is defined semantically by $[\![T\tau]\!] = \boldsymbol{T} \, [\![\tau]\!]$, where $(\boldsymbol{T}, \boldsymbol{\eta}, \boldsymbol{\mu}, \boldsymbol{t})$ is the strong monad defined in [17, 8, 19]. $\boldsymbol{T}A$ is defined as $\text{colim}_{s'} A(\_ + s') : \mathcal{I} \rightarrow \boldsymbol{Set}$. On objects, this is given by $\boldsymbol{T}As = \text{colim}_{s'} A(s + s')$, i.e., $\boldsymbol{T}As$ is the set of all equivalence classes of pairs (s', a) with s' a finite set and $a \in A(s + s')$, modulo the smallest equivalence relation $\equiv$ such that $(s', a) \equiv (s'', A(\text{id}_s + j)a)$ for every morphism $s' \xrightarrow{j} s''$ in $\mathcal{I}$. Intuitively, given a set of *names* s, elements of $\boldsymbol{T}As$ are formal expressions $(\nu s')a$ where all names in s' are bound and every name free in a is in $s + s'$—modulo the fact that $(\nu s', s'')a \equiv (\nu s')a$ for any additional set of new names s'' not free in a. We shall in fact write $(\nu s')a$ the equivalence class of (s', a), to aid intuition.

The semantics of let and val is standard [14]. Making it explicit on this particular monad, we obtain: $[\![\text{val } t]\!] \, s\rho = (\nu \emptyset) [\![t]\!] \, s\rho$ and $[\![\text{let } x \Leftarrow t \text{ in } u]\!] \, s\rho = (\nu \, s' + s'')b$, where $[\![t]\!] \, s\rho = (\nu s')a$, we assume that $\Gamma \vdash t : T\tau$ and $\Gamma, x : \tau \vdash u : T\tau'$, and where $[\![u]\!] \, (s + s')(([\![\Gamma]\!] \, (\text{inl}_{s,s'})\rho)[x := a]) = (\nu s'')b$. (Concretely, if Γ is $x_1 : \tau_1, \ldots, x_n : \tau_n$, $\rho = [x_1 := a_1, \ldots, x_n := a_n]$ where $a_i \in [\![\tau_i]\!] \, s$ for every i, then $[\![\Gamma]\!] \, (\text{inl}_{s,s'})\rho$ is $[x_1 := [\![\tau_1]\!] \, (\text{inl}_{s,s'})a_1, \ldots, x_n := [\![\tau_n]\!] \, (\text{inl}_{s,s'})a_n]$.)

The semantics of base types $b \in \Sigma$, except ν, is given by constant functors: $[\![b]\!] \, s$ is a fixed set, independent of s; e.g., $[\![\text{bool}]\!] \, s = \mathbb{B}$. The semantics of ν is $[\![\nu]\!] \, s = s$,

$[\![\nu]\!] i = i$; i.e., the names that exist at s are just the elements of s. $\textit{Set}^{\mathcal{I}}$ is a presheaf category, hence cartesian-closed [11]. This provides a semantics for λ-abstraction and applications.

Finally, the semantics of $\texttt{new} : T\nu$ is given by $[\![\texttt{new}]\!] s\rho = (\nu\{n\})n$, where n is any element not in s, and $[\![\doteq]\!]$ is defined as the only morphism in $\textit{Set}^{\mathcal{I}}$ such that $[\![\doteq xy]\!] s[x := a, y := b]$ is $\texttt{true}$ if $a = b$, and $\texttt{false}$ otherwise.

5.2   Lax Logical Relations for Monads

Given that terms now take values in some category ($\textit{Set}^{\mathcal{I}}$), not in $\textit{Set}$ as in Section 3, the proper generalization of prelogical relations is given by *lax logical relations* [16]. We introduce this notion as gently as possible.

Let Σ be the set of base types, seen as a discrete category. The simply-typed λ-calculus gives rise to the *free CCC* $\lambda(\Sigma)$ over Σ as follows: the objects of $\lambda(\Sigma)$ are typing contexts Γ, a morphism from Γ to $\Delta = y_1 : \tau_1, \ldots, y_n : \tau_n$ is a substitution $[y_1 := t_1, \ldots, y_n := t_n]$, where $\Gamma \vdash t_i : \tau_i$ $(1 \leq i \leq n)$, modulo $\beta\eta$-conversion. (In particular, Γ-environments are exactly morphisms from the terminal object, the empty context ϵ, to Γ.) Composition is substitution. Being the free CCC means that, for any CCC $\boldsymbol{C}$, for any functor $[\![\_]\!]_0$ from Σ to $\boldsymbol{C}$ (i.e., for any function $[\![\_]\!]_0$ mapping each base type in Σ to some object in $\boldsymbol{C}$), there is a unique representation $[\![\_]\!]_1$ of CCCs from $\lambda(\Sigma)$ to $\boldsymbol{C}$ such that the right diagram commutes. A representation of CCCs is any functor that preserves products and exponentials. When $\boldsymbol{C}$

$$\Sigma \xrightarrow{\subseteq} \lambda(\Sigma) \qquad (1)$$

is $\textit{Set}$, this describes all at once all the constructions $[\![\tau]\!]_1$ (denotation of types τ) and $[\![t]\!]_1$ (denotations of typed λ-terms t) as used in Section 3.

Let $\text{Subscone}_{\boldsymbol{C}}^{\mathbb{C}}$ be the *subscone* category, defined as follows. Assume $\mathbb{C}$ is another CCC, such that $\mathbb{C}$ has pullbacks. Let $|\_|$ be a functor from $\boldsymbol{C}$ to $\mathbb{C}$ that preserves finite products. Then $\text{Subscone}_{\boldsymbol{C}}^{\mathbb{C}}$ is the category whose objects are triples $\langle S, m, A \rangle$, where m is a mono $S \rightarrowtail |A|$ in $\mathbb{C}$, and whose morphisms from $\langle S, m, A \rangle$ to $\langle S', m', A' \rangle$ are pairs of morphisms $\langle u, v \rangle$ (u in $\mathbb{C}$, from S to S', and v in $\boldsymbol{C}$, from A to A'), making the obvious square commute. Noting that $\text{Subscone}_{\boldsymbol{C}}^{\mathbb{C}}$ is again a CCC (Mitchell and Scedrov [13] make this remark when $\mathbb{C}$ is $\textit{Set}$, and $|\_|$ is the global section functor $\boldsymbol{C}(1, \_)$), the following purely diagrammatic argument obtains. Assume we are given a functor from Σ to $\text{Subscone}_{\boldsymbol{C}}^{\mathbb{C}}$, i.e., a collection $\mathcal{R}_o$ of objects in $\text{Subscone}_{\boldsymbol{C}}^{\mathbb{C}}$, one for each base type o. Then there is a unique representation $\mathcal{R}$ of CCCs from $\lambda(\Sigma)$ such that the right diagram commutes. Now the crux of the argument is the following. The forgetful functor $U : \text{Subscone}_{\boldsymbol{C}}^{\mathbb{C}} \to \boldsymbol{C}$ mapping the object $\langle S, m, A \rangle$ to A and the mor-

$$\Sigma \xrightarrow{\subseteq} \lambda(\Sigma) \qquad (2)$$

phism $\langle u, v \rangle$ to v is also a representation of CCCs. It follows that $U \circ \mathcal{R}$ is a representation of CCCs again, from $\lambda(\Sigma)$ to $\boldsymbol{C}$. If $U \circ (\mathcal{R}_o)_{o \in \Sigma} = [\![\_]\!]_0$, then by the uniqueness property of $[\![\_]\!]_1$, we must have $U \circ \mathcal{R} = [\![\_]\!]_1$, i.e., diagram (3) commutes. As observed in [13], and extended to CCCs in [3], when $\mathbb{C} = \textit{Set}$, $\boldsymbol{C}$ is the product of two CCCs $\boldsymbol{A}$ and $\boldsymbol{B}$, and $|\_|$ is the functor $\boldsymbol{A}(1, \_) \times \boldsymbol{B}(1, \_)$, $(\mathcal{R}(\tau))_{\tau \text{ type}}$ behaves like a logical relation. It is really a logical relation, as we have defined it earlier, when both A and B are Set. (In this case, an object $\mathcal{R}(\tau)$ is of the form $S \hookrightarrow [\![\tau]\!]^2$, where S, up to isomorphism, is just a subset of the cartesian product of $[\![\tau]\!]$ with itself.) In case A and B are the same presheaf category $Set^{\mathcal{I}}$, $(R(\tau))_{\tau \text{ type}}$ is a Kripke logical relation with base category $\mathcal{I}$.

While the object part of functor $\mathcal{R}$, $(\mathcal{R}(\tau))_{\tau \text{ type}}$, yields logical relations (or extensions), the morphism part maps each morphism in $\boldsymbol{\lambda}(\Sigma)$, namely a typed term t modulo $\beta\eta$, of type τ, to a morphism in the subscone, i.e., a pair $\langle u, v \rangle$. The fact that diagram (3) commutes states that v is just the pair of the semantics of t in A and the semantics of t in B, and the fact that $\langle u, v \rangle$ is a morphism (saying that a certain square commutes) states that these two semantics are related by $\mathcal{R}(\tau)$: this establishes the basic lemma.

The important property to make $\mathcal{R}$ satisfy the basic lemma is just the equality in the right diagram. Logical relations are the case where $\mathcal{R}$ is a representation of CCCs, in which case, as we have seen, this diagram necessarily commutes. *Lax* logical relations are prod-uct preserving functors $\mathcal{R}$ such that Diagram (3) com-

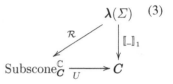

mutes [16, Section 6]. The difference is that, with lax logical relations, we do not re-quire $\mathcal{R}$ to be representations of CCCs, just product preserving functors. We say that $\mathcal{R}$ is *strict at arrow types* if and only if $\mathcal{R}$ preserves exponentials, too.

Defining lax logical relations for Moggi's monadic meta-language follows the same pattern. The monadic λ-calculus gives rise to the *free let-CCC* $\boldsymbol{Comp}(\Sigma)$ over Σ, where a let-CCC is a CCC with a strong monad. We then get Diagram (1) again, only with $\boldsymbol{\lambda}(\Sigma)$ replaced by $\boldsymbol{Comp}(\Sigma)$, $\boldsymbol{C}$ is a let-CCC, and $[\![\_]\!]_1$ is a representation of let-CCCs, i.e., a functor that preserves products, exponentials, and the monad (functor, unit, mul-tiplication, strength).

5.3   Contextual Equivalence

Defining contextual equivalence in a calculus with names is a bit tricky. First, we have to consider contexts $\mathcal{C}$ of type To ($o \in \mathrm{Obs}$), not of type o. Intuitively, contexts should be allowed to do some computations; were they of type o, they could only return values. In particular, note that contexts $\mathcal{C}$ such that $x : T\tau \vdash \mathcal{C} : o$, meant to observe computations at type τ, cannot observe anything. This is because the (let) typing rule only allows one to use computations to build other computations, never values.

Another tricky aspect is that we cannot take contexts $\mathcal{C}$ that only depend on one variable $x : \tau$. We must assume that $\mathcal{C}$ can also depend on an arbitrary set of pub-lic names. Given names $n_1, \ldots, n_m$, the only way $\mathcal{C}$ can be made to depend on them is to assume that $\mathcal{C}$ has m free variables $z_1, \ldots, z_m$ of type $\boldsymbol{\nu}$, which are mapped to $n_1, \ldots, n_m$. (It is more standard [15, 1] to consider expressions built on separate sets of variables and names, thus introducing the semantic notion of names in the syntax. It is more natural here to consider that there are variables z_l mapped, in a one-to-one way, to names n_l.) Let s_1 be any set of names containing $n_1, \ldots, n_m$, let w_1 be $\{z_1, \ldots, z_m\}$, and $w_1 \xrightarrow{i_1} s_1$ the injection mapping each z_l to n_l, $1 \le l \le m$. Write $w_1 := i_1(w_1)$ for $z_1 := n_1, \ldots, z_m := n_m$, and $\overline{w_1 : \boldsymbol{\nu}}$ for $z_1 : \boldsymbol{\nu}, \ldots, z_m : \boldsymbol{\nu}$. We shall then consider contexts $\mathcal{C}$ such that $\overline{w_1 : \boldsymbol{\nu}}, x : \tau \vdash \mathcal{C} : To$ is derivable, and evaluate

$[\![\mathcal{C}]\!] s_1[x := a, \overline{w_1 := i_1(w_1)}]$ and compare it with $[\![\mathcal{C}]\!] s_1[x := a', \overline{w_1 := i_1(w_1)}]$ to decide whether a and a' are contextually equivalent. This represents the fact that $\mathcal{C}$ is evaluated in a world where all names in s_1 have been created, and where $\mathcal{C}$ has access to all (public) names in $i(w_1)$.

This definition is not yet correct, as this requires a and a' to be in $[\![\tau]\!] s_1$, but they are in $[\![\tau]\!] s$ for some possibly different set s of names. This is repaired by considering coercion $[\![\tau]\!] k_1$, where $s \xrightarrow{k_1} s_1$ is any injection.

To sum up, say that $a, a' \in [\![\tau]\!] s$ are *contextually equivalent at* s, and write $a \approx_\tau^s a'$, if and only if, for every finite set of variables w_1, for every injections $w_1 \xrightarrow{i_1} s_1$ and $s \xrightarrow{k_1} s_1$, for every term $\mathcal{C}$ such that $\overline{w_1 : \boldsymbol{\nu}}, x : \tau \vdash \mathcal{C} : To$ is derivable ($o \in \mathrm{Obs}$), $[\![\mathcal{C}]\!] s_1[x := [\![\tau]\!] k_1(a), \overline{w_1 := i_1(w_1)}] = [\![\mathcal{C}]\!] s_1[x := [\![\tau]\!] k_1(a'), \overline{w_1 := i_1(w_1)}]$.

The notion we use here is inspired by [15, Definition 4], although it may not look so at first sight. We may simplify it a bit by noting that we lose no generality in considering that $\mathcal{C}$ has access to *all* names in s_1. Without loss of generality, we equate w_1 with s_1, and notice that $a \approx_\tau^s a'$ if and only if, for every injection $s \xrightarrow{k_1} s_1$, for every term $\mathcal{C}$ such that $\overline{s_1 : \boldsymbol{\nu}}, x : \tau \vdash \mathcal{C} : To$ is derivable ($o \in \mathrm{Obs}$), $[\![\mathcal{C}]\!] s_1[x := [\![\tau]\!] k_1(a), \overline{s_1 := s_1}] = [\![\mathcal{C}]\!] s_1[x := [\![\tau]\!] k_1(a'), \overline{s_1 := s_1}]$. (Remember we see the *variables* in s_1 as denoting the *names* in s_1 here, equating names with variables.) The use of injections between finite sets leads us naturally to switch from $\boldsymbol{Set}^{\mathcal{I}}$ to the category $\boldsymbol{Set}^{\mathcal{I}^{\rightarrow}}$, where $\mathcal{I}^{\rightarrow}$, the *arrow category* of $\mathcal{I}$, has as objects all morphisms $w \xrightarrow{i} s$ in $\mathcal{I}$, and as morphisms from $w \xrightarrow{i} s$ to $w' \xrightarrow{i'} s'$ all pairs (j, k) of morphisms such that the right diagram commutes. This is in accordance with [19], where it is noticed that

$$
\begin{array}{ccc}
w & \xrightarrow{\;\;i\;\;} & s \\
{\scriptstyle j}\downarrow & & \downarrow{\scriptstyle k} \\
w' & \xrightarrow{\;\;i'\;\;} & s'
\end{array}
\qquad (4)
$$

$\boldsymbol{Set}^{\mathcal{I}^{\rightarrow}}$ is the right category to define a Kripke logical relation (but not necessarily lax) that coincides with Pitts and Stark's on first-order types. We shall consider here the equivalent category where w is restricted to be a finite set of *variables* (and continue to call this category $\mathcal{I}^{\rightarrow}$). Objects $w \xrightarrow{i} s$ are then sets w of variables denoting those public names in s, together with an injection i. So we shall work with lax logical relations in the subscone category $\mathrm{Subscone}_{\mathbb{C}}^C$, where $C = \boldsymbol{Set}^{\mathcal{I}} \times \boldsymbol{Set}^{\mathcal{I}}$, $\mathbb{C}$ is the presheaf category $\boldsymbol{Set}^{\mathcal{I}^{\rightarrow}}$, and $|\_| : C \to \mathbb{C}$ is the composite of the binary product functor $\times : \boldsymbol{Set}^{\mathcal{I}} \times \boldsymbol{Set}^{\mathcal{I}} \to \boldsymbol{Set}^{\mathcal{I}}$ with the functor $\boldsymbol{Set}^{\mathfrak{u}} : \boldsymbol{Set}^{\mathcal{I}} \to \boldsymbol{Set}^{\mathcal{I}^{\rightarrow}}$. Here $\mathfrak{u} : \mathcal{I}^{\rightarrow} \to \mathcal{I}$ is the obvious forgetful functor that maps $w \xrightarrow{i} s$ to s. Say that a value $a \in [\![\tau]\!] s$ is *definable at* $w \xrightarrow{i} s$ if and only if there is a term t such that $\overline{w : \boldsymbol{\nu}} \vdash t : \tau$ is derivable and $a = [\![t]\!] s[\overline{w := i(w)}]$.

Definition 1. *Let* $w \xrightarrow{i} s$ *be any object of* $\mathcal{I}^{\rightarrow}$. *The value* $a, a' \in [\![\tau]\!] s$ *are said to be contextually equivalent at* $w \xrightarrow{i} s$, *written* $a \approx_\tau^{w \xrightarrow{i} s} a'$, *if and only if, for every morphism* (j_1, k_1) *from* $w \xrightarrow{i} s$ *to any object* $w_1 \xrightarrow{i_1} s_1$ *in* $\mathcal{I}^{\rightarrow}$, *for every term* $\mathcal{C}$ *such that* $\overline{w_1 : \boldsymbol{\nu}}, x : \tau \vdash \mathcal{C} : To$ ($o \in \mathrm{Obs}$) *is derivable,* $[\![\mathcal{C}]\!] s_1[x := [\![\tau]\!] k_1(a), \overline{w_1 := i_1(w_1)}] = [\![\mathcal{C}]\!] s_1[x := [\![\tau]\!] k_1(a'), \overline{w_1 := i_1(w_1)}]$. *Define the relation* $\mathcal{R}_\tau^{w \xrightarrow{i} s}$ *by:* $a \, \mathcal{R}_\tau^{w \xrightarrow{i} s} a'$ *if and only if* a *and* a' *are definable at* $w \xrightarrow{i} s$ *and* $a \approx_\tau^{w \xrightarrow{i} s} a'$.

In particular, $a \approx_\tau^s a'$ iff $a \approx_\tau^{\emptyset \to s} a'$, where $\emptyset \to s$ denotes the unique empty injection.

Note that for every value $a \in [\![\tau]\!] \, s$ definable at $w \xrightarrow{i} s$, $[\![\tau]\!] \, k(a)$ is also definable at $w' \xrightarrow{i'} s'$, whenever there is a morphism (j, k) from the former to the latter. Indeed, let $a = [\![t]\!] \, s[w := i(w)]$. Then for t' obtained from t by renaming according to j,

$$[\![\tau]\!] \, k(a) = [\![t']\!] \, s'[w' := i'(w')]. \tag{1}$$

In particular, every value $a \in [\![\tau]\!] \, s$ definable at $\emptyset \to s$, is definable at every $w \xrightarrow{i} s$.

Theorem 5. *Lax logical relations are complete for contextual equivalence in the Moggi-Stark calculus, in the strong sense that there is a lax logical relation $\mathcal{R}$ such that, for every terms u, u' such that $\overline{w : \boldsymbol{\nu}} \vdash u : \tau$ and $\overline{w : \boldsymbol{\nu}} \vdash u' : \tau$ are derivable,*
$$[\![u]\!] \, s[\overline{w := i(w)}] \approx_\tau^{w \xrightarrow{i} s} [\![u']\!] \, s[\overline{w := i(w)}] \text{ iff } [\![u]\!] \, s[\overline{w := i(w)}] \, \mathcal{R}_\tau^{w \xrightarrow{i} s} \, [\![u']\!] \, s[\overline{w := i(w)}].$$

The (non-lax) logical relation of [19] is defined on $\boldsymbol{\nu}$ by: $n \, \mathcal{R}_{\boldsymbol{\nu}}^{w \xrightarrow{i} s} \, n'$ iff $n = n' \in w$. This is exactly what the lax logical relation of Definition 1 is defined as on the $\boldsymbol{\nu}$ type:

Lemma 2. *Let $\mathcal{R}_\tau^{w \xrightarrow{i} s}$ be the logical relation of Definition 1. Then $n \, \mathcal{R}_{\boldsymbol{\nu}}^{w \xrightarrow{i} s} \, n'$ if and only if $n = n' \in i(w)$.*

To finish this section, we observe:

Lemma 3. *Assume that observation types have no junk, in the sense that every value of $[\![o]\!] \, s \, (o \in \text{Obs})$ is definable at s, for every s, equivalently at every $w \xrightarrow{i} s$. Then $\mathcal{R}_o^{w \xrightarrow{i} s}$ is equality on $[\![o]\!] \, s$, and $\mathcal{R}_{To}^{w \xrightarrow{i} s}$ is equality on $[\![To]\!] \, s$ for any observation type o.*

We almost forgot to prove soundness! It is easy to see that any lax logical relation that coincides with partial equality on types To is sound for contextual equivalence. Indeed, by the basic lemma $U \circ \mathcal{R} = [\![\_]\!]_1$, whenever $a \, \mathcal{R}_\tau^{w \xrightarrow{i} s} \, a'$, then for any $\mathcal{C}$ such that $\overline{w_1 : \boldsymbol{\nu}}, x : \tau \vdash \mathcal{C} : To \, (o \in \text{Obs})$ is derivable, for any morphism (j_1, k_1) from $w \xrightarrow{i} s$ to $w_1 \xrightarrow{i_1} s_1$, $[\![\mathcal{C}]\!] \, s_1[\overline{w_1 := i_1(w_1)}, x := [\![\tau]\!] \, k_1(a)] \, \mathcal{R}_{To}^{w_1 \xrightarrow{i_1} s_1} \, [\![\mathcal{C}]\!] \, s_1[\overline{w_1 := i_1(w_1)}, x := [\![\tau]\!] \, k_1(a')]$; so $a \approx_\tau^{w \xrightarrow{i} s} a'$.

5.4   Mixing Fresh Name Creation and Encryption

Let us get down to earth. What do we need now to get lax logical relations that are sound and complete for contextual equivalence when both fresh name creation and cryptographic primitives are involved? The answer is: just lax logical relations on $\boldsymbol{Set}^{\mathcal{I}^\rightarrow}$, as used in Section 5.3... making sure that they relate each constant itself. We have indeed been careful in being sure that our calculi were open, i.e. they can be extended to arbitrarily many new types and constants. The only requirement that the new constructs can be given a semantics in $\boldsymbol{Set}^{\mathcal{I}}$. In particular, a lax logical relation on $\boldsymbol{Set}^{\mathcal{I}^\rightarrow}$ is sound for observational equivalence in the presence of cryptographic primitives if each of the constants enc, dec, SOME, NONE, case is related to itself.

Then Theorem 5 shows that lax logical relations are complete for the Moggi-Stark calculus, which uses a name creation monad. We have in fact proved more, again because we have been particularly keen on leaving the set of types and constants open:

whatever new constants and types you allow, lax logical relations remain complete. In particular, taking enc, dec, SOME , NONE, case as new constants, we automatically get sound and complete lax logical relations for name creation *and* cryptographic primitives.

Acknowledgements. We would like to thank Michel Bidoit for having directed us to the notion of prelogical relations in the first place.

References

1. M. Abadi and A. D. Gordon. A calculus for cryptographic protocols: The spi calculus. In *Proc. 4th ACM Conference on Computer and Communications Security (CCS)*, 1997.
2. M. Abadi and A. D. Gordon. A bisimulation method for cryptographic protocols. *Nordic Journal of Computing*, 5(4), 1998.
3. M. Alimohamed. A characterization of lambda definability in categorical models of implicit polymorphism. *Theoretical Computer Science*, 146(1–2), 1995.
4. M. Boreale, R. de Nicola, and R. Pugliese. Proof techniques for cryptographic processes. In *Proc. LICS'99*. IEEE Computer Society Press, 1999.
5. J. Borgström and U. Nestmann. On bisimulations for the spi calculus. In *Proc. AMAST'02*, volume 2422 of *LNCS*. Springer, 2002.
6. H. Comon and V. Shmatikov. Is it possible to decide whether a cryptographic protocol is secure or not? *J. of Telecommunications and Information Technology*, 4, 2002.
7. D. Dolev and A. C. Yao. On the security of public key protocols. *IEEE Transactions on Information Theory*, IT-29(2), 1983.
8. J. Goubault-Larrecq, S. Lasota, and D. Nowak. Logical relations for monadic types. In *Proc. CSL'02*, volume 2471 of *LNCS*. Springer, 2002.
9. J. Goubault-Larrecq, S. Lasota, D. Nowak, and Y. Zhang. Complete lax logical relations for cryptographic lambda-calculi. Research Report, LSV, ENS de Cachan, 2004.
10. F. Honsell and D. Sannella. Pre-logical relations. In *Proc. CSL'99*, volume 1683 of *LNCS*, 1999.
11. J. Lambek and P. J. Scott. *Introduction to Higher Order Categorical Logic*, volume 7 of *Cambridge Studies in Advanced Mathematics*. Cambridge University Press, 1986.
12. J. C. Mitchell. *Foundations for Programming Languages*. MIT Press, 1985.
13. J. C. Mitchell and A. Scedrov. Notes on sconing and relators. In *Proc. CSL'93*, volume 702 of *LNCS*. Springer, 1993.
14. E. Moggi. Notions of computation and monads. *Information and Computation*, 93, 1991.
15. A. Pitts and I. Stark. Observable properties of higher order functions that dynamically create local names, or: What's *new*? In *Proc. Int. Conf. Mathematical Foundations of Computer Science (MFCS)*, volume 711 of *LNCS*. Springer, 1993.
16. G. D. Plotkin, J. Power, D. Sannella, and R. D. Tennent. Lax logical relations. In *Proc. ICALP'00*, volume 1853 of *LNCS*. Springer, 2000.
17. I. Stark. Categorical models for local names. *Lisp and Symbolic Computation*, 9(1), 1996.
18. E. Sumii and B. C. Pierce. Logical relations for encryption. In *Proc. CSFW-14*. IEEE Computer Society Press, 2001.
19. Y. Zhang and D. Nowak. Logical relations for dynamic name creation. In *Proc. CSL/KGL'03*, volume 2803 of *LNCS*. Springer, 2003.

Subtyping Union Types

Jérôme Vouillon

CNRS and Université Paris 7
Case 7014, 2 Place Jussieu, 75251 Paris Cedex 05, France
Jerome.Vouillon@pps.jussieu.fr

Abstract. Subtyping can be fairly complex for union types, due to interactions with other types, such as function types. Furthermore, these interactions turn out to depend on the calculus considered: for instance, a call-by-value calculus and a call-by-name calculus will have different possible subtyping rules. In order to abstract ourselves away from this dependence, we consider a fairly large class of calculi. This allows us to find a subtyping relation which is both robust (it is sound for all calculi) and precise (it is complete with respect to the class of calculi).

Keywords: union types, subtyping, semantics, lambda-calculus.

1 Introduction

The design of a subtyping relation for a language with a rich type system is hard. The subtyping relation should satisfy conflicting requirements. On the one hand, one would like the relation to have strong theoretical foundations, rather than being defined in an ad hoc, purely algorithmic, fashion. It is therefore tempting to base it on the semantics of the language. But, on the other hand, one should be careful not to tie it too tightly to a particular language. Especially, one should avoid accidental special cases which happen to hold only in the language considered. Indeed, the relation should be robust in order to accommodate future language extensions. It should also be simple enough so that the users can understand it, and should possess good algorithmic properties: checking whether two types are in a subtyping relation should be reasonably simple and efficient.

We should emphasize the fact that the possible subtyping relations depend on the language considered by providing some examples. Let us first give some rough intuition about types. For these examples, we take the view that well-typed terms may diverge but will evaluate without error. A term of type $\bot$ is a term that always diverges. A term of type $\top$ is a term that evaluates without error. A term of type $\tau' \to \tau$ behaves like a term of type τ once applied to a term of type τ'. A term of type $\tau \cup \tau'$ behaves as a term of type either τ or τ'. We write $\tau <: \tau'$ to mean that τ is a subtype of τ' and $\tau = \tau'$ to mean that τ and τ' are equivalent, that is, subtypes of one another. We can now present some typing relations that only hold under some conditions on the language.

- In some call-by-value languages, we can have $\top <: \bot \to \bot$. Indeed, this assertion holds when the application is strict on its right argument (for any

J. Marcinkowski and A. Tarlecki (Eds.): CSL 2004, LNCS 3210, pp. 415–429, 2004.

first argument which evaluates without error), that is, when we can apply any term e which evaluates without error to a term e' that diverges and get a term $e\,e'$ which diverges.

- In some call-by-value languages, we can have the distributivity law $(\tau_1 \cup \tau_2) \times \tau = (\tau_1 \times \tau) \cup (\tau_2 \times \tau)$. This law does not hold in a call-by-name language with non-determinism. Indeed, a term of type $(\tau_1 \cup \tau_2) \times \tau$ may well be a pair whose first component evaluates sometimes to a value of type τ_1 and sometimes to a value of type τ_2. Still, it can hold in a call-by-need language with non-determinism, as an expression is then evaluated at most once.
- In a deterministic language, union of function types $\tau \to \tau'$ obey very special subtyping rules when τ is finite (as observed by Damm [1]). The reason is that these types are isomorphic to tuple types.

On the other hand, some rules seem very robust:

- The arrow is covariant on the left and contravariant on the right: if $\tau_1 <: \tau_1'$ and $\tau_2' <: \tau_2$, then $\tau_2 \to \tau_1 <: \tau_2' \to \tau_1'$;
- Union types are least upper bounds: if $\tau <: \tau_1$ or $\tau <: \tau_2$, then $\tau <: \tau_1 \cup \tau_2$; if $\tau_1 <: \tau$ and $\tau_2 <: \tau$, then $\tau_1 \cup \tau_2 <: \tau$.

The aim of this paper is to develop a framework in which we can substantiate the above claims, and thus understand which subtyping assertions $\tau <: \tau'$ hold "by accident" (depending on some specific properties of a calculus), and which are more universal (valid for a large class of calculi).

Rather than choosing a particular calculus, we specify a broad class of calculi in a fairly abstract way. For each calculus, we interpret a type τ as a set of terms $[\![\tau]\!]$. Given a subtyping relation $<:$, defined for instance by inference rules, we can state that a subtyping assertion $\tau <: \tau'$ is *valid* when $[\![\tau]\!] \subseteq [\![\tau']\!]$. Then, a subtyping relation is *sound* when any derivable subtyping assertion is valid in all calculi. It is *complete* when every universally valid assertion can be derived. We present a relation which is both sound and complete for the class of calculi considered. Though this is not addressed in this paper, it would then be possible to study relations which are only sound under some assumptions by restricting the class of calculi.

The paper is organized as follows. The class of calculi is defined (Sect. 2) and a particular instance is given (Sect. 3). We present a simple type system, define a subtyping relation and prove the soundness and completeness of the relation (Sect. 4). We conclude by presenting related work (Sect. 5) and directions for future work (Sect. 6). Most proofs are omitted for lack of space. They are available online in an extended version of the paper [2].

2   A Class of Abstract Calculi

2.1   Informal Presentation and Definitions

We would like to study subtyping for a class of calculi with functions, pairs and constants. The first step is to associate to each type τ its semantics $[\![\tau]\!]$, that is, the set of terms of type τ. We type terms rather than values because the notion of terms is more fundamental: the notion of value depends on the language considered. Besides, it is not always possible to reduce the behavior of a term to the behavior of a set of values, especially in a call-by-name calculus. This is actually possible in the calculus of Sect. 3, but only because we made some specific choices about types.

As it turns out, it is convenient to only consider sets of terms that satisfy a given closure property: we assume given a *closure operator* on sets of terms, that is, a function $\mathcal{E} \mapsto \overline{\mathcal{E}}$ which is extensive ($\mathcal{E} \subseteq \overline{\mathcal{E}}$), idempotent ($\overline{\overline{\mathcal{E}}} = \overline{\mathcal{E}}$) and monotone. A set of terms $\mathcal{E}$ is said to be *closed* if $\mathcal{E} = \overline{\mathcal{E}}$. The idea is that the closure $\overline{\mathcal{E}}$ of a set of terms $\mathcal{E}$ is the set of terms that cannot be distinguished (as far as types are concerned) from the terms in $\mathcal{E}$. Thus, different choices of a closure operator yields different interpretation of types.

Types categorize terms according to their behavior. We should be able to use them to avoid some unsafe behavior, typically runtime errors. So, we distinguish a set $\mathcal{S}$ of *safe terms*. Dually, we define a set $\mathcal{N}$ of *neutral terms* (typically, terms that loop) as the intersection of all non-empty closed sets of terms. We call *semantic type* a closed set of terms included in $\mathcal{S}$ and including $\mathcal{N}$. We require the semantics $[\![\tau]\!]$ of a syntactic type τ to be a semantic type.

It seems really important in practice to distinguish a set of safe terms $\mathcal{S}$ from the set of all terms $\mathcal{T}$, and a set of neutral terms $\mathcal{N}$ from the least closed set $\overline{\emptyset}$. Indeed, in Sect. 3, we will have $\mathcal{S} \neq \mathcal{T}$ and $\mathcal{N} = \overline{\emptyset}$, but in [3], we have $\mathcal{S} \neq \mathcal{T}$ and $\mathcal{N} \neq \overline{\emptyset}$, and in [4], we have $\mathcal{S} = \mathcal{T}$ and $\mathcal{N} \neq \overline{\emptyset}$. Finally, in the case of *reducibility candidates* [5], one has $\mathcal{S} \neq \mathcal{T}$ and $\mathcal{N} \neq \overline{\emptyset}$ (safe terms are strongly normalizing terms, and some terms such as a variable x can be given any type).

Let us now sketch how we define the semantics of types. The idea is that we want to be able to build more complex typed terms by assembling smaller typed terms according to simple (typing) rules. For instance:

$$
\begin{array}{ccc}
\textsc{App} & \textsc{Fst} & \textsc{Snd} \\
\dfrac{e : \tau' \to \tau \quad e' : \tau'}{e\,e' : \tau} & \dfrac{e : \tau \times \tau'}{\mathbf{fst}\,e : \tau} & \dfrac{e : \tau \times \tau'}{\mathbf{snd}\,e : \tau'}
\end{array}
$$

The rules above suggest the following inclusions.

$$
[\![\tau' \to \tau]\!] \subseteq \{e \in \mathcal{S} \mid \forall e' \in [\![\tau']\!].e\,e' \in [\![\tau]\!]\}
$$
$$
[\![\tau \times \tau']\!] \subseteq \{e \in \mathcal{S} \mid \mathbf{fst}\,e \in [\![\tau]\!] \wedge \mathbf{snd}\,e \in [\![\tau']\!]\}
$$

These inclusions ensure the *soundness* of the typing rules. In order to reason about types, it is important to have a more precise characterization of their semantics. It seems therefore natural to replace these inclusions by an equality.

$$
[\![\tau' \to \tau]\!] = \{e \in \mathcal{S} \mid \forall e' \in [\![\tau']\!].e\,e' \in [\![\tau]\!]\}
$$
$$
[\![\tau \times \tau']\!] = \{e \in \mathcal{S} \mid \mathbf{fst}\,e \in [\![\tau]\!] \wedge \mathbf{snd}\,e \in [\![\tau']\!]\}
$$

But the sets $[\![\tau' \to \tau]\!]$ and $[\![\tau \times \tau']\!]$ must be semantic types. The definitions above clearly ensure that these sets are included in $\mathcal{S}$. They must also be closed and must contain $\mathcal{N}$. We cannot force this by making the sets larger, as this would violate the soundness conditions. Instead, we make more assumptions on the calculi. We say that a function is *continuous* when the inverse image of a closed set is closed, that a function is *strict* when the set $\mathcal{N}$ is included in the inverse image of $\mathcal{N}$. We can prove inductively that the sets $[\![\tau' \to \tau]\!]$ and $[\![\tau \times \tau']\!]$ are closed if $\mathcal{S}$ is closed and the functions $\mathtt{fst}$, $\mathtt{snd}$, and $e \mapsto e\,e'$ (for all terms e' in $\mathcal{S}$) are continuous. Similarly, we can prove that these sets contain $\mathcal{N}$ if $\mathcal{N} \subseteq \mathcal{S}$ and the same functions are strict. This appears more clearly if the equations above are rewritten in a more algebraic form.

$$[\![\tau' \to \tau]\!] = \mathcal{S} \cap \bigcap_{e' \in [\![\tau']\!]} \{e \mid e\,e' \in [\![\tau]\!]\}$$

$$[\![\tau \times \tau']\!] = \mathcal{S} \cap \mathtt{fst}^{-1}([\![\tau]\!]) \cap \mathtt{snd}^{-1}([\![\tau']\!])$$

It is really natural for all these functions to be strict, as they are destructors. The continuity properties may seem harder to achieve. We will see in Sect. 3.2, that it is actually straightforward to define a closure operator ensuring these properties.

Note that if $\mathcal{N} = \overline{\emptyset}$, then all continuous functions are strict. Indeed, if f is continuous, then $f^{-1}(\overline{\emptyset})$ is closed and therefore contains $\overline{\emptyset}$. Thus, if we want constant functions to be continuous, which seems reasonable, we need to have $\mathcal{N} \neq \overline{\emptyset}$.

The calculi also have constants, denoted κ. These constants are assumed to be safe. We define a singleton type κ for each constant κ. Its semantics is the least closed set of term containing the constant κ:

$$[\![\kappa]\!] = \overline{\{\kappa\}} \ .$$

2.2   Formal Specification

The class of calculi we consider are the calculi to which we can associate:

- a set of terms $\mathcal{T}$;
- a closure operator $\mathcal{E} \mapsto \overline{\mathcal{E}}$ on terms;
- a closed subset $\mathcal{S} \subseteq \mathcal{T}$ of safe terms;
- three operators:

$$\begin{aligned} \mathtt{app} &: \mathcal{T} \to \mathcal{T} \to \mathcal{T} \\ e &\mapsto e' \mapsto e\,e' \\ \mathtt{fst} &: \mathcal{T} \to \mathcal{T} \\ e &\mapsto \mathtt{fst}\,e \\ \mathtt{snd} &: \mathcal{T} \to \mathcal{T} \\ e &\mapsto \mathtt{snd}\,e \end{aligned}$$

such that $e \mapsto e\,e'$ (where $e' \in \mathcal{S}$), $\mathtt{fst}$ and $\mathtt{snd}$ are continuous and strict;
- a set of constants $\kappa \in \mathcal{S}$.

Note that we consider the closure operator as part of the calculus. Thus, two different calculi can be identical except for their closure operators. They can be understood as two (semantically) typed variants of a same untyped calculus.

2.3   Semantic Operations

We define one operation on sets of terms for each type construction we have in mind: bottom type, union of two types, function types, pair types and constant types. These operations are used to define the semantics of types in a straight-forward fashion in Sect. 4.1. Note that the semantic union $\boxed{\cup}$ of two sets of terms is not simply their union. Indeed, the union of two closed sets is usually not a closed set. In other words, there may be some terms that are in neither of the sets but cannot be distinguished from the terms in the union of both sets. Our solution is to take the least closed set containing the union. This is not just a technical point, but is actually crucial for typing a calculus with non-determinism, for which we could expect, for instance, a term to be in $\mathcal{E} \boxed{\cup} \mathcal{E}'$ if it behaves erratically either as a term in $\mathcal{E}$ or as a term in $\mathcal{E}'$.

$$
\begin{aligned}
\boxed{\bot} &= \mathcal{N} \\
\mathcal{E} \boxed{\cup} \mathcal{E}' &= \overline{\mathcal{E} \cup \mathcal{E}'} \\
\mathcal{E}' \boxed{\rightarrow} \mathcal{E} &= \{e \in \mathcal{S} \mid \forall e' \in \mathcal{E}'.e\, e' \in \mathcal{E}\} \\
\mathcal{E} \boxed{\times} \mathcal{E}' &= \{e \in \mathcal{S} \mid \mathsf{fst}\, e \in \mathcal{E} \wedge \mathsf{snd}\, e \in \mathcal{E}'\} \\
\boxed{\kappa} &= \overline{\{\kappa\}}
\end{aligned}
$$

It is clear that all these operations map semantic types to semantic types.

3   A Concrete Calculus

We present a particular instance of the class of calculi considered. This calculus is used in Sect. 4 to prove the completeness of a subtyping relation. It actually turns out to be *universal*, in the sense that a subtyping relation is complete if and only if it is complete for this particular calculus.

3.1   The Calculus

The calculus we consider is a call-by-name calculus with pairs and constants. Its main remarkable characteristics are a notion of errors, a strict **let** binder and two non-deterministic choice operators. The syntax of the calculus is given by the following grammar:

$$
\begin{array}{lll}
e ::= & x & \text{variable} \\
& \lambda x.e & \text{abstraction} \\
& e\, e & \text{application} \\
& (e, e) & \text{pair} \\
& \mathsf{fst}\, e & \text{first projection} \\
& \mathsf{snd}\, e & \text{second projection} \\
& \kappa & \text{constant} \\
& \mathsf{if}\ e = \kappa\ \mathsf{then}\ e\ \mathsf{else}\ e & \text{conditional} \\
& e \sqcup e & \text{erratic choice} \\
& e \vee e & \text{error-avoiding choice} \\
& \mathsf{let}\ x = e\ \mathsf{in}\ e & \text{strict let} \\
& \mathsf{error} & \text{error}
\end{array}
$$

The set of constants κ is supposed to be infinite. A bigstep semantics is given in Fig. 1. The values are a subgrammar of terms:

$$v ::= \lambda x.e \mid (e, e) \mid \kappa \mid \textbf{error}$$

In the reduction rules, we write $v \neq v'$ where v' describes a specific shape of values (for instance, v' is (e_1, e_2)) to mean that v is not of the same shape as v'.

VAR-ERROR

$x \Downarrow \textbf{error}$

ABS

$\lambda x.e \Downarrow \lambda x.e$

APP

$$\dfrac{e \Downarrow \lambda x.e_1 \qquad e_1[e'/x] \Downarrow v}{e\,e' \Downarrow v}$$

APP-ERROR

$$\dfrac{e \Downarrow v \qquad v \neq \lambda x.e_1}{e\,e' \Downarrow \textbf{error}}$$

PAIR

$(e_1, e_2) \Downarrow (e_1, e_2)$

FST

$$\dfrac{e \Downarrow (e_1, e_2) \qquad e_1 \Downarrow v}{\textbf{fst}\,e \Downarrow v}$$

FST-ERROR

$$\dfrac{e \Downarrow v \qquad v \neq (e_1, e_2)}{\textbf{fst}\,e \Downarrow \textbf{error}}$$

SND

$$\dfrac{e \Downarrow (e_1, e_2) \qquad e_2 \Downarrow v}{\textbf{snd}\,e \Downarrow v}$$

SND-ERROR

$$\dfrac{e \Downarrow v \qquad v \neq (e_1, e_2)}{\textbf{snd}\,e \Downarrow \textbf{error}}$$

CONSTANT

$\kappa \Downarrow \kappa$

IF-EQUAL

$$\dfrac{e \Downarrow \kappa \qquad e' \Downarrow v}{\textbf{if } e = \kappa \textbf{ then } e' \textbf{ else } e'' \Downarrow v}$$

IF-NOT-EQUAL

$$\dfrac{e \Downarrow \kappa' \qquad \kappa \neq \kappa' \qquad e'' \Downarrow v}{\textbf{if } e = \kappa \textbf{ then } e' \textbf{ else } e'' \Downarrow v}$$

IF-ERROR

$$\dfrac{e \Downarrow v \qquad v \neq \kappa'}{\textbf{if } e = \kappa \textbf{ then } e' \textbf{ else } e'' \Downarrow \textbf{error}}$$

PARA-LEFT

$$\dfrac{e \Downarrow v}{e \sqcup e' \Downarrow v}$$

PARA-RIGHT

$$\dfrac{e' \Downarrow v}{e \sqcup e' \Downarrow v}$$

CATCH-LEFT

$$\dfrac{e \Downarrow v \qquad v \neq \textbf{error}}{e \vee e' \Downarrow v}$$

CATCH-RIGHT

$$\dfrac{e' \Downarrow v \qquad v \neq \textbf{error}}{e \vee e' \Downarrow v}$$

CATCH-ERROR

$$\dfrac{e \Downarrow \textbf{error} \qquad e' \Downarrow \textbf{error}}{e \vee e' \Downarrow \textbf{error}}$$

LET

$$\dfrac{e \Downarrow v \qquad v \neq \textbf{error} \qquad e'[v/x] \Downarrow v'}{\textbf{let } x = e \textbf{ in } e' \Downarrow v'}$$

LET-ERROR

$$\dfrac{e \Downarrow \textbf{error}}{\textbf{let } x = e \textbf{ in } e' \Downarrow \textbf{error}}$$

ERROR

$\textbf{error} \Downarrow \textbf{error}$

Fig. 1. Semantics

The semantics is rather standard and unsurprising. We simply say a few words about the two non-deterministic choice operators. The first one $e \sqcup e'$ is the standard erratic operator: $e \sqcup e' \Downarrow v$ if and only if either $e \Downarrow v$ or $e \Downarrow v$. The second one $e \vee e'$ is a bit like an angelic choice operator, but instead of attempting to avoid non-termination, it attempts to avoid errors. Another way of understanding this operator is to consider it as a symmetric variant of a catch operator: it evaluates one of the terms e or e' and, if this fails, falls back to evaluating the other term. The unusual notations emphasize the fact that both operations correspond to a least upper bound, as we will see in Sect. 3.4.

We define the following diverging term:

$$\texttt{diverge} = (\lambda x.x\, x)\, (\lambda x.x\, x) \ .$$

3.2   Orthogonality

Remember that we need to specify not only a calculus but also a closure operator on sets of terms. We first present a generic way of building a closure operator. The choice of a particular closure operator is made in the next section 3.3.

A convenient way to define a closure operator on sets of terms is by *orthogonality* between terms and contexts. At this point, it does not matter what the set of contexts is. We just assume given an orthogonality relation $e \perp c$ between contexts c and terms e. Its intended meaning is that the term e behaves properly in the context c. We define the *orthogonal* of a set of terms $\mathcal{E}$ as the set of contexts in which all terms in $\mathcal{E}$ behave properly:

$$\mathcal{E}^{\perp} = \{c \,|\, \forall e \in \mathcal{E}.e \perp c\} \ .$$

Conversely, we define the *orthogonal* of a set of contexts $\mathcal{C}$ as the set of terms that behave properly in all the contexts in $\mathcal{C}$:

$$\mathcal{C}^{\perp} = \{e \,|\, \forall c \in \mathcal{C}.e \perp c\} \ .$$

These two functions define a Galois connection between sets of terms and sets of contexts. The important point here is that the composition of these two functions, which associates to a set of terms $\mathcal{E}$ its *biorthogonal* $\overline{\mathcal{E}} = \mathcal{E}^{\perp\perp}$, is a closure operator. (Dually, we can define a closure operator which associates to a set of contexts its biorthogonal $\overline{\mathcal{C}} = \mathcal{C}^{\perp\perp}$.)

Furthermore, we can rely on the following lemma to guide us in the choice of a set of contexts. Let f be a function from terms to terms, and g be a function from contexts to contexts. We say that g is an *adjoint* of f iff

$$f(e) \perp c \Leftrightarrow e \perp g(c) \ .$$

Lemma 1. *If a function f has an adjoint g, then it is continuous.*

3.3   The Closure Operator

Using the tools just developed, we can now specify the closure operator. Contexts are given by the following grammar:

$$
\begin{array}{lll}
c ::= & Id & \text{identity} \\
& c \circ F & \text{frame concatenation} \\
& c \vee c & \text{join} \\
F ::= & \_\, e & \\
& \texttt{fst}\,\_ & \\
& \texttt{snd}\,\_ & \\
& \texttt{if}\ \_ = \kappa\ \texttt{then}\ e\ \texttt{else}\ e &
\end{array}
$$

A context c can be viewed as a stack, with a weird "stack join" operation, and F can be viewed as a stack frame. Every context c and term e may be combined to generate a term denoted $c\,e$ and defined as follows (the term $F[e]$ is the term which results from replacing _ by e in the frame F):

$$
\begin{aligned}
Id\,e &= e\\
(c \circ F)\,e &= c\,(F[e])\\
(c \vee c')\,e &= \texttt{let } x = e \texttt{ in } ((c\,x) \vee (c'\,x)) \qquad \text{where } x \text{ is fresh .}
\end{aligned}
$$

A term e is safe when it does not reduce to the error. Thus, we define the set $\mathcal{S}$ by:

$$\mathcal{S} = \{e \mid \neg(e \Downarrow \texttt{error})\} \ .$$

The orthogonality relation is defined by:

$$e \perp c \text{ iff } c\,e \in \mathcal{S} \ .$$

As indicated in the previous section 3.2, this induces a closure operator on sets of terms. This is the closure operator that we choose to associate to our calculus.

The choice of this operator is crucial: it controls what can be observed by typed terms. We should therefore explain how the contexts are chosen. The identity context Id ensures that $\mathcal{S}$ is closed. The frame concatenation operation $c \circ F$ ensures that each frame is continuous (by Lemma 1). The join operation $c \vee c'$ allows for disjunctive tests. For instance, the context $(Id \circ \texttt{fst} \_) \vee (Id \circ \_ \texttt{diverge})$ will behave properly against terms which reduce to either a pair or a function, but will fail with other terms. This ensures that the closed union $\overline{\mathcal{E}} \mathbin{\overline{\sqcup}} \overline{\mathcal{E}'}$ of two semantic types $\overline{\mathcal{E}}$ and $\overline{\mathcal{E}'}$ is not "too large" (see Sect. 3.4 for a more precise characterization of this property).

3.4   Properties of the Calculus

We study some notable properties of the calculus. The completeness proof will make use of all these properties.

Terms and Values. An important property of the calculus is that the behavior of a term (as specified by the closure operator) is characterized by the behavior of the values it reduces to.

Lemma 2 (Terms and Values). *A term e is included in a closed set of terms $\overline{\mathcal{E}}$ if and only if any value v it reduces to is included in $\overline{\mathcal{E}}$.*

The contexts have been carefully chosen for the lemma 2 to hold. For instance, it does not hold if the syntax of frames is extended with a family of frames $e\,\_$. Indeed, consider the term:

$$f = \lambda x.\texttt{if } x = \kappa \texttt{ then } (\texttt{if } x = \kappa \texttt{ then diverge else error}) \texttt{ else diverge} \ .$$

We have $f\,\kappa' \in \mathcal{S}$ for all constant κ', but $f\,(\kappa \sqcup \kappa') \notin \mathcal{S}$ if the constants κ and κ' are distinct. So, if $Id \circ (f\,\_)$ is a context, then we have $\kappa' \in \{Id \circ (f\,\_)\}^{\perp}$ for all constant κ', but not $\kappa \sqcup \kappa' \in \{Id \circ (f\ \_)\}^\perp$ (when the constants κ and κ' are distinct).

Intuitively, the result holds if the evaluation of a term ce first involves the evaluation of the term e. We formalize this property by introducing a notion of linearity: we say that a function f from terms to terms is *linear* when for any term e and value v, $f\,e \Downarrow v$ if and only if there exists a value v' such that $e \Downarrow v'$ and $f\,v' \Downarrow v$. We then have the expected result.

Lemma 3 (Context Linearity). *Contexts are linear.*

Ordering of Terms and Contexts. We define the *contextual preorder* on terms by $e \leq e'$ if and only if $\overline{\{e\}} \subseteq \overline{\{e'\}}$. Likewise, we define a preorder on contexts by $c \leq c'$ if and only if $\{c\}^\perp \subseteq \{c'\}^\perp$. Note that we choose to define both preorders so that the ordering between two elements (either two terms or two contexts) derives from the inclusion ordering between the two naturally associated sets of terms. We present the relative ordering of some interesting terms and contexts. This ordering is illustrated below.

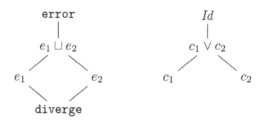

Lemma 4 (Least Upper Bounds). *For all terms e, e' and for all contexts c, c', we have: $\overline{\{e \sqcup e'\}} = \overline{\{e\}} \,\overline{\cup}\, \overline{\{e'\}}$ and $\{c \vee c'\}^\perp = \{c\}^\perp \,\overline{\cup}\, \{c'\}^\perp$. As a consequence, the term $e \sqcup e'$ is a least upper bound of the two terms e and e', and the context $c \vee c'$ is a least upper bound of the two contexts c and c'.*

Lemma 5 (Divergence). *The term* diverge *is a least term. In particular,* diverge $\in \boxed{\perp}$.

Sets of Values. We write $\mathcal{V}(\mathcal{E})$ for the set of values contained in a set of terms $\mathcal{E}$: $\mathcal{V}(\mathcal{E}) = \{v \mid v \in \mathcal{E}\}$. A direct consequence of Lemma 2 (Terms and Values) is that a closed set of terms is characterized by its values: $\overline{\mathcal{E}} = \overline{\mathcal{V}(\overline{\mathcal{E}})}$. It seems therefore natural to study some of the properties of the sets of values $\mathcal{V}(\overline{\mathcal{E}})$.

Lemma 6 (Least Semantic Type). *The least semantic type* $\boxed{\perp} = \mathcal{N}$ *does not contain any value. As a consequence, it is the least closed set of terms:* $\boxed{\perp} = \overline{\emptyset}$.

Lemma 7 (Union and Values). *The values of the closed union of two closed sets is the union of the values of each closed sets:* $\mathcal{V}(\overline{\mathcal{E}} \,\boxed{\cup}\, \overline{\mathcal{E}'}) = \mathcal{V}(\overline{\mathcal{E}}) \cup \mathcal{V}(\overline{\mathcal{E}'})$

We say that a set of terms $\mathcal{E}$ is *directed* when it is non-empty and when each pair of terms of this subset has an upper bound in this subset.

Lemma 8 (Prime when Directed). *If the set $\mathcal{V}(\overline{\mathcal{E}})$ is directed then the set $\overline{\mathcal{E}}$ is* prime, *that is, if $\overline{\mathcal{E}} \subseteq \overline{\mathcal{E}_1} \boxdot \overline{\mathcal{E}_2}$, then either $\overline{\mathcal{E}} \subseteq \overline{\mathcal{E}_1}$ or $\overline{\mathcal{E}} \subseteq \overline{\mathcal{E}_2}$.*

Instance of the Class of Calculi. We have the expected result:

Lemma 9. *The calculus is an instance of the class specified in Sect. 2.2.*

Orthogonality Functions-Arguments. Just as we defined an orthogonality relation between terms and contexts in Sect. 3.2, we can define a family of orthogonality relations between functions and arguments.

In the remainder of this section, we assume given a semantic type $\mathcal{E}_0$. We define an orthogonality relation between the elements of $\mathcal{T}$ (all terms), considered as function arguments, and the elements of $\mathcal{S}$ (safe terms), considered as functions: an argument $e' \in \mathcal{T}$ is orthogonal to a function $e \in \mathcal{S}$ when $e\,e' \in \mathcal{E}_0$. From this relation, we define the orthogonal of a set $\mathcal{E}$ of arguments by

$$\mathcal{E}^{\mathrm{fun}} = \{e \in \mathcal{S} \mid \forall e' \in \mathcal{E}.e\,e' \in \mathcal{E}_0\} = \mathcal{E} \boxdot \mathcal{E}_0$$

and the orthogonal of a set $\mathcal{E} \subseteq \mathcal{S}$ of functions by

$$\mathcal{E}^{\mathrm{arg}} = \{e' \mid \forall e \in \mathcal{E}.e\,e' \in \mathcal{E}_0\} \ .$$

The function $\mathcal{E} \mapsto \mathcal{E}^{\mathrm{fun\,arg}}$ is a closure on set of arguments.

Lemma 10 (Function Orthogonality). *The closure induced by function orthogonality is strictly finer than the closure induced by context orthogonality: for all sets of terms $\mathcal{E}$, we have*

$$\mathcal{E}^{\mathrm{fun\,arg}} \subseteq \overline{\mathcal{E}} \ ,$$

but the converse inclusion does not always hold. As a consequence,

$$\overline{\mathcal{E}}^{\mathrm{fun\,arg}} = \overline{\mathcal{E}^{\mathrm{fun\,arg}}} = \overline{\mathcal{E}}$$
$$\overline{\mathcal{E}}^{\mathrm{fun}} = \overline{\mathcal{E}} \boxdot \mathcal{E}_0$$
$$\overline{\mathcal{E}} = (\overline{\mathcal{E}} \boxdot \mathcal{E}_0)^{\mathrm{arg}} \ .$$

The key idea to prove the first inclusion is to show that for each context c there is a function $\langle c \rangle$ that behaves "similarly". This function is defined as follows.

$$\langle c \rangle = \lambda x.\mathtt{let}\ y = c\,x\ \mathtt{in}\ \mathtt{diverge}$$

It satisfies the following property.

Lemma 11 (Context as Function). *For any set of terms $\mathcal{E}$ and any context c, we have $c \in \mathcal{E}^{\perp}$ if and only if $\langle c \rangle \in \mathcal{E} \boxdot \mathcal{E}_0$.*

4   A Simple Type System

We present a simple type system and prove its soundness and completeness. These properties have been mechanically checked using the Coq proof assistant [6].

4.1   Types

The syntax of types is given by the following grammar.

$$\tau ::= \chi \qquad \text{constructed type} \qquad\qquad \chi ::= \tau \to \tau \quad \text{function type}$$
$$\bot \qquad \text{bottom type} \qquad\qquad\qquad\qquad \tau \times \tau \quad \text{pair type}$$
$$\tau \cup \tau \quad \text{union type} \qquad\qquad\qquad\qquad\quad \kappa \qquad \text{constant type}$$

The semantics $[\![\tau]\!]$ of a type τ is defined inductively on the syntax of types in a straightforward manner:

$$[\![\tau \to \tau']\!] = [\![\tau]\!] \boxed{\Rightarrow} [\![\tau']\!] \qquad\qquad [\![\bot]\!] \quad = \boxed{\bot}$$
$$[\![\tau \times \tau']\!] = [\![\tau]\!] \boxed{\times} [\![\tau']\!] \qquad\qquad [\![\tau \cup \tau']\!] = [\![\tau]\!] \boxed{\cup} [\![\tau']\!]$$
$$[\![\kappa]\!] \quad = \boxed{\kappa}$$

Clearly, the semantics $[\![\tau]\!]$ of a *syntactic type* τ is a semantic type.

4.2   Subtyping Relation

The subtyping relation $<:$ is defined inductively. The subtyping rules are given in Fig. 2. Note that the rules are almost syntax-directed: the conclusions of the rules are disjoint, except in the case of rules UNION-RIGHT-1 and UNION-RIGHT-2.

FUNCTION
$$\frac{\tau_1 <: \tau_1' \qquad \tau_2' <: \tau_2}{\tau_2 \to \tau_1 <: \tau_2' \to \tau_1'}$$

PAIR
$$\frac{\tau_1 <: \tau_1' \qquad \tau_2 <: \tau_2'}{\tau_1 \times \tau_2 <: \tau_1' \times \tau_2'}$$

CONSTANT
$$\kappa <: \kappa$$

BOTTOM
$$\bot <: \tau$$

UNION-LEFT
$$\frac{\tau <: \tau'' \qquad \tau' <: \tau''}{\tau \cup \tau' <: \tau''}$$

UNION-RIGHT-1
$$\frac{\chi <: \tau}{\chi <: \tau \cup \tau'}$$

UNION-RIGHT-2
$$\frac{\chi <: \tau'}{\chi <: \tau \cup \tau'}$$

Fig. 2. Subtyping Rules

4.3   Soundness of the Subtyping Relation

The soundness of the subtyping relation is straightforward.

Theorem 12 (Soundness). *If $\tau <: \tau'$, then $[\![\tau]\!] \subseteq [\![\tau']\!]$.*

Proof. By induction on a derivation of $\tau <: \tau'$.

- Rule FUNCTION: by covariance and contravariance of the operation $\boxed{\Rightarrow}$.
- Rule PAIR: by covariance of the operation $\boxed{\times}$.
- Rule CONSTANT: immediate.
- Rule BOTTOM: the semantic type $\boxed{\bot}$ is the least semantic type.
- Rule UNION-LEFT: by <u>induction</u> hypothesis, $[\![\tau]\!] \cup [\![\tau']\!] \subseteq [\![\tau'']\!]$; hence, as $[\![\tau'']\!]$ is closed, $[\![\tau]\!] \boxed{\cup} [\![\tau']\!] = \overline{[\![\tau]\!] \cup [\![\tau']\!]} \subseteq [\![\tau'']\!]$.
- Rule UNION-RIGHT-1: $[\![\tau]\!] \subseteq [\![\tau]\!] \boxed{\cup} [\![\tau']\!]$.
- Rule UNION-RIGHT-2: $[\![\tau']\!] \subseteq [\![\tau]\!] \boxed{\cup} [\![\tau']\!]$. $\square$

4.4   Properties of Constructed Types

Before proving the completeness of the subtyping relation $<:$, we first state some interesting properties of the semantics of constructed types.

Lemma 13 (Homogeneity). *The set of values $\mathcal{V}(\llbracket \chi \rrbracket)$ of a constructed type χ is homogeneous: $\mathcal{V}(\llbracket \tau' \to \tau \rrbracket)$ only contain functions, $\mathcal{V}(\llbracket \tau \times \tau' \rrbracket)$ only contain pairs, $\mathcal{V}(\llbracket \kappa \rrbracket)$ only contain the constant κ.*

Lemma 14 (Directed Set). *The set of values $\mathcal{V}(\llbracket \chi \rrbracket)$ is directed.*

These two lemmas are illustrated below, respectively for function types, pair types and constant types. Values are underlined. The value just above `diverge` is included in all constructed types of the corresponding kind. Given two values in $\mathcal{V}(\llbracket \chi \rrbracket)$, one of their upper bounds in $\mathcal{V}(\llbracket \chi \rrbracket)$ is given.

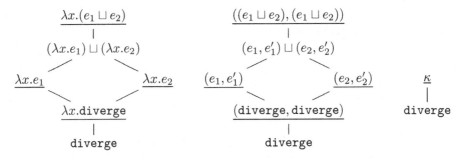

4.5   Completeness of the Subtyping Relation

We now have all the elements to prove the completeness of the subtyping relation.

Theorem 15. *If $\llbracket \tau \rrbracket \subseteq \llbracket \tau' \rrbracket$ for the calculus of Sect. 3, then $\tau <: \tau'$.*

Corollary 16 (Completeness). *If $\llbracket \tau \rrbracket \subseteq \llbracket \tau' \rrbracket$ for all calculi, then $\tau <: \tau'$.*

At several points in the proof of completeness, we need to prove an inclusion $\llbracket \tau_1 \rrbracket \subseteq \llbracket \tau_1' \rrbracket$ assuming that an inclusion between the semantics of two types built from τ_1 and τ_1' (for instance, $\llbracket \tau_1 \times \tau_2 \rrbracket \subseteq \llbracket \tau_1' \times \tau_2' \rrbracket$) holds. The proof is similar in each case. Let us call *typed transformation* a pair of a function F from types to types and a function f from terms to terms such that, for all types τ and all terms e, $e \in \llbracket \tau \rrbracket$ if and only if $f(e) \in \llbracket F(\tau) \rrbracket$. Then, it is easy to see that, if (F, f) is a typed transformation and $\llbracket F(\tau) \rrbracket \subseteq \llbracket F(\tau') \rrbracket$, then $\llbracket \tau \rrbracket \subseteq \llbracket \tau' \rrbracket$, We thus define three families of typed transformations.

Lemma 17 (Typed Transformations). *The following families of pairs of functions are typed transformations (for the calculus of Sect. 3).*

$$F_1(\tau') : \tau \mapsto \tau \times \tau' \qquad f_1 : e \mapsto (e, \mathtt{diverge})$$
$$F_2(\tau') : \tau \mapsto \tau' \times \tau \qquad f_2 : e \mapsto (\mathtt{diverge}, e)$$
$$F_3(\tau') : \tau \mapsto \tau' \to \tau \qquad f_3 : e \mapsto \lambda x.e$$

Proof (of Theorem 15). We interpret the semantics of types in the calculus defined in Sect. 3. In order to handle the contravariance of the function type, we simultaneously prove by induction on τ and τ' that if $[\![\tau]\!] \subseteq [\![\tau']\!]$ then $\tau <: \tau'$, and if $[\![\tau']\!] \subseteq [\![\tau]\!]$ then $\tau' <: \tau$. For each pair of type τ and τ', we prove that if $[\![\tau]\!] \subseteq [\![\tau']\!]$, then there exists a subtyping rule whose conclusion is $\tau <: \tau'$ and whose premises are a consequence of the induction hypothesis.

- Case $[\![\bot]\!] \subseteq [\![\tau]\!]$. By rule BOTTOM, we have $\bot <: \tau$.
- Case $[\![\tau \cup \tau']\!] \subseteq [\![\tau'']\!]$. This implies $[\![\tau]\!] \subseteq [\![\tau'']\!]$ and $[\![\tau']\!] \subseteq [\![\tau'']\!]$. Hence, by induction hypothesis, $\tau <: \tau''$ and $\tau' <: \tau''$. Finally, by rule UNION-LEFT, $\tau \cup \tau' <: \tau''$.
- Case $[\![\chi]\!] \subseteq [\![\bot]\!]$. By lemma 6 (Least Semantic Type), the set $[\![\bot]\!]$ does not contain any value. By Lemma 14 (Directed Set), $[\![\chi]\!]$ contains at least one value. Thus, this case is not possible.
- Case $[\![\chi]\!] \subseteq [\![\tau \cup \tau']\!]$. This is a direct corollary of Lemmas 14 (Directed Set) and 8 (Prime when Directed).
- Case $[\![\chi]\!] \subseteq [\![\chi']\!]$ where χ and χ' are distinct constructed types. By Lemmas 14 (Directed Set) and 13 (Homogeneity), constructed types all contain at least a value, and their values are homogeneous. Hence, $[\![\chi]\!]$ contains a value which is not in $[\![\chi']\!]$. This case is not possible.
- Case $[\![\tau_2 \to \tau_1]\!] \subseteq [\![\tau_4 \to \tau_3]\!]$. We prove that $[\![\tau_1]\!] \subseteq [\![\tau_3]\!]$ and $[\![\tau_4]\!] \subseteq [\![\tau_2]\!]$. This allow us to conclude by induction hypothesis and rule FUNCTION.
  The inclusion $[\![\tau_1]\!] \subseteq [\![\tau_3]\!]$ is a direct consequence of Lemma 17 (Typed Transformations).
  Let us prove that $[\![\tau_4]\!] \subseteq [\![\tau_2]\!]$. It is sufficient to show that $[\![\tau_2]\!]^\perp \subseteq [\![\tau_4]\!]^\perp$. Let c in $[\![\tau_2]\!]^\perp$. By Lemma 11 (Context as Function), $\langle c \rangle \in [\![\tau_2]\!] \boxminus [\![\tau_1]\!] = [\![\tau_2 \to \tau_1]\!] \subseteq [\![\tau_4 \to \tau_3]\!] = [\![\tau_4]\!] \boxminus [\![\tau_3]\!]$. Hence, by this lemma again, $c \in [\![\tau_4]\!]^\perp$.
- Case $[\![\tau_1 \times \tau_2]\!] \subseteq [\![\tau_3 \times \tau_4]\!]$. By Lemma 17 (Typed Transformations), $[\![\tau_1]\!] \subseteq [\![\tau_3]\!]$ and $[\![\tau_2]\!] \subseteq [\![\tau_4]\!]$. We conclude by induction and rule PAIR. □

The proof of the completeness theorem actually leaded us to use an orthogonality relation to define types. Indeed, for completeness to hold, we must have that, if $\tau_1 \to \tau <: \tau_2 \to \tau$, then $\tau_2 <: \tau_1$. This means that, if a term e has type τ_2 but not type τ_1, then there must exist a function e' of type $\tau_1 \to \tau$ but not $\tau_2 \to \tau$. Given that the term e has type τ_2, a natural way to prove that the function e' does not have type $\tau_2 \to \tau$ is to show that the term $e' e$ does not have type τ. So, now, for any term e of type τ_2 but not τ_1, we must be able to find a function of type $\tau_1 \to \tau$ such that the term $e' e$ does not have type τ. This must hold for any type τ_2, so the assumption that the term e has type τ_2 does not really put any constraint on the term e and it is natural to drop it. So, finally, we would like that if a term e does not have type τ_1, then there is a function e' of type $\tau_1 \to \tau$ such that $e' e$ does not have type τ. In other words, if $e \notin [\![\tau_1]\!]$, then there exists a function $e' \in [\![\tau_1]\!]^{\mathrm{fun}}$ such that e and e' are not orthogonal. That is, if a term is orthogonal to all functions in $[\![\tau_1]\!]^{\mathrm{fun}}$, then it should have type $[\![\tau_1]\!]$: the set $[\![\tau_1]\!]$ must be closed.

A noteworthy point in this discussion is that if τ is not a subtype of τ', then it is unsafe to apply a function accepting terms of type τ' to a term of type τ.

Lemma 18. *For the calculus of Sect. 3, if τ is not a subtype of τ', then there exists a term e in $[\![\tau]\!]$ and a function e' in $[\![\tau' \to \bot]\!]$ such that $e'\, e \Downarrow$ error.*

5   Related Work

This work is a continuation of our work with Melliès on semantic types [4, 3]. These two papers focus on defining types, especially recursive types, as set of terms, while we study here the subtyping relation induced by these definitions.

Defining the semantics of types as closed sets of terms is very natural. For instance, in domain theory, types can be interpreted as *ideals* [7], that is, sets that are downward closed and closed under directed limits. *Reducibility candidates* [5] are also closed sets of terms. Girard [8] reformulates the candidates as sets of terms closed by biorthogonality in his proof of cut elimination for linear logic. Meanwhile, Krivine [9, 10] has developed a comprehensive framework based on orthogonality, in order to analyze types as *specification* of terms. In semantics, Pitts [11] uses relations closed by biorthogonality to study parametric polymorphism in an operational setting.

Damm [1] studies subtyping for a deterministic calculus with recursive types with union and intersection. He takes a domain theoretic approach based on the ideal model [7]. A subtyping algorithm is specified by encoding types into tree automata and defining the subtyping relation as the inclusion of the recognized languages. The soundness and completeness of this algorithm with respect to the semantics of types is proven.

Frisch, Castagna and Benzaken [12] use an approach similar to ours to design a subtyping relation for a typed calculus with union and intersection types. They want to define the subtyping relation of this calculus in a semantic way, as the inclusion of the denotation of types. But their calculus is typed, so its semantics depends on the subtyping relation. In order to get rid of this circularity, they consider a class of calculi (called *models*). While we try to describe as large a class as possible, the authors design a class such that the subtyping relation has good properties (for instance, distributivity of union and intersection).

6   Extensions and Future Work

Polymorphism and Type Constructors. In an extended version of the paper [2], we present a refined type system with ML-style polymorphism and type constructors and we similarly prove its soundness and completeness. This is omitted here for lack of space.

Strict Pairs and Recursive Types. The type system presented here is not as rich as the type systems of XDuce [13] and CDuce [12] for two reasons. First, for the sake of simplicity, we have not considered recursive types. In previous work [4, 3], we have developed some tools to deal with them. Second, we deal with a very large class of calculi, in which some subtyping assertions such as

$(\tau_1 \cup \tau_2) \times \tau <: (\tau_1 \times \tau) \cup (\tau_2 \times \tau)$ do not hold (as hinted in the introduction). We would need to reduce the class of calculi to get a coarser subtyping relation.

Intersection Types. Intersection types are harder to handle than union types. The natural semantics for intersection types is set intersection:

$$\mathcal{E} \boxed{\cap} \mathcal{E}' = \mathcal{E} \cap \mathcal{E}' \ .$$

Then, it is clear that the dual of the subtyping rules for union types are sound. But there are other sound subtyping rules. For instance, we have $(\tau_1 \times \tau_3) \cap (\tau_2 \times \tau_4) <: (\tau_1 \cap \tau_2) \times (\tau_3 \cap \tau_4)$. Another issue is that the distributivity law $(\tau_1 \cup \tau_2) \cap \tau = (\tau_1 \cap \tau) \cup (\tau_2 \cap \tau_2)$ does not hold in general. Thus, it is not clear how union and intersection interact as far as subtyping is concerned.

References

1. Damm, F.: Subtyping with union types, intersection types and recursive types II. Research Report 2259, INRIA Rennes (1994)
2. Vouillon, J.: Subtyping union types (extended version). Manuscript; available from http://www.pps.jussieu.fr/~vouillon/publi/\#union. (2004)
3. Vouillon, J., Melliès, P.A.: Semantic types: A fresh look at the ideal model for types. In: Proceedings of the 31th ACM Conference on Principles of Programming Languages, Venezia, Italia, ACM Press (2004) 52–63
4. Melliès, P.A., Vouillon, J.: Recursive polymorphic types and parametricity in an operational framework. Preprint PPS//04/06//n°30; available from http://www.pps.jussieu.fr/~vouillon/publi/\#semtypes2. (2004)
5. Girard, J.Y.: Interprétation fonctionelle et élimination des coupures dans l'arithmétique d'ordre supérieur. Thèse de doctorat d'État, University of Paris VII (1972)
6. Coq Development Team: The Coq Proof Assistant Reference Manual – Version V7.4. (2003) Available from http://coq.inria.fr/doc/main.html.
7. MacQueen, D., Plotkin, G., Sethi, R.: An ideal model for recursive polymorphic types. Information and Control **71** (1986) 95–130
8. Girard, J.Y.: Linear logic. Theoretical Computer Science **50** (1987) 1–102
9. Danos, V., Krivine, J.L.: Disjunctive tautologies and synchronisation schemes. In: Computer Science Logic'00. Volume 1862 of Lecture Notes in Computer Science., Springer (2000) 292–301
10. Krivine, J.L.: Typed lambda-calculus in classical Zermelo-Fraenkel set theory. Archive of Mathematical Logic **40** (2001) 189–205
11. Pitts, A.M.: Parametric polymorphism and operational equivalence. Mathematical Structures in computer Science **10** (2000) 321–359
12. Frisch, A., Castagna, G., Benzaken, V.: Semantic subtyping. In: 17th IEEE Symposium on Logic in Computer Science, IEEE Computer Society Press (2002) 137–146
13. Hosoya, H., Vouillon, J., Pierce, B.C.: Regular expression types for XML. In: Proceedings of the International Conference on Functional Programming (ICFP). (2000)

Pfaffian Hybrid Systems

Margarita Korovina[1] and Nicolai Vorobjov[2]

[1] A.P.Ershov Institute of Informatics Systems, Russian Academy of Sciences
korovina@brics.dk,
http://www.brics.dk/~korovina
[2] Department of Computer Science, University of Bath, Bath BA2 7AY, England
nnv@cs.bath.ac.uk,
http://www.bath.ac.uk/~masnnv

Abstract. It is well known that in an o-minimal hybrid system the continuous and discrete components can be separated, and therefore the problem of finite bisimulation reduces to the same problem for a transition system associated with a continuous dynamical system. It was recently proved by several authors that under certain natural assumptions such finite bisimulation exists. In the paper we consider o-minimal systems defined by Pfaffian functions, either implicitly (via triangular systems of ordinary differential equations) or explicitly (by means of semi-Pfaffian maps). We give explicit upper bounds on the sizes of bisimulations as functions of formats of initial dynamical systems. We also suggest an algorithm with an elementary (doubly-exponential) upper complexity bound for computing finite bisimulations of these systems.

Introduction

We assume that the reader is familiar with the motivation and basic concepts of the theory of hybrid systems. This material can be found in collection of papers [7]. The more recent accounts are (not exclusively) [8, 3, 9].

Recall that in certain natural cases continuous and discrete components of a hybrid system can be separated. Moreover, the continuous component allows finite *bisimulation*, thus reducing the decidability questions for the original system to similar questions for a finite system. An important example having this property is the class of *o-minimal* hybrid systems, introduced in [9]. The main result of [9] is that under certain natural assumptions, o-minimal systems allow finite bisimulations. This statement was generalized in [3], where a convenient and elementary technique was developed, based on encoding of trajectories of o-minimal dynamical systems in partitioned spaces by means of words in finite alphabets. The elements of bisimulations are then encoded by *dotted* words.

In the present paper we use the technique of [3] to obtain some quantitative versions of the finite bisimulation theorems. We introduce *Pfaffian hybrid systems* which essentially reduce to *Pfaffian dynamical systems* defined by means of equations and inequalities involving *Pfaffian functions*. The latter are real

J. Marcinkowski and A. Tarlecki (Eds.): CSL 2004, LNCS 3210, pp. 430–441, 2004.
© Springer-Verlag Berlin Heidelberg 2004

analytic functions satisfying triangular systems of first order partial differential equations with polynomial coefficients. They include polynomials, real algebraic functions, and all major transcendental functions in appropriate domains. Pfaffian functions form the broadest natural class of real analytic functions for whose elements the *size* or *format* can be adequately assigned. The concept of the format can be extended to sets in $\mathbb{R}^n$ and to maps definable using Pfaffian functions.

We consider dynamical systems defined by Pfaffian functions, either implicitly (via triangular systems of ordinary differential equations) or explicitly (by means of semi-Pfaffian maps). We give explicit upper bounds on the sizes of bisimulations as functions of formats of initial dynamical systems. We also suggest an algorithm with an elementary (doubly-exponential) upper complexity bound for computing finite bisimulations of these systems.

More precisely, the outline of the paper is as follows. In Section 1 we summarize some well-known definitions and results about hybrid systems closely following [3, 9]. We also recall the *(dotted) words encoding* technique from [3]. Section 2 presents a brief digest of Pfaffian functions, upper bounds on topological complexities of semi- and sub-Pfaffian sets, and algorithms for computing their closures and cylindrical cell decompositions. In Section 3 two types of dynamical systems defined using Pfaffian functions are introduced. In Section 4 we consider dynamical systems determined by triangular systems of ordinary differential equations, and prove an upper bound on the size of its bisimulation (note that such systems may not be o-minimal in the sense of [3, 9]). In Section 5 we solve the similar problem for dynamical systems defined by explicit semi-Pfaffian maps. Finally, in Section 6 we propose an algorithm (with the usual for Pfaffian functions theory oracle) which actually computes a finite bisimulation for dynamical systems defined in Section 5. The complexity of the algorithm is doubly exponential in the format of the input system.

1   Transition Systems and Dynamical Systems

In [3, 9] it is explained how some central problems in the theory of o-minimal hybrid systems can be reduced to bisimulations of transition systems associated to o-minimal dynamical systems.

The exposition in this section closely follows [3]. The first group of definitions describes transition systems and bisimulations between the transition systems.

Definition 1. *Let Q be an arbitrary set and $\to$ be a binary relation on Q. In the context of hybrid systems theory we call Q the set of states, $\to$ the transition, and $T := (Q, \to)$ the transition system.*

Definition 2. *Given two transition systems $T_1 := (Q_1, \to_1)$ and $T_2 := (Q_2, \to_2)$ we define a* simulation *of T_1 by T_2 as a binary relation $\sim \subset Q_1 \times Q_2$ such that:*

- $\forall q_1 \in Q_1 \exists q_2 \in Q_2 (q_1 \sim q_2)$;
- $\forall q_1, q_1' \in Q_1 \forall q_2 \in Q_2 \exists q_2' ((q_1 \sim q_2 \wedge q_1 \to q_1') \Rightarrow (q_1' \sim q_2' \wedge q_2 \to q_2'))$.

Definition 3. A bisimulation between two transition systems $T_1 := (Q_1, \to_1)$ and $T_2 := (Q_2, \to_2)$ is a simulation $\sim \subset Q_1 \times Q_1$ of T_1 by T_2 such that the reciprocal relation $\sim^{-1} := \{(q_2, q_1) \in Q_2 \times Q_1 | q_1 \sim q_2\}$ is a simulation of T_2 by T_1.

Definition 4. A bisimulation between a transition system T and itself is called bisimulation on T.

Definition 5. Let $\sim$ be a bisimulation on $T = (Q, \to)$ and also an equivalence relation on Q. Let $\mathcal{P}$ be a partition of Q. We say that $\sim$ is a bisimulation with respect to $\mathcal{P}$ if any $P \in \mathcal{P}$ is a union of some equivalence classes of $\sim$.

In this paper we are concerned with estimating cardinality and computing bisimulations in the sense of Definition 5. We now give some definitions concerning dynamical systems.

Definition 6. Let $G_1 \subset \mathbb{R}^{k_1}$ and $G_2 \subset \mathbb{R}^{k_2}$ be open domains. The dynamical system is a map
$$\gamma : G_1 \times (-1, 1) \to G_2.$$

In the sequel γ will always be definable in an o-minimal structure over $\mathbb{R}$. For a given $\mathbf{x} \in G_1$ the set
$$\Gamma_{\mathbf{x}} = \{\mathbf{y} | \exists t \in (-1, 1)\, (\gamma(\mathbf{x}, t) = \mathbf{y})\} \subset G_2$$

is called the trajectory determined by $\mathbf{x}$, and the graph
$$\widehat{\Gamma}_{\mathbf{x}} = \{(t, \mathbf{y}) | \gamma(\mathbf{x}, t) = \mathbf{y}\} \subset (-1, 1) \times G_2$$

is called the integral curve determined by $\mathbf{x}$.

Definition 7. The transition system $T_\gamma = (Q, \to)$ associated to the dynamical system γ is defined as follows:

- $Q := G_2$, and
- $\mathbf{y}_1 \to \mathbf{y}_2$ for $\mathbf{y}_1, \mathbf{y}_2 \in Q$ if and only if

$$\exists \mathbf{x} \in G_1 \exists t_1, t_2 \in (-1, 1)((t_1 \leq t_2) \wedge (\gamma(\mathbf{x}, t_1) = \mathbf{y}_1) \wedge (\gamma(\mathbf{x}, t_2) = \mathbf{y}_2)).$$

We now introduce, following [3], a technique of encoding trajectories of dynamical systems by words. Let $\mathcal{P} := \{P_1, \ldots, P_s\}$ be a finite partition of $\gamma(G_1 \times (-1, 1))$ definable in the o-minimal struicture. Fix $\mathbf{x} \in G_1$. Define the set of points and open intervals in $\mathbb{R}$:

$$\mathcal{F}_{\mathbf{x}} := \{I | \ I \text{ is a point or an interval in } (-1, 1) \text{ maximal w.r.t. inclusion for the}$$
$$\text{property } \exists i \in \{1, \ldots, s\} \forall t \in I\, (\gamma(\mathbf{x}, t) \in P_i)\}.$$

Let the cardinality $|\mathcal{F}_{\mathbf{x}}| = r$ and $y_1 < \cdots < y_r$ be the set of representatives of $\mathcal{F}_{\mathbf{x}}$ such that $\gamma(\mathbf{x}, y_j) \in P_{i_j}$. Then define the word $\omega := P_{i_1} \cdots P_{i_r}$ in alphabet

$\mathcal{P}$. Informally, ω is the list of names of elements of the partition in the order they are visited by trajectory $\Gamma_{\mathbf{x}}$.

Let $\mathbf{y} \in \Gamma_{\mathbf{x}}$. Then $\mathbf{y} \in P_{i_j}$ for some $1 \le j \le r$, where P_{i_j} is a letter in ω. We represent the location of $\mathbf{y}$ on trajectory $\Gamma_{\mathbf{x}}$ by the *dotted word*

$$\dot{\omega} := P_{i_1} \cdots \dot{P}_{i_j} \cdots P_{i_r}.$$

It will be convenient to use the operation

$$\mathrm{undot}(\dot{\omega}) = \omega := P_{i_1} \cdots P_{i_j} \cdots P_{i_r}.$$

In the sequel we will always assume that a dynamical systems γ is injective. In this case there is a unique dotted word associated to a given $\mathbf{y} \in \gamma(G_1 \times (-1, 1))$.

Introduce sets of words $\Omega := \{\omega | \, \mathbf{x} \in G_1\}$, $\dot{\Omega} := \{\dot{\omega} | \, \mathbf{x} \in G_1\}$.

The following statement is an easy consequence of o-minimality.

Lemma 1. [3]   *The set Ω is finite.*

An obvious (purely combinatorial) corollary is that $\dot{\Omega}$ is also finite.

Definition 8. *The transition system $T_{\dot{\Omega}}$ is defined as follows:*

- $Q := \dot{\Omega}$, *and*
- $\dot{\omega}_1 \to \dot{\omega}_2$ *for* $\dot{\omega}_1, \dot{\omega}_2 \in Q$ *if and only if* $\omega_1 = \omega_2$ *and the dot on* $\dot{\omega}_2$ *is on the righter (or the same) position than the dot on* $\dot{\omega}_1$.

Theorem 1. [3]   *Let the o-minimal dynamical system γ be bijective, and the partition $\mathcal{P}$ be definable in the o-minimal structure. Then there is a finite bisimulation on T_γ with respect to $\mathcal{P}$.*

Proof. To prove the theorem one first shows that $T_{\dot{\Omega}}$ is a bisimulation of T_γ, and then considers the following equivalence relation $\sim$ on G_2: $\mathbf{y}_1 \sim \mathbf{y}_2$ iff for respective pre-images $(\mathbf{x}_1, t_1), (\mathbf{x}_2, t_2)$, the locations of $\mathbf{y}_1, \mathbf{y}_2$ on trajectories $\Gamma_{\mathbf{x}_1}, \Gamma_{\mathbf{x}_2}$ are described by the same dotted word $\dot{\omega}$. Then $\sim$ is the required bisimulation (see details in [3]). □

2   Pfaffian Functions and Related Sets

This section is a digest of the theory of Pfaffian functions and sets definable with Pfaffian functions. The detailed exposition can be found in the survey [4].

Definition 9. *A Pfaffian chain of the order $r \ge 0$ and degree $\alpha \ge 1$ in an open domain $G \subset \mathbb{R}^n$ is a sequence of real analytic functions $f_1, \dots, f_r$ in G satisfying differential equations*

$$\frac{\partial f_j}{\partial x_i} = g_{ij}(\mathbf{x}, f_1(\mathbf{x}), \dots, f_j(\mathbf{x})) \tag{1}$$

for $1 \leq j \leq r$, $1 \leq i \leq n$. *Here* $g_{ij}(\mathbf{x}, y_1, \ldots, y_j)$ *are polynomials in* $\mathbf{x} = (x_1, \ldots, x_n), y_1, \ldots, y_j$ *of degrees not exceeding* α. *A function*

$$f(\mathbf{x}) = P(\mathbf{x}, f_1(\mathbf{x}), \ldots, f_r(\mathbf{x})),$$

where $P(\mathbf{x}, y_1, \ldots, y_r)$ *is a polynomial of a degree not exceeding* $\beta \geq 1$, *is called a* Pfaffian function *of order* r *and degree* (α, β).

Apart from polynomials, the class of Pfaffian functions includes real algebraic functions, exponentials, logarithms, trigonometric functions, their compositions, and other major transcendental functions in appropriate domains (see [4]).

Definition 10. *A set* $X \subset \mathbb{R}^n$ *is called* semi-Pfaffian *in an open domain* $G \subset \mathbb{R}^n$ *if it consists of points in* G *satisfying a Boolean combination of some atomic equations and inequalities* $f = 0, g > 0$, *where* f, g *are Pfaffian functions having a common Pfaffian chain defined in* G. *A semi-Pfaffian set* X *is* restricted *in* G *if its topological closure lies in* G.

Definition 11. *A set* $X \subset \mathbb{R}^n$ *is called* sub-Pfaffian *in an open domain* $G \subset \mathbb{R}^n$ *if it is an image of semi-Pfaffian set under a projection into a subspace.*

In the sequel we will be dealing with the following subclass of sub-Pfaffian sets.

Definition 12. *Consider the closed cube* $[-1, 1]^{m+n}$ *in an open domain* $G \subset \mathbb{R}^{m+n}$ *and the projection map* $\pi : \mathbb{R}^{m+n} \to \mathbb{R}^n$. *A subset* $Y \subset [-1, 1]^n$ *is called* restricted sub-Pfaffian *if* $Y = \pi(X)$ *for a restricted semi-Pfaffian set* $X \subset [-1, 1]^{m+n}$.

Note that a restricted sub-Pfaffian set need not be semi-Pfaffian.

Definition 13. *Consider a semi-Pfaffian set*

$$X := \bigcup_{1 \leq i \leq M} \{\mathbf{x} \in \mathbb{R}^s | f_{i1} = \cdots = f_{iI_i}, g_{i1} > 0, \ldots, g_{iJ_i} > 0\} \subset G, \qquad (2)$$

where f_{ij}, g_{ij} *are Pfaffian functions with a common Pfaffian chain of order* r *and degree* (α, β), *defined in an open domain* G. *Its* format *is a tuple* (r, N, α, β, s), *where* $N \geq \sum_{1 \leq i \leq M}(I_i + J_i)$. *For* $s = m + n$ *and a sub-Pfaffian set* $Y \subset \mathbb{R}^n$ *such that* $Y = \pi(\overline{X})$, *its* format *is the format of* X.

We will refer to the representation of a semi-Pfaffian set in the form (2) as to *disjunctive normal form (DNF)*.

Remark 1. In this paper we are concerned with upper bounds on sizes of bisimulations and complexities of computations, as functions of the format. In the case of Pfaffian dynamical systems these sizes and complexities also depend on the domain G. So far our definitions imposed no restrictions on an open set G, thus allowing it to by arbitrarily complex and induce this complexity on the corresponding semi- and sub-Pfaffian sets. To avoid this we will always assume in the context of Pfaffian dynamical systems that G is "simple", like $\mathbb{R}^n$, or $(-1, 1)^n$.

Theorem 2. [6, 11]   *Consider a semi-Pfaffian set $X \subset G \subset \mathbb{R}^n$, where G is an open domain, represented in DNF with format (r, N, α, β, n). Then the sum of the Betti numbers (in particular, the number of connected components) of X does not exceed*

$$N^n 2^{r(r-1)/2} O(n\beta + \min\{n, r\}\alpha)^{n+r}.$$

In this paper we examine complexities of algorithms for computing bisimulations. In order to estimate the "efficiency" of a computation we need to specify more precisely a *model of computation*. As such we use a *real numbers machine* which is an analogy of a classical Turing machine but allows the exact arithmetic and comparisons on real numbers. Since we are interested only in upper complexity bounds for algorithms, we have no need in a formal definition of this model of computation (it can be found in [2]). In some of our computational problems we will need to modify the standard real numbers machine by equipping it with an *oracle* for deciding feasibility of any system of Pfaffian equations and inequalities. An oracle is a subroutine which can be used by a given algorithm any time the latter needs to check feasibility. We assume that this procedure always gives a correct answer ("true" or "false") though we do not specify how it actually works. An *elementary step* of a real numbers machine is either an arithmetic operation, or a comparison (branching) operation, or an oracle call. The *complexity* of a real numbers machine is the number of elementary steps it makes in worst case until termination, as a function of the format of the input.

Now we define cylindrical decompositions of semi- and sub-Pfaffian sets.

Definition 14. Cylindrical cell *in $[-1,1]^n$ is defined by induction as follows.*

1. *Cylindrical 0-cell in $[-1,1]^n$ is an isolated point.*
2. *Cylindrical 1-cell in $[-1,1]$ is an open interval $(a, b) \subset [-1, 1]$.*
3. *For $n \geq 2$ and $0 \leq k < n$ a cylindrical $(k+1)$-cell in $[-1,1]^n$ is either a graph of a continuous bounded function $f : C \to \mathbb{R}$, where C is a cylindrical $(k+1)$-cell in $[-1,1]^{n-1}$, or else a set of the form*

$$\{(x_1, \ldots, x_n) \in [-1, 1]^n \mid (x_1, \ldots, x_{n-1}) \in C \text{ and}$$

$$f(x_1, \ldots, x_{n-1}) < x_n < g(x_1, \ldots, x_{n-1})\},$$

where C is a cylindrical k-cell in $[-1,1]^{n-1}$, and $f, g : C \to [-1, 1]$ are continuous bounded functions such that $f(x_1, \ldots, x_{n-1}) < g(x_1, \ldots, x_{n-1})$ for all points $(x_1, \ldots, x_{n-1}) \in C$.

Definition 15. Cylindrical cell decomposition $\mathcal{D}$ *of a subset $A \subset [-1,1]^n$ is defined by induction as follows.*

1. *If $n = 1$, then $\mathcal{D}$ is a finite family of pair-wise disjoint cylindrical cells (i.e., isolated points and intervals) whose union is A.*
2. *If $n \geq 2$, then $\mathcal{D}$ is a finite family of pair-wise disjoint cylindrical cells in $[-1,1]^n$ whose union is A and there is a cylindrical cell decomposition of $\pi(A)$ such that $\pi(C)$ is its cell for each $C \in \mathcal{D}$, where $\pi : \mathbb{R}^n \to \mathbb{R}^{n-1}$ is the projection map onto the coordinate subspace of $x_1, \ldots, x_{n-1}$.*

Definition 16. *Let $B \subset A \subset [-1,1]^n$ and $\mathcal{D}$ be a cylindrical cell decomposition of A. Then $\mathcal{D}$ is* compatible *with B if for any $C \in \mathcal{D}$ we have either $C \subset B$ or $C \cap B = \emptyset$ (i.e., some subset $\mathcal{D}' \subset \mathcal{D}$ is a cylindrical cell decomposition of B).*

Definition 17. *For a given finite family $f_1, \ldots, f_N$ of Pfaffian functions f_i in an open domain G define its* consistent sign assignment *as a non-empty semi-Pfaffian set in G of the kind*

$$\{\mathbf{x} \in G \mid f_{i_1} = \cdots = f_{i_{N_1}} = 0, f_{i_{N_1}+1} > 0 \ldots, f_{i_{N_2}} > 0, f_{i_{N_2}+1} < 0, \ldots, f_{i_N} < 0\},$$

where $i_1, \ldots, i_{N_1}, \ldots, i_{N_2}, \ldots, i_N$ is a permutation of $1, \ldots, N$.

Theorem 3. [5, 10] *Let $f_1, \ldots, f_N$ be a family of Pfaffian functions in an open domain $G \subset \mathbb{R}^s$, $G \supset [-1,1]^s$ having a common Pfaffian chain of order r, and degrees (α, β). Then there is an algorithm (with the oracle) producing a cylindrical cell decomposition of $[-1,1]^s$ which is compatible with each consistent sign assignment of $f_1, \ldots, f_N$. Each cell is a sub-Pfaffian set represented as a projection of a semi-Pfaffian set in DNF. The number of cells, the components of their formats and the complexity of the algorithm are less than*

$$(\alpha + \beta N)^{r^{O(n)} 2^{O(n^2)}}.$$

3   Pfaffian Dynamical Systems

Definition 18. *A triangular system of ordinary differential equations is defined by*

$$\dot{\mathbf{x}} = \mathbf{f}(t, \mathbf{x}), \tag{3}$$

where $\mathbf{x} \in \mathbb{R}^n$, and $\mathbf{f}$ is a vector-function $\mathbf{f} = (f_1, \ldots, f_n)$ where

$$f_i \in \mathbb{R}[t, x_1, x_2, \ldots, x_i]$$

for every $1 \le i \le n$. By a solution *of (3) with* initial conditions *$(t_0, \mathbf{x}_0)$, $t_0 \in (-1, 1)$ we mean any analytic vector-function $\varphi : (-1, 1) \to \mathbb{R}^n$ such that $d\varphi/dt = \mathbf{f}(t, \varphi)$ for all $t \in (-1, 1)$ and $\varphi(t_0) = \mathbf{x}_0$.*

From the Definition 9 it follows that any solution of (3) is a vector of Pfaffian functions in the common domain $(-1, 1)$.

To any system (3) we can relate a dynamical system $\gamma : G \times (-1, 1) \to G$, where $G = I_1 \times \cdots \times I_n$, and I_i is an open interval (possibly unbounded) for all $1 \le i \le n$. More precisely, assume that for any $\mathbf{x} \in G$ the system has a solution φ with initial conditions $(\mathbf{x}, 0)$. Then $\gamma(\mathbf{x}, t) := \varphi(t)$.

Along with the dynamical systems associated with triangular systems of the kind (3) we will consider *Pfaffian dynamical systems* defined as follows.

Definition 19. *A dynamical system*

$$\gamma : G \times (-1, 1) \to G,$$

where G is open in $\mathbb{R}^n$ and γ is a map with a semi-Pfaffian graph, is called Pfaffian dynamical system.

Remark 2. Observe that the dynamical system γ associated with (3) may not be a Pfaffian dynamical system in the sense of the last definition since γ, being a Pfaffian vector-function for any fixed $\mathbf{x}$, is not necessarily a Pfaffian map for variable $\mathbf{x}$.

4   Bisimulations of Dynamical Systems Associated with a Differential Equation

Let in the system (3) the degree $\deg(f_i) < \alpha$ for any $1 \leq i \leq n$ and the associated dynamical system γ be bijective. Let $T = (G, \rightarrow)$ be the transition system associated with γ. Consider a partition $\mathcal{P} := \{P_1, \dots, P_s\}$ of G into s semi-Pfaffian sets P_j each having the format (r, N, α, β, n). Let $m := \max\{n, r\}$, $M := \max\{n, N\}$.

Theorem 4. *There is a bisimulation on T with respect to $\mathcal{P}$ consisting of*

$$s^s M^n 2^{m(m-1)/2} O(n(\alpha+\beta))^{n+m} \tag{4}$$

equivalence classes.

Proof. We use the notations and arguments of Section 1. First we estimate the length ℓ of the word ω for any $\mathbf{x} \in G$. Fix $\mathbf{x}$. Since ℓ coincides with the total number of connected components of intersections

$$P_j \cap \Gamma_{\mathbf{x}} = \{\mathbf{y} | \exists t \in (-1, 1)(\mathbf{y} = \gamma(\mathbf{x}, t) \wedge \mathbf{y} \in P_j)\}$$

for all $1 \leq j \leq s$. The semi-Pfaffian set $\widehat{P}_j := \{(\mathbf{y}, t) | \mathbf{y} = \gamma(\mathbf{x}, t) \wedge \mathbf{y} \in P_j\}$ has the format (m, M, α, β, n), thus according to Theorem 2 the number of its connected components does not exceed

$$L := M^n 2^{m(m-1)/2} O(n(\alpha + \beta))^{n+m}.$$

Since $P_j \cap \Gamma_{\mathbf{x}}$ is the projection of $\widehat{P}_j$ along t, the number of all connected components of $P_j \cap \Gamma_x$ is also less or equal to L, and the total number of connected components for all $1 \leq j \leq s$ does not exceed sL. Since the number of distinct letters in any word ω is at most s, the number of all words in the set Ω does not exceed (4). Then the cardinality of the set $\dot{\Omega}$ of all dotted words also does not exceed (4). It remains to notice that due to Theorem 1, the finite transition system $T_{\dot{\Omega}}$ is a bisimulation of T.

5   Bisimulations of Pfaffian Dynamical Systems

Consider a bijective Pfaffian dynamical system $\gamma : G \times (-1, 1) \rightarrow G$, where $G = (-1, 1)^n$, and a partition $\mathcal{P} := \{P_1, \dots, P_s\}$ of G into s semi-Pfaffian sets P_j. Let the graph of γ and each set P_j have the format (r, N, α, β, n), and all Pfaffian functions involved have the common Pfaffian chain.

Theorem 5. *There is a bisimulation on T with respect to $\mathcal{P}$ consisting of*

$$s^s N^n 2^{r(r-1)/2} O(n(\alpha+\beta))^{n+r} \tag{5}$$

equivalence classes.

Proof. A straightforward adjustment of the proof of Theorem 4.

Using the Pfaffian dependence on $\mathbf{x}$ in the case of a Pfaffian dynamical system we obtain another upper bound on the size of the bisimulation which is asymptotically better in general than the bound in Theorem 5 (except the case when n is significantly larger than the rest of the parameters).

Theorem 6. *There is a bisimulation on T with respect to $\mathcal{P}$ consisting of*

$$(\alpha + \beta s N)^{r^{O(n)} 2^{O(n^2)}} \tag{6}$$

equivalence classes.

Proof. Consider the family of Pfaffian functions in the domain

$$G \times (-1, 1) \times G$$

consisting of all functions in variables $\mathbf{x}, t, \mathbf{y}$ involved in the defining formulae for the graph of the map $\gamma : (\mathbf{x}, t) \mapsto \mathbf{y}$, and for all sets P_j considered in the latter case as functions in variables $\mathbf{y}$. According to Theorem 3, there is a cylindrical decomposition $\mathcal{D}$ for this family with respect to $(\mathbf{x}, t, \mathbf{y})$, consisting of at most (6) cylindrical cells. By the definition of cylindrical decomposition, $\mathcal{D}$ induces the cylindrical decomposition on G (equipped with coordinates $\mathbf{x}$) which we denote by $\mathcal{E}$.

We claim that for any cell $C \in \mathcal{E}$ and any two points $\mathbf{x}_1, \mathbf{x}_2 \in C$ the trajectories $\Gamma_{\mathbf{x}_1}, \Gamma_{\mathbf{x}_2} \in G$ are intersecting sets $P_1, \ldots, P_s$ in the same order (i.e., are encoded by the same word from Ω). Indeed, let $\pi : G \times (-1, 1) \times G \to G$ be the projection on G with coordinates $\mathbf{x}$. Decomposition $\mathcal{D}$ induces cylindrical decompositions $\mathcal{D}_1$ and $\mathcal{D}_2$ on $\pi^{-1}(\mathbf{x}_1)$ and $\pi^{-1}(\mathbf{x}_2)$ respectively. In particular, each of the integral curves $\widehat{\Gamma}_{\mathbf{x}_1}$ and $\widehat{\Gamma}_{\mathbf{x}_2}$ is decomposed into a sequence of alternating points and open intervals. Due to basic properties of a cylindrical decomposition, there is a natural bijection $\psi : \mathcal{D}_1 \to \mathcal{D}_2$ such that

(i) the restriction of ψ to the set of all cells in $\widehat{\Gamma}_{\mathbf{x}_1}$ is a bijection onto the set of all cells in $\widehat{\Gamma}_{\mathbf{x}_2}$;
(ii) for each $1 \leq j \leq s$ the restriction of ψ to the set of all cells in $(-1, 1) \times P_j \subset \pi^{-1}(\mathbf{x}_1)$ is a bijection onto the set of all cells in $(-1, 1) \times P_j \subset \pi^{-1}(\mathbf{x}_2)$.

It follows that if a cell $B \in \mathcal{D}_1$ and $B \subset \widehat{\Gamma}_{\mathbf{x}_1} \cap ((-1, 1) \times P_j)$ for some $1 \leq j \leq s$, then $\psi(B) \subset \widehat{\Gamma}_{\mathbf{x}_2} \cap ((-1, 1) \times P_j)$. The claim is proved.

It follows that the cardinality of Ω does not exceed the cardinality of $\mathcal{E}$ which does not exceed the cardinality of $\mathcal{D}$ which in turn is at most (6). Therefore, the cardinality of $\dot{\Omega}$ does not exceed (6), and the theorem is proved.

6   Computing Bisimulations

In this section we introduce an algorithm for computing finite bisimulations described in Theorem 6. It is sufficient to construct the set of dotted words $\dot{\Omega}$ corresponding to the bijective Pfaffian dynamical system $\gamma : G \times (-1, 1) \to G$ (with $G = (-1, 1)^n$) and a partition $\mathcal{P} := \{P_1, \ldots, P_s\}$. Since $\dot{\Omega}$ is trivially obtained from the set Ω, we will be constructing the latter set.

The algorithm applies the procedure from Theorem 3 to the family of Pfaffian functions consisting of all functions in variables $\mathbf{x}, t, \mathbf{y}$ involved in the defining formulae for the graph of the map $\gamma : (\mathbf{x}, t) \mapsto \mathbf{y}$, and for all sets P_j considered in the latter case as functions in variables $\mathbf{y}$. As a result, the algorithm produces a cell decomposition $\mathcal{D}$ which induces the cell decomposition $\mathcal{E}$ (see the proof of Theorem 6). Using the oracle, the algorithm selects the cells from $\mathcal{D}$ which are subsets of $\{(\mathbf{x}, t, \mathbf{y}) | \mathbf{y} = \gamma(\mathbf{x}, t)\}$. Denote the set of the selected cells by $\mathcal{B}$. Observe that for any fixed $\mathbf{x}' \in G$ the set $\bigcup_{B \in \mathcal{B}} B \cap \{\mathbf{x} | \mathbf{x} = \mathbf{x}'\}$ coincides with the integral curve $\widehat{\Gamma}_{\mathbf{x}'}$. Then the algorithm determines the order in which the cells $B \in \mathcal{B}$ intersected with $\{\mathbf{x} | \mathbf{x} = \mathbf{x}'\}$ appear in the trajectory $\Gamma_{\mathbf{x}'}$.

More precisely, for each pair of distinct cells $B_1, B_2 \in \mathcal{B}$ the algorithm decides, using the oracle, whether

$$\exists \mathbf{x} \exists t_1 \exists t_2 \exists \mathbf{y}_1 \exists \mathbf{y}_2 \left((\mathbf{x}, t_1, \mathbf{y}_1) \in B_1 \wedge (\mathbf{x}, t_2, \mathbf{y}_2) \in B_2 \wedge (t_1 < t_2) \right).$$

For a given $C \in \mathcal{E}$, after all pairs of cells are processed we get the ordered set of cells $B_1, \ldots, B_k$ in $\mathcal{D}$ such that for any $1 \leq i \leq k$ and any $\mathbf{x}' \in C$ the sequence of points and intervals

$$B_1 \cap \{\mathbf{x} | \mathbf{x} = \mathbf{x}'\}, \ldots, B_k \cap \{\mathbf{x} | \mathbf{x} = \mathbf{x}'\}$$

forms the integral curve $\widehat{\Gamma}_{\mathbf{x}'}$. By the definition of the cylindrical decomposition, for any pair B_i, P_j either $B_i \subset (C \times (-1, 1) \times P_j)$ or $B_i \cap (C \times (-1, 1) \times P_j) = \emptyset$. The algorithm uses the oracle to decide for every pair which of these two cases takes place. As the result, the sequence $B_1, \ldots, B_k$ becomes partitioned into subsequences of the kind

$$(B_1, \ldots, B_{k_1}), (B_{k_1+1}, \ldots, B_{k_2}), \ldots, (B_{k_{\ell-1}+1}, \ldots, B_k),$$

where for any i, $0 \leq i \leq \ell - 1$, the cells $B_{k_i+1}, \ldots, B_{k_{i+1}}$ lie in $C \times (-1, 1) \times P_{j_i}$ for some j_i, while $B_{k_i} \cap C \times (-1, 1) \times P_{j_i} = \emptyset$ and $B_{k_{i+1}+1} \cap C \times (-1, 1) \times P_{j_i} = \emptyset$. Then the word $\omega := P_{j_0} \cdots P_{j_{\ell-1}}$ corresponds to the cell C. Considering all cells in $\mathcal{E}$ the algorithm finds Ω and then $\dot{\Omega}$. This completes the description of the construction of $T_{\dot{\Omega}}$. It remains to construct the bisimulation $\sim$ on G.

As it was explained in the proof of Theorem 1, for $\mathbf{y}_1, \mathbf{y}_2 \in G$ we have $\mathbf{y}_1 \sim \mathbf{y}_2$ iff for respective pre-images $(\mathbf{x}_1, t_1), (\mathbf{x}_2, t_2)$, the locations of $\mathbf{y}_1, \mathbf{y}_2$ on trajectories $\Gamma_{\mathbf{x}_1}, \Gamma_{\mathbf{x}_2}$ are described by the same dotted word. Fix a cell $C \in \mathcal{E}$. To all points $\mathbf{x} \in C$ correspond the trajectories $\Gamma_{\mathbf{x}}$ encoded by the same word, say $\omega := P_{i_0} \cdots P_{i_q} \cdots P_{i_{\ell-1}}$. Consider a dotted word $\dot{\omega} := P_{i_0} \cdots \dot{P}_{i_q} \cdots P_{i_{\ell-1}}$.

Then all points $\mathbf{y} \in \Gamma_{\mathbf{x}}$ (for various $\mathbf{x} \in C$) whose locations are described by $\dot{\omega}$ form the set

$$A_{\dot{\omega}}(C) := \{\mathbf{y} \in G | \exists \mathbf{x} \in C \, \exists t \, ((\mathbf{x}, t, \mathbf{y}) \in (B_{k_q+1} \cup \cdots \cup B_{k_q+1}))\}.$$

Notice that $A_{\dot{\omega}}(C)$ is a sub-Pfaffian set with components of the format not exceeding (6). The equivalence relation $\sim$ is now defined by the partition

$$G = \bigcup_{\dot{\omega} \in \dot{\Omega}} A_{\dot{\omega}},$$

into disjoint classes

$$A_{\dot{\omega}} := \bigcup_{C \in \mathcal{E}} A_{\dot{\omega}}(C).$$

This completes the description of the algorithm.

A straightforward analysis shows that the complexity of the algorithm does not exceed (6), taking into account the bounds from Theorem 3.

7   Future Research

Observe that upper bounds from Theorems 5 and 6 on the size of bisimulations are *doubly exponential* in some parameters of the format of the original dynamical system. It looks feasible that there exists a *singly exponential* upper bound. The proof would require avoiding cylindrical cell decomposition technique which is intrinsically doubly exponential. Instead, it could use ideas related to those employed in effective quantifier elimination over real closed fields (see, e.g., [1]) and in recent upper bounds on topological complexity of definable sets [6].

Acknowledgements

The first author was supported in part by Grant Scientific School 2112.2003.1 and by the London Mathematical Society scheme 2 grant. A part of the work on this paper was done while she visited the Department of Computing and Software of McMaster University in Spring 2004. The second author was supported by the European RTN Network RAAG 2002–2006 (contract HPRN-CT-2001-00271).

References

1. S. Basu, R. Pollack and M.-F. Roy, *Algorithms in Real Algebraic Geometry*, Springer, Berlin-Heidelberg, 2003.
2. L. Blum, , F. Cucker, M. Shub, and S. Smale, *Complexity and Real Computation*, Springer, New York, 1997.
3. T. Brihaye, C. Michaux, C. Riviere, C. Troestler, On o-minimal hybrid systems, in: *Hybrid Systems: Computation and Control*, R. Alur, G. J. Pappas, (Eds.), LNCS, **2993**, Springer, Heidelberg, 2004, 219–233.

4. A. Gabrielov, N. Vorobjov, Complexity of computations with Pfaffian and Noetherian functions, in: *Normal Forms, Bifurcations and Finiteness Problems in Differential Equations*, Yu. Ilyashenko et al., (Eds.), NATO Science Series II, **137**, Kluwer, 2004, 211–250.

5. A. Gabrielov, N. Vorobjov, Complexity of cylindrical decompositions of sub-Pfaffian sets, *J. Pure and Appl. Algebra*, **164**, 1–2, 2001, 179–197.

6. A. Gabrielov, N. Vorobjov, Betti numbers of semialgebraic sets defined by quantifier-free formulae, to appear in: *Discrete and Computational Geometry*, 2004.

7. R.L. Grossman, et al., (Eds.), *Hybrid Systems*, LNCS, **736**, Springer, Berlin-Heidelberg, 1993, 474 p.

8. T. A. Henzinger, The theory of hybrid automata, in: *Proceedings of 11th Ann. Symp. Logic in Computer Sci.*, IEEE Press, 1996, 278–292.

9. G. Lafferriere, G.J. Pappas, S. Sastry, O-minimal hybrid systems, *Math. Control Signals Systems*, **13**, 2000, 1–21.

10. S. Pericleous, N. Vorobjov, New complexity bounds for cylindrical decompositions of sub-Pfaffian sets, in: *Discrete and Computational Geometry. Goodman-Pollack Festschrift*, B. Aronov et al. (Eds.), Springer, 2003, 673–694.

11. T. Zell, Betti numbers of semi-Pfaffian sets, *J. Pure Appl. Algebra,* **139**, 1999, 323–338.

Axioms for Delimited Continuations
in the CPS Hierarchy

Yukiyoshi Kameyama

Department of Computer Science, University of Tsukuba
Tennodai 1-1-1, Tsukuba, 305-8573, JAPAN and
Japan Science and Technology Agency
kameyama@acm.org

Abstract. A CPS translation is a syntactic translation of programs, which is useful for describing their operational behavior. By iterating the standard call-by-value CPS translation, Danvy and Filinski discovered the CPS hierarchy and proposed a family of control operators, shift and reset, that make it possible to capture successive *delimited* continuations in a CPS hierarchy.

Although shift and reset have found their applications in several areas such as partial evaluation, most studies in the literature have been devoted to the base level of the hierarchy, namely, to level-1 shift and reset. In this article, we investigate the whole family of shift and reset. We give a simple calculus with level-n shift and level-n reset for an arbitrary $n > 0$. We then give a set of equational axioms for them, and prove that these axioms are sound and complete with respect to the CPS translation. The resulting set of axioms is concise and a natural extension of those for level-1 shift and reset.

Keywords: CPS Translations, Control Operators, Delimited Continuations, Axiomatization, Type System.

1 Introduction

A CPS translation transforms a source term into continuation-passing style (CPS for short). It can be regarded as a compilation step, since it makes explicit the evaluation order of the source program and gives names to intermediate results. Another motivating fact for CPS is that it makes it possible to represent various control mechanisms, such as `callcc` in Scheme and Standard ML of New Jersey, that give programmers first-class continuations in the source language.

Logically, a CPS translation for the simply typed lambda calculus is a double negation interpretation from classical logic into minimal logic, or Friedman's A-translation [12]. The control mechanisms added to the source language can be also understood logically. For instance, Griffin [13] has revealed the Curry-Howard correspondence between the calculus with `callcc` and classical logic.

Danvy and Filinski [7,8] observed that there is room for a more refined control mechanism. By CPS translating the answer type of the standard CPS

J. Marcinkowski and A. Tarlecki (Eds.): CSL 2004, LNCS 3210, pp. 442–457, 2004.

translation, they obtained what they call a CPS hierarchy. Furthermore, they proposed a family of control operators **shift** and **reset** to abstract delimited continuations in this hierarchy. In the literature, many different control operators for delimited continuations have been proposed [10, 14, 15, 16]. In contrast to these other control operators, **shift** and **reset** are solely defined in terms of the CPS translation. In addition, they have found applications in partial evaluation [19], one-pass CPS translations [8], and normalization by evaluation [4], as well as to represent layered monads [11] and mobile computation [24].

In this article, we study a theoretical foundation of the control operators in the CPS hierarchy. Specifically, we address the problem of finding direct-style axioms for them. While these operators are used in many applications and their semantics is given by a CPS translation (be it iterated or extended), we often want to reason about source programs directly, rather than treating the image of CPS translations, since the CPS translation is sometimes said to obscure the overall structure of source programs. Also finding a good set of direct-style axioms could lead one to a better understanding of these operators.

We give a simple set of axioms consisting of only three equations for **shift** and three equations for **reset**, and then prove that this set of equations is sound and complete with respect to the iterated CPS translation. This work builds on our previous work, in which we gave a sound and complete axiomatization for level-1 **shift** and **reset** operators [18], and for level-2 [17]. Since completeness proofs of this kind often require quite a lot of calculations, we make the proof more structured by following an idea due to Sabry [22, 23] and reconstructing it in a type-theoretic setting, which further simplifies our proof.

Overview: The rest of this article is organized as follows. In Section 2, we informally introduce **shift** and **reset** and we explain their operational aspect. In Sections 3 and 4, we formally introduce the calculi with these control operators and a CPS translation for them. In Section 5 we present the axioms for control operators. In Section 6, we give a type-theoretic analysis of the CPS translation and we prove completeness. In Section 7, we conclude and mention future work.

Prerequisites: We assume that readers have some familiarity with CPS translations.

2   Control Operators in the CPS Hierarchy

We introduce the **shift** and **reset** operators through some examples.

2.1   A Simple Example

The following example uses these operators in a simple way:

$$
\begin{aligned}
3 \ + \ \langle 4 \ * \ \mathcal{S}(\lambda k. \ 5 \ + \ (k \ (k \ 2))) \rangle &= \texttt{let } k \texttt{ be } \lambda x. \langle 4 * x \rangle \\
&\quad \texttt{in } 3 \ + \ \langle 5 \ + \ (k \ (k \ 2)) \rangle \\
&= 3 \ + \ \langle 5 \ + \ \langle 4 \ * \ \langle 4 \ * \ 2 \rangle \rangle \rangle
\end{aligned}
$$

where $\langle\_\rangle$ is the reset operator and $\mathcal{S}$ is the shift operator.[1] Unlike the continuation captured by `callcc`, the continuation captured by $\mathcal{S}$ is not the whole rest of the computation (such as $3 + \langle 4 * [\ ]\rangle$), but a part which is delimited by a `reset`, that is, $\langle 4 * [\ ]\rangle$. Also it is not abortive, and thus we can compose the captured continuation with ordinary functions. When several occurrences of `reset` enclose an occurrence of `shift`, the (dynamically determined) closest one is chosen as the delimiter.

As more substantial examples, we borrow the ones by Danvy and Filinski [7].

2.2   Nondeterminism

A non-deterministic choice can be represented by backtracking in direct style using `shift` and `reset`:

$$\texttt{flip}(x) \overset{def}{=} \mathcal{S}_1(\lambda c.\ \texttt{begin}\ c(\texttt{true});\ c(\texttt{false});\ \texttt{fail}(\_)\ \texttt{end})$$

$$\texttt{fail}(x) \overset{def}{=} \mathcal{S}_1(\lambda c.\ \texttt{"no"})$$

$$\texttt{choice}(n) \overset{def}{=} \texttt{if}\ n < 1\ \texttt{then}\ \texttt{fail}(\_)$$
$$\texttt{else if}\ \texttt{flip}(\_)\ \texttt{then}\ \texttt{choice}(n-1)$$
$$\texttt{else}\ n$$

where $\_$ is a dummy value, `true`, `false` are truth values, and `begin`$\cdots$`end` is for sequencing.

To understand these programs, we CPS translate these three functions as:

$$\texttt{flip-c}(x, k) \overset{def}{=} \texttt{begin}\ k(\texttt{true});\ k(\texttt{false});\ \texttt{fail-c}(\_, k)\ \texttt{end}$$

$$\texttt{fail-c}(x, k) \overset{def}{=} \texttt{"no"}$$

$$\texttt{choice-c}(n, k) \overset{def}{=} \texttt{if}\ n < 1\ \texttt{then}\ \texttt{fail-c}(\_,\ k)$$
$$\texttt{else}\ \texttt{flip-c}(\_,\ \lambda y.\ \texttt{if}\ y\ \texttt{then}\ \texttt{choice-c}(n-1, k)$$
$$\texttt{else}\ k(n))$$

Let us consider the program $\langle\texttt{display}(\texttt{choice}(3))\rangle_1$. It is CPS translated to the program `choice-c(3, display)`, which will display 1, 2 and 3 in this order. It is easy to see that `shift` captures the current continuation, which is composable with functions (including other continuations), and that `reset` installs the identity continuation.

2.3   Collecting Successive Results

As a next step, one may wants to collect all answers generated by non-deterministic choices. This is implemented by the function `emit` defined by:

$$\texttt{emit}(n) \overset{def}{=} \mathcal{S}_1(\lambda c.\ \texttt{cons}(n,\ c(\texttt{nil})))$$

For instance, $\langle\texttt{begin}\ \texttt{emit}(1);\ \texttt{emit}(2);\ \texttt{emit}(3)\ \texttt{end}\rangle_1$ will return a list (1 2 3).

---

[1] Danvy and Filinski used the notation $\xi k.M$ for $\mathcal{S}(\lambda k.M)$.

It is then natural to expect that a combined program $\langle\text{emit}(\text{choice}(3))\rangle_1$ would work. However it does not, since the control operators in the two programs interfere. To see this, let us CPS translate emit as:

$$\text{emit-c}(n, k) \stackrel{def}{=} \text{cons}(n,\ k(\text{nil}))$$

The term $\langle\text{emit}(\text{choice}(3))\rangle_1$ is CPS translated to $\text{choice-c}(3, \lambda x.\text{emit-c}(x, \lambda a.a))$, which will generate three lists (1), (2) and (3), but never collect these answers.

A correct way of combining these programs is to make them *layered*. The continuation captured in emit should be in a higher level than that captured in choice. To achieve this, the CPS counterpart of emit should be:

$$\text{emit-c2}(n, k, \gamma) \stackrel{def}{=} \text{cons}(n,\ \gamma(k(\text{nil})))$$

where γ is a level-2 continuation. Its direct-style counterpart is:

$$\text{emit-c1}(n, k) \stackrel{def}{=} k(\mathcal{S}_1(\lambda c.\ \text{cons}(n,\ c(\text{nil}))))$$

which passes a continuation, even though it is not in CPS since the argument of k is not a trivial term. Its direct-style counterpart is:

$$\text{emit}(n) \stackrel{def}{=} \mathcal{S}_2(\lambda c.\ \text{cons}(n,\ c(\text{nil})))$$

This is the point where we need a level-2 control operator in the CPS hierarchy. Executing the term $\langle\text{emit}(\text{choice}(3))\rangle_2$ returns (1 2 3) as expected.

2.4   Summary and Conclusion

In summary, a direct-style program with level-2 control operators is CPS translated to a 1-CPS program with level-1 control operators, which is then CPS translated to a 2-CPS program with no control operators. CPS translating this program yields a real CPS program where all calls are tail calls and all subterms are trivial. The family of layered control operators thus corresponds to the CPS hierarchy.

A similar situation occurs when we perform partial evaluation of a program using shift/reset when the partial evaluator itself uses shift/reset. We refer to the reader to Asai's recent work [1, 2].

3   The Calculi with Control Operators

In this section, we define the language of our calculi, and postpone giving axioms until the next section.

The calculus we choose here is a type-free lambda calculus with control operators for delimited continuations. Later we briefly mention simply typed calculi. We consider the call-by-value evaluation order only.

We shall define calculi $\lambda\mathcal{S}_n$ and $\lambda\mathcal{C}_n$ for a natural number n. The former is a calculus with shift/reset, and the latter a calculus with $\mathcal{C}$/reset. The control operator $\mathcal{C}$ has a slightly different semantics as shift, which will be explained later.

We first assume there are infinitely many variables (written x, y, z and so on). Terms of $\lambda\mathcal{S}_n$ are type-free lambda terms augmented with control operators, and defined by:

$$\text{(terms)} \quad M, N ::= x \mid \lambda x.M \mid MN \mid \langle M\rangle_i \mid \mathcal{S}_i$$

where $1 \leq i \leq n$. Terms of $\lambda\mathcal{C}_n$ are defined in the same way with $\mathcal{S}_i$ being replaced by $\mathcal{C}_i$.

The index i denotes the level, which is conceptually the number of iterations of CPS translations that are necessary to interpret the control operator. The construct $\langle\_\rangle_i$ is level-i reset, and $\mathcal{S}_i$ is shift. Note that $\mathcal{S}_i$ (and $\mathcal{C}_i$) is a constant rather than a constructor. We use the abbreviations: $\mathcal{S}_i k.M \equiv \mathcal{S}_i(\lambda k.M)$, $\mathcal{C}_i k.M \equiv \mathcal{C}_i(\lambda k.M)$, and $\mathcal{A}_i \equiv \lambda x.\mathcal{C}_i(\lambda k.x)$. We also define $\langle M\rangle_0 \equiv M$.

A value (written V) is either a variable, λ-abstraction, or a constant ($\mathcal{S}_i$ in $\lambda\mathcal{S}_n$ and $\mathcal{C}_i$ in $\lambda\mathcal{C}_n$). Variables are bound by λ, and free and bound variables of terms are defined as usual. $FV(M)$ denotes the set of free variables in M. We identify two terms which differ only in renaming of bound variables. $M\{x := N\}$ is the result of the usual capture-avoiding substitution of N for x in M.

Contexts and evaluation contexts are defined as follows:

$$C ::= [\,] \mid CM \mid MC \mid \langle C\rangle_i$$
$$E ::= [\,] \mid EM \mid VE \mid \langle E\rangle_i$$
$$E^i ::= [\,] \mid E^i M \mid VE^i \mid \langle E^i\rangle_j \quad \text{for } j \leq i$$

E is an evaluation context in call by value, and E^i is a level-i evaluation context in which the level of reset operators enclosing the hole must be equal to or smaller than i. For example, $\langle x[\,]\rangle_2$ and $\langle x[\,]\rangle_2\langle yz\rangle_3$ are level-2 evaluation contexts. As a special case, E^0 is an evaluation context in which no reset may enclose the hole.

The operational semantics is given by the following rules (where $f \notin FV(E^{j-1})$):

$$E[\langle V\rangle_j] \rightarrow E[V]$$
$$E[\langle E^{j-1}[\mathcal{S}_j V]\rangle_j] \rightarrow E[\langle V(\lambda f.\langle E^{j-1}[f]\rangle_j)\rangle_j]$$

The first rule says that delimiting a value does nothing. The second rule shows how the $\mathcal{S}$-operator works. It captures the continuation delimited by the (dynamically determined) closest reset-operator, as the evaluation context E^{j-1} does not contain a level-j reset operator which encloses the hole. Note that the continuation captured by $\mathcal{S}_j$ is a function, whose body is enclosed by a level-j reset operator. This is an essential difference between Danvy and Filinski's shift operator and Felleisen's control operator [10].

The rule above is only a special case of the general rule given below. As we explained before, the level of the corresponding reset operator can be higher than j. Therefore a general rule for the second line is (where $j \leq i$ and $f \notin FV(E^{j-1})$):

$$E[\langle E^{j-1}[S_j V]\rangle_i] \to E[\langle V(\lambda f.\langle E^{j-1}[f]\rangle_j)\rangle_i]$$

The two operators S_i and C_i are inter-definable:

$$S_i = \lambda z.C_i(\lambda k.z(\lambda x.\langle kx\rangle_i))$$
$$C_i = \lambda z.S_i(\lambda k.z(\lambda x.S_i(\lambda d.kx)))$$

These equations are formally justified by the CPS translation in the next section.

4 CPS Translation

The CPS translation we consider is due to Danvy and Filinski [7]. It translates terms of the source calculi (λS_n or λC_n) to the type-free lambda calculus without control operators. As we explained in the introduction, their CPS translation can be thought of as a standard CPS translation followed by $n-1$ successive CPS translations of the answer type for $n > 1$. If we fix the number of iterations, then the whole translations can be expressed by a single, uncurried CPS translation, which we call an extended CPS translation. It takes n continuation parameters, each being introduced by the i-th CPS translation (for $1 \le i \le n$). This extended CPS translation gives a precise semantics to the level-i control operators (for $i < n$). The n continuation parameters are represented by the variables k_i, which we call a continuation variable of level i (for $1 \le i \le n$).

For a fixed n, and given a term M and a value V in the source calculi, we define two translations $[\![\_]\!]$ and $\_^*$, which send a term and a value to terms of type-free lambda calculus. To avoid clutter we present a $\beta\eta$-reduced version. In the following we assume $1 \le i < n$.

$$[\![V]\!] \overset{def}{=} \lambda k_1.\ k_1 V^*$$
$$[\![MN]\!] \overset{def}{=} \lambda k_1.\ [\![M]\!](\lambda m.[\![N]\!](\lambda n.mnk_1))$$
$$[\![\langle M\rangle_i]\!] \overset{def}{=} \lambda k_1.\cdots\lambda k_{i+1}.\ [\![M]\!]\theta_1\cdots\theta_i(\lambda x.\ \theta_0 x k_1 k_2\cdots k_{i+1})$$

$$x^* \overset{def}{=} x$$
$$(\lambda x.M)^* \overset{def}{=} \lambda x.[\![M]\!]$$
$$S_i{}^* \overset{def}{=} \lambda x k_1\cdots k_i.\ x(\lambda y k_1'\cdots k_{i+1}'.\theta_0 y k_1 k_2\cdots k_i(\lambda z.\theta_0 z k_1' k_2'\cdots k_{i+1}'))\theta_1\cdots\theta_i$$
$$C_i{}^* \overset{def}{=} \lambda x k_1\cdots k_i.\ x(\lambda y k_1'\cdots k_i'.\theta_0 y k_1 k_2\cdots k_i)\theta_1\cdots\theta_i$$

where $\theta_i = \lambda x k_{i+1}.k_{i+1}x$. The term θ_i can be thought of as the image of the identity continuation (the empty evaluation context [5]) of level i.

Let us briefly explain the extended CPS translation. Terms and values without control operators are translated as usual. For the term $\langle M\rangle_i$, the reset operator installs identity continuations up to level i, and composes all continuations of up to level i with the continuation of level $i + 1$. The operator $\mathcal{C}_i$ captures the current continuation up to level i, which results in $\lambda y k_1' \cdots k_i'.\theta_0 y k_1 k_2 \cdots k_i$ in this context, then applies it to the argument. It also installs instances of the identity continuation up to level i. The CPS translation of $\mathcal{S}_i$ is slightly more complex, since it captures a non-abortive delimited continuation so that we should compose $k_1', \cdots, k_n'$ with the captured continuation. Note that the result of the extended CPS translation does not depend on n (if the result is defined).

We show a few examples of the extended CPS translation and $\beta\eta$-reductions in the target calculus:

$$\begin{aligned}
[\mathcal{C}_2(\lambda f.fx)] &= \lambda k_1.(\lambda k_1.k_1 \mathcal{C}_2{}^*)(\lambda m.(\lambda k_1.k_1(\lambda f.fx)^*)(\lambda n.mnk_1)) \\
&\to_{\beta\eta} \lambda k_1.\ k_1 x \\
[\langle xy \rangle_2] &= \lambda k_1 k_2 k_3.\ [xy]\theta_1\theta_2(\lambda z.\ \theta_0 z k_1 k_2 k_3) \\
&\to_{\beta\eta} \lambda k_1 k_2 k_3.\ xy\theta_1\theta_2(\lambda z.\ k_1 z k_2 k_3)
\end{aligned}$$

The semantics of the target terms is given by the standard $\beta\eta$-equality (the type-free lambda calculus with $\beta\eta$-equality will be denoted by $\lambda_{\beta\eta}$.) Given a CPS translation and the target theory $\lambda_{\beta\eta}$, the source calculus is given a rigid semantics (CPS semantics). The fundamental question addressed in this article is, what is the equality theory that coincides with this CPS semantics. We first give an answer to this question, and then prove it.

5   Axioms of $\lambda\mathcal{S}_n$ and $\lambda\mathcal{C}_n$

We give axioms for the theories $\lambda\mathcal{S}_n$ and $\lambda\mathcal{C}_n$. The common axioms for these theories are shown in Figure 1, the specific axioms for $\lambda\mathcal{S}_n$ are in Figure 2, and the specific axioms for $\lambda\mathcal{C}_n$ are in Figure 3. In the presentation of axioms, we assume the levels of all control operators are less than n, namely, $1 \le i, j < n$. For the purpose of comparison, we list the axioms for the base level control operators $\mathcal{S}_1$ and $\langle \_ \rangle_1$ in Figure 4 which were given in our joint work with Hasegawa [18].

Recall that E^i is a level-i evaluation context, $\mathcal{S}_i k.M \equiv \mathcal{S}_i(\lambda k.M)$, $\mathcal{C}_i k.M \equiv \mathcal{C}_i(\lambda k.M)$, and $\langle M \rangle_0 \equiv M$. The last abbreviation is used when $i = 1$ in reset-lift-2 and others. (Note that $i - 1$ may be 0.)

Let us explain Figure 1. The first three axioms β_v, η_v and β_Ω are those for Moggi's computational lambda calculus, the canonical calculus in call-by-value [20]. The axiom reset-value is essentially the same as that in level-1 theory (Figure 4). The axioms reset-lift and reset-lift-2 lift a β-redex over a reset operator. The axiom β_Ω can be applied to the level-0 evaluation context only, but with the help of these axioms we can lift a β-redex over a general evaluation context. The axiom reset-lift is a natural extension of its level-1 counterpart, while no counterpart of reset-lift-2 exists in level-1 axioms. The axiom may look strange since the index j appears only in the right-hand side. We may restrict j to i in reset-lift-2 in the presence of the axiom $\langle\langle M \rangle_j \rangle_i = \langle M \rangle_i$ for $j \le i$.

$$(\lambda x.M)V = M\{x := V\} \qquad \beta_v$$
$$\lambda x.\ Vx = V \qquad \eta_v,\ \text{if}\ x \notin FV(V)$$
$$(\lambda x.E^0[x])M = E^0[M] \qquad \beta_\Omega,\ \text{if}\ x \notin FV(E^0)$$
$$\langle V \rangle_i = V \qquad \text{reset-value}$$
$$\langle (\lambda x.M)\langle N \rangle_i \rangle_j = (\lambda x.\langle M \rangle_j)\langle N \rangle_i \qquad \text{reset-lift},\ j \leq i$$
$$\langle (\lambda x.M)\langle N \rangle_{i-1} \rangle_i = \langle (\lambda x.\langle M \rangle_j)\langle N \rangle_{i-1} \rangle_i \qquad \text{reset-lift-2},\ j \leq i$$

Fig. 1. Common Axioms for $\lambda \mathcal{S}_n$ and $\lambda \mathcal{C}_n$ $(1 \leq i, j < n)$

$$\mathcal{S}_i k.\langle M \rangle_i = \mathcal{S}_i k.M \qquad \mathcal{S}\text{-reset}$$
$$\mathcal{S}_i k.k\langle M \rangle_{i-1} = \langle M \rangle_{i-1} \qquad \mathcal{S}\text{-elim, if}\ k \notin FV(M)$$
$$\langle E^j[\mathcal{S}_i k.M] \rangle_i = \langle M\{k := \lambda f.\ \langle E^j[f] \rangle_i\} \rangle_i \qquad \mathcal{S}\text{-lift}$$
$$\text{if}\ k \notin FV(E^j),\ f \notin FV(kE^j),\ \text{and}\ j < i$$

Fig. 2. Specific Axioms for $\lambda \mathcal{S}_n$ $(1 \leq i, j < n)$

$$\mathcal{C}_i k.\langle M \rangle_i = \mathcal{C}_i k.M \qquad \mathcal{C}\text{-reset}$$
$$\mathcal{C}_i k.k\langle M \rangle_{i-1} = \langle M \rangle_{i-1} \qquad \mathcal{C}\text{-elim, if}\ k \notin FV(M)$$
$$\langle E^j[\mathcal{C}_i k.M] \rangle_i = \langle M\{k := \lambda f.\ \mathcal{A}_i \langle E^j[f] \rangle_{i-1}\} \rangle_i \qquad \mathcal{C}\text{-lift}$$
$$\text{if}\ k \notin FV(E^j),\ f \notin FV(kE^j),\ \text{and}\ j < i$$

Fig. 3. Specific Axioms for $\lambda \mathcal{C}_n$ $(1 \leq i, j < n)$

Besides the common axioms, each theory has three specific axioms for $\mathcal{S}_i$ or $\mathcal{C}_i$. The axiom $\mathcal{S}$-reset is a natural extension of its level-1 counterpart, while the axiom $\mathcal{S}$-elim is not quite the same as a natural extension of its level-1 counterpart. In fact, $\mathcal{S}_i k.kN = N$ is not sound for $i > 1$. Danvy and Filinski [7] stated that the current formulation of shift/reset is not completely satisfactory since $\mathcal{S}_2 k.kN = N$ does not hold. However, by restricting N to $\langle M \rangle_i$, we have obtained a sound axiom.

The last axiom $\mathcal{S}$-lift is a natural extension of its level-1 counterpart (called reset-$\mathcal{S}$ formerly).[2] It is also a direct formulation of the operational semantics given in the earlier section, by changing reduction to equality.

We believe that the resulting axioms are simple to understand, and the soundness of these axioms is not surprising. The completeness of $\lambda \mathcal{C}_n$ may be surprising, since one may think it lacks many important equations which were included in our axiomatization of $\mathcal{C}_2$ [17], such as:

---

[2] $\mathcal{S}$-lift does not immediately subsume reset-$\mathcal{S}$, since the latter allows an arbitrary M in $\mathcal{S}_1 M$, while the former restricts it to be a function. However, this difference does not matter since we can prove $\mathcal{S}_1 M = \mathcal{S}_1 k.Mk$ for $k \notin FV(M)$ in $\lambda \mathcal{S}_n$.

$$\langle V \rangle_1 = V \qquad \text{reset-value}$$

$$\langle (\lambda x.M)\langle N \rangle_1 \rangle_1 = (\lambda x.\langle M \rangle_1)\langle N \rangle_1 \qquad \text{reset-lift}$$

$$\mathcal{S}_1 k.kM = M \qquad \mathcal{S}\text{-elim, if } k \notin FV(M)$$

$$\mathcal{S}_1 k.\langle M \rangle_1 = \mathcal{S}_1 k.M \qquad \mathcal{S}\text{-reset}$$

$$\langle E^0[\mathcal{S}_1 M] \rangle_1 = \langle M(\lambda x.\langle E^0[x] \rangle_1) \rangle_1 \qquad \text{reset-}\mathcal{S}, \text{ if } x \notin FV(E^0)$$

Fig. 4. Axioms for $\lambda_{\mathcal{S}}$ other than β_v, η_v, and β_Ω

$$\langle \langle M \rangle_i \rangle_l = \langle M \rangle_{max(l,i)} \qquad \text{reset-reset}$$

$$\langle \mathcal{C}_i k.M \rangle_j = \mathcal{C}_i k.M \qquad \text{reset-}\mathcal{C}, \text{ if } j < i$$

$$\langle \mathcal{C}_i k.M \rangle_i = \langle M\{k := \mathcal{A}_i\} \rangle_i \qquad \mathcal{C}\text{-top}$$

$$(\lambda x.\mathcal{C}_1 k.M)N = \mathcal{C}_1 k.(\lambda x.M)N \qquad \text{let-}\mathcal{C}_1, \text{ if } k \neq x$$

$$\mathcal{C}_1 M = \mathcal{C}_1 k.Mk \qquad \mathcal{C}_1\text{-fun, if } k \notin FV(M)$$

Another seemingly missing axiom is an equation for lifting $\mathcal{C}_i$ over an evaluation context such as $E[\mathcal{C}_i k.M] = M\{k := \cdots\}$. In the next theorem we show that these equations are derivable.

From now on, we will mainly investigate $\lambda \mathcal{C}_n$. After proving its completeness, we will come back to $\lambda \mathcal{S}_n$.

Theorem 1. *The equations reset-reset, reset-$\mathcal{C}$, $\mathcal{C}$-top, let-$\mathcal{C}_1$ and $\mathcal{C}_1$-fun as well as the following equations are derivable in $\lambda \mathcal{C}_n$ where $k_1, \cdots, k_i, f$ are fresh variables, and k is not bound by C in $\mathcal{C}$-abort.*

$$E^j[\mathcal{A}_i \langle M \rangle_{i-1}] = \mathcal{A}_i \langle M \rangle_{i-1} \qquad \mathcal{A}\text{-abort, if } j < i$$

$$\mathcal{C}_i k.C[E^j[kV]] = \mathcal{C}_i k.C[kV] \qquad \mathcal{C}\text{-abort, if } j < i$$

$$E^j[\mathcal{C}_i k.M] = \mathcal{C}_1 k_1. \cdots \mathcal{C}_i k_i.M\{k := N\} \qquad \text{telescope, if } j < i$$

$$\text{where } N \text{ is } \lambda f.k_i \langle k_{i-1} \cdots \langle k_1(E^j[f]) \rangle_1 \cdots \rangle_{i-1}$$

Proof. The equation reset-reset is obtained in this way. By putting $M = x$ in reset-lift, we obtain $\langle \langle N \rangle_i \rangle_j = (\lambda x.\langle x \rangle_j)\langle N \rangle_i$, and by reset-value and β_Ω, we obtain $\langle \langle N \rangle_i \rangle_j = \langle N \rangle_i$ for $j \leq i$. Similarly by putting $N = x$ in reset-lift-2, we obtain $\langle M \rangle_i = \langle \langle M \rangle_j \rangle_i$ for $j \leq i$, hence we are done.

The equation reset-$\mathcal{C}$ is obtained by putting $E^j = [\,]$ and $E^j = \langle [\,] \rangle_j$ in $\mathcal{C}$-lift and comparing the results. The equation $\mathcal{C}$-top is obtained by putting $E^j = [\,]$ in $\mathcal{C}$-lift and using reset-value and η_v.

For the remaining equations, we first derive by $\mathcal{C}$-elim and $\mathcal{C}$-reset:

$$M = \mathcal{C}_1 k_1.k_1 M = \mathcal{C}_1 k_1.\langle k_1 M \rangle_1 = \mathcal{C}_1 k_1.\mathcal{C}_2 k_2.\langle k_2 \langle k_1 M \rangle_1 \rangle_2$$

Iterating this process i-times we obtain:

$$M = \mathcal{C}_1 k_1.\mathcal{C}_2 k_2. \cdots \mathcal{C}_i k_i.\langle k_i \cdots \langle k_2 \langle k_1 M \rangle_1 \rangle_2 \cdots \rangle_i$$

By this equation and $\mathcal{C}$-lift, we obtain the following key equation (where $j < i$):

$$E^j[\mathcal{C}_i k.M] = \mathcal{C}_1 k_1.\cdots \mathcal{C}_i k_i.\langle k_i \langle \cdots \langle k_1 (E^j[\mathcal{C}_i k.M]) \rangle_1 \cdots \rangle_{i-1} \rangle_i$$
$$= \mathcal{C}_1 k_1.\cdots \mathcal{C}_i k_i.\langle M\{k := \lambda f.\mathcal{A}_i \langle k_i \langle \cdots \langle k_1 (E^j[f]) \rangle_1 \cdots \rangle_{i-1} \rangle_{i-1}\} \rangle_i$$

For $\mathcal{A}$-abort, we first note that, if $j<i$ then we can derive $E^j[(\lambda x.N)\langle M \rangle_{i-1}]=(\lambda x.E^j[N])\langle M \rangle_{i-1}$ by reset-lift and β_Ω. Also, if $k \notin FV(M)$, then the right-hand side of the key equation does not contain E^j, hence $E^j[\mathcal{C}_i k.M] = \mathcal{C}_i k.M$. Then we can compute as follows (for $j < i$):

$$E^j[\mathcal{A}_i \langle M \rangle_{i-1}] = E^j[(\lambda x.\mathcal{C}_i(\lambda k.x))\langle M \rangle_{i-1}] \qquad \text{by definition}$$
$$= (\lambda x.E^j[\mathcal{C}_i(\lambda k.x)])\langle M \rangle_{i-1} \qquad \text{by the above equation}$$
$$= (\lambda x.\mathcal{C}_i(\lambda k.x))\langle M \rangle_{i-1} \qquad \text{by the key equation}$$
$$= \mathcal{A}_i \langle M \rangle_{i-1}$$

For $\mathcal{C}$-abort we compute as follows (where $j < i$):

$$\mathcal{C}_i k.C[E^j[kV]]$$
$$= \mathcal{C}_1 k_1.\cdots \mathcal{C}_i k_i.\langle (C[E^j[kV]])\{k := \lambda f.\mathcal{A}_i \langle k_i \langle \cdots \langle k_1 f \rangle_1 \cdots \rangle_{i-1} \rangle_{i-1}\} \rangle_i$$
$$= \mathcal{C}_1 k_1.\cdots \mathcal{C}_i k_i.\langle C[E^j[\mathcal{A}_i \langle k_i \langle \cdots \langle k_1 V \rangle_1 \cdots \rangle_{i-1} \rangle_{i-1}]] \rangle_i$$
$$= \mathcal{C}_1 k_1.\cdots \mathcal{C}_i k_i.\langle C[\mathcal{A}_i \langle k_i \langle \cdots \langle k_1 V \rangle_1 \cdots \rangle_{i-1} \rangle_{i-1}]] \rangle_i \qquad \text{by } \mathcal{A}\text{-abort}$$

Since the final result does not contain E^j, we have $\mathcal{C}_i k.C[E_1^j[kV]]=\mathcal{C}_i k.C[E_2^j[kV]]$ for any level-j evaluation contexts E_1^j and E_2^j. The axiom $\mathcal{C}$-abort then follows.

Applying $\mathcal{C}$-abort to the key equation (by putting $E^j = \mathcal{A}_i \langle [\ ] \rangle_{i-1}$ in $\mathcal{C}$-abort), we obtain the telescope axiom.[3] Verification of let-$\mathcal{C}_1$ and $\mathcal{C}_1$-fun is left for the reader.

This finishes the proof of the theorem.

6   Completeness Proof

The main results of this article are that the theories $\lambda\mathcal{S}_n$ and $\lambda\mathcal{C}_n$ are sound and complete with respect to the extended CPS translation into the theory $\lambda_{\beta\eta}$. In the previous work, we proved completeness by the following strategy; (1) we analyzed the syntax of the image of CPS translation, (2) defined an inverse CPS translation, i.e., a direct-style transformation, and (3) proved that the equality is preserved through this inverse translation. The most difficult part was to find a suitable inverse translation, and we found it by trial and error. In this article our source calculus is much more complex than those in the previous studies, and therefore a better strategy is called for.

The proof method we present here is based on an idea by Sabry, who applied it to the axiomatization of a calculus with level-1 shift and *lazy* reset [22]. In this section, we develop his method in the type-theoretic framework so as to make the proof more structured.

---

[3] This is a generalized version of Murthy's telescope axiom [21].

6.1   The Target Calculus and Its Type Structure

We analyze the set of terms which contains the image of the extended CPS translation and is closed under $\beta\eta$-reductions. We call this language (under the $\beta\eta$-equality) the target language or the target calculus.

An important observation on the target calculus is that it is typed by the following type structure:

$$\text{Term}_i = \text{Cont}_{i+1} \rightarrow \text{Term}_{i+1} \quad \text{for } 0 \leq i < n$$
$$\text{Term}_n = \text{Ans}$$
$$\text{Cont}_i = \text{Value} \rightarrow \text{Term}_i \quad \text{for } 1 \leq i \leq n$$
$$\text{Value} = \text{Value} \rightarrow \text{Term}_0$$

where Ans is an arbitrary, fixed type, called the answer type [7, 9].[4] The above definition of types makes sense if we have recursive types, that is, if the recursive equation for Value has a solution. Note that if we were working in the typed setting from the beginning, that is, our source calculi were simply typed lambda calculi, we would not need recursive types for Value.

Using the type structure, terms of the target calculus can be introduced as typed terms. A typing judgment is of the form $\Gamma \vdash P : T$ where Γ is a set of variable-type pairs consisting of either $x : \text{Value}$ or $k_i : \text{Cont}_i$. As usual, a variable may occur at most once in Γ. We have the following eight type inference rules for $0 < i \leq n$:

$$\frac{\Gamma \vdash W : \text{Value} \quad \Gamma \vdash W' : \text{Value}}{\Gamma \vdash WW' : \text{Term}_0} \qquad \frac{\Gamma, k_i : \text{Cont}_i \vdash T_i : \text{Term}_i}{\Gamma \vdash \lambda k_i.T_i : \text{Term}_{i-1}}$$

$$\frac{\Gamma \vdash T_{i-1} : \text{Term}_{i-1} \quad \Gamma \vdash K_i : \text{Cont}_i}{\Gamma \vdash T_{i-1}K_i : \text{Term}_i} \qquad \frac{\Gamma \vdash K_i : \text{Cont}_i \quad \Gamma \vdash W : \text{Value}}{\Gamma \vdash K_iW : \text{Term}_i}$$

$$\frac{}{\Gamma, k_i : \text{Cont}_i \vdash k_i : \text{Cont}_i} \qquad \frac{\Gamma, x : \text{Value} \vdash T_i \; : \; \text{Term}_i}{\Gamma \vdash \lambda x.T_i \; : \; \text{Cont}_i}$$

$$\frac{}{\Gamma, x : \text{Value} \vdash x : \text{Value}} \qquad \frac{\Gamma, x : \text{Value} \vdash T_0 \; : \; \text{Term}_0}{\Gamma \vdash \lambda x.T_0 \; : \; \text{Value}}$$

where $\Gamma, x : \text{Value}$ means the set union $\Gamma \cup \{x : \text{Value}\}$.

If we can prove $\Gamma \vdash P : T$ using the typing rules above, we say that P is a term (of type T) in the target calculus. For instance, the term $\lambda k_1.\, k_1x$ (which is the $\beta\eta$-reduced term of $[\mathcal{C}_2(\lambda f.fx)]$) can be typed as follows:

$$\frac{\dfrac{}{x : \text{Value}, k_1 : \text{Cont}_1 \vdash k_1 : \text{Cont}_1} \quad \dfrac{}{x : \text{Value}, k_1 : \text{Cont}_1 \vdash x : \text{Value}}}{\dfrac{x : \text{Value}, k_1 : \text{Cont}_1 \vdash k_1x : \text{Term}_1}{x : \text{Value} \vdash \lambda k_1.\, k_1x : \text{Term}_0}}$$

For this type structure, it is not difficult to prove the following theorem.

---

[4] We can make the answer type parametric, as investigated by Thielecke [25].

Theorem 2. *(1) Let M and V be a term and a value, resp. in λC_n. Then we can derive $\Gamma \vdash [M] : \text{Term}_0$ and $\Gamma' \vdash V^* : \text{Value}$ for some Γ and Γ'.*

(2) If we can derive $\Gamma \vdash P : T$ in the target calculus, and P reduces to Q by $\beta\eta$-reductions, then we can derive $\Gamma \vdash Q : T$.

6.2   Direct-Style Translation

We define an extended direct-style translation from the target calculus to the source calculus λC_n. We first give it as a syntactic translation $\_^\dagger$ based on the type structure of target terms as follows (for $0 < i \leq n$):

$$(WW')^\dagger \overset{def}{=} W^\dagger W'^\dagger \qquad\qquad (\lambda k_i.T_i)^\dagger \overset{def}{=} \mathcal{C}_i k_i.T_i{}^\dagger$$

$$(T_{i-1}K_i)^\dagger \overset{def}{=} K_i{}^\dagger \langle T_{i-1}{}^\dagger \rangle_{i-1} \qquad (K_iW)^\dagger \overset{def}{=} K_i{}^\dagger W^\dagger$$

$$k_i{}^\dagger \overset{def}{=} k_i \qquad\qquad\qquad (\lambda x.T_i)^\dagger \overset{def}{=} \lambda x.\langle T_i{}^\dagger \rangle_i$$

$$x^\dagger \overset{def}{=} x \qquad\qquad\qquad (\lambda x.T_0)^\dagger \overset{def}{=} \lambda x.T_0{}^\dagger$$

The next theorem ensures that $\_^\dagger$ is in fact a translation from $\lambda_{\beta\eta}$ to λC_n.

Theorem 3. *The translation $\_^\dagger$ respects the $\beta\eta$-equality in the target calculus, namely, if P and Q are typable terms in the target calculus and $\lambda_{\beta\eta} \vdash P = Q$, then $\lambda C_n \vdash P^\dagger = Q^\dagger$.*

We also have that $\_^\dagger$ is really an inverse of the extended CPS translation.

Theorem 4. *If M is a term in λC_n, then $\lambda C_n \vdash [M]^\dagger = M$.*

The proofs of these theorems are not shown here due to lack of space.

Now we can prove the completeness of λC_n.

Theorem 5 (Soundness & Completeness). *Let M and N be terms in λC_n. Then we have:*

$$\lambda C_n \vdash M = N \text{ if and only if } \lambda_{\beta\eta} \vdash [M] = [N]$$

Proof. Soundness (the "only-if" direction) can be proved by calculating both sides of axioms in λC_n. For completeness (the "if" direction), suppose $\lambda_{\beta\eta} \vdash [M] = [N]$. Since $[M]$ and $[N]$ are of type T_0, we have $\lambda C_n \vdash [M]^\dagger = [N]^\dagger$ by Theorem 3. Using Theorem 4, we conclude $\lambda C_n \vdash M = N$.

6.3   Completeness of λS_n

We finally obtain the completeness of λS_n.

Theorem 6 (Soundness & Completeness). *Let M and N be terms in λS_n. Then we have:*

$$\lambda S_n \vdash M = N \text{ if and only if } \lambda_{\beta\eta} \vdash [M] = [N]$$

Proof. Soundness can be proved in the same way as λC_n. For completeness, let ϕ be a translation from terms in λS_n to terms in λC_n which replaces S_i by its "definition" in C_i given in Section 3. Similarly let ψ be a translation from λC_n to λS_n. It suffices to prove the following properties with M and M' being terms in λS_n, and N and N' being terms in λC_n:

1. $\lambda_{\beta\eta} \vdash [\![\phi(M)]\!] = [\![M]\!]$.
2. $\lambda C_n \vdash N = N'$ implies $\lambda S_n \vdash \psi(N) = \psi(N')$.
3. $\lambda S_n \vdash \psi(\phi(M)) = M$

All these properties can be proved by calculation.

6.4   Typing Source Calculi

So far we have been studying the simplest possible source calculi. Introducing type structure into the source calculi is an important problem, as most modern programming languages have a built-in type system. Another benefit of introducing types is that, in the presence of appropriate types, we can avoid the full η_v-equality ($\lambda x.Vx = V$ for $x \notin FV(V)$), which is inconsistent with the presence of basic values such as natural numbers. In order to restrict V in η_v to a functional value, we need a type system in the source calculus.

A simple choice of the typing rules of control operators would be:

$$\frac{\Gamma \vdash M : \Phi}{\Gamma \vdash \langle M \rangle_i : \Phi} \qquad \overline{\Gamma \vdash S_i : ((A \to \Phi) \to \Phi) \to A}$$

where Φ is a designated atomic type, and A is an arbitrary type. Then we can prove that the extended CPS translation preserves the typability if we add the type information to the classes $\mathbf{Term_0}$ and $\mathbf{Value}$ in the target calculus (in which case, the type structure of the target calculus does not need recursive types). All the axioms and the proof of soundness and completeness in this article go through for the simply typed case.

Introduction of types to the source calculi makes explicit the connection of the extended CPS translation and the double negation translation. If we take $n = 1$, then by the definition of types given in Section 6.1 we have $\mathbf{Term_0} = (\mathbf{Value} \to \mathbf{Ans}) \to \mathbf{Ans}$. If we take $n = 2$, then

$$\mathbf{Term_0} = (\mathbf{Value} \to (\mathbf{Value} \to \mathbf{Ans}) \to \mathbf{Ans}) \to (\mathbf{Value} \to \mathbf{Ans}) \to \mathbf{Ans}$$

Hence the type $\mathbf{Ans}$ in the $n = 1$ case (which corresponds to $\bot$ in the double negation translation) is CPS translated to $(\mathbf{Value} \to \mathbf{Ans}) \to \mathbf{Ans}$ in the $n = 2$ case. Thus, we can say the extended translation represents an iterated double-negation translation.

In the literature, Danvy and Filinski [6] and Murthy [21] have proposed more liberal type systems for **shift** and **reset**. Since these type systems are quite complicated, it is not obvious whether our axioms work for them.

7   Conclusion

In this article we have studied a family of control operators in the CPS hierarchy. In particular, we have analyzed the image of the extended CPS translation with type-theoretic machinery, and have obtained a simple set of axioms which is sound and complete for all such control operators. To our knowledge this work is the first such result about the hierarchy of delimited continuation operators. Our axioms for level-n shift/reset are a simple extension of those for level-1 shift/reset, and the axioms for level-n $\mathcal{C}$/reset are even simpler than those for level-2 $\mathcal{C}$/reset.

The control operators in the CPS hierarchy have also been investigated by Murthy [21], who gave an elaborate type system for level-n shift and reset, and also gave a set of axioms for them. The difference between his work and ours is that he only proved the soundness of the axioms and did not state completeness, and also that his set of axioms consists of many complex axioms such as the telescope axiom, while ours consists of a small number of simple axioms.

In another line of work, Danvy and Yang [9], Murthy [21], and Biernacka, Biernacki and Danvy [3] studied an operational aspect of the control operators in the CPS hierarchy by giving abstract machines for shift and reset. It seems interesting to study how our axioms fit with these abstract machines.

Future Work: Besides studying the connection to abstract machines, there are two major avenues for future work:

(1) While we have built a theoretical foundation for the control operators in the CPS hierarchy there remains a question about the application of our axioms. It is almost impossible to use them for automatic verification because they require a degree of insight. Nevertheless, besides obtaining a better understanding of control operators, we hope to use the axioms to prove the correctness of program translations such as compiler optimization.

(2) An even more fundamental question of this study is whether one needs these hierarchical control operators at all. The existence of several application programs and the correspondence between the CPS hierarchy and layered monads [11] seem to give a positive answer to this question. However, there is much room for further work.

Acknowledgments: The author would like to thank Olivier Danvy, Masahito Hasegawa, Amr Sabry and Kenichi Asai for constructive comments and discussions. He also thanks anonymous referees for many insightful comments and criticisms. This work was supported in part by Grant-in-Aid for Scientific Research No. 16500004 from Japan Society for the Promotion of Science.

References

1. K. Asai. Online Partial Evaluation for Shift and Reset. In *Proc. ACM Workshop on Partial Evaluation and Semantics-Based Program Manipulation*, pages 19–30, 2002.

2. K. Asai. Offline Partial Evaluation for Shift and Reset. In *Proc. ACM Workshop on Partial Evaluation and Semantics-Based Program Manipulation*, to appear.
3. M. Biernacka, D. Biernacki, and O. Danvy. An operational foundation for delimited continuations. In *Proc. Fourth ACM SIGPLAN Workshop on Continuations, Technical Report CSR-04-1, School of Computer Science, University of Birmingham*, pages 25–34, 2004.
4. O. Danvy. Type-directed partial evaluation. In *Proc. 23rd Symposium on Principles of Programming Languages*, pages 242–257, 1996.
5. O. Danvy. On evaluation contexts, continuations, and the rest of the computation. In *Proc. Fourth ACM SIGPLAN Workshop on Continuations, Technical Report CSR-04-1, School of Computer Science, University of Birmingham*, pages 13–23, 2004.
6. O. Danvy and A. Filinski. A functional abstraction of typed contexts. Technical Report 89/12, DIKU, University of Copenhagen, 1989.
7. O. Danvy and A. Filinski. Abstracting Control. In *Proc. 1990 ACM Conference on Lisp and Functional Programming*, pages 151–160, 1990.
8. O. Danvy and A. Filinski. Representing Control: a Study of the CPS Transformation. *Mathematical Structures in Computer Science*, 2(4):361–391, 1992.
9. O. Danvy and Z. Yang. An Operational Investigation of the CPS Hierarchy. In *Proc. 8th European Symposium on Programming*, Lecture Notes in Computer Science 1576, pages 224–242, 1999.
10. M. Felleisen. The Theory and Practice of First-Class Prompts. In *Proc. 15th Symposium on Principles of Programming Languages*, pages 180–190, 1988.
11. A. Filinski. Representing Layered Monads. In *Proc. 26th Symposium on Principles of Programming Languages*, pages 175–188, 1999.
12. H. Friedman. Classically and intuitionistically provably recursive functions. Lecture Notes in Mathematics 699, pages 21–28, 1978.
13. T. Griffin. A Formulae-as-Types Notion of Control. In *Proc. 17th Symposium on Principles of Programming Languages*, pages 47–58, 1990.
14. C. A. Gunter, D. Rémy, and J. G. Riecke. A Generalization of Exceptions and Control in ML-Like Languages. In *Proc. Functional Programming and Computer Architecture*, pages 12–23, 1995.
15. R. Hieb, R. Dybvig, and C. W. Anderson. Subcontinuations. *Lisp and Symbolic Computation*, 6:453–484, 1993.
16. Y. Kameyama. A Type-Theoretic Study on Partial Continuations. In *Proc. IFIP International Conference on Theoretical Computer Science*, Lecture Notes in Computer Science 1872, pages 489–504, 2000.
17. Y. Kameyama. Axiomatizing higher level delimited continuations. In *Proc. 3rd ACM SIGPLAN Workshop on Continuations, Technical Report 545, Computer Science Department, Indiana University*, pages 49–53, 2004.
18. Y. Kameyama and M. Hasegawa. A Sound and Complete Axiomatization of Delimited Continuations. In *Proc. ACM International Conference on Functional Programming*, pages 177–188, 2003.
19. J. Lawall and O. Danvy. Continuation-based partial evaluation. In *Proc. 1994 ACM Conference on LISP and Functional Programming*, pages 227–238, 1994.
20. E. Moggi. Computational Lambda-Calculus and Monads. In *Proc. 4th Symposium on Logic in Computer Science*, pages 14–28, 1989.
21. C. Murthy. Control operators, hierarchies, and pseudo-classical type systems: A-translation at work. In *Proc. First ACM Workshop on Continuations, Technical Report STAN-CS-92-1426, Stanford University*, pages 49–72, 1992.

22. A. Sabry. Note on Axiomatizing the Semantics of Control Operators. Technical Report CIS-TR-96-03, Dept. of Computer Science, University of Oregon, 1996.
23. A. Sabry and M. Felleisen. Reasoning about Programs in Continuation-Passing Style. *Lisp and Symbolic Computation*, 6(3-4):289–360, 1993.
24. T. Sekiguchi, T. Sakamoto, and A. Yonezawa. Portable Implementation of Continuation Operators in Imperative Languages by Exception Handling. In *Advances in Exception Handling Techniques*, Lecture Notes in Computer Science 2022, pages 217–233, 2001.
25. H. Thielecke. Answer type polymorphism in call-by-name continuation passing. In *Proc. 13th European Symposium on Programming*, Lecture Notes in Computer Science 2986, pages 279–293, 2004.

Set Constraints on Regular Terms

Paweł Rychlikowski and Tomasz Truderung*

Institute of Computer Science, Wrocaw University
{prych, tt}@ii.uni.wroc.pl

Abstract. Set constraints are a useful formalism for verifying properties of programs. Usually, they are interpreted over the universe of finite terms. However, some logic languages allow infinite regular terms, so it seems natural to consider set constraints over this domain. In the paper we show that the satisfiability problem of set constraints over regular terms is undecidable. We also show that, if each function symbol has the arity at most 1, then this problem is EXPSPACE-complete.

1 Introduction

Set constraints are inclusions between expressions denoting sets of terms. They are a natural formalism for problems that arise in program analysis, including type checking, type inference, and approximating the meaning of programs. They were used in analyzing functional [ALW94], logic [AL94], imperative [HJ94] and concurrent constraint programs [CPM99].

The most popular domain for which set constraints were considered is the Herbrand universe, i.e. the set of all finite terms constructed over a given signature. The satisfiability of such constraints was studied by many authors including N. Heintze and J. Jaffar [HJ90], A. Aiken and E. L. Wimmers [AW92], L. Bachmair, H. Ganzinger, U. Waldmann [BGW93], R. Gilleron, S. Tison and M. Tommasi [GTT93] and W. Charatonik and L. Pacholski [CP94]. Set constraints for other domains were also studied ([MGWK96], [ALW94]).

In this paper we consider a variant of set constraints, namely set constraints over the set of (finite and infinite) regular terms. This domain was first introduced in Prolog II [Col82], and now is used in many modern logic programming languages, such as SWI-Prolog [Wie03] or Eclipse [WNS97]. Classical set constraints over the Herbrand universe can be inadequate in analyzing programs written in these languages.

Infinite terms in the context of set constraints were studied by Charatonik and Podelski [CP98], who proved that, for some restricted class of set constraints, which they call co-definite set constraints, the algorithms working for the Herbrand universe also apply to infinite terms. They proved EXPTIME-completeness of the satisfiability problem of co-definite set constraints over infinite regular terms.

---

* Partially supported by Polish KBN grant 8T11C 04319.

J. Marcinkowski and A. Tarlecki (Eds.): CSL 2004, LNCS 3210, pp. 458–472, 2004.

In general, the satisfiability problem in the Herbrand universe is not equivalent to the satisfiability problem over the set of regular terms. Consider, for example, the signature containing one constant c, and one unary function symbol f. Then the set constraints $X \neq \emptyset, X = f(X)$ have no solution in the Herbrand universe, but they have a solution $X = \{f(f(f(\dots)))\}$ in the set of infinite terms. Even if we forbid negative constraints, the finite and infinite cases differ. Consider the set constraints (with the same signature) consisting of one equation $\overline{X} = f(X)$. It has a solution in the Herbrand universe, but it is not solvable in the universe of regular terms. The reason is that the regular term $t = f(f(f(\dots)))$ fulfills the equation $t = f(t)$, so the constraint implies that, in any solution, t belongs to X, if and only if t belongs to $\overline{X}$.

In this paper we show that the satisfiability problem for positive set constraints over regular terms is undecidable. The proof is by reduction of the Post Correspondence Problem. Moreover, we show that, if all function symbols have the arity less or equal to 1, this problem is EXPSPACE-complete. These are rather surprising results, since set constraints over the Herbrand universe are EXPTIME-complete in unary case and NEXPTIME-complete when we allow function symbols of any arity [AKVW93] (they stay in NEXPTIME even if we enrich the language, adding negative constraints and projections [CP94]).

2   Preliminaries

2.1   Signatures and Terms

Let $\Sigma = \Sigma_0 \cup \Sigma_1 \cup \Sigma_2 \cup \cdots$ be a signature. A function symbol from Σ_n is told to be of the arity n.

In the paper by a *term* we mean a finite or infinite tree with nodes labeled by elements from Σ. If the label of a node belongs to Σ_n, then this node has exactly n ordered sons. A term t_1 is a *subterm* of t_2, if t_1 is a subtree of t_2. A term t is *regular*, if it has only finitely many different subterms. We denote the set of all (finite and infinite) regular terms over Σ by T_Σ^R. For terms $t_1, \dots, t_n$, we define the term $f(t_1, \dots, t_n)$ as a tree t, such that the root of t is labeled by f, and the i-th son of the root is t_i, for $i = 1, \dots, n$.

In order to describe regular terms we introduce a notion of t-graphs. A *t-graph* is a tuple $\langle \Sigma, V, E \rangle$, where V is a set of vertices, Σ is a signature, and $E : V \to \Sigma \times V^*$, such that if $E(v) = \langle f, v_1 \dots v_n \rangle$, then f is of the arity n. In such a situation we say that v is labeled by f. If V is finite then we say that the t-graph $\langle \Sigma, V, E \rangle$ is finite. We write $v =_E f(v_1, \dots, v_n)$ instead of $E(v) = \langle f, v_1 \dots v_n \rangle$. We say that a vertex v_i is the *i-th son* of v, if and only if $v =_E f(v_1, \dots, v_{i-1}, v_i, v_{i+1}, \dots, v_n)$, for some f.

A regular term can be represented by a t-graph with a selected vertex, as it is shown in Figure 1. This correspondence could be defined formally in the following way: a vertex v in a t-graph $G = \langle \Sigma, V, E \rangle$ *describes* a term t, if there is a function h from the subterms of t to V such that $h(t) = v$, and, for every subterm $t' = f(s_1, \dots, s_n)$ of t, we have $h(t') =_E f(h(s_1), \dots, h(s_n))$.

For a t-graph G, by $T(G)$ we denote the set of regular terms represented by the vertices of G. We say that a t-graph is *minimal*, if each of its vertices describes a distinct regular term (in Figure 1, G_2 is minimal, whereas G_1 is not). We denote by M_Σ^R the minimal t-graph (usually infinite) such that $T(M_\Sigma^R) = T_\Sigma^R$. It is easy to see that t-graph exists, and is unique up to isomorphism.

2.2   Set Constraints

Positive set constraints are inclusions[1] of the form $E \subseteq E'$, where the expressions E and E' are given by the grammar

$$E ::= X \mid E \cap E \mid \overline{E} \mid f(E, \dots, E) \mid \bot,$$

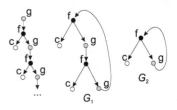

Fig. 1. A regular term (on the left) is represented by gray vertices of G_1 and G_2

where X stands for a variable from a given set, and f is a function symbol from a given signature Σ. We will use $\top$ as the abbreviation of $\overline{\bot}$, and $E_1 \cup E_2$, as the abbreviation of $\overline{\overline{E_1} \cap \overline{E_2}}$. We will also identify $\overline{\overline{E}}$ with E.

Let SC be a system of set constraints. Let Var denotes the set of variables that appear in SC, and let $\sigma : \mathsf{Var} \to 2^{T_\Sigma^R}$ be an assignment of subsets of T_Σ^R to variables in Var. Then σ in the unique way extends to a function $\hat{\sigma}$ from expressions to subsets of T_Σ^R. This extension is defined as follows: $\hat{\sigma}(X) = \sigma(X)$, for $X \in \mathsf{Var}$, $\hat{\sigma}(\bot) = \emptyset$, $\hat{\sigma}(E_1 \cap E_2) = \hat{\sigma}(E_1) \cap \hat{\sigma}(E_2)$, $\hat{\sigma}(\overline{E}) = T_\Sigma^R \setminus \hat{\sigma}(E)$, and for $f \in \Sigma_n$, we have $\hat{\sigma}(f(E_1, \dots, E_n)) = \{f(t_1, \dots, t_n) \mid t_1 \in \hat{\sigma}(E_1), \dots, t_n \in \hat{\sigma}(E_n)\}$. An assignment $\sigma : \mathsf{Var} \to 2^{T_\Sigma^R}$ is a solution of SC, if $\hat{\sigma}(E) \subseteq \hat{\sigma}(E')$, for each constraint $E \subseteq E'$ in SC. A system SC of set constraints is satisfiable, if it has a solution.

3   Automata and Set Constraints

We adapt here the definition of a t-dag automaton from [Cha99]. A *t-graph automaton* is a tuple $\langle \Sigma, Q, \Delta \rangle$, where Σ is a finite signature, Q is a finite set of states, and Δ is a set of transitions of the form $f(q_1, \dots, q_n) \mapsto q$ with $q, q_1, \dots, q_n \in Q$ and $f \in \Sigma_n$. An automaton is called *complete*, if for each $f \in \Sigma_n$, and each sequence $q_1, \dots, q_n \in Q$ there exists $q \in Q$ such that $f(q_1, \dots, q_n) \mapsto q$ belongs to Δ. An automaton $A' = \langle \Sigma, Q', \Delta' \rangle$ is an *subautomaton* of $A = \langle \Sigma, Q, \Delta \rangle$ iff $Q' \subseteq Q$, and $\Delta' \subseteq \Delta$.

A *run* of an automaton $\langle \Sigma, Q, \Delta \rangle$ on a t-graph $G = \langle V, \Sigma, E \rangle$ is a mapping ρ from V to Q such that for each $v, v_1, \dots, v_n \in V$, and $f \in \Sigma_n$, if $v =_E f(v_1, \dots, v_n)$, then Δ contains the transition $f(\rho(v_1), \dots, \rho(v_n)) \mapsto \rho(v)$. If there is a run of an automaton A on a graph G, then we say that A *accepts* G.

The following lemma states the connections between runs on finite graphs and runs on M_Σ^R.

---

[1] In so called *negative set constraints* there are also allowed negated inclusions. Such systems were analyzed for instance in [CP94].

Lemma 1. *Let Σ be a finite signature, and let A be an automaton over Σ. The following conditions are equivalent:*

(i) *A accepts M_Σ^R.*
(ii) *A accepts all finite t-graphs over Σ.*
(iii) *There exists a complete subautomaton A' of A which accepts M_Σ^R.*
(iv) *There exists a complete subautomaton A' of A which accepts all finite t-graphs over Σ.*

Proof. One can show the following implications: (ii) $\Rightarrow$ (i), (i) $\Rightarrow$ (iii), (iii) $\Rightarrow$ (iv), (iv) $\Rightarrow$ (ii). All the implications but the first one are quite straightforward. In the case of (ii) $\Rightarrow$ (i) we can use the compactness theorem for the propositional logic, as is sketched bellow. Assume that (ii) holds. We use the propositional variable p_q^v for each vertex v, and each state q of the given automata A. The intended meaning of p_q^v is "the state q is assigned to the vertex v". For each t-graph G, using these variables, it is easy to construct a set Φ_G of formulas such that Φ_G is satisfiable iff A has a run on G. Now, for any finite $\Phi' \subseteq \Phi_{M_\Sigma^R}$, one can show that there is a finite subgraph G of M_Σ^R such that $\Phi' \subseteq \Phi_G$. By the assumption, A accepts G, so the set Φ_G is satisfiable, and so is Φ'. Hence, by the compactness theorem, $\Phi_{M_\Sigma^R}$ is satisfiable, which implies that A accepts M_Σ^R.   □

Following Charatonik and Pacholski [CP94, Cha99] we define, for a system of set constraints SC, the automaton A_{SC} representing it. Let $E(SC)$ be the set of all set expressions occurring in SC together with their complements.

Definition 1. *Let SC be a system of set constraints over Σ. The automaton A_{SC} is $\langle \Sigma, Q, \Delta \rangle$, where $Q \subseteq 2^{E(SC)}$, and*

1. A subset ϕ of E(SC) is a state of A_{SC}, if
    (i) $\bot \notin \phi$,
    (ii) $E \in \phi$ iff $\overline{E} \notin \phi$,
    (iii) if $(E_1 \cap E_2) \in \phi$ then $E_1, E_2 \in \phi$,
    (iv) if $E_1 \in \phi$, $E_2 \in \phi$, and $(E_1 \cap E_2) \in$ E(SC), then $(E_1 \cap E_2) \in \phi$,
    (v) if $E \subseteq E' \in SC$, and $E \in \phi$, then $E' \in \phi$,
    (vi) if $f(E_1, \ldots, E_n) \in \phi$ and $g(E_1, \ldots, E_m) \in \phi$ then $m = n$ and $f = g$,
2. Δ is the set of transitions of the form $f(\phi_1, \ldots, \phi_n) \mapsto \phi$, where
    (i) $f \in \Sigma_n$, and $\phi_1, \ldots, \phi_n, \phi \in Q$, and
    (ii) $f(E_1, \ldots, E_n) \in \phi$ iff $E_i \in \phi_i$ for each $i = 1, \ldots, n$.

The following lemma (and its proof) is an exact 'translation' of the part of Theorem 24 from [Cha99].

Lemma 2. *Let SC be a system of positive set constraints. SC is satisfiable, if and only if A_{SC} accepts M_Σ^R.*

Proof. Suppose that σ is a solution of SC. Let t_v denotes the term described by a vertex v in M_Σ^R. Then we can define a run ρ of A_{SC} on M_Σ^R by setting $\rho(v) = \{E \in E(SC) \mid t_v \in \hat{\sigma}(E)\} \cup \{\overline{E} \mid E \in E(SC), t_v \notin \hat{\sigma}(E)\}$, for each vertex v of M_Σ^R. Conversely, if there exists a run ρ on M_Σ^R, we can define a solution σ of SC by $\sigma(X) = \{t_v \in T_\Sigma^R \mid$ there exists a vertex v of M_Σ^R such that $X \in \rho(v)\}$.   □

Notation: As the proof of Lemma 2 shows, solutions and runs express the same relation between terms and variables in different ways: (a) for a solution σ of SC, we write $t_v \in \sigma(X)$, whereas (b) for a run ρ of A_{SC}, we write $X \in \rho(v)$ (where v describes t_v).

We find it convenient to write constraints in a form which directly expresses local relations in a t-graph, using formulas which correspond to (b). So, for $f \in \Sigma_n$, we will write constraints of the form

$$\forall t_0 = f(t_1, \ldots, t_n)) : \quad \varphi, \qquad (\star)$$

where φ is a boolean combination of formulas of the form $(X \text{ in } t_i)$, and $(\overline{X} \text{ in } t_i)$ (for $0 \le i \le n$).

Constraints of the form $(\star)$ can be easily translated to ordinary set constraints: first we change all atomic formulas of the form $(E \text{ in } t_0)$ into E, and, for $1 \le i \le n$, we change $(E \text{ in } t_i)$ into $f(\top, \ldots, E, \ldots, \top)$ (with E on i-th position). Then, we replace $\wedge$ by $\cap$, $\vee$ by $\cup$ and $\neg$ by complementation, obtaining an expression S. The resulting set constraint for $(\star)$ is $f(\top, \ldots, \top) \subseteq S$.

For instance the formula $(\forall s = f(t_1, t_2) : X \text{ in } s \Rightarrow Y \text{ in } t_1)$ is translated to $f(\top, \top) \subseteq \overline{X} \cup f(Y, \top)$. This formula says that for any run ρ of A_{SC}, for all nodes v and v' such that v is labeled by f and v' is the first son of v, we have that $X \in \rho(v)$ implies $Y \in \rho(v')$.

Moreover, if a sequence of variables $\boldsymbol{X} = (X_1, \ldots, X_n)$ is supposed to code values from some finite set, then we allow to use such vectors of variables as a syntactic sugar in constraints written in the form $(\star)$, which is shown in the following example.

Example 1. The formula

$$\forall t = f(s) : (\boldsymbol{X} \text{ in } t) \ne (\boldsymbol{X} \text{ in } s) \qquad (1)$$

is an abbreviation of the formula

$$\forall t = f(s) : \quad (X_1 \text{ in } s) \wedge (\overline{X}_1 \text{ in } t) \vee (\overline{X}_1 \text{ in } s) \wedge (X_1 \text{ in } t) \vee \cdots \vee$$
$$(X_n \text{ in } s) \wedge (\overline{X}_n \text{ in } t) \vee (\overline{X}_n \text{ in } s) \wedge (X_n \text{ in } t)$$

Now, the formula (1) means that for any run ρ of the automaton for the constraint (1), for all nodes v and v', such that v' is the only son of v, and v is labeled by f, we have that the value of $\boldsymbol{X}$ in $\rho(v)$ is different than the value of $\boldsymbol{X}$ in $\rho(v')$, i.e. if we take $a_i = 1$ iff $X_i \in \rho(v)$, and $a_i = 0$ iff $\overline{X}_i \in \rho(v)$, and similarly $b_i = 1$ iff $X_i \in \rho(v')$, and $b_i = 0$ iff $\overline{X}_i \in \rho(v')$, then we have that $(a_1 \ldots a_n) \ne (b_1 \ldots b_n)$.

In a similar way we can use formulas like $\forall t = f(s) : (\boldsymbol{X} \text{ in } s) = (\boldsymbol{Y} \text{ in } t)$ or $\forall t = f(s) : (\boldsymbol{X} \text{ in } s = \boldsymbol{X} \text{ in } t + 1)$.

4   The General Case

Now we state the main result of the paper. The rest of this section is devoted to its proof.

$$\forall t = a_j(\cdot): \quad A \text{ in } t \wedge \overline{S} \text{ in } t \wedge \overline{P} \text{ in } t \tag{2}$$

$$\forall t = s_i(\cdot, \cdot): \quad \overline{A} \text{ in } t \wedge S \text{ in } t \wedge \overline{P} \text{ in } t \tag{3}$$

$$\forall t = p_i(\cdot, \cdot, \cdot): \quad \overline{A} \text{ in } t \wedge \overline{S} \text{ in } t \wedge P \text{ in } t \tag{4}$$

$$\forall t = a_j(t_1): \quad (\boldsymbol{K} \text{ in } t) = (\boldsymbol{K} \text{ in } t_1) \tag{5}$$

$$\forall t = p_i(t_1, \cdot, \cdot): \quad (\boldsymbol{K} \text{ in } t) = (\boldsymbol{K} \text{ in } t_1) \tag{6}$$

$$\forall t = a_j(t_1): \quad (E \text{ in } t) \Rightarrow (\overline{S} \text{ in } t_1) \wedge (\overline{A} \text{ in } t_1 \vee E \text{ in } t_1) \tag{7}$$

$$\forall t = p_i(t_1, t_2, t_3): \quad (E \text{ in } t) \Rightarrow (\overline{S} \text{ in } t_1) \wedge (E \text{ in } t_1 \vee \overline{P} \text{ in } t_1) \vee (\overline{A} \text{ in } t_2 \vee E \text{ in } t_2) \tag{8}$$
$$\vee (\overline{A} \text{ in } t_3 \vee E \text{ in } t_3) \vee (\boldsymbol{K} \text{ in } t) \neq (\boldsymbol{K} \text{ in } t_2) \vee (\boldsymbol{K} \text{ in } t) \neq (\boldsymbol{K} \text{ in } t_3)$$

$$\forall t = s_i(t_1, t_2): \quad (E \text{ in } t) \Rightarrow (E \text{ in } t_1) \vee (E \text{ in } t_2) \vee (\overline{P} \text{ in } t_1) \vee (\overline{A} \text{ in } t_2) \tag{9}$$

$$\forall t = s_i(t_1, t_2): \quad (H \text{ in } t) \vee (E \text{ in } t) \tag{10}$$
$$\vee ((\boldsymbol{K} \text{ in } t) \neq (\boldsymbol{K} \text{ in } t_1)) \vee ((\boldsymbol{K} \text{ in } t) \neq (\boldsymbol{K} \text{ in } t_2))$$

$$\forall t = a_j(t_1): \quad (H \text{ in } t) \Rightarrow (H \text{ in } t_1) \tag{11}$$

$$\forall t = p_i(t_1, t_2, t_3): \quad (H \text{ in } t) \Rightarrow (H \text{ in } t_1) \wedge (H \text{ in } t_2) \wedge (H \text{ in } t_3) \tag{12}$$

$$\forall t = s_i(t_1, t_2): \quad (H \text{ in } t) \Rightarrow (H \text{ in } t_1 \wedge H \text{ in } t_2) \tag{13}$$

$$\forall t = a_i(t_1): \quad (H \text{ in } t) \wedge (\boldsymbol{U} \text{ in } t) = (a_i x_2 \ldots x_m) \Rightarrow (\boldsymbol{U} \text{ in } t_1) = (x_2 \ldots x_m) \tag{14}$$

$$\forall t = a_i(\cdot): \quad (H \text{ in } t) \wedge (\boldsymbol{U} \text{ in } t) = \epsilon \Rightarrow (F \text{ in } t) \tag{15}$$

$$\forall t = p_i(t_1, t_2, t_3): \quad (H \text{ in } t) \Rightarrow (\overline{C} \text{ in } t) \vee (C \text{ in } t_1) \tag{16}$$
$$\vee ((\boldsymbol{U} \text{ in } t_2) = u_i \wedge \overline{F} \text{ in } t_1) \vee ((\boldsymbol{U} \text{ in } t_3) = v_i \wedge \overline{F} \text{ in } t_1)$$

$$\forall t = p_i(t_1, t_2, t_3): \quad (H \text{ in } t) \wedge (\overline{F} \text{ in } t) \Rightarrow (\overline{F} \text{ in } t_2 \wedge \overline{F} \text{ in } t_3) \tag{17}$$

$$\forall t = p_i(t_1, t_2, t_3): \quad (H \text{ in } t) \wedge (C \text{ in } t) \Rightarrow (\boldsymbol{V} \text{ in } t_2) \neq (\boldsymbol{V} \text{ in } t_3) \tag{18}$$

$$\forall t = s_i(t_1, t_2): \quad (H \text{ in } t) \Rightarrow (C \text{ in } t_1) \vee (\boldsymbol{U} \text{ in } t_2) = u_i \wedge \overline{F} \text{ in } t_1 \tag{19}$$
$$\vee (\boldsymbol{U} \text{ in } t_2) = v_i \wedge \overline{F} \text{ in } t_1$$

Fig. 2. Constraints Φ

Theorem 1. *Deciding whether a system of positive set constraints has a solution in the domain of regular terms is undecidable.*

Let $(u_1, v_1), \ldots, (u_N, v_N)$ be an instance of PCP over an alphabet $\Sigma = \{a_1, \ldots, a_K\}$. We can assume that the words $u_1, v_1, \ldots, u_N, v_N$ are not empty[2]. We give a system Φ of set constraints (Fig. 2) which has a solution, if and only if the given instance of PCP has no solution. The explanation of the the intended meaning of these contstraints will be postponed until Subsections 4.2 and 4.3.

The signature we use consists of the functors s_i of the arity 2, and the functors p_i of the arity 3, for each $1 \leq i \leq N$ (each s_i and p_i corresponds to the pair (u_i, v_i)), and the functors a_j of the arity 1, for each a_j belonging to Σ.

---

[2] Undecidability of PCP can be proved, if we assume that words are not empty. See e.g. the proof of undecidability of PCP in [HU79].

4.1   H-structures and Solutions of PCP

In this subsection we define a class of finite t-graphs, called *H-structures*, which can be used to code solutions of the given instance of PCP.

We say that a vertex is of the *type a*, if it is labeled by a_j, for some j. Similarly, a vertex is of the *type s* or *p*, if it is labeled by s_i, p_i respectively, for some i.

Definition 2. An *H-structure* is a t-graph which consists of one vertex x_0 of the type s, called the *root*, nodes $x_1, \ldots x_n$ of the type p, and nodes $y_1, \ldots, y_m$ of the type a such that:

(i) the first son of the root is x_1, and the second son of the root is y_1,
(ii) the only son of y_i is y_{i+1}, for $1 \leq i < m$, and the only son of y_m is x_0,
(iii) for each $1 \leq i < n$, the first son of x_i is x_{i+1}, and the first son of x_n is x_0,
(iv) for each $1 \leq i \leq n$, the second and the third son of x_i are of the type a (thus belong to $\{y_1, \ldots, y_m\}$).

An example of an H-structure is shown in Fig. 3. Let us notice that labels of vertices of the type a correspond to symbols from Σ, thus sequences of vertices of this type can code words over Σ. We formalize it in the following way: a word $w = b_1 \ldots b_n \in \Sigma^*$ *has an instance starting at a vertex* y_1, *and finished at a vertex* y, if there exists a path $y_1, \ldots, y_n$ of vertices labeled by $b_1, \ldots, b_n$, and y is the only son of y_n.

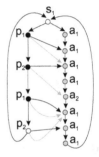

Fig. 3. An H-structure. The first sons of vertices of the type p are represented by down arrows, the second sons by gray right arrow, and the third sons by black right arrow

Let x be a vertex of the type p or s. A vertex y is said to be a *second (third) grandson of x*, if y is the second (third) son of the first son of x.

Definition 3. A vertex x labeled by p_k is *valid*, if its first son has the type p, and (a) u_k has an instance starting at the second son of x, and finished at its second grandson, and (b) v_k has an instance starting at the third son of x, and finished at its third grandson.

Similarly, a vertex x labeled by s_k is *valid*, if its first son has the type p, and (a) u_k has an instance starting at the second son of x, and finished at its second grandson, and (b) v_k has an instance starting at the second son of x, and finished at its third grandson.

If y is the first son of a valid vertex, then y is told to be *charged*.

Notice, that a charged vertex must have the type p. Now, consider the H-structure from Fig. 3, and suppose that the given instance of PCP has two pairs (a_1a_1, a_1), $(a_2, a_1a_1a_2)$. Consider the path of vertices labeled by s_1, p_1, p_2, p_1, p_2. The first three vertices of this path are valid. The charged vertices of this H-structure are the black ones.

Definition 4. *An H-structure G with the root x_0 describes a solution*, if there exist charged vertices $x_1, \ldots, x_n$ (of the type p), for $n \geq 1$, such that $x_0, x_1, \ldots, x_n$ is a path in G, and the second and the third son of x_n are the same.

Notice that the vertices $x_0, \ldots, x_{n-1}$ in the definition above must be valid, because the first son of each of them is charged.

Consider again the PCP given by the pairs $(a_1 a_1, a_1)$, $(a_2, a_1 a_1 a_2)$. The H-structure from Fig. 3 describes a solution of this PCP. The described solution is the sequence $1, 1, 2$. It is given by indices of the labels s_1, p_1, p_2 of the vertices on the path starting with the root. Note that the next vertex on this path with label p_1 (which is not valid, but is charged) is the one whose second and third sons are the same. Note also that the last vertex of the type p is of no use, and we could build a smaller H-structure which describe the same solution.

It is easy to show that the following holds.

Remark 1. The given instance of PCP has a solution, if and only if there exists an H-structure which describes a solution.

Now, we state two lemmas which constitute the major steps of the proof of Theorem 1. The proofs of these lemmas are given in separate subsections.

Lemma 3. *Let Φ be the system of set constraints from Fig. 2. A_Φ accepts all finite t-graphs, if and only if it accepts all H-structures.*

Lemma 4. *Let Φ be the system of set constraints from Fig. 2. A_Φ accepts an H-structure G, if and only if G does not describe a solution.*

Lemmas 2, 3, and 1 imply that system Φ from Fig. 2 is satisfiable, if and only if A_Φ accepts all H-structures. Consequently, by Remark 1 and Lemma 4, the given instance of PCP has no solution, if and only if the system of set constraint from Fig. 2 is satisfiable, which completes the proof of Theorem 1.

4.2   The Proof of Lemma 3

A vertex of the type p is *well-typed*, if its first son has the type p or s, and its second and third sons have the type a. A vertex of the type a is *well-typed*, if its only son has the type a or s. A vertex of the type s is *well-typed*, if its first son has the type p, and its second son has the type a. Notice that all the vertices of any H-structure are well-typed.

Now, we introduce a notion of *witness* which is intended to give an evidence that a vertex v (of the type s) is *not* a root of any H-structure, and carry some information which will be used later in the proof.

Definition 5. Let G be a t-graph, and v be a vertex of the type s. A *witness* for v has one of the following forms:

(a) $\langle v, \{v, z_1, \ldots, z_n\}\rangle$, if $v, z_1, \ldots, z_n$ is a path in G, where none of $z_1, \ldots, z_n$ has the type s, and only the last vertex of $v, z_1, \ldots, z_n$ is not well-typed.

(b) $\langle v, \{v, z_1, z_2, \dots\}\rangle$, if $v, z_1, z_2, \dots$ is an infinite path in G of well-type vertices, where none of $z_1, z_2, \dots$ has the type s.

(c) $\langle w, \{v, x_1, \dots, x_n\}\rangle$, if $v, x_1, \dots, x_n, y_1, \dots, y_m, w$ is a path in G, where the vertex $w \neq v$ has the type s; the vertices $x_1, \dots, x_n$ are well-typed, and have the type p; the vertices $y_1, \dots, y_m$ are well-typed, and have the type a.

(d) $\langle w, \emptyset\rangle$, if there is a path of the form $v, x_1, \dots, x_n, w$ in G, where the vertex $w \neq v$ has the type s; the vertices $x_1, \dots, x_n$ are well-typed, and either each of them has the type p, or each of them has the type a.

The set of all the witnesses for v is denoted by $W(v)$.

Note that a witness of the form (a) corresponds to the case, when starting with v, we can reach some vertex which is not well-typed. A witness of the form (b) corresponds to the case, when there is an infinite path of well-typed vertices of the type p or a starting with v. A witness of the form (c) or (d) corresponds to the case when starting with v, we can reach a vertex $w \neq v$ of the type s: a witness has the form (c) if the path from v to w contains vertices of the type p followed by vertices of the type a, and a witness has the form (d) if the path from v to w contains vertices only of the type p, or only of the type a.

One can check that a vertex v of the type s is the root of some H-structure, if and only if $W(v)$ is empty.

In order to prove the nontrivial implication of Lemma 3, let us assume that A_Φ accepts all H-structures. Let G be a finite t-graph, and G_0 be the subgraph of G containing exactly all the H-structures of G. Because these H-structures have disjoint sets of vertices, and each of them has an accepting run, there exists an accepting run ρ_0 of A_Φ on G_0. We will extend ρ_0 to a run ρ on G, but first we informally explain the role of some variables used in the constraints from Fig. 2:

- A, S, P – type variables. In each vertex v exactly one of these variables have to be set: A has to be set in v (i.e. $A \in \rho(v)$), if v has the type a, and so on (see (2)–(4)).

- $\boldsymbol{K}$ — vectors of variables which can code a *color*, i.e. a value from $\{\alpha, \beta, \gamma\}$. The constraints (5)–(6) guarantee that each vertex of the type a or p has the same color as its first son, thus each H-structure is colored with one color. These variables are used to detect cases related to the points (c) and (d) of Definition 5.

- E – an error flag. If set in a vertex v, it indicates that v cannot be a part of any H-structure. The constraints (7)–(9) allow us to set E in v only if either (i) E is set in some son of v of a type different than s, (ii) v is not well-typed, or (iii) v has the type p and it has a different color than its second or third son (it is related to point (c) of Definition 5).

- H – an H-structure indicator. This variable is intended to be set exactly in these vertices which are a part of some H-structure. The constraints (11)–(13) guarantees that, if H is set in some vertex, then it must be also set in all its sons. The constraint (10) guarantees that H must be set in a vertex v of the type s unless (i) E is set in v, which corresponds to the points (a),

(b) or (c) of Definition 5, or (ii) v has a different color than one of its sons, which corresponds to point the (d) of Definition 5.

Note that, if the variable H is not set in some vertex v (so v is not a part of any H-structure), then (11)–(19) are obviously satisfied in v.

As we have noticed, v of the type s is the root of some H-structure, if and only if $W(v) = \emptyset$. Let V_S denote the set of vertices of the type s from $G \setminus G_0$. For each $v \in V_S$, $W(v) \neq \emptyset$, so let us chose one witness from $W(v)$, and denote it by $f(v)$. One can assign a color $c_v \in \{\alpha, \beta, \gamma\}$ to each vertex $v \in G$ of the type s in such a way that, if $f(v) = \langle w, B \rangle$, for some $w \neq v$, then $c_v \neq c_w$, and, for $v \in G_0$, we have $c_v = (\boldsymbol{K} \text{ in } \rho_0(v))$.

Now, we extend ρ_0 to a run ρ on G. For each $v \in G \setminus G_0$, we define $\rho(v)$ as follows: We set variable H to 0. We set type variables (A, S, P) according to the type of v (e.g. $A, \overline{S}, \overline{P}$ in vertices of the type a). We set variable E to 1, if and only if, for some $v' \in V_S$ with $f(v') = \langle w, B \rangle$, the vertex $v \in B$. If $v \in V_S$ then we set $\boldsymbol{K}$ to c_v, otherwise we give v the color of its first son (when following first sons, we get into a cycle, then we can set $\boldsymbol{K}$ to α in all the vertices of this cycle). The values of the other variables do not matter. One can now show that ρ is a run on G.   □

4.3   The Proof of Lemma 4

Lemma 5. *Let G be an H-structure, and ρ be a run of A_Φ on G. For each vertex v of G, the variable H must be set in $\rho(v)$.*

Proof. It is easy to check that all the vertices of G have the same values of $\boldsymbol{K}$ in ρ. Moreover, one can check that E must not be set in $\rho(v)$, for each v in G. Thus, the only way to satisfy the constraint (10) for the root r is to set H in $\rho(r)$. So, by the constraint (11)–(13), H must be set in all the vertices of G.

Let us now describe the intended meaning of the variables used in this part of the proof. As we consider here H-structures, we assume that, according to Lemma 5, the variable H must be set in any vertex considered.

- $\boldsymbol{U}$ — a sequence of variables of the length sufficient to code a special value $\diamond$, and words over Σ not longer than l, where l is the length of the longest of the words in the given instance of PCP. It is used to check whether some word has an instance at a given place (see (14)–(15)), and so to check validity of vertices.
- $\boldsymbol{F}$ — an auxiliary variable used together with $\boldsymbol{U}$ to check validity of vertices.
- $\boldsymbol{C}$ — the constraints (14)–(16) and (19) guarantee that, in any run, $\boldsymbol{C}$ have to be set in the vertices denoted by $x_1, \ldots, x_n$ in Definition 4 (that is in the sequence of charged vertices).
- $\boldsymbol{V}$ — a vector of variables which can code one of the colors α, β, γ. It is used only in (18) which guarantees that the second and the third son of a charged vertex (a vertex with the variable C set) must not be the same,

In order to prove Lemma 4, we first show that **if an H-structure G describes a solution, then A_Φ does not accept G.**

Proof. Suppose that G describes a solution, and suppose that ρ is a run of A_Φ on G. Let $x_0, \ldots, x_n$ denote vertices according to Definition 4. Using Lemma 5, one can show that (14)–(15) have the following consequence: if a word u has an instance starting at x and finished at y, and $(\boldsymbol{U}$ in $\rho(x)) = u$, then $F \in \rho(y)$.

Using that together with (16), (17), (19) one can prove, by induction on i, that $C \in \rho(x_i)$, for $1 \leq i \leq n$. Particularly, $C \in \rho(x_n)$. By Definition 4, the second and the third sons of x_n are the same, which implies that ρ cannot fulfill (18), and contradicts the assumption that ρ is a run on G. $\square$

Now, we show that **if an H-structure G does not describe a solution, then A_Φ accepts G.**

Proof. Let $x_0, \ldots, x_n$ and $y_1, \ldots, y_m$ denote vertices according to Definition 2. For $i \in \{1, \ldots, n\}$, we define $s(i)$ and $t(i)$ in such a way that $y_{s(i)}$ is the second son of x_i, and $y_{t(i)}$ is its third son.

Let x_k be the first not valid vertex from $x_0, \ldots, x_n$ (such a vertex exists, because x_n is not valid). Notice that G does not describe a solution, and, for each $i \in \{1, \ldots, k\}$, x_i is charched, so we have $s(i) \neq t(i)$, and moreover, $s(i) < s(i{+}1)$, and $t(i) < t(i + 1)$. Using these facts, one can prove that each vertex v can be assigned a color $f(y) \in \{\alpha, \beta\}$ such that, for each $i \in \{1, \ldots, k\}$, it holds $f(y_{s(i)}) \neq f(y_{t(i)})$.

Now, we will define a function δ from the set of vertices of G to the set of values that can be coded in $\boldsymbol{U}$. If $k = n$, then let $\delta(v) = \diamond$, for each vertex v of G. Otherwise (i.e. if $k < n$), since x_k is not valid, there are two possible cases which correspond to violation of the condition (a), or the condition (b) of Definition 3. We consider the case (a), and we assume that $k \neq 0$ (in the other cases the proof proceeds similarly). Let p_j be the label of x_k, and $u_j = b_0 \ldots b_l$. Let d be the greatest natural number such that the labels of $y_{s(k)}, \ldots, y_{s(k)+d}$ are some prefix of u_j (d must not be greater than l). For $0 \leq i \leq d + 1$, let $\delta(y_{s(k)+i}) = b_i \ldots b_l$. For a vertex $v \notin \{y_{s(k)}, \ldots, y_{s(k)+d+1}\}$, let $\delta(v) = \diamond$.

Now we construct a run ρ such that:
- in each vertex v of G we set H to 1 and E to 0; the type variables we set according to the type of v, and the variables $\boldsymbol{K}$ we set to α,
- we set F to 1 in each vertex, with the exception of x_{k+1} and its second and third sons, if $k < n$,
- we set the variable C to 1 only in the vertices $x_0, \ldots, x_k$,
- in each vertex v, we set $\boldsymbol{V}$ to $f(v)$, and $\boldsymbol{U}$ to $\delta(v)$.

One can check that ρ is a run of A_Φ on G. $\square$

5   The Unary Case

In this section we consider positive set constraints which use only constants and unary function symbols. Such systems will be called *unary set constraints (USC)*. The problem of deciding whether a system of USC is satisfiable turns out to be EXPSPACE-complete, which is an immediate consequence of the following theorems.

Theorem 2. *Deciding whether a system of USC is satisfiable is in EXPSPACE.*

Theorem 3. *Deciding whether a system of USC is satisfiable is EXPSPACE-hard.*

5.1   Proof of Theorem 2

Let Σ be a signature, and $f_1, \ldots, f_n \in \Sigma$. A finite t-graph $\langle V, \Sigma, E \rangle$ with the set of vertices $V = \{v_1, \ldots, v_n\}$ is *a cycle* $f_1, \ldots, f_n$ *(over* Σ*)*, if $v_1 =_E f_1(v_n)$, and $v_i =_E f_i(v_{i-1})$, for $1 < i \leq n$.

Intuitively, cycles are the most difficult parts of a t-graph when we want to find a run. This is stated by the following lemma:

Lemma 6. *Let A be a complete automaton over Σ. A accepts all finite t-graphs over Σ, if and only if it accepts all cycles over Σ.*

Proof. Sketch. To proceed the proof in the nontrivial direction note that the connected components of a graph can be considered separately. Each such component contains at most one cycle. To find a run for a whole connected component, we first find a run on the only cycle, and then use the completeness of the automaton.

Now, for a t-graph automaton A, we define a deterministic finite automaton $\tilde{A}$ (working on finite words) in such a way that runs of A on cycles can be simulated by $\tilde{A}$.

Definition 6. *Let $A = \langle \Sigma, Q, \Delta \rangle$ be a t-graph automaton with $Q = \{q_1, \ldots, q_n\}$. We define a deterministic finite automaton $\tilde{A}$ on finite words over the alphabet Σ_1 (i.e. over the set of the unary symbols of Σ) in the following way: let $\tilde{A} = \langle \Sigma_1, \tilde{Q}, \tilde{q}_0, \tilde{\delta}, \tilde{F} \rangle$, where $\tilde{Q} = (2^Q)^n$ is the set of states, $\tilde{q}_0 = \langle \{q_1\}, \ldots, \{q_n\} \rangle$ is the initial state, and $\tilde{F} = \{ \langle Q_1, \ldots, Q_n \rangle \mid q_i \in Q_i \text{ for some } 1 \leq i \leq n \}$ is the set of accepting states. The transition function $\tilde{\delta}$ is defined by the equation*

$$\tilde{\delta}(\langle Q_1, \ldots, Q_n \rangle, f) = \langle Q'_1, \ldots, Q'_n \rangle,$$

where $Q'_i = \{q' \in Q \mid \text{there exists } q \in Q_i \text{ such that } (f(q) \mapsto q') \in \Delta \}$.

One can prove the following lemma which expresses a correspondence between automata on words and t-graph automata on cycles.

Lemma 7. *Let A be a t-graph automaton over Σ, and $f_1, \ldots, f_k \in \Sigma$. The automaton A accepts the cycle $f_1, \ldots, f_k$, iff $\tilde{A}$ accepts the word $f_1 \ldots f_k$.*

Deciding whether a deterministic finite automaton is universal (i.e. accepts all words) is NLOGSPACE complete. Knowing that, since the size of $\tilde{A}$ is $2^{O(n^2)}$, it is easy to prove that, for a t-graph automaton A, the problem of deciding whether $\tilde{A}$ is universal is in PSPACE.

Now we give the nondeterministic algorithm working in EXPSPACE, and verifying whether a given system SC of set constraints is satisfiable: *First we construct* A_{SC}. *Then we guess*[3] *a complete subautomaton* B *of* A_{SC} *and verify whether* $\tilde{B}$ *is universal. If it is, we halt with the answer "yes", otherwise halt with the answer "no".* It is easy to check that this algorithm works in EXPSPACE. Its correctness follows directly from Lemmas 1, 2, 6, 7.

5.2   Proof of Theorem 3

Suppose that M is a deterministic Turing machine which, for an input word w of length n ($w \in \{0,1\}^*$), uses space bounded by 2^N where $N = n^l$, for some integer l. We assume that the set of states $Q = \{0, \ldots, K\}$, the tape alphabet $\Gamma = \{0, \ldots, L\}$, the number 0 denotes the initial state and the blank symbol, $Q_F \subseteq Q$ is the set of accepting states, and δ is the transition function.

Without loss of generality we can assume that Q is a union of three disjoint sets: Q_L, Q_R and $\{0\}$, such that M can be in a state belonging to Q_L (Q_R) only after its head has been moved left (right respectively).[4] Thus the transition function δ can be seen as a function from $Q \times \Gamma$ to $Q \times \Gamma$.

Let $u = u_1 \ldots u_n \in \{0,1\}^*$ be the input word. We will construct a system Ψ of set constraints such that M accepts u, if and only if Ψ is not satisfiable. In set constraints Ψ we use the signature $\Sigma = \Sigma_1 = \{f_0, \ldots, f_K\}$.

Let us notice that a sequence $0, q_1, \ldots, q_n$ of states of M can be coded as the cycle $f_0, f_{q_1}, \ldots f_{q_n}$. This gives us opportunity to code computations using t-graphs (note that a sequence of states determines the position of the head). A sequence $0, q_1, \ldots, q_n$ of states of M is *accepting*, if $q_n \in Q_F$. It is *valid*, if there is a computation of M on u with these states.

We will consider a computation of M for u from the point of view of the i-th cell. A sequence $0, q_1, \ldots, q_n$ of states of M is *valid with respect to the i-th cell*, if there exists a sequence of tape symbols $a_0, a_1, \ldots, a_n$, such that (1) a_0 is the tape symbol contained in the i-th cell at the start of computation, and (2) if the position of the head of M in the j-th step is i, then $\delta_M(q_j, a_j) = \langle q_{j+1}, a_{j+1} \rangle$, otherwise $a_{j+1} = a_j$.

Note that a sequence of states is valid, if and only if it is valid with respect to the i-th cell, for all $0 \le i < 2^N$. In Figure 4 we give a system Ψ of set constraints which is solvable, if and only if M does not accept u.

In the constraints we use the following variables: $\boldsymbol{P} = P_1, \ldots, P_N$ (related to the position of the head), $\boldsymbol{A} = A_1, \ldots, A_N$ (related to the address of a cell), $\boldsymbol{X} = X_1, \ldots, X_{\lceil log_2 L \rceil}$ (related to the content of the cell pointed by A), $\boldsymbol{Q} = Q_1, \ldots, Q_{\lceil log_2 K \rceil}$ (related to a state of M). We use a special variable T which is a sign of invalid computation and variables $\boldsymbol{K}$ which are supposed to code a color (i.e. an element from $\{\alpha, \beta, \gamma\}$).

One can prove the following lemma.

Lemma 8. A_Ψ *accepts all finite t-graphs, if and only if* A_Ψ *accepts all cycles of the form* $f_0, f_{q_1}, \ldots f_{q_n}$ *where* $q_i \ne 0$, *for* $i = 1, \ldots, n$.

---

[3] We can use nondeterminism here because EXPSPACE=NEXPSPACE.

[4] We can easily transform any Turing machine, so as it meets this condition.

$$\forall t = f_0(s) : \quad (\boldsymbol{K} \text{ in } s) \neq (\boldsymbol{K} \text{ in } t) \lor \text{start}(t) \land \overline{T} \text{ in } t \land (T \text{ in } s \lor \text{nacc}(s)), \quad (20)$$

$$\forall t = f_i(s) : \quad (\boldsymbol{K} \text{ in } t) = (\boldsymbol{K} \text{ in } s) \land \qquad\qquad\qquad\qquad\qquad\quad (21)$$

$$((T \text{ in } s) \lor (\overline{T} \text{ in } t \land \text{good}_i(t, s)) \qquad \text{(for all } 0 < i \le K), \quad (22)$$
$$\lor (T \text{ in } t \land \text{bad}_i(t, s)))$$

where

$$\text{nacc}(t) \equiv \bigwedge_{i \in Q_F} (\boldsymbol{Q} \text{ in } t) \neq i, \qquad\qquad\qquad\qquad\qquad\qquad (23)$$

$$\text{start}(t) \equiv (\boldsymbol{P} \text{ in } t) = 0 \land (\boldsymbol{Q} \text{ in } t) = 0 \land \qquad\qquad\qquad\qquad (24)$$

$$((\boldsymbol{A} \text{ in } t) > n \land (\boldsymbol{X} \text{ in } t) = 0 \lor \bigvee_{1 \le i \le n} (\boldsymbol{A} \text{ in } t) = i \land (\boldsymbol{X} \text{ in } t) = u_i) \quad (25)$$

$$\text{bad}_i(t, s) \equiv (\boldsymbol{A} \text{ in } s) = (\boldsymbol{P} \text{ in } s) \land \delta_M((\boldsymbol{Q} \text{ in } s), (\boldsymbol{X} \text{ in } s)) = \langle j, v \rangle, \text{where } j \neq i \quad (26)$$

$$\text{good}_i(t, s) \equiv \text{next}_i(t, s) \land (\boldsymbol{A} \text{ in } t) = (\boldsymbol{A} \text{ in } s) \land (\boldsymbol{Q} \text{ in } t = i) \land (\Phi_i \lor \Phi_i') \quad (27)$$

$$\Phi_i \equiv (\boldsymbol{A} \text{ in } s) = (\boldsymbol{P} \text{ in } s) \land \delta_M(\boldsymbol{Q} \text{ in } s, \boldsymbol{X} \text{ in } s) = \langle i, \boldsymbol{X} \text{ in } t \rangle \quad (28)$$

$$\Phi_i' \equiv (\boldsymbol{A} \text{ in } s) \neq (\boldsymbol{P} \text{ in } s) \land (\boldsymbol{X} \text{ in } t) = (\boldsymbol{X} \text{ in } s) \quad (29)$$

$$\text{next}_i(t, s) \equiv \begin{cases} (\boldsymbol{P} \text{ in } t) = (\boldsymbol{P} \text{ in } s) + 1, & \text{if } i \in Q_R \\ (\boldsymbol{P} \text{ in } t) = (\boldsymbol{P} \text{ in } s) - 1, & \text{if } i \in Q_L \end{cases} \quad (30)$$

Fig. 4. Constraints Ψ

Now we give some intuition how A_Ψ works on t-graphs which possibly code computations of M. The main idea is as follows: successive vertices of a t-graph describe successive states of computation of M from the point of view of the k-th cell. The number k is coded in $\boldsymbol{A}$, and $\boldsymbol{X}$ represents the content of the cell within the computation. $\text{next}_i(t, s)$ describes changes of the position of the head, $\text{nacc}(t)$ says that the state coded in vertex t is not accepting. The expression $\text{good}_i(t, s)$ guarantees that consecutive vertices have proper values of $\boldsymbol{P}, \boldsymbol{A}, \boldsymbol{Q}, \boldsymbol{X}$: the value of $\boldsymbol{P}$ is incremented or decremented dependently of the state, the value of $\boldsymbol{A}$ is copied (since we still look at the same cell), and the value of $\boldsymbol{Q}$, which codes a state, changes according to the label of the node. The value of $\boldsymbol{X}$ changes only if the head of the machine M looks at the selected cell. These changes have to agree with the transition function of M. The expression $\text{bad}_i(t, s)$ says that i cannot be the next state of M, if the current state was coded in s.

The next lemma relates cycles of the form $f_0, f_{q_1}, \dots f_{q_n}$ with computations of M.

Lemma 9. A_Ψ *does not accept a cycle* $f_0, f_{q_1}, \dots f_{q_n}$ *where* $q_i \neq 0$ *for* $i = 1, \dots, n$, *if and only the sequence* $0, q_1, \dots, q_n$ *is valid and accepting.*

Lemmas 8, and 9 suffice to conclude that A_Ψ accepts all finite t-graphs iff M rejects u. So, thanks to Lemmas 1 and 2, M rejects u iff the system Ψ has a solution, which completes the proof of Theorem 3.

6   Future Works

It could be useful to find some other variants of set constraints for which the satisfiability problem over sets of regular terms is decidable. Particularly, it is worth to consider so called definite set constraints [HJ90] for which the satisfiability problem over the Herbrand universe is EXPTIME-complete [CP97].

References

[AKVW93] A. Aiken, D. Kozen, M. Vardi, and E. L. Wimmers, *The complexity of set constraints*, Proceedings of CSL 1993, Springer Verlag, 1993, pp. 1–17.

[AL94]   A. Aiken and T.K. Lakshman, *Directional type checking of logic programs*, Proceedings of the First International Static Analysis Symposium (Baudoin Le Charlier, ed.), Springer Verlag, 1994, pp. 43–60.

[ALW94]   A. Aiken, T.K. Lakshman, and E. Wimmers, *Soft typing with conditional types*, Proceedings of POPL 1994, ACM Press, 1994, pp. 163–173.

[AW92]   A. Aiken and E. Wimmers, *Solving systems of set constraints*, Proceedings of LICS 1992, IEEE Computer Society Press, 1992, pp. 329–340.

[BGW93]   L. Bachmair, H. Ganzinger, and U. Waldmann, *Set constraints are the monadic class*, Proceedings of LICS 1993, IEEE Computer Society Press, 1993, pp. 75–83.

[Cha99]   W. Charatonik, *Automata on dag representations of finite trees*, Tech. report, Max-Planck-Institut fr Informatik, 1999.

[Col82]   A. Colmerauer, *Prolog II reference manual and theoretical model*, Universite de la Mediterranee Aix-Marseille II, 1982.

[CP94]   W. Charatonik and L. Pacholski, *Set constraints with projections are in NEXPTIME*, FOCS 1994, IEEE Comp. Society Press, 1994, pp. 642–653.

[CP97]   W. Charatonik and A. Podelski, *Set constraints with intersection*, Proceedings of LICS 1997, IEEE Computer Society Press, 1997, pp. 362–372.

[CP98]   ———, *Co-definite set constraints*, Proceedings of RTA 98, Springer Verlag, 1998, pp. 211–225.

[CPM99]   W. Charatonik, A. Podelski, and M. Müller, *Set-based failure analysis for logic programs and concurrent constraint programs*, Proceedings of the European Symposium on Programming, Springer Verlag, 1999, pp. 177–192.

[GTT93]   R. Gilleron, S. Tison, and M. Tommasi, *Solving systems of set constraints with negated subset relationships*, Proceedings of FOCS 1993, IEEE Computer Society Press, 1993, pp. 372–380.

[HJ90]   N. Heintze and J. Jaffar, *A decision procedure for a class of set constraints*, Proceedings of LICS 1990, IEEE Computer Society Press, 1990, pp. 42–51.

[HJ94]   ———, *Set constraints and set-based analysis*, Proceedings of CP 1994, Springer Verlag, 1994, pp. 281–298.

[HU79]   J. Hopcroft and J. Ullman, *Introduction to automata theory, languages and computation*, Addison-Wesley, 1979.

[MGWK96] D. A. McAllester, R. Givan, C. Witty, and D. Kozen, *Tarskian set constraints*, LICS 1996, IEEE Computer Society Press, 1996, pp. 138–147.

[Wie03]   J. Wielemaker, *SWI-Prolog 5.1 reference manual*, 2003.

[WNS97]   M. Wallace, S. Novello, and J. Schimpf, *ECLiPSe: A Platform for Constraint Logic Programming*, Tech. report, IC-Parc, Imperial College, London, 1997.

Unsound Theorem Proving

Christopher Lynch

Department of Math and Computer Science
Clarkson University,* Potsdam, New York
clynch@clarkson.edu

Abstract. We discuss the benefits of complete unsound inference procedures for efficient methods of disproof. We give a framework for converting a sound and complete saturation-based inference procedure into successive unsound and complete procedures, that serve as successive approximations to the theory. The idea is to successively add new statements in such a way that the inference procedure will halt. Then the satisfiability is evaluated over a stronger theory. This gives an over-approximation to the given theory. We show how to successively compute better over-approximations. Similarly, a sound an incomplete theorem prover will give an under-approximation. In our framework, we succesively compute better over and under-approximations in this way.

We illustrate this framework with Knuth-Bendix Completion, and show that in some theories this method becomes a decision procedure. Then we illustrate the framework with a new method for the (nonground) word problem, based on Congruence Closure. We show a class where this becomes a decision procedure. Also, we show that this new inference system is interesting in its own right. Given a particular goal, in many cases we can halt the procedure at some point and say that all the equations for solving the goal have been generated already. This is generally not possible in Knuth-Bendix Completion.

1 Introduction

The major problem in automated theorem proving, that of deciding the unsatisfiability of a set of statements, is undecidable in general. This is true in first order logic and equational logic, for example. There exist sound and complete theorem provers. If a theorem prover is sound, then that means that when it gives a proof of a theorem, you are guaranteed that it is correct. If a theorem prover is complete, then when a conjecture is true, a proof is guaranteed to be found.

Automated theorem provers have been used to prove difficult theorems. But the search space is so large that in practice, more efficient incomplete sound theorem provers are often used. Then proofs can be trusted, but disproofs cannot.

Our focus in automated theorem proving is not in finding proofs for difficult mathematical theorems. Instead, we are interested in using theorem provers to

---

* This work was performed while the author was on sabbatical at Naval Research Lab.

J. Marcinkowski and A. Tarlecki (Eds.): CSL 2004, LNCS 3210, pp. 473–487, 2004.

solve verification problems. When used in that context, many of the conjectures given to a theorem prover will be false. So we would like to be able to trust a result that a conjecture is false. In that case, incomplete theorem provers are useless. Also, complete and sound theorem provers are generally not so efficient.

Therefore, we create a framework for unsound complete theorem provers for disproving conjectures. This technique is useful to weed out theorems which are obviously not true. There are simple examples where theorem provers run forever trying to solve conjectures that are trivially false to a human. In addition, we could use an unsound and complete theorem prover in combination with a sound and incomplete theorem prover to approximate a conjecture from both sides.

The framework given in this paper is for saturation theorem provers, which operate by continually inferring new statements implied by previous statements. Our framework consists of a modification to a saturation theorem prover. We modify it by adding statements that are not necessarily implied by previous statements. These potentially false statements are chosen in such a way as to force the theorem prover to halt. The effect of this is to evaluate a stronger conjecture. The procedure is complete, so that if the stronger conjecture is false, then the given conjecture is false. The theorem prover has approximated the theory with a stronger theory: an unsound approximation.

We iterate this process. We run the theorem prover again, but this time try to approximate the theory with a weaker theory than before In this way, we continually approximate the given theory.

In our framework, we have also shown that it is possible to create a weak approximation, and gradually attempt to make this approximation stronger. We iterate the construction of the two approximations, one strong and one weak. Anything found true in the weak one is true, and things found false in the strong one are false. In some cases this becomes a decision procedure.

There is another benefit to our framework that might not be so obvious. Suppose that we have a complete theorem prover that halts without proving the conjecture. Then that final saturated set is a disproof of the conjecture. Our unsound method is complete, so that it gives a disproof, although in a stronger theory. This has in effect abstracted properties from the conjecture and disproved the abstraction. We argue that in this abstraction the disproofs will generally be shorter and easier to understand than disproofs in the original theory.

For examples, we instantiate this framework with two concrete inference systems. First is Knuth-Bendix Completion [3], where we show a class of theories where this becomes a decision procedure. However, the direct purpose of this paper is not in finding new decision procedures; that is only an example of the kinds of things that can be done within this framework.

The second inference system which we use to instantiate this framework is a new inference procedure, as far as we are aware. It is based on Abstract Congruence Closure, for ground equational theories [4, 6] which we extend to non-ground theories. For this inference system, we show that it is sometimes possible to examine the set of equations during the derivation and deduce that the conjecture can never be proved from this point. This is generally impossible to do in

theorem proving, because even if the equations become large, it is always possible
that an equation can be simplified to a smaller one.

In Section 2, we introduce theorem proving derivations, and give the framework for unsound theorem proving. In Section 3, we instantiate the framework
with Knuth-Bendix Completion. In Section 4, we introduce Nonground Congruence Closure and instantiate the framework with that. We conclude the paper
with a discussion of how to apply this method to Resolution and Paramodulation inference systems, and a comparison with related work. This paper does not
contain any of the proofs. All of the proofs and all of the technical details can
be found at www.clarkson.edu/~clynch/papers/uf.ps/

2   Framework

Basic definitions of Theorem Proving Derivations are from [2, 9]. A *saturation*
inference system is an inference system that starts with some set of statements,
and uses transformation rules to create new statements and delete old ones.
Transformation rules are of the form $\Gamma \longrightarrow \Delta$, where Γ and Δ are both sets
of statements. The meaning of a transformation rule is that the statements in
Γ should be replaced by the statements of Δ. There are two kinds of transformation rules: inference rules and deletion rules. *Inference rules* are of the
form $\{C_1, \cdots, C_n\} \longrightarrow \{C_1, \cdots, C_n, C\}$. It indicates that in the presence of
$C_1, \cdots, C_n$, C should be added. We will write that inference rule in the following
notation:

$$\frac{C_1 \cdots C_n}{C}$$

Deletion rules will be of the form $\{C_1, \cdots, C_n\} \longrightarrow \{C_2, \cdots, C_n, D_1, \cdots, D_m\}$.
This means that if the statements $C_1, \cdots, C_n$ exist in the current set of state-
ments, then C_1 should be deleted and $D_1, \cdots, D_m$ added. Inference rules represent rules that **must** be performed in an inference procedure, and deletion rules
may be performed if desired.

Given a set of inference rules I and deletion rules D, an (I,D) *theorem proving
derivation* is a (possibly infinite) sequence $S_1, S_2, \cdots$ of sets of statements such
that each S_{i+1} is obtained by applying an inference rule from I or a deletion
rule from D to clauses of S_i. We define $S_\infty = \bigcup_{i \geq 1} \bigcap_{j \geq i} S_j$ The clauses in
S_∞ represent the set of *persistent* statements, i.e., the statements that are never
deleted after i for some i. Given a set of inference and deletion rules, since we
assume they are applied according to some strategy, we can assume there is one
theorem proving derivation for each set of statements.

We assume a well-founded ordering $<$ on the statements. Based on that
ordering, there is a notion of redundancy. A statement C is *redundant* in S
if there are statements $C_1, \cdots, C_n \in S$ such that each $C_i < C$ for all i, and
$C_1, \cdots, C_n \models C$. We will construct the deletion rules so that they cannot be
performed unless C_1 is redundant in $\{C_2, \cdots, C_n, D_1 \cdots D_m\}$. A set of statements
S is said to be *saturated* if the conclusion of every inference from S is either in S

or is redundant in S. A theorem proving derivation $S_1, S_2, \cdots$ is *fair* if for every inference from S_∞ with conclusion C, there exists an i such that $C \in S_i$ or C is redundant in S_i. If $S_1, S_2, \cdots$ is fair, then S_∞ is saturated.

The inference rules are *sound* if $C_1, \cdots, C_n \models C$. The deletion rules are *sound* if $C_1, \cdots, C_n \models D_i$ for all i. Inference rules are designed so that each saturated set has certain properties. The most common is the refutational property. In that case, we distinguish a new atom called $\bot$, usually called the empty clause, which indicates that a set of statements is unsatisfiable. We have the important definitions of *soundness* and *completeness* of an inference system.

Definition 1. *A set of inference rules I and deletion rules D is* sound *if for every fair theorem proving derivation $S_1, S_2, \cdots$, if $\bot \in S_\infty$ then S_1 is unsatisfiable. I is* complete *if $\bot \in S$ for every saturated and unsatisfiable S.*

It is obvious that a set of inference and deletion rules is sound if each individual inference and deletion rule is sound.

From the definitions of soundness and completeness, we see that If we can show that a sound set of inference rules produces $\bot$ from S, then S is unsatisfiable. Analogously, if we can show that a complete set of inference rules does not produce $\bot$ from S then S is satisfiable. Exhibiting the entire theorem proving derivation gives us a proof of unsatisfiability (or satisfiability) in their respective cases. Obviously, this is only possible if the theorem proving derivation is finite.

Smaller proofs are preferable. For a proof of unsatisfiability, it is only necessary to give the set of statements and inference and deletion rules that leads up to $\bot$. Therefore, it is necessary to save histories throughout the theorem proving process, or have some way of reconstructing them.

For a proof of satisfiability, we may be able to dispense with the history altogether. In that case, it is enough to exhibit just S_1 and S_∞ (if it is finite), as long as we can prove that (i) $\bot \notin S_\infty$, (ii) S_∞ is saturated, and (iii) $S_\infty \models S_1$. The first fact is trivial to show. The second fact can be easily shown for most redundancy rules used in practice by examining all the inferences from S_∞. The third fact is usually easy to determine, since S_∞ is saturated, but that depends on the inference system. For the examples in this paper, it will be easy.

Ideally, a theorem prover should be sound and complete. Then we are guaranteed that the existence or non-existence of $\bot$ determines whether a set of statements is satisfiable or not. Of course, the problem of theorem proving is, in general, undecidable for first order logic. So, in practice, theorem provers that have proved important results are not always complete. For example, the Robbins Algebra problem was proved with an incomplete theorem prover [8].

Throughout the history of automated theorem proving, until very recently, much of the emphasis has been on solving very hard theorems. A theorem proving contest is run every year at the CADE conference, with the main emphasis on proving unsatisfiability. In the past, theorem prover developers have come up with sound and incomplete methods that prove some theorems where some complete methods fail. A simple example of a sound and incomplete strategy is that strategy which discards every clause with more than a given number of symbols.

2.1   Modifying Sound and Complete Derivations

In this paper, we are interested in satisfiability. Therefore, we will develop strategies that destroy soundness but do not destroy completeness. Soundness of an inference system is implied by soundness of the inference and deletion rules. We will relax that requirement. In particular, we will allow unsound deletion rules, while still requiring the inference rules to be sound. We will keep the requirement that deletion rules only remove redundant statements. Therefore, the inference systems we consider will still be complete, but not sound.

In the rest of this section, five different ideas will be discussed. First is the idea of *Unsound Theorem Proving*. That is the idea of modifying a sound and complete theorem proving procedure so that it may be unsound, but that it remains complete and it terminates, so that it can decide satisfiability in some cases. The second well-known ideas is *Incomplete Theorem Proving* which modifies a sound complete theorem proving procedure so that it may become incomplete, but it remains sound and it terminates. This procedure can show unsatisfiability but not satisfiability. The third idea is *Iterative Unsound Theorem Proving*. This iterates Unsound Theorem Proving, with the goal of becoming more and more sound each time, thereby proving the satisfiability of more statements. The third idea is *Iterative Incomplete Theorem Proving* which iterates Incomplete Theorem Proving, with the goal of becoming more complete each time. The final idea is *Iterative Unsound and Incomplete Theorem Proving*, which simultaneously iterates Unsound and Incomplete Theorem Proving.

The idea of Unsound Theorem Proving is presented now. After each inference rule is performed, we look at the conclusion. In some cases, we will keep the conclusion. In other cases, we will perform a deletion rule where the statements $D_1, \cdots, D_m$ might not follow from previous statements. Therefore, the inference rules remain sound, but the deletion rules do not. This gives us an unsound but complete finite theorem proving derivation. We will assume that all the D_i come from a predetermined finite set F, so it will prove whether or not a larger set of statements is unsatisfiable. If this larger set is satisfiable, the original set is satisfiable. Since F is finite, this procedure must halt.

For Incomplete Theorem Proving, we also look at the conclusion of each inference. If the conclusion is not in F, then we do not add it. Therefore, we lose completeness. But since F is finite, the procedure terminates.

For Iterative Unsound Theorem Proving, we run the Unsound Theorem Proving Procedure. If this returns "unsatisfiable" to us, then we cannot be sure that the answer is correct, because of unsoundness. So we repeat the procedure with a larger set F. And this process is iterated. Iterative Incomplete Theorem Proving is similar, except that in this case we cannot trust a result of "satisfiable", so in that case we iterative Incomplete Theorem Proving for a larger value of F.

Iterative Unsound and Incomplete Theorem Proving is a combination of the two processes. We choose an F, then run the Unsound Theorem Proving procedure for that F. If it returns "unsatisfiable", we run the Incomplete Theorem Proving procedure for the same F. If that returns "satisfiable" we choose a larger value of F and iterate.

Define an F-replacement deletion rule as follows:

Definition 2. *Let F be a set of statements. A deletion rule $\{C, D_1, \cdots, D_m\} \longrightarrow \{D_1, \cdots, D_m\}$ is an F-replacement if $C \notin F$ and $D_1, \cdots, D_m \in F$.*

Non-F replacement rules will always be designed so that every clause not in F is the premise of a non-F replacement rule.

If I is a set of inference rules, and D is a set of deletion rules containing an F-replacement rule, and F is a finite set, then every (I, D) derivation is finite, because only statements from F are saved. The key idea of this paper is that a complete set of inference and deletion rules can be augmented with an F-replacement rule, so that any derivation from the augmented set of rules will halt, and if $\perp$ is not generated from S then S is satisfiable.

We describe the Iterated Unsound and Incomplete Theorem Proving Process. Let S be the set of statements for which we want to decide satisfiability. Let I and D be a sound and complete set of inference and deletion rules. Let $\hat{F} = F_1, F_2, \cdots$ be a monotonic sequence of finite sets of statements, i.e., $F_k \subseteq F_{k+1}$ for all k. We define $F_\infty = \bigcup_{k \geq 1} F_k$. For each k let I_k be a modification of I such that all inferences with a conclusion not in F_k are not performed. For each k, let D_k be D augmented with an F_k replacement rule. Then the $(I, D, \hat{F})$ *derivation from* S is the following:

1. Let $k = 1$
2. Let S_∞ be an (I, D_k) saturation of S.
3. Let T_∞ be an (I_k, D) saturation of S.
4. If $\perp \notin S_\infty$, halt and say SATISFIABLE.
5. If $\perp \in T_\infty$, halt and say UNSATISFIABLE.
6. Let $k = k + 1$
7. Go to 2

Note that steps 2 and 3 can be computed in finite time, since each F_k is finite. We will show that this process is sound. In fact we extend the definition of soundness to say that if $\perp$ is not produced or if the function returns SATISFIABLE, then S is satisfiable. The process is complete if F_∞ contains all statements.

Theorem 1. *Let I and D be a sound and complete set of inference and deletion rules. Let $\hat{F}$ be a monotonic sequence of finite sets of statements. Suppose that every (I, D) derivation from $S \subseteq F_\infty$ only produces statements in F_∞[1]. Let $S \subseteq F_\infty$. Then if the $(I, D, \hat{F})$ derivation returns SATISFIABLE (resp. UNSATISFIABLE) then S is satisfiable (resp. unsatisfiable). Also, if S is unsatisfiable, then the $(I, D, \hat{F})$ derivation returns UNSATISFIABLE.*

The proof is due to the fact that each (I, D_k) derivation is finite and complete, and each (I_k, D) derivation is finite and sound. Recall that deletion rules only remove redundant statements.

---

[1] This is true, for example, if F_∞ contains all statements.

We often consider F_∞ to be the set of all statements, then it is trivially the case that all derivations from $S \subseteq F_\infty$ only contain statements in F_∞.

We point out some of the benefits of this procedure over the (I, D) procedure. First of all, if a set of statements is satisfiable, then this procedure is more likely to give an answer. There are simple cases where sound and complete derivations will not halt. We give some examples later in the paper. Also, suppose that we have a set of satisfiable statements S, for which the (I, D) derivation is finite. It still might be better to use an $(I, D, \hat{F})$ derivation, because the proof of satisfiability might be simpler. A stronger theory may have a smaller saturated set, and therefore a smaller proof.

The $(I, D, \hat{F})$ derivation might actually become a decision procedure. We give some examples later in the paper. An example of this is when for every satisfiable set of statements S, there is a k such that $\bot$ is not in the (I, D_k) derivation.

Theorem 2. *Let (I, D) be sound and complete. Let $\hat{F}$ be a monotonic sequence of finite sets of statements such that all derivations from $S \subseteq F_\infty$ only contain statements in F_∞. Let $\hat{G}$ be a sequence such that all $G_k \subseteq F_k$. Suppose that for every satisfiable $S \subseteq G_\infty$, there is a k such that $\bot$ is not in the (I, D_k) derivation from S. Then the $(I, D, \hat{F})$ procedure is a decision procedure for all $S \in G_\infty$.*

It is not necessary to know that the $(I, D, \hat{F})$ procedure is a decision procedure in order for it to be one, whereas it is necessary to know in advance if the (I_k, D) derivation is a decision procedure in order for it to be one. We will discuss that issue further in the next section.

These ideas can also be applied to existential problems, i.e., unification problems. In that case, the (I, D_k) and the (I_k, D) derivation both produce a complete set of unifiers. One is an over-approximation, and one is an under-approximation. This could be a useful way to approximate unification.

3   Knuth-Bendix Completion

We have presented an abstract framework for using unsound theorem proving to determine satisfiability, and develop decision procedures. But that framework is not useful unless some interesting examples fit into the framework. In particular, what are the F_k in the sequence $F_1, F_2, \cdots$, and even more important, what are the values of the D_k used in the F_k replacement deletion rules.

Next we extend this framework to Knuth-Bendix Completion [3], which is an inference system over equations $s \approx t$ and disequations $s \not\approx t$. Completion consists of Inference rules **Critical Pair**, **Narrowing** and **Equation Resolution**, plus Deletion rule **Simplification**. Now we will apply our framework to Knuth-Bendix Completion. First we define the sequence $F_1, F_2, \cdots$.

Definition 3. *Given a term t, let $|t|$ be the number of non-variable symbols in t. Then F_k is the set of equations $s \approx t$ and disequations $s \not\approx t$ such that $|s| \leq k$ and $|t| \leq k$.*

Clearly F_∞ is the set of all equations and disequations. Another possibility for F_k is to let k be a limit on the depth of the terms in F_k. The F_k replacement deletion rule is to add an equation that subsumes one not in F_k:

Definition 4. *The* Unsound Subsumption *deletion rule is the rule* $\{A\} \longrightarrow \{A'\}$ *where* $A'\sigma = A$ *for some* σ, $A \notin F_k$ *and* $A' \in F_k$.

This replaces an equation or disequation A, with A' where A' is a new equation or disequation that strictly subsumes A, and $A' \in F_k$. It is easy to find such an A'. It is just necessary to replace subterms in A with variables. The best idea is to replace as few subterms as possible, so that A' is in F_k but not in F_{k-1}.

Next we show how the $(I, D, \hat{F})$ derivation becomes a decision procedure for some theories (sets of equations). For example, consider the theories E we will call *size preserving linear theories*:

Definition 5. *Let* E *be a set of equations. Then* E *is* size preserving linear *if and only if for every* $s \approx t \in E$, $|s| = |t|$ *and each variable that occurs in* $s \approx t$ *occurs exactly once in* s *and once in* t.

We prove that this property is preserved by inferences and deletions.

Lemma 1. *Any Knuth-Bendix Completion inference or deletion among size preserving linear equations results in a size preserving linear equation.*

We also show that inferences among size preserving linear equations can never result in a smaller equation. We actually prove something more general than that, because we will need it later. Let's define $|s \approx t| = min\{|s|, |t|\}$.

Lemma 2. *Let* n *be a positive integer. Let* e_1 *and* e_2 *be equations such that* $|e_1| \geq n$, *and either* $|e_2| \geq n$ *or* e_2 *is size preserving linear. Let* e_3 *be the conclusion of a Knuth Bendix Completion inference or deletion rule among* e_1 *and* e_2. *Then* $|e_3| \geq n$.

We use these results to get a decision procedure.

Definition 6. *For all* k, *define* G_k *to be the set of size preserving linear equations in* F_k, *and all ground disequations in* F_k.

The following theorem is implied by the fact that once an equation e outside of G_k is created, then any rule with e as one of its premises will have a conclusion that is not in G_k.

Theorem 3. *Let* (I, D) *be the inference and deletion Rules of KB Completion. Then* (I_k, D) *is a decision procedure for all members of* G_k.

There is a similar theorem for unsound derivations.

Theorem 4. *Let* I *and* D *be the Inference and Deletion Rules of KB Completion. Then* $(I, D, \hat{F})$ *with Unsound Subsumption is a decision procedure for all* S *in* G_∞.

As we did in the framework, we once again point out the distinction between those two theorems. For Theorem 3, it is necessary to know in advance that (I_k, D) will be a decision procedure. But for Theorem 4, $(I, D, \hat{F})$ is a decision procedure, and it is not necessary to know that in order for it to be one. For example, $(I, D, \hat{F})$ is a decision procedure for the theory $\{f(f(x)) \approx g(f(x)), h(a) \approx b\}$, whereas (I_k, D) would require a new theorem in order to turn it into a decision procedure. It is worth pointing out that Knuth-Bendix Completion will normally not halt for many size preserving linear theories, such as $\{f(f(x)) \approx g(f(x))\}$.

4   Nonground Congruence Closure

Now we extend the Abstract Congruence Closure algorithm of [6, 4] to equations with variables. That algorithm works by creating new constants representing equivalence classes of terms. In our approach, we create new function symbols in addition to new constants. The function symbols when applied to terms represent equivalence classes. So the function symbol itself represents a parametrized equivalence class. We apply the Knuth-Bendix procedure to the flattened equations. The result might not be flat. Therefore, we flatten the conclusion of the inference, and create a new function symbol. This process can go on forever, but it is still complete, if we order the new (possibly infinitely many) symbols in a well-founded way.

There are some advantages of this approach over Knuth-Bendix Completion. Equations are kept small, and inferences are easy to perform. The ordering used is trivial to calculate on flat terms. Also, rewrite chains are polynomial in the number of equations.

Unfortunately, this procedure may not halt on sets of equations where Knuth-Bendix Completion halts. However, when we apply unsound theorem proving to this method, it appears to have advantages over Knuth-Bendix Completion. In many instances of traditional theorem proving, it is possible to tell that if there was a proof we would have found it already. This corresponds to Knuth-Bendix Completion being able to determine that all future equations will be larger than a given size. As far as we know, there is no way to do that in Knuth-Bendix Completion, aside from coming up with some meta-theorem, as in the previous section, or using unsound theorem proving. In the Congruence Closure method, unsound theorem proving is not even necessary. Although we present it to strengthen this approach, and to handle additional classes of equations.

4.1   Nonground Congruence Closure

We will now define Nonground Congruence Closure. First, define the *height* $Ht(t)$ of a term t such that $Ht(x) = 0$ for all variables x, and $Ht(f(t_1, \cdots, t_n)) = 1 + max\{Ht(t_1), \cdots, Ht(t_n)\}$. Let $Vars(t)$ be the set of all variables occuring in t. The *depth* of $x \in Vars(t)$ is defined as $depth(x, x)=0$ and $depth(x, f(t_1, \cdots, t_n))= 1 + max\{depth(x, t_i, \| \ x \in Vars(t_i)\}$. Let $root(t)$ be the top symbol of t. We define *flat equations*. All equations can be flattened.

Definition 7. *An equation $s = t$ is* flat *if (i) $Ht(s) \leq 2$ and $Ht(t) \leq 1$, (ii) $Vars(t) \subseteq Vars(s)$ and t is linear[2], and (iii) if $depth(x, s) = 2$ then x only occurs once in s.*

When restricted to ground terms, this definition is the same as the usual definition of flat ground equations. Note that a term of height 1 is of the form $f(x_1, \cdots, x_n)$, where $n \geq 0$. All equations can be flattened.

We are going to consider a signature Σ, and an infinite set of new function symbols $C = \{c_1, c_2, \cdots\}$. Let $\Sigma_C = \Sigma \cup C$ be an extended signature.

Ordering: We assume a total precedence $<_p$ on the symbols, with the requirement that $arity(f) < arity(g)$ implies that $f <_p g$. Furthermore if $i < j$ and $arity(c_i) = arity(c_j)$ then $c_i <_p c_j$. This last fact guarantees that the precedence order is well-founded. From the precedence ordering, we can define an ordering $<_f$ on ground terms.

Definition 8. *Let $s = f(s_1, \cdots, s_n)$ and $t = g(t_1, \cdots, t_m)$. Then $s >_f t$ if*

1. *$|s| > |t|$, or*
2. *$|s| = |t|$ and $f >_p g$, or*
3. *$|s| = |t|$ and $f = g$ and $\{s_1, \cdots, s_n\} >_f \{t_1, \cdots, t_n\}$[3].*

The relation $<_f$ is really a well-founded monotonic ordering, and it is simple to compute on flat equations.

Theorem 5. *The $<_f$ order is well-founded, irreflexive, transitive and monotonic on ground terms.*

This ordering can be extended to a total ordering on ground terms by comparing sets of subterms lexicographically in case 3 of the definition of the order. We extend $<_f$ to nonground terms as usual: $s >_f t$ if and only if $s\sigma >_f t\sigma$ for all ground substitutions σ.

For flat terms, it is especially simple to compare sides of flat equations in the $<_f$ ordering. We simply compare the height of the terms. If the heights are equal, then we compare the precedence of the root symbols.

Theorem 6. *Let $s \approx t$ be a flat equation. Then $s >_f t$ if and only if (i) $Ht(s) > Ht(t)$ or (ii) $Ht(s) = Ht(t)$ and $root(s) >_p root(t)$.*

Note that all for all flat equations $s \approx t$, s and t can be compared using the $<_f$ ordering, except for equations of the form $f(x_1, \cdots, x_n) \approx f(y_1, \cdots, y_n)$ where $x_1, \cdots, x_n$ is a permutation of $y_1, \cdots, y_n$.

Flattening. Now we define how to flatten a set of equations E using the following rules. The resulting flattened set will be called $flat(E)$.

---

[2] Each variable occurs at most once in t.
[3] Here we mean the multiset extension of $>_f$.

1. If $u[s] \approx v \in E$ and $Ht(s) = 2$, and either $Ht(u[s]) > 2$ or $Ht(v) > 1$, then replace $u[s] \approx v$ with $u[c(x_1, \cdots, x_n)] \approx v$ and $s \approx c(x_1, \cdots, x_n)$, where $\{x_1, \cdots, x_n\} = Vars(s)$, and c is a new function symbol from C.
2. If $u \approx v \in E$ and $Ht(u) \leq 2$ and $Ht(v) = 1$ and both of the following hold:
   (a) u is not linear or $Vars(u) \not\subseteq Vars(v)$ or $Ht(u) = 2$, and
   (b) v is not linear or $Vars(v) \not\subseteq Vars(u)$
   then replace $u \approx v$ with $u \approx c(x_1, \cdots, x_n)$ and $v \approx c(x_1, \cdots, x_n)$, where $\{x_1, \cdots, x_n\} = Vars(u) \cap Vars(v)$, and c is a new function symbol from C.
3. Suppose that $u \approx v \in E$ such that $Ht(u) = 2$, $Ht(v) < 2$ and v is linear and $Vars(v) \subseteq Vars(u)$ and there is some variable x such that $depth(x, u) = 2$ and $occur(x, u) \geq 2$.

   Then let u' be the term formed from u by, for all $y \in Vars(u)$, replacing the occurrences of y with new variables $y_1, \cdots, y_{occur(y,u)}$. Define a substitution γ such that, for all y_j created in the previous sentence, $y_j \gamma = y$. Let Z be the set of all all the new y_j created. Obviously Z is the domain of γ, and $Vars(u)$ is the range. Let $z_1, \cdots, z_m$ be an enumeration of the members of Z.

   Finally, replace $u \approx v$ with $u' \approx c(z_1, \cdots, z_m)$ and $c(z_1\gamma, \cdots, z_m\gamma) \approx v$.

This procedure will always halt in a flat set of equations which is a conservative extension of the original set.

Lemma 3. *For every E, the flattening procedure halts on E, $flat(E)$ is flat, and $flat(E)$ is a conservative extension of E.*

The inference and deletion rules for Congruence Closure are the same as the inference and deletion rules for Knuth-Bendix Completion, with the addition of one flattening deletion rule that will be performed once after a non-flat equation is created by an inference or deletion rule.

Flattening:

$$\frac{u \approx v}{u \approx c(x_1, \cdots, x_n) \quad v \approx c(x_1, \cdots, x_n)}$$

where $u \approx v$ is not flat[4], $\{x_1, \cdots, x_n\} = Vars(u) \cap Vars(v)$, and c is a new function symbol from C.

Note that the result of a Critical Pair or Simplification Rule will be an equation $u \approx v$ such that $Ht(u) \leq 2$ and $Ht(v) \leq 2$.

Lemma 4. *If e_1 and e_2 are flat equations, then any inference or deletion rule among e_1 and e_2 will result in an equation $u \approx v$ such that $Ht(u) \leq 2$, $Ht(v) \leq 2$, and u contains no variable x such that $depth(x, u) = 2$ and $occur(x, u) \geq 2$, and v contains no variable x such that $depth(x, v) = 2$ and $occur(x, v) \geq 2$*

Therefore, the conclusion of Flattening will always be a flat equation, since $c(x_1, \cdots, x_n)$ is linear, and all of its variables also appear in u and v.

---

[4] For instance, because $Ht(u) = Ht(v) = 2$.

Compare equations by defining $s \approx t <_f u \approx v$ if $\{s, t\} <_f \{u, v\}$, where $<_f$ is its own multiset extension. These two new equations are smaller than the one it is replacing. For example, $c_1(x_1, \cdots, x_n) <_f u$, because $Vars(c_1(x_1, \cdots, x_n)) \subseteq Vars(u)$, $c_1(x_1, \cdots, x_n)$ is linear, and

1. $Ht(u) = 2$, or
2. $Ht(u) = 1$ and there is a variable in u that is not in $c_1(x_1, \cdots, x_n)$, so $arity(c) < arity(root(u))$, or
3. $Ht(u) = 1$ and u is not linear, so $arity(c) < arity(root(u))$.

Since the two new equations imply the replaced equation, and because the new equations are smaller, this is an instance of removing a redundant equation. Therefore, the Congruence Closure inference system is sound and complete.

Theorem 7. *The Nonground Congruence Closure inference and deletion rules are sound and complete.*

4.2   Fitting into Framework

Now that the Nonground Congruence Closure inference procedure is defined, we fit it into our framework. We have defined the inference rules so that all equations are flat, but the disequations are not necessarily flat. It would also be possible to flatten the disequations, but we chose not to approach it that way.

For unsound theorem proving, we need to define a sequence $F_1, F_2, \cdots$.

Definition 9. *Let F_k be the set of equations and disequations such that for all $s \not\approx t$ in F_k, $|s| \leq k$ and $|t| \leq k$, and the only function symbols that can appear in F_k are the function symbols of $\Sigma \cup \{c_1, \cdots, c_k\}$.*

Then the F_k replacement rule is the *Combine Equivalence Class* deletion rule.

Definition 10. Combine Equivalence Classes *is the deletion rule* $\{u = c_j(x_1, \cdots, x_n)\} \longrightarrow u = c_i(y_1, \cdots, y_m)\}$, *where* $j > k$, $i \leq k$, $arity(c_i) \leq arity(c_j)$[5]*, and* $\{y_1, \cdots, y_m\} \subseteq \{x_1, \cdots, x_n\}$.

Notice that each F_k is finite, and that the Combine Equivalence Classes rule will replace a term not in F_k with a term in F_k, assuming that A is not a disequation with too many symbols on one side. But we will use this rule in inferences where such rules are never created. The Combine Equivalence Classes rule has the effect of preventing the inference procedure from creating new function symbols at some point, which creates an unsound, complete inference procedure.

Given an equation $s \approx t \in E$, we sometimes write it as $s \rightarrow t$ if $s >_f t$. Then $\rightarrow$ represents the rewrite relation, and $\rightarrow^*$ represents its reflexive and transitive closure. We will define a size function called *minsize* on all symbols and all terms, with respect to a set of equations. The intention will be that $minsize(t, E) = min\{|s| \mid s \in T_\Sigma \text{ and } s \rightarrow^* t\}$, where $minsize(t) = |t|$ for all $t \in T_\sigma$.

---

[5] We can assume some initial constant c_i, or set of constants with small arity, if necessary so that this is always posssible.

Definition 11. *Let E be a set of equations. If x is a variable, then $minsize(x,E)=$ 0. Define $minsize(f(t_1, \cdots, t_n), E) = 1 + \Sigma_{1 \leq i \leq n} minsize(t_i, E)$. Define minsize$(c, E) = min\{minsize(t, E) \mid t \rightarrow c(x_1, \cdots, x_n) \in E\}$.*

In the long version, we show that for every term t, $minsize$ has some value, and $minsize(t, E)$ is the size of the smallest term in T_Σ which rewrites to t.

For any term t, define $maxsym(t, E) = max\{minsize(c, E) \mid c$ is a symbol in $t\}$, and define $maxsym(s \approx t, E) = min\{maxsym(s, E), maxsym(t, E)\}$. If $u \geq_f v$, then define an equation $u \approx v \in E$ to be *expanding* if $maxsym(u, E) \leq maxsym(v, E)$ and every variable of u occurs in v.

Lemma 5. *Let e_1 and e_2 be flat equations where $maxsym(e_1, E) \geq n$ and either e_2 is expanding or $maxsym(e_2, E) \geq n$. If e_3 is the conclusion of an inference between e_1 and e_2 then $maxsym(e_3, E) \geq n$.*

This theorem can be used to show that the traditional and unsound decision procedures can become decision procedures in some cases.

Lemma 6. *Let n be a number and $u \not\approx v$ be a ground disequation in T_Σ such that $|u| \leq n$ and $|v| \leq n$. Let S be a set of equations appearing in a Theorem Proving derivation from some subset of T_Σ. Let $S_n = \{s \approx t \in S \mid maxsym(s) \leq n$ and $maxsym(t) \leq n\}$. Suppose that S_n is saturated under the Nonground Congruence Closure rules, and all equations in S_n are expanding. Then $S_n \cup \{u \not\approx v\}$ is unsatisfiable if and only if $S \cup \{u \not\approx v\}$ is unsatisfiable.*

The lemma follows from the fact that for an expanding set of equations, once an equation $s \approx t$ appears with $maxsym(s) > n$ or $maxsym(t) > n$, then any descendent of that equation will also have that property.

Suppose that we are trying to prove the unsatisfiability of a set or equations and disequations. And supppose that at some point of the theorem proving derivation, we have saturated all equations of the form $s \approx t$ with $maxsym(s) \leq n$ and $maxsym(t) \leq n$. If all such equations are expanding, then (I_n, D) is a decision procedure for the word problem for any equation $u \approx v$ with $|u| \leq n$ and $|v| \leq n$. Furthermore, the $(I, D_n, \hat{F})$ procedure will be a decision procedure, even though we may not know that it is.

If we can show that some set of equations S will only create expanding equations in the saturation, then the $(I, D, \hat{F})$ procedure is a decision procedure for S. For example, we can show that it forms a decision procedure for size preserving linear theories.

Theorem 8. *The $(I, D, \hat{F})$ procedure is a decsision procedure for size preserving linear theories.*

Finally, we consider another interesting theory, where Knuth-Bendix Completion does not halt, but it is not size preserving. The theory is $\{f(g(f(x))) \approx g(f(x))\}$. If we flatten this theory, we get equations $g(f(x))=c_1(x)$ and $f(c_1(x))= c_1(x)$ (assuming a simplification). There is one inference that can be done on these two equations. Its result adds the two equations $g(c_1(x)) = c_2(x)$ and

$c_1(c_1(x)) = c_2(x)$. If we could continue this process infinitely, then for all i and j, we get $f(c_i(x)) = c_i(x)$, $g(c_i(x)) = c_{i+1}(x)$ and $c_i(c_j(x)) = c_{i+j}(x)$. Notice that $minsize(g, E) = minsize(f, E) = 1$, and $minsize(c_i, E) = i + 1$ for all i. All of the rules in the infinite saturated set are expanding, so both the unsound and traditional method will give us a decision procedure.

5   Conclusion

We have discussed the benefits of unsound theorem proving, for disproving conjectures. It can often find disproofs when traditional methods do not.

We gave a framework for unsound and complete theorem proving, which amounts to proof in a stronger theory, which can be decided. We discussed how to iterate the process to attempt to find weaker and weaker approximations, and we showed how this can be combined with a sound and incomplete theorem prover, and how to iterate them both to continually attempt to refine approximation from both sides.

We instantiated our framework with Knuth-Bendix Completion and a non-ground Congruence Closure method, based on ground Congruence Closure methods [6, 4]. Our Nonground Congruence Closure is new, as far as we know. However, it is in the same spirit as what is done in [10], which also uses the Knuth-Bendix inference rules followed by eaqer splitting of equations introducing new constant symbols, and it also has an arity-compatible precedence. The inference system of [10] was shown to terminate for standard theories. A difference is that we allow depth-2 linear variables to appear at depth one on the right hand side of rules. This means that we can capture all equational theories, but of course it makes theorem proving undecidable. We gave some evidence to indicate that our Nonground Congruence Closure be especially powerful in combination with unsound theorem proving.

We did not discuss how to instantiate the framework with clausal theorem proving methods like Resolution and Paramodulation. However, we can quickly suggest a method for unsound deletion. In the paper, we have shown how to prevent terms from becoming too large. For clauses, we must also prevent them from becoming too long. A simple method to do that is to delete some literals when a clause gets too long. There may be other more sophisticated and interesting methods.

Our work can be compared with other approximation methods. For example, [1] shows how to disprove false conjectures by translating them into second-order monadic logic. This is an unsound approximation in the same sense as our paper. In [7], an efficient approximation of E-unification is given by modifying a goal-directed inference method. Those two papers give a single approximation, using a completely different method than ours. Many goal directed theorem proving procedures and constraint solving methods could be thought of as successive unsound approximations. The paper of [5] is close in spirit to our paper. It discusses how to get successive approximations by converting first order clauses into ground clauses, and then applying a satisfiability test. When a ground solution is found, it must be verified for soundness. It also discusses other approximations besides ground clauses. We are not aware of other work besides ours which successively modifies a saturation procedure to produce strong models.

6   Acknowledgments

I would like to thank Robert Nieuwenhuis and many anonymous referees for useful suggestions on this paper.

References

1. Serge Autexier and Carsten Schurmann. Disproving false conjectures. In *LPAR*, volume 2850 of *Lecture Notes in Computer Science*, pages 33–48. Springer, September 2003.
2. L. Bachmair and H. Ganzinger. Resolution theorem proving. In A. Robinson and A. Voronkov, editors, *Handbook of Automated Reasoning*, volume I, chapter 2, pages 19–99. Elsevier Science, 2001.
3. Leo Bachmair, Nachum Dershowitz, and David Plaisted. *Completion without Failure*, volume II. Academic Press, 1989.
4. Leo Bachmair and Ashish Tiwari. Abstract congruence closure and specializations. In David McAllester, editor, *Automated Deduction — CADE-17*, volume 1831 of *Lecture Notes in Artificial Intelligence*, pages 64–78, Pittsburgh, PA, jun 2000. Springer-Verlag.
5. Harald Ganzinger and Konstantin Korovin. New directions in instantiation-based theorem proving. In *IEEE Symposium on Logic in Computer Science*, pages 55–64, Ottawa, Ont., jun 2003. IEEE.
6. Deepak Kapur. Shostaks congruence closure as completion. In *International Conference on Rewriting Techniques and Applications*, volume 1232 of *LNCS*, pages 23–37, Bacelona Spain, jun 1997. Springer-Verlag.
7. Christopher Lynch and Barbara Morawska. Approximating e-unification. In *15th Annual Workshop on Unification Theory*, Siena, Italy, 2001.
8. William McCune. Solution of the robbins problem. *Journal of Automated Reasoning*, 19(3):263–276, 1997.
9. R. Nieuwenhuis and A. Rubio. Paramodulation-based theorem proving. In A. Robinson and A. Voronkov, editors, *Handbook of Automated Reasoning*, volume I, chapter 7, pages 371–443. Elsevier Science, 2001.
10. Robert Nieuwenhuis. Complexity analysis by basic paramodulation. *Information and Computation*, 147:1–21, 1998.

A Space Efficient Implementation of a Tableau Calculus for a Logic with a Constructive Negation

Alessandro Avellone[1], Camillo Fiorentini[2],
Guido Fiorino[1], and Ugo Moscato[1]

[1] Dipartimento di Metodi Quantitativi per le Scienze Economiche Aziendali,
Università Milano-Bicocca, piazza dell'Ateneo Nuovo, 1, 20126 Milano
{alessandro.avellone, guido.fiorino, ugo.moscato}@unimib.it
[2] Dipartimento di Scienze dell'Informazione,
Università degli Studi di Milano, via Comelico, 39, 20135 Milano, Italy
fiorenti@dsi.unimi.it

Abstract. A tableau calculus for a logic with constructive negation and an implementation of the related decision procedure is presented. This logic is an extension of Nelson logic and it has been used in the framework of program verification and timing analysis of combinatorial circuits. The decision procedure is tailored to shrink the search space of proofs and it is proved correct by using a semantical technique. It has been implemented in C++ language.

1 Introduction

Since the works of Nelson [11] and Thomason [14], logics with *constructive nega-tion* ($\sim$) have been deeply investigated in the literature. Unlike intuitionistic negation ($\neg$), where $\neg A$ is understood as "A implies falsehood", the meaning of $\sim A$ is defined according to the structure of A, where the notion of falsity of atomic formula is as primitive as the concept of its truth (for a thorough discussion about constructive negation see [17]). Nelson logic $\mathbf{N}$ extends intuitionistic logic by adding a constructive negation. Accordingly, both positive and negative information has a constructive nature; indeed, $\mathbf{N}$ enjoys *disjunction property* (if a formula $A \vee B$ belongs to $\mathbf{N}$, then either A or B belongs to $\mathbf{N}$) and its negative counterpart, namely *constructible falsity* (if $\sim (A \wedge B)$ belongs to $\mathbf{N}$, then either $\sim A$ or $\sim B$ belongs to $\mathbf{N}$). Beyond $\mathbf{N}$, many other logical systems with the same constructive features have been studied; for a comprehensive picture, we refer the reader to [7], where sequent calculi and Kripke semantics of predicate con-structive logics with constructive negation are presented. The interest in such logics has been increased thanks to their applications in Computer Science; first of all, we mention the relevance of constructive negation in logic programming and in knowledge representation (see, e.g., [12,13]).

In this paper we focus on a particular propositional logic with constructive negation, namely the propositional fragment of the logic $\mathbf{E}$ introduced in [10] (where $\mathbf{E}$ stays for "effective"). Instead of two negations, only the constructive

J. Marcinkowski and A. Tarlecki (Eds.): CSL 2004, LNCS 3210, pp. 488–502, 2004.

negation is used, but a new unary logical operator ($\square$) is introduced to represent *classical truth* inside **E**: a formula $\square A$ belongs to **E** if and only if A is classically valid. In the classification of [7], **E** coincides with the logic **N3o**, namely the logic obtained by adding to Nelson logic the *potential omniscience axiom* $\neg\neg(A \vee \sim A)$. The interaction between constructive falsity and classical truth provides a powerful environment where one can embed classical reasoning in a constructive setting, and this has a fruitful impact for Computer Science. Two recent trends encourage the research in this direction: in [3] it is described a framework based on logic **E** oriented to verification of computer programs; in [5] formal proofs of **E** are used to extract information about the propagation delays of signals in a combinatorial circuit (timing analysis).

In this context, the main contribution of the paper is to supply a space efficient tool to generate proofs of **E**. Firstly, we present a tableau calculus for **E**. Differently from the calculi presented in [1, 7], we aim to avoid what might produce inefficiency in the proof search task. Along the lines of [4, 9, 16], we avoid duplications of formulas: when a rule is applied to a formula A, A must not occur in the obtained configuration. The rules for the formulas $A \vee B$ and $\sim(A \wedge B)$ are defined according to their constructive meaning. A peculiar feature of our calculus is the combination of constructive and non-constructive tools; indeed, in particular configurations, we are allowed to continue a proof by using the rules for classical logic. The more expensive task in proof search strategy is due to backtracking. Typically, if one fails to build a closed proof table, one has to restore some old configuration and try the application of a different rule. In our implementation we reduce this kind of backtracking and, using a semantical argumentation, we show that the backtracking can be actually limited to few rules.

We have implemented in C++ language a decision procedure for the logic **E** based on the proof search strategy. The program is available at `http://www.dimequant.unimib.it/elogic/index.html`.

2   The Logic E

We consider the propositional language $\mathcal{L}$ based on a denumerable set of *propositional variables* and the logical constants $\sim, \wedge, \vee, \rightarrow$ and $\square$. We denote with $p, q, \ldots$ propositional variables and with $A, B, \ldots$ arbitrary formulas. We write $A \leftrightarrow B$ as an abbreviation of $(A \rightarrow B) \wedge (B \rightarrow A)$. A *literal* is any formula of the kind p or $\sim p$, where p is a propositional variable. We denote with **Int** the set of intuitionistic valid formulas of the propositional language $\mathcal{L}_{\mathbf{Int}}$ having as logical constants $\neg, \wedge, \vee, \rightarrow$; **Cl** denotes the set of classically valid formulas of $\mathcal{L}$, where $\square$ has to be trivially understood as the identity operator (namely, $\square A$ is equivalent to A).

The logic **E** (in the predicate language) has been introduced in [10], where both a natural deduction calculus and a Kripke semantics is provided. In this section we outline some results presented in [10]. The logic **E** can be axiomatized by adding to the positive axioms of **Int** (see, for instance, [15]) the following axioms which characterize $\sim$ as a *constructive negation* and $\square$ as an operator to represent *classical truth*:

(E1). $\sim (A \wedge B) \leftrightarrow (\sim A \vee \sim B)$
(E2). $\sim (A \vee B) \leftrightarrow (\sim A \wedge \sim B)$
(E3). $\sim (A \to B) \leftrightarrow (A \wedge \sim B)$
(E4). $\sim\sim A \leftrightarrow A$
(E5). $A \wedge \sim A \to B$
(E6). $(\square A \wedge \square \sim A) \to B$
(E7). $(\sim A \to B \wedge \sim B) \to \square A$
(E8). $(A \to B \wedge \sim B) \to \sim \square A$

Clearly, **E** is contained in **Cl**. Constructive negation (also called *strong negation*) is weaker, with respect to provability, than classical negation; as a matter of fact, the classical tautologies $\sim (A \wedge \sim A)$, $(A \to B) \to (\sim B \to \sim A)$ and $(A \to B) \to \sim A \vee B$ do not belong to **E**. Moreover, unlike intuitionistic negation, constructive negation satisfies the principle of *constructible falsity (cf)*, which is the negative counterpart of *disjunction property (dp)*. This means that:

(cf). $\sim (A \wedge B) \in \mathbf{E}$ implies $\sim A \in \mathbf{E}$ or $\sim B \in \mathbf{E}$;
(dp). $A \vee B \in \mathbf{E}$ implies $A \in \mathbf{E}$ or $B \in \mathbf{E}$.

The $\square$ operator allows us represent *classical truth* inside **E**; indeed:

(ct). $\square A \in \mathbf{E}$ if and only if $A \in \mathbf{Cl}$.

Intuitionistic validity can be represented inside **E** by means of a translation map $\mathcal{T}$ defined on formulas of $\mathcal{L}_{\mathbf{Int}}$. As a matter of fact, let us define:

$$\mathcal{T}(p) = p, \text{ with } p \text{ a propositional variable;}$$
$$\mathcal{T}(A \oplus B) = \mathcal{T}(A) \oplus \mathcal{T}(B), \text{ with } \oplus \in \{\wedge, \vee, \to\};$$
$$\mathcal{T}(\neg A) = \square \sim A.$$

Then:

(int). $A \in \mathbf{Int}$ if and only if $\mathcal{T}(A) \in \mathbf{E}$.

We point out that in the literature logics with both intuitionistic and constructive negation have been investigated (see, e.g., [7, 11, 14, 17]). The logic **E**, provided we define $\square A$ as $\neg\neg A$, coincides with the logic **N3o** of [7], namely, the logic obtained by adding to Nelson logic **N3** the *potential omniscience axiom* $\neg\neg(A \vee \sim A)$. In [10] it is also presented the logic $\mathbf{E}^*$, which is maximal among the logics containing **E** and satisfying (dp), (cf) and (ct).

To treat constructive negation, we introduce a Kripke semantics equivalent to the one in [7, 10]. We denote with $\langle P, \leq \rangle$ a *poset* (partially ordered set), where P is a nonempty set and $\leq$ is a partial ordering between elements of P; $\langle P, \leq, \rho \rangle$ means that ρ is the minimum element of $\langle P, \leq \rangle$. We call *final element* of $\langle P, \leq \rangle$ any $\phi \in P$ that is maximal in $\langle P, \leq \rangle$ (that is, for every $\alpha \in P$, $\phi \leq \alpha$ implies $\phi = \alpha$). Given $\alpha \in P$, $\mathrm{Fin}(\alpha)$ denotes the set of final elements ϕ of $\langle P, \leq \rangle$

such that $\alpha \leq \phi$. Without loss of generality, we assume that, for every $\alpha \in P$, $\mathrm{Fin}(\alpha) \neq \emptyset$. A *Kripke model* for $\mathcal{L}$ is a structure $\underline{K} = \langle P, \leq, \rho, \Vdash \rangle$, where $\langle P, \leq, \rho \rangle$ is a poset and $\Vdash$ (the *forcing relation*) is a binary relation between elements α of P and literals l of $\mathcal{L}$ such that:

(K1). $\alpha \Vdash l$ and $\alpha \leq \beta$ implies $\beta \Vdash l$;
(K2). For every propositional variable p, it is not true that $\alpha \Vdash p$ and $\alpha \Vdash \sim p$;
(K3). For every final element ϕ of $\underline{K}$ and every propositional variable p, $\phi \Vdash p$ or $\phi \Vdash \sim p$.

The forcing relation is extended in a standard way to arbitrary formulas of $\mathcal{L}$ as follows:

1. $\alpha \Vdash A \wedge B$ iff $\alpha \Vdash A$ and $\alpha \Vdash B$;
2. $\alpha \Vdash A \vee B$ iff $\alpha \Vdash A$ or $\alpha \Vdash B$;
3. $\alpha \Vdash A \rightarrow B$ iff, for every $\beta \in P$ such that $\alpha \leq \beta$, $\beta \Vdash A$ implies $\beta \Vdash B$;
4. $\alpha \Vdash \Box A$ iff, for every $\phi \in \mathrm{Fin}(\alpha)$, $\phi \Vdash A$;
5. $\alpha \Vdash \sim (A \wedge B)$ iff $\alpha \Vdash \sim A$ or $\alpha \Vdash \sim B$;
6. $\alpha \Vdash \sim (A \vee B)$ iff $\alpha \Vdash \sim A$ and $\alpha \Vdash \sim B$;
7. $\alpha \Vdash \sim (A \rightarrow B)$ iff $\alpha \Vdash A$ and $\alpha \Vdash \sim B$;
8. $\alpha \Vdash \sim \Box A$ iff, for every $\phi \in \mathrm{Fin}(\alpha)$, $\phi \Vdash \sim A$;
9. $\alpha \Vdash \sim\sim A$ iff $\alpha \Vdash A$.

We write $\alpha \nVdash A$ to mean that $\alpha \Vdash A$ does not hold. It is easy to check that properties (K1), (K2) and (K3) hold for arbitrary formulas as well. In this generalized formulation, (K1) is the usual monotonicity property of forcing relation, (K3) states that a final element ϕ of $\underline{K}$ behaves like a classical interpretation. Note that a classical interpretation $\mathcal{I}$ can be seen as a Kripke model having $\mathcal{I}$ as the only element and forcing relation defined in the obvious way.

A formula A *is valid in a Kripke model* $\underline{K}$ if and only if $\alpha \Vdash A$ for all elements α of $\underline{K}$. As proved in [10], **E** coincides with the set of formulas valid in all Kripke models.

3   The Tableau Calculus

The major contribute of this paper is the definition of a tableau calculus **Tab** for **E**. As far as we know, no tableau calculus for this logic has been presented in the literature. The object language of the calculus is based on the signs **T** and **F**. A *signed formula* (*sf* for short) is an expression of the form **T**A or **F**A, where A is any formula; a **T**-formula is a sf with sign **T**, whereas an **F**-formula is a sf with sign **F**. The rules of **Tab** are in Tables 1-3. The meaning of the signs **T** and **F** is explained by the notion of *realizability*. Let $\underline{K} = \langle P, \leq, \Vdash \rangle$ be a Kripke model, let $\alpha \in P$, let A be a formula and let S be a set of sfs. We say that:

- $\alpha \triangleright \mathbf{T}A$ (α *realizes* A) iff $\alpha \Vdash A$;
- $\alpha \triangleright \mathbf{F}A$ iff $\alpha \nVdash A$;
- $\alpha \triangleright S$ iff, for every $H \in S$, $\alpha \triangleright H$.

We say that S is *realizable* iff there exists an element α of some model $\underline{K}$ such that $\alpha \triangleright S$. A *configuration* is an expression of the form $S_1 \mid \ldots \mid S_n$ where, for all $i = 1, \ldots, n$, S_i is a set of sfs. In the rules of the calculus, we denote with $S, H_1, \ldots, H_m$ the set $S \cup \{H_1, \ldots, H_m\}$ and with S_T the set of **T**-formulas of S. Every rule is applied to a signed formula of a configuration $S_1 \mid \ldots \mid S_i \mid \ldots S_n$; e.g., the notation $S, \mathbf{T}(A \wedge B)$ points out that the rule $\mathbf{T}\wedge$ is applied to the formula $\mathbf{T}(A \wedge B)$ of the set $S \cup \{\mathbf{T}(A \wedge B)\}$, where S is possibly empty; the schema

$$\frac{S_1 \mid \ldots \mid S, \mathbf{T}(A \wedge B) \mid \ldots \mid S_n}{S_1 \mid \ldots \mid S, \mathbf{T}A, \mathbf{T}B \mid \ldots \mid S_n} \mathbf{T}\wedge$$

illustrates an application of the rule $\mathbf{T}\wedge$. In every rule we distinguish two parts: the *premise*, that is the configuration above the line, and the *conclusion*, that is the configuration below the line. Differently from the calculi for logics with

Table 1.

$$\frac{S, \mathbf{T}(A \wedge B)}{S, \mathbf{T}A, \mathbf{T}B} \mathbf{T}\wedge \qquad \frac{S, \mathbf{F}(A \wedge B)}{S, \mathbf{F}A \mid S, \mathbf{F}B} \mathbf{F}\wedge$$

$$\frac{S, \mathbf{T}(A \vee B)}{S, \mathbf{T}A \mid S, \mathbf{T}B} \mathbf{T}\vee \qquad \frac{S, \mathbf{F}(A \vee B)}{S, \mathbf{F}A} \mathbf{F}\vee_1 \qquad \frac{S, \mathbf{F}(A \vee B)}{S, \mathbf{F}B} \mathbf{F}\vee_2$$

$$\frac{S, \mathbf{F}(A \rightarrow B)}{S_T, \mathbf{T}A, \mathbf{F}B} \mathbf{F}\rightarrow$$

$$\frac{S, \mathbf{T} \sim (A \wedge B)}{S, \mathbf{T} \sim A \mid S, \mathbf{T} \sim B} \mathbf{T}\sim\wedge \qquad \frac{S, \mathbf{F} \sim (A \wedge B)}{S, \mathbf{F} \sim A} \mathbf{F}\sim\wedge_1 \qquad \frac{S, \mathbf{F} \sim (A \wedge B)}{S, \mathbf{F} \sim B} \mathbf{F}\sim\wedge_2$$

$$\frac{S, \mathbf{T} \sim (A \vee B)}{S, \mathbf{T} \sim A, \mathbf{T} \sim B} \mathbf{T}\sim\vee \qquad \frac{S, \mathbf{F} \sim (A \vee B)}{S, \mathbf{F} \sim A \mid S, \mathbf{F} \sim B} \mathbf{F}\sim\vee$$

$$\frac{S, \mathbf{T} \sim (A \rightarrow B)}{S, \mathbf{T}A, \mathbf{T} \sim B} \mathbf{T}\sim\rightarrow \qquad \frac{S, \mathbf{F} \sim (A \rightarrow B)}{S, \mathbf{F}A \mid S, \mathbf{F} \sim B} \mathbf{F}\sim\rightarrow$$

$$\frac{S, \mathbf{T} \sim\sim A}{S, \mathbf{T}A} \mathbf{T}\sim\sim \qquad \frac{S, \mathbf{F} \sim\sim A}{S, \mathbf{F}A} \mathbf{F}\sim\sim$$

strong negation presented in [1, 7], we are interested in a calculus oriented to an efficient implementation. First of all, we aim to avoid duplications, thus to treat $\mathbf{T}(A \rightarrow B)$ we need several rules according to the structure of A (see [4, 16]). Moreover, to further reduce the depth of the proofs, the rules

$$\frac{S, \mathbf{T}((A \vee B) \rightarrow C)}{S, \mathbf{T}(A \rightarrow C), \mathbf{T}(B \rightarrow C)} \mathbf{T}\rightarrow\vee \qquad \frac{S, \mathbf{T}((A \rightarrow B) \rightarrow C)}{S_T, \mathbf{F}(A \rightarrow B), \mathbf{T}(B \rightarrow C) \mid S, \mathbf{T}C} \mathbf{T}\rightarrow\rightarrow$$

of [4, 16] are rewritten as in Table 2, where the new propositional variable p avoids the repetition of C in the former rule and of B in the latter rule (see [6, 8]).

Table 2.

$$\frac{S, \mathbf{T}A, \mathbf{T}(A \to B)}{S, \mathbf{T}A, \mathbf{T}B} \mathbf{T}{\to}$$

$$\frac{S, \mathbf{T}((A \wedge B) \to C)}{S, \mathbf{T}(A \to (B \to C))} \mathbf{T}{\to}\wedge \qquad \frac{S, \mathbf{T}(\sim(A \wedge B) \to C)}{S, \mathbf{T}(\sim A \to p), \mathbf{T}(\sim B \to p), \mathbf{T}(p \to C)} \mathbf{T}{\to}\sim\wedge$$

$$\frac{S, \mathbf{T}((A \vee B) \to C)}{S, \mathbf{T}(A \to p), \mathbf{T}(B \to p), \mathbf{T}(p \to C)} \mathbf{T}{\to}\vee \qquad \frac{S, \mathbf{T}(\sim(A \vee B) \to C)}{S, \mathbf{T}(\sim A \to (\sim B \to C))} \mathbf{T}{\to}\sim\vee$$

$$\frac{S, \mathbf{T}((A \to B) \to C)}{S_T, \mathbf{T}A, \mathbf{F}p, \mathbf{T}(B \to p), \mathbf{T}(p \to C) \mid S, \mathbf{T}C} \mathbf{T}{\to}{\to} \qquad \frac{S, \mathbf{T}(\sim(A \to B) \to C)}{S, \mathbf{T}(A \to (\sim B \to C))} \mathbf{T}{\to}\sim{\to}$$

$$\frac{S, \mathbf{T}(\sim\sim A \to B)}{S, \mathbf{T}(A \to B)} \mathbf{T}{\to}\sim\sim$$

where the propositional variable p in $\mathbf{T}{\to}\sim\wedge$, $\mathbf{T}\to\vee$ and $\mathbf{T}{\to}{\to}$ is new

As already mentioned in the introduction, in some configurations we are allowed to apply to a set S the rules of a tableau calculus for **Cl**. To mark these sets, we use the notation $[S]_{\mathbf{Cl}}$ and we say that $[S]_{\mathbf{Cl}}$ is a *classical set*; intuitively, the signs $\mathbf{T}$ and $\mathbf{F}$ occurring in $[S]_{\mathbf{Cl}}$ have to be understood in a classical way (see the rules of Table 3).

Table 3.

$$\frac{S, \mathbf{T}\square A}{[S_T, \mathbf{T}A]_{\mathbf{Cl}}} \mathbf{T}\square \qquad \frac{S, \mathbf{F}\square A}{[S_T, \mathbf{T} \sim A]_{\mathbf{Cl}}} \mathbf{F}\square$$

$$\frac{S, \mathbf{T} \sim \square A}{[S_T, \mathbf{T} \sim A]_{\mathbf{Cl}}} \mathbf{T}{\sim}\square \qquad \frac{S, \mathbf{F} \sim \square A}{[S_T, \mathbf{T}A]_{\mathbf{Cl}}} \mathbf{F}{\sim}\square$$

$$\frac{S, \mathbf{T}(\square A \to B)}{[S_T, \mathbf{T} \sim A]_{\mathbf{Cl}} \mid S, \mathbf{T}B} \mathbf{T}{\to}\square \qquad \frac{S, \mathbf{T}(\sim\square A \to B)}{[S_T, \mathbf{T}A]_{\mathbf{Cl}} \mid S, \mathbf{T}B} \mathbf{T}{\to}\sim\square$$

A set S of sfs is *contradictory* iff one of the following conditions holds:

1. $\mathbf{T}A \in S$ and $\mathbf{F}A \in S$;
2. $\mathbf{T}A \in S$ and $\mathbf{T} \sim A \in S$;
3. S is a classical set and S is not **Cl**-consistent.

It is immediate to prove that:

Proposition 1. *Let $\underline{K} = \langle P, \leq, \Vdash \rangle$ be a Kripke model, let $\alpha \in P$ and let S be a contradictory set.*

1. *If S is not a classical set, then $\alpha \rhd S$ does not hold.*
2. *If S is a classical set and α is a final element of $\underline{K}$, then $\alpha \rhd S$ does not hold.*

A *proof table* for a set S is a finite sequence of configurations $\Gamma_1, \ldots, \Gamma_n$, where Γ_1 is the set S and the configuration Γ_{i+1} is obtained from $\Gamma_i = S_1 \mid \ldots \mid S_m$ by applying a rule to a non-contradictory set S_i. A *closed proof table* is a proof table $\Gamma_1, \ldots, \Gamma_n$ where all the sets in the last configuration are contradictory. We point out that to check that a classical set is contradictory, we can use any classical tableau calculus (extended in a trivial way to the language $\mathcal{L}$). Closed proof tables are the proofs of our calculus **Tab**. A set S is provable in **Tab** iff there exists a closed proof table for S; *A is provable in* **Tab** iff there exists a closed proof table for $\{\mathbf{F}A\}$. We remark that the rules of the calculus do not increase the number of **F**-formulas in a set. In particular, if the set in the first configuration of a proof table contains an **F**-formula at most, then every set occurring in the proof table contains an **F**-formula at most. Note that the rule $\mathbf{F} \to$ applied to a set S of this kind is invertible. On the other hand, the rules $\mathbf{F}\vee_i$ and $\mathbf{F} \sim \wedge_i$ are non-invertible, but they capture the constructive meaning of disjunction and negation.

Our aim is to exhibit an "efficient" sound and complete proof search strategy for closed proof tables of **Tab**. We begin by proving that **Tab** is sound for **E**. The main step consists in showing that the rules of **Tab** preserve realizability. It is easy to prove that:

Lemma 1. *Let $\underline{K} = \langle P, \leq, \Vdash \rangle$ be a Kripke model, let $\alpha \in P$ and let R be a rule of the calculus having S as premise and S_1 or $S_1 \mid S_2$ as consequence. If $\alpha \rhd S$, then there is $\beta \in P$ and $i \in \{1, 2\}$ such that $\alpha \leq \beta$ and $\beta \rhd S_i$. Moreover, if S_i is a classical set, then β is a final element of $\underline{K}$.*

From the above lemma we deduce that, if A does not belong to **E**, then no closed proof table for $\{\mathbf{F}A\}$ can exist. Indeed, let $\underline{K} = \langle P, \leq, \Vdash \rangle$ be a model such that A is not valid in $\underline{K}$ and let us assume that there exists a closed proof table $\Gamma_1, \ldots, \Gamma_n$ for $\{\mathbf{F}A\}$. Since $\underline{K}$ realizes $\{\mathbf{F}A\}$, by the previous lemma $\underline{K}$ realizes a set S of Γ_n, moreover if S a classical set, then S is realized in a final element of $\underline{K}$. This contradicts Proposition 1. It follows that A is not provable in **Tab**, hence:

Theorem 1 (Soundness). *If A is provable in* **Tab**, *then A belongs to* **E**.

In the following sections we prove that every formula of **E** is provable in **Tab** (Completeness Theorem).

4   The Proof Search Strategy

In this section we describe a procedure TAB which, given a set S of sfs, searches for a closed proof table for S. The main issue is to reduce backtracking in proof search. In our calculus the rules requiring backtracking are:

$$\mathbf{F}\vee_i \, , \, \mathbf{F}\sim\wedge_i \, , \, \mathbf{T}\rightarrow\rightarrow, \mathbf{T}\rightarrow\Box \, , \, \mathbf{T}\rightarrow\sim\Box \, , \, \mathbf{T}\Box \, , \, \mathbf{T}\sim\Box \quad (i=1,2)$$

Since we consider sets of sfs having an $\mathbf{F}$-formula at most, the rules $\mathbf{F}\rightarrow$, $\mathbf{F}\Box$ and $\mathbf{F}\sim\Box$ are invertible, thus they do not require backtracking. Moreover, if S_T satisfies some properties (see Definition 1 in the next section), also the rules $\mathbf{F}\vee_i$, $\mathbf{F}\sim\wedge_i$, $\mathbf{T}\Box$ and $\mathbf{T}\sim\Box$ are invertible, as proved in Lemma 2 (see [2] for a thorough discussion).

To describe our procedure we introduce some classes $\mathcal{C}_j$ to identify sfs with the same behaviour:

$\mathcal{C}_1 = \{\, \mathbf{F}\Box A, \, \mathbf{F}\sim\Box A \,\};$
$\mathcal{C}_2 = \{\, \mathbf{T}(A \wedge B), \, \mathbf{F}(A \rightarrow B), \, \mathbf{T}\sim(A \vee B), \, \mathbf{T}\sim(A \rightarrow B), \, \mathbf{T}\sim\sim A, \mathbf{F}\sim\sim A,$
$\quad \mathbf{T}((A \wedge B) \rightarrow C), \, \mathbf{T}(\sim(A \wedge B) \rightarrow C), \, \mathbf{T}((A \vee B) \rightarrow C),$
$\quad \mathbf{T}(\sim(A \vee B) \rightarrow C), \, \mathbf{T}(\sim(A \rightarrow B) \rightarrow C), \, \mathbf{T}(\sim\sim A \rightarrow B) \,\};$
$\mathcal{C}_3 = \{\, \mathbf{F}(A \wedge B), \, \mathbf{T}(A \vee B), \, \mathbf{T}\sim(A \wedge B), \, \mathbf{F}\sim(A \vee B), \, \mathbf{F}\sim(A \rightarrow B) \,\};$
$\mathcal{C}_4 = \{\, \mathbf{T}(\Box A \rightarrow B), \, \mathbf{T}(\sim\Box A \rightarrow B) \,\};$
$\mathcal{C}_5 = \{\, \mathbf{T}((A \rightarrow B) \rightarrow C) \,\};$
$\mathcal{C}_6 = \{\, \mathbf{F}(A \vee B), \, \mathbf{F}\sim(A \wedge B) \,\};$
$\mathcal{C}_7 = \{\, \mathbf{T}\Box A, \, \mathbf{T}\sim\Box A \,\}.$

We describe a recursive procedure $\text{TAB}(S, \texttt{applyAll})$ that, given a set S of sfs containing at most an $\mathbf{F}$-formula and a boolean value $\texttt{applyAll}$, returns either a closed proof table for S or NULL if S is realizable (hence, no closed proof table for S can exist). The role of $\texttt{applyAll}$ will be clarified in the next section; here we only point out that, when $\texttt{applyAll}$ is $\texttt{false}$, we do not apply any rule to signed formulas in $S \cap (\mathcal{C}_4 \cup \mathcal{C}_5)$ (see line 29 of the procedure). We assume to have a subroutine $\text{TABCL}(S)$ that, given a set of sfs S, searches for a classical closed proof table for S. If a proof is found, $\text{TABCL}(S)$ returns $[S]_{\mathbf{Cl}}$, otherwise it returns NULL (this means that S is Cl-consistent). Let S be a set of sfs, let $H \in S$ and let S_1 or $S_1 \mid S_2$ the configuration obtained by applying to S the rule $Rule(H)$ corresponding to H (when $H \in \mathcal{C}_6$, we write $Rule_1(H)$ or $Rule_2(H)$ to identify the rule). If Tab_1 and Tab_2 are closed proof tables for S_1 and S_2 respectively, then $\dfrac{S}{Tab_1} Rule(H)$ or $\dfrac{S}{Tab_1 \mid Tab_2} Rule(H)$ denotes the closed proof table for S defined in the obvious way. Moreover, $\mathcal{R}_i(H)$ $(i = 1, 2)$ denotes the set containing the sfs of S_i which replace H. For instance:

$\mathcal{R}_1(\mathbf{T}(A \wedge B)) = \{\, \mathbf{T}A, \mathbf{T}B \,\};$
$\mathcal{R}_1(\mathbf{T}(A \vee B)) = \{\, \mathbf{T}A \,\} \, ; \, \mathcal{R}_2(\mathbf{T}(A \vee B)) = \{\mathbf{T}B\};$
$\mathcal{R}_1(\mathbf{T}((A \rightarrow B) \rightarrow C)) = \{\, \mathbf{T}A, \mathbf{F}p, \mathbf{T}(B \rightarrow p), \mathbf{T}(p \rightarrow C) \,\};$
$\mathcal{R}_2(\mathbf{T}((A \rightarrow B) \rightarrow C)) = \{\mathbf{T}C\}.$

The pseudo-code for TAB is the following:

FUNCTION TAB(S, applyAll)

1   if (($\mathbf{T}A, \mathbf{F}A \in S$) OR ($\mathbf{T}A, \mathbf{T} \sim A \in S$))
2   then return S;
3   if ($S \cap \mathcal{C}_1 \neq \emptyset$)
4   then Let H be the $\mathbf{F}$-formula of S;
5            $Tab_1 \leftarrow$ TABCL($S_T \cup R_1(H)$);
6   if ($Tab_1 \neq$ NULL)
7   then return $\dfrac{S}{Tab_1}$ $Rule(H)$;
8   else return NULL ;
9   if ($\mathbf{T}A, \mathbf{T}(A \to B) \in S$)
10   then $Tab_1 \leftarrow$ TAB($(S \setminus \mathbf{T}(A \to B)) \cup \{\mathbf{T}B\}$, true);
11   if ($Tab_1 \neq$ NULL)
12   then return $\dfrac{S}{Tab_1}$ $\mathbf{T}\to$;
13   else return NULL ;
14 if ($S \cap \mathcal{C}_2 \neq \emptyset$)
15   then Let $H \in S \cap \mathcal{C}_2$;
16           $Tab_1 \leftarrow$ TAB($(S \setminus \{H\}) \cup \mathcal{R}_1(H)$, true);
17   if ($Tab_1 \neq$ NULL)
18   then return $\dfrac{S}{Tab_1}$ $Rule(H)$;
19   else return NULL ;
20 if ($S \cap \mathcal{C}_3 \neq \emptyset$)
21   then Let $H \in S \cap \mathcal{C}_3$;
22           $Tab_1 \leftarrow$ TAB($(S \setminus \{H\}) \cup \mathcal{R}_1(H)$, true);
23   if ($Tab_1 \neq$ NULL)
24   then $Tab_2 \leftarrow$ TAB($(S \setminus \{H\}) \cup \mathcal{R}_2(H)$, true);
25   if ($Tab_2 \neq$ NULL)
26   then return $\dfrac{S}{Tab_1 \mid Tab_2}$ $Rule(H)$;
27   else return NULL ;
28   else return NULL ;
29 if (applyAll AND ($S \cap (\mathcal{C}_4 \cup \mathcal{C}_5) \neq \emptyset$))
30   then for ($H \in (S \cap (\mathcal{C}_4 \cup \mathcal{C}_5))$)
31   do $Tab_2 \leftarrow$ TAB($(S \setminus \{H\}) \cup \mathcal{R}_2(H)$, true);
32   if ($Tab_2 =$ NULL)
33   then return NULL ;
34   if ($H \in \mathcal{C}_4$)
35   then $Tab_1 \leftarrow$ TABCL($(S_T \setminus \{H\}) \cup \mathcal{R}_1(H)$);
36   else $Tab_1 \leftarrow$ TAB($(S_T \setminus \{H\}) \cup \mathcal{R}_1(H)$, true);
37   if ($Tab_1 \neq$ NULL)
38   then return $\dfrac{S}{Tab_1 \mid Tab_2}$ $Rule(H)$;
39 if ($S \cap \mathcal{C}_6 \neq \emptyset$)
40   then Let H be the $\mathbf{F}$-formula of S;
41           $Tab_1 \leftarrow$ TAB($S_T \cup \mathcal{R}_1(H)$, false);
42   if ($Tab_1 \neq$ NULL)

43   **then return** $\dfrac{S}{Tab_1}$ $Rule_1(H)$;
44   **else** $Tab_2 \leftarrow$ TAB$(S_T \cup \mathcal{R}_2(H),$ **false**);
45   **if** $(Tab_2 \neq$ NULL $)$
46   **then return** $\dfrac{S}{Tab_2}$ $Rule_2(H)$;
47   **else return** NULL ;
48   **if** $(\ (S \cap (\mathcal{C}_4 \cup \mathcal{C}_5) = \emptyset)$ AND $(S \cap \mathcal{C}_7 \neq \emptyset)\ )$
49   **then** Let $H \in S \cap \mathcal{C}_7$;
50              $Tab_1 \leftarrow$ TABCL$((S \setminus \{H\}) \cup \mathcal{R}_1(H))$;
51   **if** $(Tab_1 \neq$ NULL $)$
52   **then return** $\dfrac{S}{Tab_1}$ $Rule(H)$;
53   **else return** NULL ;
54 **return** NULL ;

We remark that, when one of the **if** conditions at lines 1, 3, 9, 14, 20, 39 and 48 is matched, the corresponding **then** instruction is executed and the procedure ends returning a value. This means that, independently of the choice of H, no backtracking is needed. On the contrary, in the **for** instruction at line 30 it might be necessary to try the application of a rule to all the formulas H in $S \cap (\mathcal{C}_4 \cup \mathcal{C}_5)$ and possibly to continue in line 39. We emphasize that to implement $\mathbf{F}\vee_i$, $\mathbf{F} \sim \wedge_i$, $\mathbf{T}\square$ and $\mathbf{T} \sim \square$ without backtracking, it is essential to apply these rules after having tried the application of all the other rules (see the proof of Proposition 2 in the next section).

Example 1. Let us consider the set of signed formulas
$$S = \{\, \mathbf{T}((a \rightarrow (b \rightarrow c)) \rightarrow \square d)\, , \ \mathbf{F}(a \vee \square((b \rightarrow c) \rightarrow d))\, \}$$

To search for a closed proof table for S, we call TAB(S, true).

(1). Since $\mathbf{T}((a \rightarrow (b \rightarrow c)) \rightarrow \square d) \in \mathcal{C}_5$ and `applyAll` is `true`, the condition in the **if** statement of line 29 is matched. This means that the procedure tries to apply $\mathbf{T} \rightarrow\rightarrow$ to S, therefore closed proof tables for the sets
$$S_1 = \{\, \mathbf{T}a\, , \ \mathbf{T}((b \rightarrow c) \rightarrow p)\, , \ \mathbf{T}(p \rightarrow \square d)\, , \ \mathbf{F}p\, \}$$
$$S_2 = \{\, \mathbf{T}\square d\, , \ \mathbf{F}(a \vee \square((b \rightarrow c) \rightarrow d))\, \}$$
are searched.

(2). The call TAB(S_2, true) of line 31 is executed in order to build a closed proof table for S_2.

(3). The call TAB(S_3, false) of line 41 is executed, where $S_3 = \{\mathbf{T}\square d, \mathbf{F}a\}$, which corresponds to the application of $\mathbf{F}\vee_1$ to S_2.

(4). The application of $\mathbf{T}\square$ to S_3 is tried by the call TABCL$(\{\mathbf{T}d\})$ (line 50). The NULL value is returned (indeed, $\mathbf{T}d$ is Cl-consistent) and also TAB(S_3, false) fails (line 53 is executed and NULL is returned).

(5). The execution of TAB(S_2, true) continues in line 44 with the computation of TAB(S_4, false), where
$$S_4 = \{\, \mathbf{T}\square d\, , \ \mathbf{F}\square((b \rightarrow c) \rightarrow d)\, \}$$
namely, $\mathbf{F}\vee_2$ is applied to S_2.

(6). The call $\text{TABCL}(\{\mathbf{T}\Box d, \mathbf{T} \sim ((b \to c) \to d)\})$ of line 5 is executed ($\mathbf{F}\Box$ is applied to S_4) and a classical proof table is found. Thus, both the calls in (5) and (2) succeed and the closed proof table Tab_2 is built as follows:

$$\cfrac{\cfrac{\mathbf{T}\Box d, \ \mathbf{F}(a \vee \Box((b \to c) \to d))}{\cfrac{\mathbf{T}\Box d, \ \mathbf{F}\Box((b \to c) \to d)}{[\mathbf{T}\Box d, \ \mathbf{T} \sim ((b \to c) \to d)]_{\text{Cl}}} \mathbf{T}\Box}{} \mathbf{F}\vee_2}$$

(7). Now, the computation of $\text{TAB}(S, \mathtt{true})$ continues with the call $\text{TAB}(S_1, \mathtt{true})$ (line 36) in order to build a closed proof table for S_1.

(8). The condition in the **if** statement of line 29 is matched, thus the **for** loop in line 30 is executed. This means that it is tried the application of the rule $\mathbf{T} \to\to$ to S_1 for *all* the signed formulas of the kind $\mathbf{T}((H_1 \to H_2) \to H_3)$. We have *two* signed formulas of this kind and it is easy to check that in both cases the search for a closed proof table fails. It follows that no closed proof table for S_1 can be built; nevertheless, $\text{TAB}(S, \mathtt{true})$ does not fail, but the computation continues with the statements after line 38.

(9). The condition in the **if** statement of line 39 is satisfied, thus the computation continues with the call $\text{TAB}(S_5, \mathtt{false})$ of line 41, where

$$S_5 = \{\mathbf{T}((a \to (b \to c)) \to \Box d), \ \mathbf{F}a\}$$

(it corresponds to apply $\mathbf{F}\vee_1$ to S). The procedure immediately fails. As a matter of fact, the application of $\mathbf{T} \to\to$ is not tried since the value of applyAll is false, and no other **if** statement can be executed; hence, line 54 is executed and NULL is returned.

(10). The call $\text{TAB}(S_6, \mathtt{false})$ of line 44, where

$$S_6 = \{\mathbf{T}((a \to (b \to c)) \to \Box d), \ \mathbf{F}\Box((b \to c) \to d)\}$$

is executed (it corresponds to apply $\mathbf{F}\vee_2$ to S).

(11). Since the value of applyAll is false, the instructions inside the **if** statement of line 29 are not executed (namely, the application of $\mathbf{T} \to\to$ is not tried), but the call $\text{TABCL}(S_7)$ in line 5 is executed ($\mathbf{F}\Box$ is applied to S_6), where

$$S_7 = \{\mathbf{T}((a \to (b \to c)) \to \Box d), \ \mathbf{T} \sim ((b \to c) \to d)\}$$

The procedure succeeds in finding out a classical closed proof table for S_7; accordingly, both $\text{TAB}(S_6, \mathtt{false})$ and $\text{TAB}(S, \mathtt{true})$ succeed and the returned closed table for S is:

$$\cfrac{\cfrac{\mathbf{T}((a \to (b \to c)) \to \Box d), \ \mathbf{F}((a \vee \Box((b \to c) \to d)))}{\cfrac{\mathbf{T}((a \to (b \to c)) \to \Box d), \ \mathbf{F}\Box((b \to c) \to d)}{[\mathbf{T}((a \to (b \to c)) \to \Box d), \ \mathbf{T} \sim ((b \to c) \to d)]_{\text{Cl}}} \mathbf{F}\Box}{} \mathbf{F}\vee_2}$$

Example 2. Let us consider the set

$$S = \{ \mathbf{T}((p \vee q) \vee r) \,,\, \mathbf{F}q \}$$

To build a closed proof table for S, we call $\mathrm{TAB}(S, \texttt{true})$. Line 22 is executed and $\mathrm{TAB}(S_1, \texttt{true})$ is called, where $S_1 = \{\mathbf{T}(p \vee q), \mathbf{F}q\}$ ($\mathbf{T}\vee$ is applied to S). Again, line 22 is executed and $\mathrm{TAB}(S_2, \texttt{true})$ is called, where $S_2 = \{\mathbf{T}p, \mathbf{F}q\}$. Now, no condition associated with the **if** statements is matched, hence $\mathrm{TAB}(S_2, \texttt{true})$ immediately fails (line 54 is executed and NULL is returned). This implies that $\mathrm{TAB}(S_1, \texttt{true})$ and $\mathrm{TAB}(S, \texttt{true})$ immediately fail (indeed, in both cases line 28 is executed and NULL is returned) and no proof table for S is found.

To prove the termination of TAB and the Completeness Theorem we define the function dg as follows:

- if l is a literal, then $\mathrm{dg}(l) = 0$;
- $\mathrm{dg}(A \wedge B) = \mathrm{dg}(A) + \mathrm{dg}(B) + 2$;
- $\mathrm{dg}(A \vee B) = \mathrm{dg}(A) + \mathrm{dg}(B) + 3$;
- $\mathrm{dg}(A \to B) = \mathrm{dg}(A) + \mathrm{dg}(B) +$ (number of implications occurring in A) $+ 1$;
- $\mathrm{dg}(\sim A) = \mathrm{dg}(A) + 1$;
- $\mathrm{dg}(\Box A) = \mathrm{dg}(A)$;
- if S is a set of sfs, we set $\mathrm{dg}(S) = \sum_{H \in S} \mathrm{dg}(H)$.

It is easy to check that, if S is a set of sfs and S' is obtained from S by an application of a rule of **Tab**, then $\mathrm{dg}(S') < \mathrm{dg}(S)$. Using this fact, it is immediate to prove that TAB always terminates.

Remark 1. Along the lines of [6], it is possible to prove that the depth of every proof table of **Tab** is linearly bounded in the proved formula. This property implies the space efficiency of TAB (see the discussion in Section 6).

5   Completeness

We prove that, when the call $\mathrm{TAB}(S, \texttt{applyAll})$ returns NULL, S is realizable and we can actually build a countermodel for S (namely, a model $\underline{K} = \langle P, \leq, \rho, \Vdash \rangle$ such that $\rho \rhd S$). To justify the lack of backtracking in rules $\mathbf{F}\vee_i$, $\mathbf{F} \sim \wedge_i$, $\mathbf{T}\Box$ and $\mathbf{T} \sim \Box$ we introduce the notion of $\to$-realizability.

Definition 1. *A set S_T of $\mathbf{T}$-formulas is $\to$-realizable iff the following holds:*

1. $S_T \subseteq \{\mathbf{T}l \mid l$ *is a literal*$\} \cup \mathcal{C}_4 \cup \mathcal{C}_5 \cup \mathcal{C}_7$;
2. *For all* $\mathbf{T}(l \to B) \in S_T$, *with* l *a literal,* $\mathbf{T}l \notin S_T$;
3. *For all* $\mathbf{T}(\Box A \to B) \in S_T$, *the set* $(S_T \setminus \{\mathbf{T}(\Box A \to B)\}) \cup \{\mathbf{T} \sim A\}$ *is realizable;*
4. *For all* $\mathbf{T}(\sim \Box A \to B) \in S_T$, *the set* $(S_T \setminus \{\mathbf{T}(\sim \Box A \to B)\}) \cup \{\mathbf{T}A\}$ *is realizable;*
5. *For all* $\mathbf{T}((A \to B) \to C) \in S_T$, *the set* $(S_T \setminus \{\mathbf{T}((A \to B) \to C)\}) \cup \{\mathbf{T}A, \mathbf{F}p, \mathbf{T}(B \to p), \mathbf{T}(p \to C)\}$, *where p is new, is realizable.*

Remark 2. If S_T is $\rightarrow$-realizable and $S_T \cap (\mathcal{C}_4 \cup \mathcal{C}_5) \neq \emptyset$, then S_T is realizable as well.

Using a semantical construction on Kripke models, in [2] it is proved that:

Lemma 2. *Let S_T be $\rightarrow$-realizable. Then:*

(i). $S_T, \mathbf{F}(A \vee B)$ is realizable iff both $S_T, \mathbf{F}A$ and $S_T, \mathbf{F}B$ are realizable.
(ii). $S_T, \mathbf{F} \sim (A \wedge B)$ is realizable iff both $S_T, \mathbf{F} \sim A$ and $S_T, \mathbf{F} \sim B$ are realizable.
Moreover, if l is a literal and $\mathbf{T}l \notin S_T$, then:
(iii). $S_T, \mathbf{F}l$ is realizable iff S_T is $\mathbf{Cl}$-consistent.

We say that the call $\text{TAB}(S, \texttt{applyAll})$ is *sound* iff $\texttt{applyAll}$ is $\texttt{true}$ or S_T is $\rightarrow$-realizable.

Proposition 2. *Let S be a set of sfs containing at most one $\mathbf{F}$-formula, let $\text{TAB}(S, \texttt{applyAll})$ be a sound call and suppose that $\text{TAB}(S, \texttt{applyAll})$ returns the $\texttt{NULL}$ value. Then, there is a Kripke model $\underline{K} = \langle P, \leq, \rho, \Vdash \rangle$ such that $\rho \triangleright S$.*

Proof. Let us assume, by induction hypothesis, that the proposition holds for all sets S' such that $\text{dg}(S') < \text{dg}(S)$. We prove that the proposition holds for S by inspecting all the possible cases where the procedure returns the $\texttt{NULL}$ value. We show some significant cases (the whole proof is in [2]).

The instruction at line 33 has been executed.
Let us assume, for instance, that, at line 31, the call $\text{TAB}(S', \texttt{true})$ has been executed, with $S' = (S \setminus \{\mathbf{T}(\square A \rightarrow B)\}) \cup \{\mathbf{T}B\}$. By induction hypothesis there exists a Kripke model $\underline{K} = \langle P, \leq, \rho, \Vdash \rangle$ such that $\rho \triangleright S'$, hence $\rho \triangleright S$.

Remark 3. If none of the conditions at lines 1, 3, 9, 14, 20 and 29 holds, we claim that S_T is $\rightarrow$-realizable. Indeed, if the second parameter of the function is $\texttt{false}$, this follows by the hypothesis of the proposition. Otherwise, using the induction hypothesis, one can easily check that S_T satisfies the definition of $\rightarrow$-realizability. In particular, the realizability of $(S_T \setminus \{H\}) \cup \mathcal{R}_1(H)$, for $H \in \mathcal{C}_4 \cup \mathcal{C}_5$, follows by the fact that the procedure has not terminated inside the **for** instruction at line 30, since the value of Tab_1 is $\texttt{NULL}$.

The instruction at line 47 has been executed.
Suppose that $S = S_T \cup \{\mathbf{F}(A \vee B)\}$ and let $S_A = S_T \cup \{\mathbf{F}A\}$, $S_B = S_T \cup \{\mathbf{F}B\}$. Then, both the call $\text{TAB}(S_A, \texttt{false})$ and $\text{TAB}(S_B, \texttt{false})$ have returned the $\texttt{NULL}$ value. Note that both calls are sound (indeed, $(S_A)_T = (S_B)_T = S_T$ and S_T is $\rightarrow$-realizable) thus, by induction hypothesis, both $S_T, \mathbf{F}A$ and $S_T, \mathbf{F}B$ are realizable. By Lemma 2(i), S is realizable. The case $S = S_T \cup \{\mathbf{F} \sim (A \wedge B)\}$ is similar. $\square$

By the above proposition, it immediately follows the Completeness Theorem.

Theorem 2 (Completeness). *If A belongs to $\mathbf{E}$, then $\text{TAB}(\{\mathbf{F}A\}, \texttt{true})$ returns a closed proof table for $\mathbf{F}A$.*

As a consequence of the above theorem, we have a trivial proof of properties (cf), (dp) and (ct) of $\mathbf{E}$ stated in Section 2

6   Implementation of the Decision Procedure

We have implemented a decision procedure based on TAB and TABCL. The implementation uses the signs $\mathbf{T}, \mathbf{F}, \mathbf{T_c}$ and $\mathbf{F_c}$. The signs $\mathbf{T}$ and $\mathbf{F}$ are used as in TAB. When the procedure TABCL is called, the signs of the formulas are turned into $\mathbf{T_c}$ (classical truth) and $\mathbf{F_c}$ (classical falsity). To treat formulas with signs $\mathbf{T_c}$ and $\mathbf{F_c}$ rules of a classical tableau calculus are used. Since the rules for Classical logic are invertible backtracking is not required. To reduce the number of nodes in the classical tableau the rules having one set in the conclusion are applied first. Since signed formulas $\mathbf{T}H$ and $\mathbf{T_c}H$, where H is an axiom of $\mathbf{Cl}$, are not needed to close a proof table, they are deleted any time they appear in a configuration. Whenever a rule is applied, the condition in Line 1 is checked as follows. If the rule related to H is applied to "$(S \setminus \{H\}), H$", the consistence of the resulting set S_i is checked considering every formula in $\mathcal{R}_i(H)$: every $\mathbf{T}$-formula in $\mathcal{R}_i(H)$ is checked against the $\mathbf{F}$-formula and the $\mathbf{T}$-formulas of $S_i \setminus \mathcal{R}_i(H)$. If $\mathcal{R}_i(H)$ contains the $\mathbf{F}$-formula, then it is checked against the $\mathbf{T}$-formulas of $S_i \setminus \mathcal{R}_i(H)$. The implementation proceeds in a similar way for the signs $\mathbf{T_c}$ and $\mathbf{F_c}$.

TAB is implemented as an iterative procedure. The implementation uses a stack to take into account two different levels of backtracking. The former level of backtracking, related to the **for** statement in line 30, is used to explore the search space of the proof table. The latter level of backtracking, related to lines 24 and 36, is used to visit with a depth-first strategy a single proof table to determine if it is closed. The stack has, at most, as many elements as the longest branch of the deepest proof table in the search space. Every element of the stack contains the sets of formulas of the nodes in the branch the procedure is visiting and two integers denoting, respectively, which formula of the set has been used to get the subsequent set and if the right subtree of the node has already been visited. By Remark 1, the stack has a number of elements linearly bounded in the length of the formula to be proved. Moreover, the number of symbols in each node of every proof table is linearly bounded in the length of the formula to be proved. Thus, the implementation is $O(n^2)$-SPACE. Finally, the iteration in line 30 is implemented to apply the rules of $\mathcal{C}_4$ first, since the first set in the conclusion gives rise to a classical set of formulas.

7   Conclusion and Future Work

We have provided a tableau calculus for the logic $\mathbf{E}$ and the related decision procedure that minimizes the backtracking. The implementation has been developed in C++ language. [3] Since we are interested in using the logic $\mathbf{E}$ in the field of timing analysis, we plan to extend our program in order to extract timing information from proofs of $\mathbf{E}$, to implement the algorithms described in [5].

---

[3] The program is available at `http://www.dimequant.unimib.it/elogic/index.html`.

References

1. S. Akama. Tableaux for logic programming with strong negation. In *Automated reasoning with analytic tableaux and related methods (Pont-à-Mousson, 1997)*, Lecture Notes in Artificial Intelligence, pages 31–42. Springer, Berlin, 1997.
2. A. Avellone, C. Fiorentini, G. Fiorino, and U. Moscato. An efficient implementation of a tableau calculus for a logic with a constructive negation. Technical Report 83, Dipartimento di Metodi Quantitativi per le Scienze Economiche Aziendali, Università Milano-Bicocca., 2004. Available at http://homes.dsi.unimi.it/~fiorenti.
3. M. Benini. *Verification and Analysis of Programs in a Constructive Environment*. PhD thesis, Dipartimento di Scienze dell'Informazione, Università degli Studi di Milano, Italy, 1999.
4. R. Dyckhoff. Contraction-free sequent calculi for intuitionistic logic. *Journal of Symbolic Logic*, 57(3):795–807, 1992.
5. M. Ferrari, C. Fiorentini, and M. Ornaghi. Extracting exact time bounds from logical proofs. In A. Pettorossi, editor, *Logic Based Program Synthesis and Transformation, 11th International Workshop, LOPSTR 2001, Selected Papers*, volume 2372 of *Lecture Notes in Computer Science*, pages 245–265. Springer-Verlag, 2002.
6. G. Fiorino. Space-efficient decision procedures for three interpolable propositional intermediate logics. *J. Logic Comput.*, 12(6):955–992, 2002.
7. I. Hasuo and R. Kashima. Kripke completeness of first-order constructive logics with strong negation. *IGPL*, 11(6):615–646, 2003.
8. J. Hudelmaier. An $O(n \log n)$-SPACE decision procedure for intuitionistic propositional logic. *Journal of Logic and Computation*, 3(1):63–75, 1993.
9. P. Miglioli, U. Moscato, and M. Ornaghi. Avoiding duplications in tableau systems for intuitionistic logic and Kuroda logic. *Logic Journal of the IGPL*, 5(1):145–167, 1997.
10. P. Miglioli, U. Moscato, M. Ornaghi, and G. Usberti. A constructivism based on classical truth. *Notre Dame J. Formal Logic*, 30(1):67–90, 1989.
11. D. Nelson. Constructible falsity. *J. Symbolic Logic*, 14:16–26, 1949.
12. D. Pearce. Reasoning with negative information. II. Hard negation, strong negation and logic programs. In *Nonclassical logics and information processing (Berlin, 1990)*, volume 619 of *Lecture Notes in Comput. Sci.*, pages 63–79. Springer, Berlin, 1992.
13. D. Pearce and G. Wagner. Logic programming with strong negation. In *Extensions of logic programming (Tübingen, 1989)*, volume 475 of *Lecture Notes in Comput. Sci.*, pages 311–326. Springer, Berlin, 1991.
14. R.H. Thomason. A semantical study of constructible falsity. *Z. Math. Logik Grundlagen Math.*, 15:247–257, 1969.
15. A.S. Troelstra and H. Schwichtenberg. *Basic Proof Theory*, volume 43 of *Cambridge Tracts in Theoretical Computer Science*. Cambridge University Press, 1996.
16. N. N. Vorob'ev. A new algorithm of derivability in a constructive calculus of statements. In *Sixteen papers on logic and algebra*, volume 94 of *American Mathematical Society Translations, Series 2*, pages 37–71. American Mathematical Society, Providence, R.I., 1970.
17. H. Wansing. *The logic of information structures*, volume 681 of *Lecture Notes in Computer Science*. Springer-Verlag, Berlin, 1993. Lecture Notes in Artificial Intelligence.

Automated Generation of Analytic Calculi for Logics with Linearity

Agata Ciabattoni[*]

Institut für Diskrete Mathematik und Geometrie
Research group for Computational Logic TU Wien, Austria
agata@logic.at

Abstract. We show how to automatically generate analytic hypersequent calculi for a large class of logics containing the linearity axiom (lin) $(A \supset B) \lor (B \supset A)$ starting from existing (single-conclusion) cut-free sequent calculi for the corresponding logics without (lin). As a corollary, we define an analytic calculus for Strict Monoidal T-norm based Logic **SMTL**.

1 Introduction

A central task of logic in computer science is to provide *automated generation* of suitable *analytic calculi* for a wide range of non-classical logics. By analytic calculi we mean calculi in which the proof search proceeds by step-wise decomposition of the formula to be proved. The most famous examples of such calculi are the Gentzen sequent calculus **LK** and its single-conclusion version **LJ** for classical and intuitionistic logic respectively. Cut-free "Gentzen-style" calculi serve as a basis for automated deduction, and allow the extraction of important implicit information from proofs such as numerical bounds and programs in proof-style.

The presence of the linearity axiom (lin) $(A \supset B) \lor (B \supset A)$ in the Hilbert-style axiomatization of a logic ensures a total ordering among the elements of its intended models (e.g., Kripke structures, truth-value interpretations). Several logics have been defined adding (lin) to well known systems. E.g., all fuzzy logics based on t-norm[1] connectives [12] – a prominent example being Gödel logic[2] [11, 8, 19] which arises by extending intuitionistic logic **IL** with (lin). Weaker logics such as Monoidal T-norm based Logic **MTL** [9] – the logical counterpart of left continuous t-norms and their residua – or both versions of Urquhart's **C** [21], have also been defined adding (lin) to suitable contraction-free versions of **IL**.

In this paper we show how to automatically generate analytic Gentzen style calculi for a large class of logics containing (lin). To this end we consider a natural generalization of sequent calculi: hypersequent calculi. Hypersequent calculi arise

---

[*] Work Supported by C. Bühler-Habilitations-Stipendium H191-N04, from the Austrian Science Fund (FWF).

[1] T-norms are the main tool in fuzzy logic to combine vague information.

[2] Gödel logic is also known as Dummett's **LC** [8] or Intuitionistic Fuzzy Logic [19].

J. Marcinkowski and A. Tarlecki (Eds.): CSL 2004, LNCS 3210, pp. 503–517, 2004.

by extending Gentzen calculi to refer to whole contexts of sequents instead of single sequents. They are particularly suitable for dealing with logics including (*lin*). Indeed, as shown by Avron in [2], this axiom can be enforced in **LJ**, once one embeds sequents into hypersequents and adds suitable rules to manipulate the additional layer of structure. In particular, the crucial rule added to **LJ** is the communication rule (*com*). This design resulted in an analytic calculus for Gödel logic. The same methodology was used e.g. in [6, 5] to introduce analytic hypersequent calculi for some basic fuzzy logics, including **MTL** and Urquhart's **C**, arising by adding (*lin*) to suitable contraction-free versions of **IL**.

Here we generalize these results showing that (*com*) can be viewed, in fact, as a *transfer principle* that translates (single-conclusion) cut-free sequent calculi for a *large class of* logics that do not satisfy (*lin*) into cut-free hypersequent calculi for the corresponding logics with (*lin*). This will give us the means to derive systematically analytic deduction methods for logics whose Hilbert-style axiomatizations contain (*lin*), starting from existing analytic calculi for the corresponding logics without (*lin*). To do this,

- we first introduce a general cut-elimination method for sequent calculi (*cut-elimination by substitutions*) that can be easily transferred to the hypersequent level. Sufficient conditions a calculus has to satisfy in order to admit cut-elimination by substitution are also provided. Among other things, these conditions render our cut-elimination procedure easier to verify than "ad hoc" procedures. (The verification of *unstructured* cut-elimination procedures for hypersequent calculi has been shown to be problematic in the literature.)
- We characterize *which* logics admit this transfer principle, providing some general conditions (on their sequent calculi/Hilbert-style systems) they have to satisfy both at the propositional and at the first-order level.
- As an easy corollary of the transfer principle we define an analytic hypersequent calculus for Strict Monoidal T-norm based Logic **SMTL** [9] – the logic of left-continuous *t*-norms satisfying the pseudo-complementation property.

2   Sequent and Hypersequent Calculi

The aim of this section is to settle the (hyper)sequent calculi we will deal with. We start by recalling some basic definitions in order to fix the notation and terminology we shall use throughout the paper.

The sequent calculus was introduced by Gentzen [10] in 1934 (see [18] or [20] for a detailed overview). Gentzen sequents are expressions of the form $\Gamma \Rightarrow \Delta$ where Γ and Δ are finite sequences of formulas, respectively called the antecedent and succedent of the sequent. If in a sequent calculus, succedents of all sequents contain at most one formula, the calculus is said to be *single-conclusion*.

In general, in a sequent calculus there are *axioms* (or initial sequents) and inference *rules*. The latter are divided into structural rules, logical rules and cut.

In each logical rule, the introduced formula and the corresponding auxiliary formula(s) are called *principal formula* and *active formula(s)*, respectively. We will refer to the remaining formulas in logical rules as well as to the formulas that remain unchanged in structural rules as (internal) *contexts*.

We call *additive* a multi-premises rule whose contexts in its premises are the same. If those contexts are different and simply merged in the conclusion, the rule is said to be *multiplicative*.

Recall that the structural rules introduced by Gentzen are exchange, weakening and contraction, with single-conclusion versions:

$$\frac{\Gamma, B, A, \Gamma' \Rightarrow C}{\Gamma, A, B, \Gamma' \Rightarrow C} \ (e) \qquad \frac{\Gamma, A, A \Rightarrow C}{\Gamma, A \Rightarrow C} \ (c) \qquad \frac{\Gamma \Rightarrow C}{\Gamma, A \Rightarrow C} \ (w, l) \qquad \frac{\Gamma \Rightarrow}{\Gamma \Rightarrow C} \ (w, r)$$

As is well known, their presence or absence determines completely different systems. For instance, a sequent formulation $\mathbf{ScFL}_{ew}$ for Full Lambek calculus with exchange and weakenings[3] $\mathbf{FL}_{ew}$ is obtained by eliminating (c) from the $\mathbf{LJ}$ sequent calculus for $\mathbf{IL}$ see [13]. This entails the splitting of the connective "and" of $\mathbf{IL}$, into (the additive version) $\wedge$ and (the multiplicative version) $\odot$.

Further structural rules can be defined. Here below are some examples of weaker forms of contraction i.e. *weak contraction* and *n-contraction*:

$$\frac{\Gamma, A, A \Rightarrow}{\Gamma, A \Rightarrow} \ (wc) \qquad \frac{\Gamma, A^n \Rightarrow C}{\Gamma, A^{n-1} \Rightarrow C} \ (nc)$$

where A^k stands for $A, \ldots, A$, k times.

A *derivation* in a sequent calculus is a labelled finite tree with a single root (called *end sequent*), with axioms at the top nodes, and each node-label connected with the label of the (immediate) successor nodes (if any) according to one of the rules. We refer to those connections as (correct) *inferences*.

Definition 1. *We call any propositional single-conclusion sequent calculus* standard *when it satisfies the following conditions:*

1. *antecedents of each sequent are multisets of formulas (or, equivalently, the calculus contains rule (e));*
2. *axioms have the form $A \Rightarrow A$ or $\bot \Rightarrow$;*
3. *each logical rule*
   (a) *has left and right versions, according to the side of the sequent it modifies;*
   (b) *introduces only one connective at a time;*
   (c) *has no side conditions limiting its application (besides, possibly, a condition saying that succedents of some sequents are empty)*
   (d) *has active formulas that are immediate subformulas of the principal formula;*
4. *the cut rule is multiplicative, i.e., it has the form*

$$\frac{\Gamma \Rightarrow A \quad A, \Gamma' \Rightarrow C}{\Gamma, \Gamma' \Rightarrow C} \ (cut)$$

5. *structural rules do not mention any connective.*

---

[3] $\mathbf{FL}_{ew}$ also coincides with the exponential-free fragment of affine Intuitionistic Linear Logic $\mathbf{ILL}$, i.e. $\mathbf{ILL}$ with weakenings.

Definition 2. *We call a standard sequent calculus containing the rules for quantifiers of Gentzen* **LJ** *calculus for* **IL**, *a* first-order standard sequent calculus.

Henceforth we will only consider (first-order) standard sequent calculi.

Hypersequent calculi were introduced in [1] and [14]. They are a natural generalization of Gentzen sequent calculi.

Definition 3. *A* hypersequent *is a multiset* $\Gamma_1 \Rightarrow \Pi_1 \mid \dots \mid \Gamma_n \Rightarrow \Pi_n$ *where, for all* $i = 1, \dots n$, $\Gamma_i \Rightarrow \Pi_i$ *is a Gentzen sequent.* $\Gamma_i \Rightarrow \Pi_i$ *is called a* component *of the hypersequent. A hypersequent is called* single-conclusion *if so are its components.*

The symbol "$\mid$" is intended to denote disjunction at the meta-level.

Like ordinary sequent calculi, hypersequent calculi consist of initial hypersequents (i.e., axioms) as well as logical, structural rules and cut. Axioms, logical rules and cut are essentially the same as in sequent calculi. The only difference is the presence of a *side hypersequent*, denoted by G, representing a (possibly empty) hypersequent. E.g. the hypersequent version of the **LJ** rules $(\supset, r), (\vee, r)_{1,2}$ and $(\vee, l)$ are[4] respectively:

$$\frac{G \mid \Gamma, A \Rightarrow B}{G \mid \Gamma \Rightarrow A \supset B} \; (\supset, r) \qquad \frac{G \mid \Gamma \Rightarrow A_i}{G \mid \Gamma \Rightarrow A_1 \vee A_2} \; (\vee, r)_i \qquad \frac{G \mid \Gamma, A \Rightarrow C \quad G \mid \Gamma, B \Rightarrow C}{G \mid \Gamma, A \vee B \Rightarrow C} \; (\vee, l)$$

Structural rules are divided into *internal* and *external rules*. The internal structural rules deal with formulas within components. They are the same as in ordinary sequent calculi. The external structural rules manipulate whole components of a hypersequent. Examples of this kind of rules are external weakening (ew) and external contraction (ec):

$$\frac{G}{G \mid \Gamma \Rightarrow A} \; (ew) \qquad\qquad \frac{G \mid \Gamma \Rightarrow A \mid \Gamma \Rightarrow A}{G \mid \Gamma \Rightarrow A} \; (ec)$$

Let **Sc** be any sequent calculus. We refer to its *hypersequent version* **HSc** as the calculus containing axioms and rules of **Sc** augmented with side hypersequents and in addition (ew) and (ec). (Note that **HSc** has the same expressive power as **Sc**.) However, in hypersequent calculi it is possible to define *additional external structural rules* which simultaneously act on several components of one or more hypersequents. It is this type of rule which increases the expressive power of hypersequent calculi compared to ordinary sequent calculi. A remarkable example of this kind of rules is Avron's communication rule [2]:

$$\frac{G \mid \Gamma, \Gamma' \Rightarrow A \quad G \mid \Gamma_1, \Gamma_1' \Rightarrow A'}{G \mid \Gamma, \Gamma_1 \Rightarrow A \mid \Gamma', \Gamma_1' \Rightarrow A'} \; (com)$$

Adding this rule to **HLJ** yields an analytic calculus for Gödel logic [11].

---

[4] We will use the same notation both for sequent and hypersequent rules. However, the context will always provide the relevant information.

The hypersequent version of the quantifier rules we will consider are:

$$\frac{G \mid A(t), \Gamma \Rightarrow B}{G \mid (\forall x)A(x), \Gamma \Rightarrow B} \; (\forall, l) \qquad\qquad \frac{G \mid \Gamma \Rightarrow A(a)}{G \mid \Gamma \Rightarrow (\forall x)A(x)} \; (\forall, r)$$

$$\frac{G \mid A(a), \Gamma \Rightarrow B}{G \mid (\exists x)A(x), \Gamma \Rightarrow B} \; (\exists, l) \qquad\qquad \frac{G \mid \Gamma \Rightarrow A(t)}{G \mid \Gamma \Rightarrow (\exists x)A(x)} \; (\exists, r)$$

where the eigenvariable condition in $(\exists, l)$ and $(\forall, r)$ has to apply to the whole hypersequent conclusion of the rule, i.e., the free variable a must not occur in the lower *hypersequent*. Indeed, in hypersequent calculi with (com), if one requires the weaker condition that a must not occur (only) in the lower *sequent*, then $\exists x F(x) \Rightarrow \forall x F(x)$ turns out to be derivable.

Definition 4. *We call a single-conclusion hypersequent calculus satisfying the conditions of Definition 1 (and containing the above quantifier rules) a (first-order) standard hypersequent calculus.*

Let **HS** be any sequent or hypersequent calculus. In the following we write $d, S' \vdash_{\mathbf{HS}} S$ if d is a derivation in **HS** of the (hyper)sequent S from the assumption S', i.e. a labelled tree whose nodes are applications of rules of **HS** and whose leaves are either S' or axioms.

Definition 5. *The length $|d|$ of a derivation d in **HS** is (the maximal number of inference rules) + 1 occurring on any branch of d. The complexity $|A|$ of a formula A is defined as the number of occurrences of its connectives and quantifiers. The cut-rank $\rho(d)$ of d is (the maximal complexity of cut-formulas in d) + 1. ($\rho(d) = 0$ if d is cut free).*

3   Cut-Elimination by Substitutions

Cut-elimination is one of the most important procedures in logic. The removal of cuts corresponds to the elimination of "lemmas" from derivations. This renders a derivation *analytic*, in the sense that all formulas occurring in the derivation are subformulae of the formula to be proved.

Here we prove that if a standard (first-order) sequent calculus **Sc** admits cut-elimination, **HSc** + (com) i.e. its hypersequent version with in addition (com), admits cut-elimination too. For this purpose, we introduce a cut-elimination method for sequent calculi (*cut-elimination by substitutions*) that can be easily transferred to the hypersequent level (and in particular to the corresponding hypersequent calculi with (com)).

We start discussing which of, and how, the main cut-elimination methods for sequent calculi can be used in hypersequent context. Recall that Gentzen's cut-elimination method proceeds by eliminating a *uppermost cut* in a derivation by a double induction on the complexity c of the cut formula (+1) and on the sum l of the lengths of its left and right derivations. In his original proof of the cut-elimination theorem for sequent calculus [10], Gentzen met the following problem: If the cut formula is derived by (c), the permutation of cut with (c) does not necessarily move the cut higher up in the derivation. To solve this problem, he introduced the mix rule – a derivable generalization of cut.

In hypersequent calculi a similar problem arises when one tries to permute cut with (ec). (Note that the solution proposed in [6], i.e., to proceed by induction on $(\#(ec), c, l)$ where $\#(ec)$ is the number of applications of (ec) in a derivation, does not work.) In analogy with Gentzen's solution, a way to overcome the problem due to (ec) is to introduce suitable "ad hoc" (derivable) generalizations of the mix rule for each hypersequent calculus. These rules should allow certain cuts to be reduced *in parallel*. E.g. to prove cut-elimination in the hypersequent calculus for propositional Gödel logic, Avron used the following induction hypothesis [2] (generalized mix rule):

If $H \mid \Gamma_1 \Rightarrow A \mid \ldots \mid \Gamma_n \Rightarrow A$ and $H \mid \Sigma_1, A^{n_1} \Rightarrow B_1 \mid \ldots \mid \Sigma_k, A^{n_k} \Rightarrow B_k$ are cut-free provable, so is $H \mid \Gamma, \Sigma_1 \Rightarrow B_1 \mid \ldots \mid \Gamma, \Sigma_k \Rightarrow B_k$, where $\Gamma = \Gamma_1, \ldots, \Gamma_n$ and A^{n_l} stands for $A, \ldots, A$, n_l times.

However, this generalized mix rule does not work for calculi not admitting, e.g., (c) or (w). (Note that to shift upward a cut in which a component $\Gamma_i \Rightarrow A$, with $i \in \{1, \ldots, n\}$, is derived by (ec) or (ew), one needs to use rules (c) and (w), respectively).

A different cut-elimination method for sequent calculus was introduced by Schütte-Tait [15, 17]. This proceeds by eliminating a *largest cut* in a derivation (w.r.t. the number of connectives and quantifiers). The main feature of this method is that a cut with a non-atomic cut formula is not shifted upward but simply reduced (i.e., replaced by smaller cuts) using the inversion(s) of the premises of the original cut (see, e.g. [16]). This renders the presence of (ec) unproblematic once one uses this method in hypersequent calculi. Proofs of cut-elimination à la Schütte-Tait for the hypersequent calculi for (first-order) Gödel logic and **MTL** can be found, e.g., in [3, 5]. There in fact to eliminate a cut with a non-atomic cut formula only *one* premise of this cut is inverted and used to replace the cut by smaller ones exactly in the place(s) in which the cut formula (of the remaining premise of the cut) is introduced.

However, cut-elimination à la Schütte-Tait cannot be straightforwardly transferred from a sequent to the corresponding hypersequent calculus. Moreover, demanding the invertibility (even) of (only) one of the premises of cuts seems to be a rather strong condition. Indeed, there do exist (hyper)sequent calculi in which cuts are eliminable but in which none of the premises of a cut is invertible. An example of such a calculus is obtained by replacing the right rule introducing $\land$ in the $\mathbf{ScFL}_{ew}$ calculus for $\mathbf{FL}_{ew}$ by the following rules:

$$\frac{\Gamma \Rightarrow A_1 \quad \Gamma', A_1 \Rightarrow A_2}{\Gamma, \Gamma' \Rightarrow A_1 \land A_2} \; (\land, r)_1 \qquad \frac{\Gamma \Rightarrow A_2 \quad \Gamma', A_2 \Rightarrow A_1}{\Gamma, \Gamma' \Rightarrow A_1 \land A_2} \; (\land, r)_2$$

This calculus admits cut-elimination (e.g., using Gentzen's method, see [5]) but neither of the premises of a cut with cut formula $A \land B$ can be inverted in the usual way.

In the proof of Theorem 1 below, we introduce *cut-elimination by substitutions*. This proceeds by eliminating a *largest uppermost* cut in a derivation. The idea behind this method is to eliminate a cut via suitable substitutions in the derivations $d_0 \vdash_{\mathbf{Sc}} \Sigma \Rightarrow A$ and $d_1 \vdash_{\mathbf{Sc}} \Gamma, A \Rightarrow C$ of its premises. We substitute all the occurrences of the cut formula. When we do this we have also to replace all the subproofs of d_0 and d_1 ending in an inference whose principal formula is an occurrence of the cut-formula. This requires us to trace up the occurrences of the cut formula through d_0 and d_1. For this purpose we use below the notion of *decoration of a formula A in a (hyper)sequent derivation d*. This essentially amounts to the (marked) derivation obtained by following up and marking in d all occurrences of the considered formula A starting from the end sequent of d: if at some stage any marked occurrence of A –indicated by A^*– is multiplied by a certain (internal or external) structural rule we mark and trace up all these occurrences of the formula from the premise(s). In outline, two cases can occur.

- If the cut formula (A^*) was not introduced by *any* logical (or quantifier) rule in d_0 (respectively d_1), the cut is replaced by the derivation d_0 (respectively d_1) in which one substitutes all A^* by Γ and C (respectively Σ) ($\star$).
- Suppose A^* was introduced by some logical (or quantifier) rules in d_0 and d_1. The required derivation is obtained from d_0 and d_1 by replacing all A^*s via suitable substitutions ($\star$), and replacing the inferences which introduced A^* with suitable cuts on subformulas of A ($\star\star$).

The applicability of cut elimination by substitutions relies on the fact that the considered (standard) sequent calculus satisfies ($\star\star$) and ($\star$), namely, its rules allow the replacement of cuts by smaller ones (i.e. logical and quantifier rules are *reductive*) and they lead to correct inferences once one uniformly replaces any formula in their premises and (some occurrences of this formula in their) conclusions by multisets of formulas (i.e., rules are *substitutive*). The latter condition can be equivalently expressed as: the rules allow any cut to be shifted upward replacing the cut formula in their premises by the contexts of the remaining premise of the cut.

Before introducing the formal definition of reductive and substitutive rules let us consider the following explanatory example:

Example 1. The contraction rule (c) is substitutive. Indeed the sequents obtained by replacing any formula $X \in \Gamma$ (or by replacing A) with a multiset Σ in its conclusion, can be derived by applying (c) to the sequent $\Gamma, A, A \Rightarrow C$ after having replaced $X \in \Gamma$ (or the two occurrences of A) with Σ. Moreover, the sequent $\Gamma, A, \Sigma \Rightarrow D$, obtained by substituting C in the conclusion of (c) with Σ and D, can be derived by applying (c) to $\Gamma, A, A \Rightarrow C$ in which one carries out the same substitution. By contrast, the n-contraction rule (nc) is not substitutive. Indeed e.g. the sequent $\Gamma, A^{n-2}, \Sigma \Rightarrow C$, obtained by substituting one occurrence of A with Σ in its conclusion cannot be derived by applying (nc) to $\Gamma, \Sigma^n \Rightarrow C$.

Definition 6. *Let* **HS** *be any standard (hyper)sequent calculus.*

*We call its (logical or quantifier) rules $\{(\star, r)_1, \ldots, (\star, r)_n\}$ and $\{(\star, l)_1,$ $\ldots, (\star, l)_m\}$ for introducing a connective (or a quantifier) $\star$ reductive, whenever the sequent obtained via (cut) on the principal formula of the conclusions of $(\star, l)_i$ and $(\star, r)_j$ (for each $i = 1, \ldots, m$ and $j = 1, \ldots m$) can be derived from their premises using (cut) and the structural rules of **HS**. Any **HS**-rule ($n \geq 1$ and $C \neq C'$)*

$$\frac{(G \mid \Gamma'_1 \Rightarrow C'_1 \mid)\Gamma_1 \Rightarrow C_1 \ldots \quad \ldots (G \mid \Gamma'_n \Rightarrow C'_n \mid)\Gamma_n \Rightarrow C_n}{(G \mid \Gamma' \Rightarrow C' \mid)\Gamma \Rightarrow C} \ (R)$$

is said to be substitutive *whenever the following conditions hold:*

1. *Let X be any formula that is not principal in (R) occurring in Γ (or Γ') and let H be the (hyper)sequent arising by replacing some occurrences of X in Γ or Γ' with any multiset of formulas Σ. H can be derived using only (R) and the structural rules of **HS** from the premises of (R) with Σ uniformly substituted for every occurrence of X in each Γ_i and Γ'_i ($i = 1, \ldots, n$).*
2. *If C (respectively C') is neither empty nor principal in (R), the (hyper)sequent $(G \mid \Gamma' \Rightarrow C' \mid)\Sigma, \Gamma \Rightarrow D$ (respectively $(G \mid \Gamma', \Sigma \Rightarrow D) \mid \Gamma \Rightarrow C$), for any Σ and D, is derivable only using (R) and the structural rules of **HS** from the premises of (R) with $\Gamma_i^{(')}, \Sigma \Rightarrow D$ uniformly substituted for each $\Gamma_i^{(')} \Rightarrow C_i^{(')}$ in which $C_i^{(')} = C$ (respectively $C_i^{(')} = C'$).*

Let $d(s)$ and $H(s)$ denote the results of substituting the term s for all free occurrences of x in the derivation $d(x)$ and in the (hyper)sequent $H(x)$.

Lemma 1 (Substitution Lemma). *Let **HS** be any standard first-order (hyper)sequent calculus. If $d(x) \vdash_{\mathbf{HS}} H(x)$, then $d(s) \vdash_{\mathbf{HS}} H(s)$, with $|d(s)| = |d(x)|$ and $\rho(d(s)) = \rho(d(x))$, where s only contains variables that do not occur in $d(x)$.*

Using the above lemma one can show

Lemma 2. *The (hyper)sequent rules $(\forall, \lhd)$ and $(\exists, \lhd)$, with $\lhd \in \{l, r\}$, are substitutive in any standard first-order (hyper)sequent calculus.*

Theorem 1. *Any standard (first-order) sequent calculus **Sc** in which (a) logical rules are reductive and (b) rules are substitutive, admits cut-elimination.*

Proof. Let $d \vdash_{\mathbf{Sc}} S$, with $\rho(d) > 0$. The proof proceeds by induction on the pair $(\rho(d), \#\rho(d))$, where $\#\rho(d)$ is the number of cuts in d with cut-rank $\rho(d)$. Suppose $\rho(d) = |A| + 1$ and let

$$d_0 \vdash_{\mathbf{Sc}} \Sigma \Rightarrow A \quad \text{and} \quad d_1 \vdash_{\mathbf{Sc}} \Gamma, A \Rightarrow C$$

be the premises of the uppermost cut in d with cut-formula A. We can find a derivation $d' \vdash_{\mathbf{Sc}} \Gamma, \Sigma \Rightarrow C$ with $\rho(d') < \rho(d)$. Hence, replacing in d the subderivation ending in this largest uppermost cut by d', results in a derivation $\bar{d}$ such that either $\rho(\bar{d}) < \rho(d)$ or $\#\rho(\bar{d}) = \#\rho(d) - 1$. Two cases can occur:

1. The cut-formula A is not introduced by any logical (or quantifier) inference in d_0 or d_1. Assume first that this is the case in d_1. We consider the decoration of A in d_1 starting from $d_1 \vdash_{\mathbf{Sc}} \Gamma, A^* \Rightarrow C$. We then substitute A^* everywhere in d_1 by Σ. Let us call d_1^* the obtained labelled tree. Since A is not introduced by any logical (or quantifier) inference in d_1 and $\mathbf{Sc}$ is a (first-order) standard sequent calculus whose rules are substitutive, all the inferences in d_1^* are correct (upon adding some structural inferences, if needed). Note that if A^* originates in an axiom $A^* \Rightarrow A$, this is transformed into $\Sigma \Rightarrow A$. Hence d_1^* is a derivation in $\mathbf{Sc}$ and either $d_1^*, \Sigma \Rightarrow A \vdash_{\mathbf{Sc}} \Gamma, \Sigma \Rightarrow C$ or $d_1^* \vdash_{\mathbf{Sc}} \Gamma, \Sigma \Rightarrow C$. A derivation $d' \vdash_{\mathbf{Sc}} \Gamma, \Sigma \Rightarrow C$ with $\rho(d') < \rho(d)$ is thus obtained by replacing d_0 and d_1 in d by (the juxtaposition of d_0 and) d_1^*. The case where A is not introduced by any logical (or quantifier) inference in d_0 is symmetric. Here we consider the decoration of A in d_0 starting from $d_0 \vdash_{\mathbf{Sc}} \Sigma \Rightarrow A^*$ and we substitute in d_0 each sequent of the form $\Pi \Rightarrow A^*$ with $\Pi, \Gamma \Rightarrow C$ possibly adding suitable structural inferences, if needed. The rest of the proof proceeds (similarly) as above.

2. The cut-formula A is introduced by logical (or quantifier) inferences both in d_0 and d_1. Let us consider the decoration of A in d_0 and d_1 starting from $d_0 \vdash_{\mathbf{Sc}} \Sigma \Rightarrow A^*$ and $d_1 \vdash_{\mathbf{Sc}} \Gamma, A^* \Rightarrow C$ respectively. Suppose $A = \star(A_1, \ldots A_p)$, where $\star$ is any connective, or $A = \forall x B(x)$. Let $\Sigma_1 \Rightarrow A^*, \ldots, \Sigma_n \Rightarrow A^*$ and $\Gamma_1, A^* \Rightarrow C_1 \ldots \Gamma_m, A^* \Rightarrow C_m$ be the conclusions of the logical (or $\forall$) inferences introducing A^* in d_0 and d_1. We first replace A^* with Σ_1 everywhere in d_1. Note that the resulting tree is not a derivation anymore. However, since the rules of (first-order) $\mathbf{Sc}$ are substitutive, all the inferences – except those that introduced A^* in d_1 – are correct (upon adding some structural inferences, if needed). These incorrect inferences have the following form (assume w.l.o.g. that $(\star, l)$ is a one-premise rule)

$$\begin{array}{c} \vdots\, d_1' \\ \dfrac{\Gamma_1', A_l, \ldots A_t \Rightarrow B_1'}{\Gamma_1, \Sigma_1 \Rightarrow B_1}\,{}_{(\star,l)} \end{array}$$

We replace them by cut(s) with $d_1' \vdash_{\mathbf{Sc}} \Gamma_1', A_l, \ldots A_t \Rightarrow B_1'$ and the premise(s) of the inference rule introducing A^* in d_0, with conclusion $\Sigma_1 \Rightarrow A^*$, (previously applying the Substitution Lemma and), adding some structural inferences, if needed. We call the resulting tree d_{1_1}. Note that if d_1 also contains axioms $A^* \Rightarrow A$, these are transformed into sequents $\Sigma_1 \Rightarrow A$ in d_{1_1}. These are simply replaced by the subderivation of d_0 ending in $\Sigma_1 \Rightarrow A$. Since the rules of $\mathbf{Sc}$ are reductive, $d_{1_1}^*$ is a derivation in $\mathbf{Sc}$. Moreover, it is easy to check that $d_{1_1}^* \vdash_{\mathbf{Sc}} \Gamma, \Sigma_1 \Rightarrow C$. Similarly, we can obtain derivations $d_{1_2}^*, \ldots d_{1_n}^*$ of $\Gamma, \Sigma_2 \Rightarrow C, \ldots \Gamma, \Sigma_n \Rightarrow C$, with $\rho(d_{1_i}^*) < \rho(d)$, for $i = 1, \ldots, n$. This is not yet what we were looking for. Let us substitute in (the decorated version of) d_0 each sequent of the form $\Pi \Rightarrow A^*$ with $\Pi, \Gamma \Rightarrow C$, possibly adding suitable structural inferences, if needed. (If d_0 also contains axioms $A \Rightarrow A^*$, these are replaced by the derivation d_1). As before, the resulting tree is not a derivation anymore and the only incorrect inferences are those which introduced A^* that now have the form (assume w.l.o.g. that $(\star, r)$ is a one-premise rule)

$$\vdots$$
$$\frac{\Sigma_i' \Rightarrow A_k}{\Sigma_i, \Gamma \Rightarrow C} \,\, {\scriptstyle (\star,r)}$$

To correct these inferences we replace the whole subtree ending in $\Sigma_i, \Gamma \Rightarrow C$ with the derivation $d_{1_i}^*$ obtained before. Iterating this procedure for all the n inferences introducing A^* in d_0, leads to the required derivation $d' \vdash_{\mathbf{Sc}} \Gamma, \Sigma \Rightarrow B$ with $\rho(d') < \rho(d)$.

If $A = \exists x B(x)$, the proof proceeds as above exchanging, however, the role of d_0 and d_1. This way, one can replace the incorrect $(\exists, r)$ inferences by introducing $(\exists x B(x))^*$ with a cut from their premises and the premises of the $(\exists, l)$ inferences introducing $(\exists x B(x))^*$ in d_1, previously applying the Substitution Lemma to the latter.

Cut-elimination by substitutions can be easily used in hypersequent calculi. First note that (ew) and (ec) are substitutive in any hypersequent calculus.

Theorem 2. *Any (first-order) standard hypersequent calculus* **HL** *in which (a) logical rules are reductive and (b) rules are substitutive, admits cut-elimination.*

Proof. Let $d \vdash_{\mathbf{HL}} H$, with $\rho(d) = |A| + 1$ and let $d_0 \vdash_{\mathbf{HL}} G \mid \Sigma \Rightarrow A$ and $d_1 \vdash_{\mathbf{HL}} G \mid \Gamma, A \Rightarrow C$ be the premises of the uppermost cut in d with cut-formula A. We show that we can find a derivation $d' \vdash_{\mathbf{HL}} G \mid \Gamma, \Sigma \Rightarrow C$ with $\rho(d') < \rho(d)$. The proof proceeds by induction on $(\rho(d), \#\rho(d))$. We sketch below the (few) additional steps – w.r.t. those outlined in the proof of Theorem 1 – needed to cope with side hypersequents.

1. The cut-formula A is not introduced by any logical (or quantifier) inference in d_0 or d_1. Assume w.l.o.g. that this is the case in d_1. We first add G to all the hypersequents in d_1 and for each newly generated hypersequent $G \mid B \Rightarrow B$ or $G \mid \bot \Rightarrow$ (if any), we add an application of (ew) to recover the original axiom $B \Rightarrow B$ or $\bot \Rightarrow$ of d_1. The remaining steps are as in the proof of Theorem 1. The required derivation is finally obtained by applying (ec) to d_1^*.

2. The cut-formula A is introduced by logical (or quantifier) inferences both in d_0 and d_1. Let $G_1 \mid \Sigma_1 \Rightarrow A^*, \ldots, G_n \mid \Sigma_n \Rightarrow A^*$ (and $H_1 \mid \Gamma_1, A^* \Rightarrow C_1 \ldots H_m \mid \Gamma_m, A^* \Rightarrow C_m$) be the conclusions of the logical (or quantifier) inferences introducing A^* in d_0 and d_1, respectively. Assume, w.l.o.g., $A = \star(A_1, \ldots A_p)$ or $A = \forall x B(x)$. We first add G_i to all the hypersequents in d_1 and we add applications of (ew) to recover the original axioms of d_1, if needed. Following the same steps as in the proof of Theorem 1, we obtain the derivations $d_{1_i}^* \vdash_{\mathbf{HL}} G_i \mid G \mid \Gamma, \Sigma_i \Rightarrow C$, for $i = 1, \ldots, n$. We now first add G to all the hypersequents in d_0 and we then proceed as in the proof of Theorem 1. This leads to $d'' \vdash_{\mathbf{HL}} G \mid G \mid \Gamma, \Sigma \Rightarrow B$. The required derivation is finally obtained by applying (ec) to d''.

Corollary 1. *Let* **Sc** *be a standard (first-order) sequent calculus in which (a) logical rules are reductive and (b) rules are substitutive.* **HSc** + *(com) admits cut-elimination.*

Proof. It is easy to verify that **HSc** with in addition *(com)* satisfies conditions (a) and (b) too. The claim follows by Theorem 2.

4   Transfer Principle

Let **Sc** be a (first-order) standard sequent calculus that admits cut-elimination by substitutions. Here we show that if **Sc** (or, equivalently, the formalized logic **L**) is "expressive enough", then **HSc** + *(com)* is an analytic calculus for **L**+ axiom schemata $(A \supset B) \vee (B \supset A)$ (+, in the first-order case, $\forall x(P(x) \vee Q) \supset (\forall x P(x) \vee Q)$, where x does not occur free in Q).

Henceforth we assume logics to be specified by Hilbert-style systems. A logic **L** is identified with the set of its provable formulas. By a first-order logic **L** we mean a Hilbert system whose rules are *modus ponens* and *generalization* and whose axioms for quantifiers are those of first-order intuitionistic logic.

In order to interpret (hyper)sequents into the language of the considered logics, we assume these contain a disjunction connective $\vee$, an implication $\supset$ and the constant $\bot$. Since sequents (respectively hypersequents) are multisets of formulas (respectively sequents), we assume $\vee$ is commutative and $\supset$ satisfies exchange (i.e. $(A \supset (B \supset C)) \supset (B \supset (A \supset C))$). Moreover, $\bot \supset A$ belongs to the provable formulas.

Definition 7. *Let* $A_1, \ldots, A_n \Rightarrow B$ *be a sequent. Its* generic interpretation $\mathcal{I}$ *is defined as follows:*

$$\mathcal{I}(\Rightarrow B) := B$$
$$\mathcal{I}(A_1, \ldots, A_n \Rightarrow B) := (A_1 \supset \ldots \supset (A_n \supset B) \ldots)$$
$$\mathcal{I}(A_1, \ldots, A_n \Rightarrow) := (A_1 \supset \ldots \supset (A_n \supset \bot) \ldots)$$

Let G *be the hypersequent* $S_1 \mid \cdots \mid S_n$. *Then its* generic interpretation $\mathcal{I}(G)$ *is defined as* $\mathcal{I}(S_1) \vee \ldots \vee \mathcal{I}(S_n)$.

Definition 8. *A (Hyper)sequent rule*

$$\frac{S_1 \quad \ldots \quad S_n}{S_0} \; (r) \qquad \text{with } n \geq 1$$

is sound *for a Hilbert style system* **L**, *if whenever* **L** *derives the generic interpretations of its premises,* **L** *derives the generic interpretation of its conclusion too.* (r) *is* strongly sound *for* **L** *if* **L** *derives the formula* $\mathcal{I}(S_1) \supset (\ldots (\mathcal{I}(S_n) \supset \mathcal{I}(S_0)) \ldots)$. *A (hyper)sequent calculus* **HL** *is called* sound *(resp. strongly sound) for* **L** *if all the axioms and rules of* **HL** *are sound (resp. strongly sound) for* **L**. **HL** *is called* complete *for* **L** *if for all formulas* A *derivable in* **L**, *the (hyper)sequent* $\Rightarrow A$ *is derivable in* **HL**.

Lemma 3. *Let* **Sc** *be a standard sequent calculus in which the* **LJ** *rules* $(\supset, r)$, $(\vee, r)_{1,2}$, $(\vee, l)$ *as well as the rule*

$$\frac{\Gamma \Rightarrow A \quad \Gamma', B \Rightarrow C}{\Gamma, \Gamma', A \supset B \Rightarrow C} \; (\supset, l)$$

are derivable. If **Sc** *is strongly sound and complete for* **L** *then the following properties hold:*

1. $(A \supset B) \supset ((B \supset C) \supset (A \supset C)) \in \mathbf{L}$
2. $A \supset (G \vee A) \in \mathbf{L}$
3. *If* $A \supset B \in \mathbf{L}$ *then* $(H \vee A) \supset (H \vee B) \in \mathbf{L}$,
4. $(A \vee A) \supset A \in \mathbf{L}$
5. *If* $A \in \mathbf{L}$, $B \in \mathbf{L}$ *and* $A \supset X \vee B \supset X \in \mathbf{L}$, *then* $X \in \mathbf{L}$.
6. *If* $A \supset B \in \mathbf{L}$ *and* $C \supset D \in \mathbf{L}$ *then* $(A \vee C) \supset (B \vee D) \in \mathbf{L}$,
7. *If* $(A \supset B) \vee H \in \mathbf{L}$ *and* $A \in \mathbf{L}$, *then* $B \vee H \in \mathbf{L}$,
8. *If* $A \vee B \in \mathbf{L}$ *and* $A \supset X \in \mathbf{L}$, *then* $X \vee B \in \mathbf{L}$,
9. *If* $A \supset (B \supset C) \in \mathbf{L}$, $A \vee H$, $B \vee H \in \mathbf{L}$, *then* $C \vee H \in \mathbf{L}$.
10. *If* $A_1 \supset (A_2 \supset \ldots (A_n \supset B) \ldots)) \in \mathbf{L}$ *and* $A_i \vee H \in \mathbf{L}$, *for each* $i = 1, \ldots n$, *then* $B \vee H \in \mathbf{L}$.

Proof. 3. By Property 2, $B \supset (H \vee B) \in \mathbf{L}$, hence by Property 1 and modus ponens, $A \supset (H \vee B) \in \mathbf{L}$. Since $H \supset (H \vee B) \in \mathbf{L}$, follows that $(H \vee A) \supset (H \vee B) \in \mathbf{L}$.

5. From $A \in \mathbf{L}$ and $B \in \mathbf{L}$ follows $(A \supset X \vee B \supset X) \supset X \in \mathbf{L}$. The claim follows by modus ponens.

7. From $A \in \mathbf{L}$ we get $[(A \supset B) \vee H] \supset B \vee H$. The claim follows by modus ponens.

9. By Property 3, $(B \supset C) \supset [(B \vee H) \supset (C \vee H)] \in \mathbf{L}$. By Property 1 and modus ponens follows $A \supset [(B \vee H) \supset (C \vee H)] \in \mathbf{L}$. By Property 3 and modus ponens we get $(A \vee H) \supset [(B \vee H) \supset (C \vee H) \vee H] \in \mathbf{L}$. By modus ponens we obtain $[(B \vee H) \supset (C \vee H)] \vee H$ and by Property 7 $(C \vee H) \vee H \in \mathbf{L}$. The claim follows since $[(C \vee H) \vee H] \supset (C \vee H) \in \mathbf{L}$.

10. Follows by repetedely applying Properties 3, 7 and 9.

Theorem 3. *Let* **Sc** *be a standard sequent calculus in which the rules* $(\supset, r)$, $(\supset, l), (\vee, r)_{1,2}, (\vee, l)$ *are derivable. If* **Sc** *is strongly sound and complete for* **L**, *then* **HSc** $+ (com)$ *is sound and complete for*

$$\mathbf{L} + (A \supset B) \vee (B \supset A)$$

Proof. *(Soundness)* The soundness of logical and internal structural rules of **HSc** follows by the strongly soundness of **Sc** w.r.t. **L** together with Property 10. The soundness of (ec) is ensured by Properties 3 and 4, while that of (ew) follows by Property 2. For *(com)* we can argue as follows: Assume $\mathcal{I}(\Gamma, \Gamma' \Rightarrow A) \vee H \in \mathbf{L}$ and $\mathcal{I}(\Gamma_1, \Gamma_1' \Rightarrow A') \vee H \in \mathbf{L}$. We show that

$$(*) \quad \mathcal{I}(\Gamma, \Gamma_1 \Rightarrow A) \vee \mathcal{I}(\Gamma', \Gamma_1' \Rightarrow A') \vee H \in \mathbf{L}$$

Indeed, let the notation $[\Sigma]$, where $\Sigma = \Sigma_1, \ldots \Sigma_n$, stand for $[(\Sigma_1 \supset (\ldots (\Sigma_{n-1} \supset \Sigma_n) \ldots))$. We have

$$([\Gamma_1] \supset [\Gamma']) \supset (\mathcal{I}(\Gamma, \Gamma' \Rightarrow A) \supset \mathcal{I}(\Gamma, \Gamma_1 \Rightarrow A)) \quad \text{and}$$

$$([\Gamma'] \supset [\Gamma_1]) \supset (\mathcal{I}(\Gamma_1, \Gamma_1' \Rightarrow A') \supset \mathcal{I}(\Gamma', \Gamma_1' \Rightarrow A')).$$

By Properties 2, 3, 1 and modus ponens follow

$$(\mathcal{I}(\Gamma, \Gamma' \Rightarrow A) \supset \mathcal{I}(\Gamma, \Gamma_1 \Rightarrow A)) \supset ((\mathcal{I}(\Gamma, \Gamma' \Rightarrow A) \vee H) \supset (*)) \in \mathbf{L} \quad \text{and}$$

$$(\mathcal{I}(\Gamma_1, \Gamma_1' \Rightarrow A') \supset \mathcal{I}(\Gamma', \Gamma_1' \Rightarrow A')) \supset ((\mathcal{I}(\Gamma_1, \Gamma_1' \Rightarrow A') \vee H) \supset (*)) \in \mathbf{L}$$

By Properties 1, 6 and axiom $(A \supset B) \vee (B \supset A)$ we get

$$((\mathcal{I}(\Gamma, \Gamma' \Rightarrow A) \vee H) \supset (*)) \vee ((\mathcal{I}(\Gamma_1, \Gamma_1' \Rightarrow A') \vee H) \supset (*)) \in \mathbf{L}$$

the claim follows by Property 5.

(*Completeness*) Since **Sc** (and hence **HSc**) is complete for **L**, the claim follows by the derivability of the linearity axiom in **HSc** + (*com*):

$$\frac{\dfrac{A \Rightarrow A \quad B \Rightarrow B}{A \Rightarrow B \mid B \Rightarrow A} \; {\scriptstyle (\text{com})}}{\dfrac{\Rightarrow A \supset B \mid \Rightarrow B \supset A}{\dfrac{\Rightarrow (A \supset B) \vee (B \supset A) \mid \Rightarrow (A \supset B) \vee (B \supset A)}{\Rightarrow (A \supset B) \vee (B \supset A)} \; {\scriptstyle (\text{ec})}} \; {\scriptstyle 2\text{x}(\vee_1,\text{r})}} \; {\scriptstyle 2\text{x}(\supset,\text{r})}$$

Corollary 2 (Transfer Principle). *Let* **Sc** *be a standard sequent calculus whose logical rules are reductive and all its rules are substitutive and in which the rules* $(\supset, r), (\vee, r)_{1,2}, (\vee, l)$ *and* $(\supset, l)$ *are derivable. If* **Sc** *is strongly sound and complete for* **L** *then* **HSc** + (*com*) *is an analytic calculus sound and complete for* $\mathbf{L} + (A \supset B) + (B \supset A)$.

If **Sc** contains quantifier rules, this result does not hold anymore. E.g. in **LJ** the rules $(\supset, r), (\vee, r)_{i: i=1,2}, (\vee, l)$ and $(\supset, l)$ are derivable. However the calculus obtained by adding (*com*) to the hypersequent version of **LJ** is *not* sound for first-order **IL** with the linearity axiom. (This logic, introduced by Corsi in [7], is semantically characterized by linearly ordered Kripke frames.) Indeed in this calculus one can derive the shifting law of universal quantifiers w.r.t. $\vee$, i.e., $(\vee\forall) \; \forall x(P(x) \vee Q) \supset (\forall x P(x) \vee Q)$, where x does not occur free in Q. This law, that forces the domains of the corresponding Kripke models to be constant, is not valid in Corsi's logic. In fact, **HLJ** + (*com*) turns out to be sound and complete for first-order Gödel logic [4] – whose axiomatization is obtained by adding $(\vee\forall)$ to Corsi's logic. As the theorem below shows, this is not by chance, but follows a general principle (note that $(\vee\forall)$ is needed to prove the soundness of the hypersequent rule $(\forall, r)$).

Theorem 4. *Let* **Sc** *be a standard sequent calculus in which the rules* $(\supset, r)$, $(\vee, r)_{1,2}, (\vee, l)$ *and* $(\supset, l)$ *are derivable. If (propositional)* **Sc** *is strongly sound and complete for* **L**, *then first-order* **HSc** + (*com*) *is sound and complete for*

$$\text{first-order} \quad \mathbf{L} + (A \supset B) \vee (B \supset A) + (\vee\forall)$$

Proof. (*Soundness*) By Theorem 3 it is enough to prove the soundness of the hypersequent rules for quantifiers w.r.t. **L**. The cases $(\forall, l)$ and $(\exists, r)$ are easy. For $(\forall, r)$ we may argue as follows: If $\mathcal{I}(G) \vee \mathcal{I}(\Gamma \Rightarrow A(a)) \in \mathbf{L}$, $\forall x (\mathcal{I}(G) \vee \mathcal{I}(\Gamma \Rightarrow A(x))) \in \mathbf{L}$ too. Since a did not occur in $\mathcal{I}(G)$ or in $\mathcal{I}(\Gamma \Rightarrow A(a))$, we may now assume that x does not either. Hence $\mathcal{I}(G) \vee \forall x \mathcal{I}(\Gamma \Rightarrow A(x)) \in \mathbf{L} + (\vee\forall)$. The result follows by Property 8 since $\forall x \mathcal{I}(\Gamma \Rightarrow A(x)) \supset \mathcal{I}(\Gamma \Rightarrow \forall x A(x)) \in \mathbf{L}$. The soundness of $(\exists, l)$ can be proved in a similar way.

(*Completeness*) Since the generalization rule is a particular case of $(\forall, r)$, by Theorem 3 it is enough to prove that $\vdash_{\mathbf{HSc}+(com)} \Rightarrow (\vee\forall)$. Indeed

$$
\cfrac{
A(a) \Rightarrow A(a) \quad
\cfrac{
\cfrac{
\cfrac{
\cfrac{
\cfrac{A(a) \Rightarrow A(a) \quad B \Rightarrow B}{B \Rightarrow A(a) \mid A(a) \Rightarrow B}(com) \quad B \Rightarrow B
}{A(a) \vee B \Rightarrow A(a) \mid A(a) \vee B \Rightarrow B} 2x(\vee,l)+(ew)s
}{\forall x(A(x) \vee B) \Rightarrow A(a) \mid \forall x(A(x) \vee B) \Rightarrow B} 2x(\forall,l)
}{\forall x(A(x) \vee B) \Rightarrow \forall x A(x) \mid \forall x(A(x) \vee B) \Rightarrow B}(\forall,r)
}{\forall x(A(x) \vee B) \Rightarrow \forall x A(x) \vee B \mid \forall x(A(x) \vee B) \Rightarrow \forall x A(x) \vee B} 2x(\vee,r)
}
{\forall x(A(x) \vee B) \Rightarrow \forall x A(x) \vee B}(ec)
}{\Rightarrow \forall x(A(x) \vee B) \supset (\forall x A(x) \vee B)}(\supset,r)
$$

Corollary 3 (Transfer Principle). *Let* **Sc** *be a standard first-order sequent calculus whose logical rules are reductive and rules are substitutive and in which the rules* $(\supset, r), (\vee, r)_{1,2}, (\vee, l)$ *and* $(\supset, l)$ *are derivable. If* **Sc** *is strongly sound and complete for* **L** *then first-order* **HSc** $+ (com)$ *is an analytic calculus sound and complete for first-order* **L** $+ (lin) + (\vee\forall)$.

5   SMTL: A Case Study

As an easy corollary of the transfer principle introduced above, we define here an analytic calculus for Strict Monoidal T-norm based Logic **SMTL**. This logic was defined in [9] by adding axioms $((A \supset \bot) \wedge A) \supset \bot$ and (lin) to $\mathbf{FL_{ew}}$. **SMTL** turns out to be the logic based on left-continuous t-norms satisfying the pseudo-complementation property. To the best of our knowledge no analytic calculi have been provided for **SMTL** so far.

Proposition 1. $\mathbf{ScFL_{ew}} + (wc)$ *is strongly sound and complete for* $\mathbf{FL_{ew}}$ *extended with* $((A \supset \bot) \wedge A) \supset \bot$.

Proof. (*Soundness*) $\vdash_{\mathbf{ScFL_{ew}}} \mathcal{I}(\Gamma, A, A \Rightarrow), ((A \supset \bot) \wedge A) \supset \bot \Rightarrow \mathcal{I}(\Gamma, A \Rightarrow)$. Hence the claim follows by the strongly soundness of $\mathbf{ScFL_{ew}}$ w.r.t. $\mathbf{FL_{ew}}$ ([13]) and axiom $((A \supset \bot) \wedge A) \supset \bot$.

(*Completeness*) By the completeness of $\mathbf{ScFL_{ew}}$ w.r.t. $\mathbf{FL_{ew}}$ it is enough to check that $\vdash_{\mathbf{ScFL_{ew}}+(wc)} ((A \supset \bot) \wedge A) \supset \bot$. This is straightforward.

Corollary 4. *The hypersequent version of* $\mathbf{ScFL_{ew}} + (wc)$ *with in addition* (*com*) *is an analytic calculus for* **SMTL**.

Proof. $\mathbf{ScFL_{ew}} + (wc)$ is a standard sequent calculus in which the rules $(\supset, r), (\vee, r)_{1,2}, (\vee, l)$ and $(\supset, l)$ are derivable. Moreover its rules are reductive and substitutive. The claim follows by Proposition 1 and Corollary 2.

References

1. A. Avron. A constructive analysis of RM. *J. of Symbolic Logic*, vol. 52. pp. 939-951. 1987.
2. A. Avron. Hypersequents, logical consequence and intermediate logics for concurrency. *Annals of Mathematics and Artificial Intelligence*, 4. pp. 225–248. 1991.
3. M. Baaz, A. Ciabattoni. A Schütte-Tait style cut-elimination proof for first-order Gödel logic. In *Proc. of Automated Reasoning with Analytic Tableaux and Related Methods (Tableaux'2002)*. LNAI 2381, pp. 23-33. 2002.
4. M. Baaz, A. Ciabattoni, C.G. Fermüller. Hypersequent Calculi for Gödel Logics — a Survey. *J. of Logic and Computation*. vol. 13. pp. 1-27. 2003.
5. M. Baaz, A. Ciabattoni, F. Montagna. Analytic Calculi for Monoidal T-norm Based Logic. *Fundamenta Informaticae* vol. 59(4), pp. 315-332. 2004.
6. M. Baaz, A. Ciabattoni, C.G. Fermüller and H. Veith. Proof Theory of Fuzzy Logics: Urquhart's C and Related Logics. In: Proc. of *MFCS'98*. LNCS 1450, pp. 203-212. 1998.
7. G. Corsi. Completeness theorem for Dummett's LC quantified and some of its extension. *Studia Logica*, vol. 51, pp. 317-335. 1992.
8. M. Dummett. A propositional calculus with denumerable matrix. *J. Symbolic Logic*, vol. 24. pp. 97–106. 1959.
9. F. Esteva, J. Gispert, L. Godo, F. Montagna. On the Standard and Rational Completeness of some Axiomatic Extensions of the Monoidal T-norm Logic *Studia Logica*. vol. 71. pp. 393-420, 2002.
10. G. Gentzen. Untersuchungen über das logische Schliessen I, II. *Mathematische Zeitschrift*, 39. pp. 176–210, 405–431. 1934.
11. K. Gödel. Zum Intuitionistischen Aussagenkalkül. *Ergebnisse eines mathematischen Kolloquiums*, vol. 4, pp. 34-38. 1933.
12. P. Hájek. *Metamathematics of Fuzzy Logic*. Kluwer, 1998.
13. H. Ono, Y. Komori. Logics without the contraction rule. *J. of Symbolic Logic*, vol. 50. 169–201. 1985.
14. G. Pottinger. Uniform, cut-free formulation of T,S$_4$ and S$_5$, (abstract). *J. of Symbolic Logic*, vol. 48, p. 900. 1983.
15. K. Schütte. *Beweistheorie*. Springer Verlag. 1960.
16. H. Schwichtenberg. Proof Theory: Some applications of Cut-Elimination, in: *Handbook of Mathematical Logic* (J. Barwise, Ed.), North-Holland, 868–894. 1977.
17. W.W. Tait. Normal derivability in classical logic. In *The Syntax and Semantics of infinitary Languages*, LNM 72, 204–236. 1968.
18. G. Takeuti. *Proof Theory*. North-Holland, Amsterdam, 2nd edition, 1987.
19. G. Takeuti, T. Titani. Intuitionistic fuzzy logic and intuitionistic fuzzy set theory, *J. of Symbolic Logic*, vol. 49, 851–866. 1984.
20. A.S. Troelstra and H. Schwichtenberg. Basic Proof Theory. Cambridge University Press. 1996.
21. A. Urquhart. Basic Many-valued Logic. *Handbook of Philosophical Logic*. Vol III, D.M. Gabbay and F. Guenthner eds. First and Second edition. 1984 and 2000.

Author Index

Lecture Notes in Computer Science

For information about Vols. 1–3112

please contact your bookseller or Springer